High-Energy Aspects of Solar Flares

A. Gordon Emslie • Brian R. Dennis • Robert P. Lin •
Hugh S. Hudson
Editors

High-Energy Aspects of Solar Flares

Previously published in *Space Science Reviews* Volume 159,
Issues 1–4, 2011

Springer

Editors
A. Gordon Emslie
Dept. of Physics and Astronomy
Western Kentucky University
1906 College Heights Blvd, #11077
Bowling Green, KY 42101-1077
USA

Robert P. Lin
Space Sciences Lab
University of California
7 Gauss Way
Berkeley, CA 94720-7450
USA

Brian R. Dennis
Solar Physics Laboratory
Heliophysics Science Division
NASA Goddard Space Flight Center
Greenbelt, MD
USA

Hugh S. Hudson
Space Sciences Lab
University of California
7 Gauss Way
Berkeley, CA 94720-7450
USA

ISBN 978-1-4614-3072-8
Springer New York Dordrecht Heidelberg London

Library of Congress Control Number: 2011945429

Printed on acid-free paper

Springer is part of Springer Science+Business Media (www.springer.com)

Contents

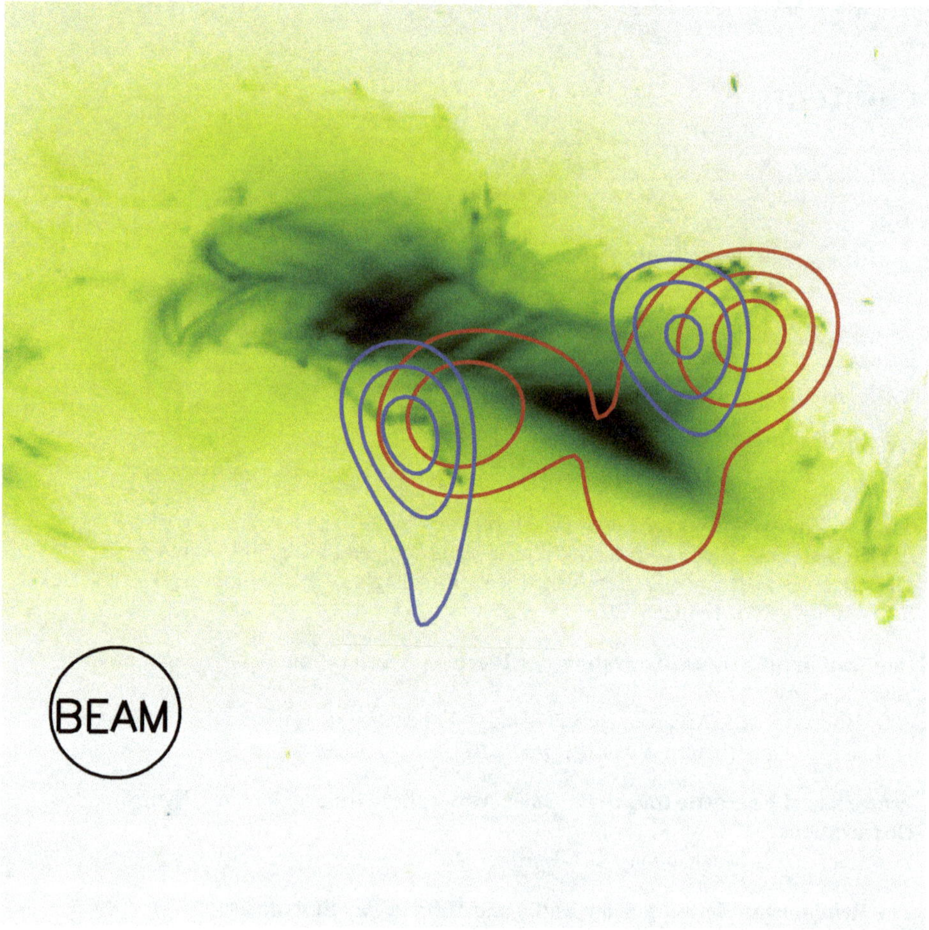

RHESSI images of 200–300 keV electron-bremsstrahlung emission (*blue contours*) and the ionassociated 2.218–2.228 MeV neutron-capture γ-ray line emission (*red* 50, 70 and 90% *contours*). These are overlaid on a simultaneous TRACE 195 Å image of flare SOL2001-10-28. The image captures, for the first time, the double-footpoint structure of the nuclear emission from ions accelerated in coronal magnetic loops to tens of MeV. The neutron-capture line imaging is based on the two RHESSI subcollimators (6 and 9) whose grids are thick enough to stop γ-rays. They provide angular resolution of 35″ FWHM as shown in the beam circle (from Hurford et al. 2005)

Space Sci Rev (2011) 159:1–2
DOI 10.1007/s11214-011-9815-7

Preface

A. Gordon Emslie · Brian R. Dennis · Hugh Hudson · Robert P. Lin

Published online: 13 August 2011

The Reuven Ramaty High Energy Solar Spectroscopic Imager (*RHESSI*) was launched on February 5, 2002. During the years after launch, we organized a series of workshops to share ideas and results from the analysis of the new *RHESSI* observations, and to facilitate those analyses. The workshops sought to integrate the unique high-energy perspective of *RHESSI* with data from observations at many wavelengths. At the same time we wanted the workshops to spur modeling and the development of basic theory. Some of the workshops also provided training and familiarization with the analysis of *RHESSI* data using the project's software. This volume summarizes all of this work through 2010. Broadly speaking, each paper in this volume represents the work of a topical group within the workshop series, with the authors comprising the contributors to the work of that group. To this we have added an introductory chapter and a concluding summary article.

To ensure accuracy, we have tried to restrict the references to articles that, at the time of writing, had actually been published or accepted for publication prior to our cutoff date in 2010 (but some 2011 items slipped in anyway). The work presented essentially covers solar activity from Cycle 23 and should serve as reference material from the current Cycle 24, during which *RHESSI* continues to operate well.

A.G. Emslie
Department of Physics & Astronomy, Western Kentucky University, Bowling Green, KY 42101, USA

B.R. Dennis
Solar Physics Laboratoty (Code 671), NASA Goddard Space Flight Center, Greenbelt, MD 20771, USA

H. Hudson (✉)
University of California-Berkeley, Berkeley, CA, USA
e-mail: hhudson@ssl.berkeley.edu

R.P. Lin
Physics Department and Space Sciences Laboratory, US Berkeley, Berkeley, CA 94720, USA

R.P. Lin
School of Space Research, Kyung Hee University, Seoul, Korea

To the extent possible, we have used a consistent notation throughout, invoking the new IAU naming convention for solar targets whereby a flare is uniquely referenced by its date and *GOES* peak time (e.g., SOL2002-07-23T00:35). A glossary of acronyms is in the introductory chapter, separate bibliographies are provided for each chapter, and a detailed book-wide subject index is at the end.

The intended audience for this volume includes active workers in the field of high-energy solar physics, scientists working in other areas who may be surprised by how much can be gained from a more careful study of the Sun, and graduate students entering (or considering entry) into this field. The goal, therefore, is to provide both quality of scientific content and pedagogical value. To this end, each paper was reviewed by both one senior member of the community and one younger scientist.

We thank the group leaders for their tireless efforts. We are grateful to the referees of these papers, and also to Amir Caspi for editorial help with the entire volume. Finally we thank UC Berkeley, NASA's Goddard Space Flight Center, and the Paul Scherrer Institute in Switzerland for the development of the spacecraft, its instrumentation, and the software. The international team of *RHESSI* Co-Investigators continues to play a fundamental role in the interpretation of the data, as do the many participants from outside the core team itself. As *RHESSI* enters its second solar maximum with unique hard X-ray and γ-ray observational capabilities, we hope that this volume can help lead to a better understanding of many discoveries yet to come.

24 May 2011

Space Sci Rev (2011) 159:3–17
DOI 10.1007/s11214-011-9802-z

Overview of the Volume

B.R. Dennis · A.G. Emslie · H.S. Hudson

Received: 3 May 2010 / Accepted: 15 June 2010 / Published online: 23 July 2011

Abstract In this introductory chapter, we provide a brief summary of the successes and remaining challenges in understanding the solar flare phenomenon and its attendant implications for particle acceleration mechanisms in astrophysical plasmas. We also provide a brief overview of the contents of the other chapters in this volume, with particular reference to the well-observed flare of 2002 July 23.

Keywords Sun: flares · Sun: X-rays · Sun: acceleration · Sun: energetic particles

1 Historical Perspective

This volume of *Space Science Reviews* contains a comprehensive review of our current understanding of the high energy aspects of solar flares. It is written with the same philosophy as the book on solar flares (Sturrock 1980b) that grew out of the *Skylab* workshops. The nine chapters are intended to display the accumulated wisdom of the many scientists who have attended the eleven *RHESSI* (*Reuven Ramaty High Energy Solar Spectroscopic Imager*) science workshops to date. *RHESSI* is a NASA Small Explorer satellite launched in February 2002. The intent was to cover the relevant published literature into 2010, and this succeeded as illustrated in Fig. 1. Results are summarized from hard X-ray and γ-ray observations in Solar Cycle 23 and the complementary observations at other wavelengths that have provided information on the same flares and the often-associated coronal mass ejections (CMEs). We

B.R. Dennis (✉)
Code 671, NASA Goddard Space Flight Center, Greenbelt, MD 20771, USA
e-mail: brian.r.dennis@nasa.gov

A.G. Emslie
Western Kentucky University, Bowling Green, KY 42101, USA
e-mail: emslieg@wku.edu

H.S. Hudson
Space Sciences Laboratory, UC Berkeley, Berkeley, CA 94720, USA
e-mail: hhudson@ssl.berkeley.edu

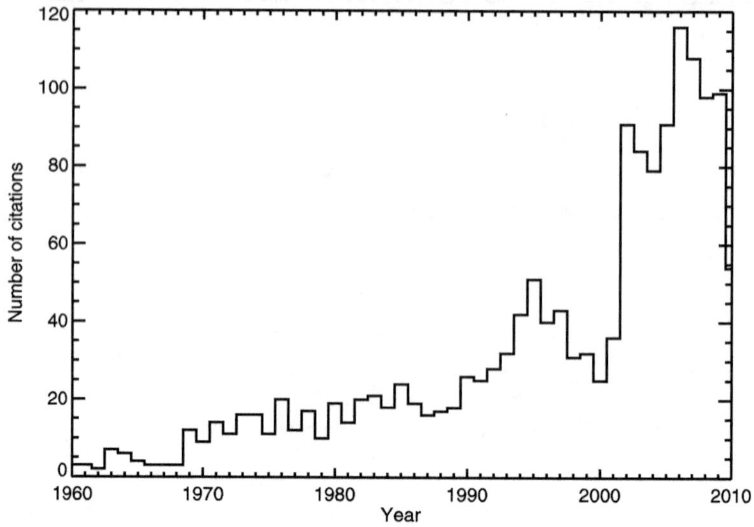

Fig. 1 The distribution by year of the 1,619 (non-unique) citations in Chaps. 1–8 back to 1960; the 20 citations prior to that included papers by Alfvén, Carrington, and Giovanelli among other pioneers. The peak in 2002 coincides with the publication of the initial *RHESSI* results and the peak at around 1995 may be attributable to *Yohkoh* results

anticipate that this volume will be a comprehensive reference of our current state of knowledge and relevant published literature up to the onset of renewed solar activity in Solar Cycle 24.

Although great progress has been made in understanding the solar flare phenomenon since the *Skylab* report, the basic concepts were nevertheless well established at that time from the extensive ground-based observations and early space missions. As summarized by Sturrock (1980a), the basic picture of a flare involved the sudden explosive release of the "free" magnetic energy of a current-carrying magnetic field in the corona. During the flare, the energy would be released by altering (or even destroying) the currents to convert the field to a lower-energy (or even current-free) form. Various energy release mechanisms were considered, including magnetic reconnection, but then, as now, it was recognized that the plasma processes are very complicated and no definitive conclusions could be made on which specific processes were involved.

The acceleration of particles was considered as posing a primary requirement for any flare model. It was recognized though that there was a serious electron "number problem," in that the number of electrons required to explain the measured hard X-ray fluxes was a substantial fraction of the electron content of the corresponding coronal region before the flare. Also, the total power in the 10–100 keV electrons (some 2×10^{29} erg s^{-1} in SOL1972-08-04T06:25 or SOL1972-08-04T06:25)[1] and the total energy contained in these electrons assuming thick-target interactions was known to be an unexpectedly large fraction of the total flare energy (Ramaty et al. 1980). Indeed, Lin and Hudson (1976) had shown that the ~10 to 100 keV electrons "constitute the bulk of the flare energy," perhaps as high as 10 to 50%. This issue remains one of the key problems in understanding the mechanism of

[1] In this volume we identify individual flares using the IAU naming convention; see Solar Physics **263**(1) (2010); doi:10.1007/s11207-010-9553-0.

energy release in solar flares as described in Chaps. (3, 7, and 8) of this volume (Holman et al. 2011; Kontar et al. 2011; Zharkova et al. 2011).

The thermal or nonthermal origin of the hard X-ray emission was an ongoing debate in the *Skylab* workshops (Sturrock 1980b). This debate is still not fully resolved (White et al. 2011) but strong support for the nonthermal electron-beam model came with the hard X-ray imaging of the *Solar Maximum Mission (SMM)*, *Hinotori*, and *Yohkoh* satellites. The Hard X-ray Telescope (HXT) on *Yohkoh* produced images of a great many flares (Sato et al. 1999), with the advantage of simultaneous microwave observations (e.g., Bastian et al. 1998; Hannah et al. 2011). The X-ray images revealed the prevalence of double footpoint sources and the near-simultaneity of their light curves to within a second (Sakao et al. 1996). The stronger X-ray source and the weaker radio source both tend to be located in the weaker magnetic field region, consistent with an electron beam model where magnetic mirroring is significant (e.g., Bastian et al. 1998).

The acceleration of ions was also recognized from the γ-ray observations on *OSO-3* (Chupp et al. 1973) and *HEAO-1* (Hudson et al. 1980) but their numbers and energy content were not well understood. It required observations made with the Gamma Ray Spectrometer (Forrest et al. 1980) on *SMM* in the 1980s to reveal that the ions could be accelerated nearly simultaneously with the electrons (Forrest and Chupp 1983) and that the total energy in ions above 1 MeV/nucleon could be comparable to, or even exceed, the total energy in electrons above 20 keV (Ramaty et al. 1995). Similar to electrons (Holman et al. 2011), the total energy content in ions depends to a significant extent on an accurate determination of the low-energy end of the accelerated spectrum, and considerable progress has been made on this front (Vilmer et al. 2011).

The biggest discrepancy between our current understanding of the flare phenomenon and the *Skylab* ideas is in the relationship between flares and coronal mass ejections (CMEs). In the *Skylab* era, CMEs were referred to as coronal transients, and it could still be questioned if they were "incidental to flares or whether they reveal something fundamental about the energy release process" (Rust et al. 1980). It was not until the "solar flare myth" was described by Gosling (1993) that many realized that the CMEs are the major cause of solar effects at the Earth, not the flares as such. This was controversial (Hudson et al. 1995), and with modern data we now understand that flares and energetic CMEs are, in fact, intimately related and may have comparable energy content (Emslie et al. 2004a, 2005) in major events. It is now clear that the largest and fastest CMEs, which have the greatest effect on space weather and pose the greatest danger to satellites, are mostly (possibly always) associated with large flares of comparable total energy. Indeed, the origins and energy sources of flares and CMEs are so intimately entwined that it is impossible to explain one without understanding the other. Thus, if we are to understand these phenomena and develop predictive capabilities, it is imperative that we consider them as interrelated phenomena.

2 Review of Flare Models

It was established very early on in flare studies that the source of the energy released in a flare is in current-carrying (i.e., "non-potential") magnetic fields. Not only are flares invariably connected with magnetism in the photosphere, but an examination of the various candidate sources of energy reveals magnetic energy to be the only plausible contender (Tandberg-Hanssen and Emslie 1988). Release of the energy stored in the twisted magnetic field configuration can proceed through a process termed *magnetic reconnection*, in which the connectivity of the magnetic field redefines itself. However, it is far less clear—and this is what to a large extent drives flare research—how the energy can be built up *without* reconnection dissipating it immediately, and what "trigger" initiates the flare process. The energy can build up for many hours, or even days, without significant dissipation of the energy prior to the flare onset.

Sturrock (1980a) provided a concise, yet thorough, review of the various reconnection scenarios that existed through the end of the 1970s. In fairness, it must be conceded that most of these scenarios are still valid, although many have fallen out of favor. Perhaps the most significant "casualty" is the model of Gold and Hoyle (1960), which invoked the self-attraction of magnetic loops carrying parallel, or near-parallel, currents, leading to an energy release at their mutual interface. While this laboratory analogy is rather appealing, it unfortunately fails to recognize that it is not currents that attract *per se*, but rather one current interacts with the magnetic field produced by the other. Thus, in the global force-free field appropriate to the low-β solar corona, the current density $\mathbf{J}$ is always nearly parallel to the local magnetic field $\mathbf{B}$, and no $\mathbf{J} \times \mathbf{B}$ force exists. Independent magnetic flux loops therefore have no particular attraction to (or repulsion from) each other, and some external influence (such as a photospheric velocity field; e.g. Heyvaerts et al. 1977) must be postulated in order to drive them together.

Basic Sweet-Parker reconnection (e.g., Sweet 1969), has also fallen out of favor because it is perceived to develop too slowly. In its place Petschek reconnection, which is both much faster and is characterized by energy release at reconnection-driven shocks (Petschek 1964) rather than at the point of reconnection, has appeared in many descriptions. The standard "CSHKP" (Carmichael 1964; Sturrock 1966; Hirayama 1974; Kopp and Pneuman 1976) model, suggested by the growth of soft X-ray "loop prominence systems" accompanying the increasing separation of Hα "ribbons," invokes a reconnection site near the apex of the loop system. This injects energy (e.g., as flows) into the underlying loops. Although this model does not naturally account for the impulsive phase of a flare, nor the particle acceleration, a great deal of effort has been expended in quantitative modeling of such structures (density of loop-top source, standing shocks in loop legs, etc.). Aspects of these problems are discussed in Zharkova et al. (2011).

The planning of the *RHESSI* workshop series addressed the need to consider the physics of magnetic reconnection, and the concomitant acceleration of electrons and ions, *in the context of the observations*. For example, it was recognized early on that "test-particle" approaches to particle acceleration, in which the electric and magnetic fields are *prescribed* and constant in time, do not adequately take into account the fact that the mass, momentum, energy, and electrical current carried by the large number of accelerated particles necessary to account for flare observations must have a major feedback on both the electrodynamic and magnetohydrodynamic environments of the acceleration region. A major emphasis of the theory team (Zharkova et al. 2011) at the *RHESSI* workshops was the development of acceleration models that explicitly take these nonlinear aspects of the process into account.

3 Challenges for Simple Acceleration Models

The impulsive phase of a solar flare is characterized, in part, by the emission of a copious flux of hard X-rays (photon energy $\varepsilon \gtrsim 10$ keV). It is generally accepted that these hard X-rays are produced by collisional bremsstrahlung (free-free emission) when accelerated electrons encounter ambient protons and heavier ions in the solar atmosphere, although other emission mechanisms have also been considered (Kontar et al. 2011). The amount of electron energy required to produce these hard X-rays depends on the model used to characterize the interaction of the accelerated electrons with the target. A lower limit is given by a *collisional thick-target* interpretation (e.g., Brown 1971), in which all of the electron energy is absorbed in the target only through Coulomb collisions with ambient particles (primarily electrons). In this interpretation, the ratio of electron power to emitted hard X-ray power is of order 10^5 (see (4.4) below), and this gives an order-of-magnitude estimate for the rate of electron acceleration in the flare (Holman et al. 2011). For a large (e.g., *GOES* X-class) flare, the required rate of acceleration of electrons above 20 keV can exceed 10^{37} s^{-1}, a large number with interesting consequences (e.g., Miller et al. 1997), as we now discuss.

First, the number of electrons $\mathcal{N}$ confined in the coronal portion of a flare loop is simply the number density n (cm^{-3}) multiplied by the volume V (cm^3). Inserting typical values $n \approx 10^{11}$ and $V \approx 10^{27}$ gives $\mathcal{N} \approx 10^{38}$, so that the acceleration process would deplete the store of available electrons in 10 s or so, significantly less than the observed duration of the flare. This simple calculation therefore rules out models in which all the electrons to be accelerated are stored in a coronal volume prior to the onset of the flare. If the acceleration site is indeed in the corona, this requires instead that the electrons "recycle" multiple times or that fresh electrons enter from the chromosphere. Given that electrons in the solar atmosphere have gyroradii of order a centimeter and so are strongly tied to the guiding magnetic field lines, this presents formidable difficulties for current closure of the accelerated electron streams. Although solutions to this problem have been offered (Emslie and Hénoux 1995), they require the synergistic interaction of a large number ($\sim 10^{10}$ or so) of separate acceleration regions, and the origin (not to mention stability) of such a system has yet to be adequately explored.

Second, the accelerated electrons carry with them an associated *electrical current* of some 10^{18} A (3×10^{27} statamps). In steady-state, such a current, if assumed to propagate unidirectionally in a flux tube of radius 10^9 cm, gives rise, via Ampère's Law, to a magnetic field $B \approx 2 \times 10^8$ Gauss). Not only is such a high magnetic field completely untenable on observational grounds, the associated energy density $B^2/8\pi \approx 10^{15}$ erg cm^{-3}, for a total energy content $\int (B^2/8\pi)\, dV$ of some 10^{42} ergs, some ten orders of magnitude larger than the energy content in the electron beam that ostensibly produces the field, even for a 1000-second beam.

Third, a current of this magnitude cannot appear instantaneously. The self-inductance $\mathcal{L}$ of a structure scales with ℓ, the characteristic dimension. For a solar flare loop of typical dimensions, we find $\mathcal{L} \approx 10$ H. Hence, to initiate a current $I \approx 10^{18}$ A in a time $\tau \approx 10$ s requires a voltage $V \approx \mathcal{L}I/\tau \approx 10^{18}$ V, which is *fourteen* orders of magnitude higher than the typical energy of the accelerated electrons.

Such considerations have led various authors (e.g., Knight and Sturrock 1977; Emslie 1980; Brown and Bingham 1984; Spicer and Sudan 1984; Larosa and Emslie 1989; van den Oord 1990; Zharkova et al. 1995) to consider models in which cospatial return currents locally neutralize the beam current. However, such models can only produce beam-current neutralization in the *propagation* region, and considerable analysis has been performed on the details of the beam/return current interaction in this propagation region (van Oss and van den Oord 1995). In the acceleration region itself, return-current electrons would have to flow in a direction counter to the applied electromotive force and so cannot neutralize the unacceptably large currents therein. Although significant progress on this issue has been made recently (Zharkova et al. 2011), a satisfactory resolution of this issue has yet to be offered. A self-consistent electrodynamic theory would require a description of current closure in the acceleration region as well as in the beam itself. This difficulty has led various authors to reject acceleration models featuring large-scale electric fields in favor either of acceleration by very large electric fields in localized current sheets, or via acceleration in (stochastic) MHD or plasma waves.

The problems imposed by the return current become more severe as the beam becomes more intense. It has become increasingly clear directly from *RHESSI* imaging (e.g., Dennis and Pernak 2009) that the hypothetical electron beam would occupy only a small area. Closely related emissions such as UV and white light are often unresolved at even higher resolution and suggest areas substantially smaller than 10^{16} cm^2 (Fletcher et al. 2011). Accordingly, alternatives that replace electron-beam energy transport with Poynting fluxes have been proposed (Emslie and Sturrock 1982; Haerendel 2006; Fletcher and Hudson 2008; Haerendel 2009).

4 Importance of Hard X-Rays and γ-Rays as Diagnostics of Accelerated Particles

It is important to emphasize that the energy released as hard X-rays and γ-rays is—in and of itself—a negligible component of that released in the flare. The importance of this radiation lies not in its energy content *per se*, but rather in the energy in accelerated particles required to produce this diagnostic radiation (Holman et al. 2011; Vilmer et al. 2011).

The process of hard X-ray emission is very inefficient. In order to produce a photon by bremsstrahlung, an electron must suffer a near-direct collision on an ambient ion. Most electrons instead lose their energy in a large number of small-angle scatterings off ambient electrons, and do not contribute to the bremsstrahlung yield. We may compare the energy emitted through bremsstrahlung to that suffered in Coulomb collisions by comparing the cross-sections for the two processes. The nonrelativistic differential cross-section (cm^2 per unit photon energy) for free-free emission of a photon of energy ε by an electron of energy E may be, to order-of-magnitude, approximated by the Kramers form

$$\sigma(\varepsilon, E) \approx \alpha \frac{r_o^2 mc^2}{\varepsilon E}, \tag{4.1}$$

where $\alpha \approx 1/137$ is the fine structure constant, $mc^2 = 511$ keV is the electron rest mass, and r_0 is the classical radius of the electron. From this it follows that the cross-section (cm^2 keV) for energy loss through bremsstrahlung by an electron of energy E is

$$\sigma_E^B = \int_0^E \varepsilon\sigma(\varepsilon, E)\, dE \approx \alpha r_o^2 mc^2. \tag{4.2}$$

By contrast, the cross-section for Coulomb energy loss (cm^2 keV) by an electron of energy E is (Brown 1972; Emslie 1978)

$$\sigma_E^C(\varepsilon, E) = \frac{2\pi e^4 \Lambda}{E} \approx \Lambda \frac{r_o^2 (mc^2)^2}{E}, \tag{4.3}$$

where Λ is the Coulomb logarithm, ε is the electron charge, and we have used $r_o = e^2/mc^2$. Taking the ratio

$$\eta = \frac{\sigma_E^B}{\sigma_E^C} \approx \frac{\alpha}{\Lambda} \frac{E}{mc^2} \approx 4 \times 10^{-4} \frac{E}{mc^2} \tag{4.4}$$

gives the energetic efficiency of the bremsstrahlung process relative to Coulomb collisions. For $E \approx 20$ keV, $\eta \approx 1.5 \times 10^{-5}$, i.e., for each erg of bremsstrahlung, $\gtrsim 10^5$ ergs of energy in electrons are required.

Gamma-rays in solar flares (Vilmer et al. 2011) are produced principally by the interaction of accelerated protons and heavier ions with nuclei in the ambient atmosphere, although electron-ion bremsstrahlung from accelerated electrons can also contribute. Unlike hard X-ray emission, not all γ-ray emission is prompt—in particular, the capture of neutrons onto ambient protons to create the 2.223 MeV deuterium-formation line can take several minutes because of the need to reduce the momentum of the neutrons to a value where the cross-section for recombination is sufficiently high. The low speed of the captured neutrons also leads to the extremely small spectral width of the 2.223 MeV line; by contrast, most γ-ray lines (e.g., the prompt nuclear de-excitation lines of ^{12}C at 4.4 MeV and ^{16}O at 6.1 MeV) have both narrow and broad spectral profiles depending on whether the heavy ions are the target or the projectile in interactions with accelerated or ambient protons, respectively.

RHESSI not only provides γ-ray spectra with unprecedented spectral resolution (Smith et al. 2003) but also, on a few occasions, *images* of the γ-ray line emission—in particular in the 2.223 MeV line. Interestingly, the locations of the hard X-ray and γ-ray sources are *not* coincident (Hurford et al. 2003, 2006), indicating a preferential acceleration of electrons vs. protons in different substructures within the flare volume (cf. Emslie et al. 2004b).

Some time after the publication of the *Skylab* volume, studies of observations from the *SMM* Gamma-Ray Spectrometer (GRS) (Share and Murphy 1995) led Ramaty et al. (1995) to realize that the spectra of the accelerated ions could remain steep down to energies as low as 1 MeV, and hence that the *energy content* of accelerated ions in solar flares, as revealed by their γ-ray emission, could rival that of the accelerated electrons, hitherto thought to dominate the energy budget of accelerated particles. Further study of the relative partitioning of energy between accelerated electrons and ions was carried out by Emslie et al. (2004a, 2005), who reached a similar conclusion. Further discussion of the partitioning of flare energy amongst its constituent parts can be found in Fletcher et al. (2011), Holman et al. (2011), and Vilmer et al. (2011).

5 *RHESSI* Design and Capabilities

5.1 Operations

RHESSI (Lin et al. 2002) uses nine cooled and segmented germanium detectors to achieve high-resolution X-ray and γ-ray spectroscopy across the full energy range from 3 keV to 17 MeV (Smith et al. 2002). The FWHM energy resolution increases from $\sim$1 keV at the lowest energies to $\sim$10 keV at the highest. This has proved adequate to detect the iron-line complex at $\sim$6.7 keV, measure the steep hard X-ray continuum spectra with relative flux accuracies as fine as 1%, measure the width of the positron-annihilation line at 511 keV as it varies with time during a flare, and resolve all of the narrow nuclear γ-ray lines except for the intrinsically narrow neutron-capture line at 2.223 MeV. A bi-grid tungsten collimator over each detector modulates the incident photon flux as the spacecraft rotates at $\sim$15 rpm to provide the temporal information needed for the Fourier-transform technique that is used to reconstruct the X-ray and γ-ray images (Hurford et al. 2002). Imaging is possible at all energies up to about 1 MeV, with an angular resolution of $\sim 2''$ (FWHM) up to $\sim$100 keV increasing to $\sim 20''$ at 1 MeV. When the count rates are sufficiently high, images can be made with a cadence as short as 4 s. In addition, images can also be made in the neutron-capture γ-ray line at 2.223 MeV in the relatively few flares when the total number of photons detected in this line is sufficiently high (several thousand; Hurford et al. 2006). The field of view is $\sim 1^\circ$, and so a flare can be imaged no matter where on the visible disk it occurs. *RHESSI* has an effective sensitive area that reaches ∼60 cm^2 at 100 keV. Two thin aluminum disks can be automatically moved above each detector to attenuate intense soft X-ray fluxes so that *RHESSI* can operate with minimal detector saturation and pulse pile-up over a wide dynamic range in flux level. This allows coverage of both faint microflares (with the attenuators removed) and the most powerful flares (with both attenuators in place over each detector).

RHESSI was launched on 2002 February 5 and has been in operation almost continuously since 2002 February 11, with brief intervals of pointing away from the Sun for observations of the Crab Nebula and solar global emission (the quiet Sun as a star). The *GOES* class X4.8 flare SOL2002-07-23T00:35 yielded the first γ-ray emission lines detected by *RHESSI* (see Shih et al. 2009 for a summary of all γ-ray events seen with *RHESSI* as of the Cycle 23/Cycle 24 solar minimum). By the time of writing, there have been two successful *anneals* of the germanium detectors, in November 2007 and again in April 2010. These are month-long procedures needed to restore sensitive volume and energy resolution that become degraded by the accumulation of radiation damage.

5.2 Collaborations

Because solar flares and other forms of activity are defined by particle acceleration and extreme heating, *RHESSI* observations are at the heart of many broader studies. Instruments on other spacecraft and at ground-based observatories around the world have been active participants in providing the magnetic, thermal, and dynamic context in which the X-ray and γ-ray sources are produced. In addition, microwave observations of the gyrosynchrotron emission provide additional information on the accelerated electrons themselves. Coronagraph observations of CMEs, and *in situ* particle-and-field measurements in the near-Earth environment, also provide information that can be used to establish the links between these related phenomena and any associated flares. A partial list of all collaborating space-based observatories with further information can be found on the Max Millennium Web site at http://solar.physics.montana.edu/max_millennium/obs/SBO.html. They include the *ACE, Cluster, CORONAS, GOES, INTEGRAL, SOHO, TRACE*, and *WIND* spacecraft. More recently, complementary observations have been made with the newer solar missions including *Hinode*, *STEREO*, *SDO*, and *Fermi*. They have provided X-ray, EUV, UV, optical, and *in situ* particle-and-field measurements relevant to the many events recorded by *RHESSI*. Thanks to the daily email messages and the coordinating efforts of the Max Millennium program, many collaborative observing campaigns have been conducted to maximize the overlap of the various observatory programs.

6 Outline of This Volume

The contents of this volume center on *RHESSI* capabilities and research flowing from the *RHESSI* data, but this embraces most of flare physics because of the dominating importance of the high-energy processes. The current volume consists of a series of articles, each representing the work of multiple authors. The purpose of each article is to present a review of the pertinent subject matter, linking the results of a variety of published works into a coherent whole, which we hope is useful both for the reader who wishes an overview of contemporary knowledge in the area, and for the experienced researcher to view results in context. Each article presents mid-length reviews of the literature, and provides a comprehensive reference list for the reader who seeks more detailed developments or information.

Taken as a whole, the Editors hope that this volume will be a useful successor for the *Skylab* workshop volume (Sturrock 1980b), and we hope that Zharkova et al. (2011) in particular will play the same role for solar-flare particle acceleration as reviewed earlier by Miller et al. (1997).

Although each of the articles presents a somewhat different aspect of solar flare research, they are all inter-related and should be read in this context. To illustrate the inter-relationship of these articles, we note that much attention has been paid to the characteristics of the first γ-ray line flare detected with *RHESSI* mentioned above (SOL2002-07-23T00:35) This event forms the basis for much of the discussion in each article, namely:

- The temporal, spatial and spectral properties of the intense hard X-ray radiation from this flare are discussed in Kontar et al. (2011); this includes a discussion of the temporal evolution of the hard X-ray spectrum, both for the flare as a whole and for subregions (e.g., coronal sources, chromospheric footpoints) observed within the active region. Using an appropriate cross-section for hard X-ray production, regularized spectral inversion of the observed hard X-ray spectrum then yields the volume-averaged mean source electron spectrum for the event.
- A useful check on the electron spectrum comes from observations of deka-GHz radio emission (White et al. 2011); this radio emission is believed to be produced by the high-energy tail of the same ensemble of electrons that produces the deka-keV hard X-ray emission, but the inferred electron spectra are intriguingly different.
- As discussed in Holman et al. (2011), combining the mean source electron spectrum with an appropriate electron transport model then leads to the *accelerated* electron spectrum. Analysis of this accelerated spectrum (with particular attention to the low-energy end where, due to the typically steep spectra involved, most of the particle energy resides) then yields information on the total energy in the accelerated electrons.
- Vilmer et al. (2011) discuss various aspects of the γ-ray emission from this event. It includes a comparison of the time profiles and spatial locations of hard X-ray and γ-ray sources. The intensity and Doppler shifts of the γ-ray lines are also presented along with the information that these measurements provide on the number and angular distribution of accelerated ions, and even on the magnetic field geometry in the active region.
- Results from all photon energy ranges (hard X-ray, soft X-ray, γ-ray, optical and EUV) are synthesized into a global picture of the energetics of this flare in Fletcher et al. (2011).
- The position of this flare within the statistical ensemble of all flare events detected with *RHESSI*, extending from B-class microflares to large X-class events, is provided in Hannah et al. (2011).
- Zharkova et al. (2011) review the implications of these observational results for theoretical models of particle acceleration and transport in flare plasmas, and in the broader field of acceleration in astrophysical sources.
- Lin (2011) summarizes these results and presents prospects for future research directions.

Acknowledgements H. Hudson was supported by NASA under contract NAS5-98033 for RHESSI. A.G. Emslie was supported by NASA Grant NNX10AT78G.

Appendix: Glossary of Acronyms Used in the Monograph

3DP 3-D Particles, onboard *WIND*

AAS American Astronomical Society
ACE *Advanced Composition Explorer*

ACRIM	Active Cavity Radiometer Irradiance Monitor, onboard *SMM*
AIA	Atmospheric Imaging Assembly, onboard *SDO*
ARTB	active-region transient brightening
ATST	Advanced Technology Solar Telescope
BATSE	Burst And Transient Source Experiment, onboard *CGRO*
BBSO	Big Bear Solar Observatory
BCS	Bent or Bragg Crystal Spectrometer, onboard *SMM* or *Yohkoh*
CA	cellular automaton
CCD	charge-coupled device
CDF	cumulative distribution function
CDS	Coronal Diagnostic Spectrometer, onboard *SoHO*
CGRO	*Compton Gamma Ray Observatory*
CME	coronal mass ejection
COMPTEL	Imaging Compton Telescope, onboard *CGRO*
CoMP	Coronal Multi-channel Polarimeter
CORONAS	*Complex ORbital ObservatioNs of the Active Sun* satellite series
CoSMO	Coronal Solar Magnetism Observatory
CSHKP	Carmichael, Sturrock, Hirayama, Kopp & Pneuman flare model
DC	direct current
DEM	differential mission measure
DOI	Digital Object Identifier
DR	diffusion region
EDF	empirical distribution function
EIS	EUV Imaging Spectrometer, onboard *Hinode*
EIT	Extreme ultraviolet Imaging Telescope, onboard *SoHO*
EM	emission measure
ENA	energetic neutral atom
EP	erupting prominence
ESA	European Space Agency
EST	European Solar Telescope
EUV	extreme ultraviolet
EVE	EUV Variability Experiment, onboard *SDO*
FAL	Fontenla, Avrett, & Loeser solar atmospheric model
FASR	Frequency Agile Solar Radiotelescope
FIP	first ionization potential
FMSS	fast-mode standing shock
FOXSI	Focusing Optics hard X-ray Spectrometer Imager
FP	footpoint
FWHM	full width at half maximum
GBM	Gamma-ray Burst Monitor, onboard *Fermi*
GeDs	germanium detectors
GEM	Geospace Environment Modeling
GLE	ground-level event

GOES	*Geostationary Operational Environmental Satellite*
GONG	Global Oscillation Network Group
GRANAT	*Gamma Rentgenovskii Astronomicheskii Nauchni ApparaT*
GRB	gamma-ray burst
GRIPS	*Gamma-Ray Imaging Polarimeter for Solar flares*
GRS	Gamma Ray Spectrometer, onboard *SMM*
GSAT-2	Geostationary Satellite launched by the Indian Space Research Organization on May 8, 2003
GSFC	[NASA] Goddard Space Flight Center
HEAO	*High Energy Astrophysical Observatory* satellite series
HMI	Helioseismic and Magnetic Imager, onboard *SDO*
HXIS	Hard X-ray Imaging Spectrometer, onboard *SMM*
HXR	hard X-ray
HXRBS	Hard X-Ray Burst Spectrometer, onboard *SMM*
HXRS	Hard X-ray Spectrometer, onboard *MTI*
HXT	Hard X-ray Telescope, onboard *Yohkoh*
IAU	International Astronomical Union
ICE	*International Cometary Explorer*, a.k.a *ISEE-3*
ICME	interplanetary coronal mass ejection
IDL	Interactive Data Language
IGY	International Geophysical Year
INTEGRAL	*INTErnational Gamma-Ray Astrophysics Laboratory*
IR	infrared
ISEE	*International Sun-Earth Explorer* satellite series
KOSMA	Kölner Observatorium für SubMillimeter Astronomie
KS	Kolmogorov-Smirnov statistic
LASCO	Large Angle and Spectrometric Coronagraph, onboard *SoHO*
LDE	long-decay event
LHDI	lower hybrid drift instability
LOFAR	Low Frequency Array for Radio Astronomy
LOS	line-of-sight
LPF	large proton flare
LTE	local thermal equilibrium
MDI	Michelson Doppler Imager, onboard *SoHO*
MEKAL	Mewe-Kaastra-Liedahl atomic code
MEM	maximum entropy method
MESSENGER	*MErcury Surface, Space ENvironment, GEochemistry, and Ranging* mission
MHD	magnetohydrodynamic(s)
MISS	Multichannel Infrared Solar Spectrograph, at PMO
MSFC	[NASA] Marshall Space Flight Center
MTI	*Multi-Thermal Imager*
NASA	National Aeronautics and Space Administration

NOAA	National Oceanic and Atmospheric Administration
NoRH	Nobeyama Radioheliograph
NoRP	Nobeyama Radio Polarimeters
NRBH	nonrelativistic Bethe-Heitler (bremsstrahlung cross-section)
NRH	Nançay Radioheliograph
OSO	*Orbiting Solar Observatory* satellite series
OSPEX	Object SPectral EXecutive (IDL-based spectral analysis software)
OSSE	Oriented Scintillation Spectrometer Experiment, onboard *CGRO*
OVSA	Owens Valley Solar Array
PA	position angle
PASJ	Publications of the Astronomical Society of Japan
PDF	probability density function
PFL	post-flare loop
PHEBUS	Payload for High Energy BUrst Spectroscopy, onboard *GRANAT*
PIC	particle-in-cell
PIL	polarity inversion line
PMO	Purple Mountain solar Observatory
POLAR	*POLAR* spacecraft, not an acronym
PVO	*Pioneer Venus Orbiter*
QPP	quasi-periodic pulsations
RCS	reconnecting current sheet
RESIK	REntgenovsky Spektrometr s Izognutymi Kristalami, onboard *CORONAS-F*
RHESSI	*Reuven Ramaty High Energy Solar Spectroscopic Imager*
RMC	rotation modulation collimator
SC	(*RHESSI*) subcollimators
SDO	*Solar Dynamics Observatory*
SEE	*Solar Eruptive Events*
SEP	solar energetic particle
SFU	solar flux unit (10^{-22} W m^{-2} Hz^{-1})
SHH	soft-hard-harder (temporal behavior of spectral index)
SHS	soft-hard-soft (temporal behavior of spectral index)
SIGMA	Systéme d'Imagerie Gamma à Masque Aléatoire, instrument onboard *GRANAT*
SMART	Hida Solar Magnetic Activity Research Telescope
SMM	*Solar Maximum Mission*
SMSS	slow-mode standing shock
SOC	self-organized criticality
SOHO or *SoHO*	*Solar and Heliospheric Observatory*
SOLIS	Synoptic Optical Long-term Investigations of the Sun magnetograph
SONG	SOlar Neutrons and Gamma-rays instrument, onboard *CORONAS-F*
SORCE	*Solar Radiation and Climate Experiment*
SOT	Solar Optical Telescope, onboard *Hinode*

SOXS	SOlar X-ray Spectrometer, onboard *GSAT-2*
SPI	SPectrometer on Integral, onboard *INTEGRAL*
SPR-N	Solar Spectropolarimeter, onboard *CORONAS-F*
SSRT	Siberian Solar Radio Telescope
SST	Solar Submillimeter Telescope
ST	Hubble Space Telescope
STEREO	*Solar TErrestrial RElations Observatory*
SUMER	Solar Ultraviolet Measurements of Emitted Radiation instrument, onboard *SoHO*
SXI	Soft X-ray Imager, onboard *GOES*
SXR	soft X-ray
SXT	Soft or Solar X-ray Telescope, onboard *Yohkoh* or *Hinotori*, respectively
THEMIS	*Time History of Events and Macroscale Interactions during Substorms*
TIM	Total Irradiance Monitor, onboard *SORCE*
TOF	time-of-flight
TRACE	*Transition Region and Coronal Dynamics Explorer*
TS	termination shock
TSI	total solar irradiance
UV	ultraviolet
UVCS	Ultraviolet Coronagraph Spectrometer, onboard *SoHO*
VAL	Vernazza, Avrett, & Loeser solar atmospheric model
VLA	Very Large Array
VUV	vacuum ultraviolet
WATCH	Wide Angle Telescope for Cosmic Hard X-rays; onboard *GRANAT*
WAVES	Not an acronym; instrument onboard *WIND*
WBS	Wide Band Spectrometer, onboard *Yohkoh*
WIND	Spacecraft; not an acronym
WL	white light
XBP	X-ray bright point
XRT	X-Ray Telescope, onboard *Hinode*
XUV	X-ray/EUV/UV

References

T.S. Bastian, A.O. Benz, D.E. Gary, Annu. Rev. Astron. Astrophys. **36**, 131 (1998)
J.C. Brown, Sol. Phys. **18**, 489 (1971)
J.C. Brown, Sol. Phys. **26**, 441 (1972)
J.C. Brown, R. Bingham, Astron. Astrophys. **131**, L11 (1984)
H. Carmichael, NASA Spec. Publ. **50**, 451 (1964)
E.L. Chupp, D.J. Forrest, P.R. Higbie, A.N. Suri, C. Tsai, P.P. Dunphy, Nature **241**, 333 (1973)
B.R. Dennis, R.L. Pernak, Astrophys. J. **698**, 2131 (2009)
A.G. Emslie, Astrophys. J. **224**, 241 (1978)
A.G. Emslie, Astrophys. J. **235**, 1055 (1980)
A.G. Emslie, B.R. Dennis, G.D. Holman, H.S. Hudson, J. Geophys. Res. **110**, 11103 (2005)
A.G. Emslie, J.-C. Hénoux, Astrophys. J. **446**, 371 (1995)

A.G. Emslie, H. Kucharek, B.R. Dennis, N. Gopalswamy, G.D. Holman, G.H. Share, A. Vourlidas, T.G. Forbes, P.T. Gallagher, G.M. Mason, T.R. Metcalf, R.A. Mewaldt, R.J. Murphy, R.A. Schwartz, T.H. Zurbuchen, J. Geophys. Res. **109**, 10104 (2004a)
A.G. Emslie, J.A. Miller, J.C. Brown, Astrophys. J. Lett. **602**, L69 (2004b)
A.G. Emslie, P.A. Sturrock, Sol. Phys. **80**, 99 (1982)
L. Fletcher, H.S. Hudson, Astrophys. J. **675**, 1645 (2008)
L. Fletcher et al., Space Sci. Rev. (2011). doi:10.1007/s11214-010-9701-8
D.J. Forrest, E.L. Chupp, Nature **305**, 291 (1983)
D.J. Forrest, E.L. Chupp, J.M. Ryan, M.L. Cherry, I.U. Gleske, C. Reppin, K. Pinkau, E. Rieger, G. Kanbach, R.L. Kinzer, Sol. Phys. **65**, 15 (1980)
T. Gold, F. Hoyle, Mon. Not. R. Astron. Soc. **120**, 89 (1960)
J.T. Gosling, J. Geophys. Res. **98**, 18937 (1993)
G. Haerendel, Space Sci. Rev. **124**, 317 (2006)
G. Haerendel, Astrophys. J. **707**, 903 (2009)
I. Hannah et al., Space Sci. Rev. (2011)
J. Heyvaerts, E.R. Priest, D.M. Rust, Astrophys. J. **216**, 123 (1977)
T. Hirayama, Sol. Phys. **34**, 323 (1974)
G. Holman et al., Space Sci. Rev. (2011). doi:10.1007/s11214-010-9680-9
H. Hudson, B. Haisch, K.T. Strong, J. Geophys. Res. **100**, 3473 (1995)
H.S. Hudson, T. Bai, D.E. Gruber, J.L. Matteson, P.L. Nolan, L.E. Peterson, Astrophys. J. Lett. **236**, L91 (1980)
G.J. Hurford, S. Krucker, R.P. Lin, R.A. Schwartz, G.H. Share, D.M. Smith, Astrophys. J. Lett. **644**, L93 (2006)
G.J. Hurford, E.J. Schmahl, R.A. Schwartz, A.J. Conway, M.J. Aschwanden, A. Csillaghy, B.R. Dennis, C. Johns-Krull, S. Krucker, R.P. Lin, J. McTiernan, T.R. Metcalf, J. Sato, D.M. Smith, Sol. Phys. **210**, 61 (2002)
G.J. Hurford, R.A. Schwartz, S. Krucker, R.P. Lin, D.M. Smith, N. Vilmer, Astrophys. J. Lett. **595**, L77 (2003)
J.W. Knight, P.A. Sturrock, Astrophys. J. **218**, 306 (1977)
E. Kontar et al., Space Sci. Rev. (2011). doi:10.1007/s11214-011-9804-x
R.A. Kopp, G.W. Pneuman, Sol. Phys. **50**, 85 (1976)
T.N. Larosa, A.G. Emslie, Sol. Phys. **120**, 343 (1989)
R. Lin, Space Sci. Rev. (2011). doi:10.1007/s11214-011-9801-0
R.P. Lin, B.R. Dennis, G.J. Hurford, D.M. Smith, A. Zehnder, P.R. Harvey, D.W. Curtis, D. Pankow, P. Turin, M. Bester, A. Csillaghy, M. Lewis, N. Madden, H.F. van Beek, M. Appleby, T. Raudorf, J. McTiernan, R. Ramaty, E. Schmahl, R. Schwartz, S. Krucker, R. Abiad, T. Quinn, P. Berg, M. Hashii, R. Sterling, R. Jackson, R. Pratt, R.D. Campbell, D. Malone, D. Landis, C.P. Barrington-Leigh, S. Slassi-Sennou, C. Cork, D. Clark, D. Amato, L. Orwig, R. Boyle, I.S. Banks, K. Shirey, A.K. Tolbert, D. Zarro, F. Snow, K. Thomsen, R. Henneck, A. Mchedlishvili, P. Ming, M. Fivian, J. Jordan, R. Wanner, J. Crubb, J. Preble, M. Matranga, A. Benz, H. Hudson, R.C. Canfield, G.D. Holman, C. Crannell, T. Kosugi, A.G. Emslie, N. Vilmer, J.C. Brown, C. Johns-Krull, M. Aschwanden, T. Metcalf, A. Conway, Sol. Phys. **210**, 3 (2002)
R.P. Lin, H.S. Hudson, Sol. Phys. **50**, 153 (1976)
J.A. Miller, P.J. Cargill, A.G. Emslie, G.D. Holman, B.R. Dennis, T.N. LaRosa, R.M. Winglee, S.G. Benka, S. Tsuneta, J. Geophys. Res. **102**, 14631 (1997)
H.E. Petschek, NASA Spec. Publ. **50**, 425 (1964)
R. Ramaty, N. Mandzhavidze, B. Kozlovsky, R.J. Murphy, Astrophys. J. Lett. **455**, L193+ (1995)
R. Ramaty, C. Paizis, S.A. Colgate, G.A. Dulk, P. Hoyng, J.W. Knight, R.P. Lin, D.B. Melrose, F. Orrall, P.R. Shapiro, in *Skylab Solar Workshop II*, ed. by P.A. Sturrock (1980), pp. 117–185
D.M. Rust, E. Hildner, R.T. Hansen, M. Dryer, A.N. McClymont, S.M.P. McKenna-Lawlor, D.J. McLean, E.J. Schmahl, R.S. Steinolfson, E. Tandberg-Hanssen, in *Skylab Solar Workshop II*, ed. by P.A. Sturrock (1980), pp. 273–339
T. Sakao, T. Kosugi, S. Masuda, K. Yaji, M. Inda-Koide, K. Makishima, Adv. Space Res. **17**, 67 (1996)
J. Sato, T. Kosugi, K. Makishima, Publ. Astron. Soc. Jpn. **51**, 127 (1999)
G.H. Share, R.J. Murphy, Astrophys. J. **452**, 933 (1995)
A.Y. Shih, R.P. Lin, D.M. Smith, Astrophys. J. Lett. **698**, L152 (2009)
D.M. Smith, R.P. Lin, P. Turin, D.W. Curtis, J.H. Primbsch, R.D. Campbell, R. Abiad, P. Schroeder, C.P. Cork, E.L. Hull, D.A. Landis, N.W. Madden, D. Malone, R.H. Pehl, T. Raudorf, P. Sangsingkeow, R. Boyle, I.S. Banks, K. Shirey, R. Schwartz, Sol. Phys. **210**, 33 (2002)
D.M. Smith, G.H. Share, R.J. Murphy, R.A. Schwartz, A.Y. Shih, R.P. Lin, Astrophys. J. Lett. **595**, L81 (2003)

D.S. Spicer, R.N. Sudan, Astrophys. J. **280**, 448 (1984)
P.A. Sturrock, Nature **211**, 695 (1966)
P.A. Sturrock, in *Skylab Solar Workshop II*, ed. by P.A. Sturrock (1980a), pp. 411–449
P.A. Sturrock (ed.), *Solar Flares: A Monograph From Skylab Solar Workshop II* (1980b)
P.A. Sweet, Annu. Rev. Astron. Astrophys. **7**, 149 (1969)
E. Tandberg-Hanssen, A.G. Emslie, *The Physics of Solar Flares* (Cambridge University Press, Cambridge, 1988), p. 286
G.H.J. van den Oord, Astron. Astrophys. **234**, 496 (1990)
R.F. van Oss, G.H.J. van den Oord, Astron. Astrophys. **299**, 297 (1995)
N. Vilmer et al., Space Sci. Rev. (2011). doi:10.1007/s11214-010-9728-x
S. White et al., Space Sci. Rev. (2011). doi:10.1007/s11214-010-9708-1
V.V. Zharkova, J.C. Brown, D.V. Syniavskii, Astron. Astrophys. **304**, 284 (1995)
V.V. Zharkova et al., Space Sci. Rev. (2011). doi:10.1007/s11214-010-9803-y

Space Sci Rev (2011) 159:19–106
DOI 10.1007/s11214-010-9701-8

An Observational Overview of Solar Flares

L. Fletcher · B.R. Dennis · H.S. Hudson · S. Krucker · K. Phillips · A. Veronig · M. Battaglia · L. Bone · A. Caspi · Q. Chen · P. Gallagher · P.T. Grigis · H. Ji · W. Liu · R.O. Milligan · M. Temmer

Received: 19 August 2009 / Accepted: 8 October 2010 / Published online: 11 August 2011

Abstract We present an overview of solar flares and associated phenomena, drawing upon a wide range of observational data primarily from the *RHESSI* era. Following an introductory discussion and overview of the status of observational capabilities, the article is split into topical sections which deal with different areas of flare phenomena (footpoints and ribbons, coronal sources, relationship to coronal mass ejections) and their interconnections. We also discuss flare soft X-ray spectroscopy and the energetics of the process. The emphasis is to describe the observations from multiple points of view, while bearing in mind the models that link them to each other and to theory. The present theoretical and observational understanding of solar flares is far from complete, so we conclude with a brief discussion of models, and a list of missing but important observations.

Keywords Sun

L. Fletcher (✉) · M. Battaglia
School of Physics and Astronomy, Astronomy and Astrophysics Group, University of Glasgow, Glasgow G12 8QQ, UK
e-mail: lyndsay.fletcher@glasgow.ac.uk

B.R. Dennis · W. Liu
NASA Goddard Space Flight Center, Greenbelt, MD, USA

R.O. Milligan
School of Mathematics and Physics, Queen's University Belfast, Belfast BT7 1NN, UK

H.S. Hudson · S. Krucker · A. Caspi
Space Sciences Laboratory, U.C. Berkeley, Berkeley, CA 94720-7450, USA

K. Phillips · L. Bone
Mullard Space Science Laboratory, Holmbury St. Mary, Dorking RH5 CNT, UK

A. Veronig
Institute for Geophysics, Astrophysics and Meteorology, University of Graz, Graz 8010, Austria

M. Temmer
Institute of Physics, University of Graz, Universitaetsplatz 5, 8010 Graz, Austria

1 The Multi-wavelength Flare

Solar flares are the most powerful magnetic events in the solar system. In tens of minutes they can release in excess of 10^{32} erg of energy. They emit radiation across the entire electromagnetic spectrum, from radio to γ-rays, and are also intimately associated with the acceleration of particles into interplanetary space and with coronal mass ejections. The flare results from the rapid release of energy previously stored as inductive magnetic fields due to electrical currents flowing into the corona. The total flare energy is comparable with the amount of magnetic "free" energy (usually defined as the energy stored in the magnetic field that is over and above the energy of the potential magnetic field defined by the same boundaries) inferred to be available in the magnetic active regions (i.e., the coronal connections of a sunspot group) where most flares take place. The magnetic free energy is hard to evaluate from observations, depending as it does on the magnetic vector magnetic field, but in the few cases in which this has been possible (e.g., Metcalf et al. 1995, 2005; Schrijver et al. 2008; Jing et al. 2008), and it is found that the free energy is comparable with that of large flares. Furthermore, the energy budget is difficult to explain from other possible coronal or chromospheric energy sources (e.g., Hudson 2007). So we can conclude that conversion of stored magnetic energy is at the heart of the flare process.

The term "flare" is normally taken to refer specifically to the electromagnetic radiation of this whole magnetically-driven event, which embodies a significant fraction of the total energy liberated. The total energy released varies from event to event, with many more small events than large events. The distribution of the number of flares as a function of their peak energy, or their total energy, or their duration, is approximately a power law, the gradient of which is a critical factor in understanding the contribution of flare-like heating events to the overall energy budget of the solar corona (e.g., Crosby et al. 1993; Hannah et al. 2011). The primary way of classifying the "importance" of a flare is via its soft X-ray (SXR) flux at 1–8 Å, as measured by *GOES* (the *Geostationary Orbiting Environmental Satellites*). Flares are classified into X, M, C, B and A flares, with X corresponding to *GOES* flux in excess of 10^{-4} W m^{-2} at Earth, and successive classifications decreasing in decades. Table 1 puts the X-ray class of a flare in the context of the older classification based on Hα area (data from Thomas and Teske 1971).

Q. Chen
Stanford University Physics Department, Varian Building, 382 Via Pueblo Mall, Stanford, CA 94305, USA

P. Gallagher
School of Physics, Trinity College Dublin, Dublin 2, Ireland

P.T. Grigis
Harvard-Smithsonian Center for Astrophysics, 60 Garden Street, Cambridge, MA 02138, USA

H. Ji
Purple Mountain Observatory, 2 W. Beijing Rd, Nanjing 210008, China

H. Ji
Big Bear Solar Observatory, New Jersey Institute of Technology, 40386 North Shore Lane, Big Bear City, CA 92314, USA

W. Liu
Stanford-Lockheed Institute for Space Research, 466 Via Ortega, Cypress Hall, Stanford, CA 94305-4085, USA

Table 1 *GOES* and Hα classifications

GOES class	EM[a] (cm^{-3})	Hα class	Hα area (sq. degrees)
X10	10^{51}	4	24.7
X	10^{50}	3	12.4
M	10^{49}	2	5.1
C	10^{48}	1	2.0
B	10^{47}	S	<2.0
A	10^{46}	S	<2.0

[a]Soft X-ray emission measure (approx.)

The majority of the radiative flare energy emerges at visible and ultraviolet wavelengths (Woods et al. 2006). Where a bolometric measurement is possible, i.e., in the most energetic flares, we find that the radiated optical luminosity is comparable to the kinetic energy of the coronal mass ejection (see Sect. 6.2), and also to the energy of the accelerated electrons as inferred from the hard X-radiation (HXR) (Fletcher et al. 2007) under the assumptions of the collisional thick-target model (Brown 1971). A lot of emphasis has been placed on hard X-rays (HXRs) in understanding the flare energization process—see Holman et al. (2011)—despite the fact that energetically they represent only a small fraction of the total radiation. However, as HXRs result mainly from the well-understood bremsstrahlung radiation process, and the sources are optically thin, it is relatively straightforward to interpret them. The HXR emission is thus a powerful diagnostic for flare electrons, compared to longer wavelength, optically thick radiation, and the measurement of flare HXRs has been a primary goal of the *Reuven Ramaty High Energy Solar Spectroscopic Imager* (*RHESSI*, Lin et al. 2002). However, hard X-rays alone give only a restricted view of the overall configuration, development and energetics of a flare, and of its relationship to accompanying dynamical processes. The aim of this article is therefore to set the HXRs in the context of the multi-wavelength flare, to give an up-to-date observational picture, and to provide context for subsequent articles in this volume. This article focuses on the radiative flare, and discusses the coronal mass ejection and solar energetic particles only in association. It is not intended as a comprehensive historical review, although selected historical observations appear. The following recent reviews of observations of flare and related phenomena, including theory, are also recommended for further reading: Priest et al. (2002), Aschwanden (2002), Benz (2008), Krucker et al. (2008a), Vršnak and Cliver (2008), and Schrijver (2009).

As a preview of the introductory sections of this article, Fig. 1 (from Qiu et al. 2009) sketches out the temporal and spatial evolution of a well-observed major flare, SOL2001-10-19T01:05 (X1.6).

1.1 Flare development

Flares—of all sizes—tend strongly to occur in magnetic active regions and are associated with strong magnetic fields in the neighborhood of magnetic polarity inversion lines (dividing lines between regions of positive and negative vertical component of the photospheric magnetic field, sometimes also called neutral lines). A small fraction of flares do occur in so-called "spotless" regions (Dodson and Hedeman 1970; Martin 1980), and large-scale filament eruptions with flare-like properties can happen anywhere on the quiet Sun (e.g., Harvey and Sheeley 1986). It is not yet possible to predict the time or location of a solar flare. Several statistical studies have attempted to identify active-region magnetic properties that are correlated with active region flare productivity, or can even act as solar flare forecasters. The best indicators of flare productivity are those known in the "lore" of flare observers regarding the

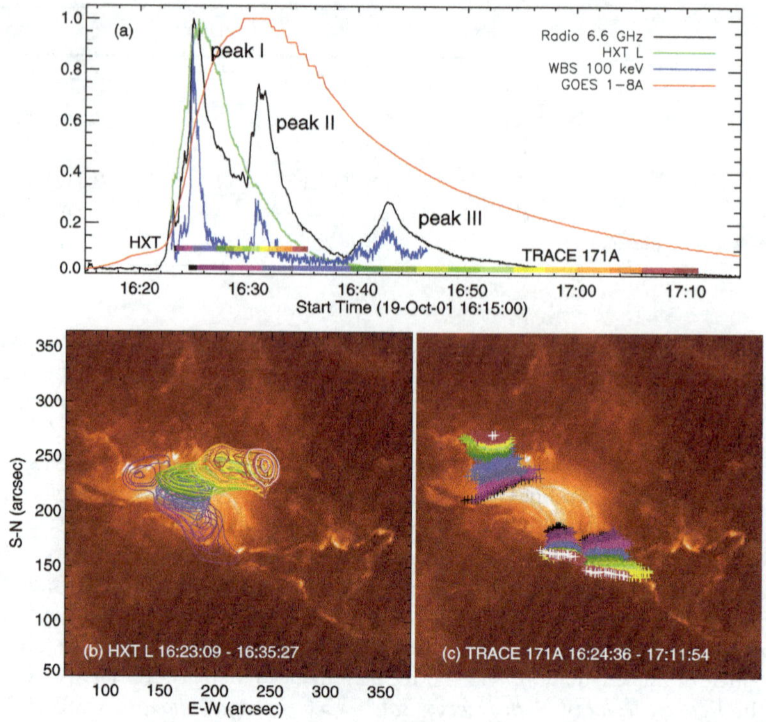

Fig. 1 Time evolution of the flare SOL2001-10-19T01:05 (X1.6) in multiple X-ray, EUV, and radio wavelengths (from Qiu et al. 2009). The *impulsive phase* is best characterized by the hard X-ray light curve (*blue*), in the 100 keV band of *Yohkoh* Wide Band Spectrometer (WBS) or the shortest-wavelength radio emissions (*black*, at 6.6 GHz), from the Owens Valley Solar Array. The *GOES* lightcurve (*red*) shows the *gradual phase* well. The *two panels at the bottom* show *TRACE* 171 Å images defining the flare arcade and its footpoints; *left*, with hard X-ray contours (14–23 keV); *right*, with EUV footpoint locations color-coded by time in the *upper panel*. The *black line* shows the microwave emission at 6.6 GHz, the electron gyrofrequency for a 2,400 G magnetic field. Reproduced by permission of the AAS

size and complexity of the sunspots in the active region. The largest flares occur in "delta-spot" regions, which have two umbrae within a single penumbra (Zirin and Liggett 1987; McIntosh 1990). They follow the rate of evolution of the active region (Schrijver et al. 2005) and require the presence of strong magnetic gradients (Hagyard et al. 1984). An example of a flare-productive active region, AR10486, is shown in Fig. 2. The magnetic properties of flaring regions are encapsulated in various studies of the photospheric field which find higher flare probabilities in regions with high total photospheric magnetic flux, excess magnetic energy, long polarity inversion lines with a strong, highly variable distribution of shear along their length (Cui et al. 2006; Leka and Barnes 2007) and a high fractal dimension of the photospheric field (McAteer et al. 2005). Incorporating information on the evolution of observed photospheric parameters, the rate of change of the strongest photospheric magnetic twists in the region, is the best predictor of a flare (Leka and Barnes 2003). However, in general the photospheric properties alone appear to offer poor predictive capabilities, and it appears likely that parameters of the coronal magnetic configuration offer a better prospect. For example, a high degree of complexity—expressed in parameters such as the number of topologically distinct regions (e.g., Cui et al. 2006; Barnes and Leka 2006), and the "effective connected magnetic field" (Georgoulis and Rust

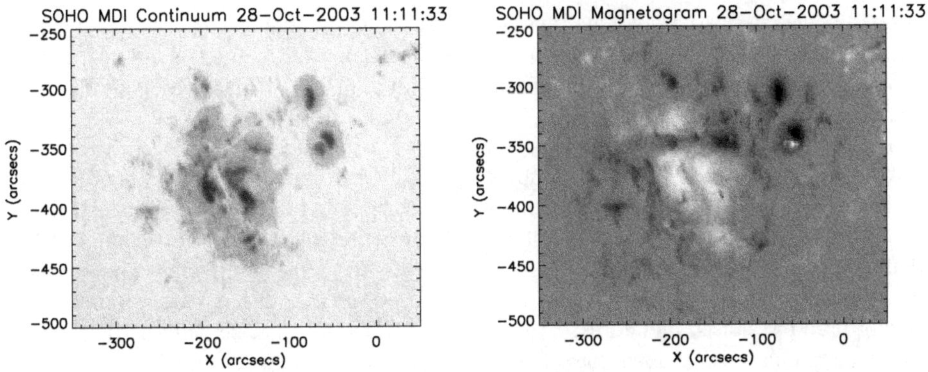

Fig. 2 Continuum (*left*) and magnetogram (*right*) images of NOAA Active Region 10486, which was the seat of many major flares in late October/early November 2003, including SOL2003-10-29T20:49 (X10.0), SOL2003-11-03T09:55 (X3.9), and SOL2003-11-04T19:53 (X17.4), the most powerful *GOES* flare on record. The δ configuration is visible in the positive polarity region of this group, with three major umbrae (those for X in the range $[-140, -200]$ and Y in the range $[-370, -410]$) within one penumbra. Also visible in the magnetogram are oppositely-colored inclusions within the main spot areas, e.g., at $[-140, -390]$. These occur as a result of disruption of the magnetically-sensitive spectral line because of the strong disturbance to the atmosphere at the time of the flare. Such features are transient and unrelated to actual magnetic changes. However, non-reversing magnetic changes at the time of flares are observed; these are discussed in Sect. 3.6

2007)—shows promise, as do estimators of the global non-potentiality of the magnetic field such as the flux-normalized field twist (e.g., Falconer et al. 2002). Aside from those methods based on magnetic field information, a Bayesian approach using past history of flare occurrence in an active region has also been proposed (Wheatland 2004).

One final indicator of approaching flare activity that is worth mentioning is filament activation. A filament is a narrow concentration of dense, cool material ($n_e \approx 10^{12}$ cm^{-3}, $T \approx 10^4$ K) which overlies and runs parallel to a magnetic neutral line. The filament material is supported in the atmosphere by a strongly-sheared magnetic field in the corona, primarily oriented along the neutral line, and it exhibits substantial plasma flows parallel to this. Quiescent filaments are visible in absorption in Hα against the disk (and emission at the limb, against the background sky), and also in absorption in the *Transition Region and Coronal Explorer* (*TRACE*) extreme ultraviolet (EUV) lines. Prior to a flare, such features are often observed to start rising slowly. Just prior to their eruption, brightenings in part or all of the filament may also be observed along its length, in wavelengths from Hα to EUV. Filaments and their activation/eruption are interesting in the context of flare development as their involvement in the earliest phase of the eruption, their low altitude in the corona, and their strong concentration along the region of sheared field all point towards the flare initiation, and possibly a large part of the pre-flare energy storage taking place within 1–2 $\times$ 10^4 km above the photosphere.

1.2 Terminology

Over the decades, the task of describing the appearance and time evolution of a flare led to a flowering of classifications and related vocabulary, which evolved along with our understanding of the range of flare and flare-related phenomena. This included much to-ing and fro-ing over the direction of causality between flare and CME phenomena (see Cliver 1995, for a historical overview). The time evolution of a flare based on observations is now normally characterized by two main phases, these being the impulsive phase and the gradual phase. The rough division of the time-history of emission into an abrupt and a more slowly-varying phase has been recognized through the history of flare observations in different wavelengths, including Hα (Ellison 1946), microwaves (Wild et al. 1963), and HXRs (Kane 1969; Kane and Donnelly 1971). "Spotless" flares, mentioned above, tend to have less rapid rises and less significant impulsive phases than flares in normal active regions. Some authors have also classified events as either "gradual flares" or "impulsive flares" based on their time characteristics within a single wavelength range, for example SXRs (Pallavicini et al. 1977) or HXRs (Ohki et al. 1983; Bai 1986) emissions. The notion of the "impulsive flare," i.e., one with no gradual phase has more-or-less faded from parlance among flare physicists, though it persists in discussions of solar energetic particle events. However, exclusively "gradual flares" do occur; such events have essentially no impulsive component and show only a more slowly-varying X-ray component at relatively low peak temperatures—so low, indeed, as to be undetectable by *GOES* (see Hudson et al. 1995, for example). These are often associated with filament eruptions and have properties similar to "long decay events" (LDEs) that tend to be accompanied by CMEs (Kahler 1977; Sheeley et al. 1983), but which typically do have impulsive phases.

1.3 Time Evolution

The time evolution of a flare is characterized by different timescales visible at different wavelengths. In the SXR range, following a rise phase lasting a few minutes, evolution is slow. The return of this thermal emission to its pre-event levels is a smooth decay that can last for hours. At the opposite extreme, the lower-frequency end of the radio spectrum exhibits bursts known as type III bursts, with a high brightness temperature, rapid drift rate (frequency decreasing with time) and duration as short as tens of milliseconds (Aschwanden et al. 1995a). These are most normally seen in the decimetric regime and far below, but can also be found in the microwaves. Observations of X-rays at tens of keV and above, with the BATSE detector (on the *Compton Gamma Ray Observatory, CGRO*) exhibit pulses with widths of hundreds of milliseconds to a second or two (see Aschwanden et al. 1995b, for details). This fast structure is sometimes apparently superposed on a slower, large amplitude variation (though the slow component could be the superposition of many rapid variations). It has not proved possible to detect these short pulses with *RHESSI*. However with both *Yohkoh*/HXT and *RHESSI*, the phenomenon of HXR quasi-periodic pulsations has become very clear (Foullon et al. 2005; Ofman and Sui 2006; Li and Gan 2008; Inglis and Nakariakov 2009). These are modulations in the X-ray intensity, in the few to tens of keV range with a modulation depth of up to 90%, and observed periods of tens of seconds to a few minutes. The periods observed demand an explanation in terms of MHD timescales; for example, a sausage mode could alter the magnetic field in the loop leading to variations in trapped particle precipitation rates (Inglis and Nakariakov 2009), or perturbations to the accelerator itself (Ofman and Sui 2006). Optical flare variations are typically abrupt, tracking the HXR lightcurves well at energies of tens of keV in the rise phase, but sometimes having a slower decay. UV and EUV also show mixed impulsive and slow variations—major HXR spikes are reflected in the lightcurves, but superposed on a slowly-varying background.

In the HXR impulsive phase, a well-known temporal pattern is the so-called *soft-hard-soft* spectral pattern—namely that the spectral index of the non-thermal part of the photon spectrum becomes harder as the non-thermal flux increases. This is a pattern that holds universally across individual bursts in the impulsive phase of a flare, both in the footpoint regions and the looptop regions (e.g., Battaglia and Benz 2006; see also Fig. 10), and on timescales down to a few seconds. In many flares it also represents the evolution of the entire event at lower time resolution, as the HXR flux increases and then decreases again. However there is a separate class of events—the *soft-hard-harder* events—in which the spectrum continues to become harder throughout the duration of the event (Silva et al. 2000). This is a property of gradual HXR events (e.g., Cliver et al. 1986; Saldanha et al. 2008), and in at least some events may reflect the presence of long-lived high energy coronal sources (Krucker et al. 2008b). It has been interpreted both as a consequence of energy-dependent particle losses from a coronal trap (with low-energy particles being scattered out of the trap before high-energy particles), and as evidence of continued injection of high energy particles after the flare impulsive phase.

The final time-behavior mentioned here is the *Neupert effect*—the phenomenon that in many flares the time integral of the non-thermal emissions tracks the time profile of the thermal emissions. The underlying reason for this is that the mostly thermal coronal plasma has a much longer energy loss timescale than the chromosphere, and "integrates" the energy deposited there, presumably from "evaporation" (Sect. 3.7) caused by more impulsive lower-atmosphere energy input. This was first discovered as a delay between the peaks of SXR and microwave emissions (Neupert 1968), but has since become more commonly associated with the HXR/SXR time profiles; essentially the HXRs show the (non-thermal) energy release as it happens, and the SXRs show the part of it that winds up in the corona as high-temperature thermal plasma. The fidelity with which the Neupert effect holds depends on the wavelength ranges chosen to test it, with higher temperature thermal emissions showing a better relationship (McTiernan et al. 1999). The Neupert timing relationship was found to hold in 80% of 66 large events studied using the *GOES* 1–8 Å channel and the *SMM*/HXRBS 26–41 keV channel (Dennis and Zarro 1993). A statistical study of the timing of the SXR peak compared to the HXR impulsive phase in more than 1000 events, using *GOES* 1–8 Å and *CGRO*/BATSE HXR counts at 25–50 and 50–100 keV, found that 50% of events were consistent with Neupert-like timing behavior, 25% were inconsistent, with SXR emission peaking substantially after the end of the HXR emission, and the remaining 25% were unclear (Veronig et al. 2002a). In those flares consistent with Neupert timing there was also a strong linear correlation between the HXR fluence (time-integrated counts) and SXR peak flux, as expected. It has been speculated (Li and Gan 2006) that the flares which do not show the expected Neupert timing behavior may occur if the SXR peak only when plasma "evaporated" from the chromosphere (see Sect. 3.7) reaches the top of the flaring loops (Reale and Peres 1995), which introduces a longer time lag for larger flare loops. But the level of disagreement has prompted further investigations of individual flares, in which the beam power inferred from HXR and the power required to explain the SXR flux and spectrum were tested for consistency (Veronig et al. 2005). Under straightforward model assumptions these two powers were found not to correlate well in time, which may suggest that there is energy input other than by non-thermal electrons during some phases of the flare (see counter-examples from Ning 2008). Of course, plasma flows may also lead to heating via compression that might not match the non-thermal signatures so well (e.g., Caspi and Lin 2010). Violating one particular model assumption, that of a constant value low-energy cutoff, could (if the cutoff varies through the flare in the right way) lead to better agreement (Sui et al. 2005; Veronig et al. 2005). At the present time, the discrepancies between observed and theoretical Neupert effects are probably within the observational limits, but hints of different physics from the standard ideas that hot thermal emission is coronal and non-thermal emission mostly chromospheric should of course not be ignored.

Cross-correlation analysis has indicated a delay of the SXR time derivative by $\sim$10 s relative to the HXR flux (Liu et al. 2006). This delay can be interpreted as the hydrodynamic timescale for the redistribution of energy deposited by non-thermal electrons, consistent with the results of hydrodynamic simulations (e.g., Mariska et al. 1989; Li et al. 1993; Liu 2008; Zharkova et al. 2011).

1.4 Preflare Evolution

Preceding the impulsive phase, there may be initial signs of activity termed the *pre-flare phase*. This term covers both pre-flare activity, which refers to the very earliest stages of the flare before the impulsive phase radiation is detectable, and the *flare precursor* events, which are small-scale brightenings in UV to SXR wavelengths happening some tens of minutes before the flare. Spatially unresolved lightcurve data such as that obtained with *GOES* may be misleading in this respect, since apparent flare precursors can originate from distant active regions, whereas actual flare precursors may fail to be visible against the integrated intensity from the disk. However with spatially resolved observations it is apparent that flare precursors often do occur in the neighborhood of, but not usually at exactly the same location as, the site in which the majority of the flare radiation will subsequently occur (Fárník et al. 1996, 2003; Fárník and Savy 1998; Warren and Warshall 2001). Some authors have explicitly linked pre-flare brightenings with the destabilization of the magnetic structure that will lead to a CME (e.g., Harrison et al. 1985; Sterling and Moore 2005) or a filament eruption (Fárník et al. 2003; Chifor et al. 2007; Liu et al. 2009d) although coronagraphic studies of the former kind are plagued by the lack of knowledge of the CME launch time, since substantial parts of the event occur unseen below the coronagraph occulting disk. The onset of pre-flare activity, in the form of weak SXR emission, precedes the onset of impulsive HXR emission by around 3 minutes in the vast majority of flares, regardless of their total energy or duration (Veronig et al. 2002b). Spectral line broadening has been observed in the pre-flare phase (Harra et al. 2001; Harra et al. 2009) starting minutes to hours before the impulsive phase, consistent with non-thermal effects such as plasma turbulence.

1.5 The Impulsive Phase

The primary energy release occurs during the impulsive phase. This phase of flare activity lasts from tens of seconds to tens of minutes and is characterized by HXRs, γ-rays, non-thermal (synchrotron) microwaves and white-light continuum emission, indicating strong acceleration of both electrons and ions. These radiations are accompanied also by strong enhancements in chromospheric line and continuum emission, ultraviolet and extreme ultraviolet radiation, and bulk plasma upflows in the EUV and SXRs at speeds on the order of 100 $\mathrm{km\,s^{-1}}$ coupled with downflows in cooler lines such as Hα (Zarro et al. 1988; Milligan et al. 2006a). The impulsive-phase radiation is concentrated at the chromospheric endpoints of the magnetic field involved in the flare; indeed prior to the availability of the EUV and SXR imaging that has led to a shift of focus to the coronal aspects of flares, the strong lower atmosphere signatures led to the term "chromospheric eruption" or "chromospheric flare" being used in early studies (Giovanelli 1948) even as the relationship of the radiation increase to flows and to processes higher in the atmosphere and at the Earth were becoming apparent. Even now, it is clear that while the magnetic drama takes place in the corona, the dominant radiative energy of a flare, from both non-thermal and thermal particles, is from the lower atmosphere (e.g., Chupp et al. 1973; Hudson 1972; Kane et al. 1979b; Lemaire et al. 2004; McIntosh and Donnelly 1972; Woodgate et al. 1983). So somehow the flare energy must be transported into the chromosphere, there to be dissipated by radiation and flows. There is unfortunately only very

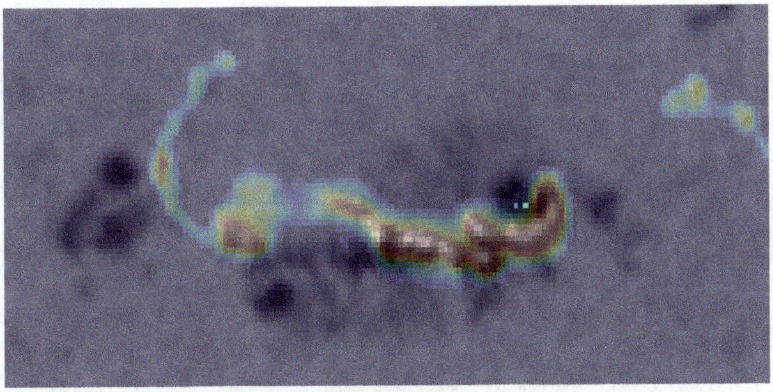

Fig. 3 SOL2002-10-04T05:38 (M4.0), with *TRACE* UV (1700 Å) ribbons in color overlaid on *TRACE* continuum in *black and white*. The flare *white-light* sources are visible as small white patches, within the UV ribbons, and having a considerably smaller area. The field of view is 135″ by 65″, pixel size 0.5″. Here the main neutral line lies between the concave-up UV feature under the sunspot *on the right*, and the more linear feature *below it*

sparse knowledge of the optical and UV properties of the impulsive phase, and this is a substantial shortcoming and missed opportunity. Rich diagnostic information, particularly in spectroscopic data, informs our understanding of the quiet-Sun chromosphere, and to have such data available also for flaring regions would substantially improve our knowledge of the results of flare energy deposition in these layers, and even give insights into the flare particle acceleration problem.

The magnetic reconfiguration that allows the rapid release of stored magnetic energy in a flare is generally agreed to occur somewhere above the chromosphere in the (low-β) corona. The main theoretical argument for a coronal energy release is that the corona provides adequate volume for storing the energy required for a flare. Observationally, coronal manifestations such as large SXR and Hα flare loops, HXR looptop sources, and coronal mass ejections have almost universally been interpreted in a framework involving large-scale coronal magnetic reconnection. Non-linear force-free reconstructions of the magnetic field find that the energy is concentrated low in the corona in a newly-emerged active region, and can be sufficient for flaring activity (Régnier and Priest 2007; Schrijver et al. 2008) whereas in an older decaying active region it is stored higher in the corona. It has proved marginally possible to detect differences between the free magnetic energy before and after a flare or CME event (Metcalf et al. 2002; Bleybel et al. 2002; Schrijver et al. 2008). More quantitative reconstructions of the coronal magnetic field which can track the actual redistribution of coronal magnetic energy during a flare are just beginning to appear (Jing et al. 2008; Thalmann and Wiegelmann 2008).

The impulsive-phase flare signatures in the lower solar atmosphere are termed "footpoints" (originally HXR) or "ribbons" (originally Hα) and are now detectable in a wide range of wavelengths. They are interpreted as the chromospheric ends of the coronal magnetic field structures involved with the flare energization at a given instant. The impulsive-phase reconnection of the coronal magnetic field is not visible in coronal signatures with current instruments, though its effects certainly may be. These effects include EUV and SXR flare loops and cusp-shaped structures that appear in the gradual phase (see Sect. 1.6), and impulsive phase supra-arcade downflows (Asai et al. 2004b; Khan et al. 2007), interpreted as the "dipolarization" of newly-reconnected field. These correspond extremely well in time with impulsive phase HXR bursts. Flare footpoints occur on

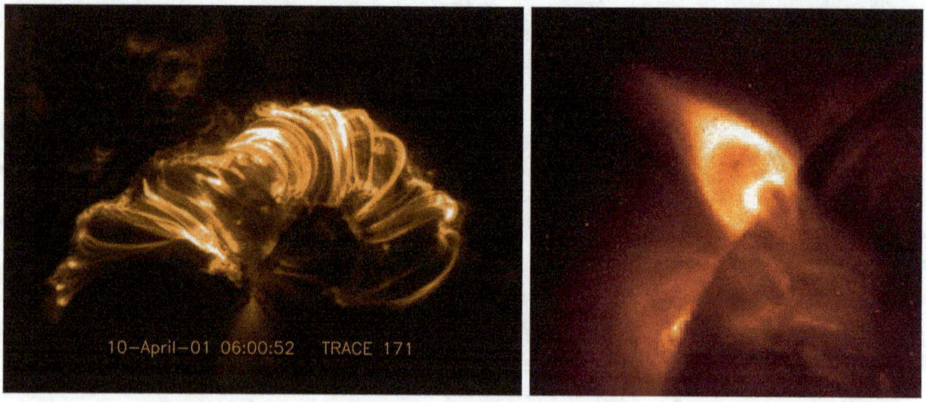

Fig. 4 *Left*: a beautiful flare arcade (SOL2001-04-10T05:26, X2.3) seen here in the 171 Å channel of *TRACE*, revealing plasma at ~1 MK emitting in lines of Fe IX/X. *Right*: a large post-flare cusp structure observed several hours after the impulsive peak of SOL1999-03-18T08:31 (M3.3) by *Yohkoh*/SXT and reported by Yokoyama et al. (2001). The temperature in this structure is 3–4 MK

either side of the magnetic polarity inversion line, as illustrated in Fig. 3. Early in the impulsive phase they tend to be concentrated around this line, and move with respect to it as the flare evolves. In the later phase of a flare, when strong Hα and UV ribbons are visible, the ribbons tend to move outward from the polarity inversion line, but in the impulsive phase, both ribbon and HXR footpoint motion is more complex, sometimes showing parallel or approaching motions. This is discussed in more detail in Sect. 3.4.

The importance of flare ribbons and footpoints in marking regions of changing magnetic connectivity is now well established (e.g., Mandrini et al. 1991). The large-scale reconnection model in two dimensions (e.g., Kopp and Pneuman 1976) was originally developed to explain the spreading Hα ribbons and the Hα arcade that appears in the gradual phase of a flare. The outer edges of the ribbons show the chromospheric projection of the interface (magnetic separatrix surface) between the post-reconnection ("post-flare") arcade fields, and the field that is yet to be reconnected. The importance to flare energy release of the three-dimensional connectivity of the solar magnetic field around sunspots was known early on (Giovanelli 1948; Sweet 1969; Syrovatskii 1981) and early observational associations were made between topological structures and observable chromospheric Hα features by, e.g., Gorbachev and Somov (1989), Mandrini et al. (1991), and Démoulin et al. (1992). Interpreting the chromospheric features within this framework, it becomes possible to establish some of the global properties of the magnetic field and its evolution, such as the different magnetic domains, the amount and rate of magnetic flux transfer during flare events, and also—under the assumption of two-dimensional translational symmetry—the convective electric field ($\mathbf{v} \times \mathbf{B}$) of the magnetic reconnection.

1.6 The Gradual Phase

During the gradual phase, identified by its slowly decaying SXR and microwave signatures, the effects of the flare on the corona become apparent. Loops and loop arcades emitting in SXRs and EUV form and appear to grow (Fig. 4), filled (it is usually assumed) by chromospheric plasma forced to expand into the corona as the chromosphere is rapidly heated by particle energy deposition or thermal conduction. This expansion is known as *chromospheric evaporation* (see Sect. 3.7). The gas pressure of these flaring coronal loops can increase from ~0.1 dyne cm^{-2}, typical of the quiet corona, to $>10^3$ dyne cm^{-2}, as shown by semi-empirical models based on radiative transfer theory in chromospheric lines (e.g., Machado et al. 1980). The pressure increase is mainly due to the growth of density in the loops, but the new coronal material is also at flare temperatures (10–20 MK), as opposed to pre-flare coronal temperatures (1–3 MK). The loop arcades show a gradient in temperature, with the outermost loops (i.e., those corresponding to the outer edges of the ribbons) being the hottest (Forbes and Acton 1996). The hottest outer loops sometimes exhibit a "cusp" (Fig. 4), consistent with the shape of the field that would be expected below a coronal current sheet; this cusp is most pronounced later in the gradual phase of the flare (Tsuneta et al. 1992; Yokoyama et al. 2001; Hara et al. 2008). This is circumstantial evidence for coronal magnetic reconnection during this phase. Later on, as the corona cools, the arcade becomes visible in lower temperature emissions, including EUV and Hα (Schmieder et al. 1995). Cooling occurs by both conduction and radiation, depending on the flare loop length and plasma parameters (Culhane et al. 1970; Cargill et al. 1995; Aschwanden and Alexander 2001). Observational studies tend to find that early on the hottest plasma cools for a few minutes by conduction (e.g., Culhane et al. 1994) which may then be followed by dominant radiative cooling (e.g., Aschwanden and Alexander 2001; Vršnak et al. 2006). Models have been formulated which also take into account thermal energy redistribution throughout the loop due to conduction, and the gentle chromospheric evaporation that results (Cargill et al. 1995). As the loop plasma cools it begins to drain under gravity, and Hα downflows ("coronal rain") become visible along the legs of the arcade. The plasma upflows and downflows have been detected spectroscopically in a number of events (e.g., Zarro and Lemen 1988; Czaykowska et al. 1999; Brosius 2003).

Loop arcades observed at a particular wavelength appear to grow upwards and outwards in time. This can be seen in Hα (Švestka et al. 1987), microwaves (Li and Gan 2005), EUV, SXRs and even HXRs (e.g., Gallagher et al. 2002). In the common interpretation, the point of magnetic reconnection moves slowly upwards in the solar corona as the gradual phase proceeds. Successive shells of reconnected loops fill with hot plasma expanding from the chromospheric footpoints, a pattern often described as the "CSHKP model" after some of its major contributors (H. Carmichael, P. Sturrock, T. Hirayama, R. Kopp, and J. Pneuman). As new loops or their footpoints brighten in a particular wavelength, their angle with respect to the magnetic polarity inversion line may increase, as reported since *Skylab* in many events (e.g., Sakurai et al. 1992; Sakao 1994; Masuda et al. 2001; Aschwanden and Alexander 2001; Su et al. 2006; Schmahl et al. 2006; Ji et al. 2008; Liu et al. 2009b). This could be interpreted as due to reconnection starting in highly-sheared field anchored close to the polarity inversion line, and progressing to less-sheared field further from it. Figure 5 illustrates this process.

The gradual phase may last several hours, depending on the magnitude of the flare. In many events, the cooling timescales point towards an additional energy source during the gradual phase, which could come from ongoing slow reconnection and its associated heating (MacCombie and Rust 1979; Forbes et al. 1989), either above the flare arcade or conceivably also between individual tangled strands of the arcade as they shrink down. The arcade appears to grow in scale as newer and higher-altitude loops appear, while individual loops physically shrink with time (Švestka et al. 1982; Forbes and Acton 1996). Multi-strand loop modeling (Hori et al. 1998; Warren et al. 2002; Reeves and Warren 2002) may explain cooling time profiles, and also the fact that the observed velocity characteristics of evaporative upflows tend to be smaller than those predicted by radiation hydrodynamic simulations.

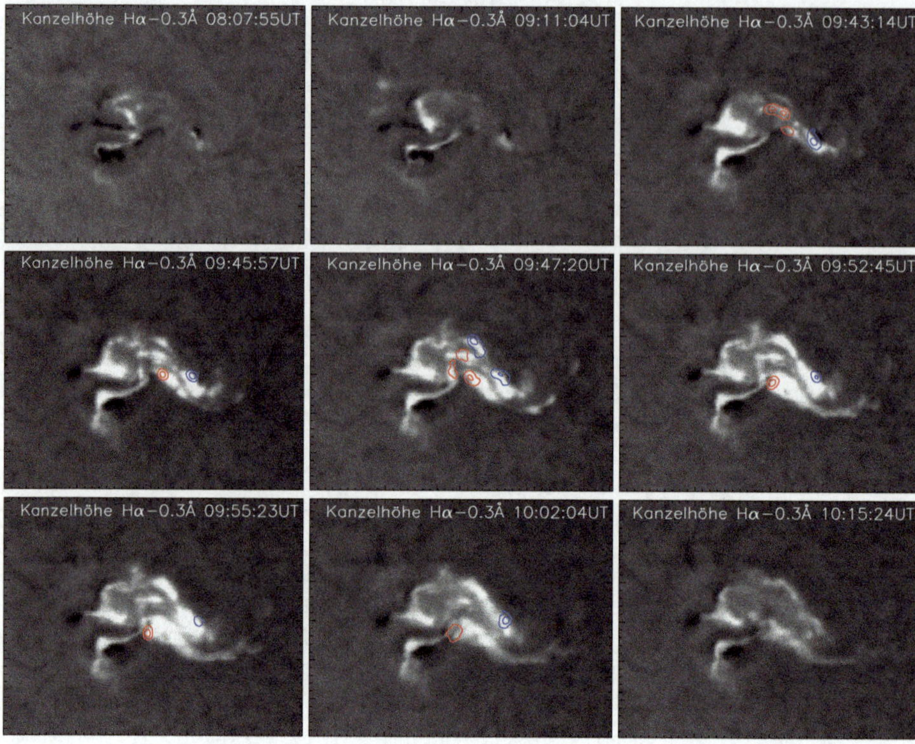

Fig. 5 Sequence of Hα–0.3 Å filtergrams taken at Kanzelhöhe Observatory during SOL2005-01-17T09:52 (X3.8) in a $400'' \times 300''$ field of view. The contours are co-temporal *RHESSI* 30–100 keV images, and the color indicates the polarity of the underlying photospheric magnetic field (*red*: positive, *blue*: negative). Before 09:35 UT, no *RHESSI* data were available. During certain flare periods, up to five individual HXR footpoint sources were observed simultaneously, located on different Hα flare ribbons, but two main footpoints prevailed during the overall flare impulsive phase. Note the general counterclockwise rotation of the line connecting the opposite-polarity footpoints, an observation often interpreted as a reduction in shear. Adapted from Temmer et al. (2007). Reproduced by permission of the AAS

The gradual phase and the standard magnetic-reconnection model nicely link the arcade formation with the occurrence of a filament eruption and a CME; this would create the (approximately) oppositely-directed field lines that then reconnect (Sect. 5). Further evidence for ongoing magnetic coronal reconnection comes from the supra-arcade downflows (McKenzie and Hudson 1999; Sheeley et al. 2004) observed mainly during the gradual phase and apparently moving at a fraction of the Alfvén speed. As we have seen already, such downflows also occur in the impulsive phase of a flare. The magnetic geometry in which the reconnection and shrinkage occur may be considerably more complex in the impulsive phase.

1.7 The Magnetic Field

How well do we understand the structure of the magnetic field, which underlies all aspects of solar activity? There are routine measurements of the Zeeman splitting in the solar photosphere (see Sect. 2.3), especially for the line-of-sight component of the field. We can identify the magnetic environment of flare occurrence morphologically—a large, rapidly-forming sunspot group containing a "delta spot" certainly has a higher probability of flaring (e.g., Zirin and Liggett 1987). At the photospheric level, the polarity inversion line ("PIL" or "neutral line") plays an important role morphologically. This line by definition must separate the footpoints of coronal loops, such as flare loops. By definition, the field lies tangent to the plane of the photosphere at the neutral line, and where it is concave up (the "bald patch" configuration) it does not lead to an overlying arcade, but still may play an important role in flare development (Titov et al. 1993).

Unfortunately the field in the photosphere, which can be observed comprehensively, maps into the corona in an elusive manner (e.g., Schrijver et al. 2008). Many competing mathematical and MHD-based approaches have been applied to the problem of extrapolating the photospheric knowledge into the corona (Schrijver et al. 2006), but several substantial problems prevent these techniques from being quantitatively persuasive. From the perspective of flare research, one major objective (for example) would be to characterize the "non-potential energy" resulting from current systems linking the corona and the solar interior. This energy exceeds the minimum level derivable from a "potential field" representation, i.e., one without embedded currents. This should decrease when a flare happens, but the results of all analyses to date have been uncertain on this point (e.g., Metcalf et al. 2002).

The direct measurement of the field via line-splitting observations in an active-region corona has met extreme difficulties, and accordingly the results at the present time have great uncertainties. In principle one needs to measure a full vector function $\mathbf{B}(\mathbf{r})$ at each point $\mathbf{r}$ in the coronal volume. There is progress on this formidable task (e.g., Tomczyk et al. 2008). Radio techniques also have great promise to determine at least $|\mathbf{B}|$, since the Larmor frequency and its harmonics have clearly identifiable and precise signatures (e.g., Bastian et al. 1998). In general the most exact knowledge of at least the plane-of-the-sky projection of $\mathbf{B}(\mathbf{r})$ comes from high-resolution imaging in coronal emissions (visible, UV, EUV, X-ray), since they often have fine striations that can be interpreted as due to the alignment of the field. A potential-field interpretation of the intensity in the corona above a large sunspot can be approximated by that of a simple current loop lying in the photosphere: $B(z)/B_0 = (z^2+1)^{-3/2}$, where z is the height above the umbra in units of the umbral radius. For a radius of 10^4 km and an umbral field of 3500 G (e.g., Cox 2000) this formula gives axial fields of 300–1200 G at coronal altitudes of 1–2 $\times$ 10^4 km. For reference, the current in the loop has a magnitude of 5.5×10^{12} A.

When a solar flare happens, the observed photospheric field changes in a stepwise manner (Wang 1992; Sudol and Harvey 2005). This would be expected from the extraction of energy by its dipolarization following large-scale reconnection, or generally just to reduce $\int (B^2/8\pi)\, dV$ (Hudson 2000); see Sect. 3.6.

2 Status of Observational Techniques

Flare emissions have been detected all the way from about 10^{-10} eV (30 kHz, a typical plasma frequency in the solar wind near the Earth) out to some hundreds of MeV (the pion decay spectrum). This whole vast spectrum, in principle, could be broadly observed with sensitive remote-sensing instruments, and in stereo. We could thus aspire to the observation of a many-dimensional data cube: x, y, z, λ, polarization, and time for starters, and even the directional components (limb-darkening function) of the emitted radiation. Clearly the observations to date have only begun to scratch the surface of this potential wealth of material. In each parameter there is an implied sampling capability—temporal cadence, signal-to-noise ratio, contrast, and scattered light considerations provide much further diversity. The limits on the available capabilities of course relate to somewhat intangible matters such as community preferences, technical feasibility, and cost. We note that solar observers tend (for several reasons) to want to study the most powerful events, whereas non-solar astronomers may strive for sensitivity instead. Unfortunately, this leaves a huge range of solar parameter space unobserved by the best instrumentation. In the following we briefly review the capabilities of the available multiwavelength observations existing at present, broken down by wavelength.

2.1 Radio

Radio observations broadly reveal both non-thermal and thermal emissions via several mechanisms and from a broad range of phenomena occurring anywhere above the photosphere (e.g., Bastian et al. 1998). For flares the most important of these are cm-wave gyrosynchrotron radiation from high-energy electrons (the extension, to a few MeV, of the energetically important electrons responsible for the impulsive-phase HXR emissions), and meter-wave plasma-frequency emissions that show many dynamical processes in the corona.

The past two decades have seen the improvement and deployment of several instruments. The major dedicated cm/mm-wave observatory is the radioheliograph at Nobeyama, Japan (e.g., Koshiishi et al. 1994) which images at fixed frequencies of 17 and 34 GHz with a beam size of $10''$ at 17 GHz. This is accompanied by a non-imaging radio polarimeter operating at nine fixed frequencies between 1 and 80 GHz. The Owens Valley Solar Array (OVSA) (Hurford et al. 1984; Gary and Hurford 1999) has a smaller number of dishes but superior frequency coverage, operating at 86 fixed frequencies from 1 to 18 GHz. This allows the use of "frequency synthesis" (i.e., the interpretation of the measured (u, v)-plane[1] in terms of a model source spectrum) to augment the coverage in the (u, v)-plane to a certain extent.

At decimeter wavelengths the most productive facility is the Nançay radioheliograph (e.g., Avignon et al. 1989), typically operating at five fixed frequencies in the 150–450 MHz range. Joint *RHESSI*–radio observations have also been made with a number of dynamic spectrographs spanning a wide range of frequencies, such as Ondrejov at 0.1–4.2 GHz (Jiřička et al. 1993), Huairou at 2.6–3.8 GHz (Fu et al. 2004), the Phoenix-2 instrument at 0.1–4 GHz (Messmer et al. 1999), Tremsdorf at 40–800 MHz (Mann et al. 1992), the Greenbank Solar Radio Burst Spectrometer at 10–110 MHz (White et al. 2005) and Hiraiso at 25–2500 MHz (Kondo et al. 1994). There are plans for an ambitious new solar radio interferometer, the Frequency Agile Solar Radio Telescope (FASR; Bastian 2003) to cover the range 0.1 to 24 GHz, while low-frequency solar imaging and spectroscopy is planned with the Low Frequency Array for Radio Astronomy (LOFAR, Bastian 2004); see Claßen et al. (2003) and Benz et al. (2005). At submillimeter wavelengths a dedicated site now exists at El Leoncito, Argentina, where the Solar Submillimeter Telescope observes at frequencies of a few hundred GHz (Kaufmann 2003).

2.2 Infrared

The solar infrared spectrum extends from visible wavelengths out to the 10 μm band accessible to ground-based observations, the mid-infrared accessible only from space, and the submm-THz range again accessible from the ground. This huge region contains thermal free-free continuum and coronal line emission, plus other possible contributors at the

[1] The radio astronomers' "(u, v)-plane" is the map of observed spatial frequencies; Fourier transformation of measurements at the observed points in the (u, v)-plane yields the source image projected on the plane of the sky.

longest wavelengths. The IR has some advantages, for example, in terms of improved seeing and in magnetic sensitivity (e.g., Debi Prasad 1998); on the other hand the diffraction limit becomes severe for non-interferometric imagers.

No dedicated solar infrared observatories at wavelengths much longward of 1 μm exist at present, although developments are underway in imaging at 10 μm (e.g., Hudson 1975), with flares already having been detected (Melo et al. 2006). Specific observations on the 1 m McMath Solar Telescope make use of the mid-infrared to study photospheric magnetic fields (Moran et al. 2007), and at the 0.76 m Dunn Solar Telescope to probe flare activity at the "opacity minimum" spectral region around 1.5 μm (Xu et al. 2004). Direct observations of the coronal magnetic field are also becoming possible (Lin et al. 2004; Tomczyk et al. 2008). Full disk observations are made daily in the Ca II 8542 Å and He I 10830 Å lines by the SOLIS synoptic telescope (and by earlier Kitt Peak telescopes), and flare spectroscopic observations at these wavelengths are provided by the Multichannel Infrared Solar Spectrograph (MISS; Li et al. 2002) at Purple Mountain Solar Observatory (PMO), with a spectral resolution on the order of 1 Å.

2.3 Optical

2.3.1 General

High-resolution observations have become much more commonplace from ground-based observatories. This results both from the availability of large apertures but especially from active control of the optics to ameliorate the effects of atmospheric seeing; in addition there are advanced post-processing techniques such as speckle reconstruction (von der Luehe 1993). Specific telescopes include the Dunn Solar Telescope, the Swedish Solar Observatory on La Palma, the German Vacuum Tower Telescope on Tenerife, and the Dutch Open Telescope. The best of these observations have substantially better spatial resolution than contemporaneous telescopes in space, but without the consistently perfect seeing conditions.

Several flexible optical/IR instruments operate in user mode at ground-based observatories (e.g., Cauzzi et al. 2008). With such instruments one can generate multidimensional data cubes incorporating spatial, temporal, spectral, and polarimetric measurements. Being ground-based observatories, they may operate with state-of-the-art adaptive optics and, having large apertures, obtain high resolution while running at high cadence. They also have minimal constraints on data bandwidth and can flexibly study a variety of spectral features. Future instrumentation is emphasizing larger apertures, higher frame rates (Jess et al. 2007), and the implementation of high-order adaptive optics (e.g., Rimmele et al. 2003; Denker et al. 2007). Major new facilities under study include the Advanced Technology Solar Telescope (ATST) and the European Solar Telescope (EST).

Optical observations from space have included the white-light telescope on *Yohkoh*, which despite being short-lived, observed several solar flares in the Fraunhofer G band (Matthews et al. 2003). In the *RHESSI* era, the Michelson Doppler Imager instrument on the *Solar and Heliospheric Observatory (SOHO)* (MDI; Scherrer et al. 1995) provides white-light full disk images, and the *TRACE* satellite also has a "white-light" channel with a broad response (~1000–8000 Å) which was successfully used in the study of solar flares at high cadence (Hudson et al. 2006). However, the first full large-aperture optical solar observatory in space, on board *Hinode* (Shimizu 2002), was not launched until 2006. This telescope provides diffraction-limited observations with a 50 cm primary and includes both narrow-band filter imaging and spectrographic imaging, the latter permitting vector magnetograms (e.g., Lites et al. 2001).

2.3.2 Hα

The 3 → 2 Balmer transition line of neutral hydrogen, centered at 6563 Å, is historically important in flare observations and continues to be observed and studied for its rich diagnostic capability. Many tens of observatories worldwide are engaged in monitoring the Sun in Hα for flares, making both full- and partial-disk observations. The US National Oceanic and Atmospheric Administration (NOAA) maintains a database of Hα flares stretching back decades. Currently five stations provide Hα data for this: Kanzelhöhe (Austria), Learmonth (Australia), Holloman Air Force Base (USA), San Vito (Italy) and Kharkov (Ukraine). The major Hα instruments with facility for high cadence and high resolution observations, which have been used in connection with *RHESSI* observations, are at Kanzelhöhe (Otruba and Pötzi 2003), the 65 cm telescope at Big Bear Solar Observatory (removed from service in 2006, and being replaced with a 1.6 m instrument), and the Hα Fine Structure Telescope at Purple Mountain Observatory. The Global High-Resolution H-alpha Network consists of the stations at Big Bear, Kanzelhöhe, Catania, Meudon, Pic du Midi, Huairou, Yunnan, and Mauna Loa. Spectroheliographs making Hα observations (other than daily images) include those on the THEMIS telescope (Mein 2002) and at Hida Observatory.

2.3.3 Magnetographic

It is difficult to measure directly the magnetic field in the solar corona. Photospheric glare makes observations against the disk almost inconceivable except perhaps at radio wavelengths; above the limb one has the high temperatures of the corona and line-of-sight confusion to contend with. Photospheric magnetic fields are typically observed in a weak Fraunhofer line, e.g., Ni I 6758 Å for MDI.

Nevertheless, precise knowledge of the vector magnetic field in the corona is fundamental to our understanding of flares and CMEs, and much effort has been expended in extrapolating near-photospheric measurements into the corona. To understand the coronal field, measurements would ideally be done above the solar chromosphere, where the magnetic field approaches the force-free condition and can therefore, if it remains static, be used as a reliable lower boundary for mathematical extrapolations into the corona. Special observational facilities exist for the observation of the solar magnetic field in the photosphere via spectropolarimetric techniques, with perhaps the earliest serious work done at the Crimean Observatory (Severny 1964). These facilities have different objectives, ranging from characterization of "sun-as-a-star" average line-of-sight field components (e.g., Wilcox and Hundhausen 1983), to full vector measurements at the highest possible resolution and at different heights in the solar atmosphere. The latter data provide the indispensable boundary condition for the increasingly ambitious efforts to extrapolate the structure of the coronal field in three dimensions (e.g., Schrijver et al. 2006). Ground-based magnetographs include the early MSFC Vector Magnetograph (Hagyard et al. 1982), the Mees Imaging Vector Magnetograph (Mickey et al. 1996), Huairou (Ai and Hu 1986), the Hida Solar Magnetic Activity Research Telescope (SMART; Ueno et al. 2004), the BBSO Digital Vector Magnetograph (Spirock et al. 2001), the GONG Network, including magnetographs located in 6 sites around the world (Harvey et al. 1996), and SOLIS (Keller et al. 2003). The *Hinode* spectropolarimeter, which uses the Fe I lines at 6302.1 and 6302.5 Å, is the only high-resolution vector magnetograph in space, while the HMI instrument on the *Solar Dynamics Observatory* now provides high-cadence, full-disk photospheric vector magnetograms at a $1''$ spatial resolution (Graham et al. 2003).

The coronal magnetic field, in spite of frequent small-scale activity, normally is considered to be force-free. Thus a simple extrapolation of any photospheric magnetic field, the (non-force-free) source of the stresses that drive these currents, incorporates systematic errors. Accordingly several observatories have begun systematic observations of the chromospheric magnetic field in Na I 5896 Å (Metcalf et al. 1995) or another chromospheric line such as Hβ. The upper chromosphere should be more force-free and these observations should enable more accurate force-free extrapolations.

2.4 UV–EUV–X-ray Wavelengths

From the UV into the SXR domain (up to a few keV)[2] one can use specialized astronomical techniques involving normal-incidence or, at the shortest wavelengths, grazing-incidence mirrors. Such observations need to be made in space because of the large opacity of the Earth's atmosphere. Indeed, solar observations at these previously inaccessible wavelengths were a part of the early history of space astronomy, leading to *Skylab*. The following paragraphs summarize some of the observations concurrent with *RHESSI*.

2.4.1 Ultraviolet

For spectroscopy, SUMER[3] (Wilhelm et al. 1995) has provided excellent data but with limited solar flare coverage for technical reasons. *TRACE* has provided excellent imaging observations but with poor spectral resolution (Handy et al. 1999). The *TRACE* UV bands includes a narrow ($\sim$30 Å) band centered on strong lines of C IV, plus broader UV channels covering Lyman-α and the UV continuum to around 2500 Å, and a "white light" channel covering the full range of spectral response of the Lumogen-coated charge-coupled device (CCD).[4] Nevertheless this rich spectral domain, the most important for studying the chromosphere and transition region, still remains underexploited; for example, there are no comprehensive observations yet even of Lyman-α, the strongest line.

2.4.2 EUV

TRACE has provided excellent EUV observations at the standard normal-incidence narrow spectral bands centered at 171, 195, and 284 Å, corresponding to transitions of Fe IX/X, Fe XII (and Fe XXIV in flares), and Fe XV. Such observations, pioneered by the EIT[5] instrument (Delaboudinière et al. 1995), now continue stereoscopically (Kaiser et al. 2007) on the *STEREO* spacecraft (Howard et al. 2008). In addition to these broad-band imaging instruments, stigmatic slit spectrographs have flown on *SOHO* (310–380 Å and 520–630 Å; Harrison et al. 1995), and now on *Hinode* (170–210 Å and 250–290 Å; Culhane et al. 2007). These provide high resolution spectroscopy enabling detailed diagnostics for plasma density, temperature, velocity and abundance. Currently the AIA[6] instrument (Rochus et al. 2004) on *SDO* provides comprehensive coverage of this type, with 10 s cadence, many passbands, and whole-Sun imaging with negligible gaps in time coverage.

[2]Roughly, the UV and vacuum UV extend through the Lyman continuum; at shorter wavelengths to about 44 Å, the carbon K-edge, it would be EUV, and still shorter, soft X-rays.

[3]The Solar UV Measurements of Emitted Radiation instrument on *SOHO*.

[4]Lumogen is a proprietary coating that extends the UV response of the CCD.

[5]Extreme-EUV Imaging Telescope on *SOHO*.

[6]Atmospheric Imaging Assembly.

2.4.3 Soft X-rays

The SXR flux from a solar flare has become the definitive flare observable via the *GOES* flare classifications, and photometric SXR observations have continued in an unbroken photometric sequence since the 1970s. High-resolution imaging began with grazing-incidence telescopes on rockets and then (with film readout) on *Skylab*. In 1991 *Yohkoh* was launched with its SXT imager (Tsuneta et al. 1991) which could image at spatial resolutions down to 5″, and now *Hinode* carries the improved XRT instrument (Golub et al. 2007). The temperature coverage of XRT is around 1–30 MK, and multiple filter combinations permit a temperature discrimination as good as $\log T_e = 0.2$.

In terms of energy, the end of the soft X-ray range and the beginning of the hard X-ray range is a somewhat arbitrary matter, which varies from author to author in solar physics, and also from field to field in astrophysics. Roughly speaking, energies of 0.1–10 keV are counted as "soft" X-rays in solar physics, and energies above about 20 keV are "hard." Between these two, the description "soft" or "hard" can depend on whether the spectrum looks thermal or non-thermal.

2.4.4 Spectroscopy

High-resolution spectroscopy (from infrared to the γ-rays) provides the closest approach to learning the state of the flaring plasma via remote sensing. Generally these techniques provide information about the electron energy distribution function, but extracting information about the electron angular distribution is extremely difficult, even adopting the common assumption of azimuthal isotropy with respect to the magnetic field. The SXR emission lines have been particularly productive in flare research, with instruments flown on many spacecraft—most recently the *Yohkoh* Bent Crystal Spectrometer (Lang et al. 1992) and its follow-on RESIK[7] (Sylwester et al. 2005). The latter observes in narrow bands around the principal X-ray emission lines of Si, S, Cl, Ar, and K in the 3.1–6.6 Å range, with access to other features at higher grating orders.

At longer wavelengths the definitive observations have come from stigmatic slit spectrographs, which image in one dimension and thus multiplex the second spatial dimension in time. The definitive instruments in this category are not particularly optimized for flare observations but have produced many interesting results. These are currently SUMER (Wilhelm et al. 1995), which covers the VUV range 500–1610 Å, and EIS (Culhane et al. 2007) which covers the XUV range 170–290 Å (in two bands). This latter range conveniently overlaps the standard spectral selections for normal-incidence imagers such as those on *SOHO* and *TRACE* (see above).

The *Solar Dynamics Observatory* (SDO) now also provides EUV spectral irradiance measurements for the Sun as a star via its EVE[8] instrument (Woods et al. 2010).

2.4.5 Hard X-rays and γ-rays

At energies above a few keV, focusing optics have until recently been prohibitively difficult. Accordingly simple counter spectrometers provided the main source of information. HXR time series from such detectors provide important information about the flare impulsive

[7] *Rentgenovsky Spektrometr s Izognutymi Kristalami.*

[8] EUV Variability Experiment.

phase (and other epochs of particle acceleration) even without imaging, and many instruments have contributed to this. At present *RHESSI* and *Fermi*, through its GBM[9] instrument (Schwartz et al. 2010) provide this information routinely. New spectroscopic data extending into the γ-ray range are available from other instruments, including *INTEGRAL* (e.g., Gros et al. 2004).

Over the years 1991–2001 *Yohkoh* provided systematic HXR imaging for the first time, using non-focusing optics. The HXT instrument (Kosugi et al. 1991) thus anticipated some of *RHESSI*'s results in HXRs, providing four energy channels over ~15–92 keV. True imaging spectroscopy for hard X-rays and γ-rays began in 2002 with *RHESSI* (Lin et al. 2002). The single *RHESSI* instrument consists of nine high-purity germanium detectors, segmented into front and rear segments for sensitive and simultaneous hard X-ray and γ-ray measurements. These detectors have resolutions as good as about 1 keV (FWHM) and a range 3 keV to 17 MeV (Smith et al. 2002). Each of the detectors has a bigrid modulation collimator with different parameters (thickness and angular resolution) (Hurford et al. 2002), giving imaging (at a basic 4-s time resolution governed by spacecraft rotation) down to the arcsec range.

2.5 Particles and Fields

2.5.1 Solar Energetic Particles

The Sun emits charged particles, neutrons, and energetic neutral atoms as a part of flare/CME physical processes, and these (as well as the solar wind) provide independent samples of material or accelerated particles with which to compare the remote-sensing observations. Ideally one would be able to compare the separate populations of energetic electrons and ions detected *in situ* with those inferred from the *RHESSI* HXRs and γ-ray observations. The "multiwavelength" flare observations in a sense include the direct detection of high-energy particles emitted both promptly and with delays by solar flares and CMEs. These SEPs (Solar Energetic Particles) have a rich history of observation from space, and have also been termed "solar cosmic rays." Many spacecraft have observed SEPs in the past, and the current flotilla includes *ACE* (*Advanced Composition Explorer*), *SOHO*, the two *STEREO* spacecraft, *Ulysses*, *WIND*, and other missions both in deep space and in near-Earth space. The relativistic "ground level event" solar particles can also be detected by neutron monitors at the Earth (e.g., Cliver 2006), either as neutron secondaries or directly in a few cases.

2.5.2 Neutrons and ENAs

Solar neutrons, produced mainly in (p, p), (p, α), and (α, p) reactions (Hua and Lingenfelter 1987), can propagate into interplanetary space and even arrive at the Earth's surface if at high enough energies. The natural decay time of a free neutron is about 866 s.[10] An array of neutron monitors and neutron telescopes (e.g., Muraki et al. 2007) can detect both direct neutrons and secondary neutrons induced by charged particles creating cascade showers. The largest advances in our understanding of solar neutrons will come from instruments flown within a few tenths of an AU of the Sun, within the decay-time horizon for lower-energy neutrons that do not survive to one AU.

[9] Gamma-ray Burst Monitor.

[10] A free neutron decays, into a proton, and electron, and an antineutrino.

Neutron-decay products (energetic protons, with a characteristic spectrum) can also be detected in interplanetary space (Evenson et al. 1983, 1990), and in principle fast electrons as well (Dröge et al. 1996). A further channel for flare study has recently surfaced: one event (SOL2006-12-05T10:35, X9) has been found to have emitted detectable levels of energetic neutral hydrogen atoms (Mewaldt et al. 2009) in the 1.8–5 MeV range.

2.5.3 *Interplanetary Coronal Mass Ejections*

Finally, the low-energy ejecta of an ICME (Interplanetary Coronal Mass Ejection) may contain a large total energy and accelerate particles copiously (see Sect. 6.2), as well as provide information via their magnetic fields (Burlaga et al. 1982; Bothmer and Schwenn 1998).

2.6 Summary

From the foregoing one can recognize that our observational power has increased considerably during solar cycles 22 and 23, and in fact provides some information in almost all wavelength bands. We list some of the dedicated whole-Sun instruments in Table 2 for reference. In spite of all of the data available, it would be quite misleading to imagine that any of the material is definitive, and we return to discuss the missing things in Sect. 7.2. The lack of imaging spectroscopy in the optical (other than a few prominent lines) and the UV is especially painful, and the THz and γ-ray regimes are very under-exploited. In general one can recognize that non-solar observational facilities (e.g., NASA's Great Observatories such as *Chandra*) have highly desirable characteristics (resolution, sensitivity, background levels) that are not at present achieved by dedicated solar instruments. In addition, the exciting era of stereoscopic observations has now begun, but with only a limited instrument complement. Clearly stereoscopic observations, including an out-of-the ecliptic capability, would be of great value at all wavelengths; for example, at HXR or radio bands such a capability would make it possible to study the anisotropic radiation patterns of flare emissions, as well as the three-dimensional structures of their sources.

3 Footpoints and Ribbons

3.1 Overview

Traditionally we call the HXR brightenings, often observed as paired compact sources, *footpoints*, and the elongated structures originally found in Hα images *ribbons*. In fact both features can be observed at many wavelengths; the ribbons are prominent in the UV and EUV (observed, for example, by *TRACE*—Warren and Warshall 2001; Saba et al. 2006). Footpoint sources tend to be distinguished by their strong HXR and/or white light emission. Either way, these features occur pairwise separated by a magnetic polarity inversion line in the photosphere, which separates the two polarities of the vertical magnetic field. Regions connected by coronal magnetic fields are called *conjugate points*. Figures 1, 5, and 6 show the context.

The principal physical distinction between footpoint and ribbon morphologies may reflect the idea that the footpoints are the direct result of non-thermal processes (traditionally interpreted in terms of the thick-target model, in which electrons lose their energy collisionally in the chromosphere), whereas the ribbons also show excitation which could be due to thermal conduction from the overlying coronal arcade, or from weak particle precipitation.

Table 2 Full-disk solar data sources[a]

Observatory	Observation	Dates	URL
I. Radio			
Nançay	Decimetric imaging	1997–present	http://bass2000.bagn.obs-mip.fr/Tarbes/
Nobeyama	17 & 35 GHz imaging	1992–present	http://solar.nro.nao.ac.jp/norh/
SSRT (Irkutsk)	5.73 GHz imaging	2000–present	http://ssrt.iszf.irk.ru/
II. Infrared			
Kitt Peak	8542 Å, 10830 Å imaging	2007–present	http://solis.nso.edu/
III. Optical			
GONG	Quasi-continuum images	1995–present	http://gong.nso.edu/
GONG	LOS magnetic field	1995–present	http://gong.nso.edu/
Global Hα	Hα Imaging	2000–present	http://swrl.njit.edu/ghn_web/
SDO	HMI	2010–present	http://jsoc.stanford.edu/
SOHO/MDI	Quasi-continuum images	1995–2011	http://soi.stanford.edu/
SOHO/MDI	LOS magnetic field	1995–2011	http://soi.stanford.edu/
SOLIS	Vector magnetic field	2008–present	http://solis.nso.edu/
IV. UV/EUV			
SOHO/EIT	171, 195, 284, 304 Å imaging	1995–2011	http://umbra.nascom.nasa.gov/eit/
STEREO/EUVI	171, 195, 284, 304 Å imaging	2006–present	http://secchi.lmsal.com/EUVI/
SDO	AIA	2010–present	http://jsoc.stanford.edu/
V. Soft X-ray			
Yohkoh/SXT	~1 keV imaging	1991–2001	http://solar.physics.montana.edu/ylegacy/
GOES-12/SXI	~1 keV imaging	2001–2007	http://www.swpc.noaa.gov/sxi/index.html
GOES-13/SXI	~1 keV imaging	2006	http://www.swpc.noaa.gov/sxi/index.html
Hinode/XRT	~1 keV imaging	2006–present	http://sdc.uio.no/sdc/welcome
GOES-14/SXI	~1 keV imaging	2009–present	http://www.swpc.noaa.gov/sxi/index.html
VI. Hard X-ray/γ-ray			
Yohkoh/HXT	~13–91 keV imaging	1991–2001	http://solar.physics.montana.edu/ylegacy/
RHESSI	3 keV–17 MeV imaging spectroscopy	2002–present	http://hesperia.gsfc.nasa.gov/hessi/

[a]This is only a selection, emphasizing stable full-disk observations; please note that there are many more resources with limited fields of view and/or coverage

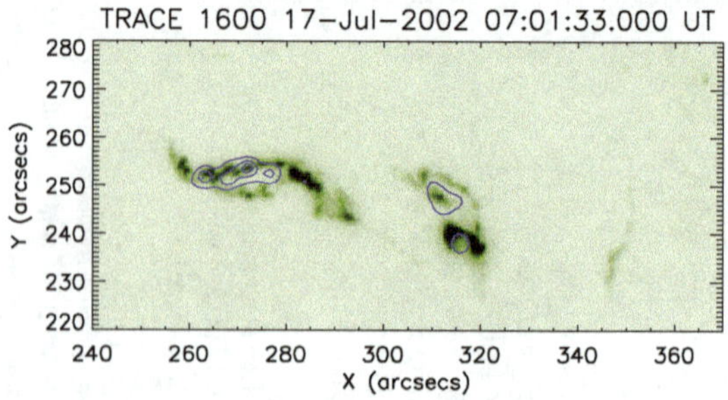

Fig. 6 *RHESSI* and *TRACE* observation of SOL2002-07-17T07:13 (M8.5), showing UV ribbons made up of multiple small sources, and simultaneous *RHESSI* 25–50 keV footpoints superposed (*blue contours*, at 25%, 50% and 75% of maximum). The *RHESSI* contours were made using the Pixon algorithm (Metcalf et al. 1996), including Grid 1 which is capable of showing spatial detail at 2.3″ scales. The *RHESSI* sources correspond well to the ribbons overall, but the ribbons are significantly more extended

Careful observation and modeling of individual flare events demonstrates that the location and evolution of the ribbons is clearly related to the magnetic topology in which the event occurs. In particular, flare ribbons reflect the projection of the separatrix or quasi-separatrix structure where flux transfer between magnetic domains is occurring (e.g., Mandrini et al. 1991; Démoulin et al. 1997) (see Sect. 3.5 for illustrations and further discussion). In Hα and the UV/EUV, the flare ribbons have a tendency to spread systematically outwards from the magnetic polarity inversion line, with their appearance becoming more ordered as the flare progresses.

On the other hand, the HXR footpoint sources frequently move along the Hα or UV/EUV flare ribbons, rather than away from the polarity inversion line, and are found at locations distinguished by their high magnetic field strengths (Temmer et al. 2007) or their high magnetic flux transfer rates (Liu et al. 2008; Fletcher 2009). HXR footpoints, generated by electron-ion bremsstrahlung of non-thermal electrons in the chromospheric plasma, are well-correlated in space and time with white-light sources, confirming them as locations of intense energy deposition (e.g., Hudson et al. 2006). Though it was for a long time believed that white-light emission is a "big flare" phenomenon, targeted observations at high cadence and high spectral resolution seem to suggest that even the smallest flares have a white-light counterpart. The generation mechanism for the white-light emission is not known, one issue being that if the white-light emission is a photospheric or lower chromospheric phenomenon this places perhaps unreasonably strong demands on beam excitation models. Observations in the infrared at 1.56 μm, corresponding to the wavelength at which the solar opacity is a minimum, also suggest strongly that the deep atmosphere is involved (Xu et al. 2004). In EUV, the counterparts to the HXR footpoints are often observed as particularly intense brightenings, and in *TRACE* observations this leads to characteristic diffraction patterns caused by the metal grids supporting the EUV filters, such that even if the CCD is saturated at the footpoint pixel some limits to its intensity can be calculated (Lin et al. 2001; Gburek et al. 2006; Mrozek et al. 2007).

One of the significant observational discoveries by *RHESSI* has been γ-ray footpoints in a small number of solar flares, imaged in the 2.223 MeV neutron capture line (Hurford et al. 2003, 2006). The 2.223 MeV line is produced by the radiative capture on hydrogen of a thermalized neutron. The neutron itself results from an earlier interaction by a primary energetic ion undergoing inelastic scattering resulting in a neutron-emitting isotope (Hua and Lingenfelter 1987). It takes a substantial column depth to slow the initially fast neutrons, so the 2.223 MeV radiation presumably forms in the dense lower atmosphere.

The *RHESSI* 2.223 MeV γ-ray observations should determine the centroid location of the final neutron captures to within about an arcsecond (Hurford et al. 2003). These observations are difficult to make, involving long integrations because of the low intensity of the radiation, and the high energy photons are modulated only by the thickest *RHESSI* grids, providing limited spatial information. The four flares that have been imaged in this way so far show single or double (in one case) γ-ray image components at roughly, but not exactly, the same location as the HXR footpoints. In three out of the four events, there is a statistically significant displacement between the 2.223 MeV sources and the 200–300 keV sources. This is dealt with in Vilmer et al. (2011).

Strong evidence for the impact of solar flares on the deep atmosphere—the lower chromosphere or photosphere—is also present in two other signatures. Flare-induced seismic waves ripple out across the photosphere as detected by helioseismic techniques (Kosovichev and Zharkova 1998), and strong changes occur in the photospheric magnetic field, most readily visible in the line-of-sight component. The changes are essentially simultaneous with the flare (e.g., Sudol and Harvey 2005) and with emission at the 1.56 μm "opacity minimum" height (Xu et al. 2004). There is no question that the effects of the flare reach deep into the solar atmosphere.

The behavior of the chromospheric footpoint plasma during the impulsive phase remains ill-understood theoretically. At the same time it is in one sense the most important flare problem, since as we will describe in Sect. 6.2 the white-light and UV continuum in this phase may dominate the radiated energy of a flare. The problem has been repeatedly treated in the 1-D "radiation hydrodynamics" approach (Kostiuk and Pikelner 1975) in which gas dynamics is coupled with some treatment of radiative transfer. Recent developments in this area (Allred et al. 2005) include an elementary formulation of beam heating, along with ionization calculations, line and continuum radiative transfer, and 1-D hydrodynamics. However the limitations of even this advanced treatment (one dimension; no self-consistent treatment of particle or wave energy transport) strongly suggest a need for additional development. The coupling of radiation and matter is decisive for the radiation signatures, of course, and also is necessary to understand the upwards mass flow and the seismic signatures. Self-consistent modeling should also include wave energy transport (Emslie and Sturrock 1982; Fletcher and Hudson 2008).

3.2 Timing

HXR sources can fluctuate rapidly, on time scales below 1 s (e.g., Kiplinger 1995). Such short timescale variations could reflect transport or acceleration processes. For reference, a 30 keV electron at $v = 0.3c$ would move $\sim 10^4$ km in of order 0.1 s. The timing measurements are especially interesting in comparison with radio measurements, which show myriad fine structures in their spectrograms. Figure 7 shows a representative example of a decimeter-wavelength dynamic spectrum (unflitered and frequency-filtered) compared with HXRs (Aschwanden et al. 1995a), in which detailed correlation between spikes is seen. It is more difficult to interpret the decimeter bursts because of their non-linear generation mechanisms; the timescale for the development of the kinetic instability related to the beam transport time (the "bump-on-tail" instability) is important. In some flares the decimeter wavelength/HXR correlations are very good, but not in all—possibly due to the complexities of the plasma radiation generation and absorption/Razin suppression (Bastian et al. 1998). At centimeter wavelengths there can be a much more precise relationship.

Sakao et al. (1996) have used HXR timing measurements to establish conjugacy (see Sect. 3.1) statistically, drawing the inference that a coronal particle acceleration region sends

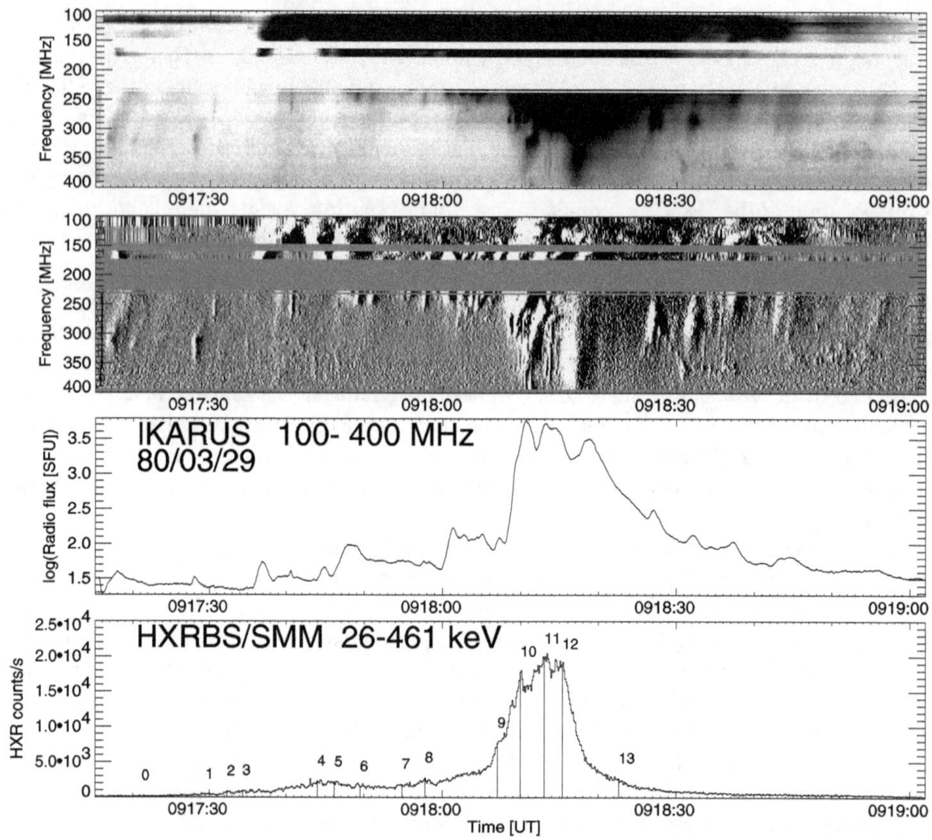

Fig. 7 Comparison of decimeter-wave and HXR time variations (Aschwanden et al. 1995a); data from the Ikarus spectrograph (Perrenoud 1982) and the *SMM*/HXRBS instrument (Orwig et al. 1980), respectively. The *upper two panels* show the radio spectrogram for SOL1980-03-29T09:18, with the *lower one* filtered to show temporal gradients; the *bottom two panels* show the radio flux integrated over 100–400 MHz and the HXR flux. Reproduced by permission of the AAS

bremsstrahlung-producing fast electrons simultaneously (to a few tenths of a second) into the two footpoint regions. At present there is no analysis of *RHESSI* data along these lines. Aschwanden et al. (1995b) have extended HXR timing analysis in an effort to establish time-of-flight delays between HXRs at different energies. Their decomposition of the time series of the Masuda flare (SOL1992-01-13T17:25, M2.0; Masuda et al. 1994), to name only one example, is consistent with signatures of loop-top injection superposed on a slowly varying envelope; note though that the interpretation of such a decomposition is not unique (Brown et al. 1998).

3.3 Morphology of Flare Footpoints and Ribbons

There are distinctions between the impulsive (early) and gradual (late) phase behavior of flare ribbons and footpoints. In the gradual phase of a flare, the emission tends to be in two roughly parallel ribbons, visible particularly well in Hα and UV (see, for example, Fig. 8). These tend to separate slowly from one another and from the magnetic polarity inversion line as the flare proceeds, although the HXR footpoints can have more complicated

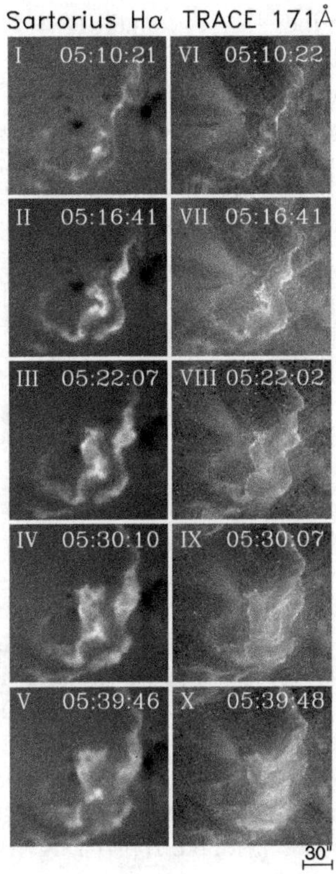

Fig. 8 The flare SOL2001-04-10T05:26 (X2.3) (Asai et al. 2003) in Hα, showing chromospheric emission, and *TRACE* 171 Å showing a mixture of transition-region ribbons and loops. This sequence shows clearly the spreading of the flare ribbons as the flare progresses. This event had an extended impulsive phase consisting of several HXR spikes; the cross diffraction patterns at the footpoints are characteristic of *TRACE* impulsive phase EUV observations. The thickening of the Hα ribbons is clear, and where these thickened inner portions map to the *TRACE* emission in panel VII, there is a "moss"-like appearance. Reproduced by permission of the AAS

apparent motions as discussed below. The roughly parallel expansion of the ribbons led to the standard CHKSP magnetic reconnection model. In the gradual phase of the flare, footpoint emission is thought to be powered primarily by electron beams, but may instead result from thermal conduction from the overlying hot loops (see the analysis of Czaykowska et al. 2001). Flare ribbons in Hα can be some tens of arcseconds wide, particularly in the gradual phase, and show internal structure (also visible in some cases in UV). The gradual phase ribbon morphology in EUV appears similar to a patterning associated with hot, high pressure coronal loops (in this case flare loops). In such structures conduction from the loops causes plasmas at transition-region or coronal temperature to appear at chromospheric heights. The resulting pattern of spatially-intermittent and dynamic hot plasma and spicules is known as "moss" (Berger et al. 1999). The leading edges of the ribbons, illuminated in EUV, are considerably narrower (e.g., Asai et al. 2003), and are also where the brightest Hα emission (Švestka et al. 1982) and the HXR footpoint sources are found in the impulsive phase. Hard X-ray bremsstrahlung is not normally a feature of the ribbon-associated gradual phase, although late-phase non-thermal emission with different morphologies is now known to be common (Krucker et al. 2008a).

In the flare impulsive phase the picture is more complicated. For example, more than two UV or Hα ribbons may be seen in magnetically quadrupolar flares (e.g., Tang 1985; Su et al. 2007), as well as a small number of bright footpoints visible in white-light, UV, HXR or Hα kernels. In general the HXR sources are confined to localized areas situated on the outer edges of the elongated flare ribbons observed in UV and Hα (Fig. 5) and are predominantly associated with bright Hα/UV kernels (Asai et al. 2002, 2004a; Krucker et al. 2005. Often there are two main HXR footpoints, but sometimes there are more (e.g., Fletcher and Hudson 2002; Lin et al. 2003; Temmer et al. 2007). The HXR emission is direct confirmation of the presence of non-thermal electrons in the lower chromosphere, which also heat and (further) ionize the chromospheric plasma as they stop collisionally.

A basic property of flare footpoints is that they are compact. In wavelengths from the infrared through the extreme ultraviolet, the dimension of the brightest flare footpoints matches the minimum scale resolvable by the instrumentation used. Only with the launch of *Hinode* do we have hints that we are approaching the basic scale for the optical flare kernels. Sub-arcsecond structure has been detected in optical flare sources which are seen to consist of a bright emission core with a FWHM of around 500 km (corresponding to an area of around 10^{16} cm^2), surrounded by a diffuse halo of emission having greater extent (Isobe et al. 2007). This diffuse halo is interpreted as radiation from the core backscattered by deeper atmospheric layers. Arcsecond-scale widths (diffraction limited) for flare ribbons were also observed in the infrared at 1.56 μm (Xu et al. 2004). Spatial resolution at UV/EUV wavelengths is not so high, but *TRACE* UV observations of flare kernels in the 1600 Å band, where not saturated, are consistent with them being on the scale of the telescope point spread function (see images in e.g., Warren and Warshall 2001; Alexander and Coyner 2006). The size scale of HXR images is harder to quantify, but in some flares *RHESSI* imaging reveals HXR footpoint sizes comparable with the resolution capability of the finest grids (2.3 arcseconds FWHM; e.g., Fletcher et al. 2007; Schmahl et al. 2007; Kontar et al. 2008; Dennis and Pernak 2009), corresponding to an area on the order of 10^{17} cm^2. The smaller optical sizes may imply that the optical excitation is taking place deeper down, in a converging magnetic field, or it may mean that the HXR footpoints simply have not been resolved yet at the best *RHESSI* resolution.

During the *Yohkoh* era, the *Yohkoh*/SXT observed impulsive SXR footpoint sources indicating heating of upper chromosphere or transition region plasmas to around 10 MK (McTiernan et al. 1993; Hudson et al. 1994; Mrozek and Tomczak 2004). The SXR emission is not consistent with an extrapolation of the bremsstrahlung power law spectrum to low energies, and originates in the chromosphere. Thus it appears to correspond to strong heating of the chromosphere (Hudson et al. 1994). In beam-driven radiative hydrodynamic simulations, to achieve such temperatures at the appropriate heights in the atmosphere requires beam energy fluxes on the order of $10^{11}\ \mathrm{erg\,cm^{-2}\,s^{-1}}$, or alternatively a somewhat lower electron flux for hundreds of seconds (Allred et al. 2005).

Although impulsive-phase HXRs and optical emissions are well correlated in space and in time, the relationship between HXR and UV/EUV emission is not so clear. There are often pre-flare brightenings in UV which remain bright during the flare, and the HXR footpoints occur only at locations which were not bright in UV before the flare (Warren and Warshall 2001). There are good temporal UV/HXR correlations during the flare, but as noted the UV ribbons are more extended than the HXR footpoints (Alexander and Coyner 2006). So there are evidently only a few locations in the flare magnetic field which are involved in the acceleration of a large number of non-thermal particles. Looking next at the UV/EUV and HXR sources which are at the same locations during the flare itself, there is a relatively good correlation between the *TRACE* 1600 Å channel flux and the *Yohkoh*/HXT 33–53 keV flux, and a weak anticorrelation between the *TRACE* UV (1600 Å) and *Yohkoh* HXR spectral index (Mrozek et al. 2007). An anticorrelation would be expected due to the plasma heating in the upper chromospheric levels produced by an electron beam in which low energies (with short collisional stopping depths) dominate. The *TRACE* 1600 Å channel is rather broad in wavelength, so it is not clear which lines or continua dominate may be dominating the UV flux. This may explain in part the relative weakness of this correlation. However, the

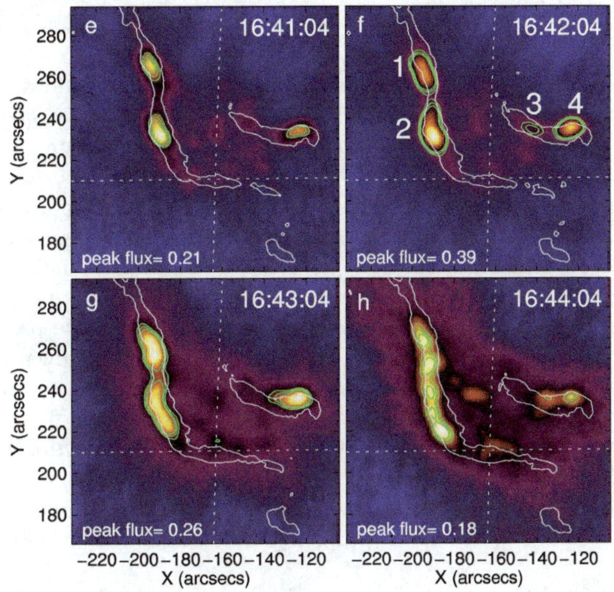

Fig. 9 Time sequence of *RHESSI* 25–50 keV HXR images together with *TRACE* 1600 Å image contours (*white lines*) showing the evolution of the HXR source emission from localized footpoints to ribbon-like emission. Flare SOL2005-05-13T16:57 (M8.0), adapted from Liu et al. (2007). Reproduced by permission of the AAS

weak correlation may well indicate the importance of heating mechanisms other than beam heating (e.g., thermal conduction) in producing EUV footpoints. The study has not yet been repeated using the better spectroscopic capabilities offered by *RHESSI*.

The dominant impulsive-phase HXR morphology is small number of isolated footpoints, with elongated HXR flare ribbons rarely observed. The first report of HXR ribbons was in *Yohkoh* HXT observations of SOL2000-07-14T10:24 (X5.7) (Masuda et al. 2001). In the well-observed SOL2005-05-13T16:57 (M8.0) flare, the *RHESSI* HXR sources evolve from footpoints concentrated in strong magnetic field areas along the *TRACE* UV ribbons in the HXR rising phase to ribbon-like HXR source structures closely matching the UV flare ribbon morphology after the HXR peak (see Fig. 9). The simpler structure of the HXR ribbons presumably reflects more closely the pattern of electron acceleration along the flare loop arcade during its formation (Liu et al. 2007, 2008). Figure 9 suggests that *RHESSI* has sufficient resolution to resolve discrete footpoint features that would be interpreted as multiple simultaneous footpoint brightenings. We should note that the appearance of elongated HXR ribbon-like features might also result from source motions during the relatively long HXR integration times (>4 s for *RHESSI* imaging).

3.4 Motions of Ribbons and Footpoints

In the standard model, as flare reconnection proceeds, different elements of the magnetic field move into the reconnection region. This leads to the expectation that the Hα/UV flare ribbons or footpoint sources move. In many flares the footpoints appear to move away from one another and from the magnetic polarity inversion line as the flare loop system grows (for recent studies see, e.g., Fletcher and Hudson 2002; Qiu et al. 2002; Krucker et al. 2003; Asai et al. 2004a; Veronig et al. 2006; Miklenic et al. 2007; Temmer et al. 2007). The flare ribbons are understood to somehow map to the energy release site in solar flares, and the movement of the ribbons and kernels across the photosphere, and their relationship to magnetic fields, are important means by which the magnetic reconnection process can be explored. For example, under the assumption of magnetic flux conservation, the progress of the flare sources across the magnetic flux of the photosphere can be used to measure the magnetic flux transfer rate.

However, much more complex HXR footpoint motions than the straightforward separating ribbon motions mentioned above are observed. In *Yohkoh*/HXT data only about 13% of HXR flares exhibit HXR footpoint motions corresponding strictly to separation with respect to the polarity inversion line (Bogachev et al. 2005), and are more likely to have a component of motion along the ribbon direction. They can also approach one another. Both converging and separating footpoints travel at tens of kilometers per second. A recent statistical study of footpoint motions in 27 *RHESSI* X- and M-class flares by Yang et al. (2009) has found that parallel/antiparallel motions are more likely during the SXR rise phase than during the flare peak, where separating motions become more prominent. Many individual examples confirm that the HXR footpoint motions do not always agree with the standard-model predictions of separating footpoints (e.g., Fletcher and Hudson 2002; Krucker et al. 2003; Grigis and Benz 2005; Ji et al. 2006). As in the 2-D flare model, the interpretation of the footpoint motion is still in terms of magnetic reconnection, but in a complex magnetic field. The complicated footpoint motions are likely to be linked to the projection(s) of the locus of reconnection. Particular examples of this can be found in sheared arcade models (Somov et al. 2002), and the "slip-running" reconnection model (Aulanier et al. 2006).

3.4.1 Footpoint Motion Parallel to the Magnetic Polarity Inversion Line

An example of footpoint motion parallel to the magnetic inversion line is shown in Fig. 10. The *SOHO*/EIT image of SOL2002-11-09T13:23 (M4.9) shows a postflare arcade, with the centroid position of the *RHESSI* HXR footpoints superimposed. As the event progresses, in several emission spikes, the HXR footpoint pairs move along the arcade. Source motion specific to the emission spikes shows up as deviations from the overall trend of the footpoint motions, and are decomposed into parallel and perpendicular components relative to this trend. Footpoint motion is directed parallel to the ribbons and is smooth at these scales, in contrast with the bursty evolution of the HXR flux. There is no evidence for a systematic trend in outward perpendicular displacement, or of discontinuities during the transitions from one spike to the next. The emission spikes originate at different sources along the arcade. The overall picture is that the HXR footpoint motion may be a consequence of a moving trigger, possibly caused by an asymmetric eruption of a filament (Tripathi et al. 2006), or a "domino effect" where energy release in part of the field triggers activity in its neighbors.

3.4.2 Converging Footpoints

In its first couple of minutes, SOL2002-11-09T13:23 (M4.9) also exhibits a convergence of the flare footpoints, explained in terms of the sequential activation of a flare arcade which varies in width along its length (Grigis and Benz 2005). A different kind of converging footpoint motion, corresponding to footpoints traveling anti-parallel to one another and along the ribbons, was seen earlier with *Yohkoh*/HXT (Bogachev et al. 2005), and now frequently in *RHESSI* (e.g., Fletcher and Hudson 2002). However, a new feature of such events became apparent in *RHESSI* observations showing the accompanying HXR coronal source. A number of events observed in HXRs, Hα and UV/EUV showed approaching footpoints in the early impulsive phase, accompanied by a projected downward motion of the coronal HXR source, and followed by separation of the footpoints and a projected rise in the coronal source (Fig. 11). These include SOL2002-03-14T01:44 (M5.7), SOL2003-10-29T20:49 (X10.0) (Zhou et al. 2008; Ji et al. 2008;

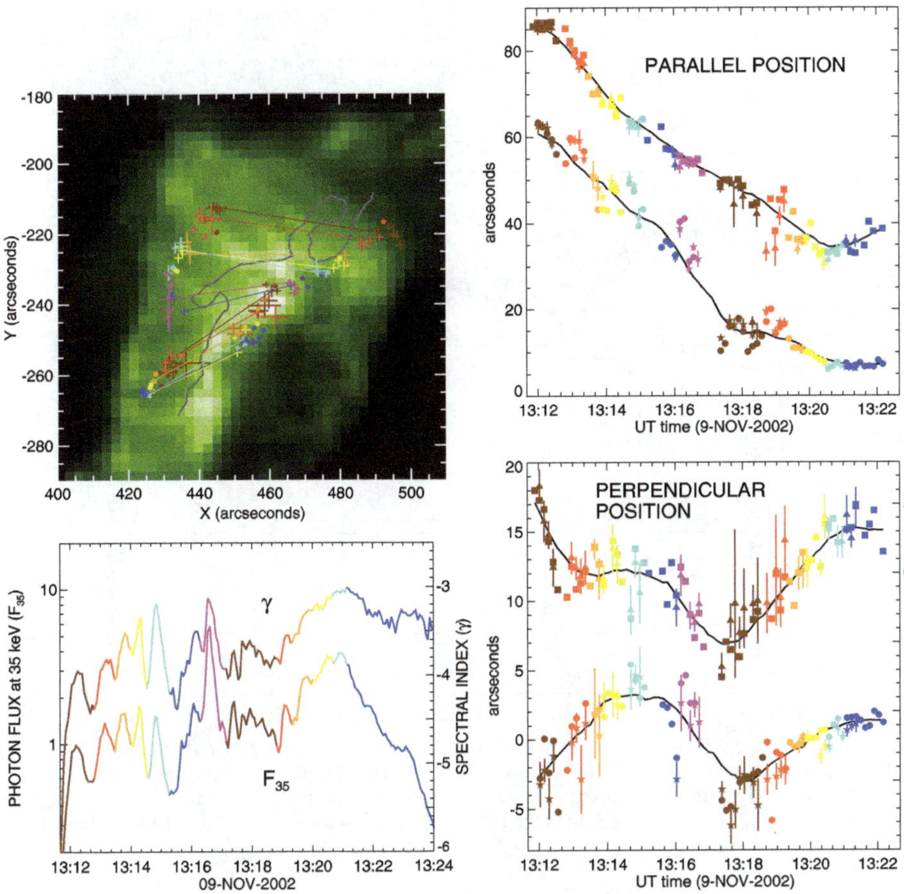

Fig. 10 *Top left*: *SOHO*/EIT 195 Å image of post-flare loops with the *RHESSI* HXR source positions superimposed for SOL2002-11-09T13:23 (M4.9). The positions of the 20–50 keV sources from the CLEAN images are represented by *crosses* with arm lengths equal to the errors, positions from the PIXON images are given by *circles*. Simultaneous footpoints are connected and color coded according to the time intervals defined in the *bottom part*. The neutral line is shown in *gray*. *Bottom left*: Time evolution of the flux and spectral index. *Right*: Time evolution of the source positions relative to the trend lines. *Triangles and stars with error bars* refer to values derived using CLEAN, *squares and circles* using PIXON, for the western and eastern footpoints, respectively. Reproduced by permission of the AAS

Liu et al. 2009b), SOL2002-09-09T17:52 (M2.1), and SOL2004-11-01T03:22 (M1.1) (Ji et al. 2004, 2006). The initial converging motion has also been noted in UV/EUV ribbons (Zhou et al. 2008). The footpoint convergence phase lasts for a few minutes, and both footpoints and coronal sources move with a projected speed of some tens of $\mathrm{km\,s^{-1}}$. Liu et al. (2009b) found an example of simultaneous height decrease and footpoint convergence. In the cases of antiparallel footpoint convergence, the empirical shear (determined from the angle relative to the neutral line made by the line joining footpoints) may decrease no matter whether the footpoints move inward or outward with respect to the polarity inversion line.

The downward motion of the coronal sources is consistent with the extraction of magnetic energy from the field (Hudson 2000). It has been interpreted in two related ways—as an initial shrinkage of the field in a 2-D "collapsing trap" immediately fol-

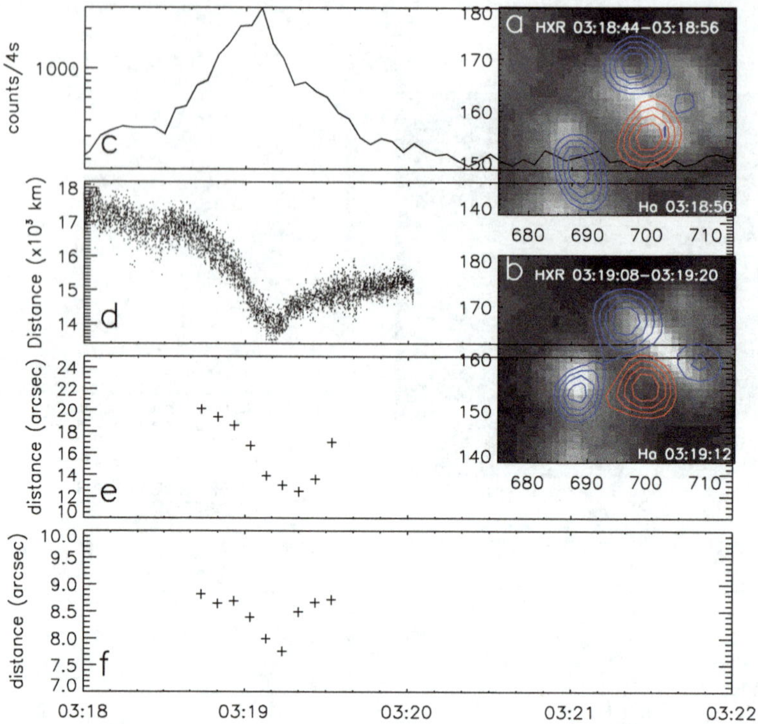

Fig. 11 Converging footpoints in SOL2004-11-01T03:22 (M1.1): (**a**)–(**b**) *RHESSI* HXR contours overlaid on H_α—0.5 Å filtergrams taken at PMO, with the *blue contours* at 25–60 keV and the *red ones* at 3–6 keV, (**c**) *RHESSI* 50–100 keV time profiles, (**d**) distance between the two conjugate Hα kernels or HXR conjugate footpoint sources, and (**e**) projected height of the *RHESSI* loop-top source. From Ji et al. (2006). Reproduced by permission of the AAS

lowing reconnection (e.g., Karlický and Kosugi 2004; Veronig et al. 2006) and as the consequence of the relaxation of shear in a 3-D arcade model (see also Ji et al. 2007; Somov et al. 2002). In the latter model, the less-sheared field reaches a lower altitude in the corona than more-sheared field, and its relaxation can in principle explain both the decrease in coronal source altitude and the converging footpoint motion. We discuss the relationship between ribbon motions and coronal dynamics extensively in Sect. 4.6 below.

3.5 Flare Footpoints and the Magnetic Environment

It has long been known that the magnetic structure of the solar corona is reflected in the distribution and evolution of flare footpoint sources. The simplest example of this is the straightforward mapping between the pre- and post-reconnected field in the 2-D standard model, and the spreading Hα ribbons. Even in more magnetically complex configurations, in principle each X-ray footpoint, or white-light/UV kernel, maps via the coronal field to a conjugate counterpart. In practice this has been difficult to demonstrate quantitatively. The flare impulsive phase is characterized by complex magnetic geometries, and recent years have also seen great advances in breaking down the active region coronal field into its topological elements—separatrix and quasi-separatrix layers, separator field lines and null-points. This remains an area of intense theoretical activity, as part of an overall effort to understand how magnetic reconnection takes place in three dimensions. The hope is that

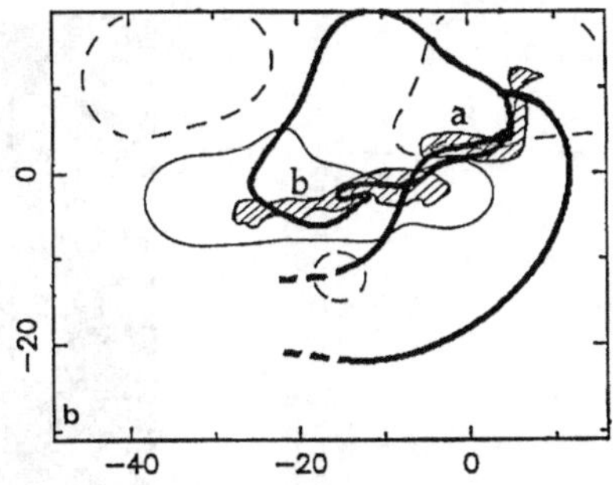

Fig. 12 Portions of the photospheric projection of separatrix structures from a linear force-free magnetic extrapolation are shown by Mandrini et al. (1995) to correspond to the locations of flare Hα ribbons. Contours show ± 400 G levels of magnetic field strength. Note, the correspondence between separatrix and ribbon positions is not found for a potential model. The Hα ribbons are found by Mandrini et al. (1995) to correspond to the regions of highest current density in their model

the observed evolution of footpoints can aid in this overall goal. During the flare impulsive phase, where the HXR and WL footpoints are typically observed, the magnetic geometry is not readily interpreted from EUV loop observations (as is the case in the gradual phase). The magnetic field is presumably stressed, and therefore the relatively straightforward potential field extrapolations may provide a misleading picture of the overall coronal structure (though as we see below they have been used to explore certain aspects of the flare geometry). Since the *Skylab* era (Zirin and Tanaka 1973; Pallavicini et al. 1975) it has been clear that later loops in the gradual phase of a flare look more potential-like, and make a larger angle to the neutral line, and this is often glibly taken as evidence for the reduction of shear expected to reduce the stored magnetic energy. Clearly it is not that simple; the observed changing pattern of shear is determined by the amount of shear in the pre-flare field as a function of distance from the neutral line before the flare, as well as its reduction as a function of time during the flare. The distribution of magnetic shear within the flaring volume reflects the paths taken by coronal current systems, about which we have little knowledge.

3.5.1 Footpoints and Magnetic Topology

The earliest observational studies demonstrated that the locations of Hα ribbons could be explained (with a suitable arrangement of magnetic charges,[11] extrapolated in a potential approximation) as the intersection of coronal separatrix surfaces with the photospheric boundary (Gorbachev and Somov 1989; Mandrini et al. 1991; Longcope 1996). An example of this is shown in Fig. 12 (Mandrini et al. 1995). Many flares have been modeled using similar approaches with increasing degrees of complexity—for example, incorporating linear force-free fields (Démoulin et al. 1994) and non-linear force-free fields (Régnier et al. 2002; Régnier and Canfield 2006; Schrijver et al. 2008), and using increasingly precise representations of the photospheric field (Barnes et al. 2005) and chromospheric magnetic fields (Metcalf et al. 2005) as input. Overall the correlations between the photospheric mappings of separatrices, quasi-separatrices and separators, and the observed location of flare footpoints are convincing, but certain aspects still evade a clear explanation. These mappings

[11] This is "magnetic charge topology," in which an array of fictitious magnetic monopoles is used as a best fit to a photospheric magnetogram, as a basis for 3-D potential-field extrapolations (Baum and Bratenahl 1980; Démoulin et al. 1993; Longcope 1996).

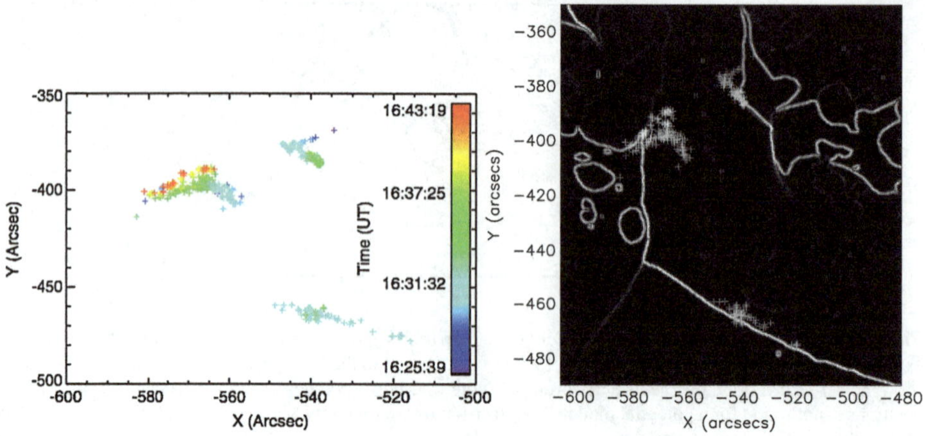

Fig. 13 *Left:* positions of the HXR footpoints in SOL2001-08-25T16:45 (X5.3) (Metcalf et al. 2003), with time color-coded *on the right*; data from the *Yohkoh*/HXT instrument (Sakao 1994). *Right:* locations of the separatrix surfaces projected onto the photospheric plane, with the shading representing the magnitude of the separatrix discontinuity. The *upper footpoints* illustrate the complex motions along the flare ribbons, while the *lower points* show rapid motion. The authors showed that this motion coincided with the photospheric intersection of a magnetic separatrix structure. Reproduced by permission of the AAS

have not yet allowed us to fully understand the reason why the HXR or WL footpoints are few and compact, while the Hα ribbons are extended, nor the properties of the reconnection that determine the motions of the footpoints. But they are providing us with fascinating clues.

Prior to *RHESSI*, a small number of studies had been carried out in which the relationship between magnetic field and HXR footpoint location was investigated. As well as the investigations into footpoint asymmetry described below, there was some early work on HXR sources in the context of magnetic topology. For example, Fig. 13 shows the *Yohkoh* HXR and *TRACE* WL source motions observed in SOL2001-08-25T16:45 (X5.3), compared to the projections of coronal separatrix structures onto the photosphere (Metcalf et al. 2003). This figure reveals strong resemblances between observed and calculated features; in particular the lower source moves extremely rapidly almost in coincidence with one of the separatrix intersections, in a manner suggesting the "slip-running" reconnection model of Aulanier et al. (2006). Evidence was also found for a coronal null and reconnection of external field through a separatrix "dome" (Fletcher et al. 2001). *Yohkoh*-era observations also showed that HXR sources tended to avoid sites of high vertical current density, preferentially occurring adjacent to them (Li et al. 1997).

RHESSI data have led to more studies relating HXR footpoint behavior and magnetic fields. For example, in SOL2005-01-17T09:52 (X3.8), which exhibits four Hα ribbons and corresponding HXR footpoint sources, the two strongest and long-lived Hα kernels and HXR footpoints are observed to tend to avoid the strongest fields, and move approximately along iso-Gauss contours, along the border between the sunspot's umbra and penumbra (Temmer et al. 2007). Also in this event, the magnetic reconnection rates derived for flare ribbon locations showing HXR footpoints are higher (by two orders of magnitude) than those in flare ribbon locations not showing HXRs. The strongest HXR sources were preferentially located in those regions of the ribbons with the strongest magnetic field, although this cannot readily be seen from Fig. 14. Similar results were obtained from *Yohkoh*/HXT studies (Asai et al. 2002, 2004a).

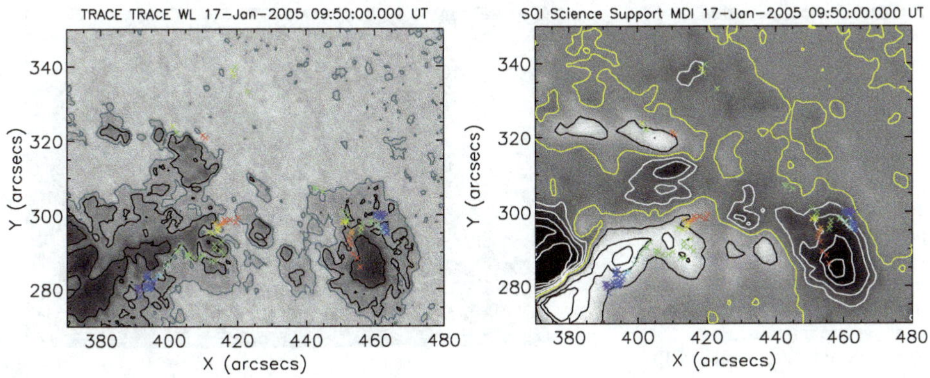

Fig. 14 *RHESSI* hard X-ray source centroids overlaid on a *TRACE* white-light image (*left*) and MDI magnetogram (*right*) for the X4 flare SOL2005-01-17T09:52. The subsequent occurrence of the HXR sources from 09:43:20 to 10:04:10 UT is color-coded *from red to blue*. *Left: TRACE white-light contours* roughly outline the umbrae as well as inner and outer penumbrae of the sunspots. *Right:* Isogauss lines at −2000, −1600, −1300, and −600 G (*white contours*) and +600, +1300, and +1500 G (*black contours*). The *yellow line* marks the magnetic inversion line. Adapted from Temmer et al. (2007). Reproduced by permission of the AAS

For three major *RHESSI* flares, it has been demonstrated (Des Jardins et al. 2009) that the path of HXR footpoints corresponds to a particular type of topological structure, a subset of the photospheric spine lines identified in magnetic charge topology models (see Longcope 2005, and the discussion in Sect. 3.5). These are lines that join two magnetic sources of the same sign ("charge") via a magnetic null. The implication is that the footpoint movement reflects the changing length of the separator joining the nulls on the two spine lines, as the coronal reconnection proceeds and the reconnection region moves.

The huge differences derived in the local energy release rates for flare ribbon locations with/without HXR footpoints, in combination with the limited dynamic range of present HXR instruments (of order 10:1) can explain the different flare morphologies typically observed in HXRs (compact footpoints) and Hα/UV (extended ribbons). However, it is still implied that a large fraction of the electrons is accelerated into spatially confined subsystems of magnetic loops as outlined by the HXR footpoints, and only a minor fraction goes into the large flare arcade outlined by the Hα/UV ribbons and EUV postflare loops (Temmer et al. 2007).

Although white light is an important indicator of the locations of strongest energy input (e.g., Neidig 1989), and although it is substantially simpler to image than HXRs, we do not have adequate systematic observations. In particular the white-light footpoint motion has only rarely been studied. SOL2001-08-25T16:45 (X5.3) (Metcalf et al. 2003) is one such example. The SOL2002-09-30T01:50 (M2.1) white-light flare studied by Chen and Ding (2006) is another. It showed systematic footpoint motion in the white-light continuum, following roughly that of the corresponding HXR source. Footpoints at both wavelengths zigzag back and forth, primarily parallel to the magnetic neutral line (see Fig. 15) in a manner which may be explained by the particular magnetic configuration in the flaring region.

3.5.2 Footpoint Asymmetry

A basic prediction of flare models that invoke electron acceleration in the corona and precipitation to the chromosphere is that regions with stronger magnetic field convergence (i.e.,

Fig. 15 Footpoint motion history in the white-light continuum (*pluses*) and 12–25 keV HXR emission (*diamonds*) superposed on the MDI magnetogram in the *white-light flare* SOL2002-09-30T01:50 (M2.1). The magnetic neutral line is plotted as the *dashed line* (Chen and Ding 2006). The points generally describe a clockwise motion, and cover a time interval of about five minutes. Reproduced by permission of the AAS

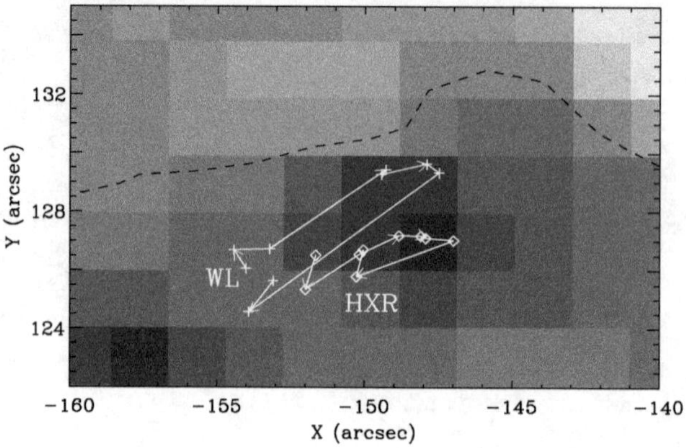

a stronger chromospheric or photospheric magnetic field) should be locations of weaker HXR footpoint sources, because the higher mirror ratio leads to a larger fraction of accelerated electrons mirroring before they reach the thick-target footpoints. The ratio of brightness in footpoint pairs should thus be inversely correlated with the ratio of magnetic field strengths at the location of those footpoints. This tendency was demonstrated systematically in early analyses of a small number of double-footpoint flares observed with *Yohkoh*/HXT, using both line-of-sight (e.g., Sakao et al. 1994) and vector fields (Li et al. 1997), but later work revealed counter-examples. Stronger HXR footpoints were found in stronger magnetic field regions in at least one-third of 32 flares examined (e.g., Goff et al. 2004). A detailed study by Liu et al. (2009b) of *RHESSI* footpoint pairs in SOL2003-10-29T20:49 (X10.0) showed the expected relationship in the first few minutes of the event, but thereafter it disappeared. Furthermore, although the sign of the correlation was as expected early in the event, the magnitude was not consistent with simple predictions of the magnetic mirroring model. They also found that collisional losses due to asymmetric column densities from the looptop (assumed to be the acceleration region) to the footpoints alone cannot explain the totality of the observed HXR fluxes and spectra. This is consistent with the result of a statistical study of *RHESSI* footpoint asymmetry carried out by Saint-Hilaire et al. (2008), though these authors did not examine the footpoint magnetic fields. As Liu et al. (2009c) suggest, more detailed modeling including mirroring, collisional losses, and other particle transport effects (such as nonuniform target ionization, relativistic beaming, photospheric albedo, and return currents) may provide a resolution to the above discrepancies (see Fig. 16). An alternative investigation of the footpoint asymmetry intrinsic to the acceleration process has been pursued (McClements and Alexander 2005) to explain the observations of Alexander and Metcalf (2002).

Most of the microwave emission in flares is gyrosynchrotron from non-thermal electrons, the intensity of which depends on the electron energy and on the ambient magnetic field. Thus the stronger-field footpoint should correspond to stronger microwave sources (and weaker HXR footpoints), giving a complementary view of footpoint asymmetry. This relationship has been found to hold in some recent investigations (Kundu et al. 1995; Wang et al. 1995), but the modeling presents the complicated problem of understanding the microwave absorption. Without understanding absorption it is not possible to get at information about the microwave footpoints, and so any conclusions drawn from HXR and microwave comparisons are premature.

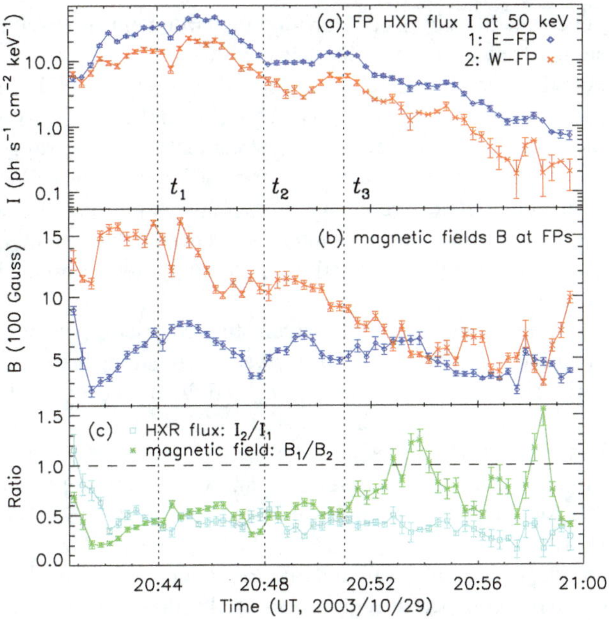

Fig. 16 Time profiles of X-ray and magnetic field parameters of the conjugate footpoints in SOL2003-10-29T20:49 (X10.0). (**a**) HXR fluxes at 50 keV of the eastern footpoint (E-FP, *blue diamonds*) and western footpoint (W-FP, *red crosses*) obtained from power-law fits in the 50–150 keV range. (**b**) *SOHO*/MDI magnetic field strengths registered at the two footpoints. (**c**) Ratios of the 50 keV fluxes (W-to-E) and magnetic fields (E-to-W) of the two footpoints. The expected correlation between these two ratios only holds for the first half of the flare duration (from Liu et al. 2009b). Reproduced by permission of the AAS

3.5.3 Inference of the Properties of Magnetic Reconnection

The product of the footpoint apparent speed and the line-of-sight magnetic field, expressed as a flux transfer rate $\dot{\Phi}$, has been used as a measure of the coronal reconnection rate by various authors (e.g., Qiu et al. 2002; Fletcher et al. 2004; Jing et al. 2005; Temmer et al. 2007). The quantity $\dot{\Phi}$ can be estimated from the observations as (e.g., Forbes and Lin 2000):

$$\dot{\Phi} = \frac{\partial}{\partial t} \int B_{\mathrm{ph}}\, da, \tag{1}$$

where da denotes the ribbon area and B_{ph} the normal component of the photospheric magnetic field.

In the gradual phase of large two-ribbon flares where the magnetic configuration is well-approximated by a 2- or 2.5-D field (i.e., no significant shear or twist component of the field), the coronal reconnection rate is also equal to an equivalent electric field E_c (Poletto and Kopp 1986; Forbes and Lin 2000), i.e.,

$$E_c = v_{\mathrm{fp}} B_{\mathrm{ph}}, \tag{2}$$

where E_c is the convective electric field at the magnetic reconnection site, v_{fp} the observed speed of the apparent HXR footpoint or Hα/UV flare ribbon and B_{ph} the vertical component of the photospheric magnetic field. Moreover the energy release rate equals the Poynting flux into the current sheet (e.g., Lee et al. 2006). Estimates of the flux can be obtained from the motions of footpoints and the line-of-sight magnetic field strength at the flare footpoints (e.g., Isobe et al. 2002; Asai et al. 2004a; Temmer et al. 2007).

In the impulsive phase, which observationally is far from two-dimensional, the relationship between footpoint motion, magnetic field and flux transfer rate should be preserved (this follows from magnetic flux conservation) but the coronal reconnection electric field is not so readily obtained. Nor, in an environment of strong twist and shear, will there be a straightforward relationship between $\dot{\Phi}$, the Poynting flux, and the energy release rate. We note that a study with a prescribed 3-D coronal field (Hesse et al. 2005) showed a relationship between the instantaneous reconnected magnetic flux at a field line and the "field-line-integrated" parallel electric field along that field line, suggesting that the endpoints of field lines with high values of this electric field correspond to locations of chromospheric excitation. Although the relationship between $\dot{\Phi}$ and the parallel field is determined by the coronal magnetic configuration, which is not generally known, this analytic work can give us some confidence that $\dot{\Phi}$ calculated from footpoint motions in the impulsive phase is a meaningful quantity.

Observationally there are interesting correlations between $\dot{\Phi}$ and properties of the footpoint radiation. Several studies reveal correlations between the HXR flux evolution and the derived reconnection quantities E_c and $\dot{\Phi}$, and also with the speed of the footpoint separation or flare loop growth (Qiu et al. 2002; Fletcher and Hudson 2002; Krucker et al. 2003; Asai et al. 2004a; Liu et al. 2004; Krucker et al. 2005; Veronig et al. 2006; Saba et al. 2006; Miklenic et al. 2007; Temmer et al. 2007). Figure 17 shows the result for SOL2003-10-29T20:49 (X10.0), with derived v_{fp}, $v_{\mathrm{fp}}B_{\mathrm{ph}}$ and $v_{\mathrm{fp}}B^2_{\mathrm{ph}}$ curves correlated with the *RHESSI* HXR flux (Krucker et al. 2005). In this flare the HXR flux exponentially correlates with the magnetic field strength at the footpoints, which may scale with the field strength in the coronal reconnection region (Liu et al. 2009b). Figure 18 shows the relationship in time of $\dot{\Phi}(t)$ derived from SOL2003-11-18T08:31 (M3.9). Each of the strongest three *RHESSI* HXR peaks is well reflected in the derived $\dot{\Phi}(t)$ time profiles but shifted in time by 1–2 min (for a discussion of this effect see Miklenic et al. 2007). The correlation in time and space between locations of high $v_{\mathrm{fp}}B_{\mathrm{ph}}$ and footpoint intensity has also been demonstrated in detail using *TRACE* UV footpoints (Fletcher 2009).

Interestingly, in the rare example of the HXR ribbon flare SOL2005-05-13T16:57 (M8.0), which has at first glance quasi-2-D properties, there is a better correlation between the HXR intensity and the derived local magnetic reconnection rate and energy release rate when there are only a few isolated HXR footpoints, than when the ribbon-like HXR emission appears (Jing et al. 2007; Liu et al. 2008).

Further substantial progress in determining the reconnected flux and Poynting flux will be extremely difficult. It will require not only the measurement of the chromospheric vector magnetic field (to assess time-dependent field perturbations at the boundaries) but also microwave observations of gyroresonance emission leading to improved knowledge of the coronal magnetic field. Using multi-frequency microwave observations, such as are planned with FASR (Bastian 2003), isosurfaces of magnetic field strength can be computed. Coupled with plasma flow information from EUV spectroscopic diagnostics, "before and after" changes of the field around a flare would give an independent view of the energy extracted from the field, as well as some information about how it moves through the configuration. Direct observations of the coronal magnetic field at the limb using infrared Zeeman splitting (Lin et al. 2000) are proceeding now with the CoMP instrument (Tomczyk et al. 2008). The Hanle effect in the UV (Raouafi et al. 2009) is also utilized.

3.6 Excitation of the Deep Atmosphere

As is well known, the first recorded observation of a solar flare (Carrington 1859) was in the optical or "white light" wavelength range. Since such a flare is visible over and above the bright photospheric radiation (6.27×10^{10} erg cm^{-2} s^{-1} in the quiet Sun), roughly doubling it, this emission is a significant component in the flare energy budget. A small number of

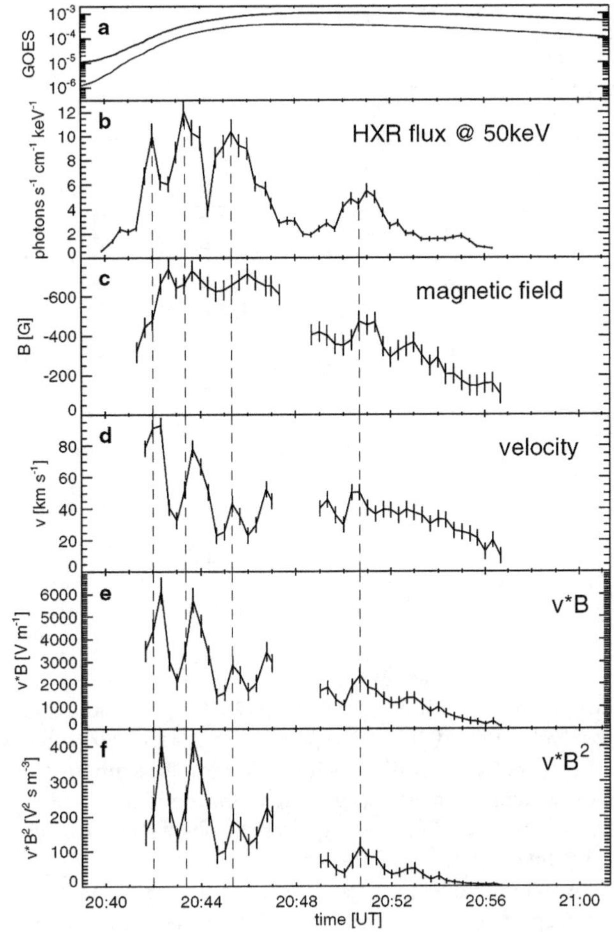

Fig. 17 Magnetic reconnection analysis of SOL2003-10-29T20:49 (X10.0). Time series of: (**a**) *GOES* SXR flux, (**b**) *RHESSI* HXR flux at 50 keV of the Eastern footpoint source, (**c**) photospheric magnetic field B at the instantaneous footpoint location, (**d**) velocity v of the HXR footpoint source, (**e**) product vB (magnitude of convective electric field E_c), (**f**) product vB^2 (measure of the energy release rate $\dot{W}$) (from Krucker et al. 2005). Reprinted by permission from Advances in Space Research

direct measurements of the flare total irradiance now exist for large flares (Woods et al. 2006; Kretzschmar 2008), in which the total radiant energy of the flare is measured to be a few times 10^{31} to 10^{32} erg. White-light emission can also be present in relatively weak flares, down to low *GOES* C class (Matthews et al. 2003; Hudson et al. 2006; Jess et al. 2008).

The generation of flare optical radiation (IR/visual/UV continuum) is not yet well explained, and it may be that there are different processes operating in different flares. Where spectroscopic observations are available, observed white-light flares have been split into two types (Machado et al. 1986; Neidig 1989). Type I flares[12] show intense and broad Balmer lines and Balmer and Paschen edges (resulting from recombination), and are thought to occur in a heated chromosphere. Type II flares, much less frequent, do not show these features and may arise from enhanced H^- continuum. The location at which the Type II flare radiation is produced is not known. Generally it is hard to see how the deep layers of the photosphere could be directly excited by electron beams without requiring rather unreasonable electron energy budgets (Aboudarham and Hénoux 1986). Excitation by

[12] Type I and Type II as used here should not be confused with the meter-wave radio bursts (e.g., Wild et al. 1963), nor with the spicules (De Pontieu et al. 2007).

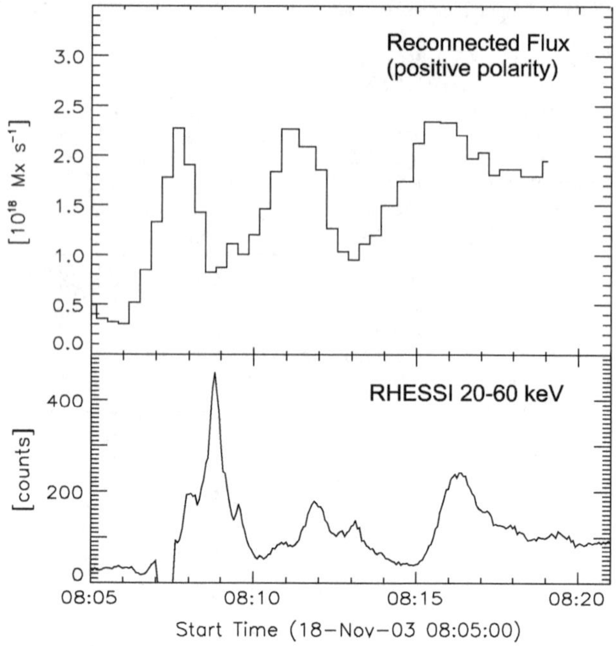

Fig. 18 Magnetic flux change rate $\dot{\Phi}(t)$ together with the *RHESSI* 20–60 keV HXR flux for SOL2003-11-18T08:31 (M3.9). Adapted from Miklenic et al. (2007)

proton beams with energies of 10–20 MeV has also been proposed (e.g., Švestka 1970; Machado et al. 1978), since protons have a greater penetration depth than electrons of the same velocity. A small velocity dispersion is required to focus the energy deposition over a narrow range of depths, as well as to find agreement with proton fluxes at ~ 30 MeV implied by γ-ray observations. Energetic electrons would also need to be present to account for the HXR flux.

A popular model for the Type I events, which does not require electron beams to reach the deepest layers of the chromosphere, is the "radiative backwarming" model. In this model energy is deposited in the upper chromosphere generating a strong Balmer-Paschen continuum by recombination, which warms lower levels (Metcalf et al. 1990). This idea is close to the original suggestion of Hudson (1972) to bypass the complicated problems of radiative transfer with the "specific ionization" approximation, which implies secondary ionizations. Hudson (1972) and Aboudarham and Hénoux (1986) also note that the non-thermal ionization of hydrogen would also strongly enhance the continua. Non-LTE simulations suggest that a purely chromospheric temperature rise may be insufficient to produce the continuum intensity enhancements seen (Ding and Fang 1996) and an enhancement near the temperature minimum region may still be necessary. Evidence for the effect of energy deposition was found in the white-light flare SOL2002-09-29T06:39 (M2.6) (Ding et al. 2003; Chen and Ding 2005). This event had two HXR footpoints, one with weaker HXR emission but stronger white-light continuum emission, and a relatively weak, centrally-reversed Hα profile (see e.g., Švestka 1966, for a discussion of flares at optical wavelengths). This profile indicates that at the weaker HXR footpoint the atmosphere had not been fully heated, and under such conditions it is possible that an electron beam could effectively penetrate the chromosphere and produce the observed continuum emission via radiative backwarming. By contrast, the local atmosphere at the other footpoint had been appreciably heated, producing a high coronal pressure (Canfield et al. 1984). Electrons would thus be prevented from penetrating into the deeper atmosphere.

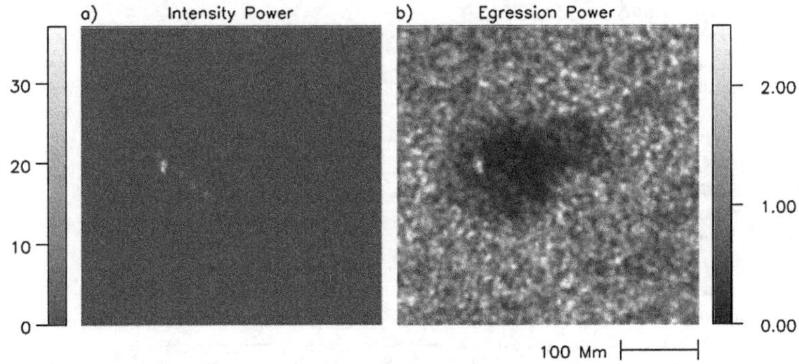

Fig. 19 Signatures of SOL2003-10-29T20:49 (X10.0) filtered in the 5–7 mHz band, i.e. at frequencies above most of the p-mode power. *Left:* intensity; *right:* "egression power," showing the source of the seismic waves observed from this flare. The main seismic source is within the area of the large sunspot. From Lindsey and Donea (2008)

Direct evidence for excitation of the deep atmosphere during a flare comes from the flare seismic waves ("sunquakes") first observed by Kosovichev and Zharkova (1998), and thereafter in several other flares of M and X class.

These waves refract through layers deep in the convection zone and appear as surface ripples, traveling at apparent speeds of only some tens of $\mathrm{km\,s^{-1}}$, far from the flare site. Seismic waves have been observed from several flares, but they are still comparatively rare—most large flares do not produce detectable wave amplitudes (Donea and Lindsey 2005). However, the energy required to produce the disturbances is found to be small, on the order of 10^{-4} of the total flare energy. The sources of this seismic emission can be located holographically (Fig. 19), and a large proportion of the seismic sources are located within the penumbrae of sunspots. The sources appear to coincide with the HXR footpoints and white-light flare kernels (Martínez-Oliveros et al. 2008), and move with them (Kosovichev 2006). They do not appear to be so strongly associated with the γ-ray sources—i.e., the accelerated ions (Kosovichev 2007)—but the γ-ray imaging is much inferior to the HXR imaging. Flare seismic waves may be associated with downwards-moving material in the MDI Doppler data. Theory suggests that a shock could be produced by intense heating of the chromosphere by an electron beam, and initially it was proposed that the waves resulted from this shock impacting on the photosphere. However, the momentum required to produce the seismic disturbance is substantially higher than that observed directly in the plasma downflows, and the shock propagation time to the photosphere is inconsistent with observations of the seismic pulse onset versus the HXR peak (Zharkova 2008, and references therein). It is also likely that such shocks would be radiatively damped before reaching the chromosphere (Lindsey and Donea 2008). Thus another method for delivering momentum into the deep photosphere may be required. Proton beams have been proposed (Zharkova and Zharkov 2007).

An alternative view is that the seismic waves are launched by a "jerk" of the magnetic field, caused by field re-organization in the corona, imparting momentum to the ions and collisionally-coupled neutrals at the photosphere (Hudson et al. 2008). The jerk is the Lorentz force imparted at the photosphere and is thus capable of launching an interior seismic wave. In the standard reconnection flare model, the perturbation producing the jerk would originate in the corona and propagate as a wave into the photosphere (e.g., Fletcher and Hudson 2008). Initial analyses cast doubt on the viability of this mechanism (Martínez-Oliveros and Donea 2009), but it is difficult to disentangle this mechanism and

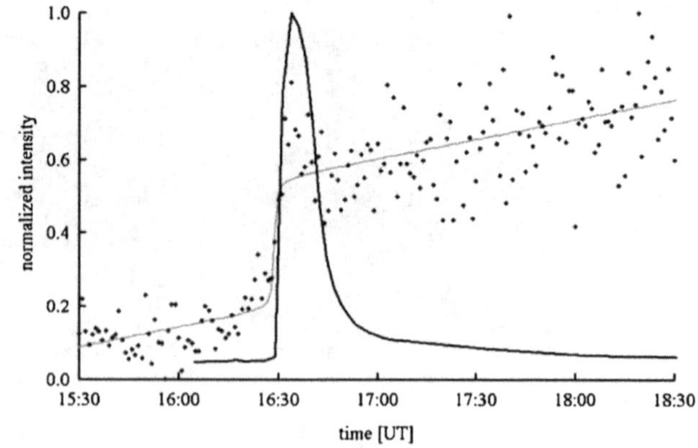

Fig. 20 *Points and gray line* show the normalized time variations of the line-of-sight magnetic field measured by GONG at a point, during SOL2001-08-25T16:45 (X5.3). The line is the *TRACE* 1600 Å intensity measured co-spatially with the changing magnetic field, to within around 20″. The line-of-sight magnetic field changes abruptly at the time of the flare impulsive phase (from Sudol and Harvey 2005). Reproduced by permission of the AAS

the others proposed: the original idea of a hydrodynamic shock (Kostiuk and Pikelner 1975; Kosovichev and Zharkova 1998), and the more recent discussion of direct photospheric heating, e.g., through radiative backwarming (e.g., Lindsey and Donea 2008).

The Lorentz-force jerk is consistent with the non-reversible magnetic field changes in the line-of-sight magnetic field observed in many flares. Early observations showed permanent changes in the vector field around the magnetic neutral line (Wang et al. 1994), in one case the observations being separated by only a few minutes before and after the flare. Similar observations for a flare on the limb found variations in the line-of-sight magnetic field (i.e., the component tangential to the photosphere for a limb flare), the importance of this being that the line-of-sight component does not suffer from the 180° directional ambiguity of the vector field (Cameron and Sammis 1999). Further work on irreversible changes to the line-of sight magnetic field followed using *SOHO*/MDI (Kosovichev 2006) and they have now been confirmed to occur in all large flares, usually close in time to the flare impulsive phase (Sudol and Harvey 2005), and close spatially to the HXR footpoint sources. Figure 20 shows an example of such field changes, the typical magnitude of which is 100–200 G (or on the order of 10% of the photospheric field in the region). In many cases, the field changes are also associated with visible evolution in the sunspot, particularly to a disappearance of a part of the penumbra (Anwar et al. 1993; Wang et al. 2004; Liu et al. 2005).

Changes of this magnitude at the photosphere imply that the overlying field undergoing rearrangement—presumably in the low corona—must be strong, and energy transported by magnetic disturbances propagating through the chromosphere to the photosphere may be an important component in the flare chromosphere energization (Emslie and Sturrock 1982; Fletcher and Hudson 2008). An interesting aspect of such observations is that many flares show an *increase* in the observational shear along the magnetic polarity inversion line (Wang et al. 1994), counter to what would be expected in a scenario in which the active region free energy should decrease to power the flare. However, *Hinode*/SOT observations of a flare show that below about 8000 km above the photosphere the shear increases after a flare, whereas above this altitude it decreases (Jing et al. 2008). It is possible that the shear increase close to the polarity inversion line is associated with flux emergence, as part of the ongoing build-up of magnetic free energy in a repeatedly flaring region.

3.7 Chromospheric Evaporation

The arcades of loops characteristic of the gradual phase of solar flares are filled with hot, dense plasma, usually interpreted as chromospheric plasma which expands to a new equilibrium following chromospheric heating in the impulsive phase. This process is termed chromospheric evaporation. It has been studied for almost 40 years, since it was first proposed to explain the delay between the peaks of SXR (gradual) and microwave (impulsive) emissions (Neupert 1968). Latterly, the Neupert effect usually refers to the often-observed relationship between time-integrated HXR flux and the SXR flux. The Neupert effect gives indirect evidence for chromospheric evaporation; more direct support comes from observations of blueshifted emission of high-temperature plasma, often correlated with impulsive HXR bursts as discussed further below. Early on, spatially-unresolved 300–400 $\mathrm{km\,s^{-1}}$ upflows in resonance lines of Ca XIX and Fe XXV were observed using instruments on board the *Solar Maximum Mission* (Doschek et al. 1980; Antonucci and Dennis 1983; Zarro et al. 1988), and since then confirmed in spatially-resolved observations with *SOHO*/CDS[13] (e.g., Czaykowska et al. 1999; Teriaca et al. 2003; Brosius and Phillips 2004; Del Zanna et al. 2006) and more recently *Hinode*/EIS (Milligan 2008). It has also been suggested that the heating of quiescent active-region loops is not actually "coronal heating" at all, but happens in the chromosphere and stems from the principles of flare-induced chromospheric evaporation (Aschwanden et al. 2007).

SXR spectra of flares often show a dominant stationary component as well as the upflow. This is a puzzle for evaporation theory in a single loop excitation, but it could be explained by a filamentary structure, in which many sub-resolution magnetic loops are activated in succession, each having such a small cross-section that it produces undetectably small amounts of emission. Emission would then be detected only after some time, when a number of these loops are emitting together, and the evaporated plasma in each has come to rest at the looptop (Doschek and Warren 2005). It has also been argued that hot dense plasma exists in the flare corona in advance of the flare impulsive phase (Feldman 1990; Caspi and Lin 2010). It is usually assumed that the hot emission comes from the flare corona, but recently, stationary Fe XXIV emission has been detected at loop footpoints (Milligan and Dennis 2009). There are also observations which suggest that the upflows do not reach high into the corona, and that the coronal density increase occurs as a result of compression (Feldman et al. 2004; Caspi and Lin 2010). An adiabatic compression would in fact create a *negative* microwave flare in the free-free continuum because the free-free emissivity scales as $T^{-0.5}$. On the contrary, the observations show a good correlation between the radio and X-ray continuum emission measures, so a negative flare is contrary to the observations (e.g., Kundu 1965; Hudson and Ohki 1972); see White et al. (2011) for more detail. Note that negative microwave bursts do occur, but they can be explained by intervening absorptions (Covington 1973).

Early in the flare the chromosphere is heated rapidly and impulsively, primarily by energetic electrons which lose energy collisionally in the chromosphere. Thermal conduction from the corona may also play a role in heating the chromospheric plasma, particularly in pre-impulsive (Battaglia et al. 2009) or gradual (e.g., Zarro and Lemen 1988) phases. Liu et al. (2009c) also suggest that conduction may play an important role in the impulsive phase for a flare with a substantial low-energy component. The heated atmosphere can radiate or conduct away the energy, and can also expand upwards and downwards. Whether this evaporation is gentle or explosive depends on the energy deposition rate by accelerated electrons as treated in the 1-D radiation hydrodynamics calculations (Fisher et al. 1985a, 1985b, 1985c, Abbett and Hawley 1999). For energy input rates of less than $\sim 3\times 10^{10}$ $\mathrm{erg\,cm^{-2}\,s^{-1}}$, theory suggests that the evaporation is gentle, with upward plasma flows at several tens of

[13] Coronal Diagnostic Spectrometer.

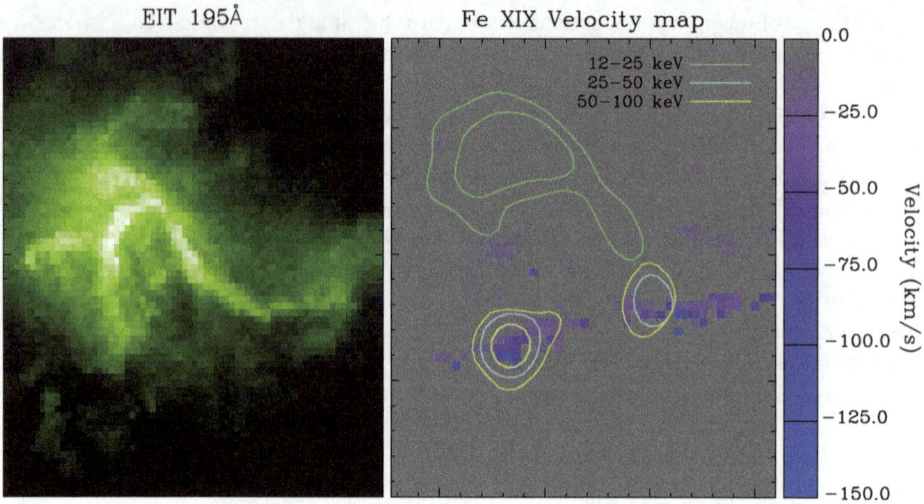

Fig. 21 *Left:* EIT image of a flare observed by *SOHO*/EIT showing a loop-like structure and possibly an extended EUV ribbon to the west. *Right:* the event was also observed by *RHESSI* and the *SOHO* Coronal Diagnostic Spectrometer. The *RHESSI* high energy footpoints (*yellow contours*) coincide with regions of upflow in Fe XIX of formation temperature ($\log T_e = 6.9$), with speeds of up to 150 $\mathrm{km\,s^{-1}}$. This provides evidence for explosive evaporation driven by electron heating. Reproduced by permission of the AAS

kilometers per second. Gentle evaporation can also be conductively driven. At high non-thermal electron rates ($> 3 \times 10^{10}$ $\mathrm{erg\,cm^{-2}\,s^{-1}}$), the chromosphere is unable to radiate at a sufficient rate and consequently expands rapidly. This condition is met when the heating time-scale is less than the hydrodynamic expansion time-scale:

$$\frac{3kT}{Q} < \frac{L_0}{c_s} \tag{3}$$

where Q is the flare heating rate per particle, T is the final temperature of the heated plasma, c_s is the corresponding sound speed, and L_0 is the length-scale of the flaring region. If this condition holds, the heated chromospheric plasma expands upward at hundreds of $\mathrm{km\,s^{-1}}$ in a process known as "explosive" evaporation. The overpressure of the flare plasma relative to the underlying chromosphere causes cooler, more dense material to recoil downward at tens of $\mathrm{km\,s^{-1}}$ (known as "chromospheric condensation"). In addition to the magnitude of the energy deposition rate, Fisher et al. (1985a) also stipulate that the direction of flows in the transition region/upper chromosphere determines whether the evaporation is gentle or explosive.

3.7.1 Explosive Evaporation

Spatially-resolved spectroscopic EUV observations of explosive upflows during the impulsive phase of a flare are relatively rare, since they require the spectrometer slit to be located at the flare footpoints exactly at the time of strong energy deposition. However, this was managed in SOL2003-06-10T14:36 (M2.2). *SOHO*/CDS was used to detect simultaneous strongly blueshifted ($\sim$250 $\mathrm{km\,s^{-1}}$) Fe XIX emission (peak formation temperature $\log T_e = 6.9$; Mazzotta et al. 1998) from footpoints, supportive of explosive evaporation, along with weakly redshifted He I ($\log T_e = 4.5$) and O V ($\log T_e = 5.4$) emission (Milligan et al. 2006a). Coordinated *RHESSI* imaging confirmed that the origin of these flows was at the flare HXR footpoints (see, for example, Fig 21). The combination of images and spectra from *RHESSI* also allows estimates to be made of the energy flux contained in the non-thermal electrons (in $\mathrm{erg\,cm^{-2}\,s^{-1}}$) in order to make a direct comparison with the predictions of theory, under the assumption of the collisional thick-target model. Subject to uncertainties in the area of the unresolved HXR footpoints, this event revealed an electron energy flux greater than $4 \times 10^{10}\ \mathrm{erg\,cm^{-2}\,s^{-1}}$, in agreement with the predictions of Fisher et al. (1985a).

Other observations without HXRs have instead used the time derivative of the *GOES* flux (assuming the Neupert effect) to identify the impulsive phase (Teriaca et al. 2006; Del Zanna et al. 2006). Spectroscopy from CDS at high time cadence (10 s) of a "flare-like transient" has also been obtained (Brosius and Holman 2007). All authors report line profiles in hot lines (e.g., Si XII with peak formation temperature $\log T_e = 6.3$, Fe XVI at $\log T_e = 6.4$ and Fe XIX at $\log T_e = 6.9$) consistent with upflows on the order of 100–200 $\mathrm{km\,s^{-1}}$. In the transition lines of O V at $\log T_e = 5.4$ and He I at $\log T_e = 4.5$ downflows of a few tens of $\mathrm{km\,s^{-1}}$ were most commonly observed, though upflows are also reported (Del Zanna et al. 2006). Recent observations also confirm the cospatial downflows in Hα spectra and upflows in Ca XIX (first observed by Wuelser et al. 1994) of a few $\mathrm{km\,s^{-1}}$ (Teriaca et al. 2006). These can be used to determine the expected momentum balance between the evaporated and condensing material (Canfield et al. 1987).

3.7.2 Gentle Evaporation

Plasma flows attributed to gentle evaporation driven by thermal conduction from the overlying hot corona have frequently been observed during the gradual/decay phase of flares, after the non-thermal beam heating has ceased and upflows are sustained by the thermal conduction fronts set up by the steep temperature gradients (e.g., Schmieder et al. 1987; Zarro and Lemen 1988; Czaykowska et al. 1999, 2001). However, gentle evaporation due to a weak non-thermal electron flux has only recently been observed by *SOHO*/CDS in conjunction with *RHESSI* in SOL2002-07-15T11:55 (C9.1) (Milligan et al. 2006b). Doppler shifts of lines formed at a range of temperatures showed upflows of $\lesssim$100 $\mathrm{km\,s^{-1}}$ at all temperatures, consistent with gentle evaporation. The upflow velocity of the Fe XIX material was a factor of two higher when the electron energy flux was an order of magnitude greater (Milligan et al. 2006a) (see Fig. 22). The absence of any redshifted lines supports the hypothesis that only a large flux of electrons is capable of driving the downflows in the transition region and upper chromosphere associated with explosive evaporation.

The improved spatial, spectral and temporal capabilities of the Extreme Ultraviolet Imaging Spectrometer (EIS) onboard *Hinode* have also been used to study chromospheric evaporation using emission lines formed over a broad range of temperatures (Milligan and Dennis 2009). During the impulsive phase of SOL2007-12-14T14:16 (C1.1), blueshifted emission (coincident with *RHESSI* HXR emission) was observed in six emission lines (Fe XIV–XXIV) formed over the temperature range $\log T_e = 6.3$–7.2. These upflows were found to scale with temperature over the range 8–18 MK, reaching speeds of >250 $\mathrm{km\,s^{-1}}$ in the Fe XXIV line. This dependence on temperature exists as chromospheric material, heated to a range of different temperatures by a distribution of electron energies, will be subject to different pressure gradients relative to the overlying corona and therefore rise at different rates. A new finding was that cospatial material formed at temperatures from $\log T_e = 4.7$–6.2 was redshifted by several tens of $\mathrm{km\,s^{-1}}$.

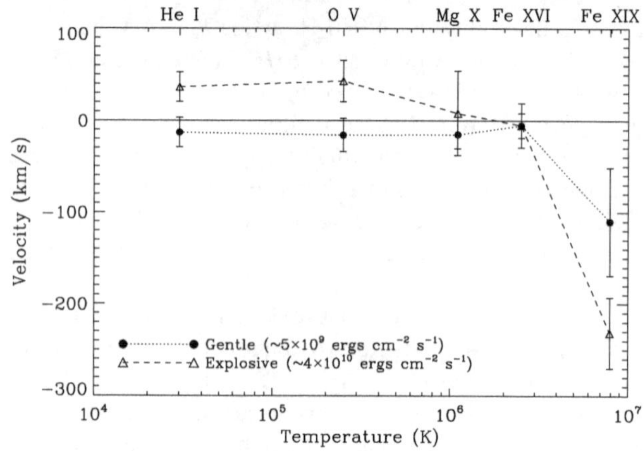

Fig. 22 Plasma velocity as a function of temperature for each of the five emission lines observed using CDS during the impulsive phases of two flares, plotted against their characteristic temperatures. Positive velocities indicate downflows (redshifts), while negative values indicate upflows. The data points plotted with *filled circles* denote the case of gentle evaporation, while the *open triangles* illustrate the case of explosive evaporation. From Milligan et al. (2006b). Reproduced by permission of the AAS

At this time the origin of these higher-temperature downflows is unclear and presents a challenge for current evaporation models. They may be related to the spectrum of the electron beam, with a soft spectrum resulting in energy deposition higher in the atmosphere (and downflowing transition-region plasma) and a hard spectrum resulting in energy deposition low in the atmosphere (leading to upflowing transition region plasma). However, a recent model by Liu et al. (2009c) suggests that this may be a result of sustained chromospheric heating, rather than a single heating burst as is used in most modeling.

3.7.3 *Imaging of Evaporation*

Direct imaging observations of the expected fronts of multi-million K upflowing plasma in the act of filling the loops are rather hard to come by (Doschek et al. 1996). One reported observation with *Yohkoh*/SXT (Silva et al. 1997) has features suggestive of upwards plasma flows at the rather slow speed of around 60 $\mathrm{km\,s^{-1}}$ in the impulsive phase. Evidence for evaporative upflows has been claimed in *RHESSI* observations of SOL2003-11-13T05:01 (Liu et al. 2006). This event has a pair of footpoint/loop leg sources in the range 12 to 30 keV (see Fig. 23), which converge into a single source near the center of the loop at a speed of some hundreds of km $\mathrm{s^{-1}}$, perhaps as much as 10^3 km $\mathrm{s^{-1}}$. The emission centroids in this event shift systematically toward the footpoints with increasing energies up to $\sim$70 keV, and the upward source motion occurs first at low energies and progresses to higher energies. The source motion is thus also consistent with the behavior expected in a coronal thick target (see Sect. 4.1). This is accompanied by an increase in density at the looptop (see also Jin and Ding 2008 who observe the appearance of a coronal loop at high energies).

Finally, a recent observation of a weak B1.7-class flare using *RHESSI* and *Hinode*/EIS found hot ($\sim$2 MK) upflowing plasma at one footpoint, but hot downflowing plasma at the other (Milligan 2008), whereas both theory and previous observations at lower spatial resolution have only shown upflows at this temperature. An interpretation in terms of a siphon-like flow is inconsistent with the apparent filling of the flare loop observed with *Hinode*/XRT.

4 Coronal Sources

The presence of coronal hard X-ray sources was first inferred in disk-occulted events (flares with HXR footpoints behind the solar limb) using data from HXR spectrometers on the OSO-5 and OSO-7 satellites (Frost and Dennis 1971; Hudson 1978). The radial height above

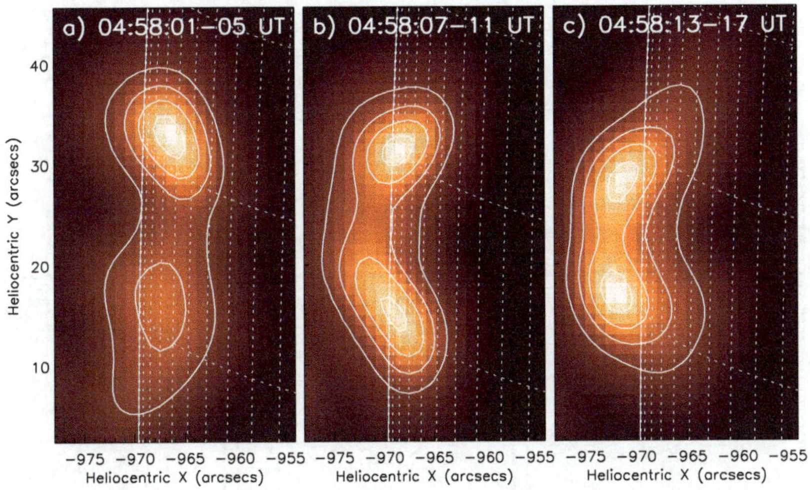

Fig. 23 *RHESSI* hard X-ray images of SOL2003-11-13T05:01 (M1.6) at 12–15 keV integrated over three 4 s time intervals in sequence. The contours are at 30%, 50%, 70% and 90% of the image maximum. The two sources rapidly move from the footpoints toward the looptop. This is consistent with the flows of hot plasma in chromospheric evaporation, and the Neupert effect is present in this flare as expected (adapted from Liu et al. 2006). Reproduced by permission of the AAS

the likely active regions in these cases could be estimated at more than 2×10^4 km. This placed them above the mean heights of flare SXR sources (Catalano and van Allen 1973; van Beek et al. 1981). Similar HXR source altitudes were inferred using observations from multiple spacecraft (Kane et al. 1979a). The earliest HXR imaging (from *Hinotori*) directly showed coronal source energies up to at least 25 keV (Takakura et al. 1983).

The launch of the *Yohkoh* satellite led to several further reports of coronal emission. Observations of an impulsive, and by implication, non-thermal coronal component, with energies up to the 33–53 keV band, were first made by using *Yohkoh*/HXT (Masuda et al. 1994). More *Yohkoh*/HXT coronal sources were subsequently found, though few with quite the same remarkable properties as the "Masuda flare." HXR coronal sources were demonstrated to exhibit both gradual and impulsive HXR characters, with the impulsive spikes being more energetic and having harder spectra (Tomczak 2001).

The *RHESSI* observations have substantially added to the literature on coronal HXR sources, and it is one area in which *RHESSI* has made a tremendous impact. This body of new observational work has recently been summarized, along with an overview of the theoretical ideas, by Krucker et al. (2008a). This paper should be consulted for more detail on this rich and relatively new field than can be provided below.

4.1 Thick-Target Looptop Sources

Coronal densities are generally too low to produce HXR emission efficiently, which is why footpoint sources normally dominate the images. The observation of the Masuda source inspired the suggestion that it could be explained by a partially thick coronal target (Wheatland and Melrose 1995). Subsequent observations with *RHESSI* (Veronig and Brown 2004) uncovered a new type of coronal HXR emission embedded in coronal loops, rather than "above the loop top" as in the Masuda source. Several events could be interpreted as "coronal thick target" sources (Fig. 24), identifiable by their lack of compact

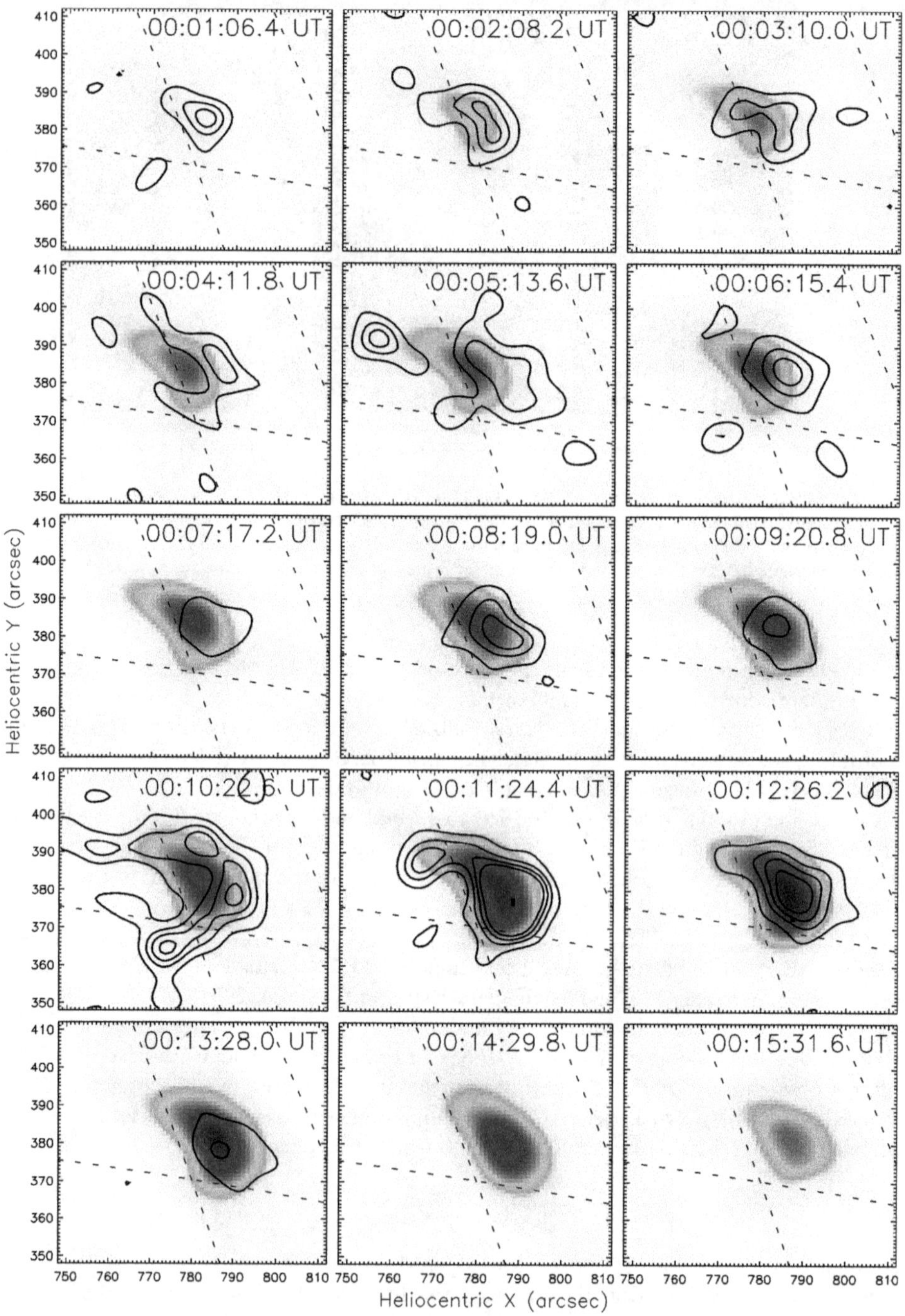

Fig. 24 *RHESSI* images of a thick-target looptop source in SOL2002-04-15T00:15 (M3.7) (Veronig and Brown 2004): images 6–12 keV, contours 25–50 keV with minimum level 0.17 times the maximum in each frame. The HXR contours show little evidence for footpoint brightening. Reproduced by permission of the AAS

footpoint HXR sources. In the first case, SOL2002-04-15T00:15 (M3.7), the spectrum observed with *RHESSI* was rather steep ($\gamma \geq 6$) and the column density N_{loop}, estimated using *GOES*, was high ($N_{loop} \geq 10^{20}$ cm^{-2}), leading to the interpretation that plasma in the corona was dense enough to act as a thick target. This can happen when the stopping energy $E_{loop} = \sqrt{3KN} \approx 8.8\sqrt{N_{19}}$ keV is greater than the energy of an electron, where N_{19} is the column density in 10^{19} cm^2. A 25 keV electron has a range (stopping column density) of about 10^{19} cm^{-2} (Brown 1973). In the Veronig-Brown flare such a column density appears already to have been present within the loop at the onset of the flare, and during the flare it increased to several times 10^{20} cm^{-2} which allowed electrons up to 50 keV to be fully stopped.

A high pre-flare coronal density is puzzling, because if the material is at coronal temperatures and static, the resulting high equilibrium pressure will make it a bright X-ray source. This consideration would not apply to a prominence or a loop structure at intermediate temperatures. Having a high density at the start of a flare is also a difficult problem for the standard flare model, which envisions the opening of the field prior to the reconnection. An earlier flare in the same region could create an enhanced coronal column density by evaporation in a loop that subsequently flares again (Veronig et al. 2005), or, similarly, slow pre-flare heating could lead to evaporation into a system which then becomes unstable. However, at least in the standard 2-D scenario, electrons would propagate down a different set of field lines from the closed, post-reconnection loops onto which the plasma has already evaporated, precluding the scenario in which energetic electrons are accelerated onto dense loops as a result of subsequent reconnection. A more complicated magnetic geometry or a less direct link between reconnection and acceleration is needed.

The SOL2002-04-15T00:15 (M3.7) "coronal thick target" event could also be observed via Nobeyama microwave imaging at 17 GHz. These data include a circularly polarized component (Veronig et al. 2005; Bone et al. 2007), establishing the presence of non-thermal electrons, visible in the whole loop through unpolarized thermal emission and consistent with the high density needed for the coronal thick target (Bone et al. 2007). A rather non-standard explanation is that the previously flaring dense loops could become unstable due to a high-beta instability resulting in a second flare (Shibasaki 2001).

The idea of a coronal thick target might extend to a seemingly separate class of events, namely the "soft-hard-harder" coronal HXR sources discussed below in Sect. 4.5. This possibility might require the stable trapping of high-energy electrons in a theoretically unstable loss-cone distribution (Wentzel 1976). Other than high density and magnetic trapping, turbulence or plasma waves (e.g., Benz 1977; Miller et al. 1996; Petrosian and Liu 2004), generated as a consequence of magnetic reconnection, provide an alternative mechanism that could accelerate electrons and, at the same time, confine them to the region near the looptop.

4.2 X-ray Observations Suggesting Coronal Current Sheets

RHESSI observations revealed the existence of double coronal X-ray sources, interpreted in terms of the current sheet expected in the large-scale reconnection model.

In a series of flares that occurred during April 2002, a coronal X-ray source was observed above the flare loops and was detectable to about 20 keV (Sui and Holman 2003; Sui et al. 2004). This source was initially stationary before moving outwards at around 300 km s^{-1} (Sui and Holman 2003). *RHESSI* images typically show a height dependence on energy in the flare loops, with the higher energy X-ray sources located above the lower energy sources, whereas here the lower-energy sources are located above the higher-energy sources (e.g.,

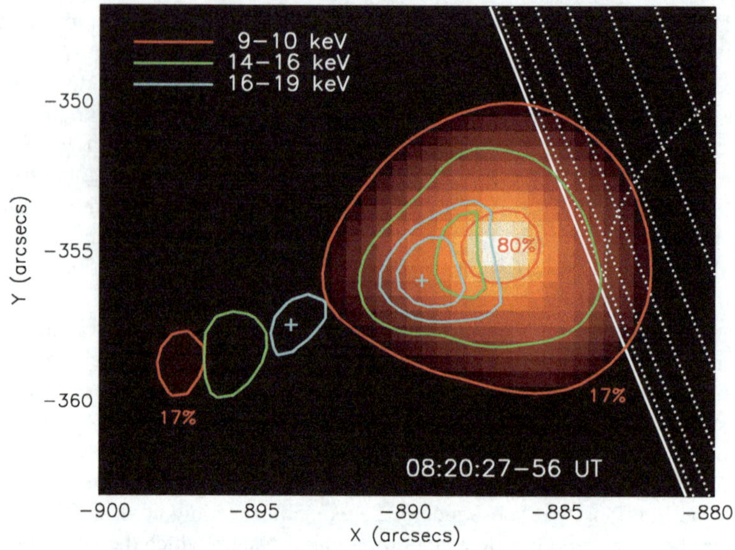

Fig. 25 *RHESSI* soft X-ray observations (Liu et al. 2008b) of a double coronal SXR configuration in SOL2002-04-30T08:22 (M1.3) suggesting the presence of a current sheet, as originally discovered by Sui and Holman (2003). The separation between coronal and loop-top sources decreases as a function of photon energy. The separation between the 16–19 keV source centroids (+signs) was found to be $(4.6 \pm 0.3)''$ in this case. Reproduced by permission of the AAS

Fig. 25). These observations matched the theoretical expectations for signatures of a current sheet formed between the top of the flare loops and the coronal source, although the velocity of the coronal source may be somewhat too low to be consistent with Alfvénic ejection speeds from the upper end of a large-scale current sheet (the same comment applies to the speed of supra-arcade downflows). For one of these events EUV spectroscopic observations were available which showed high-speed, high-temperature plasma flows near the inferred current sheet; these observations were interpreted as reconnection outflows (Wang et al. 2007). Similar but less prominent double coronal X-ray sources were reported by Veronig et al. (2006) in SOL2003-11-0T309:55 (X3.9) and by Li and Gan (2007) in the occulted flare SOL2002-11-02T06:07 (C3.9).

Imaging spectroscopy of a double coronal source was carried out for SOL2002-04-30T08:22 (M1.3) (Fig. 26; see Liu et al. 2004). The HXR footpoints were occulted by the limb, and this facilitated imaging the otherwise relatively faint coronal sources at energies up to ∼40 keV. The two coronal sources, both visible for about 12 minutes, had similar light curves and power-law spectra above ∼20 keV, suggesting production by similar populations of non-thermal electrons possibly energized by a common acceleration mechanism. At low energies ($\lesssim$20 keV), both sources were dominated by thermal emission, and the lower coronal source had a larger emission measure but a lower temperature, suggesting that the different magnetic connectivity above and below the current sheet could lead to different plasma densities. In addition, the trend of the energy-dependent source structure (Sui and Holman 2003) visible at thermal energies showed a reversal above ∼25 keV (see Fig. 26), with the two sources being further away from each other at higher, non-thermal energies. A possible explanation is the larger stopping distances, from the acceleration site, of the higher energy electrons. The above two properties were also found in SOL2003-04-24T15:53 (C8.2) (Liu et al. 2009; see their Fig. 5).

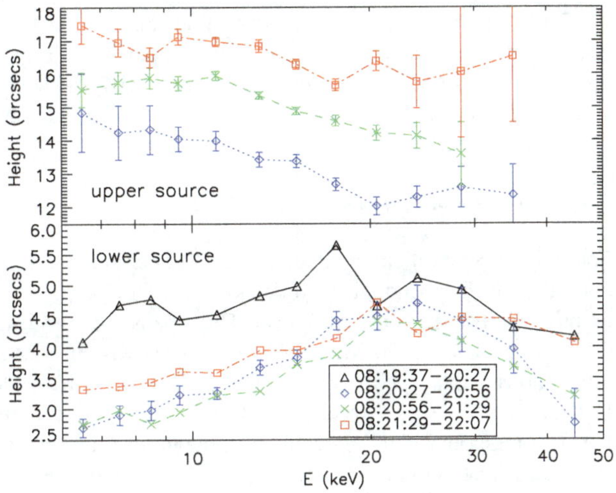

Fig. 26 Height above the limb of the centroids for the upper and lower coronal sources in SOL2002-04-30T08:22 (M1.3) plotted as a function of energy for four consecutive time intervals. Note the reversal of the trend from low-energy thermal to high-energy non-thermal regime (from Liu et al. 2008b). Reproduced by permission of the AAS

4.3 Early-Phase Coronal Sources

In its first γ-ray flare, SOL2002-07-23T00:35 (X4.8), *RHESSI* also discovered a remarkable new type of coronal HXR source. The observations showed a coronal HXR predecessor to the main impulsive phase of a flare (Lin et al. 2003; Asai et al. 2006). The HXR spectrum of these early coronal sources extends directly down to low energies (<10 keV) with power-law indices of around 5, but the spectra also show the characteristic Fe emission feature at 6.7 keV (e.g., Phillips 2004), establishing that some high-temperature background plasma also exists in (or near) the source (Caspi and Lin 2010). While the time evolution of the thermal component is gradual, the emission at higher energies shows time variations of tens of seconds' duration suggesting that the two components are produced by different emission mechanisms. X-ray spectral fitting of the high energy component shows that either a non-thermal model (i.e., broken power-law spectrum) or a multi-thermal model (temperatures up to $\sim$100 MK are needed) can represent the spectra well. However the microwave observations (Asai et al. 2006) favor the non-thermal alternative (see White et al. 2011). This interpretation suggests strong coronal magnetic fields (around 200 G) at relatively high altitudes ($> 2 \times 10^4$ km).

Despite the ambiguous continuum models, the Fe and Fe/Ni line emission can also be used to constrain the thermal plasma parameters. During the impulsive and decay phases of the flare, the fluxes of the two line complexes and their ratio are correlated with the continuum temperature; by assuming that the same relationship holds during this pre-impulsive phase, the observed line fluxes and ratio (which can be accurately measured by *RHESSI*) thus provide upper and lower limits on the temperature and emission measure of the thermal plasma. During the peak of the pre-impulsive phase, the line observations constrain the thermal component temperature to be between $\sim$29 and $\sim$37 MK; a cooler component with temperature between $\sim$21 and $\sim$18 MK (respectively) is also required to fit the SXR spectrum. Consequently, the low-energy cutoff of the non-thermal electrons is at least as low as $\sim$20 and $\sim$27 keV, respectively (Caspi and Lin 2010).

Strong coronal magnetic fields are also supported by observations of the thermal plasma later in the flare. At the time of peak temperature of SOL2002-07-23T00:35 (X4.8), the thermal energy density of the super-hot component ($T > 50$ MK) was $\sim$4800 erg cm^{-3}, suggesting a coronal field strength exceeding $\sim$350 G to contain it. We note that a survey of 37 M- and X-class flares shows that strong coronal fields (>200 G) are required in all X-class flares, which were also invariably super-hot (Caspi 2010) in nature. All three HXR-dominated pre-impulsive sources observed by RHESSI were X-class flares, and this suggests an intimate link between the existence of super-hot plasma, strong coronal fields, and the existence of such pre-impulsive HXR sources.

4.4 High Coronal Sources

For flares occurring more than about 20° behind the solar limb, the occultation should normally be deep enough to hide not only the footpoint sources, but also the main flare loops as well (e.g., Tomczak 2001). This opens the possibility of observing emissions from the high corona (~200 Mm above flare site), and in fact the early non-imaging observations showed that such events really do happen (Frost and Dennis 1971; Hudson 1978; Hudson et al. 1982), if rarely. HXR emissions from a flare occulted by 40° as seen from Earth, corresponding to an occultation height of roughly a third of a solar radius, have been reported (Kane et al. 1992). Despite this large occultation height, HXR emissions were observed up to 80 keV, with a rather hard spectrum ($\gamma < 3.5$). Another high coronal event observed by *Yohkoh* (Hudson et al. 2001) revealed rapid outward source motions with accompanying microwave emission. The event morphology suggested filament eruption and CME occurrence. An early HXR stereo observation using multiple spacecraft (Kane et al. 1979a) showed that the HXR emissions from the high corona can occur during the impulsive phase of the flare simultaneously with the HXR footpoint emissions, and the source size might at the same time be large (of order 200″).

The *RHESSI* observations (Krucker et al. 2007) of high coronal events all tend to have similar time profiles. They show a fast rise and a slower exponential decay, as illustrated in Fig. 27. The exponential decay is surprisingly constant, lasting sometimes several minutes without significant deviation, and the photon spectrum exhibits progressive spectral hardening (hence the "soft-hard-harder" morphology—Cliver et al. 1986; Kiplinger 1995; Grigis and Benz 2008). The decay suggests that collisional losses—without further acceleration—dominate. Density estimates of the ambient plasma support this; these allow estimates of collisional loss timescales of 25 keV electrons comparable with those measured. While the flare-accelerated electrons in the high corona are only a small fraction (0.1%) of the total number of accelerated electrons in the flare (Kane et al. 1992), the relative number of energetic electrons (>10 keV) in the high coronal source may be of order 10% of the thermal electrons.

4.5 Gradual Late-Phase Sources

Gradual late-phase sources are characterized by flat HXR spectra (power-law index $\gamma \approx 2$), gradual time profiles, low microwave peak frequencies, anomalously weak SXR emission, and association with coronal radio bursts. The prototype event SOL1969-03-30T02:47 (Frost and Dennis 1971) occurred in an active region known circumstantially to have been some distance behind the solar limb (Palmer and Smerd 1972) so that the HXRs visible from the Earth-orbiting *OSO-5* spacecraft probably originated from relatively high in the corona.

A "soft-hard-harder" pattern of spectral evolution characterizes many long-duration HXR events (Cliver et al. 1986; Kiplinger 1995; Grigis and Benz 2008). This pattern differs from the otherwise ubiquitous "soft-hard-soft" pattern associated with the impulsive phase (Parks and Winckler 1969; Hudson and Fárník 2002; Grigis and Benz 2004). From a non-imaging perspective, the HXR spectrum of such a source consists of a gradual, continuously hardening component plus a series of spikes with soft-hard-soft evolution. Often these spikes

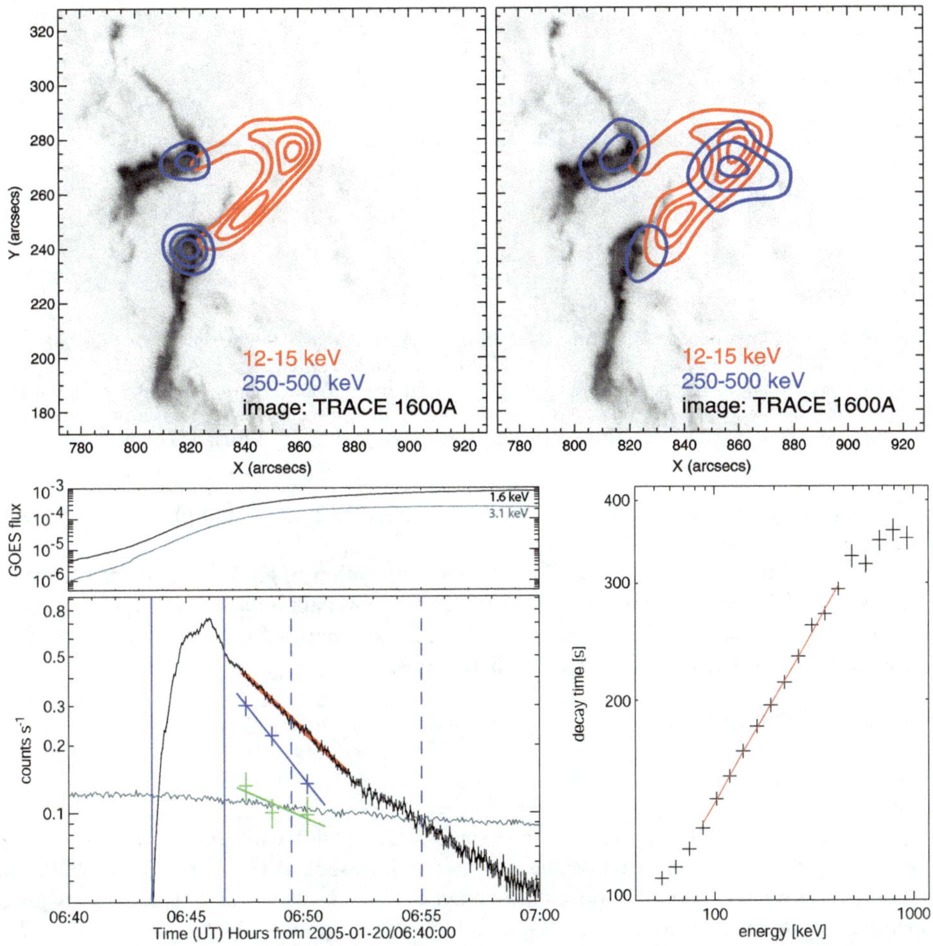

Fig. 27 *Upper: RHESSI* hard X-ray observations at 12–15 keV (*red*) and >250 keV (*blue*) from SOL2005-01-20T07:01 (X7.1) (Krucker et al. 2008a), showing a strong high-energy coronal hard X-ray source. The background *TRACE* image shows the flare ribbons as observed in the UV by *TRACE*, and the contour levels are at 30–90% of the image maxima. *Lower: at left, the light curves*, with *GOES* soft X-rays *at the top* and *RHESSI* hard X-rays *at the bottom*. The latter show three points in the time histories of the looptop and footpoint sources at >250 keV, while the *full curve* shows the total. The footpoints decay more rapidly. The *plot at lower right* shows the decay time as a function of photon energy. The *vertical lines at lower left* show the integration times for the images

become more gradual as the event develops (Saldanha et al. 2008); see also the early non-imaging observations from the *TD-1A* spacecraft (Hoyng et al. 1976). Figure 28 illustrates this.

The physics of the gradual-phase coronal HXR sources with their "soft-hard-harder" temporal development remains to be worked out. There is every reason at present to suppose that the observed spectral flattenings and exponential-law time decays can be explained by some combination of trapping and collisions, but important theoretical work involving wave-particle interactions, specifically loss-cone instabilities, and large-scale magnetic restructurings remains to be done. A possibly interesting theoretical aspect of these sources is

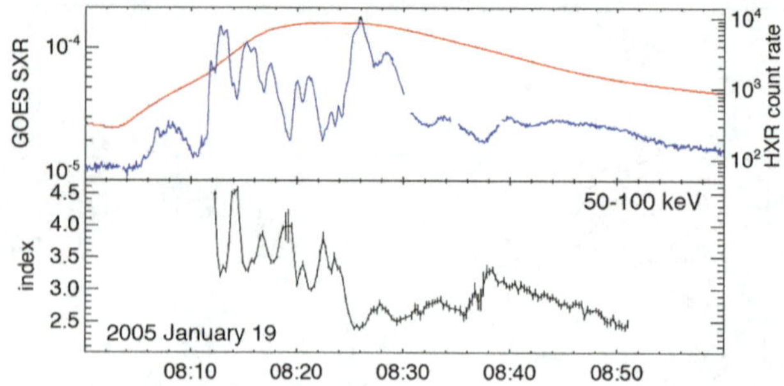

Fig. 28 Recent illustration of the soft-hard-harder spectral evolution in the late phase of SOL2005-01-19T08:22 (X1.3) (adapted from Saldanha et al. 2008). Note the anticorrelation between HXR flux (*upper, blue line*; *the red line* shows the *GOES* SXR flux) and HXR spectral index (*lower*) at the beginning of the event, during the more impulsive variations. Towards the end the variability has longer time scales and the spectral index systematically diminishes (gradual hardening). Reproduced by permission of the AAS

the idea that the non-thermal particles could dominate the background plasma component energetically (e.g., Hudson et al. 2001).

Because of their long duration, a thin-target explanation might be imagined, such that the HXR emission would come predominantly from a coronal trap. However it now appears (Qiu et al. 2004) that these late sources emit hard X-rays mainly from footpoints, at least at energies below 100 keV, as with ordinary flare loops.

4.6 Looptop Source Motions

4.6.1 Upwards

A flare arcade gradually develops to larger and larger scales with time, a process which has long been observed in chromospheric emission lines such as Hα and in SXRs. *RHESSI* observes this phenomenon in a different manner because of its uniquely sensitive imaging spectroscopy in the 3–20 keV range. Figure 29 shows this graphically for one of the first X-class long-duration flares observed by *RHESSI* (Gallagher et al. 2002). The "shrinkage" expected (Švestka et al. 1987; Forbes and Acton 1996) from the standard reconnection model is clearly visible, in the sense that the higher-energy (higher-temperature) X-ray sources, identifiable with loop tops, lie systematically at higher altitudes than the EUV sources that they presumably evolve into.

4.6.2 Downwards

Observations of hot flare loops made by *RHESSI* in X-rays show that in many cases the loops contract downward during the early, most explosive part of the flare before the apparent outward expansion is observed. The downward motion of the X-ray loop-top centroid early in the impulsive phase of solar flares has only recently been recognized. This new observation may have been missed with the *Yohkoh* observations because of coverage biases induced by the operation of its flare mode, which initiated hard X-ray spectral observations only at a soft X-ray flux level typically corresponding to a low C-class flare. The first reported observation, during the rising phase of SOL2002-04-15T00:15 (M3.7), showed shrinking of the underlying HXR flaring loop at $\sim$9 km s^{-1}(Sui and Holman 2003). Several further reported events confirm this pattern (e.g., Liu et al. 2004; Sui et al. 2004; Veronig et al. 2006; Joshi et al. 2007).

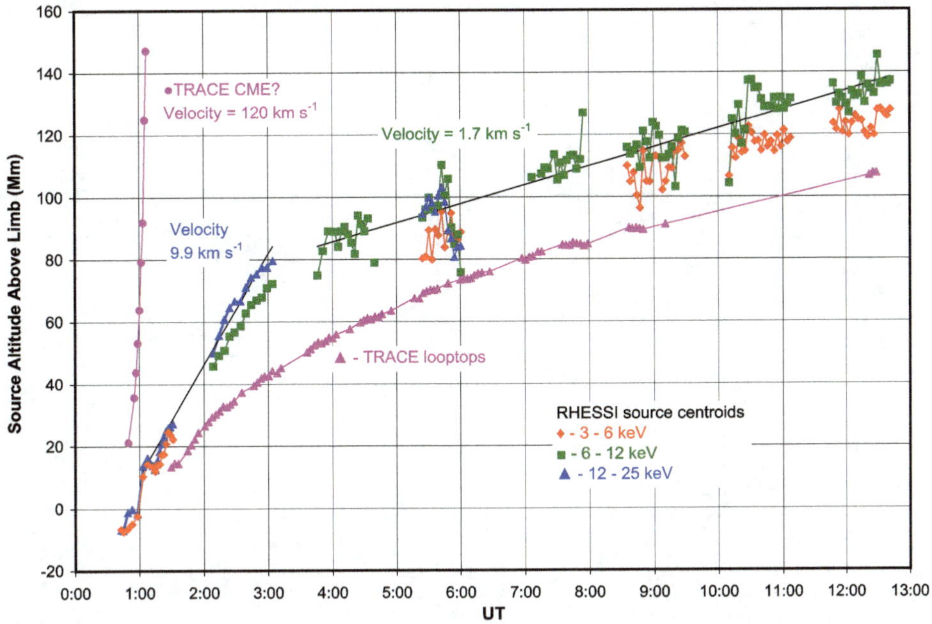

Fig. 29 Observations of SOL2002-04-21T01:51 (X1.5), an event at the extreme W limb (adapted from Gallagher et al. 2002). The plot shows apparent altitude above the solar limb as a function of time for features seen in *RHESSI* and *TRACE* images. The slower motions are related to the growth of the arcade, and the *RHESSI* data clearly show higher temperatures at larger altitudes. The rapid motions at the beginning are associated with the CME (Gallagher et al. 2003)

In all events, the looptop sources of the flares at higher X-ray energy bands were located at higher altitudes and showed higher downward velocities than at lower energies. For example, in SOL2003-11-03T09:55 (X3.9) (Liu et al. 2004; Veronig et al. 2006) the mean downward velocities range from 45 km s^{-1} in the 25–30 keV band to 14 km s^{-1} in the *RHESSI* 10–15 keV band, and in SXR observations from the *GOES* Soft X-ray Imager (SXI) the looptop altitude decreased at 12 km s^{-1}, in agreement with the general trend (Veronig et al. 2006). This trend of lower speed at lower energies is also carried through the wavelength range to loops that are both shrinking and cooling, visible in EUV and Hα (Vršnak et al. 2006). An interesting phenomenon that follows the energy-dependent looptop source velocity is the anti-correlation between the HXR flux and the separation between emission centroids of the looptop sources at different energies as shown in Fig. 30 (Liu et al. 2004; Liu et al. 2008b). This has been interpreted as the looptop source being more spatially homogeneous during the HXR peaks, as a consequence of the rapid energy release. It is now known that the early downwards motion of coronal HXR sources is a common characteristic of flares.

The explanation for the converging motion of conjugate footpoints and the simultaneous descending of looptop sources is still in a preliminary stage of development. As a basic consequence of the extraction of excess energy from the magnetic field the contraction of flaring loops at the initial phase of solar flares should occur as an "implosion" (Hudson 2000; Janse and Low 2007; Liu et al. 2009), a process consistent with the observations mentioned in the previous paragraph (see also Sect. 3.4). In the framework of a reconnection model with sheared magnetic field lines, the contraction might be caused by the relaxation

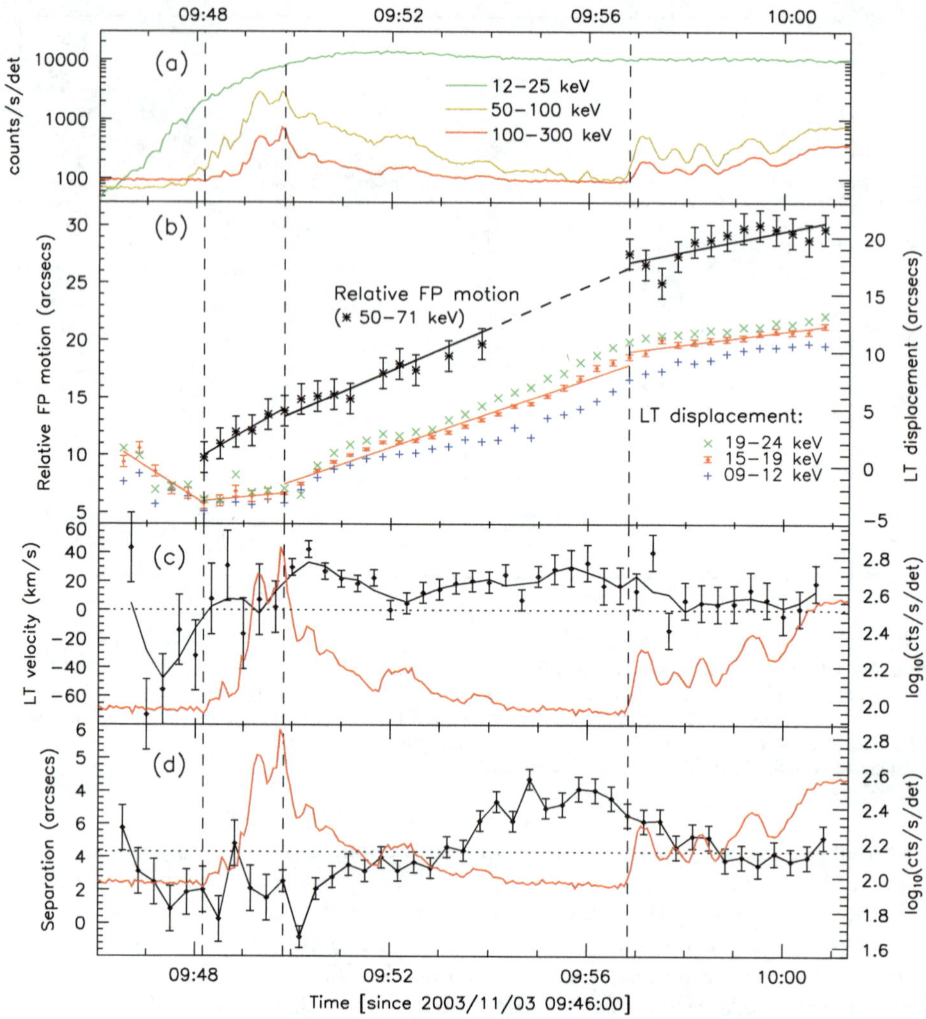

Fig. 30 Looptop and footpoint source motions in SOL2003-11-03T09:55 (X3.9) and the energy-dependent source structure. (**a**) *RHESSI* light curves. (**b**) Projected height of the looptop centroid (*right scale*) and the separation of the two footpoints (*left scale*), which both increase at comparable speeds. Note the early downward looptop motion. (**c**) Velocity of the looptop at 15–19 keV, with the values smoothed over 1-minute intervals shown as the *dark line*. The *red curve* here and in panel (**d**) is the logarithm of the 100–300 keV count rate (right scale). (**d**) Separation of the looptop centroids at 19–24 keV and 9–12 keV, which is in anti-correlation with the HXR count rate (from Liu et al. 2004). Reproduced by permission of the AAS

of highly sheared magnetic field after magnetic reconnection. In a force-free arcade in which the magnetic field strength decreases exponentially with height, the dissipation of magnetic energy in a flaring region could lead to a decrease in the scale height of the magnetic field and thus a shortening of the field lines (Ji et al. 2007). Therefore, in the initial phase of flares, the contraction caused by the relaxation of highly-sheared core magnetic field may dominate over the apparent expansion of the hot loops which occurs as a result of reconnection taking place at higher and higher altitudes. In such a case it is interesting to note that the remaining shear may indicate stresses in the field remaining even after the flare has occurred.

5 Flare Relationships to Coronal Mass Ejections, Dimmings, and Particles

5.1 Overview

The large-scale behavior of the solar corona during a flare or CME disruption has many observational consequences. Traditionally the corona itself has only been observable as Thomson-scattered photospheric light, but well above the photosphere so as to avoid its glare. Thus most of the low corona, with high density and strong magnetic field, was unobservable. Now with SXR and EUV imaging one can study essentially the same structures via these thermal emissions (adding to what one can learn from radio techniques; see White et al. 2011, for further information).

A flare and/or CME disrupts the coronal structure in ways that are not completely clear yet, and in the process of this large-scale dynamics there is powerful particle acceleration. It is impossible in this short review to do justice to the vast number of studies of CMEs both at the Sun and in interplanetary space, and so we focus here on how the flare, the CME and the related particle acceleration fit together observationally, a process sometimes likened to the fable of the blind man and the elephant (e.g., Hudson and Cliver 2001; Cliver and Hudson 2002, where much more extensive and amusing reviews of the flare/CME relationship can also be found).

A flare/CME event marks the conversion of stored magnetic energy into various other forms that propagate through the solar atmosphere and into interplanetary space. Traditionally the brightening is associated with the "flare" and the motions with the "CME" (Cliver 1995), but the physics of either phenomenon seems to require both motions and brightenings. Ultimately, all of the energy extracted from the magnetic field appears either as energy associated with the CME, or as enhanced radiative output of the Sun. The dominant term of the CME energy appears to be its kinetic energy, which can be estimated from LASCO images (Vourlidas et al. 2000). The dominant energy product of a flare, of course, is the transient excess it makes in the solar luminosity. We discuss this in detail in Sect. 6.2. One immediately evident property of the flare/CME combination is the comparative sizes of the flaring region and the associated CME, illustrated quite startlingly in Fig. 31.

5.2 Flare Energy Release and CME Dynamics

From reconnection models it is supposed that the CME kinematics and the energy release of the associated flare are closely related. The degree of association between flares and CMEs had always been problematic, largely because of the poor coverage of the low corona provided by coronagraphs. Almost half of the CMEs, for example, originate on the far side of the Sun, for which no low-coronal observations had been available until the advent of *STEREO*. Nevertheless the X-ray observations of coronal dimmings (Hudson et al. 1996) and EUV observations of compact CME sources (Dere et al. 1997a) had made it clear that there was often a very tight relationship between flares and CMEs.

Further studies revealed a close correlation between the CME acceleration and the derivative of the flare SXR flux, taken as a proxy for the flare energy release (Zhang et al. 2001, 2004; Maričić et al. 2007). For some well-observed CME/flare events, further direct evidence was recently provided showing a very close synchronization between the CME acceleration and the *RHESSI* HXR flux. For example, in the SOL2005-01-17T09:52 (X3.8) flare/CME event and in SOL2006-07-06T08:36 (M2.5) (Temmer et al. 2008), the use of *GOES*/SXI and *TRACE* running difference images showed that the CME impulsive acceleration in the low corona and the flare energy release (deduced from the *RHESSI* HXR flux)

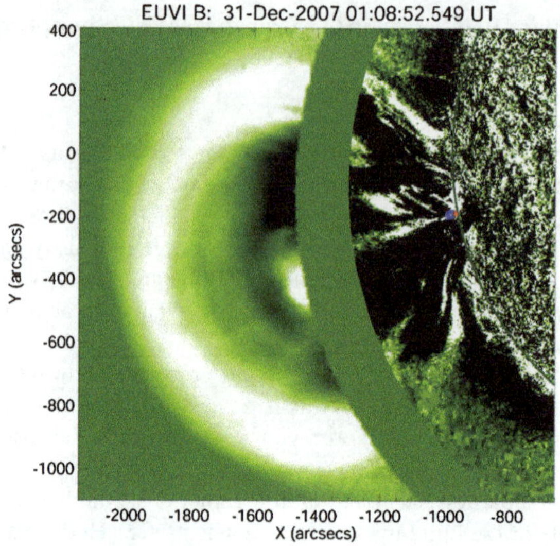

Fig. 31 A combined *RHESSI*/*STEREO* image of SOL2007-12-31T01:11 (C8.3). It is composed of a difference image of the solar disk and inner corona, which illustrates well the loss of coronal plasma referred to as the "dimming," *RHESSI* sources (*red and blue contours*) and the CME bright front, dark following cavity and inner bright core. As is apparent, a CME is a very much more spatially extensive phenomenon than is the flare (image courtesy S. Krucker)

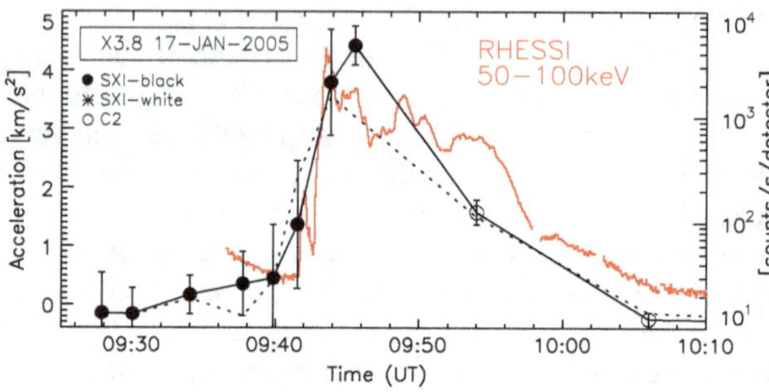

Fig. 32 Flare HXR flux (*RHESSI* 30–100 keV) and CME acceleration profile derived from *GOES*/SXI SXR images and *SOHO*/LASCO coronagraph images for SOL2005-01-17T09:52 (X3.8). Note the close synchronization (Temmer et al. 2008). Reproduced by permission of the AAS

are closely synchronized, and peak simultaneously within ±3 min, i.e., within the CME measurement uncertainties (see Fig. 32). Such correlations provide strong evidence that the CME large-scale acceleration and the flare particle acceleration are intimately connected phenomena, reflecting the rapid extraction of energy from the reconfiguring field both below and above the coronal reconnection region. The effect of the magnetic boundary conditions (i.e., with field being line-tied to the photosphere below the coronal reconnection region, and able to expand relatively freely above) obviously has a substantial impact on the energetically dominant terms in the flare versus the CME.

5.3 Large-Scale Waves

The restructuring of the large-scale coronal magnetic field implicit in a flare and a CME can be considered as a magnetic impulse, and the magnetic field that permeates the surrounding corona and deeper atmosphere guarantees that large-scale waves will ripple away from the site (for a recent review of coronal waves see Vršnak and Cliver 2008). Because of dispersion, any of the global waves will shock and dissipate their energy in non-thermal effects, such as the acceleration of few-keV electrons in a radio type II burst. The waves may contain large energies, as discussed in Sect. 6.3.6. In general an analysis of the structure and timing of these waves may provide key clues to the nature of the energy release by helping to define its geometry.

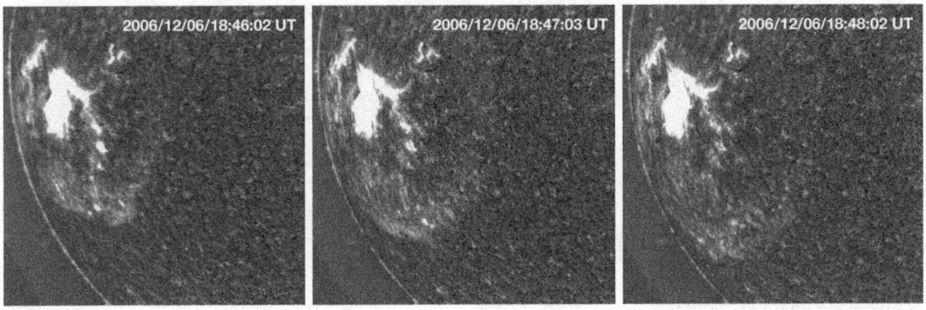

Fig. 33 Moreton wave observed in Hα at Sacramento Peak in SOL2006-12-06T18:47 (X6.5). Courtesy K. S. Balasubramaniam

The existence of large-scale coronal shock waves has in fact been known since the early interpretation of the "slow drift" or metric type II radio bursts (e.g., Wild et al. 1963). These were then linked to the Moreton waves, which are observed in chromospheric Hα signatures (Athay and Moreton 1961), by the theory of a weak fast-mode MHD shock (Uchida 1974). Figure 33 shows a well-observed example of a Moreton wave, which typically travels across the chromosphere at 1000 km s^{-1}. Even before this time "flare ejecta" or "driver gas," now known as an ICME (for Interplanetary CME), had been identified as the cause of the "storm sudden commencement" geomagnetic effect. This is the abrupt onset of a magnetic storm resulting from the compressive interaction of an interplanetary shock wave driven by what we now term an ICME (e.g., Caroubalos 1964) with the magnetosphere Now in addition to this well-known coronal and chromospheric evidence of large-scale waves, we can add the "EIT wave" observations (Moses et al. 1997; Thompson et al. 1998). These EUV perturbations take the form of an expanding wave front most clearly visible in difference images. In a small subset of the EIT waves one can make an identification with the Moreton wave/type II phenomenon, but for the most part they have clearly different properties (Biesecker et al. 2002). Their interpretation in terms of X-ray, EUV or white light dimming (or depletion) (e.g., Hudson et al. 1996; Sterling and Hudson 1997) is complicated because of the temperature sensitivity of the EIT response. The EIT signature is presumably a mixture of true depletion, simple waves, and large-scale restructurings of the field as required by the CME.

The similarity of the radio signatures (metric type II for the flare-associated wave, and interplanetary type II for the CME-driven wave) have led to much recent discussion (see Pick et al. 2006, for a recent comprehensive survey) regarding the distinction, if any, between the meter-wave and the interplanetary shock signatures. The radio spectrograms at long and short wavelengths each have complex signatures, and have been traditionally made in disjoint spectral bands. It has thus been common (e.g., Cliver et al. 2004) to speculate that a common mechanism, specifically the CME bow wave, could explain all of the large-scale wave observations. Indeed, the coronagraphic observations show image evidence, in many cases, for CME-driven shocks in the middle corona (Vourlidas et al. 2003; Ontiveros and Vourlidas 2009). However there is no clear evidence for continuity in the radio signatures, using new observations in the 1–14 MHz range that separates the traditional ground-based and interplanetary observations (Cane and Erickson 2005). The distinction between a flare origin and a CME origin has also become more difficult to make now that improved data have established tighter relationships between flares and CMEs in both point of origin and timing (see Fig. 32). The coronagraph data show the importance of the flanks of

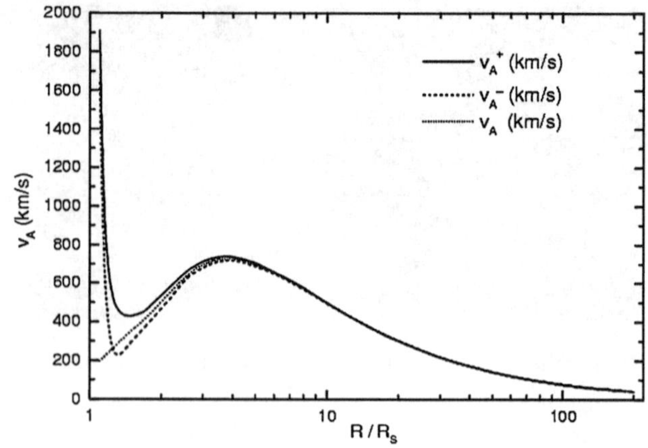

Fig. 34 Model variation of Alfvén speed with height in the corona, showing the local minimum above an active region (Mann et al. 2003). Here the *dotted line* shows the quiet Sun, modeled as a simple dipole at Sun center, and the active region as a vertical dipole near the surface. The *solid and dashed lines* show the cases where these two components have parallel or antiparallel radial contributions

the CME-driven wave, both directly and also indirectly via the excitation of streamers that the flanks intersect. SXR observations may show the earliest signatures of the large-scale wave disturbance (Khan and Aurass 2002; Hudson et al. 2003).

5.4 Solar Energetic Particles and Particle Acceleration

Shock waves can be efficient accelerators of high-energy particles, and there is clear (though indirect) evidence for shock acceleration of SEPs (e.g., Reames 1999). Indeed, the energy ending up in energetic particles can be a substantial fraction of the CME kinetic energy (see Sect. 6.3.6). The shock acceleration of SEPs probably takes place at some distance from the event origin (Kahler 1994). This would be consistent with the idea that the shock condition does not develop immediately, presuming that larger Mach numbers correspond to more efficient acceleration. The existence of a minimum in the Alfvén speed in the middle corona (Mann et al. 2003) allows the CME-driven disturbance to attain a higher Mach number even as its absolute speed may be constant or even decreasing. Figure 34 shows a model view of the coronal Alfvén speed (Mann et al. 2003).

A more directly flare-related acceleration of SEP ions may also occur, and the fact that SEP occurrence is strongly associated with a "soft-hard-harder" spectral evolution (Kiplinger 1995; Grayson et al. 2009) may support this more direct connection. Solar electron events detected in interplanetary space have a strong association with metric-decimetric type III bursts (e.g., Lin 1970). Here the starting frequency of the radio emission, taken to be the plasma frequency or its first harmonic, points to relatively high densities (the lower corona) consistent with a process physically close to the flare site.

6 Physical Properties of Flares

6.1 X-ray Spectroscopy

The emission spectra of flares, especially in the SXR range, convey the most direct information obtainable by remote sensing of the flaring coronal plasmas. *RHESSI* touches on this domain via its capability for measuring the thermal free-free and free-bound continua, as well as to detect the K-shell emission lines of highly ionized Fe around 6.7 keV. These spectral features appear commonly in a wide variety of astrophysical sources, such as active galactic nuclei, stellar flares, and supernova remnants. The parameters thus available include the electron temperature T_e, the emission measure $n_e n_i V$, and information about elemental abundances given adequate models of the plasma physics and the atomic physics. *RHESSI* data, for example, can determine the abundance ratio Fe/H from the equivalent width of the Fe-line feature at ∼6.7 keV (a measure of the ratio of the Fe-line flux to the continuum flux at the Fe-line energy). Note that the Fe feature actually consists of many lines in the K-shell energy range, as described by Phillips (2004). The solar Fe K-shell spectral feature typically does not exhibit the ∼6.4 keV emission of Fe in low ionization states. This is commonly observed elsewhere in the Universe, for example, in reflection spectra from accretion disks, but is largely absent in the solar spectra (in favor of the ∼6.7 keV emission band due to high-temperature plasmas).

6.1.1 Thermal Continua Observed by RHESSI

The flare thermal spectrum observed by *RHESSI* in the ∼3–20 keV energy range consists of free-free (bremsstrahlung) and free-bound (recombination) continuum emission. The contributions made by these continua vary with energy and T_e. In general, thermal free-free radiation is predominant at lower energies and higher temperatures, as was evident in early calculations (Culhane 1969; Gronenschild and Mewe 1978). Newer continuum calculations are included in the Chianti atomic database and software package (Dere et al. 1997b; Young et al. 2003), but the dependence on energy and temperature is nearly the same as the earlier work. There is now, however, the recognition that the abundances of some elements important for free-bound emission are enhanced in the corona, giving rise to enhanced free-bound emission. For flare temperatures between 10 and 20 MK and with coronal element abundances, the cross-over energy where free-free and free-bound continua fluxes are equal is at the lower end of the range that *RHESSI* observes, so both free-free and free-bound continua are important (e.g., White et al. 2005b). (The two-photon continuum, due to the de-excitation of metastable levels in H-like and He-like ions, is much less important.)

6.1.2 The Fe-Line and Fe/Ni-Line Features Observed by RHESSI

As well as the continuous emission, *RHESSI* observes two line features at ∼6.7 keV and ∼8 keV, known as the Fe-line and Fe/Ni-line complexes. They are composed of numerous individual spectral lines emitted mainly by He-like Fe XXV ions and dielectronic satellite lines emitted by mainly Li-like Fe XXIV and lower Fe ions, with a small contribution from highly ionized Ni lines to the ∼8 keV feature. In coronal ionization equilibrium, these ions are expected to be abundant at temperatures above ∼10 MK, and in confirmation of this, the Fe-line feature is evident in *RHESSI* spectra with $T_e \gtrsim 10$ MK. Both line features are conspicuous for $T_e \gtrsim 20$ MK. The spectral resolution of the *RHESSI* detectors at these energies is ∼1 keV FWHM, which is insufficient for resolving the line structure of the Fe-line and Fe/Ni-line features. Nevertheless, *RHESSI* has the advantages of covering a much broader energy range than previous high-resolution crystal spectrometers (which have only covered the immediate vicinity of the ∼6.7 keV lines) and by directly observing the continuum—some crystal spectrometers have had a strong background due to crystal fluorescence, which can obscure the true flare continuum. A valuable diagnostic means for studying the hot component of the solar flare plasma is thus available (Phillips 2004).

Temperature, line equivalent width and abundance analysis has been carried out for SXRs ($\gtrsim$5 keV) spectra taken during 27 flares observed by *RHESSI*, with *GOES* class between C3 and X8 (Phillips et al. 2006). The measured spectra were fitted with model spectra consisting of a continuum (isothermal free-free plus free-bound emission) and lines at ~6.7 keV to characterize the Fe-line feature and at ~8 keV for the Fe/Ni-line feature. Figure 35 shows an example of this during SOL2003-04-26T03:06 (M2.1), at a time when *RHESSI* was in its single-attenuator state.[14] The temperatures T_e in this analysis ranges between 11 MK and 29 MK (Phillips et al. 2006).

The equivalent widths of the Fe-line and Fe/Ni-line features were also derived. Many flares lasted long enough to allow repeated measurements of particularly the Fe-line equivalent width, mostly during the flare decay phase. All the measurements were made with the thin *RHESSI* attenuator in place. An example of measured equivalent width variations during SOL2002-05-31T00:16 (M2.4) is shown in Fig. 36. Similar calculations were done for the ~8 keV Fe/Ni-line feature. Both are based on a coronal Fe/H abundance ratio of 1.26×10^{-4} (Feldman and Laming 2000), or 4 times the photospheric value. These are compared with theoretical calculations of the equivalent width vs. temperature. There is a general agreement in the trend for both the Fe-line and Fe/Ni-line equivalent widths. This indicates that the coronal Fe abundance is appropriate for this flare, though there is a systematic displacement of the points towards higher temperatures. This may be due to the multithermal nature of the flare plasma, or to the presence of non-thermal effects, or to instrumental effects at high photon count rates. They may also be due to incorrect atomic rates used in the calculation of the He-like Fe XXV ion fractions in ionization equilibrium calculations, since for most of the temperature range shown the fraction of Fe XXV ions is small ($\lesssim 0.3$) where the uncertainties are greatest.

The measured Fe/H ratios for up to 22 of the 27 flares when in the first *RHESSI* attenuator state were found to be consistent with a Fe/H abundance ratio between 0.8 and 1.0 times the coronal value. No large-scale time variations in the Fe/H abundance are apparent, as were derived for Ca/H abundance ratios from the BCS instrument on *SMM* (Sylwester et al. 1984; Sylwester et al. 1998). Ratios measured in the thickest *RHESSI* attenuator state were up to nearly a factor of two higher than the theoretical curve, but for these there was generally a poor spectral fit. The best agreement of measured equivalent widths are for spectra taken with *RHESSI* in its first attenuator state during flare decay stages, for which spectra were apparently more nearly isothermal than near the flare peak and rise stages.

6.1.3 Ratio of Fe-Line to Fe/Ni-Line in RHESSI Spectra

The Fe-line feature at ~6.7 keV in *RHESSI* spectra is made up of Fe XXV lines and Fe XXIV satellite lines, both emitted as a result of transitions like $1s-2p$ or $1s-2s$, whereas most (~85%) of the Fe/Ni-line feature at ~8 keV is made up of Fe XXV lines and Fe XXIV satellites with $1s-np$ ($n \geqslant 3$) transitions (the remaining 15% being due to highly ionized Ni lines). The flux ratio of the Fe-line to the Fe/Ni-line features should therefore be sensitive to T_e because of the different excitation energies of the transitions, and so T_e could be derived independently of the temperature from the continuum emission which might be contaminated with non-thermal emission near the flare impulsive stage. The agreement of the observed and calculated equivalent widths of both line features (Fig. 36) indicates that for most *RHESSI* spectra the continuum temperature (which is plotted as the abscissa in Fig. 36) describes both equivalent widths well, so that the temperature from the ratio of the line features is nearly equal to the continuum temperature.

[14]*RHESSI* achieves great dynamic range via two attenuators, which incrementally (in four combinations, but in actual practice only three) cut off the intense low-energy fluxes of major events and also allow the sensitive detection of weak microflares (Lin et al. 2002).

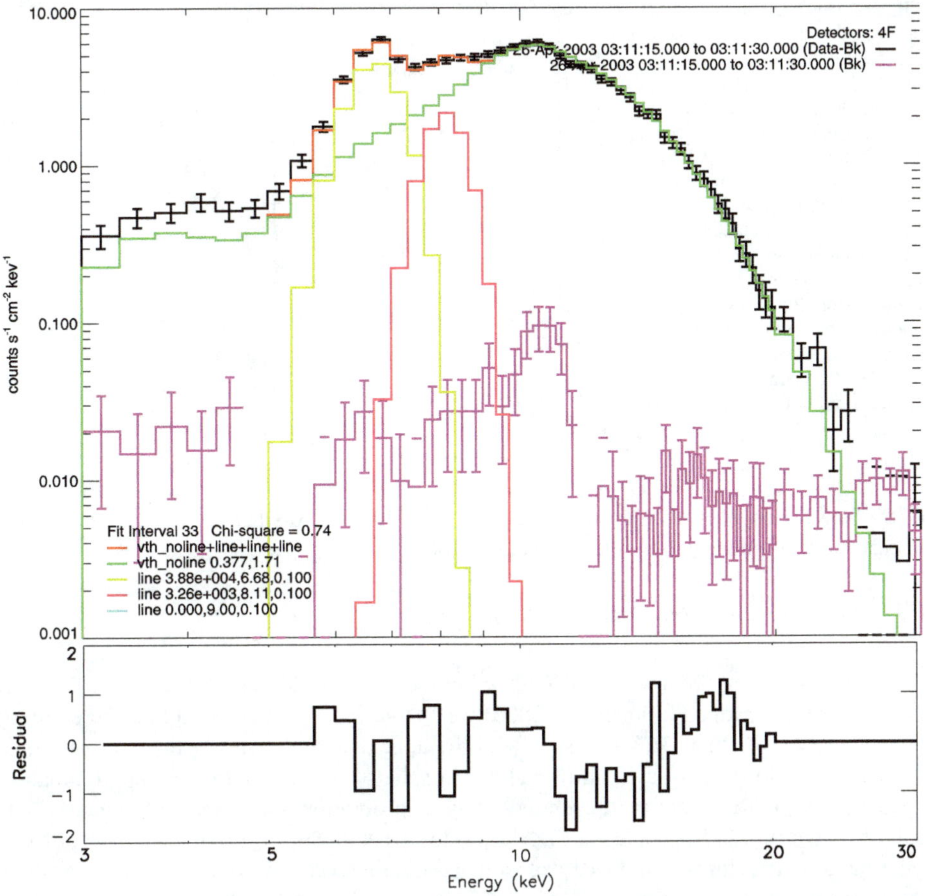

Fig. 35 *Top:* measured and modeled count rate spectra in the energy interval 3–30 keV for *RHESSI* detector 4 shortly after the peak of SOL2003-04-26T03:06 (M2.1). The measured background-subtracted spectrum in the interval 03:11:15-03:11:30 UT is the *black histogram* with $\pm 1\sigma$ uncertainties in each energy bin. The background spectrum is the *purple histogram with error bars* at count rates between $\sim$0.03 and 0.1 counts $(\mathrm{cm}^2\,\mathrm{s\,keV})^{-1}$. The green histogram shows the thermal continuum calculated with the MEKAL atomic code (Meyer 1985) and folded through the spectral response matrix of the *RHESSI* instrument. The histograms with *yellow and magenta lines* are two Gaussian line features representing the Fe-line feature ($\sim$6.7 keV) and the Fe/Ni-line feature ($\sim$8 keV) respectively, while the *red histogram* represents the total model. The fit range was 5.7 to 20 keV, and the reduced χ^2 of the fit was 0.74. *Bottom:* residuals in the fit range plotted as the number of standard deviations

For SOL2002-07-23T00:35 (X4.8), the continuum spectrum is well-fit by two thermal components throughout the impulsive and decay phases. However, the Chianti predictions (cf. Phillips 2004) of the Fe and Fe/Ni line fluxes and ratio based on the observed temperatures and emission measures of the two thermal components are significantly larger than the measured line values—by, on average, $\sim$55%, $\sim$20%, and $\sim$34%, respectively, with larger deviations at lower continuum temperatures (Caspi and Lin 2010). These discrepancies suggest against the interpretation of non-thermal excitation for the lines (as that would produce fluxes in excess of the predictions); while abundance variations could explain the deviations

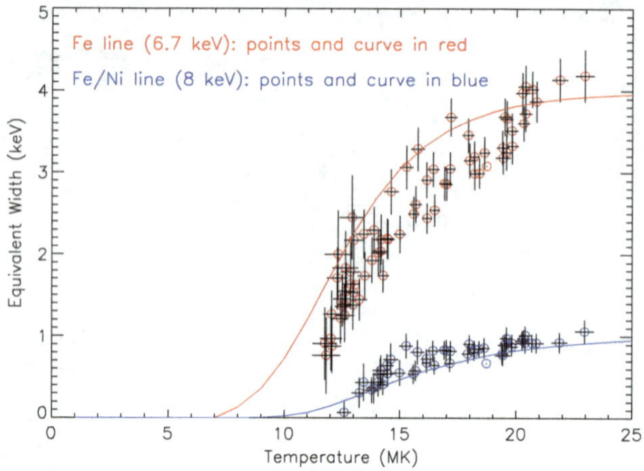

Fig. 36 Measured and predicted equivalent widths (see text) as a function of T_e for SOL2002-05-31T00:16 (M2.4). The *red upper points* are for the Fe-line feature with $\pm 1\sigma$ *error bars* obtained from spectral fits similar to the one shown in Fig. 35. The *blue lower points* show the Fe/Ni-line feature. The corresponding *solid curves* show the theoretical dependence, updated from Phillips (2004). Atomic data used is from the Chianti database (v. 5.1) (Dere et al. 1997b; Young et al. 2003). (Data from Phillips et al. 2006)

of the line fluxes, the variation in the ratio is more likely explained by ionization-fraction uncertainties as suggested by Phillips et al. (2006).

6.1.4 Non-isothermal Effects in RHESSI Spectra

As indicated earlier, the analysis of Fe-line emission in *RHESSI* spectra to derive the flare Fe/H abundance ratio (Phillips et al. 2006) was based on the assumption that the emitting plasma was isothermal. This conflicts with the appearance of flares with multiple loops, in which each loop may have a different T_e. In spite of this the isothermal approximation appears to apply to many flares, even including long-duration flares observed spectroscopically from *Yohkoh* in S XV and Ca XIX lines (Phillips et al. 2005). Such flares have obvious multiple-loop structures. The success of isothermal fits also applies to *RHESSI* spectra at low energies, for which good fits to spectra obtained even with the *RHESSI* attenuators in place are generally achieved. For the rise phase of many flares, however, fits to *RHESSI* spectra are not so satisfactory, even for spectra which appear not to have any non-thermal component in the low-energy continuum. Inspection of *RHESSI* images and the higher-resolution images from *TRACE* and *SOHO*/EIT instrument at the developing stages of flares shows that many individual loop structures contribute to the total flare emission, perhaps with different temperatures.

Analysis of the differential mission measure (DEM) is therefore required, or at least *RHESSI* imaging spectroscopy, for these initial flare stages Although the parameters of the DEM are difficult to determine from *RHESSI* spectra alone, some progress can be made using simple forms for the temperature structure of the developing flare plasma, such as DEM $= T_e^{-\alpha}$ or $\exp(-T_e/T_0)$ where α or T_0 characterize the emitting plasma at any particular time. A more physically based description would involve a set of nearly isothermal components reflecting the multiple loops each in its own cooling equilibrium, so that the DEM parameters α or T_0 would characterize the distribution of the components needed. However, Caspi and Lin (2010) showed that for at least one flare (SOL2002-07-23T00:35), the spectrum is well represented by two isothermal components, and a DEM analysis is consistent with a bimodal DEM. Cursory analysis of other X-class flares reveals a similar bimodal structure. A simple exponential or power-law DEM is therefore not necessarily a reasonable approximation.

6.1.5 Non-thermal Excitation of Continuum and Line Emission in *RHESSI* Spectra

It has been commonly assumed that the non-thermal electrons accelerated during the flare impulsive stage produce bremsstrahlung (free-free radiation) as they interact with the chromospheric or coronal material. The free-free emission is basically due to hydrogen, with small contributions from He atoms or ions, reflecting the solar composition. Free-bound radiation from the non-thermal electron continuum has until recently been neglected, on the grounds that high-energy electrons are less likely to recombine on ions than they are to emit free-free radiation. This has now been questioned (Brown and Mallik 2008, 2009), who argue that for Fe particularly the free-bound radiation may be very important for hot flare sources such as limb flares in which the footpoints are occulted.

This little-studied effect could considerably alter the emission for such flares, and should be taken into account where it is detectable. This mechanism is fundamentally different from free-free emission in that the emitted X-ray energy maps one-to-one with the energy of the parent electron, rather than as an integral over the entire distribution. This means that its spectral features in principle can be interpreted much more directly.

As well as non-thermal effects on the flare continuum, non-thermal electrons may give rise to excitation of lines such as those making up the Fe-line feature at $\sim$6.7 keV. Excitation could occur by the ionization of K-shell electrons in near-neutral Fe, with re-arrangement of the Fe atoms and emission of Auger electrons (67%) or photons (33%), resulting in the Kα or Kβ lines (inner-shell transitions $1s$–$2p$ and $1s$–$3p$ respectively). An energy of at least 7.1 keV is required for the removal of the K-shell electron in each case. Observations of these lines (at 6.4 keV for Kα and 7.1 keV for Kβ) could provide a diagnostic for a non-thermal electron distribution that has sharp cut-off energy E_0, since the lines would not be observed for $E_0 > 7.1$ keV but would if $E_0 < 7.1$ keV (Phillips and Neupert 1973). In practice a sharp cut-off would be quickly smoothed out by interaction of the lower-energy electrons in the distribution with ambient plasma. Most of the observed Fe Kα line emission is due to fluorescence of neutral Fe in the photosphere for disk flares (Bai 1979; Parmar et al. 1984) and not to K-shell ionization by electrons. However, an intriguingly marginal case was observed with the BCS instrument on *SMM* in which excitation by non-thermal electrons might have been significant (Emslie et al. 1986).

6.1.6 Summary

RHESSI was intended primarily as a probe of the non-thermal emission spectra of flares at high energies, but observation of low-energy ($\sim$3–20 keV) flare spectra has yielded important information. Results include the derivation of electron temperature evolution during the peak and decay stages of flares from the thermal continuum based on an isothermal assumption, though simple approximations to the temperature distribution have also been used. The continuum is theoretically due to free-free and free-bound radiation in comparable amounts in the *RHESSI* energy range. The two line features, the Fe-line (at $\sim$6.7 keV) and Fe/Ni-line ($\sim$8 keV) features, enable the abundance of Fe relative to H to be determined from their fluxes relative to nearby continuum emission. Analyses of spectra during the peak and decay phases of flares suggest a coronal value of Fe/H, i.e., one that is larger than the photospheric value by a factor 2–4. This is confirmed by measurements from the SOXS instrument on the *GSAT-2* spacecraft (Jain et al. 2006). Additional temperature information is offered by the flux ratio of the two line features: generally the temperatures derived are similar to those obtained from the energy dependence of the thermal continuum. Nonthermal effects are currently being investigated particularly for the Fe-line and Fe/Ni-line features, and could be a sensitive probe of the low-energy cut-off energy in the non-thermal electron distribution if one exists.

6.2 Flare Energetics

In the matter of constructing the overall energetics of a flare and its associated mass ejection, the requirement for multi-wavelength observations is clear. Direct measurements of the total radiative output are available for only a few flares, notably SOL2003-10-28T11:10 (X17.2) from the Total Irradiance Monitor on the *SORCE* spacecraft (Woods et al. 2004). While such measurements are the most accurate for estimating the total energy released in an event, information about the nature of the energy release process itself can be acquired only through analysis of the partition of energy amongst the various components such as energetic particles and thermal plasma that are present as the flare proceeds.

The first attempts to estimate the total irradiance excess from a flare were made with the Active Cavity Radiometer Irradiance Monitor (ACRIM) on board the *Solar Maximum Mission*. These observations, unfortunately, yielded only upper limits (Hudson and Willson 1983). The TIM instrument currently flying on *SORCE* has now made definite observations (Woods et al. 2004, 2006), and these have proven to be a key factor in our new ability to characterize the partition of energy since they provide a direct measure of the total flare radiation. These results may improve with time if filtering can be developed to reduce the TIM background fluctuations. However the best way to measure this important parameter sensitively could be to have imaging bolometric measurements, which would avoid the large background fluctuations due to p-modes and convective motions in the rest of the photosphere.

Considerable constraints on the energy release processes follow from a consideration of the partition of the released energy between accelerated particles, radiation, heated plasma, and ejected solar material. But any exercise of this sort must be done rather carefully in order not to "double count" energy terms that are directly related to each other, e.g., energy in accelerated electrons that is used to produce thermal plasma (Emslie et al. 2005) or radiation. This requires distinguishing amongst "primary" components of energy (e.g., the magnetic field), "intermediate" components (e.g., accelerated particles and thermal plasma), and "final" components (e.g., kinetic energy of ejecta, radiant energy in various wavebands), and recognizing the overlap of these components.

In the *RHESSI* era, the partitioning of energy in two well-observed solar flare/CME events was carried out using data from a variety of missions including *RHESSI, ACE, SOHO* and *GOES*. This study yielded the result that "flare radiant energy and CME mechanical energy are the same order of magnitude." The SXR flare (from 1.5 keV) appears to contain substantially less than about 10% of the total radiant energy (Emslie et al. 2005). The impulsive-phase radiation appears to dominate the flare luminosity. Both the SEPs and the impulsive phase acceleration contain a substantial fraction of the total energy (Emslie et al. 2005), as described below.

6.3 Energetics of Two Large *RHESSI* Flares

A number of previous studies have examined the energy budget of a limited number of energy components in certain flares. The radiative energy budget of SOL1973-02-05 was evaluated by Canfield et al. (1980), but in the absence of HXRs or γ-ray observations for this event the role of energetic particles in the event could not be assessed. The X-ray and γ-ray observations of several flares, including the major γ-ray flare SOL1972-08-04, were used to show that the kinetic energy of the accelerated electrons constituted a surprisingly large fraction of the total flare energy, perhaps as high as 10 to 50% of the $\sim 10^{32}$ erg released during the flare (Lin and Hudson 1976). Two flares within the same active region

Table 3 Flare and CME energy budgets[a]

	21 April 2002	23 July 2002
Primary energy		
Magnetic	32.3 ± 0.3	32.3 ± 0.3
Flare		
Intermediate energies		
Electrons ($> E_{\min}$)	31.3 ± 0.5	31.3 ± 0.5
Ions (> 1 MeV nucleon^{-1})	< 31.6	31.9 ± 0.5
Thermal plasma ($T > 5$ MK)	$31.1^{+0.4}_{-1.0}$	$30.4^{+0.4}_{-1.0}$
Radiant energy		
From *GOES* plasma	31.3 ± 0.3	31.0 ± 0.3
Assuming $L_{total}/L_X = 100$	32.2 ± 0.3	32.2 ± 0.3
CME		
Kinetic	32.3 ± 0.3	32.3 ± 0.3
Gravitational potential	30.7 ± 0.3	31.1 ± 0.3
Energetic particles at 1 AU	31.5 ± 0.6	< 30

[a]Tabulated values are $\log_{10}$ of the energies in erg of the different components given in Table 1 of Emslie et al. (2005)

on 1980 August 31 provided the energy content in thermal plasma, non-thermal electrons, and hydrodynamic mass motions of non-ejected material (Strong et al. 1984), while the energy content in radiative, thermal, non-thermal electron, and non-CME associated plasma ejected was calculated for SOL2002-02-26T10:27 (C9.6) (Saint-Hilaire and Benz 2002). *RHESSI* X-ray observations were used determine the energy in accelerated electrons and in the hot plasma for nine medium-sized flares (*GOES* class C6 to M8), with the conclusion that despite the large uncertainties, the energies in these two components were of the same magnitude in each case (Saint-Hilaire and Benz 2005).

The energetics of two X-class flares (SOL2002-04-21T01:51 and SOL2002-07-23T00:35) have been analyzed in a very comprehensive study made possible by overlapping observations at a variety of wavelengths (Emslie et al. 2004, 2005). SOL2002-04-21T01:51 was a long-lived SXR event which occurred near the west limb; SOL2002-07-23T00:35 was much more impulsive, a strong emitter of HXRs and γ-rays (see Lin et al. 2003), and was located near the east limb at S13E72. Observations were used from instruments on the *ACE*, *SOHO*, and *RHESSI* to provide quantitative estimates of the energy contents of (1) the coronal mass ejection, (2) the thermal plasma at the Sun, (3) the accelerated electrons producing hard X-rays, (4) the accelerated ions producing gamma rays, and (5) the solar energetic particles accelerated by the outward eruptive disturbance/CME. The detailed energy budget for these two events, including the CME kinetic and potential energies and the energy in the SEPs at 1 AU, is reproduced in Table 3.

6.3.1 Magnetic Energy

The total magnetic energy available for conversion into other forms (flare and CME) can be estimated in principle from the extrapolation of the field observed at the photosphere. There are many uncertainties with such a procedure. The flares providing the information in Table 3 were each at the limb and therefore not amenable to the extrapolation in any case. Thus we view the uncertainty of 0.3 dex quoted in the table as highly optimistic.

6.3.2 Energy in the CME

To derive the energy in the CME one must first derive a coronal density distribution from the excess brightness (due to Thomson scattering of photospheric light) in coronagraphic white-light images (e.g., from the LASCO C2 and C3 coronagraphs on *SOHO*; Brueckner et al. 1995). These observations permit an estimate of the mass distribution of the CME (Poland et al. 1981; Vourlidas et al. 2000, 2002). Then, the flow patterns of the plasma during the ejection must be determined from the projected images as a function of time to derive the velocity field of the CME material. Finally these properties can be used to find the potential (U_Φ) energy, kinetic (U_K) energy, and enthalpy associated with the CME. For SOL2002-04-21T01:51 and SOL2002-07-23T00:35, the potential energies obtained were $U_\Phi = 10^{30.7}$ and $10^{31.1}$ erg, respectively, and kinetic energies $U_K = 10^{32.3}$ and $10^{32.0}$ erg. These are unusual CME events. Their large kinetic energies place both of them in the top 1% of all observed CMEs for the period 1996–2000 (Vourlidas et al. 2002). In both cases the gravitational potential energy is $\lesssim$10% of the total energy contained in the CME. Note that the magnetic energy of a CME, thought to be its dominant term, is almost impossible to assess observationally.

6.3.3 Thermal Energy

The thermal energy of the heated plasma is obtained from the temperature T_0 (K) and emission measure $EM = \int_V n_e^2\, dV$ (cm^{-3}) for the thermal portion of the overall spectral fit to the HXR data (see, e.g., Holman et al. 2003). Here n_e is the electron density (cm^{-3}) and V is the emitting volume (cm^3). Account must be taken of the filling factor f, equal to the ratio of the emitting volume to the apparent volume (V_{ap}) as determined with an imaging instrument having limited spatial resolution. The thermal energy content of the plasma is then given by

$$U_{th} = 3n_e k T_0 f V_{ap} \simeq 3kT_0\sqrt{EM \times fV_{ap}} = 4.14 \times 10^{-16} T_0 \sqrt{EMfV_{ap}}\ \text{erg}, \qquad (4)$$

where k is the Boltzmann constant and V_{ap} the source volume, estimated from the area information contained in the *RHESSI* observations assuming $V_{ap} = A^{3/2}$. For the common assumption of a filling factor of unity ($f = 1$), this is an upper limit for the instantaneous thermal energy. It also can be taken as a lower limit to the total thermal energy since it does not account for the cooling of the plasma prior to a given time, nor for any heating at later times. Each of these contributions could add perhaps a factor of two to the total thermal energy. Application of (4) to SOL2002-04-21T01:51 and SOL2002-07-23T00:35 yielded values of $U_{th} = 10^{31.3}$ erg and $10^{31.1}$ erg, respectively. An estimate of the total radiated energy can be rather straightforwardly obtained from the *GOES* SXR data, simply by integrating the product of the emission measure and the optically thin radiative loss function (Cox and Tucker 1969) (for coronal abundances) over the duration of the flare (Emslie et al. 2005). This exercise gives values of $U_R \sim 10^{31.3}$ erg for the SOL2002-04-21T01:51 flare and $10^{31.0}$ erg for SOL2002-07-23T00:35. Note that no knowledge of the source volume, density, or filling factor is required to make this calculation; hence the good agreement between these values and those obtained immediately above suggests that the simplifying assumption of unity filling factor is not unreasonable. In particular, the volume filling factor for the soft-X-ray-emitting plasma cannot be too small (< 0.01), otherwise the plasma energy calculated using the *RHESSI* source areas would be significantly below the estimate obtained from the *GOES* data.

A separate physical argument also suggests that the filling factor cannot be too small. At the time of the peak temperature, the energy density in the thermal plasma, assuming

$f = 1$, already requires coronal field strengths exceeding ~350 G to contain the plasma; this number increases as $f^{-1/2}$ so the filling factor must be no smaller than ~0.01 unless the coronal field can significantly exceed ~1000 G. Caspi (2010) found that this same argument holds for essentially all X-class flares, as they all require coronal field strengths exceeding ~220–460 G. This again suggests that a filling factor in the range 0.1–1 is a reasonable assumption for super-hot, X-class flares.

6.3.4 Accelerated Electrons

The energy in accelerated electrons can be determined from applying a thick-target model see (see Holman et al. 2011) to the measured HXR spectrum in order to obtain the injected spectrum $F_0(E_0)$ (electrons s^{-1} keV^{-1}) and calculating the corresponding injected power from

$$U_e = A_i \int_{E_{min}}^{\infty} E_0 F_0(E_0)\, dE_0, \tag{5}$$

where E_{min} is the lowest particle energy in the non-thermal component of the electron distribution and A_i is the injection area (which, however, cancels in the determination of $F_0(E_0)$). The accumulated energy in non-thermal electrons is then obtained by integrating the injected electron power over time. The thermal spectrum is typically dominant at low energies, so the largest value of E_{min} consistent with an acceptable fit to the spatially integrated spectral data, is chosen; the energies obtained are necessarily lower limits. The values of U_e thus determined were $U_e = 10^{31.5}$ erg for SOL2002-07-23T00:35 and $10^{31.3}$ erg for SOL2002-04-21T01:51. These results are higher than the corresponding values of $10^{31.3}$ erg and $10^{31.1}$ erg for the energy contained in the thermal plasma U_{th}. This result is reinforced by the wide lower error bar on U_{th} caused by the uncertain filling factor f and the fact that U_e may be an *under*estimate. This suggests that much of the electron power is radiated in other wavelengths, such as optical and EUV (see below).

6.3.5 Accelerated Ions

As explained in Vilmer et al. (2011), accelerated ions are also energetically important in energetic solar flares with significant emission above ~300 keV (Ramaty and Mandzhavidze 2000). The primary ions undergo nuclear collisions and thereby produce γ-ray lines and continua of various kinds, for example, by direct de-excitation following inelastic scattering to produce lines mainly in the ~1–10 MeV range (e.g., Ramaty et al. 1979). These are broad enough for *RHESSI* to resolve. There is also a highly-broadened set of lines (a "pseudo-continuum" because of overlaps) resulting from α-particles and higher-Z ions striking ambient nuclei. The flux in the highly-broadened component is typically >3 times that in the moderately-broadened component. The threshold energies for producing all of these nuclear lines are $\gtrsim$2.5 MeV, and so the spectrum below that energy is unknown observationally. Assuming a flat spectrum below 2.5 MeV, a lower limit of (1.0–4.0) × 10^{30} erg of energy and an upper limit of (1.2–120) × 10^{32} erg (assuming a power-law spectrum down to 0.1 MeV) was found for protons in SOL2002-07-23T00:35 (X4.8) (Lin et al. 2003). Protons and heavier ions together range from $U_i \simeq$ (6–24) × 10^{30} erg to (7–700) × 10^{32} erg. For SOL2002-04-21T01:51 (X1.5), for which no significant γ-ray line emission was produced, the upper limit is in the range $U_i \simeq 4.0 \times 10^{30}$ erg to 1.2×10^{34} erg, depending on the spectral model used.

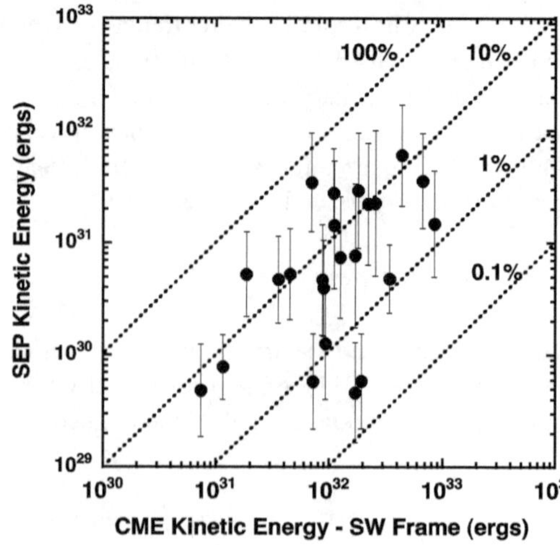

Fig. 37 Comparison of SEP energy with CME kinetic energy for a sample of events, adapted from Mewaldt et al. (2008). There is no suggestion of a correlation, but these estimates do imply that the SEPs contain a large fraction of the total CME kinetic energy. Reproduced by permission of the AIP

6.3.6 Solar Energetic Particles

SEPs accelerated at the flare site and/or at shocks driven by the CME represent another significant contribution to the global energy budget. SOL2002-07-23T00:35 (X4.8) occurred near the East limb of the Sun (S13E72) and, as is typical for east-limb events, was apparently not magnetically well-connected to Earth. As a result, near-Earth spacecraft such as *ACE* and *GOES* did not observe significant SEP fluxes that could be traced to this event. On the other hand, SOL2002-04-21T01:51 at S14W84 was relatively well connected to Earth, and indeed, a strong interplanetary shock (Mach number $M_A = 3.7$) was observed some two days later, at ∼04:15 UT on April 23. That the intensities of the SEPs coincided well with the time of shock arrival indicated that acceleration was taking place locally at the shocks. Integrating the energy spectra over energy and solid angle covered by the shock (typically $\sim\pi$ sr) gives the total particle energy incident at 1 AU as $U_p \approx 2.8 \times 10^{31}$ erg), which is a significant fraction (∼15%) of the CME kinetic energy ($\sim 1.8 \times 10^{32}$ erg). This implies that shock acceleration must be relatively efficient. Indeed, a broader survey of SEP events shows that this result holds commonly, though not invariably (Mewaldt et al. 2008). Figure 37 illustrates this.

6.3.7 Future Improvements

The measurements in Table 3 all have sizeable error bars. How can these measurements be refined? We tackle this line-by-line in the table.

(i) Estimating the primary magnetic energy reliably is a difficult problem, but its solution will certainly involve measurements of the vector magnetic field at the "top" of the chromosphere, with spatial resolution adequate to capture the field close to the magnetic polarity inversion line—scales below one arcsecond. This must be accompanied by fast and robust mathematical methods for the non-linear force-free field extrapolations and/or by direct measurements of the vector field in the corona, e.g., by CoMP or similar instruments.

(ii) Table 2 identifies "intermediate energies," reservoirs between the fundamental magnetic field and the true losses (radiation and ejecta). These table entries are themselves model-dependent, for example, in the cooling time scale of the thermal plasma. Detailed EUV-SXR imaging spectroscopy of flare arcades can sharpen these estimates substantially. The current uncertainties on the non-thermal electron energy budget derived from the collisional thick-target model (which is the model requiring the smallest amount of non-thermal particle energy) lie wholly in a lack of knowledge of the "low energy cutoff" of the electron spectrum, if there is one. One theoretical possibility for deducing this information, is to find edges in the X-ray spectrum produced by free-bound radiation (Brown and Mallik 2008; Brown et al. 2010). No hard X-ray edge structure has yet been reported. The ion (or proton) component is also very poorly constrained, and improving this will require very much better γ-ray spectroscopy, with much improved signal-to-noise ratio. We also have no knowledge at all of lower energy protons (of a few tens or hundreds of keV), though diagnostics such as Doppler-shifted Lyman-α (emitted by charge-exchanging low-energy protons (Orrall and Zirker 1976) and the (disputed) observations of Hα impact polarization (Vogt and Hénoux 1999) might help us in this direction.

(iii) Substantial improvement in CME energetics is now being obtained from the *STEREO* observations. The major uncertainty now, as before, is the lack of knowledge of the magnetic field and its dynamics. Table 2 does not even list this item though it is the dominant one.

(iv) The sparse sampling of SEP fluxes, and the difficulty of bookkeeping their distribution in the heliosphere, are some of the issues. Data from nearer the Sun will help, as will improvements in our understanding of their sources. The observation of energetic neutral atoms via charge exchange (Mewaldt et al. 2009) and neutrons, again from inner-heliospheric vantage points if possible, will be a major step forward. Because of the sampling issues, though, this will likely remain in the domain of statistics rather than precise measurement.

7 Summary: Models and Observations

7.1 Flare Model Constraints

Where within this mass of data can we identify the decisive observations that will choose one model over another? It is a daunting task to find "a flare model" that is capable of explaining all observed phenomena in all events. As we have seen a flare can be geometrically complicated and have many associated temporal components, and its effects appear across a large dynamic range of physical parameters. The extend from the solar interior to beyond one AU. Furthermore, not all flares exhibit the same behavior and many deviate sharply from the "standard" two-ribbon eruptive flare scenario that has framed much of our thinking for decades. For example, some flares do not have associated eruptions, and many have multiple ribbons instead of just two. Some flares continue accelerating electrons well into their X-ray decay phase, some have associated γ rays but many—perhaps most—do not.

Despite having orders of magnitude more data than was available at the time that the "standard" scenario was first formulated, our knowledge is still incomplete, and this incompleteness forces us to invoke cartoons (see http://solarmuri.ssl.berkeley.edu/~hhudson/cartoons/ for a compilation) to relate one observable to another, and to suggest cause and effect. They extend analytical or computational models with some degree of refinement tend to deal separately with different aspects of the flare; e.g., the magnetic configuration and how it evolves, electron acceleration, expansion of the chromosphere, production of radiation of one type or another. These various parts are then stitched together into cartoons—also loosely called models—but without the details of any individual aspect yet being fully worked out.

So, what are the "top-level" components to a flare model? Firstly, we must understand how a particular coronal magnetic configuration can become loaded with or emerge with sufficient stored magnetic energy. We must understand how, after a period of stability, the configuration becomes unstable in such a way as to produce a dramatic energy release. The conversion of stored energy into the various forms which we infer observationally is a—perhaps the—central aspect of flare models, which still defies detailed explanation. A fourth model element—undoubtedly also related to the initial magnetic configuration—must address the relationship between the localized radiation burst that is the flare, and the coronal mass ejection. Each of these top-level model components has many sub-components. For example, considering one element of the energy conversion problem, even after 150 years we do not know the origin of flare optical emission. Several alternative theoretical scenarios exist for this part alone, and likewise for the tens of other observed and inferred flare phenomena. Where within this mass of models can we find the decisive predictions?

Let us confine further discussion to the model elements identified as top-level above, and try and identify some ways in which observations support or refute the various options. This is necessarily a very abbreviated discussion as a full comparison between the multifarious models and the observations would require many more pages.

7.1.1 Energy Storage

The storage of energy presents an interesting problem. If magnetic reconnection, releasing stored energy, can readily happen in a flare, or indeed in coronal heating, why does it not happen right away? This may come down to the configuration of the magnetic field. Observationally, some progress has been made in understanding energy storage using field extrapolations from observed photospheric vector magnetic fields but these remain problematic, not least because the photosphere is not force-free, in contradiction to the basic mathematical assumptions of the extrapolations. The suggestion from the existing extrapolations, at least of newer active regions, is that free energy storage occurs low down in the atmosphere, close to the polarity inversion line (Régnier and Priest 2007; Schrijver et al. 2008). Unfortunately, present observations are barely able to resolve the photospheric vector magnetic field, let alone the chromospheric field, at the arcsecond scales required, and the reconstruction techniques are temperamental, so these conclusions should be treated with caution. Nonetheless, many flares also start with their footpoints close to a magnetic neutral line. This often involves the activation of low-lying active region filaments overlying strong sheared polarity inversion lines. This suggests that the properties of what is known as the "core field" in filament models—the strong, twisted or sheared field supporting filament material—are also core to understanding the early phase of flares, and models capable of producing such configurations, whether by shearing or by emergence of structures with concentrated twist, are very relevant. To that end, we will profit from paying close attention to active-region filament magnetic-field observations, such as those of Wiehr and Stellmacher (1991) or Kuckein et al. (2009), as well as anything that can be learned from coronal field diagnostics in the microwave, IR and UV regimes (Sect. 3.5). Observations diagnosing the typical properties in or near the reconnection region, such as the field strength and connectivity, number density, temperature and velocity, are necessary input to impulsive phase reconnection and acceleration or heating models. Unfortunately the bulk properties readily observable with spectroscopic remote sensing may not be sufficient to understand the conditions that lead to what is in large part a plasma kinetic process (i.e., the decoupling of particles from the magnetic field and from each other). We should be prepared to learn what we can, in terms of the microphysics, from *in situ* observations in the solar wind or terrestrial environment.

7.1.2 Instability and Field Rearrangement

The evolution of a stable, energy-loaded configuration towards an unstable one, and the release of the stored energy, are related problems. Both require an understanding of the conditions under which magnetic reconnection will set in—or not—and this is a vast field of study in itself. One view is that the rate of magnetic reconnection is determined primarily by the plasma microphysical conditions. Another is that reconnection is determined by the magnetohydrodynamic state, and the microphysics somehow adapts to keep up with what the field dictates. But with neither field measurements nor detailed information on non-equilibrium plasma conditions in the corona where flare reconnection is thought to take place, we know little for sure about the reconnection physics. Observations of plasma inflow in the late phase of flares allows estimates to be made of the reconnection rate, which turns out to be compatible with fast reconnection (Yokoyama et al. 2001; Narukage and Shibata 2006). But the rate alone is not very informative, and the late phase of a flare is a less challenging environment in terms of the reconnection and rate of energy conversion required.

The standard flare reconnection model has basically been in development since Giovanelli (1948). Note that Giovanelli's sketches of the magnetic scenarios for flares mainly emphasized the behavior of current systems, rather than flux transfer; as late as 1963 Parker (1963) could describe magnetic-field "annihilation" as a "presently popular belief" to explain solar flares, but that "There is very little in the observations to support such views." With Petschek (1964) augmenting the earlier work of Giovanelli, Cowling, Dungey, Sweet, and Parker himself, the case became much stronger. Nowadays it would be difficult to discuss flare phenomenology without appealing to magnetic reconnection in one form or another, although all of the abundant evidence for it is necessarily indirect. But once again, observational constraints on model geometry are few. On the disk, observationally-grounded extrapolations of the magnetic field are suggestive of one topology or another, sometimes backed up by the shape of the field outlined in EUV. For example, there is evidence for coronal nulls in flaring active regions (Aulanier et al. 2000; Fletcher et al. 2001; Masson et al. 2009). On the limb, double coronal sources are sometimes observed with a temperature structure consistent with outflows from a reconnecting structure, and a vertical displacement between the two sources suggestive of a vertically-extended current sheet (Sui and Holman 2003). Though not strictly flare activity, the magnetic and EUV coronal evolution during a flux emergence event is consistent with a model involving separator reconnection (Longcope et al. 2005). During the impulsive phase, the portion of the field outlined by hot plasma can look very asymmetric and disordered, but later on in the EUV/SXR arcade it looks rather symmetric and quasi-2-D, like the standard cartoon. So all topologies are possible, and perhaps flare magnetic systems evolve through different topologies. However, returning to Parker's discussion of "annihilation," until we know whether we are dealing with reconnection or flux transfer at a neutral point, a neutral line, a separator, a current sheet or indeed a volume filled with many small current sheets it is not possible to say whether the energy released is dominated by field annihilation (i.e., dissipation of antiparallel components of **B** within the reconnecting structure) or field relaxation following reconnection ("dipolarization"). For example, reconnection at a null dissipates essentially no magnetic field because the reconnection volume is so small—energy conversion happens elsewhere in the system—but this would not necessarily apply to a volume filled with many small current-sheet structures.

7.1.3 Energy Conversion

The sudden transformation of energy contained in the solar magnetic field is the heart of the flare problem. As mentioned above, magnetic reconnection itself does not transform much energy, but may facilitate large-scale restructurings that do. The common ingredient in all flares, by any definition of their sudden appearance, is the impulsive phase. We identify this as the timing of the energetically fundamental non-thermal process of particle acceleration as originally observed at radio wavelengths, and then in X-rays and γ-rays as well. It is this impulsive phase, which we can associate with all of the dominant energetic processes of a flare (the white-light/UV continuum emission, large-scale wave generation, CME eruption, and powerful particle acceleration), that is not captured in the many cartoons showing different versions of the standard flare model.

The dominant models hold that particle acceleration is an entirely coronal process, with HXR footpoints produced by accelerated coronal electrons which precipitate, possibly having escaped a magnetic trap formed by the expansion of magnetic field into the corona. The presence of coronal hard X-ray sources (Krucker et al. 2008a) as well as radio bursts provides ample evidence that non-thermal electron populations are present in the corona during the flare impulsive phase, with high fractions of all electrons present being accelerated. In one case in particular, the data suggest that essentially all coronal electrons present in the source are non-thermal (Krucker et al. 2010), with minimal remaining thermal distribution. Details of the acceleration mechanism are obscure, and observations in the optical to SXR regime which deal primarily with thermal (i.e., "processed" energy) are of limited help beyond suggesting the geometry and evolution of the environment in which acceleration occurs. Observational suggestions of a current sheet provide one possible environment, but if one is to produce the electron fluxes necessary to explain the HXR footpoints the required current sheet dimensions are rather unfeasible, unless the Alfvén speed is high (the number of electrons accelerated per second being limited to what can be advected into the sheet each second). We have evidence of magnetic field relaxation (Sui and Holman 2003) consistent with shrinkage beneath a reconnection region, in which betatron acceleration would occur (Somov and Kosugi 1997). But though this process can accelerate a good number of electrons, the energy increases achieved are modest, requiring a suprathermal population to start with (Giuliani and Neukirch 2005; Karlický and Bárta 2006). Non-thermal line broadening in soft X-ray (e.g., Ranns et al. 2000) and EUV lines (Imada et al. 2008; Hara et al. 2008) may be interpreted as evidence for plasma turbulence, a central ingredient in many electron acceleration models, though our observations are at a spatial scale far larger than that at which plasma wave energy can be effectively dissipated by electron acceleration.

At some level, coronal electron acceleration is straightforward even though the details are unknown. A certain amount of energy dumped into the coronal plasma in the form of plasma turbulence, for example, must be shared between the coronal particles resulting in a mean energy per particle. If that mean energy is high enough that the electrons are collisionless on timescales of interest, a non-thermal distribution of some description must result. Even in the exceptional coronal source studied by Krucker et al. (2010) the numbers and energies are plausible. More challenging are the chromospheric HXR sources interpreted as due to escaping accelerated coronal electrons. The standard collisional thick-target electron beam model requires electron fluxes up to a few $\times 10^{35}$ electrons per second (Holman et al. 2003) leaving the corona. It is well known that this places strong demands on the coronal electron population, which amounts to perhaps 10^{36} in the volume above a flare region, and which would require replenishing during the course of the flare. Replenishing of the coronal acceleration volume by a counterstreaming return current generated by the plasma in which the electrons propagate (of equal and opposite electron flux) has been postulated, but it was pointed out relatively early on there would be problems with beam stability (Brown and Melrose 1977), and this remains the case. White-light footpoint areas, which we might take as a proxy for the electron beam areas (HXR footpoint areas being hard to measure reliably) imply a beam flux and a return current speed in excess of what can propagate stably through the corona. At least according to our present theoretical understanding, the beam can travel stably only if the density of the loop in which it moves is large—on the order of 10^{11} cm^{-3} (see, e.g., van den Oord 1990, for details). Though this is possible later in the flare (once evaporation has started) there is little imaging or spectroscopic evidence for such loop densities before the flare (except in coronal thick target loop flares of Veronig and Brown 2004, which do not show footpoints). Other exceptions might be found in the pre-flare phase of SOL2002-07-23T00:35 (X4.8), as discussed by Lin et al. (2002) and Caspi and Lin (2010). Despite this uncertainty the coronal electron beam model has been accepted for decades, but it is clear that renewed theoretical effort must be dedicated to understanding the propagation of a dense electron beam through the corona. Some alternatives to this model have also recently been proposed. Fletcher and Hudson (2008) have introduced a model in which electrons producing the HXR footpoints are wholly accelerated in the chromosphere, and Brown et al. (2009) discuss the chromospheric re-acceleration of a small number of originally coronal electrons so that their photon yield per electron is increased, reducing the electron number and flux requirement. Overall, an instantaneous emission measure for the non-thermal electrons of as much as 10^{46} cm^{-3} is required to explain chromospheric HXR footpoints (Hoyng et al. 1976) which is achievable in the chromosphere, though still demanding. For example, if the flare footpoint area were around 10^{17} cm^{-2}, then assuming a chromospheric slab of thickness 10^{8} cm would require on average 3×10^{10} electrons cm^{-3} to be accelerated—on average about a tenth of the total electron population (bound or unbound) in the top 1000 km of the VAL-C chromospheric model (Vernazza et al. 1981).

7.1.4 Flares and CMEs

At present it seems clear that the standard flare scenario of large-scale reconnection in magnetic fields stretched by an eruption must explain a great deal of flare phenomenology. Probably Hirayama (1974) provided the first clear 3-D visualization of how this might feasibly happen, although the ideas certainly had been available long before this seminal paper appeared. In fact, the Hirayama work described a filament eruption (we would associate it with a CME nowadays), rather than a flare as such. Furthermore, we recognize that it is not the mass of the filament that is important in the overall dynamics, rather the evolution of the magnetic field which carries the filament mass along. The magnetic free energy released as the outwards ejection of mass is therefore at least as large as the CME kinetic energy estimated from the visible material. The CME is of course just another manifestation of the re-arrangement of coronal magnetic fields, but in an environment in which the perturbation can expand relatively freely into the corona above the reconnection region. This provides an interesting contrast to the situation below the reconnection region, where the magnetic rearrangement of strong fields releases a similar amount of energy into a small, confined volume of very low beta plasma, resulting primarily in non-thermal particles and heat, rather than mass motions. A basic calculation shows that 10^{31} erg released into a volume of 10^{27} cm^3 at a mean density of say 10^{11} cm^3 results in a mean energy per particle of 60 keV. Of course radiative and conductive losses reduce the instantaneous value, but it does make reasonably clear that flares are almost certain to be efficient producers of non-thermal populations.

The rise of the active region filament heralding the imminent onset of a flare identifies the importance of the MHD instability—which will go on to produce the CME—in the whole flare/CME combination. But the extremely good correspondence in time between HXR bursts and both filament lift-off (Maričić et al. 2007; Temmer et al. 2008) and coronal "supra-arcade downflows" (Asai et al. 2004b) (interpreted as retracting flux tubes formed in reconnection behind a CME) does not really permit us to say whether one is the "cause" and the other the "effect."

A further pioneering paper by Gold and Hoyle (1960) had already described a large-scale reconnection scenario that did not require a prior eruption, and many flares, even including some as powerful as the low X class (Wang and Zhang 2007) events, do not result from the open fields stretched out by a CME. Therefore models which permit the release of stored magnetic energy without field opening are also necessary—such as those proposing internal reconnections in active region core field (Gibson and Fan 2006), or reconnection without full field opening (DeVore and Antiochos 2008; Masson et al. 2009).

A final observational link which deserves modeling attention is the association of solar energetic particles with "soft-hard-harder" X-ray spectral evolution and fast CMEs (Grayson et al. 2009). The SHH evolution is clearly a property of a solar acceleration process which operates long after the CME has left, and long after the flare impulsive phase. It is apparently unique to flares exhibiting CMEs. Perhaps slow reconnection behind the departing CME, or dipolarization of reconnected fields which have been greatly stretched by the process, plays a part in the ongoing acceleration.

7.2 Future Observational Progress

Here we list a few important areas in which observations should be improved.

1. *UV/EUV imaging spectroscopy.* It is a major embarrassment to solar physics that we often turn to stellar observations to learn how to fill in missing "details" from the solar data. One such "detail" is the spectroscopy and morphology of Ly-α in flares (Rubio da Costa et al. 2009). Ly-α is a primary radiating component, rich in diagnostic information about the chromosphere. In general the visible/UV continuum contains the majority of flare radiated energy and yet we have few good observations of it (e.g., Neidig 1989).
2. *Sensitive high-energy observations. RHESSI* has made it abundantly clear that the key non-thermal processes involved in the disruption of coronal plasmas (i.e., flares and CMEs) can readily be detected even in the tenuous middle corona. There is a vast parameter space awaiting sensitive instruments.
3. *Microwave/meter-wave imaging spectroscopy.* Solar radio astronomy has not had the benefit yet of broadband observations in this key domain, or of radio imaging at more than a few frequencies. We know it to contain emission and absorption features of great diagnostic significance, as well as giving insight into the 3-D structure of the coronal magnetic field.
4. *Neutral particle emissions.* The detection of neutrons and energetic neutral atoms from solar flares is in its infancy but holds great promise for understanding the behavior of accelerated ions in the virtually unknown domain below a few MeV. Neutron detectors placed at a few tenths of an AU will be of great value.

5. *Coronal seismology.* The wave population—background and transient—is another means whereby the coronal magnetic structure can be probed. This can be observed by high-resolution imaging and imaging spectroscopy of the global corona. The first steps in this area are now emerging from ground-based and *Hinode* observations, and the Solar Dynamics Observatory will provide a comprehensive imaging view because of its large telemetry bandwidth.

8 Conclusions

We have reviewed flare observations in a broad sense, touching on related phenomena and models that attempt to describe the overall process. The multifarious observations across the broad spectrum of phenomena each help us to characterize the equilibrium change in the corona and chromosphere that we call a flare, and it should be clear that the multiwavelength approach is crucial in flare studies. It tells us where the flare energy starts and where it ends up, and something about the intermediate steps. It also provides some geometrical and diagnostic information about the flare magnetic environment, at different levels in the atmosphere, and how and when this changes as the flare proceeds. This big picture cannot be reached using one spectral region on its own. The multiwavelength observations have many detailed applications as we try to understand specific mechanisms that are at work in various phases and regions of the flare development. Some of the mechanisms are at the stage of recent discovery and have only the sketchiest understanding at present. The coming decade will see a flood of multi-wavelength data, mastery of which will require the development of new analysis techniques, such as fast image processing and feature recognition. It will be clear to the reader that much of the observational evidence presented here is based on the detailed analysis of small numbers of flares, and even basic statistical studies are rather few. But a comprehensive understanding of the flare phenomenon will require a blend of both approaches—i.e., the collection, sifting, comparison and assimilation of detailed properties of large samples of events. We look forward to the challenge.

Acknowledgements The authors would collectively like to acknowledge the work by the many instrument teams and software teams, whose sustained efforts over the years have made this kind of multiwavelength analysis possible. The chief architects of and major contributors to the Solarsoft library deserve particular thanks (Sam Freeland, Richard Schwartz, William Thompson, Kim Tolbert and Dominic Zarro). L.F. would like to acknowledge financial support from the UK STFC (Rolling grant ST/F0026371), the EU's SOLAIRE Research and Training Network (MTRN-CT-2006-035484) and L.F. and M.B. also acknowledge the support of the Leverhulme Foundation (Grant F/00 179/AY). The work of H. Ji was supported by NSFC 10833007W. W. Liu was partly supported by an appointment to the NASA Postdoctoral Program at Goddard Space Flight Center, administered by Oak Ridge Associated Universities through a contract with NASA. Partial support to W. Liu's work was also provided by *Hinode* SOT contract NNM07AA01C. A. Caspi, H. Hudson, and S. Krucker were supported by NASA under contract NAS5-98033 and grants NAG5-12878 and NNX08AJ18G. A.V. and M.T. acknowledge the Austrian Science Fund (FWF) projects no. P20867-N16 and P20145-N16. We are grateful to several insitutions for granting permission for the use of previously published figures. In addition to those noted in figure captions, we also acknowledge the following figures from A&A, reproduced with permission (c) ESO: Fig. 12 (Mandrini et al. 1995, 303, 927); Fig. 18 (Miklenic et al. 2007, 461, 697); Fig. 34 (Mann et al. 2003, 400, 329). The following figures are reproduced with kind permission from Springer Science+Business Media: Fig. 19 (Lindsey and Donea 2008); Fig. 27 (Krucker et al. 2008a); and Fig. 29 (Gallagher et al. 2002).

References

W.P. Abbett, S.L. Hawley, Astrophys. J. **521**, 906 (1999). doi:10.1086/307576
J. Aboudarham, J.C. Hénoux, Astron. Astrophys. **156**, 73 (1986)

G.-X. Ai, Y.-F. Hu, Publ. Beijing Astron. Obs. **8**, 1 (1986)
D. Alexander, A.J. Coyner, Astrophys. J. **640**, 505 (2006). doi:10.1086/500076
D. Alexander, T.R. Metcalf, Sol. Phys. **210**, 323 (2002). doi:10.1023/A:1022457413628
J.C. Allred, S.L. Hawley, W.P. Abbett, M. Carlsson, Astrophys. J. **630**, 573 (2005). doi:10.1086/431751. arXiv:astro-ph/0507335
E. Antonucci, B.R. Dennis, Sol. Phys. **86**, 67 (1983)
B. Anwar, L.W. Acton, H.S. Hudson, M. Makita, A.N. McClymont, S. Tsuneta, Sol. Phys. **147**, 287 (1993). doi:10.1007/BF00690719
A. Asai, S. Masuda, T. Yokoyama, M. Shimojo, H. Isobe, H. Kurokawa, K. Shibata, Astrophys. J. Lett. **578**, L91 (2002). doi:10.1086/344566. arXiv:astro-ph/0209106
A. Asai, T.T. Ishii, H. Kurokawa, T. Yokoyama, M. Shimojo, Astrophys. J. **586**, 624 (2003). doi:10.1086/367694
A. Asai, T. Yokoyama, M. Shimojo, S. Masuda, H. Kurokawa, K. Shibata, Astrophys. J. **611**, 557 (2004a). doi:10.1086/422159
A. Asai, T. Yokoyama, M. Shimojo, K. Shibata, Astrophys. J. Lett. **605**, L77 (2004b). doi:10.1086/420768
A. Asai, H. Nakajima, M. Shimojo, S.M. White, H.S. Hudson, R.P. Lin, Publ. Astron. Soc. Jpn. **58**, L1 (2006)
M.J. Aschwanden, Space Sci. Rev. **101**, 1 (2002). doi:10.1023/A:1019712124366
M.J. Aschwanden, D. Alexander, Sol. Phys. **204**, 91 (2001). doi:10.1023/A:1014257826116
M.J. Aschwanden, A.O. Benz, B.R. Dennis, R.A. Schwartz, Astrophys. J. **455**, 347 (1995a). doi:10.1086/176582
M.J. Aschwanden, R.A. Schwartz, D.M. Alt, Astrophys. J. **447**, 923 (1995b). doi:10.1086/175930
M.J. Aschwanden, A. Winebarger, D. Tsiklauri, H. Peter, Astrophys. J. **659**, 1673 (2007). doi:10.1086/513070
R.G. Athay, G.E. Moreton, Astrophys. J. **133**, 935 (1961). doi:10.1086/147098
G. Aulanier, E.E. DeLuca, S.K. Antiochos, R.A. McMullen, L. Golub, Astrophys. J. **540**, 1126 (2000). doi:10.1086/309376
G. Aulanier, E. Pariat, P. Démoulin, C.R. Devore, Sol. Phys. **238**, 347 (2006). doi:10.1007/s11207-006-0230-2
Y. Avignon, J. Bonmartin, A. Bouteille, B. Clavelier, E. Hulot, Sol. Phys. **120**, 193 (1989)
T. Bai, Sol. Phys. **62**, 113 (1979). doi:10.1007/BF00150138
T. Bai, Astrophys. J. **308**, 912 (1986). doi:10.1086/164561
G. Barnes, K.D. Leka, Astrophys. J. **646**, 1303 (2006). doi:10.1086/504960
G. Barnes, D.W. Longcope, K.D. Leka, Astrophys. J. **629**, 561 (2005). doi:10.1086/431175
T.S. Bastian, Adv. Space Res. **32**, 2705 (2003). doi:10.1016/S0273-1177(03)90454-6
T.S. Bastian, Planet. Space Sci. **52**, 1381 (2004). doi:10.1016/j.pss.2004.09.015
T.S. Bastian, A.O. Benz, D.E. Gary, Annu. Rev. Astron. Astrophys. **36**, 131 (1998). doi:10.1146/annurev.astro.36.1.131
M. Battaglia, A.O. Benz, Astron. Astrophys. **456**, 751 (2006). doi:10.1051/0004-6361:20065233. arXiv:astro-ph/0606353
M. Battaglia, L. Fletcher, A.O. Benz, Astron. Astrophys. **498**, 891 (2009). doi:10.1051/0004-6361/200811196. arXiv:0903.2754
P.J. Baum, A. Bratenahl, Sol. Phys. **67**, 245 (1980). doi:10.1007/BF00149805
A.O. Benz, Astrophys. J. **211**, 270 (1977)
A.O. Benz, Living Rev. Sol. Phys. **5**, 1 (2008)
A.O. Benz, P.C. Grigis, A. Csillaghy, P. Saint-Hilaire, Sol. Phys. **226**, 121 (2005). doi:10.1007/s11207-005-5254-5. astro-ph/0410436
T.E. Berger, B. de Pontieu, C.J. Schrijver, A.M. Title, Astrophys. J. Lett. **519**, L97 (1999). doi:10.1086/312088
D.A. Biesecker, D.C. Myers, B.J. Thompson, D.M. Hammer, A. Vourlidas, Astrophys. J. **569**, 1009 (2002). doi:10.1086/339402
A. Bleybel, T. Amari, L. van Driel-Gesztelyi, K.D. Leka, Astron. Astrophys. **395**, 685 (2002). doi:10.1051/0004-6361:20021332
S.A. Bogachev, B.V. Somov, T. Kosugi, T. Sakao, Astrophys. J. **630**, 561 (2005). doi:10.1086/431918
L. Bone, J.C. Brown, L. Fletcher, A. Veronig, S. White, Astron. Astrophys. **446**, 339 (2007). doi:10.1051/0004-6361:20020947
V. Bothmer, R. Schwenn, Ann. Geophys. **16**, 1 (1998)
J.W. Brosius, Astrophys. J. **586**, 1417 (2003). doi:10.1086/367958
J.W. Brosius, G.D. Holman, Astrophys. J. Lett. **659**, L73 (2007). doi:10.1086/516629
J.W. Brosius, K.J.H. Phillips, Astrophys. J. **613**, 580 (2004). doi:10.1086/422873
J.C. Brown, Sol. Phys. **18**, 489 (1971)
J.C. Brown, Sol. Phys. **31**, 143 (1973). doi:10.1007/BF00156080

J.C. Brown, P.C.V. Mallik, Astron. Astrophys. **481**, 507 (2008). doi:10.1051/0004-6361:20078103. arXiv:0706.2823

J.C. Brown, P.C.V. Mallik, Astrophys. J. Lett. **697**, L6 (2009). doi:10.1088/0004-637X/697/1/L6

J.C. Brown, D.B. Melrose, Sol. Phys. **52**, 117 (1977). doi:10.1007/BF00935795

J.C. Brown, A.J. Conway, M.J. Aschwanden, Astrophys. J. **509**, 911 (1998). doi:10.1086/306522

J.C. Brown, R. Turkmani, E.P. Kontar, A.L. MacKinnon, L. Vlahos, Astron. Astrophys. **508**, 993 (2009). doi:10.1051/0004-6361/200913145. arXiv:0909.4243

J.C. Brown, P.C.V. Mallik, N.R. Badnell, Astron. Astrophys. **515**, C1+ (2010). doi:10.1051/0004-6361:20078103e

G.E. Brueckner, R.A. Howard, M.J. Koomen, C.M. Korendyke, D.J. Michels, J.D. Moses, D.G. Socker, K.P. Dere, P.L. Lamy, A. Llebaria, M.V. Bout, R. Schwenn, G.M. Simnett, D.K. Bedford, C.J. Eyles, Sol. Phys. **162**, 357 (1995)

L.F. Burlaga, L. Klein, N.R. Sheeley Jr., D.J. Michels, R.A. Howard, M.J. Koomen, R. Schwenn, H. Rosenbauer, Geophys. Res. Lett. **9**, 1317 (1982). doi:10.1029/GL009i012p01317

R. Cameron, I. Sammis, Astrophys. J. Lett. **525**, L61 (1999). doi:10.1086/312328

H.V. Cane, W.C. Erickson, Astrophys. J. **623**, 1180 (2005). doi:10.1086/428820

R.C. Canfield, C.-C. Cheng, K.P. Dere, G.A. Dulk, D.J. McLean, E.J. Schmahl, R.D. Robinson Jr., S.A. Schoolman, in *Skylab Solar Workshop II*, ed. by P.A. Sturrock (1980), pp. 451–469

R.C. Canfield, T.A. Gunkler, P.J. Ricchiazzi, Astrophys. J. **282**, 296 (1984). doi:10.1086/162203

R.C. Canfield, T.R. Metcalf, K.T. Strong, D.M. Zarro, Nature **326**, 165 (1987). doi:10.1038/326165a0

P.J. Cargill, J.T. Mariska, S.K. Antiochos, Astrophys. J. **439**, 1034 (1995). doi:10.1086/175240

C. Caroubalos, Ann. Astrophys. **27**, 333 (1964)

R.C. Carrington, Mon. Not. R. Astron. Soc. **20**, 13 (1859)

A. Caspi, Super-hot (T > 30 MK) thermal plasma in solar flares. Doctoral dissertation, University of California, Berkeley, Department of Physics (2010). URL http://sprg.ssl.berkeley.edu/~cepheid/.eprints/Caspi_Dissertation.pdf

A. Caspi, R.P. Lin, Astrophys. J. Lett. **725**, L161 (2010). doi:10.1088/2041-8205/725/2/L161

C.P. Catalano, J.A. van Allen, Astrophys. J. **185**, 335 (1973). doi:10.1086/152420

G. Cauzzi, K.P. Reardon, H. Uitenbroek, F. Cavallini, A. Falchi, R. Falciani, K. Janssen, T. Rimmele, A. Vecchio, F. Wöger, Astron. Astrophys. **480**, 515 (2008). doi:10.1051/0004-6361:20078642. arXiv:0709.2417

Q.R. Chen, M.D. Ding, Astrophys. J. **618**, 537 (2005). doi:10.1086/425856. arXiv:astro-ph/0412171

Q.R. Chen, M.D. Ding, Astrophys. J. **641**, 1217 (2006). doi:10.1086/500635. arXiv:astro-ph/0512496

C. Chifor, D. Tripathi, H.E. Mason, B.R. Dennis, Astron. Astrophys. **472**, 967 (2007). doi:10.1051/0004-6361:20077771

E.L. Chupp, D.J. Forrest, P.R. Higbie, A.N. Suri, C. Tsai, P.P. Dunphy, Nature **241**, 333 (1973)

H.T. Claßen, G. Mann, A. Klassen, H. Aurass, Astron. Astrophys. **409**, 309 (2003). doi:10.1051/0004-6361:20031097

E.W. Cliver, Sol. Phys. **157**, 285 (1995)

E.W. Cliver, Astrophys. J. **639**, 1206 (2006). doi:10.1086/499765

E.W. Cliver, H.S. Hudson, J. Atmos. Sol.-Terr. Phys. **64**, 231 (2002). doi:10.1016/S1364-6826(01)00086-4

E.W. Cliver, B.R. Dennis, A.L. Kiplinger, S.R. Kane, D.F. Neidig, N.R. Sheeley, M.J. Koomen, Astrophys. J. **305**, 920 (1986)

E.W. Cliver, N.V. Nitta, B.J. Thompson, J. Zhang, Sol. Phys. **225**, 105 (2004). doi:10.1007/s11207-004-3258-1

A.E. Covington, Sol. Phys. **33**, 439 (1973). doi:10.1007/BF00152431

A.N. Cox, Allen's astrophysical quantities. ADS: http://adsabs.harvard.edu/abs/2000asqu.book.....C, ISBN: 0387987460 (2000)

D.P. Cox, W.H. Tucker, Astrophys. J. **157**, 1157 (1969)

N.B. Crosby, M.J. Aschwanden, B.R. Dennis, Sol. Phys. **143**, 275 (1993)

Y. Cui, R. Li, L. Zhang, Y. He, H. Wang, Sol. Phys. **237**, 45 (2006). doi:10.1007/s11207-006-0077-6

J.L. Culhane, Mon. Not. R. Astron. Soc. **144**, 375 (1969)

J.L. Culhane, J.F. Vesecky, K.J.H. Phillips, Sol. Phys. **15**, 394 (1970). doi:10.1007/BF00151847

J.L. Culhane, A.T. Phillips, M. Inda-Koide, T. Kosugi, A. Fludra, H. Kurokawa, K. Makishima, C.D. Pike, T. Sakao, T. Sakurai, Sol. Phys. **153**, 307 (1994)

J.L. Culhane, L.K. Harra, A.M. James, K. Al-Janabi, L.J. Bradley, R.A. Chaudry, K. Rees, J.A. Tandy, P. Thomas, M.C.R. Whillock, B. Winter, G.A. Doschek, C.M. Korendyke, C.M. Brown, S. Myers, J. Mariska, J. Seely, J. Lang, B.J. Kent, B.M. Shaughnessy, P.R. Young, G.M. Simnett, C.M. Castelli, S. Mahmoud, H. Mapson-Menard, B.J. Probyn, R.J. Thomas, J. Davila, K. Dere, D. Windt, J. Shea, R. Hagood, R. Moye, H. Hara, T. Watanabe, K. Matsuzaki, T. Kosugi, V. Hansteen, Ø. Wikstol, Sol. Phys. **243**, 19 (2007). doi:10.1007/s01007-007-0293-1

A. Czaykowska, B. de Pontieu, D. Alexander, G. Rank, Astrophys. J. Lett. **521**, L75 (1999). doi:10.1086/312176

A. Czaykowska, D. Alexander, B. De Pontieu, Astrophys. J. **552**, 849 (2001). doi:10.1086/320553

B. De Pontieu, S. McIntosh, V.H. Hansteen, M. Carlsson, C.J. Schrijver, T.D. Tarbell, A.M. Title, R.A. Shine, Y. Suematsu, S. Tsuneta, Y. Katsukawa, K. Ichimoto, T. Shimizu, S. Nagata, Publ. Astron. Soc. Jpn. **59**, 655 (2007). arXiv:0710.2934

C. Debi Prasad, Bull. Astron. Soc. India **26**, 253 (1998)

G. Del Zanna, A. Berlicki, B. Schmieder, H.E. Mason, Sol. Phys. **234**, 95 (2006). doi:10.1007/s11207-006-0016-6

J.-P. Delaboudinière, G.E. Artzner, J. Brunaud, A.H. Gabriel, J.F. Hochedez, F. Millier, X.Y. Song, B. Au, K.P. Dere, R.A. Howard, R. Kreplin, D.J. Michels, J.D. Moses, J.M. Defise, C. Jamar, P. Rochus, J.P. Chauvineau, J.P. Marioge, R.C. Catura, J.R. Lemen, L. Shing, R.A. Stern, J.B. Gurman, W.M. Neupert, A. Maucherat, F. Clette, P. Cugnon, E.L. van Dessel, Sol. Phys. **162**, 291 (1995). doi:10.1007/BF00733432

P. Démoulin, J.C. Hénoux, C.H. Mandrini, Sol. Phys. **139**, 105 (1992)

P. Démoulin, L. van Driel-Gesztelyi, B. Schmieder, J.C. Hénoux, G. Csepura, M.J. Hagyard, Astron. Astrophys. **271**, 292 (1993)

P. Démoulin, C.H. Mandrini, M.G. Rovira, J.C. Hénoux, M.E. Machado, Sol. Phys. **150**, 221 (1994). doi:10.1007/BF00712887

P. Démoulin, L.G. Bagala, C.H. Mandrini, J.C. Hénoux, M.G. Rovira, Astron. Astrophys. **325**, 305 (1997)

C. Denker, A. Tritschler, T.R. Rimmele, K. Richards, S.L. Hegwer, F. Wöger, Publ. Astron. Soc. Pac. **119**, 170 (2007). doi:10.1086/512493

B.R. Dennis, R.L. Pernak, Astrophys. J. **698**, 2131 (2009). doi:10.1088/0004-637X/698/2/2131

B.R. Dennis, D.M. Zarro, Sol. Phys. **146**, 177 (1993)

K.P. Dere, E. Landi, H.E. Mason, B.C. Monsignori Fossi, P.R. Young, Astron. Astrophys. Suppl. **125**, 149 (1997b)

K.P. Dere, G.E. Brueckner, R.A. Howard, M.J. Koomen, C.M. Korendyke, R.W. Kreplin, D.J. Michels, J.D. Moses, N.E. Moulton, D.G. Socker, O.C. St. Cyr, J.P. Delaboudinière, G.E. Artzner, J. Brunaud, A.H. Gabriel, J.F. Hochedez, F. Millier, X.Y. Song, J.P. Chauvineau, J.P. Marioge, J.M. Defise, C. Jamar, P. Rochus, R.C. Catura, J.R. Lemen, J.B. Gurman, W. Neupert, F. Clette, P. Cugnon, E.L. van Dessel, P.L. Lamy, A. Llebaria, R. Schwenn, G.M. Simnett, Sol. Phys. **175**, 601 (1997a). doi:10.1023/A:1004907307376

A. Des Jardins, R. Canfield, D. Longcope, C. Fordyce, S. Waitukaitis, Astrophys. J. **693**, 1628 (2009). doi:10.1088/0004-637X/693/2/1628

C.R. DeVore, S.K. Antiochos, Astrophys. J. **680**, 740 (2008). doi:10.1086/588011

M.D. Ding, C. Fang, Astron. Astrophys. **314**, 643 (1996)

M.D. Ding, Q.R. Chen, J.P. Li, P.F. Chen, Astrophys. J. **598**, 683 (2003). doi:10.1086/378877. arXiv:astro-ph/0308085

H.W. Dodson, E.R. Hedeman, Sol. Phys. **13**, 401 (1970). doi:10.1007/BF00153560

A.-C. Donea, C. Lindsey, Astrophys. J. **630**, 1168 (2005). doi:10.1086/432155

G.A. Doschek, H.P. Warren, Astrophys. J. **629**, 1150 (2005). doi:10.1086/431920

G.A. Doschek, U. Feldman, R.W. Kreplin, L. Cohen, Astrophys. J. **239**, 725 (1980). doi:10.1086/158158

G.A. Doschek, J.T. Mariska, T. Sakao, Astrophys. J. **459**, 823 (1996). doi:10.1086/176946

W. Dröge, D. Ruffolo, B. Klecker, Astrophys. J. Lett. **464**, L87+ (1996). doi:10.1086/310093. arXiv:astro-ph/9604019

M.A. Ellison, Mon. Not. R. Astron. Soc. **106**, 500 (1946)

A.G. Emslie, P.A. Sturrock, Sol. Phys. **80**, 99 (1982). doi:10.1007/BF00153426

A.G. Emslie, K.J.H. Phillips, B.R. Dennis, Sol. Phys. **103**, 89 (1986)

A.G. Emslie, H. Kucharek, B.R. Dennis, N. Gopalswamy, G.D. Holman, G.H. Share, A. Vourlidas, T.G. Forbes, P.T. Gallagher, G.M. Mason, T.R. Metcalf, R.A. Mewaldt, R.J. Murphy, R.A. Schwartz, T.H. Zurbuchen, J. Geophys. Res. (Space) Phys. **109**, 10104 (2004). doi:10.1029/2004JA010571

A.G. Emslie, B.R. Dennis, G.D. Holman, H.S. Hudson, J. Geophys. Res. (Space) Phys. **110**, 11103 (2005). doi:10.1029/2005JA011305

P. Evenson, P. Meyer, K.R. Pyle, Astrophys. J. **274**, 875 (1983). doi:10.1086/161500

P. Evenson, R. Kroeger, P. Meyer, D. Reames, Astrophys. Suppl. **73**, 273 (1990). doi:10.1086/191462

D.A. Falconer, R.L. Moore, G.A. Gary, Astrophys. J. **569**, 1016 (2002). doi:10.1086/339161

F. Fárník, S.K. Savy, Sol. Phys. **183**, 339 (1998)

F. Fárník, H. Hudson, T. Watanabe, Sol. Phys. **165**, 169 (1996). doi:10.1007/BF00149096

F. Fárník, H.S. Hudson, M. Karlický, T. Kosugi, Astron. Astrophys. **399**, 1159 (2003). doi:10.1051/0004-6361:20021852

U. Feldman, Astrophys. J. **364**, 322 (1990). doi:10.1086/169414

U. Feldman, J.M. Laming, Phys. Scr. **61**, 222 (2000)

U. Feldman, I. Dammasch, E. Landi, G.A. Doschek, Astrophys. J. **609**, 439 (2004). doi:10.1086/420964

G.H. Fisher, R.C. Canfield, A.N. McClymont, Astrophys. J. **289**, 434 (1985a). doi:10.1086/162903

G.H. Fisher, R.C. Canfield, A.N. McClymont, Astrophys. J. **289**, 425 (1985b), doi:10.1086/162902

G.H. Fisher, R.C. Canfield, A.N. McClymont, Astrophys. J. **289**, 414 (1985c). doi:10.1086/162901

L. Fletcher, Astron. Astrophys. **493**, 241 (2009). doi:10.1051/0004-6361:20077972

L. Fletcher, H.S. Hudson, Sol. Phys. **210**, 307 (2002). doi:10.1023/A:1022479610710

L. Fletcher, H.S. Hudson, Astrophys. J. **675**, 1645 (2008). doi:10.1086/527044. 0712.3452

L. Fletcher, T.R. Metcalf, D. Alexander, D.S. Brown, L.A. Ryder, Astrophys. J. **554**, 451 (2001)

L. Fletcher, J.A. Pollock, H.E. Potts, Sol. Phys. **222**, 279 (2004). doi:10.1023/B:SOLA.0000043580.89730.4d

L. Fletcher, I.G. Hannah, H.S. Hudson, T.R. Metcalf, Astrophys. J. **656**, 1187 (2007). doi:10.1086/510446

T.G. Forbes, L.W. Acton, Astrophys. J. **459**, 330 (1996). doi:10.1086/176896

T.G. Forbes, J. Lin, J. Atmos. Sol.-Terr. Phys. **62**, 1499 (2000)

T.G. Forbes, J.M. Malherbe, E.R. Priest, Sol. Phys. **120**, 285 (1989)

C. Foullon, E. Verwichte, V.M. Nakariakov, L. Fletcher, Astron. Astrophys. **440**, L59 (2005). doi:10.1051/0004-6361:200500169

K.J. Frost, B.R. Dennis, Astrophys. J. **165**, 655 (1971)

Q. Fu, H. Ji, Z. Qin, Z. Xu, Z. Xia, H. Wu, Y. Liu, Y. Yan, G. Huang, Z. Chen, Z. Jin, Q. Yao, C. Cheng, F. Xu, M. Wang, L. Pei, S. Chen, G. Yang, C. Tan, S. Shi, Sol. Phys. **222**, 167 (2004). doi:10.1023/B:SOLA.0000036876.14446.dd

P.T. Gallagher, B.R. Dennis, S. Krucker, R.A. Schwartz, A.K. Tolbert, Sol. Phys. **210**, 341 (2002). doi:10.1023/A:1022422019779

P.T. Gallagher, G.R. Lawrence, B.R. Dennis, Astrophys. J. Lett. **588**, L53 (2003). doi:10.1086/375504

D.E. Gary, G.J. Hurford, in *Solar Physics with Radio Observations; Proc. Nobeyama Symp. (NRO Rep. 479)*, ed. by T.S. Bastian, N. Gopalswamy, K. Shibasaki (1999), p. 429

S. Gburek, J. Sylwester, P. Martens, Sol. Phys. **239**, 531 (2006). doi:10.1007/s11207-006-1994-0

M.K. Georgoulis, D.M. Rust, Astrophys. J. Lett. **661**, L109 (2007). doi:10.1086/518718

S.E. Gibson, Y. Fan, Astrophys. J. Lett. **637**, L65 (2006). doi:10.1086/500452

R.G. Giovanelli, Mon. Not. R. Astron. Soc. **108**, 163 (1948)

P. Giuliani, T. Neukirch, P. Wood, Astrophys. J. **635**, 636 (2005)

C.P. Goff, S.A. Matthews, L. van Driel-Gesztelyi, L.K. Harra, Astron. Astrophys. **423**, 363 (2004). doi:10.1051/0004-6361:20040392

T. Gold, F. Hoyle, Mon. Not. R. Astron. Soc. **120**, 89 (1960)

L. Golub, E. Deluca, G. Austin, J. Bookbinder, D. Caldwell, P. Cheimets, J. Cirtain, M. Cosmo, P. Reid, A. Sette, M. Weber, T. Sakao, R. Kano, K. Shibasaki, H. Hara, S. Tsuneta, K. Kumagai, T. Tamura, M. Shimojo, J. McCracken, J. Carpenter, H. Haight, R. Siler, E. Wright, J. Tucker, H. Rutledge, M. Barbera, G. Peres, S. Varisco, Sol. Phys. **243**, 63 (2007). doi:10.1007/s11207-007-0182-1

V.S. Gorbachev, B.V. Somov, Sov. Astron. **33**, 57 (1989)

J.D. Graham, A. Norton, A. López Ariste, B. Lites, H. Socas-Navarro, S. Tomczyk, in *Astronomical Society of the Pacific Conference Series*, ed. by J. Trujillo-Bueno, J. Sanchez Almeida. Astronomical Society of the Pacific Conference Series, vol. 307 (2003), p. 131

J.A. Grayson, S. Krucker, R.P. Lin, Astrophys. J. **707**, 1588 (2009). doi:10.1088/0004-637X/707/2/1588

P.C. Grigis, A.O. Benz, Astron. Astrophys. **426**, 1093 (2004). doi:10.1051/0004-6361:20041367. astro-ph/0407431

P.C. Grigis, A.O. Benz, Astrophys. J. Lett. **625**, L143 (2005). doi:10.1086/431147. arXiv:astro-ph/0504436

P.C. Grigis, A.O. Benz, Astrophys. J. **683**, 1180 (2008). doi:10.1086/589826. 0708.2472

E.H.B.M. Gronenschild, R. Mewe, Astron. Astrophys. Suppl. **32**, 283 (1978)

M. Gros, V. Tatischeff, J. Kiener, B. Cordier, C. Chapuis, G. Weidenspointner, G. Vedrenne, A. von Kienlin, R. Diehl, A. Bykov, M.N. Méndez, in *5th INTEGRAL Workshop on the INTEGRAL Universe*, ed. by V. Schoenfelder, G. Lichti, C. Winkler. ESA Special Publication, vol. 552 (2004), p. 669

M.J. Hagyard, N.P. Cumings, E.A. West, J.E. Smith, Sol. Phys. **80**, 33 (1982). doi:10.1007/BF00153422

M.J. Hagyard, D. Teuber, E.A. West, J.B. Smith, Sol. Phys. **91**, 115 (1984)

B.N. Handy, L.W. Acton, C.C. Kankelborg, C.J. Wolfson, D.J. Akin, M.E. Bruner, R. Caravalho, R.C. Catura, R. Chevalier, D.W. Duncan, C.G. Edwards, C.N. Feinstein, S.L. Freeland, F.M. Friedlaender, C.H. Hoffmann, N.E. Hurlburt, B.K. Jurcevich, N.L. Katz, G.A. Kelly, J.R. Lemen, M. Levay, R.W. Lindgren, D.P. Mathur, S.B. Meyer, S.J. Morrison, M.D. Morrison, R.W. Nightingale, T.P. Pope, R.A. Rehse, C.J. Schrijver, R.A. Shine, L. Shing, K.T. Strong, T.D. Tarbell, A.M. Title, D.D. Torgerson, L. Golub, J.A. Bookbinder, D. Caldwell, P.N. Cheimets, W.N. Davis, E.E. Deluca, R.A. McMullen, H.P. Warren, D. Amato, R. Fisher, H. Maldonado, C. Parkinson, Sol. Phys. **187**, 229 (1999)

I. Hannah et al. Space Sci. Rev. (2011, this issue). doi:10.1007/s11214-010-9705-4

H. Hara, T. Watanabe, K. Matsuzaki, L.K. Harra, J.L. Culhane, P. Cargill, J.T. Mariska, G.A. Doschek, Publ. Astron. Soc. Jpn. **60**, 275 (2008)

L.K. Harra, S.A. Matthews, J.L. Culhane, Astrophys. J. Lett. **549**, L245 (2001). doi:10.1086/319163

L.K. Harra, D.R. Williams, A.J. Wallace, T. Magara, H. Hara, S. Tsuneta, A.C. Sterling, G.A. Doschek, Astrophys. J. Lett. **691**, L99 (2009). doi:10.1088/0004-637X/691/2/L99

R.A. Harrison, P.W. Waggett, R.D. Bentley, K.J.H. Phillips, M. Bruner, M. Dryer, G.M. Simnett, Sol. Phys. **97**, 387 (1985)

R.A. Harrison, E.C. Sawyer, M.K. Carter, A.M. Cruise, R.M. Cutler, A. Fludra, R.W. Hayes, B.J. Kent, J. Lang, D.J. Parker, J. Payne, C.D. Pike, S.C. Peskett, A.G. Richards, J.L. Culhane, K. Norman, A.A. Breeveld, E.R. Breeveld, K.F.A. Janabi, A.J. McCalden, J.H. Parkinson, D.G. Self, P.D. Thomas, A.I. Poland, R.J. Thomas, W.T. Thompson, O. Kjeldseth-Moe, P. Brekke, J. Karud, P. Maltby, B. Aschenbach, H. Brauninger, M. Kuhne, J. Hollandt, O.H.W. Siegmund, M.C.E. Huber, A.H. Gabriel, H.E. Mason, B.J.I. Bromage, Sol. Phys. **162**, 233 (1995)

K. Harvey, N. Sheeley Jr., J. Harvey, in *Solar-Terrestrial Predictions*, ed. by P.A. Simon, G. Heckman, M.A. Shea (1986), p. 198

J.W. Harvey, F. Hill, R. Hubbard, J.R. Kennedy, J.W. Leibacher, J.A. Pintar, P.A. Gilman, R.W. Noyes, A.M. Title, J. Toomre, R.K. Ulrich, A. Bhatnagar, J.A. Kennewell, W. Marquette, J. Patrón, O. Saá, E. Yasukawa, Science **272**, 1284 (1996). doi:10.1126/science.272.5266.1284

M. Hesse, T.G. Forbes, J. Birn, Astrophys. J. **631**, 1227 (2005). doi:10.1086/432677

T. Hirayama, Sol. Phys. **34**, 323 (1974). doi:10.1007/BF00153671

G.D. Holman, L. Sui, R.A. Schwartz, A.G. Emslie, Astrophys. J. Lett. **595**, L97 (2003). doi:10.1086/378488

G. Holman et al., Space Sci. Rev. (2011, this issue). doi:10.1007/s11214-010-9680-9

K. Hori, T. Yokoyama, T. Kosugi, K. Shibata, Astrophys. J. **500**, 492 (1998). doi:10.1086/305725

R.A. Howard, J.D. Moses, A. Vourlidas, J.S. Newmark, D.G. Socker, S.P. Plunkett, C.M. Korendyke, J.W. Cook, A. Hurley, J.M. Davila, W.T. Thompson, O.C. St Cyr, E. Mentzell, K. Mehalick, J.R. Lemen, J.P. Wuelser, D.W. Duncan, T.D. Tarbell, C.J. Wolfson, A. Moore, R.A. Harrison, N.R. Waltham, J. Lang, C.J. Davis, C.J. Eyles, H. Mapson-Menard, G.M. Simnett, J.P. Halain, J.M. Defise, E. Mazy, P. Rochus, R. Mercier, M.F. Ravet, F. Delmotte, F. Auchere, J.P. Delaboudinière, V. Bothmer, W. Deutsch, D. Wang, N. Rich, S. Cooper, V. Stephens, G. Maahs, R. Baugh, D. McMullin, T. Carter, Space Sci. Rev. **136**, 67 (2008). doi:10.1007/s11214-008-9341-4

P. Hoyng, H.F. van Beek, J.C. Brown, Sol. Phys. **48**, 197 (1976)

X.-M. Hua, R.E. Lingenfelter, Sol. Phys. **107**, 351 (1987). doi:10.1007/BF00152031

H.S. Hudson, Sol. Phys. **24**, 414 (1972)

H.S. Hudson, Sol. Phys. **45**, 69 (1975). doi:10.1007/BF00152217

H.S. Hudson, Astrophys. J. **224**, 235 (1978). doi:10.1086/156370

H.S. Hudson, Astrophys. J. Lett. **531**, L75 (2000). doi:10.1086/312516

H.S. Hudson, in *The Physics of Chromospheric Plasmas*. Astronomical Society of the Pacific Conference Series, vol. 368, ed. by P. Heinzel, J. Dorotovič, R.J. Rutten (2007), p. 365

H.S. Hudson, E.W. Cliver, J. Geophys. Res. **106**, 25199 (2001). doi:10.1029/2000JA004026

H.S. Hudson, F. Fárník, in *ESA SP-506: Solar Variability: From Core to Outer Frontiers* ed. by A. Wilson (2002), pp. 261–264

H.S. Hudson, K. Ohki, Sol. Phys. **23**, 155 (1972)

H.S. Hudson, R.C. Willson, Sol. Phys. **86**, 123 (1983)

H.S. Hudson, R.P. Lin, R.T. Stewart, Sol. Phys. **75**, 245 (1982)

H.S. Hudson, K.T. Strong, B.R. Dennis, D. Zarro, M. Inda, T. Kosugi, T. Sakao, Astrophys. J. Lett. **422**, L25 (1994). doi:10.1086/187203

H. Hudson, B. Haisch, K.T. Strong, J. Geophys. Res. **100**, 3473 (1995). doi:10.1029/94JA02710

H.S. Hudson, L.W. Acton, S.L. Freeland, Astrophys. J. **470**, 629 (1996). doi:10.1086/177894

H.S. Hudson, T. Kosugi, N.V. Nitta, M. Shimojo, Astrophys. J. Lett. **561**, L211 (2001). doi:10.1086/324760

H.S. Hudson, J.I. Khan, J.R. Lemen, N.V. Nitta, Y. Uchida, Sol. Phys. **212**, 121 (2003). doi:10.1023/A:1022904125479

H.S. Hudson, C.J. Wolfson, T.R. Metcalf, Sol. Phys. **234**, 79 (2006). doi:10.1007/s11207-006-0056-y

H.S. Hudson, G.H. Fisher, B.T. Welsch, in *Subsurface and Atmospheric Influences on Solar Activity* Astronomical Society of the Pacific Conference Series, vol. 383, ed. by R. Howe, R.W. Komm, K.S. Balasubramaniam, G.J.D. Petrie (2008), p. 221

G.J. Hurford, R.B. Read, H. Zirin, Sol. Phys. **94**, 413 (1984). doi:10.1007/BF00151327

G.J. Hurford, E.J. Schmahl, R.A. Schwartz, A.J. Conway, M.J. Aschwanden, A. Csillaghy, B.R. Dennis, C. Johns-Krull, S. Krucker, R.P. Lin, J. McTiernan, T.R. Metcalf, J. Sato, D.M. Smith, Sol. Phys. **210**, 61 (2002). doi:10.1023/A:1022436213688

G.J. Hurford, R.A. Schwartz, S. Krucker, R.P. Lin, D.M. Smith, N. Vilmer, Astrophys. J. Lett. **595**, L77 (2003). doi:10.1086/378179

G.J. Hurford, S. Krucker, R.P. Lin, R.A. Schwartz, G.H. Share, D.M. Smith, Astrophys. J. Lett. **644**, L93 (2006). doi:10.1086/505329
S. Imada, H. Hara, T. Watanabe, A. Asai, T. Minoshima, L.K. Harra, J.T. Mariska, Astrophys. J. Lett. **679**, L155 (2008). doi:10.1086/589444
A.R. Inglis, V.M. Nakariakov, Astron. Astrophys. **493**, 259 (2009). doi:10.1051/0004-6361:200810473
H. Isobe, T. Yokoyama, M. Shimojo, T. Morimoto, H. Kozu, S. Eto, N. Narukage, K. Shibata, Astrophys. J. **566**, 528 (2002). doi:10.1086/324777
H. Isobe, M. Kubo, T. Minoshima, K. Ichimoto, Y. Katsukawa, T.D. Tarbell, S. Tsuneta, T.E. Berger, B. Lites, S. Nagata, T. Shimizu, R.A. Shine, Y. Suematsu, A.M. Title, Publ. Astron. Soc. Jpn. **59**, 807 (2007). 0711.3946
R. Jain, A.K. Pradhan, V. Joshi, K.J. Shah, J.J. Trivedi, S.L. Kayasth, V.M. Shah, M.R. Deshpande, Sol. Phys. **239**, 217 (2006). doi:10.1007/s11207-006-0066-9. arXiv:astro-ph/0612356
Å.M. Janse, B.C. Low, Astron. Astrophys. **472**, 957 (2007). doi:10.1051/0004-6361:20077532
D.B. Jess, A. Andić, M. Mathioudakis, D.S. Bloomfield, F.P. Keenan, Astron. Astrophys. **473**, 943 (2007). doi:10.1051/0004-6361:20077142. arXiv:0707.2716
D.B. Jess, M. Mathioudakis, P.J. Crockett, F.P. Keenan, Astrophys. J. Lett. **688**, L119 (2008). doi:10.1086/595588. 0810.1443
H. Ji, H. Wang, P.R. Goode, Y. Jiang, V. Yurchyshyn, Astrophys. J. Lett. **607**, L55 (2004). doi:10.1086/421550
H. Ji, G. Huang, H. Wang, T. Zhou, Y. Li, Y. Zhang, M. Song, Astrophys. J. Lett. **636**, L173 (2006). doi:10.1086/500203
H. Ji, G. Huang, H. Wang, Astrophys. J. **660**, 893 (2007). doi:10.1086/513017
H. Ji, H. Wang, C. Liu, B.R. Dennis, Astrophys. J. **680**, 734 (2008). doi:10.1086/587138
M. Jin, M. Ding, Publ. Astron. Soc. Jpn. **60**, 835 (2008)
J. Jing, J. Qiu, J. Lin, M. Qu, Y. Xu, H. Wang, Astrophys. J. **620**, 1085 (2005). doi:10.1086/427165
J. Jing, J. Lee, C. Liu, D.E. Gary, H. Wang, Astrophys. J. Lett. **664**, L127 (2007). doi:10.1086/520812
J. Jing, T. Wiegelmann, Y. Suematsu, M. Kubo, H. Wang, Astrophys. J. Lett. **676**, L81 (2008). doi:10.1086/587058
K. Jiřička, M. Karlicky, O. Kepka, A. Tlamicha, Sol. Phys. **147**, 203 (1993). doi:10.1007/BF00675495
B. Joshi, P.K. Manoharan, A.M. Veronig, P. Pant, K. Pandey, Sol. Phys. **242**, 143 (2007). doi:10.1007/s11207-007-0275-x. arXiv:astro-ph/0701368
S. Kahler, Astrophys. J. **214**, 891 (1977)
S. Kahler, Astrophys. J. **428**, 837 (1994). doi:10.1086/174292
M.L. Kaiser, T.A. Kucera, J.M. Davila, O.C. St. Cyr, M. Guhathakurta, E. Christian, Space Science Reviews 198 (2007). doi:10.1007/s11214-007-9277-0
S.R. Kane, Astrophys. J. Lett. **157**, L139+ (1969). doi:10.1086/180403
S.R. Kane, R.F. Donnelly, Astrophys. J. **164**, 151 (1971)
S.R. Kane, K.A. Anderson, W.D. Evans, R.W. Klebesadel, J. Laros, Astrophys. J. Lett. **233**, L151 (1979a). doi:10.1086/183095
S.R. Kane, K.J. Frost, R.F. Donnelly, Astrophys. J. **234**, 669 (1979b). doi:10.1086/157544
S.R. Kane, J. McTiernan, J. Loran, E.E. Fenimore, R.W. Klebesadel, J.G. Laros, Astrophys. J. **390**, 687 (1992). doi:10.1086/171320
M. Karlický, M. Bárta, Astrophys. J. **647**, 1472 (2006). doi:10.1086/505460
M. Karlický, T. Kosugi, Astron. Astrophys. **419**, 1159 (2004). doi:10.1051/0004-6361:20034323
P. Kaufmann, in *Energy Conversion and Particle Acceleration in the Solar Corona*, ed. by L. Klein. Lecture Notes in Physics, vol. 612 (Springer, Berlin, 2003), pp. 294–313
C.U. Keller, J.W. Harvey, M.S. Giampapa, in *Society of Photo-Optical Instrumentation Engineers (SPIE) Conference Series*, ed. by S.L. Keil, S.V. Avakyan. Society of Photo-Optical Instrumentation Engineers (SPIE) Conference Series, vol. 4853 (2003), pp. 194–204
J.I. Khan, H. Aurass, Astron. Astrophys. **383**, 1018 (2002). doi:10.1051/0004-6361:20011707
J.I. Khan, H.M. Bain, L. Fletcher, Astron. Astrophys. **475**, 333 (2007). doi:10.1051/0004-6361:20077894
A.L. Kiplinger, Astrophys. J. **453**, 973 (1995). doi:10.1086/176457
T. Kondo, T. Isobe, S. Igi, S.-I. Watari, M. Tokumaru, Commun. Res. Lab. Rev. **40**, 85 (1994)
E.P. Kontar, I.G. Hannah, A.L. MacKinnon, Astron. Astrophys. **489**, L57 (2008). doi:10.1051/0004-6361:200810719. arXiv:0808.3334
R.A. Kopp, G.W. Pneuman, Sol. Phys. **50**, 85 (1976)
H. Koshiishi, S. Enome, H. Nakajima, K. Shibasaki, M. Nishio, T. Takano, Y. Hanaoka, C. Torii, H. Sekiguchi, S. Kawashima, T. Bushimata, N. Shinohara, Y. Irimajiri, Y. Shiomi, Publ. Astron. Soc. Jpn. **46**, L33 (1994)
A.G. Kosovichev, Sol. Phys. **238**, 1 (2006). doi:10.1007/s11207-006-0190-6. arXiv:astro-ph/0601006
A.G. Kosovichev, Astrophys. J. Lett. **670**, L65 (2007). doi:10.1086/524036. arXiv:0710.0757

A.G. Kosovichev, V.V. Zharkova, Nature **393**, 317 (1998). doi:10.1038/30629
N.D. Kostiuk, S.B. Pikel'ner, Sov. Astron. **18**, 590 (1975)
T. Kosugi, S. Masuda, K. Makishima, M. Inda, T. Murakami, T. Dotani, Y. Ogawara, T. Sakao, K. Kai, H. Nakajima, Sol. Phys. **136**, 17 (1991)
M. Kretzschmar, in *37th COSPAR Scientific Assembly*. COSPAR, Plenary Meeting, vol. 37 (2008), p. 1617
S. Krucker, G.J. Hurford, R.P. Lin, Astrophys. J. Lett. **595**, L103 (2003). doi:10.1086/378840
S. Krucker, M.D. Fivian, R.P. Lin, Adv. Space Res. **35**, 1707 (2005). doi:10.1016/j.asr.2005.05.054
S. Krucker, S.M. White, R.P. Lin, Astrophys. J. Lett. **669**, L49 (2007). doi:10.1086/523759
S. Krucker, M. Battaglia, P.J. Cargill, L. Fletcher, H.S. Hudson, A.L. MacKinnon, S. Masuda, L. Sui, M. Tomczak, A.L. Veronig, L. Vlahos, S.M. White, Astron. Astrophys. Rev. **16**, 155 (2008a). doi:10.1007/s00159-008-0014-9
S. Krucker, G.J. Hurford, A.L. MacKinnon, A.Y. Shih, R.P. Lin, Astrophys. J. Lett. **678**, L63 (2008b), doi:10.1086/588381
S. Krucker, H.S. Hudson, L. Glesener, S.M. White, S. Masuda, J. Wuelser, R.P. Lin, Astrophys. J. **714**, 1108 (2010). doi:10.1088/0004-637X/714/2/1108
C. Kuckein, R. Centeno, V. Martínez Pillet, R. Casini, R. Manso Sainz, T. Shimizu, Astron. Astrophys. **501**, 1113 (2009). doi:10.1051/0004-6361/200911800. arXiv:0904.4876
M.R. Kundu, *Solar Radio Astronomy* (Interscience, New York, 1965)
M.R. Kundu, N. Nitta, S.M. White, K. Shibasaki, S. Enome, T. Sakao, T. Kosugi, T. Sakurai, Astrophys. J. **454**, 522 (1995). doi:10.1086/176503
J. Lang, R.D. Bentley, C.M. Brown, J.L. Culhane, G.A. Doschek, T. Watanabe, E. Hiei, R.D. Deslattes, A. Fludra, P.R. Guttridge, J.E. Magraw, J. Payne, C.D. Pike, M.W. Trow, Publ. Astron. Soc. Jpn. **44**, L55 (1992)
J. Lee, D.E. Gary, G.S. Choe, Astrophys. J. **647**, 638 (2006). doi:10.1086/505416
K.D. Leka, G. Barnes, Astrophys. J. **595**, 1277 (2003). doi:10.1086/377511
K.D. Leka, G. Barnes, Astrophys. J. **656**, 1173 (2007). doi:10.1086/510282
P. Lemaire, P. Gouttebroze, J.-C. Vial, W. Curdt, U. Schühle, K. Wilhelm, Astron. Astrophys. **418**, 737 (2004). doi:10.1051/0004-6361:20034405
Y.P. Li, W.Q. Gan, Astrophys. J. Lett. **629**, L137 (2005). doi:10.1086/444617
Y.P. Li, W.Q. Gan, Astrophys. J. Lett. **652**, L61 (2006). doi:10.1086/509879
Y.P. Li, W.Q. Gan, Adv. Space Res. **39**, 1389 (2007). doi:10.1016/j.asr.2007.04.009
Y.P. Li, W.Q. Gan, Sol. Phys. **247**, 77 (2008). doi:10.1007/s11207-007-9092-5
P. Li, A.G. Emslie, J.T. Mariska, Astrophys. J. **417**, 313 (1993). doi:10.1086/173314
J. Li, T.R. Metcalf, R.C. Canfield, J.-P. Wuelser, T. Kosugi, Astrophys. J. **482**, 490 (1997). doi:10.1086/304131
H. Li, J.-Q. You, Q.-D. Wu, X.-F. Yu, Chin. Phys. Lett. **19**, 742 (2002). doi:10.1088/0256-307X/19/5/342
R.P. Lin, Sol. Phys. **15**, 453 (1970)
R.P. Lin, H.S. Hudson, Sol. Phys. **50**, 153 (1976)
H. Lin, M.J. Penn, S. Tomczyk, Astrophys. J. Lett. **541**, L83 (2000). doi:10.1086/312900
A.C. Lin, R.W. Nightingale, T.D. Tarbell, Sol. Phys. **198**, 385 (2001). doi:10.1023/A:1005213527766
R.P. Lin, B.R. Dennis, G.J. Hurford, D.M. Smith, A. Zehnder, P.R. Harvey, D.W. Curtis, D. Pankow, P. Turin, M. Bester, A. Csillaghy, M. Lewis, N. Madden, H.F. van Beek, M. Appleby, T. Raudorf, J. McTiernan, R. Ramaty, E. Schmahl, R. Schwartz, S. Krucker, R. Abiad, T. Quinn, P. Berg, M. Hashii, R. Sterling, R. Jackson, R. Pratt, R.D. Campbell, D. Malone, D. Landis, C.P. Barrington-Leigh, S. Slassi-Sennou, C. Cork, D. Clark, D. Amato, L. Orwig, R. Boyle, I.S. Banks, K. Shirey, A.K. Tolbert, D. Zarro, F. Snow, K. Thomsen, R. Henneck, A. Mchedlishvili, P. Ming, M. Fivian, J. Jordan, R. Wanner, J. Crubb, J. Preble, M. Matranga, A. Benz, H. Hudson, R.C. Canfield, G.D. Holman, C. Crannell, T. Kosugi, A.G. Emslie, N. Vilmer, J.C. Brown, C. Johns-Krull, M. Aschwanden, T. Metcalf, A. Conway, Sol. Phys. **210**, 3 (2002). doi:10.1023/A:1022428818870
R.P. Lin, S. Krucker, G.J. Hurford, D.M. Smith, H.S. Hudson, G.D. Holman, R.A. Schwartz, B.R. Dennis, G.H. Share, R.J. Murphy, A.G. Emslie, C. Johns-Krull, N. Vilmer, Astrophys. J. Lett. **595**, L69 (2003). doi:10.1086/378932
H. Lin, J.R. Kuhn, R. Coulter, Astrophys. J. Lett. **613**, L177 (2004). doi:10.1086/425217
R.P. Lin et al., Solar Phys. **210** (2002)
C. Lindsey, A.-C. Donea, Sol. Phys. **251**, 627 (2008). doi:10.1007/s11207-008-9140-9
B.W. Lites, D.F. Elmore, K.V. Streander, in *Advanced Solar Polarimetry—Theory, Observation, and Instrumentation*, ed. by M. Sigwarth. Astronomical Society of the Pacific Conference Series, vol. 236 (2001), p. 33

W. Liu, *Solar Flares as Natural Particle Accelerators: A High-energy View from X-ray Observations and Theoretical Models* (VDM Verlag Dr, Saarbrücken, 2008)
W. Liu, Y.W. Jiang, S. Liu, V. Petrosian, Astrophys. J. Lett. **611**, L53 (2004). doi:10.1086/423371. arXiv: astro-ph/0401381
C. Liu, N. Deng, Y. Liu, D. Falconer, P.R. Goode, C. Denker, H. Wang, Astrophys. J. **622**, 722 (2005). doi:10. 1086/427868
W. Liu, S. Liu, Y.W. Jiang, V. Petrosian, Astrophys. J. **649**, 1124 (2006). doi:10.1086/506268. arXiv:astro-ph/ 0603510
C. Liu, J. Lee, D.E. Gary, H. Wang, Astrophys. J. Lett. **658**, L127 (2007). doi:10.1086/513739. arXiv: astro-ph/0702326
C. Liu, J. Lee, J. Jing, D.E. Gary, H. Wang, Astrophys. J. Lett. **672**, L69 (2008). doi:10.1086/525849. 0711.1370
W. Liu, V. Petrosian, B.R. Dennis, Y.W. Jiang, Astrophys. J. **676**, 704 (2008b). doi:10.1086/527538. arXiv:0709.1963
R. Liu, H. Wang, D. Alexander, Astrophys. J. **696**, 121 (2009). doi:10.1088/0004-637X/696/1/121
W. Liu, V. Petrosian, B.R. Dennis, G.D. Holman, Astrophys. J. **693**, 847 (2009b). doi:10.1088/0004-637X/ 693/1/847. arXiv:0805.1055
W. Liu, V. Petrosian, J.T. Mariska, ArXiv e-prints (2009c). arXiv:0906.2449
W. Liu, T.-J. Wang, B.R. Dennis, G.D. Holman, Astrophys. J. **698**, 632 (2009d). doi:10.1088/0004-637X/ 698/1/632. arXiv:0902.1805
D.W. Longcope, Sol. Phys. **169**, 91 (1996). doi:10.1007/BF00153836
D.W. Longcope, Living Rev. Sol. Phys. **2**, 7 (2005)
D.W. Longcope, D.E. McKenzie, J. Cirtain, J. Scott, Astrophys. J. **630**, 596 (2005). doi:10.1086/432039
W.J. MacCombie, D.M. Rust, Sol. Phys. **61**, 69 (1979). doi:10.1007/BF00155447
M.E. Machado, A.G. Emslie, J.C. Brown, Sol. Phys. **58**, 363 (1978). doi:10.1007/BF00157282
M.E. Machado, E.H. Avrett, J.E. Vernazza, R.W. Noyes, Astrophys. J. **242**, 336 (1980). doi:10.1086/158467
M.E. Machado, E.H. Avrett, R. Falciani, C. Fang, L. Gesztelyi, J.-C. Hénoux, E. Hiei, D.F. Neidig, D.M. Rust, P. Sotirovski, Z. Svestka, H. Zirin, in *The Lower Atmosphere of Solar Flares* (1986), pp. 483–488
C.H. Mandrini, P. Démoulin, J.C. Hénoux, M.E. Machado, Astron. Astrophys. **250**, 541 (1991)
C.H. Mandrini, P. Démoulin, M.G. Rovira, J.-F. de La Beaujardiere, J.C. Hénoux, Astron. Astrophys. **303**, 927 (1995)
G. Mann, H. Aurass, W. Voigt, J. Paschke, in *Coronal Streamers, Coronal Loops, and Coronal and Solar Wind Composition, Proceedings of the First SOHO Workshop, ESA SP-348*, ed. by V. Domingo, A. Poland, J. Mariska (1992), p. 129
G. Mann, A. Klassen, H. Aurass, H.-T. Classen, Astron. Astrophys. **400**, 329 (2003). doi:10.1051/0004-6361: 20021593
D. Maričić, B. Vršnak, A.L. Stanger, A.M. Veronig, M. Temmer, D. Roša, Sol. Phys. **241**, 99 (2007). doi:10.1007/s11207-007-0291-x
J.T. Mariska, A.G. Emslie, P. Li, Numerical simulations of impulsively heated solar flares. Astrophys. J. **341**, 1067–1074 (1989). doi:10.1086/167564
S.F. Martin, Sol. Phys. **68**, 217 (1980)
J.C. Martínez-Oliveros, A.-C. Donea, Mon. Not. R. Astron. Soc. **395**, L39 (2009). doi:10.1111/ j.1745-3933.2009.00637.x. arXiv:0902.3856
J.C. Martínez-Oliveros, A.-C. Donea, P.S. Cally, H. Moradi, Mon. Not. R. Astron. Soc. **389**, 1905 (2008). doi:10.1111/j.1365-2966.2008.13722.x. arXiv:0807.3783
S. Masson, E. Pariat, G. Aulanier, C.J. Schrijver, Astrophys. J. **700**, 559 (2009). doi:10.1088/0004-637X/ 700/1/559
S. Masuda, T. Kosugi, H. Hara, S. Tsuneta, Y. Ogawara, Nature **371**, 495 (1994). doi:10.1038/371495a0
S. Masuda, T. Kosugi, H.S. Hudson, Sol. Phys. **204**, 55 (2001)
S.A. Matthews, L. van Driel-Gesztelyi, H.S. Hudson, N.V. Nitta, Astron. Astrophys. **409**, 1107 (2003). doi:10.1051/0004-6361:20031187
P. Mazzotta, G. Mazzitelli, S. Colafrancesco, N. Vittorio, Astron. Astrophys. Suppl. **133**, 403 (1998). doi:10.1051/aas:1998330. arXiv:astro-ph/9806391
R.T.J. McAteer, P.T. Gallagher, J. Ireland, Astrophys. J. **631**, 628 (2005). doi:10.1086/432412
K.G. McClements, D. Alexander, Astrophys. J. **619**, 1153 (2005). doi:10.1086/426581
P.S. McIntosh, Sol. Phys. **125**, 251 (1990)
P.S. McIntosh, R.F. Donnelly, Sol. Phys. **23**, 444 (1972). doi:10.1007/BF00148107
D.E. McKenzie, H.S. Hudson, Astrophys. J. Lett. **519**, L93 (1999). doi:10.1086/312110
J.M. McTiernan, S.R. Kane, J.M. Loran, J.R. Lemen, L.W. Acton, H. Hara, S. Tsuneta, T. Kosugi, Astrophys. J. Lett. **416**, L91+ (1993). doi:10.1086/187078
J.M. McTiernan, G.H. Fisher, P. Li, Astrophys. J. **514**, 472 (1999). doi:10.1086/306924

P. Mein, Astron. Astrophys. **381**, 271 (2002). doi:10.1051/0004-6361:20011418

A.M. Melo, P. Kaufmann, A.S. Kudaka, J.-P. Raulin, R. Marcon, A. Marun, P. Pereyra, H. Levato, Publ. Astron. Soc. Pac. **118**, 1558 (2006). doi:10.1086/509267

P. Messmer, A.O. Benz, C. Monstein, Sol. Phys. **187**, 335 (1999). doi:10.1023/A:1005194314845

T.R. Metcalf, R.C. Canfield, E.H. Avrett, F.T. Metcalf, Astrophys. J. **350**, 463 (1990). doi:10.1086/168400

T.R. Metcalf, L. Jiao, A.N. McClymont, R.C. Canfield, H. Uitenbroek, Astrophys. J. **439**, 474 (1995). doi:10.1086/175188

T.R. Metcalf, H.S. Hudson, T. Kosugi, R.C. Puetter, R.K. Pina, Astrophys. J. **466**, 585 (1996). doi:10.1086/177533

T.R. Metcalf, D.L. Mickey, B.J. Labonte, L.A. Ryder, in *Multi-wavelength Observations of Coronal Structure and Dynamics*, ed. by P.C.H. Martens, D. Cauffman (2002), p. 249

T.R. Metcalf, D. Alexander, H.S. Hudson, D.W. Longcope, Astrophys. J. **595**, 483 (2003). doi:10.1086/377217

T.R. Metcalf, K.D. Leka, D.L. Mickey, Astrophys. J. Lett. **623**, L53 (2005). doi:10.1086/429961

R.A. Mewaldt, C.M.S. Cohen, J. Giacalone, G.M. Mason, E.E. Chollet, M.I. Desai, D.K. Haggerty, M.D. Looper, R.S. Selesnick, A. Vourlidas, in *American Institute of Physics Conference Series*, ed. by G. Li, Q. Hu, O. Verkhoglyadova, G.P. Zank, R.P. Lin, J. Luhmann. American Institute of Physics Conference Series, vol. 1039 (2008), pp. 111–117. doi:10.1063/1.2982431

R.A. Mewaldt, R.A. Leske, E.C. Stone, A.F. Barghouty, A.W. Labrador, C.M.S. Cohen, A.C. Cummings, A.J. Davis, T.T. von Rosenvinge, M.E. Wiedenbeck, Astrophys. J. Lett. **693**, L11 (2009). doi:10.1088/0004-637X/693/1/L11

J.-P. Meyer, Astrophys. Suppl. **57**, 173 (1985). doi:10.1086/191001

D.L. Mickey, R.C. Canfield, B.J. Labonte, K.D. Leka, M.F. Waterson, H.M. Weber, Sol. Phys. **168**, 229 (1996). doi:10.1007/BF00148052

C.H. Miklenic, A.M. Veronig, B. Vršnak, A. Hanslmeier, Astron. Astrophys. **461**, 697 (2007). doi:10.1051/0004-6361:20065751

J.A. Miller, T.N. Larosa, R.L. Moore, Astrophys. J. **461**, 445 (1996). doi:10.1086/177072

R.O. Milligan, Astrophys. J. Lett. **680**, L157 (2008). doi:10.1086/589856

R.O. Milligan, B.R. Dennis, Astrophys. J. **699**, 968 (2009). doi:10.1088/0004-637X/699/2/968. arXiv:0905.1669

R.O. Milligan, P.T. Gallagher, M. Mathioudakis, D.S. Bloomfield, F.P. Keenan, R.A. Schwartz, Astrophys. J. Lett. **638**, L117 (2006a). doi:10.1086/500555. arXiv:astro-ph/0509664

R.O. Milligan, P.T. Gallagher, M. Mathioudakis, F.P. Keenan, Astrophys. J. Lett. **642**, L169 (2006b). doi:10.1086/504592. arXiv:astro-ph/0603652

T.G. Moran, D.E. Jennings, L.D. Deming, G.H. McCabe, P.V. Sada, R.J. Boyle, Sol. Phys. **241**, 213 (2007). doi:10.1007/s11207-007-0271-1

D. Moses, F. Clette, J.-P. Delaboudinière, G.E. Artzner, M. Bougnet, J. Brunaud, C. Carabetian, A.H. Gabriel, J.F. Hochedez, F. Millier, X.Y. Song, B. Au, K.P. Dere, R.A. Howard, R. Kreplin, D.J. Michels, J.M. Defise, C. Jamar, P. Rochus, J.P. Chauvineau, J.P. Marioge, R.C. Catura, J.R. Lemen, L. Shing, R.A. Stern, J.B. Gurman, W.M. Neupert, J. Newmark, B. Thompson, A. Maucherat, F. Portier-Fozzani, D. Berghmans, P. Cugnon, E.L. van Dessel, J.R. Gabryl, Sol. Phys. **175**, 571 (1997). doi:10.1023/A:1004902913117

T. Mrozek, M. Tomczak, Astron. Astrophys. **415**, 377 (2004). doi:10.1051/0004-6361:20034598

T. Mrozek, M. Tomczak, S. Gburek, Astron. Astrophys. **472**, 945 (2007). doi:10.1051/0004-6361:20077652

Y. Muraki, H. Tsuchiya, K. Fujiki, S. Masuda, Y. Matsubara, H. Menjyo, T. Sako, K. Watanabe, M. Ohnishi, A. Shiomi, M. Takita, T. Yuda, Y. Katayose, N. Hotta, S. Ozawa, T. Sakurai, Y.H. Tan, J.L. Zhang, Astropart. Phys. **28**, 119 (2007). doi:10.1016/j.astropartphys.2007.04.012

N. Narukage, K. Shibata, Astrophys. J. **637**, 1122 (2006). doi:10.1086/498492

D.F. Neidig, Sol. Phys. **121**, 261 (1989). doi:10.1007/BF00161699

W.M. Neupert, Astrophys. J. **153**, L59 (1968)

Z. Ning, Sol. Phys. **248**, 99 (2008). doi:10.1007/s11207-008-9124-9

L. Ofman, L. Sui, Astrophys. J. Lett. **644**, L149 (2006). doi:10.1086/505622

K. Ohki, T. Takakura, S. Tsuneta, N. Nitta, Sol. Phys. **86**, 301 (1983)

V. Ontiveros, A. Vourlidas, Astrophys. J. **693**, 267 (2009). doi:10.1088/0004-637X/693/1/267. arXiv:0811.3743

F.Q. Orrall, J.B. Zirker, Astrophys. J. **208**, 618 (1976). doi:10.1086/154642

L.E. Orwig, K.J. Frost, B.R. Dennis, Sol. Phys. **65**, 25 (1980). doi:10.1007/BF00151382

W. Otruba, W. Pötzi, Hvar Obs. Bull. **27**, 189 (2003)

R. Pallavicini, G.S. Vaiana, S.W. Kahler, A.S. Krieger, Sol. Phys. **45**, 411 (1975)

R. Pallavicini, S. Serio, G.S. Vaiana, Astrophys. J. **216**, 108 (1977)

I.D. Palmer, S.F. Smerd, Sol. Phys. **26**, 460 (1972)

E.N. Parker, Astrophys. Suppl. **8**, 177 (1963). doi:10.1086/190087
G.K. Parks, J.R. Winckler, Astrophys. J. **155**, L117 (1969)
A.N. Parmar, J.L. Culhane, C.G. Rapley, C.J. Wolfson, L.W. Acton, K.J.H. Phillips, B.R. Dennis, Astrophys. J. **279**, 866 (1984). doi:10.1086/161957
M.R. Perrenoud, Sol. Phys. **81**, 197 (1982). doi:10.1007/BF00151991
V. Petrosian, S. Liu, Astrophys. J. **610**, 550 (2004). doi:10.1086/421486. arXiv:astro-ph/0401585
H.E. Petschek, in *The Physics of Solar Flares*, ed. by W.N. Hess (1964), p. 425
K.J.H. Phillips, Astrophys. J. **605**, 921 (2004). doi:10.1086/382523
K.J.H. Phillips, W.M. Neupert, Sol. Phys. **32**, 209 (1973). doi:10.1007/BF00152737
K.J.H. Phillips, U. Feldman, L.K. Harra, Astrophys. J. **634**, 641 (2005). doi:10.1086/491693
K.J.H. Phillips, C. Chifor, B.R. Dennis, Astrophys. J. **647**, 1480 (2006). doi:10.1086/505518. arXiv:astro-ph/0607309
M. Pick, T.G. Forbes, G. Mann, H.V. Cane, J. Chen, A. Ciaravella, H. Cremades, R.A. Howard, H.S. Hudson, A. Klassen, K.L. Klein, M.A. Lee, J.A. Linker, D. Maia, Z. Mikić, J.C. Raymond, M.J. Reiner, G.M. Simnett, N. Srivastava, D. Tripathi, R. Vainio, A. Vourlidas, J. Zhang, T.H. Zurbuchen, N.R. Sheeley, C. Marqué, Space Sci. Rev. **123**, 341 (2006). doi:10.1007/s11214-006-9021-1
A.I. Poland, R.A. Howard, M.J. Koomen, D.J. Michels, N.R. Sheeley Jr., Sol. Phys. **69**, 169 (1981)
G. Poletto, R.A. Kopp, in *The Lower Atmosphere of Solar Flares; Proceedings of the Solar Maximum Mission Symposium* (1986), pp. 453–465. URL http://adsabs.harvard.edu/cgi-bin/nph-bib_query?bibcode=1986lasf.symp..453P&db_key=AST
E.R. Priest, T.G. Forbes, Astron. Astrophys. Rev. **10**, 313 (2002)
J. Qiu, J. Lee, D.E. Gary, H. Wang, Astrophys. J. **565**, 1335 (2002)
J. Qiu, J. Lee, D.E. Gary, Astrophys. J. **603**, 335 (2004). doi:10.1086/381353
J. Qiu, D.E. Gary, G.D. Fleishman, Sol. Phys. **255**, 107 (2009). doi:10.1007/s11207-009-9316-y
R. Ramaty, N. Mandzhavidze, in *Highly Energetic Physical Processes and Mechanisms for Emission from Astrophysical Plasmas*, ed. by P.C.H. Martens, S. Tsuruta, M.A. Weber. IAU Symposium, vol. 195 (2000), p. 123
R. Ramaty, B. Kozlovsky, R.E. Lingenfelter, Astrophys. Suppl. **40**, 487 (1979). doi:10.1086/190596
N.D.R. Ranns, S.A. Matthews, L.K. Harra, J.L. Culhane, Astron. Astrophys. **364**, 859 (2000)
N. Raouafi, S.K. Solanki, T. Wiegelmann, in *Astronomical Society of the Pacific Conference Series*, ed. by S.V. Berdyugina, K.N. Nagendra, R. Ramelli. Astronomical Society of the Pacific Conference Series, vol. 405 (2009), p. 429
F. Reale, G. Peres, Astron. Astrophys. **299**, 225 (1995)
D.V. Reames, Space Sci. Rev. **90**, 413 (1999). doi:10.1023/A:1005105831781
K.K. Reeves, H.P. Warren, Astrophys. J. **578**, 590 (2002). doi:10.1086/342310
S. Régnier, R.C. Canfield, Astron. Astrophys. **451**, 319 (2006). doi:10.1051/0004-6361:20054171
S. Régnier, E.R. Priest, Astron. Astrophys. **468**, 701 (2007). doi:10.1051/0004-6361:20077318. arXiv:astro-ph/0703756
S. Régnier, T. Amari, E. Kersalé, Astron. Astrophys. **392**, 1119 (2002). doi:10.1051/0004-6361:20020993
T.R. Rimmele, K. Richards, S.L. Hegwer, D. Ren, S. Fletcher, S. Gregory, L.V. Didkovsky, C.J. Denker, W. Marquette, J. Marino, P.R. Goode, in *Society of Photo-Optical Instrumentation Engineers (SPIE) Conference Series*, ed. by P.L. Wizinowich, D. Bonaccini. Society of Photo-Optical Instrumentation Engineers (SPIE) Conference Series, vol. 4839 (2003), pp. 635–646. doi:10.1117/12.457018
P.L. Rochus, J. Defise, J. Halain, C.A.J. Jamar, E. Mazy, L. Rossi, T. Thibert, F. Clette, P. Cugnon, D. Berghmans, J. Hochedez, J. Delaboudiniére, F. Auchere, R. Mercier, M. Ravet, F. Delmotte, M. Idir, U.H. Schuehle, V. Bothmer, S. Fineschi, R.A. Howard, J.D. Moses, J.S. Newmark, in *Society of Photo-Optical Instrumentation Engineers (SPIE) Conference Series*, ed. by S. Fineschi, M.A. Gummin. Society of Photo-Optical Instrumentation Engineers (SPIE) Conference Series, vol. 5171 (2004), pp. 53–64. doi:10.1117/12.503964
F. Rubio da Costa, L. Fletcher, N. Labrosse, F. Zuccarello, Astron. Astrophys. **507**, 1005 (2009). doi:10.1051/0004-6361/200912651. 0909.4705
J.L.R. Saba, T. Gaeng, T.D. Tarbell, Astrophys. J. **641**, 1197 (2006). doi:10.1086/500631
P. Saint-Hilaire, A.O. Benz, Sol. Phys. **210**, 287 (2002). doi:10.1023/A:1022478300679. arXiv:astro-ph/0210023
P. Saint-Hilaire, A.O. Benz, Astron. Astrophys. **435**, 743 (2005). doi:10.1051/0004-6361:20041918
P. Saint-Hilaire, S. Krucker, R.P. Lin, Sol. Phys. **250**, 53 (2008). doi:10.1007/s11207-008-9193-9
T. Sakao, Characteristics of solar flare hard X-ray sources as revealed with the Hard X-ray Telescope aboard the Yohkoh satellite. Ph.D. thesis, University of Tokyo (1994)
T. Sakao, T. Kosugi, S. Masuda, K. Yaji, M. Inda-Koide, K. Makishima, in *Proceedings of Kofu Symposium* (1994), pp. 169–172

T. Sakao, T. Kosugi, S. Masuda, K. Yaji, M. Inda-Koide, K. Makishima, Adv. Space Res. **17**, 67 (1996). doi:10.1016/0273-1177(95)00544-O

T. Sakurai, K. Shibata, K. Ichimoto, S. Tsuneta, L.W. Acton, Publ. Astron. Soc. Jpn. **44**, L123 (1992)

R. Saldanha, S. Krucker, R.P. Lin, Astrophys. J. **673**, 1169 (2008). doi:10.1086/524929

P.H. Scherrer, R.S. Bogart, R.I. Bush, J.T. Hoeksema, A.G. Kosovichev, J. Schou, W. Rosenberg, L. Springer, T.D. Tarbell, A. Title, C.J. Wolfson, I. Zayer, MDI Engineering Team, Sol. Phys. **162**, 129 (1995). doi:10.1007/BF00733429

E.J. Schmahl, M.R. Kundu, V.I. Garaimov, Astrophys. J. **643**, 1271 (2006). doi:10.1086/503151

E.J. Schmahl, R.L. Pernak, G.J. Hurford, J. Lee, S. Bong, Sol. Phys. **240**, 241 (2007). doi:10.1007/s11207-007-0263-1

B. Schmieder, T.G. Forbes, J.M. Malherbe, M.E. Machado, Astrophys. J. **317**, 956 (1987). doi:10.1086/165344

B. Schmieder, P. Heinzel, J.E. Wiik, J. Lemen, B. Anwar, P. Kotrc, E. Hiei, Sol. Phys. **156**, 337 (1995)

C.J. Schrijver, Adv. Space Res. **43**, 739 (2009). doi:10.1016/j.asr.2008.11.004. arXiv:0811.0787

C.J. Schrijver, M.L. DeRosa, A.M. Title, T.R. Metcalf, Astrophys. J. **628**, 501 (2005). doi:10.1086/430733

C.J. Schrijver, M.L. Derosa, T.R. Metcalf, Y. Liu, J. McTiernan, S. Régnier, G. Valori, M.S. Wheatland, T. Wiegelmann, Sol. Phys. **235**, 161 (2006). doi:10.1007/s11207-006-0068-7

C.J. Schrijver, M.L. DeRosa, T. Metcalf, G. Barnes, B. Lites, T. Tarbell, J. McTiernan, G. Valori, T. Wiegelmann, M.S. Wheatland, T. Amari, G. Aulanier, P. Démoulin, M. Fuhrmann, K. Kusano, S. Régnier, J.K. Thalmann, Astrophys. J. **675**, 1637 (2008). doi:10.1086/527413. arXiv:0712.0023

R.A. Schwartz, B. Dennis, A.K. Tolbert, R. Murphy, G. Share, G. Fishman, M. Briggs, F. Longo, R. Diehl, R. Wijers, in *American Astronomical Society Meeting Abstracts #216*. Bulletin of the American Astronomical Society, vol. 41 (2010), p. 900

A. Severny, Space Sci. Rev. **3**, 451 (1964). doi:10.1007/BF00214468

N.R. Sheeley Jr., R.A. Howard, M.J. Koomen, D.J. Michels, Astrophys. J. **272**, 349 (1983). doi:10.1086/161298

N.R. Sheeley Jr., H.P. Warren, Y.-M. Wang, Astrophys. J. **616**, 1224 (2004). doi:10.1086/425126

K. Shibasaki, Astrophys. J. **557**, 326 (2001). doi:10.1086/321651

T. Shimizu, Adv. Space Res. **29**, 2009 (2002)

A.V.R. Silva, H. Wang, D.E. Gary, N. Nitta, H. Zirin, Astrophys. J. **481**, 978 (1997). doi:10.1086/304076

A.V.R. Silva, H. Wang, D.E. Gary, Astrophys. J. **545**, 1116 (2000). doi:10.1086/317822

D.M. Smith, R.P. Lin, P. Turin, D.W. Curtis, J.H. Primbsch, R.D. Campbell, R. Abiad, P. Schroeder, C.P. Cork, E.L. Hull, D.A. Landis, N.W. Madden, D. Malone, R.H. Pehl, T. Raudorf, P. Sangsingkeow, R. Boyle, I.S. Banks, K. Shirey, R. Schwartz, Sol. Phys. **210**, 33 (2002). doi:10.1023/A:1022400716414

B.V. Somov, T. Kosugi, Astrophys. J. **485**, 859 (1997)

B.V. Somov, T. Kosugi, H.S. Hudson, T. Sakao, S. Masuda, Astrophys. J. **579**, 863 (2002). doi:10.1086/342842

T. Spirock, C. Denker, H. Chen, J. Chae, J. Qiu, J. Varsik, H. Wang, P.R. Goode, W. Marquette, in *Advanced Solar Polarimetry—Theory, Observation, and Instrumentation*, ed. by M. Sigwarth. Astronomical Society of the Pacific Conference Series, vol. 236 (2001), p. 65

A.C. Sterling, H.S. Hudson, Astrophys. J. Lett. **491**, L55+ (1997). doi:10.1086/311043

A.C. Sterling, R.L. Moore, Astrophys. J. **630**, 1148 (2005). doi:10.1086/432044

K.T. Strong, A.O. Benz, B.R. Dennis, A.I. Poland, J.W. Leibacher, R. Mewe, J. Schrijver, G. Simnett, J.B. Smith Jr., J. Sylwester, Sol. Phys. **91**, 325 (1984)

Y.N. Su, L. Golub, A.A. van Ballegooijen, M. Gros, Sol. Phys. **236**, 325 (2006). doi:10.1007/s11207-006-0039-z

Y. Su, L. Golub, A.A. Van Ballegooijen, Astrophys. J. **655**, 606 (2007). doi:10.1086/510065

J.J. Sudol, J.W. Harvey, Astrophys. J. **635**, 647 (2005). doi:10.1086/497361

L. Sui, G.D. Holman, Astrophys. J. Lett. **596**, L251 (2003). doi:10.1086/379343

L. Sui, G.D. Holman, B.R. Dennis, Astrophys. J. **612**, 546 (2004). doi:10.1086/422515

L. Sui, G.D. Holman, B.R. Dennis, Astrophys. J. **626**, 1102 (2005). doi:10.1086/430086

Z. Švestka, Space Sci. Rev. **5**, 388 (1966). doi:10.1007/BF02653250

Z. Švestka, Sol. Phys. **13**, 471 (1970). doi:10.1007/BF00153567

Z. Švestka, P. Hoyng, W. van Tend, A. Boelee, C. de Jager, R.T. Stewart, L.W. Acton, E.C. Bruner, A.H. Gabriel, C.G. Rapley, C. de Jager, H. LaFleur, G. Nelson, G.M. Simnett, H.F. van Beek, W.J. Wagner, Sol. Phys. **75**, 305 (1982). doi:10.1007/BF00153479

Z.F. Švestka, J.M. Fontenla, M.E. Machado, S.F. Martin, D.F. Neidig, Sol. Phys. **108**, 237 (1987)

P.A. Sweet, Annu. Rev. Astron. Astrophys. **7**, 149 (1969). doi:10.1146/annurev.aa.07.090169.001053

J. Sylwester, J.R. Lemen, R. Mewe, Nature **310**, 665 (1984). doi:10.1038/310665a0

J. Sylwester, J.R. Lemen, R.D. Bentley, A. Fludra, M.-C. Zolcinski, Astrophys. J. **501**, 397 (1998). doi:10.1086/305785

J. Sylwester, I. Gaicki, Z. Kordylewski, M. Kowaliński, S. Nowak, S. Płocieniak, M. Siarkowski, B. Sylwester, W. Trzebiński, J. Bakała, J.L. Culhane, M. Whyndham, R.D. Bentley, P.R. Guttridge, K.J.H. Phillips, J. Lang, C.M. Brown, G.A. Doschek, V.D. Kuznetsov, V.N. Oraevsky, A.I. Stepanov, D.V. Lisin, Sol. Phys. **226**, 45 (2005). doi:10.1007/s11207-005-6392-5

S.I. Syrovatskii, Annu. Rev. Astron. Astrophys. **19**, 163 (1981). doi:10.1146/annurev.aa.19.090181.001115

T. Takakura, S. Tsuneta, N. Nitta, K. Makishima, T. Murakami, Y. Ogawara, M. Oda, K. Ohki, S. Miyamoto, Astrophys. J. Lett. **270**, L83 (1983). doi:10.1086/184075

F. Tang, Sol. Phys. **102**, 131 (1985)

M. Temmer, A.M. Veronig, B. Vršnak, C. Miklenic, Astrophys. J. **654**, 665 (2007). doi:10.1086/509634

M. Temmer, A.M. Veronig, B. Vršnak, J. Rybák, P. Gömöry, S. Stoiser, D. Maričić, Astrophys. J. Lett. **673**, L95 (2008). doi:10.1086/527414

L. Teriaca, A. Falchi, G. Cauzzi, R. Falciani, L.A. Smaldone, V. Andretta, Astrophys. J. **588**, 596 (2003). doi:10.1086/373946

L. Teriaca, A. Falchi, R. Falciani, G. Cauzzi, L. Maltagliati, Astron. Astrophys. **455**, 1123 (2006). doi:10.1051/0004-6361:20065065. arXiv:0903.0232

J.K. Thalmann, T. Wiegelmann, Astron. Astrophys. **484**, 495 (2008). doi:10.1051/0004-6361:200809508

R.J. Thomas, R.G. Teske, Sol. Phys. **16**, 431 (1971). doi:10.1007/BF00162486

B.J. Thompson, S.P. Plunkett, J.B. Gurman, J.S. Newmark, O.C. St. Cyr, D.J. Michels, Geophys. Res. Lett. **25**, 2465 (1998). doi:10.1029/98GL50429

V.S. Titov, E.R. Priest, P. Démoulin, Astron. Astrophys. **276**, 564 (1993)

M. Tomczak, Astron. Astrophys. **366**, 294 (2001). doi:10.1051/0004-6361:20000204

S. Tomczyk, G.L. Card, T. Darnell, D.F. Elmore, R. Lull, P.G. Nelson, K.V. Streander, J. Burkepile, R. Casini, P.G. Judge, Sol. Phys. **247**, 411 (2008). doi:10.1007/s11207-007-9103-6

D. Tripathi, H. Isobe, H.E. Mason, Astron. Astrophys. **453**, 1111 (2006). doi:10.1051/0004-6361:20064993

S. Tsuneta, L. Acton, M. Bruner, J. Lemen, W. Brown, R. Caravalho, R. Catura, S. Freeland, B. Jurcevich, J. Owens, Sol. Phys. **136**, 37 (1991)

S. Tsuneta, H. Hara, T. Shimizu, L.W. Acton, K.T. Strong, H.S. Hudson, Y. Ogawara, Publ. Astron. Soc. Jpn. **44**, L63 (1992)

Y. Uchida, Sol. Phys. **39**, 431 (1974)

S. Ueno, S. Nagata, R. Kitai, H. Kurokawa, in *The Solar-B Mission and the Forefront of Solar Physics*. ASP Conference Series, vol. 325 (2004), p. 319

H.F. van Beek, C. de Jager, A. Schadee, Z. Svestka, A. Boelee, A. Duijveman, M. Galama, R. Hoekstra, P. Hoyng, R. Fryer, Astrophys. J. Lett. **244**, L157 (1981). doi:10.1086/183502

G.H.J. van den Oord, Astron. Astrophys. **234**, 496 (1990)

J.E. Vernazza, E.H. Avrett, R. Loeser, Astrophys. Suppl. **45**, 635 (1981). doi:10.1086/190731

A.M. Veronig, J.C. Brown, Astrophys. J. Lett. **603**, L117 (2004). doi:10.1086/383199

A. Veronig, B. Vršnak, B.R. Dennis, M. Temmer, A. Hanslmeier, J. Magdalenić, Astron. Astrophys. **392**, 699 (2002a). doi:10.1051/0004-6361:20020947. arXiv:astro-ph/0207217

A. Veronig, B. Vršnak, M. Temmer, A. Hanslmeier, Sol. Phys. **208**, 297 (2002b). arXiv:astro-ph/0208088

A.M. Veronig, J.C. Brown, B.R. Dennis, R.A. Schwartz, L. Sui, A.K. Tolbert, Astrophys. J. **621**, 482 (2005). doi:10.1086/427274

A.M. Veronig, M. Karlický, B. Vršnak, M. Temmer, J. Magdalenić, B.R. Dennis, W. Otruba, W. Pötzi, Astron. Astrophys. **446**, 675 (2006). doi:10.1051/0004-6361:20053112

N. Vilmer et al., Space Sci. Rev. (2011, this issue). doi:10.1007/s11214-010-9728-x

E. Vogt, J. Hénoux, Astron. Astrophys. **349**, 283 (1999)

O. von der Luehe, Astron. Astrophys. **268**, 374 (1993)

A. Vourlidas, P. Subramanian, K.P. Dere, R.A. Howard, Astrophys. J. **534**, 456 (2000). doi:10.1086/308747

A. Vourlidas, D. Buzasi, R.A. Howard, E. Esfandiari, in *ESA Special Publication*, ed. by J. Kuijpers. ESA Special Publication, vol. 506 (2002), pp. 91–94

A. Vourlidas, S.T. Wu, A.H. Wang, P. Subramanian, R.A. Howard, Astrophys. J. **598**, 1392 (2003). doi:10.1086/379098. arXiv:astro-ph/0308367

B. Vršnak, E.W. Cliver, Sol. Phys. **253**, 215 (2008). doi:10.1007/s11207-008-9241-5

B. Vršnak, M. Temmer, A. Veronig, M. Karlický, J. Lin, Sol. Phys. **234**, 273 (2006). doi:10.1007/s11207-006-0093-6

H. Wang, Sol. Phys. **140**, 85 (1992)

Y. Wang, J. Zhang, Astrophys. J. **665**, 1428 (2007). doi:10.1086/519765. 0808.2976

H. Wang, M.W. Ewell Jr., H. Zirin, G. Ai, Astrophys. J. **424**, 436 (1994). doi:10.1086/173901

H. Wang, D.E. Gary, H. Zirin, R.A. Schwartz, T. Sakao, T. Kosugi, K. Shibata, Astrophys. J. **453**, 505 (1995). doi:10.1086/176411

H. Wang, C. Liu, J. Qiu, N. Deng, P.R. Goode, C. Denker, Astrophys. J. Lett. **601**, L195 (2004). doi:10.1086/382188

T. Wang, L. Sui, J. Qiu, Astrophys. J. Lett. **661**, L207 (2007). doi:10.1086/519004

H.P. Warren, A.D. Warshall, Astrophys. J. Lett. **560**, L87 (2001). doi:10.1086/324060

H.P. Warren, A.R. Winebarger, P.S. Hamilton, Astrophys. J. Lett. **579**, L41 (2002). doi:10.1086/344921

D.G. Wentzel, Astrophys. J. **208**, 595 (1976)

M.S. Wheatland, Astrophys. J. **609**, 1134 (2004). doi:10.1086/421261. arXiv:astro-ph/0403613

M.S. Wheatland, D.B. Melrose, Sol. Phys. **158**, 283 (1995)

S.M. White, T.S. Bastian, R. Bradley, C. Parashare, L. Wye, in *Astronomical Society of the Pacific Conference Series*, ed. by N. Kassim, M. Perez, W. Junor, P. Henning. Astronomical Society of the Pacific Conference Series, vol. 345 (2005), p. 176

S.M. White, R.J. Thomas, R.A. Schwartz, Sol. Phys. **227**, 231 (2005b). doi:10.1007/s11207-005-2445-z

S. White et al., Space Sci. Rev. (2011, this issue). doi:10.1007/s11214-010-9708-1

E. Wiehr, G. Stellmacher, Astron. Astrophys. **247**, 379 (1991)

J.M. Wilcox, A.J. Hundhausen, J. Geophys. Res. **88**, 8095 (1983). doi:10.1029/JA088iA10p08095

J.P. Wild, S.F. Smerd, A.A. Weiss, Annu. Rev. Astron. Astrophys. **1**, 291 (1963). doi:10.1146/annurev.aa.01.090163.001451

K. Wilhelm, W. Curdt, E. Marsch, U. Schühle, P. Lemaire, A. Gabriel, J.-C. Vial, M. Grewing, M.C.E. Huber, S.D. Jordan, A.I. Poland, R.J. Thomas, M. Kühne, J.G. Timothy, D.M. Hassler, O.H.W. Siegmund, Sol. Phys. **162**, 189 (1995). doi:10.1007/BF00733430

B.E. Woodgate, R.A. Shine, A.I. Poland, L.E. Orwig, Astrophys. J. **265**, 530 (1983). doi:10.1086/160696

T.N. Woods, F.G. Eparvier, J. Fontenla, J. Harder, G. Kopp, W.E. McClintock, G. Rottman, B. Smiley, M. Snow, Geophys. Res. Lett. **31**, 10802 (2004). doi:10.1029/2004GL019571

T.N. Woods, G. Kopp, P.C. Chamberlin, J. Geophys. Res. (Space) Phys. **111**, 10 (2006). doi:10.1029/2005JA011507

T.N. Woods, F.G. Eparvier, R. Hock, A.R. Jones, D. Woodraska, D. Judge, L. Didkovsky, J. Lean, J. Mariska, H. Warren, D. McMullin, P. Chamberlin, G. Berthiaume, S. Bailey, T. Fuller-Rowell, J. Sojka, W.K. Tobiska, R. Viereck, Solar Phys. 3 (2010). doi:10.1007/s11207-009-9487-6

J.-P. Wuelser, R.C. Canfield, L.W. Acton, J.L. Culhane, A. Phillips, A. Fludra, T. Sakao, S. Masuda et al., Astrophys. J. **424**, 459 (1994). doi:10.1086/173903

Y. Xu, W. Cao, C. Liu, G. Yang, J. Qiu, J. Jing, C. Denker, H. Wang, Astrophys. J. Lett. **607**, L131 (2004). doi:10.1086/422099

Y.-H. Yang, C.Z. Cheng, S. Krucker, R.P. Lin, W.H. Ip, Astrophys. J. **693**, 132 (2009). doi:10.1088/0004-637X/693/1/132

T. Yokoyama, K. Akita, T. Morimoto, K. Inoue, J. Newmark, Astrophys. J. Lett. **546**, L69 (2001). doi:10.1086/318053

P.R. Young, G. Del Zanna, E. Landi, K.P. Dere, H.E. Mason, M. Landini, Astrophys. Suppl. **144**, 135 (2003). doi:10.1086/344365. arXiv:astro-ph/0209493

D.M. Zarro, J.R. Lemen, Astrophys. J. **329**, 456 (1988). doi:10.1086/166391

D.M. Zarro, R.C. Canfield, T.R. Metcalf, K.T. Strong, Astrophys. J. **324**, 582 (1988). doi:10.1086/165919

J. Zhang, K.P. Dere, R.A. Howard, M.R. Kundu, S.M. White, Astrophys. J. **559**, 452 (2001). doi:10.1086/322405

J. Zhang, K.P. Dere, R.A. Howard, A. Vourlidas, Astrophys. J. **604**, 420 (2004). doi:10.1086/381725

V.V. Zharkova, Sol. Phys. **251**, 641 (2008). doi:10.1007/s11207-008-9216-6

V.V. Zharkova, S.I. Zharkov, Astrophys. J. **664**, 573 (2007). doi:10.1086/518731. arXiv:0802.2787

V.V. Zharkova et al., Space Sci. Rev. (2011, this issue). doi:10.1007/s11214-011-9803-y

T. Zhou, H. Ji, G. Huang, Adv. Space Res. **41**, 1195 (2008). doi:10.1016/j.asr.2007.04.095

H. Zirin, M.A. Liggett, Sol. Phys. **113**, 267 (1987)

H. Zirin, K. Tanaka, Sol. Phys. **32**, 173 (1973). doi:10.1007/BF00152736

Space Sci Rev (2011) 159:107–166
DOI 10.1007/s11214-010-9680-9

Implications of X-ray Observations for Electron Acceleration and Propagation in Solar Flares

G.D. Holman · M.J. Aschwanden · H. Aurass · M. Battaglia · P.C. Grigis · E.P. Kontar · W. Liu · P. Saint-Hilaire · V.V. Zharkova

Received: 8 April 2008 / Accepted: 21 July 2010 / Published online: 12 August 2011

Abstract High-energy X-rays and γ-rays from solar flares were discovered just over fifty years ago. Since that time, the standard for the interpretation of spatially integrated flare X-ray spectra at energies above several tens of keV has been the collisional thick-target model. After the launch of the *Reuven Ramaty High Energy Solar Spectroscopic Imager* (*RHESSI*) in early 2002, X-ray spectra and images have been of sufficient quality to allow a greater focus on the energetic electrons responsible for the X-ray emission, including their origin and their interactions with the flare plasma and magnetic field. The result has been new insights into the flaring process, as well as more quantitative models for both electron acceleration and propagation, and for the flare environment with which the electrons interact.

G.D. Holman (✉) · W. Liu
NASA Goddard Space Flight Center, Code 671, Greenbelt, MD 20771, USA
e-mail: Gordon.D.Holman@nasa.gov

M.J. Aschwanden
Lockheed Martin Advanced Technology Center, Solar and Astrophysics Laboratory, Organization ADBS, Building 252, 3251 Hanover Street, Palo Alto, CA 94304, USA

H. Aurass
Astrophysikalisches Institut Potsdam, Potsdam, Germany

M. Battaglia
Department of Physics & Astronomy, University of Glasgow, Glasgow, G12 8QQ, Scotland, UK

P.C. Grigis
Harvard-Smithsonian Center for Astrophysics P-148, 60 Garden St., Cambridge, MA 02138, USA

E.P. Kontar
Department of Physics and Astronomy, University of Glasgow, Kelvin Building, Glasgow, G12 8QQ, UK

P. Saint-Hilaire
Space Sciences Lab, UC Berkeley, Berkeley, CA, USA

V.V. Zharkova
Department of Computing and Mathematics, University of Bradford, Bradford, BD7 1DP, UK

In this article we review our current understanding of electron acceleration, energy loss, and propagation in flares. Implications of these new results for the collisional thick-target model, for general flare models, and for future flare studies are discussed.

Keywords Sun: flares · Sun: X-rays · Gamma rays · Sun: radio radiation

1 Introduction

A primary characteristic of solar flares is the acceleration of electrons to high, suprathermal energies. These electrons are observed directly in interplanetary space, and indirectly at the Sun through the X-ray, γ-ray, and radio emissions they emit (Hudson and Ryan 1995). Understanding how these electrons are produced and how they evolve is fundamental to obtaining an understanding of energy release in flares. Therefore, one of the principal goals of solar flare research is to determine when, where, and how these electrons are accelerated to suprathermal energies, and what happens to them after they are accelerated to these high energies.

A major challenge to obtaining an understanding of electron acceleration in flares is that the location where they are accelerated is not necessarily where they are most easily observed. The flare-accelerated electrons that escape the Sun are not directly observed until they reach the instruments in space capable of detecting them, usually located at the distance of the Earth from the Sun. The properties of these electrons are easily modified during their long journey from the flaring region to the detecting instruments (e.g., Agueda et al. 2009). Distinguishing flare-accelerated electrons from electrons accelerated in interplanetary shock waves is also difficult (Kahler 2007).

The electrons that are observed at the Sun through their X-ray or γ-ray emissions radiate most intensely where the density of the ambient plasma is highest (see Sect. 2). Therefore, the radiation from electrons in and near the acceleration region may not be intense enough to be observable. Although these radiating electrons are much closer to the acceleration region than those detected in interplanetary space, their properties can still be significantly modified as they propagate to the denser regions where they are observed. The radio emission from the accelerated electrons also depends on the plasma environment, especially the magnetic field strength for the gyrosynchrotron radiation observed from flares (Bastian et al. 1998). Therefore, determining when, where, and how the electrons were accelerated requires a substantial amount of plausible reasoning.

Here we focus primarily on the X-ray emission from the accelerated electrons. Interplanetary electrons and low-energy emissions are addressed by Fletcher et al. (2011), while the γ-ray emission is addressed by Vilmer et al. (2011), and the radio by White et al. (2011). The X-rays are predominantly *electron-ion bremsstrahlung* (free-free radiation), emitted when the accelerated electrons scatter off ions in the ambient thermal plasma. Issues related to the emission mechanism and deducing the properties of the emitting electrons from the X-ray observations are primarily addressed by Kontar et al. (2011). Here we address the interpretation of the X-ray observations in terms of flare models, and consider the implications of the observations for the acceleration process, energy release in flares, and electron propagation. Specific models for particle acceleration and energy release in flares are addressed by Zharkova et al. (2011).

The accelerated electrons interact with both ambient electrons and ions, but lose most of their energy through *electron-electron Coulomb collisions*. Consequently, the brightest X-ray sources are associated with high collisional energy losses. These losses in turn change the energy distribution of the radiating electrons. When the accelerated electrons lose their suprathermal energy to the ambient plasma as they radiate, the source region is called a *thick target*. Electrons streaming downward into the higher densities in lower regions of the solar atmosphere, or trapped long enough in lower density regions, will emit thick-target X-rays. Hence, thick-target models are important to understanding the origin and evolution of accelerated electrons in flares. Thick-target X-ray emission is addressed in Sect. 2.

The total energy carried by accelerated electrons is important to assessing acceleration models, especially considering that these electrons carry a significant fraction of the total energy released in flares. Also, the energy carried by electrons that escape the acceleration region is deposited elsewhere, primarily to heat the plasma in the thick-target source regions. The X-ray flux from flares falls off rapidly with increasing photon energy, indicating that the number of radiating electrons decreases rapidly with increasing electron energy. Therefore, the energy carried by the accelerated electrons is sensitive to the value of the low-energy cutoff to the electron distribution. The determination of this low-energy cutoff and the total energy in the accelerated electrons is addressed in Sect. 3.

In the standard thick-target model, the target plasma is assumed to be fully ionized. If the target ionization is not uniform, so that the accelerated electrons stream down to cooler plasma that is partially ionized or un-ionized, the X-ray spectrum is modified. This is addressed in Sect. 4.

Observations of the radiation from hot flare plasma have shown this plasma to primarily be confined to magnetic loops or arcades of magnetic loops (cf. Aschwanden 2004). The observations also indicate that the heating of this plasma and particle acceleration initially occur in the corona above these hot loops (see Sect. 12.2 and Fletcher et al. 2011). When the density structure in these loops is typical of active region loops, or at least not highly enhanced above those densities, the highest intensity, thick-target X-ray emission will be from the footpoints of the loops, as is most often observed to be the case. If accelerated electrons alone, unaccompanied by neutralizing ions, stream down the legs of the loop from the acceleration region to the footpoints, they will drive a co-spatial return current in the ambient plasma to neutralize the high current associated with the downward-streaming electrons. We refer to this primarily downward-streaming distribution of energetic (suprathermal) electrons as an electron beam. The electric field associated with the return current will decelerate electrons in the beam, which can in turn modify the X-ray spectrum from the accelerated electrons. This is addressed in Sect. 5.

Both the primary beam of accelerated electrons and the return current can become unstable and drive the growth of waves in the ambient plasma. These waves can, in turn, interact with the electron beam and return current, altering the energy and angular distributions of the energetic electrons. These plasma instabilities are discussed in Sect. 6.

The collisional energy loss rate is greater for lower energy electrons. Therefore, for suprathermal electrons streaming downward to the footpoints of a loop, the footpoint X-ray sources observed at lower energies should be at a higher altitude than footpoint sources observed at higher X-ray energies. The height dispersion of these sources provides information about the height distribution of the plasma density in the footpoints. The spatial resolution of the *Reuven Ramaty High Energy Solar Spectroscopic Imager* (*RHESSI*—see Lin et al. 2002) has made such a study possible. This is described in Sect. 7. *RHESSI* has observed X-ray sources that move downward from the loop top and then upward from the footpoints during some flares. This source evolution in time is also discussed in Sect. 7.

If electrons of all energies are simultaneously injected, the footpoint X-ray emission from the slower, lower energy electrons should appear after that from faster, higher energy electrons. The length of this time delay provides an important test for the height of the acceleration region. Longer time delays can result from magnetic trapping of the electrons. The evolution of the thermal plasma in flares can also exhibit time delays associated with the balance between heating and cooling processes. These various time delays and the information they provide are addressed in Sect. 8.

An important diagnostic of electron acceleration and propagation in flares is the time evolution of the X-ray spectrum during flares. In most flares, the X-ray spectrum becomes harder (flatter, smaller spectral index) and then softer (steeper, larger spectral index) as the X-ray flux evolves from low to high intensity and then back to low intensity. There are notable exceptions to this pattern, however. Spectral evolution is addressed in Sect. 9.

One of the most important results from the *Yohkoh* mission is the discovery in some flares of a hard (high energy) X-ray source above the top of the thermal (low energy) X-ray loops. This, together with the *Yohkoh* observations of cusps at the top of flare X-ray loops, provided strong evidence that energy release occurs in the corona above the hot X-ray loops (for some flares, at least). Although several models have been proposed, the origin of these "above-the-looptop" hard X-ray sources is not well understood. We need to determine how their properties and evolution compare to the more intense footpoint hard X-ray sources. These issues are addressed in Sect. 10.

Radio observations provide another view of accelerated electrons and related flare phenomena. Although radio emission and its relationship to flare X-rays are primarily addressed in White et al. (2011), some intriguing radio observations that bear upon electron acceleration in flares are presented in Sect. 11.

RHESSI observations of flare X-ray emission have led to substantial progress, but many questions remain unanswered. Part of the progress is that the questions are different from those that were asked less than a decade ago. The primary context for interpreting the X-ray emission from suprathermal electrons is still the thick-target model, but the ultimate goal is to understand how the electrons are accelerated. In Sect. 12 we summarize and discuss the implications of the X-ray observations for the thick-target model and electron acceleration mechanisms, and highlight some of the questions that remain to be answered. Implications of these questions for future flare studies are discussed.

2 Thin- and Thick-Target X-ray Emission

As was summarized in Sect. 1, the electron-ion bremsstrahlung X-rays from a beam of accelerated electrons will be most intense where the density of target ions is highest, as well as where the flux of accelerated electrons is high. The local emission (emissivity) at position $\mathbf{r}$ of photons of energy ε by electrons of energy E is given by the plasma ion density, $n(\mathbf{r})$ ions cm^{-3}, times the electron-beam flux density distribution, $F(E,\mathbf{r})$ electrons cm^{-2} s^{-1} keV^{-1}, times the differential electron-ion bremsstrahlung cross-section, $Q(\varepsilon, E)$ cm^2 keV^{-1}. For simplicity, we do not consider here the angular distribution of the beam electrons or of the emitted photons, topics addressed in Kontar et al. (2011).

The emissivity of the radiation at energy ε from all the electrons in the beam is obtained by integrating over all contributing electron energies, which is all electron energies above the photon energy. The photon flux emitted per unit energy is obtained by integrating over the emitting source volume (V) or, for an imaged source, along the line of sight through the source region. Finally, assuming isotropic emission, the observed spatially integrated flux density of photons of energy ε at the X-ray detector, $I(\varepsilon)$ photons cm^{-2} s^{-1} keV^{-1}, is simply the flux divided by the geometrical dilution factor $4\pi R^2$, where R is the distance to the X-ray detector. Thus,

$$I(\varepsilon) = \frac{1}{4\pi R^2} \int_V \int_\varepsilon^\infty n(\mathbf{r})\, F(E,\mathbf{r})\, Q(\varepsilon, E)\, dE\, dV. \tag{2.1}$$

We refer to $I(\varepsilon)$ as the X-ray flux spectrum, or simply the X-ray spectrum. The spectrum obtained directly from an X-ray detector is generally a spectrum of counts versus energy loss in the detector, which must be converted to an X-ray flux spectrum by correcting for the detector response as a function of photon energy (see, for example, Smith et al. 2002).

Besides increasing the X-ray emission, a high plasma density also means increased Coulomb energy losses for the beam electrons. In the solar plasma, the bremsstrahlung losses are small compared to the collisional losses to the plasma electrons. For a fully ionized plasma and beam electron speeds much greater than the mean speed of the thermal electrons, the (nonrelativistic) energy loss rate is

$$dE/dt = -(K/E)\, n_e(\mathbf{r})\, \mathrm{v}(E), \tag{2.2}$$

where $K = 2\pi e^4\, \Lambda_{ee}$, Λ_{ee} is the Coulomb logarithm for electron-electron collisions, e is the electron charge, $n_e(\mathbf{r})$ is the plasma electron number density, and $\mathrm{v}(E)$ is the speed of the electron (see Brown 1971; Emslie 1978). The coefficient K is usually taken to be constant, although Λ_{ee} depends weakly on the electron energy and plasma density or magnetic field strength, typically falling in the range 20–30 for X-ray-emitting electrons. Numerically, the energy loss rate in keV s^{-1} or erg s^{-1}, with E in keV is

$$K = 3.00 \times 10^{-18} \left(\frac{\Lambda_{ee}}{23}\right) \mathrm{keV^2\,cm^2} = 4.80 \times 10^{-27} \left(\frac{\Lambda_{ee}}{23}\right) \mathrm{erg\,keV\,cm^2}. \tag{2.3}$$

Here and in equations to follow, notation such as $(\Lambda_{ee}/23)$ is used to show the scaling of a computed constant (here, K) with an independent variable and the numerical value taken for the independent variable ($\Lambda_{ee} = 23$ in (2.3)). If the plasma is not fully ionized, K also depends on the ionization state (see Sect. 4.1).

Noting that $\mathrm{v}\,dt = dz$, (2.2) can be simplified to $dE/dN_e = -K/E$, where $dN_e(z) = n_e(z)\,dz$ and $N_e(z)$ (cm^{-2}) is the plasma electron *column density*. (Here we treat this as a one-dimensional system and do not distinguish between the total electron velocity and the velocity component parallel to the magnetic field.) Hence, the evolution of an electron's energy with column density is simply

$$E^2 = E_0^2 - 2K N_e, \tag{2.4}$$

where E_0 is the initial energy of the electron where it is injected into the target region. For example, a 1 keV electron loses all of its energy over a column density of $1/(2K) = 1.7 \times 10^{17}$ cm^{-2} (for $\Lambda_{ee} = 23$). A 25 keV electron loses 20% of its energy over a column density of 3.8×10^{19} cm^{-2}.

If energy losses are not significant within a spatially unresolved X-ray source region, the emission is called *thin-target*. If, on the other hand, the non-thermal electrons lose all their suprathermal energy within the spatially unresolved source during the observational integration time, the emission is called *thick-target*. We call a model that assumes these energy losses are from Coulomb collisions ((2.2) and (2.3)) a *collisional thick-target model*. Collisional thick-target models have been applied to flare X-ray/γ-ray emission since the discovery of this emission in 1958 (Peterson and Winckler 1958, 1959).

The maximum information that can be obtained about the accelerated electrons from an X-ray spectrum alone is contained in the *mean electron flux distribution*, the plasma-density-weighted, target-averaged electron flux density distribution (Brown et al. 2003; Kontar et al.

2011). The mean electron flux distribution is defined as

$$\bar{F}(E) = \frac{1}{\bar{n}V} \int_V n(\mathbf{r})\, F(E, \mathbf{r})\, dV \text{ electrons cm}^{-2}\,\text{s}^{-1}\,\text{keV}^{-1}, \tag{2.5}$$

where $\bar{n}$ and V are the mean plasma density and volume of the emitting region. As can be seen from (2.1), the product $\bar{n}V\bar{F}$ can, in principle, be deduced with only a knowledge of the bremsstrahlung cross-section, $Q(\varepsilon, E)$. Additional information is required to determine if the X-ray emission is thin-target, thick-target, or something in between. The flux distribution of the emitting electrons and the mean electron flux distribution are equivalent for a homogeneous, thin-target source region.

Equation (2.1) gives the observed X-ray flux in terms of the accelerated electron flux density distribution throughout the source. However, we are interested in the electron distribution injected into the source region, $F_0(E_0, \mathbf{r_0})$, since that is the distribution produced by the unknown acceleration mechanism, including any modifications during propagation to the source region. To obtain this, we need to know how to relate $F(E, \mathbf{r})$ at all locations within the source region to $F_0(E_0, \mathbf{r_0})$. Since we are interested in the X-rays from a spatially integrated, thick-target source region, the most direct approach is to first compute the bremsstrahlung photon yield from a single electron of energy E_0, $\nu(\varepsilon, E_0)$ photons keV^{-1} per electron (Brown 1971). As long as the observational integration time is longer than the time required for the electrons to radiate all photons of energy ε (i.e., longer than the time required for energy losses to reduce all electron energies to less than ε), the thick-target X-ray spectrum is then given by

$$I_{thick}(\varepsilon) = \frac{1}{4\pi R^2} \int_\varepsilon^\infty \mathscr{F}_0(E_0)\, \nu(\varepsilon, E_0)\, dE_0, \tag{2.6}$$

where $\mathscr{F}_0(E_0)$ is the electron beam flux distribution (electrons s^{-1} keV^{-1}). $\mathscr{F}_0(E_0)$ is the integral of $F_0(\mathbf{r_0}, E_0)$ over the area at the injection site through which the electrons stream into the thick-target region.

The rate at which an electron of energy E radiates bremsstrahlung photons of energy ε is $n(\mathbf{r})\text{v}(E)Q(\varepsilon, E)$. The photon yield is obtained by integrating this over time. Since the electrons are losing energy at the rate dE/dt, the time integration can be replaced by an integration over energy from the initial electron energy E_0 to the lowest energy capable of radiating a photon of energy ε:

$$\nu(\varepsilon, E_0) = \int_{E_0}^{\varepsilon} \frac{n(\mathbf{r})\, \text{v}(E)\, Q(\varepsilon, E)\, dE}{dE/dt}. \tag{2.7}$$

Using (2.2) for dE/dt, (2.6) becomes

$$I_{thick}(\varepsilon) = \frac{1}{4\pi R^2} \frac{1}{\overline{\overline{Z}}K} \int_{E_0=\varepsilon}^{\infty} \mathscr{F}_0(E_0) \int_{E=\varepsilon}^{E_0} E\, Q(\varepsilon, E)\, dE\, dE_0, \tag{2.8}$$

where we have used the relationship $n_e = \overline{Z}n$, where $\overline{Z} \simeq 1.1$ is the ion-species-number-density-weighted (or, equivalently, relative-ion-abundance-weighted) mean atomic number of the target plasma.

By (2.8), the thick-target X-ray flux spectrum does not depend on the plasma density. However, the plasma must be dense enough for the emission to be thick-target, i.e., dense enough for all the electrons to be thermalized in the observation time interval. Integration of

(2.2) shows that this typically implies a plasma density $\gtrsim 10^{11}$–10^{12} cm^{-3} for an observational integration time of 1 s (see Sects. 10.4 and 12.1 for more about this). This condition is well satisfied at loop footpoints.

Observed non-thermal X-ray spectra from solar flares can usually be well fitted with a model photon spectrum that is either a single or a double power-law. For a single power-law electron flux distribution of the form $\mathcal{F}(E) = AE^{-\delta}$, the photon spectrum is also well approximated by the power-law form $I(\varepsilon) = I_0\varepsilon^{-\gamma}$. The relationship between the electron and photon spectral indices δ and γ can most easily be obtained from (2.1) and (2.8) using the Kramers approximation or the nonrelativistic Bethe-Heitler (NRBH) bremsstrahlung cross section (see Koch and Motz 1959 for bremsstrahlung cross-sections). The NRBH cross-section is given by:

$$Q_{NRBH}(\varepsilon, E) = \frac{\overline{Z^2}Q_0}{\varepsilon E}\ln\left(\frac{1+\sqrt{1-\varepsilon/E}}{1-\sqrt{1-\varepsilon/E}}\right)\ \text{cm}^2\,\text{keV}^{-1}, \tag{2.9}$$

where $Q_0 = 7.90 \times 10^{-25}$ cm^2 keV and $\overline{Z^2} \simeq 1.4$ is the ion-species-number-density-weighted mean square atomic number of the target plasma. The Kramers approximation to this cross-section is (2.9) without the logarithmic term. The bremsstrahlung cross-section is zero for $\varepsilon > E$, since an electron cannot radiate a photon that is more energetic than the electron. Analytic expressions for the photon flux from both a uniform thin-target source and a thick-target source can be obtained with the Kramers and the NRBH cross-sections when the electron flux distribution has the single-power-law form (Brown 1971; Tandberg-Hanssen and Emslie 1988). The thin-target result also generalizes to the photon flux from a single-power-law mean electron flux distribution.

For a uniform thin-target source and $\mathcal{F}(E) = AE^{-\delta}$,

$$I_{thin}(\varepsilon) = 3.93 \times 10^{-52}\left(\frac{1\ \text{AU}}{R}\right)^2\left(\frac{\overline{Z^2}}{1.4}\right)NA\beta_{tn}(\delta)\varepsilon^{-(\delta+1)}, \tag{2.10}$$

giving $\gamma_{thin} = \delta + 1$. The photon energy ε is in keV and N is the ion column density. The power-law-index-dependent coefficient for the NRBH cross-section is

$$\beta_{tn}(\delta) = \frac{B(\delta, 1/2)}{\delta}, \tag{2.11}$$

where $B(x, y)$ is the standard Beta function. (In the Kramers approximation, $\beta_{tn}(\delta) = 1/\delta$.) Typical values for the ion column density and A, the differential electron flux at 1 keV, are $N = 10^{18}$–10^{20} cm^{-2} and $A = 10^{34}$–10^{38} electrons s^{-1} keV^{-1}.

For a thick-target source region,

$$I_{thick}(\varepsilon) = 1.17 \times 10^{-34}\left(\frac{1\ \text{AU}}{R}\right)^2\left(\frac{\overline{Z^2}/\overline{Z}}{1.25}\right)\left(\frac{23}{\Lambda_{ee}}\right)A\beta_{tk}(\delta)\varepsilon^{-(\delta-1)}, \tag{2.12}$$

giving $\gamma_{thick} = \delta - 1$. The power-law-index-dependent coefficient for the NRBH cross-section is

$$\beta_{tk}(\delta) = \frac{B(\delta-2, 1/2)}{(\delta-1)(\delta-2)}. \tag{2.13}$$

In the Kramers approximation, $\beta_{tk}(\delta) = 1/[(\delta-1)(\delta-2)]$.

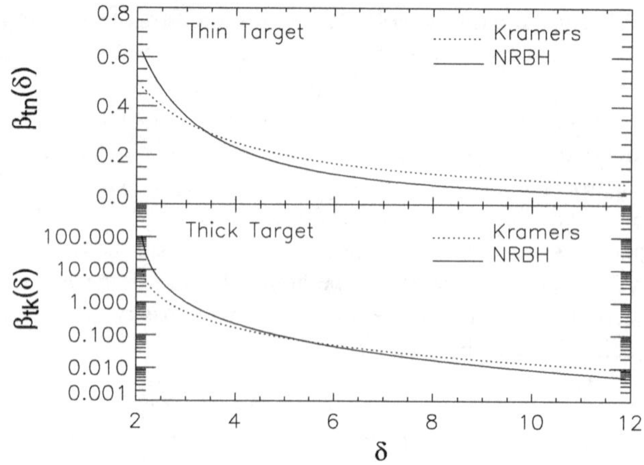

Fig. 1 The power-law-index-dependent terms $\beta_{tn}(\delta)$ (2.11) and $\beta_{tk}(\delta)$ (2.13) in the analytic expressions for the thin-target (2.10) and the thick-target (2.12) photon flux from a power-law electron flux distribution. The *solid curves* are for the nonrelativistic Bethe-Heitler bremsstrahlung cross-section and the *dotted curves* are for the Kramers cross-section. Note that the β axis is linear for the thin-target coefficient, and logarithmic for the thick-target coefficient

The coefficients $\beta_{tn}(\delta)$ and $\beta_{tk}(\delta)$ for both the NRBH and the Kramers cross-sections are plotted as a function of δ in Fig. 1. The Kramers and NRBH results are equal for thin-target emission when $\delta \simeq 3.4$, and for thick-target emission when $\delta \simeq 5.4$. For the plotted range of δ, the Kramers approximation can differ from the NRBH result by over 90%. For δ in the range 3–10, the Kramers result can differ from the NRBH result by as much as 76% and 57% for thin- and thick-target emission, respectively.

It is important to recognize that the above power-law relationships are only valid if the electron *flux density distribution*, $F(\mathbf{r}, E)$ electrons $\mathrm{cm}^{-2}\,\mathrm{s}^{-1}\,\mathrm{keV}^{-1}$, or the electron *flux distribution*, $\mathscr{F}(E)$ electrons $\mathrm{s}^{-1}\,\mathrm{keV}^{-1}$, is assumed to have a power-law energy dependence. It is sometimes convenient to work with the electron *density distribution*, $f(\mathbf{r}, E)$ (electrons $\mathrm{cm}^{-3}\,\mathrm{keV}^{-1}$), rather than the flux density distribution, especially when considering thin-target emission alone or comparing X-ray spectra with radio spectra. The flux density and density distributions are related through $F(\mathbf{r}, E) = f(\mathbf{r}, E)\mathrm{v}(E)$. If the electron density distribution rather than the flux or flux density distribution is assumed to have a power-law index δ', so that $f(\mathbf{r}, E) \propto E^{-\delta'}$, the relationships between this power-law index and the photon spectral index become $\gamma_{thin} = \delta' + 0.5$ and $\gamma_{thick} = \delta' - 1.5$.

The simple power-law relationships are *not valid* if there is a break or a cutoff in the electron distribution at an energy less than ~2 orders of magnitude above the photon energies of interest. Since all electrons with energies above a given photon energy ε contribute to the bremsstrahlung at that photon energy, for the power-law relationships to be valid the break energy must be high enough that the deficit (or excess) of electrons above the break energy does not significantly affect the photon flux at energy ε. The power-law relationship is typically not accurate until photon energies one to two orders of magnitude below the break energy, depending on the steepness of the power-law electron distribution (see Figs. 9 and 10 of Holman 2003). Thus, for example, these relationships are not correct for the lower power-law index of a double power-law fit to a photon spectrum at photon energies within about an order of magnitude below the break energy in the double power-law electron distribution. Equation (2.1) or (2.8) can be used to numerically compute the X-ray spectrum from an arbitrary flux distribution in electron energy.

When electrons with kinetic energies approaching or exceeding 511 keV significantly contribute to the radiation, the relativistic Bethe-Heitler bremsstrahlung cross-section (Eq. 3BN of Koch and Motz 1959) or a close approximation (Haug 1997) must be used. Haug (1997) has shown that the maximum error in the NRBH cross-section relative to the relativistic Bethe-Heitler cross section becomes greater than 10% at electron energies of 30 keV and above. Numerical computations using the relativistic Bethe-Heitler cross-section have been incorporated into the *RHESSI* spectral analysis software (OSPEX) for both thin- and thick-target emission from, in the most general case, a broken-power-law electron flux distribution with both low- and high-energy cutoffs (the functions labeled "thin" and "thick" using the IDL programs `brm_bremspec.pro` and `brm_bremthick.pro`—see Holman 2003). Faster versions of these programs are now available in OSPEX and are currently labeled "thin2" and "thick2" and use the IDL programs `brm2_thintarget.pro` and `brm2_thicktarget.pro`.

The analytic results based on the NRBH cross-section have been generalized to a broken-power-law electron flux distribution with cutoffs by Brown et al. (2008). They find a maximum error of 35% relative to results obtained with the relativistic Bethe-Heitler cross-section for the range of parameters they consider. These results provide the fastest method for obtaining thin- and thick-target fits to X-ray spectra in the *RHESSI* spectral analysis software, where they are labeled "photon_thin" and "photon_thick" and use the IDL programs `f_photon_thin.pro` and `f_photon_thick.pro`.

3 Low-Energy Cutoffs and the Energy in Non-thermal Electrons

One of the most important aspects of the distribution of accelerated electrons is the low-energy cutoff. The acceleration of charged particles out of the thermal plasma typically involves a competition between the collisions that keeps the particles thermalized and the acceleration mechanism. The particles are accelerated out of the tail of the thermal distribution, down to the lowest particle energy for which the acceleration mechanism can overcome the collisional force. Thus, the value of the low-energy cutoff can provide information about the force of the acceleration mechanism. More generally, as discussed below, the electron distribution must have a low-energy cutoff (1) so that the number and energy flux of electrons is finite and reasonable, and (2) because electrons with energies that are not well above the thermal energy of the plasma through which they propagate will be rapidly thermalized. Knowledge of the low-energy cutoff and its evolution during a flare is critical to determining the energy flux and energy in non-thermal electrons and, ultimately, the efficiency of the acceleration process.

3.1 Why Do We Need to Determine the Low-Energy Cutoff of Non-thermal Electron Distributions?

An important feature of the basic thick-target model is that the photon spectrum $I(\varepsilon)$ is directly determined by the injected electron flux distribution $\mathscr{F}_0(E_0)$. As can be seen from (2.8), no additional parameters such as source density or volume need to be determined. Consequently, by integrating over all electron energies, we can also determine the total flux of non-thermal electrons, N_{nth} electrons s^{-1}, the power in non-thermal electrons, P_{nth} erg s^{-1}, and, integrating over time, the total number of, and energy in, non-thermal electrons.

The total non-thermal electron number flux and power are computed as follows:

$$N_{nth} = \int_{E_c}^{+\infty} \mathscr{F}_0(E_0)\,dE_0 = \frac{A}{\delta-1} {E_c}^{-\delta+1} \text{ electrons s}^{-1} \tag{3.1}$$

$$P_{nth} = \kappa_E \int_{E_c}^{+\infty} E_0 \cdot \mathscr{F}_0(E_0)\,dE_0 = \frac{\kappa_E A}{\delta-2} {E_c}^{-\delta+2} \text{ erg s}^{-1} \tag{3.2}$$

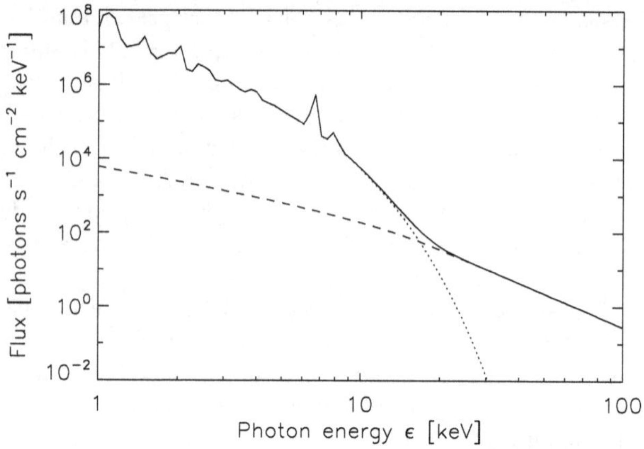

Fig. 2 Typical full-Sun flare spectrum. *Dashed:* Nonthermal thick-target spectrum from an accelerated electron distribution with $\delta = 4$, and a low-energy cutoff of 20 keV. *Dotted:* Thermal spectrum, from a plasma with temperature $T = 20$ MK and emission measure $EM = 10^{49}\ \mathrm{cm}^{-3}$. *Solid:* Total radiated spectrum. The multiple peaks in the thermal spectrum are from spectral lines, as observed by an instrument with ∼1 keV spectral resolution

The last expression in each equation is the result for a power-law electron flux distribution of the form $\mathscr{F}_0(E_0) = A \cdot E_0^{-\delta}$. The constant $\kappa_E = 1.60 \times 10^{-9}$ is the conversion from keV to erg. E_c is a low-energy cutoff to the electron flux distribution. These expressions are valid and finite for $\delta > 2$ and $E_c > 0$. We call this form of low-energy cutoff a *sharp low-energy cutoff*. An electron distribution that continues below a transition energy E_c that has a positive slope, is flat, or in general has a spectral index $\delta_{low} < 1$ also provides finite electron and energy fluxes, but these fluxes are somewhat higher than those associated with the sharp low-energy cutoff.

For this single-power-law electron flux distribution with a sharp low-energy cutoff, the non-thermal *power* (erg s^{-1}), and ultimately the non-thermal *energy* (erg), from the power-law electron flux distribution depends on only three parameters: δ, A, and E_c. Observations indicate that δ is greater than 2 (Dennis 1985; Lin and Schwartz 1987; Winglee et al. 1991; Holman et al. 2003). Hence, were $E_c = 0$, the integral would yield an infinite value, a decidedly unphysical result! Therefore, the power-law electron distribution cannot extend all the way to zero energy with the same or steeper slope, and some form of *low-energy cutoff* in the accelerated electron spectrum must be present. As we will see, the determination of the energy at which this cutoff occurs is not a straightforward process, but it is the single most important parameter to determine (as the other two are generally more straightforward to determine—see Sect. 2 and Kontar et al. 2011). For example, with $\delta = 4$ (typical during the peak time of strong flares), a factor of 2 error in E_c yields a factor of 4 error in P_{nth}. For larger δ (as found in small flares, or rise/decay phases of large flares), such an error quickly leads to an order of magnitude (or even greater) difference in the injected power P_{nth} and in the total energy in the non-thermal electrons accelerated during the flare!

3.2 Why Is the Low-Energy Cutoff Difficult to Determine?

The essence of the problem in many flare spectra is summarized in Fig. 2: the non-thermal power-law is well-observed above ∼20 keV, but any revealing features that it might possess at lower energies, such as a low-energy cutoff, are masked by the thermal emission.

Even if a spectrum does show a flattening at low energies that could be the result of a low-energy cutoff, other mechanisms that could produce the flattening must be ruled out (see Sect. 3.4). The low-energy cutoff has the characteristic feature, determined by the photon energy dependence of the bremsstrahlung cross-section (see (2.9)), that the X-ray spectrum eventually approaches a spectral index of $\gamma \approx 1$ at low energies (cf. Holman 2003). It is

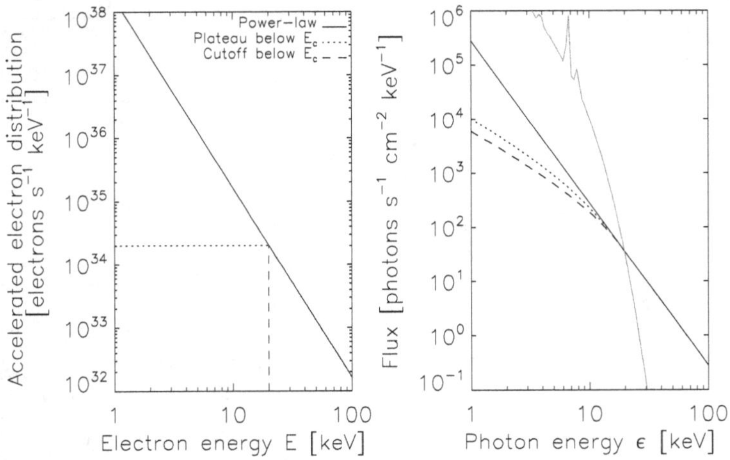

Fig. 3 Different shapes of low-energy cutoff in the injected electron distribution (*left*) lead to slightly different photon spectra (*right*). The cutoff/turnover electron energy is $E_c = 20$ keV. The *thin curve in the right panel* demonstrates how the cutoff can be masked by emission from thermal plasma. See also Holman (2003) for a thorough discussion of bremsstrahlung spectra generated from electron power-laws with cutoff

currently impossible, however, to observe a flare spectrum to low enough photon energies to see that it does indeed become this flat. Generally we can only hope to rule out the other mechanisms based on additional data and detailed spectral fits.

3.3 What Is the Shape of the Low-Energy Cutoff, and How Does It Impact the Photon Spectrum and P_{nth}?

Bremsstrahlung photon spectra are obtained from convolution integrals over the electron flux distribution ((2.1) and (2.8)). Hence, features in an electron distribution are smoothed out in the resulting photon spectrum (see, e.g., Brown et al. 2006).

As can be seen in Fig. 3, both a sharp cutoff at E_c and a "turnover" (defined here to be a constant $F_0(E)$ below E_c, a "plateau") in the injected electron distribution lead to similar thick-target photon spectra. This subtle difference is difficult to discriminate observationally, and the problem is compounded by the dominance of the thermal component at low energies.

A sharp cutoff would lead to plasma instabilities that should theoretically flatten the distribution around and below the cutoff within microseconds (see Sect. 6). On the other hand, the electron flux distribution below the cutoff must be flatter than E^{-1}, as demonstrated by (3.1), or the total electron number flux would be infinite. Having a constant value for the distribution below E_c (turnover case) seems like a reasonable middle ground and approximates a quasilinearly relaxed electron distribution (Sect. 6; Krall and Trivelpiece 1973, Chap. 10). Coulomb collisional losses, on the other hand, yield an electron distribution that increases linearly at low energies (see Fig. 4), leading to a photon spectrum between the sharp cutoff case and the turnover case.

Notice that the photon spectra actually flatten gradually to the spectral index of 1 at low energies from the spectral index of $\gamma = \delta + 1$ at E_c and higher energies. Below E_c, it is *not* a power-law. Fitting a double power-law model photon spectrum, and using the break (i.e., kink) energy as the low-energy cutoff typically leads to a large error in E_c (e.g., Gan et al. 2001; Saint-Hilaire and Benz 2005), and hence to an even larger error in P_{nth}.

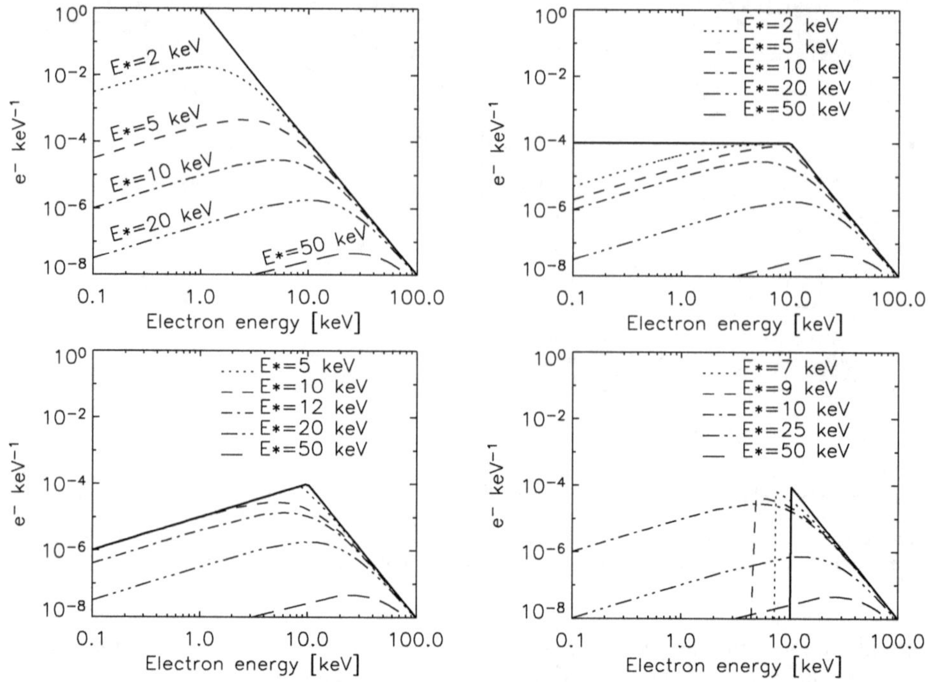

Fig. 4 The *four plots* show the Coulomb-collisional evolution with column density of an injected electron distribution (*thick, solid line*). For the simple power-law case (*upper left*), the low-energy end of the distribution becomes linear, and the peak of the distribution is found at $E_{peak} = E_*/\sqrt{\delta}$, where δ is the injected distribution power-law spectral index ($\delta = 4$ in the plots), and $E_* = \sqrt{2K \cdot N_*}$ is the initial energy that electrons must possess in order not to be fully stopped by a column density N_* (2.4). When a low-energy cutoff is present, the peak of the distribution is seen to first decrease in energy until E_* exceeds the cutoff energy (from Saint-Hilaire 2005).

In terms of the energetics, Saint-Hilaire and Benz (2005) have shown that the choice of an exact shape for the low-energy cutoff as a model is not dramatically important. For a fixed cutoff energy, from (3.2) it can be shown that the ratio of the power in the turnover model to the power in the sharp cutoff model without the flat component below the cutoff energy is $\delta/2$. In obtaining spectral fits, however, the turnover model gives higher cutoff energies than the sharp cutoff model. Using simulations, Saint-Hilaire and Benz (2005) found that assuming either a sharp cutoff model or a turnover model led to differences in P_{nth} typically less than $\sim$20%. Hence, the sharp cutoff, being the simplest, is the model of choice for computing flare energetics. Nevertheless, knowing the shape of the low-energy cutoff would not only yield more accurate non-thermal energy estimates, but would be a source of information on the acceleration mechanism and/or propagation effects.

Spectral inversion methods have recently been developed for deducing the *mean electron flux distribution* (2.5) from X-ray spectra (Johns and Lin 1992; Brown et al. 2003, 2006). A spectral "dip" has been found just above the presumed thermal component in some deduced mean electron flux distributions that may be associated with a low-energy cutoff (e.g., Piana et al. 2003). In the collisional thick-target model, the slope of the high-energy "wall" of this dip should be linear or flatter, with a linear slope indicating the absence of emitting electrons in the injected electron distribution at the energies displaying this slope. Kontar and Brown (2006a) have found evidence for slopes that are steeper than linear, but their spectra were not corrected for photospheric albedo (see Sect. 3.5). Finding and understanding these dips is a crucial element for gaining an understanding of the low-energy properties of flare electron distributions (see Kontar et al. 2011).

Emslie (2003) has pointed out that the non-thermal electron distribution could seamlessly merge into the thermal distribution, removing the need for a low-energy cutoff. As was shown by Holman et al. (2003) for SOL2002-07-23T00:35 (X4.8), however, merger of the electron distribution into the typically derived $\sim$10–30 MK thermal flare plasma generally implies an exceptionally high energy in non-thermal electrons. Thus, for a more likely energy content, a higher low-energy cutoff or a hotter plasma would need to be present in the target region. Any emission from this additional "hot core," because of its much lower emission measure, is likely to be masked by the usual $\sim$10–30 MK thermal emission. This merger of the non-thermal electron distribution into the thermal tail in the target region does not remove the need for a low-energy cutoff in the electrons that escape the acceleration region, however.

This section has dealt with the shape of the low-energy cutoff under the assumptions that the X-ray photon spectra are not altered by other mechanisms and that the bremsstrahlung emission is isotropic. The next section lists the important caveats to these assumptions, and their possible influence in the determination of the low-energy cutoff to the electron flux distribution.

3.4 Important Caveats

As previously discussed, apparently minor features in the bremsstrahlung photon spectrum can have substantial implications for the mean electron flux and, consequently, the injected electron distribution. This means that unknown or poorly-understood processes that alter the injected electron distribution (propagation effects, for example) or the photon spectrum (including instrumental effects) can lead to significant errors in the determination of the low-energy cutoff. Known processes that affect the determination of the low-energy cutoff are enumerated below.

1. Detector pulse pileup effects (Smith et al. 2002), if not properly corrected for, can introduce a flattening of the spectrum toward lower energies that simulates the flattening resulting from a low-energy cutoff.
2. The contribution of Compton back-scattered photons (photospheric albedo) to the measured X-ray spectrum can simulate the spectral flattening produced by a low-energy cutoff. Kašparová et al. (2005) have shown that the dip in a spectrum from SOL2002-08-20T08:25 (M3.4) becomes statistically insignificant when the spectrum is corrected for photospheric albedo (also see Kontar et al. 2011). Kašparová et al. (2007) show that spectra in the 15–20 keV energy band tend to be flatter near disk center when albedo from isotropically emitted photons is not taken into account, further demonstrating the importance of correcting for photospheric albedo.
3. The assumed differential cross-section and electron energy loss rate can influence the results (for a discussion of this, see Saint-Hilaire and Benz 2005). In some circumstances, a contribution from recombination radiation may significantly change the results (Brown et al. 2010; also see Kontar et al. 2011).
4. Anisotropies in the electron beam directivity and the bremsstrahlung differential cross-section can significantly alter the X-ray spectrum (Massone et al. 2004).
5. Non-uniform target ionization (the fact that the chromosphere's ionization state varies with depth, see Sect. 4) can introduce a spectral break that may be confused with the break associated with a low-energy cutoff.

6. Energy losses associated with a return current produce a low-energy flattening of the X-ray spectrum (Sect. 5). This is a low-energy "cutoff" in the electron distribution injected into the thick target, but it is produced between the acceleration region and the emitting source region.
7. A non-power-law distribution of injected electrons or significant evolution of the injected electron distribution during the observational integration time could affect the deduced value of the low-energy cutoff.

For all the above reasons, the value of the low-energy cutoff in the injected electron flux distribution has not been determined with any degree of certainty except perhaps in a few special cases. Even less is known about the shape of the low-energy cutoff. The consensus in the solar physics community for now is to assume the simplest case, a sharp low-energy cutoff. Existing studies, presented in the next section, tend to support the adequacy of this assumption for the purposes of estimating the total power and energy in the accelerated electrons.

3.5 Determinations of E_c and Electron Energy Content from Flare Data

Before *RHESSI*, instruments did not cover well (if at all) the ~10–40 keV photon energies where the transition from thermal emission to non-thermal emission usually occurs. Researchers typically assumed an arbitrary low-energy cutoff at a value at or below the instrument's observing range (one would talk of the "injected power in electrons above E_c keV" instead of the total non-thermal power P_{nth}). An exception is Nitta et al. (1990). They argued that spectral flattening observed in two flares with the *Solar Maximum Mission* and *Hinotori* indicated a cutoff energy of $\gtrsim$50 keV. Also, Gan et al. (2001) interpreted spectral breaks at ~80 keV in *Compton Gamma-Ray Observatory* (*CGRO*) flare spectra as the low-energy cutoff in estimating flare energetics, resulting in rather small values for the non-thermal energy in the analyzed flares. The relatively low resolution of the spectra from these instruments prevented the quantitative evaluation of any spectral flattening toward lower energies, however.

The only high-resolution flare spectral data before the launch of *RHESSI* was the balloon data of Lin et al. (1981) for SOL1980-06-27T16:17 (M6.7) along with ~25 microflares observed during the same balloon flight. Benka and Holman (1994) applied a direct electric field electron acceleration model to the SOL1980-06-27 flare data. They derived, along with other model-related parameters, the time evolution of the critical energy above which runaway acceleration occurs—the model equivalent to the low-energy cutoff. They found this critical energy to range from ~20–40 keV.

It is now possible in most cases to obtain a meaningful upper limit on E_c, thanks to *RHESSI*'s high-spectral-resolution coverage of the 10–40 keV energy range and beyond. Holman et al. (2003), Emslie et al. (2004), and Saint-Hilaire and Benz (2005), in determining the low-energy cutoff, obtained the "highest value for E_c that still fits the data." In many solar flare spectra, because of the dominance of radiation from thermal plasma at low energies, a range of values for E_c fit the data equally well, up to a certain critical energy, above which the χ^2 goodness-of-fit parameter becomes unacceptably large. The low-energy cutoff is taken to be equal to this critical value. This upper limit on the cutoff energy results in a lower limit for the non-thermal power and energy. The results obtained for the maximum value of E_c were typically in the 15–45 keV range, although late in the development of SOL2002-07-23T00:35 (X4.8) some values as high as ~80 keV were obtained for E_c. The minimum non-thermal energies thus determined were comparable to or somewhat larger than the calculated thermal energies.

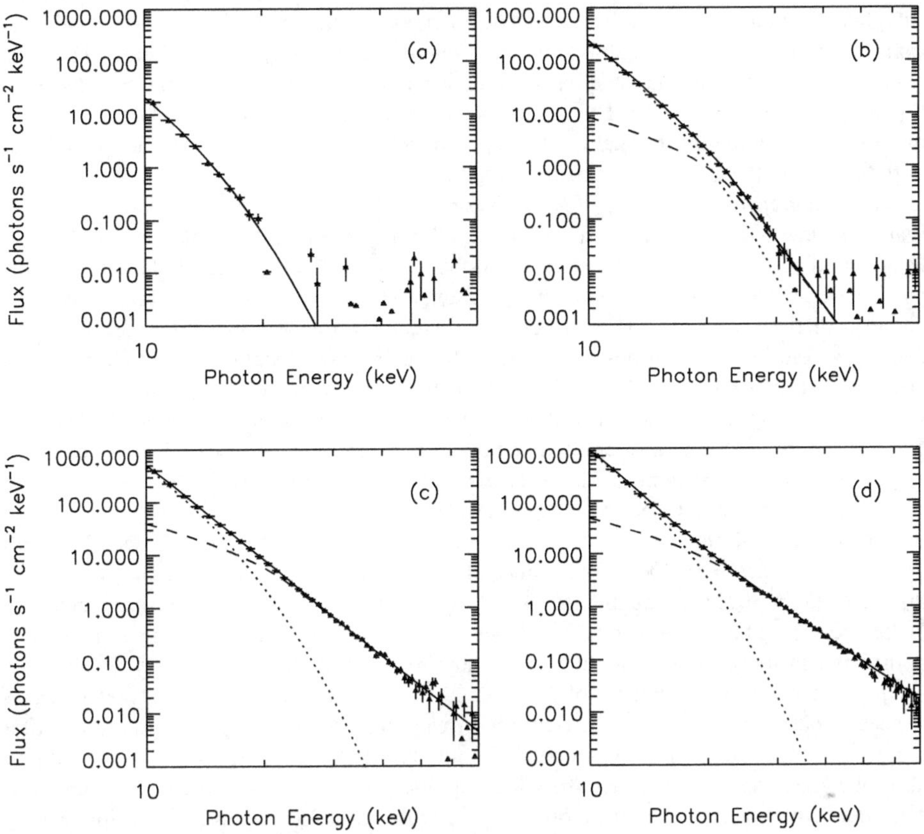

Fig. 5 *RHESSI* spatially integrated spectra in four time intervals during SOL2002-04-15T03:55 (M1.2). (**a**) Spectrum at 23:06:20–23:06:40 UT (early rise phase). (**b**) Spectrum at 23:09:00-23:09:20 UT (just before impulsive phase). (**c**) Spectrum at 23:10:00–23:10:20 UT (soon after the impulsive rise). (**d**) Spectrum at 23:11:00–23:11:20 UT (at the hard X-ray peak). The plus signs with error bars represent the spectral data. The *lines* represent model spectral fits: the *dashed lines* are non-thermal thick-target bremsstrahlung, the *dotted lines* are thermal bremsstrahlung, and the *solid lines* are the summation of the two (from Sui et al. 2005a). Reproduced by permission of the AAS

One of the best determinations of the low-energy cutoff so far was obtained by Sui et al. (2005a). They complemented the spatially-integrated spectral data for the SOL2002-04-15T03:55 (M1.2) limb flare with imaging and lightcurve information. Four spectra from this flare are shown in Fig. 5. The earliest spectrum, before the impulsive rise of the higher energy X-rays, was well fitted with an isothermal model. The last spectrum, from the time of the hard X-ray peak, clearly shows a thermal component below ~20 keV. Of particular interest is the second spectrum, showing both thermal and non-thermal fit components. As a consequence of the flattening of the isothermal component at low energies, the low-energy cutoff to the non-thermal component cannot extend to arbitrarily low energies without exceeding the observed emission. This places a tight constraint on the value of the low-energy cutoff. The additional requirement that the time evolution of the derived temperature and emission measure of the thermal component be smooth and continuous throughout the flare constrains the value at other times. Applying the collisional thick-target model with a power-law distribution of injected electrons, they found the best cutoff value to be $E_c = 24 \pm 2$ keV (roughly constant throughout the flare). The energy associated with these non-thermal electrons was found to be comparable to the peak energy in the X-ray-emitting thermal plasma, but an order of magnitude greater than the kinetic energy of the associated coronal mass ejection (CME) (Sui et al. 2005b). This contrasts with results obtained for large flares, where the minimum energy in non-thermal electrons is typically found to be less than or on the order of the energy in the CME (e.g., Emslie et al. 2004).

The importance of correcting for the distortion of spectra by albedo was revealed by a search for low-energy cutoffs in a sample of 177 flares with relatively flat spectra ($\gamma \leq 4$) between 15 and 20 keV (Kontar et al. 2008a). Spectra can be significantly flattened by the presence of albedo photons in this energy range. The X-ray spectra, integrated over the duration of the impulsive phase of the flares, were inverted to obtain the corresponding mean electron flux distributions. Eighteen of the flares showed significant dips in the mean electron flux distribution in the 13–19 keV electron energy range that might be associated with a low-energy cutoff (see Sect. 3.3). However, when the X-ray spectra were corrected for albedo from isotropically emitted X-rays, all of the dips disappeared. Therefore, the authors concluded that none of these flare electron distributions had a low-energy cutoff above 12 keV, the lowest electron energy in their analysis.

Low-energy cutoffs were identified in the spectra of a sample of *early impulsive flares* observed by *RHESSI* in 2002 (Sui et al. 2007). Early impulsive flares are flares in which the >25 keV hard X-ray flux increase is delayed by less than 30 s after the flux increase at lower energies. The pre-impulsive-phase heating of plasma to X-ray-emitting temperatures is minimal in these flares, allowing the nonthermal part of the spectrum to be observed to lower energies. In the sample of 33 flares, 9 showed spectral flattening at low energies in spectra obtained throughout the duration of each flare with a 4 s integration time. After correcting for the albedo from isotropically emitted X-rays, the flattening in 3 of the 9 flares, all near Sun center, disappeared. The flattening that persisted in the remaining 6 flares was consistent with that produced by a low-energy cutoff. The values derived for the low-energy cutoff ranged from 15 to 50 keV. The authors found the evolution of the spectral break and the corresponding low-energy cutoff in these flares to be correlated with the non-thermal hard X-ray flux. Further studies are needed to assess the significance of this correlation.

Low-energy cutoffs with values exceeding 100 keV were identified in the spectra of the large flare SOL2005-01-19T08:22 (X1.3) (Warmuth et al. 2009). The hard X-ray light curve of this flare consisted of multiple peaks that have been interpreted as quasi-periodic oscillations driven by either magnetoacoustic oscillations in a nearby loop (Nakariakov et al. 2006) or by super-Alfvénic beams in the vicinity of the reconnection region (Ofman and Sui 2006). The high low-energy cutoffs were found in the last major peak of the series of hard X-ray peaks. Unlike the earlier peaks, this peak was also unusual in that it was not accompanied by the Neupert effect (see Sect. 8.3), consistent with the high values of the low-energy cutoff, and it exhibited soft-hard-harder rather than soft-hard-soft spectral evolution (see Sect. 9.1). A change in the character of the observed radio emission and movement of one of the two hard X-ray footpoints into a region of stronger photospheric magnetic field were also observed at the time of this peak. These changes suggest a strong connection between large-scale flare evolution and electron acceleration.

4 Nonuniform Ionization in the Thick-Target Region

In the interpretation of hard X-ray (HXR) spectra in terms of the thick-target model, one effect which has been largely ignored until recently is that of varying ionization along the path of the thick-target beam. As first discussed by Brown (1973), the decrease of ionization with depth in the solar atmosphere reduces long-range collisional energy losses. This enhances the HXR bremsstrahlung efficiency there, elevating the high energy end of the HXR spectrum by a factor of up to 2.8 above that for a fully ionized target. The net result is that a power-law electron spectrum of index δ produces a photon spectrum of index $\gamma = \delta - 1$ at low and high energies (see (2.12)), but with $\gamma < \delta - 1$ in between. The upward break, where the spectrum begins to flatten toward higher energies, occurs at fairly low energies, probably masked in measured spectra by the tail of the thermal component. The downward knee, where the spectrum steepens again to $\gamma = \delta - 1$, occurs in the few deka-keV range, depending on the column depth of the transition zone. Thus, the measured X-ray spectrum may show a flattening similar to that expected for a low-energy cutoff in the electron distribution.

4.1 Electron Energy Losses and X-ray Emission in a Nonuniformly Ionized Plasma

The collisional energy-loss cross-section $Q_c(E)$ is dependent on the ionization of the background medium. Flare-accelerated electron beams can propagate in the fully ionized corona as well as in the partially ionized transition region and chromosphere. Following Hayakawa and Kitao (1956) and Brown (1973), the cross-section $Q_c(E)$ can be written for a hydrogen plasma ionization fraction x

$$Q_c(E) = \frac{2\pi e^4}{E^2}(x\Lambda_{ee} + (1-x)\Lambda_{eH}) = \frac{2\pi e^4}{E^2}\Lambda(x+\lambda), \tag{4.1}$$

where e is the electronic charge, Λ_{ee} the electron-electron logarithm for fully ionized media and Λ_{eH} is an effective Coulomb logarithm for electron-hydrogen atom collisions. Numerically $\Lambda_{ee} \simeq 20$ and $\Lambda_{eH} \simeq 7.1$, so $\Lambda = \Lambda_{ee} - \Lambda_{eH} \simeq 12.9$ and $\lambda = \Lambda_{eH}/\Lambda \simeq 0.55$.

Then, in a hydrogen target of ionization level $x(N)$ at column density $N(z)$, the energy loss equation for electron energy E is (cf. (2.2))

$$\frac{dE}{dN} = -\frac{2\pi e^4 \Lambda}{E}(\lambda + x(N)) = -\frac{K'}{E}(\lambda + x(N)), \tag{4.2}$$

where $K' = 2\pi e^4 \Lambda = (\Lambda/\Lambda_{ee})K \simeq 0.65K$.

The energy loss of a given particle with initial energy E_0 depends on the column density $N(z) = \int_0^z n(z')\,dz'$, so the electron energy at a given distance z from the injection site can be written $E^2 = E_0^2 - 2K'M(N(z))$ (cf. (2.4)), where

$$M(N(z)) = \int_0^{N(z)} (\lambda + x(N'))\,dN' \tag{4.3}$$

is the "effective" ionization-weighted collisional column density.

The fractional atmospheric ionization x as a function of column density N (cm^{-2}) changes from 1 to near 0 over a small spatial range in the solar atmosphere. Therefore, to lowest order, $x(N)$ can be approximated by a step function $x(N) = 1$ for $N < N_*$, and $x(N) = 0$ for $N \geq N_*$. This gives $M(N) = (\lambda + 1)N$ for $N < N_*$ and $M(N) = N_* + \lambda N$ for $N \geq N_*$. Electrons injected into the target with energies less than $E_* = \sqrt{2K'(\lambda+1)N_*} = \sqrt{2KN_*}$ experience energy losses and emit X-rays in the fully ionized plasma with $x = 1$, as in the standard thick-target model. Electrons injected with energies higher than E_* lose part of their energy and partially emit X-rays in the un-ionized ($x = 0$), or, more generally, partially ionized plasma.

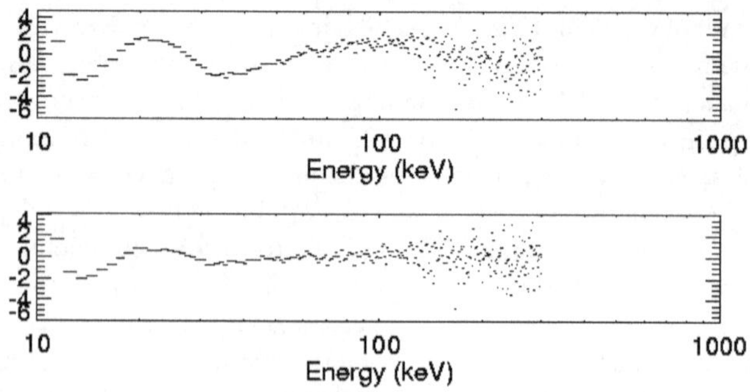

Fig. 6 Photon spectrum residuals, normalized by the statistical error for the spectral fit, for the time interval 00:30:00–00:30:20 UT, 2002-July-23, for (*upper panel*) an isothermal Maxwellian plus a power-law and (*lower panel*) an isothermal Maxwellian plus the nonuniform ionization spectrum with $\delta = 4.24$ and $E_* = 53$ keV (from Kontar et al. 2003). Reproduced by permission of the AAS

We can deduce the properties of the X-ray spectrum by substituting (4.2) into (2.7) (with $dN = nv\,dt$) and comparing $I_{thick}(\varepsilon)$ from (2.6) with $I_{thick}(\varepsilon)$ from (2.8). We see that for the nonuniformly ionized case the denominator in the inner integral now contains $\lambda + x(N)$ and K is replaced with K'. In the step-function model for $x(N)$, photon energies greater than or equal to $\varepsilon_* = E_*$ are emitted by electrons in the un-ionized plasma with $E \geq E_*$. Since $\lambda + x(N)$ has the constant value λ, the thick-target power-law spectrum is obtained (for injected power-law spectrum), but the numerical coefficient contains $K'\lambda = 2\pi e^4 \Lambda_{eH}$ instead of K. At photon energies far enough below ε_* that the contribution from electrons with $E \geq E_*$ is negligible, $\lambda + x(N) = \lambda + 1$ and the numerical coefficient contains $(\lambda + 1)K' = K$. The usual thick-target spectral shape and numerical coefficient are recovered. The ratio of the amplitude of the high-energy power-law spectrum to the low-energy power-law spectrum is $(\lambda + 1)/\lambda \simeq 2.8$. The photon energy $\varepsilon_*(\mathrm{keV}) \simeq 2.3 \times 10^{-9}\sqrt{\mathrm{N}_*(\mathrm{cm}^{-2})}$, between where the photon spectrum flattens below the high-energy power law and above the low-energy power law, determines the value of the column density where the plasma ionization fraction drops from 1 to 0.

4.2 Application to Flare X-ray Spectra

The step-function nonuniform ionization model was used by Kontar et al. (2002, 2003) to fit photon spectra from five flares. They assume a single power-law distribution of injected electrons with power-law index δ and approximate the bremsstrahlung cross-section with the Kramers cross-section. First, they fit the spectra to the sum of a thermal Maxwellian at a single temperature T plus a single power law of index γ. For SOL2002-07-23T00:35 (X4.8) (Kontar et al. 2003) they limit themselves to deviations from a power law in the non-thermal component of the spectrum above ~40 keV. The top panel of Fig. 6 shows an example of such deviations, which represent significant deviations from the power-law fit. These deviations are much reduced by replacing the power law with the spectrum from the nonuniform ionization model, with the minimum rms residuals obtained for values of $\delta = 4.24$ and $E_* = 53$ keV (Fig. 6, bottom panel). The corresponding minimum (reduced) χ^2 value obtained for the best fit to the X-ray spectrum (10–130 keV) dropped from 1.4 for the power-law fit to 0.8 for the nonuniform ionization fit. There are still significant residuals present in the range from 10 to 30 keV; these might be due to photospheric albedo or the assumption of a single-temperature thermal component.

By assuming that the main spectral feature observed in a hard X-ray spectrum is due to the increased bremsstrahlung efficiency of the un-ionized chromosphere, allowance for nonuniform target ionization offers an elegant direct explanation for the shape of the observed hard X-ray spectrum and provides a measure of the location of the transition region.

Table 1 Best fit nonuniformly ionized target model parameters for a single power-law $\mathscr{F}_0(E_0)$, the equivalent N_* (energy range 20–100 keV), and the ratio of $\chi^2_{nonuni}/\chi^2_{uni}$ (from Kontar et al. 2002)

Date	Time, UT	kT(keV)	δ	E_* (keV)	N_* (cm^2)	$\chi^2_{nonuni}/\chi^2_{uni}$
20 Feb 2002	11:06	1.47	5.29	37.4	2.7×10^{20}	0.032
17 Mar 2002	19:27	1.27	4.99	24.4	1.1×10^{20}	0.047
31 May 2002	00:06	2.02	4.15	56.2	6.1×10^{20}	0.041
1 Jun 2002	03:53	1.45	4.46	21.0	8.4×10^{19}	0.055

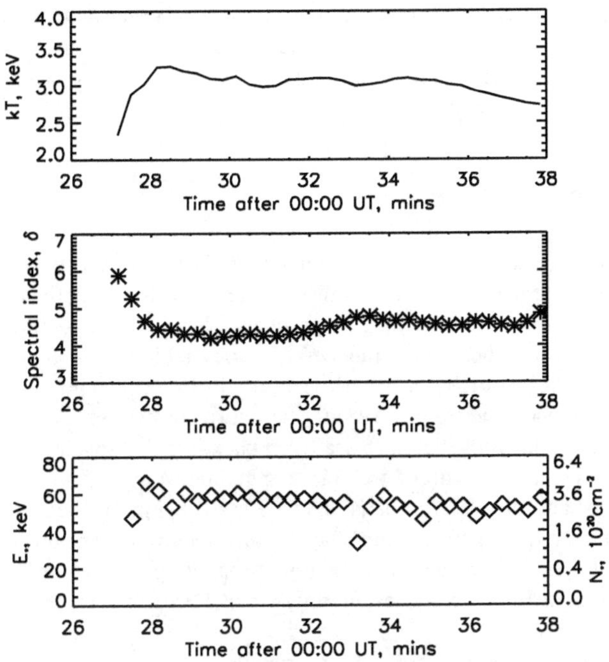

Fig. 7 Variation of kT, δ, E_*, and N_* throughout SOL2002-07-23T00:35 (X4.8) (Kontar et al. 2003). Reproduced by permission of the AAS. The variation of other parameters, such as emission measure, can be found in Holman et al. (2003) and Caspi and Lin (2010)

Table 1 shows the best fit parameters derived for the four flare spectra analyzed by Kontar et al. (2002). The last column shows the ratio of the minimum χ^2 value obtained from the nonuniform ionization fit to the minimum χ^2 value obtained from a uniform ionization (single power-law) fit to the non-isothermal part of the spectrum. The nonuniform ionization model fits clearly provide substantially better fits than single power-law fits.

Values of the fit parameters kT (keV), δ and E_* as a function of time for SOL2002-07-23T00:35 (X4.8), together with the corresponding value of $N_*(\text{cm}^{-2}) \simeq 1.9 \times 10^{17} E_*$ (keV)^2 were obtained by Kontar et al. (2003). The results (Fig. 7) demonstrate that the thermal plasma temperature rises quickly to a value $\simeq$3 keV and decreases fairly slowly thereafter. The injected electron flux spectral index δ follows a general "soft-hard-soft" trend and qualitatively agrees with the time history of the simple best-fit power-law index γ (Holman et al. 2003). E_* rises quickly during the first minute or so from ~40 keV to ~70 keV near the flare peak and thereafter declines rather slowly. The corresponding values of N_* are ~2–5 × 10^{20} cm^{-2}.

The essential results of these studies are that (1) for a single power-law electron injection spectrum, the expression for bremsstrahlung emission from a nonuniformly-ionized target provides a significantly better fit to observed spectra than the expression for a uniform target; and (2) the value of E_* (and correspondingly N_*) varies with time.

An upper limit on the degree of spectral flattening $\Delta\gamma$ that can result from nonuniform ionization was derived by Su et al. (2009). They applied this upper limit to spectra from a sample of 20 flares observed by *RHESSI* in the period 2002 through 2004. They found that 15 of the 20 flare spectra required a downward spectral break at low energies and for each of these 15 spectra derived the difference $\Delta\gamma$ of the best-fit power-law spectral indices above and below the break. A Monte Carlo method was used to determine the 95% confidence interval for each of the derived values of $\Delta\gamma$. Taking the value of $\Delta\gamma$ to be incompatible with nonuniform ionization if the 95% confidence interval fell above the derived upper limit, Su et al. (2009) found that six of the flare spectra could not be explained by nonuniform ionization alone. Thus, for these six flares some other cause such as a low-energy cutoff or return-current-associated energy losses (Sect. 5) must be at least partially responsible for the spectral flattening.

5 Return Current Losses

The thick-target model assumes that a beam of electrons is injected at the top of a loop and "precipitates" downwards in the solar atmosphere. Unless accompanied by an equal flux of positively charged particles, these electrons constitute a current and must create a significant self-induced electric field that in turn drives a co-spatial return current for compensation (Hoyng et al. 1976; Knight and Sturrock 1977; Emslie 1980; Diakonov and Somov 1988). The return current consists of ambient electrons, plus any primary electrons that have scattered back into the upward direction. By this means we have a full electric circuit of precipitating and returning electrons that keeps the whole system neutral and the electron beam stable against being pinched off by the self-generated magnetic field required by Ampère's law for an unneutralized beam current. However, the self-induced electric field results in a potential drop along the path of the electron beam that decelerates and, therefore, removes energy from the beam electrons.

5.1 The Return Current Electric Field

The initial formation of the beam/return-current system has been studied by van den Oord (1990) and references therein. We assume here that the system has time to reach a quasi-steady state. Van den Oord finds this time scale to be on the order of the thermal electron-ion collision time. This time scale is typically less than or much less than one second, depending on the temperature and density of the ambient plasma. In numerical simulations by Siversky and Zharkova (2009), times to reach a steady state after injection ranged from 0.07 s to 0.2 s, depending on the initial beam parameters.

The self-induced electric field strength at a given location z along the beam and the flare loop, $\mathcal{E}(z)$, is determined by the current density associated with the electron beam, $j(z)$, and the local conductivity of the loop plasma, $\sigma(z)$, through Ohm's law: $\mathcal{E}(z) = j(z)/\sigma(z)$. Relating the current density to the density distribution function of the precipitating electrons, $f(z, E, \theta)$, where E is the electron energy and θ is the electron pitch angle, gives

$$\mathcal{E}(z) = \frac{2\sqrt{2}\pi}{\sigma(z)} \frac{e}{\sqrt{m_e}} \int_0^1 \int_0^\infty f(z, E, \theta)\sqrt{E}\mu\, dE\, d\mu. \tag{5.1}$$

Here μ is the cosine of the pitch angle and e and m_e are the electron charge and mass, respectively. The self-induced electric field strength $\mathscr{E}(z)$ depends on the local distribution of the beam electrons, which in turn depends on the electric field already experienced by the beam as well as any Coulomb energy losses and pitch-angle scattering that may have significantly altered the beam. It also depends on the local plasma temperature (and, to a lesser extent, density) through $\sigma(z)$, which can, in turn, be altered by the interaction of the beam with the loop plasma (i.e., local heating and "chromospheric evaporation"). Therefore, determination of the self-induced electric field and its impact on the precipitating electrons generally requires self-consistent modeling of the coupled beam/plasma system.

Such models have been computed by Zharkova and Gordovskyy (2005, 2006). They numerically integrate the time-dependent Fokker-Planck equation to obtain the self-induced electric field strength and electron distribution function along a model flare loop. The injected electron beam was assumed to have a single power-law energy distribution in the energy range from $E_{low} = 8$ keV to $E_{upp} = 384$ keV and a normal (Gaussian) distribution in pitch-angle cosine μ with half-width dispersion $\Delta\mu = 0.2$ about $\mu = 1$.

The model computations show that the strength of the self-induced electric field is nearly constant at upper coronal levels and rapidly decreases with depth (column density) in the lower corona and transition region. The rapidity of the decrease depends on the beam flux spectral index. It is steeper for softer beams ($\delta = 5$–7) than for harder ones ($\delta = 3$). The strength of the electric field is higher for a higher injected beam energy flux density ($\mathrm{erg\,cm^{-2}\,s^{-1}}$), and the distance from the injection point over which the electric field strength is highest (and nearly constant) decreases with increasing beam flux density.

5.2 Impact on Hard X-ray Spectra

Deceleration of the precipitating beam by the electric field most significantly affects the lower energy electrons (<100 keV), since the fraction of the original particle energy lost to the electric field is greater for lower energy electrons. This leads to flattening of the electron distribution function towards the lower energies and, therefore, flattening of the photon spectrum.

Photon spectra computed from kinetic solutions that include return current energy losses and collisional energy losses and scattering are shown in Fig. 8 (a) and (b). Low- and high-energy spectral indices and their dependence on the power-law index of the injected electron distribution and on the injected beam energy flux density are shown in Fig. 8 (c) and (d). The difference between the high-energy and low-energy spectral indices is seen to increase with both the beam energy flux density and the injected electron power-law index δ. The low-energy index is found to be less than 2 for δ as high as 5 when the energy flux density is as high as $10^{12}\ \mathrm{erg\,cm^{-2}\,s^{-1}}$.

5.3 Observational Evidence for the Presence of the Return Current

We have seen that return current energy losses can introduce curvature into a spectrum, possibly explaining the "break" often seen in observed flare X-ray spectra. A difficulty in directly testing this explanation is that the thick-target model provides the power (energy flux) in the electron beam ($\mathrm{erg\,s^{-1}}$), but not the energy flux density ($\mathrm{erg\,cm^{-2}\,s^{-1}}$). X-ray images provide information about the area of the target, but this is typically an upper limit on the area. Even if the source area does appear to be well determined, the electron beam can be filamented so that it does not fill the entire area (the filling factor is less than 1). Also, if only an upper limit on the low-energy cutoff to the electron distribution is known,

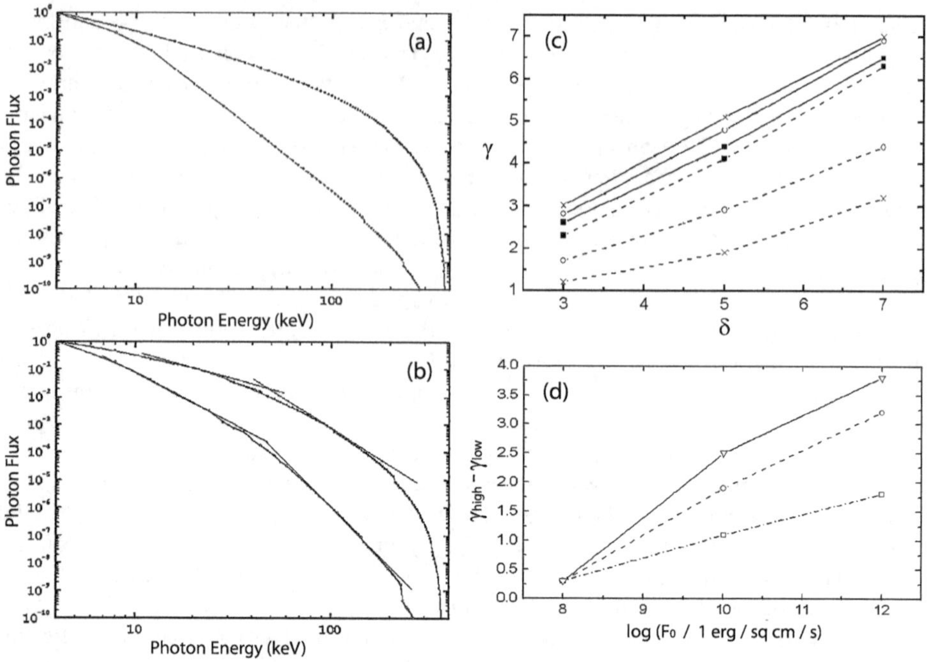

Fig. 8 (**a**) Photon spectra computed from full kinetic solutions including return current losses and collisional losses and scattering. The top spectrum is for an injected single-power-law electron flux distribution between 8 keV and 384 keV with an index of $\delta = 3$, and the bottom spectrum is for $\delta = 7$. The injected electron energy flux density is 10^8 erg cm^{-2} s^{-1}. (**b**) Same as (**a**), but for an injected energy flux density of 10^{12} erg cm^{-2} s^{-1}. The tangent lines at 20 and 100 keV demonstrate the determination of the low-energy and high-energy power-law spectral indices γ_{low} and γ_{high}. (**c**) The photon spectral indices γ_{low} (*dashed lines*) and γ_{high} (*solid lines*) vs. δ for an injected energy flux density of 10^8 (*squares*), 10^{10} (*circles*), and 10^{12} erg cm^{-2} s^{-1} (crosses). (**d**) $\gamma_{high}-\gamma_{low}$ vs. the log of the injected electron energy flux density for δ equal to 3 (*bottom curve*, *squares*), 5 (*middle curve*, *circles*), and 7 (*top curve*, *triangles*) (from Zharkova and Gordovskyy 2006). Reproduced by permission of the AAS

as described in Sect. 3.5, the energy flux density may be higher. Therefore, the observations typically only give a lower limit on the beam energy flux density.

The non-thermal hard X-ray flux is proportional to the electron beam flux density, but the return-current energy losses are also proportional to the beam flux density. As a consequence, Emslie (1980) concluded that the flux density of the non-thermal X-ray emission from a flare cannot exceed on the order of 10^{-15} cm^{-2} s^{-1} above 20 keV. Alexander and Daou (2007) have deduced the photon flux density from non-thermal electrons in a sample of 10 flares ranging from *GOES* class M1.8 to X17. They find that the non-thermal photon flux density does not monotonically increase with the thermal energy flux, but levels off (saturates) as the thermal energy flux becomes high. They argue that this saturation most likely results from the growing importance of return-current energy losses as the electron beam flux increases to high values in the larger flares. They find that the highest non-thermal photon flux densities agree with an upper limit computed by Emslie.

A correlation between the X-ray flux and spectral break energy was found by Sui et al. (2007) in their study of X-ray spectra in early impulsive flares (see Sect. 3.5). They point out that the increasing impact of return current energy losses on higher energy electrons as the electron beam energy flux density increases could be an explanation for this correlation.

Battaglia and Benz (2008) studied two flares with non-thermal coronal hard X-ray sources for which the difference between the measured photon spectral index at the footpoints and the spectral index of the coronal source was greater than two, the value expected for coronal thin-target emission and footpoint thick-target emission from a single power-law electron distribution (see Sect. 10.2). They argue that return-current losses between the coronal and footpoint source regions are most likely responsible for the large difference between the spectral indices.

The return current can also affect the spectral line emission from flares. Evidence for the presence of the return current at the chromospheric level from observations of the linear polarization of the hydrogen Hα and Hβ lines has been presented by Hénoux and Karlický (2003). Dzifčáková and Karlický (2008) have shown that the presence of a return current in the corona may have a distinguishable impact on the relative intensities of spectral lines emitted from the corona.

6 Beam-Plasma and Current Instabilities

Interaction of the accelerated electrons with plasma turbulence as they stream toward the thick-target emission region can modify the electron distribution. In this section we briefly discuss a likely source of such turbulence: that generated by the electron beam itself. If the beam is or becomes unstable to driving the growth of plasma waves, these waves can interact with the beam and modify its energy and/or pitch angle distribution until the instability is removed or the wave growth is stabilized. The return current associated with the beam (Sect. 5) can also become unstable, resulting in greater energy loss from the beam. Beam-plasma and return-current instabilities in solar flares have been reviewed by Melrose (1990) and Benz (2002).

A sharp, low-energy cutoff to an electron beam or, more generally, a positive slope in the beam electron energy distribution is well-known to generate the growth of electrostatic plasma waves (Langmuir waves). The characteristic time scale for the growth of these waves is on the order of $[(N_b/n)\omega_{pe}]^{-1}$, where ω_{pe} is the electron plasma frequency and N_b/n the ratio of the density of unstable electrons in the beam to the plasma density. This is on the order of microseconds for a typical coronal loop plasma density and $N_b/n \approx 10^{-3}$. This plasma instability is often referred to as the bump-on-tail instability. The result is that on a somewhat longer but comparable time scale electrons from the unstable region of the electron distribution lose energy to the waves until the sharp cutoff is flattened so that the distribution no longer drives the rapid growth of the waves. Therefore, the electron energy distribution below the low-energy cutoff is likely to rapidly become flat or nearly flat (sufficiently flat to stabilize the instability) after the electrons escape the acceleration region (see Chaps. 9 and 10 of Krall and Trivelpiece 1973).

A recent simulation of the bump-on-tail instability for solar flare conditions, including Coulomb collisions and wave damping as the electrons propagate into an increasingly dense plasma, has been carried out by Hannah et al. (2009). The authors compute the mean electron flux distribution in their model flare atmosphere after injecting a power-law distribution with a sharp low-energy cutoff. They find a mean electron flux distribution with no dip (see Sect. 3.3) and a slightly negative slope below the original cutoff energy with a spectral index δ between 0 and 1.

A beam for which the mean electron velocity parallel to the magnetic field substantially exceeds the mean perpendicular velocity can drive the growth of waves that resonantly interact with the beam. When the electron gyrofrequency exceeds the plasma frequency, these waves are electrostatic and primarily scatter the electrons in pitch angle. Generally known as the anomalous Doppler resonance instability, this instability tends to isotropize the beam electrons. Holman et al. (1982) showed that under solar flare conditions this instability can grow and rapidly isotropize the beam electrons in less than a millisecond. They found that electrons at both the low- and high-energy ends of the distribution may remain unscattered, however, because of wave damping. This could result in up to two breaks in the emitted X-ray spectrum. On the other hand, Vlahos and Rowland (1984) have argued that non-thermal tails will form in the ambient plasma and stabilize the anomalous Doppler resonance instability by suppressing the growth of the plasma waves.

Electrons streaming into a converging magnetic field can develop a loss-cone distribution, with a deficit of electrons at small pitch angles. Both classical Coulomb collisions and loss-cone instabilities can relax this distribution by scattering electrons into the loss cone or extracting energy from the component of the electron velocities perpendicular to the magnetic field (e.g., Aschwanden 1990). One loss-cone instability, the electron-cyclotron or gyrosynchrotron maser, produces coherent radiation observable at radio frequencies (Holman et al. 1980; Melrose and Dulk 1982).

The return current associated with the streaming electrons becomes unstable to the ion-acoustic instability when its drift speed exceeds a value on the order of the ion sound speed. The excited ion sound waves enhance the plasma resistivity, increasing the electric field strength associated with the return current, the heating of the plasma by the current, and the energy loss from the electron beam.

It has been argued that rapid plasma heating and particle acceleration in the corona should result in the expansion of hot plasma down the legs of flare loops at the ion sound speed, confined behind a collisionless ion-acoustic conduction front (Brown et al. 1979). Electrons with speeds greater than about three times the electron thermal speed would be able to stream ahead of the conduction front. This scenario has not been observationally verified, but the observational signature may be confused by the chromospheric evaporation produced by the high-energy particles streaming ahead of the conduction front.

Rowland and Vlahos (1985) argued that if the electron beam is unstable to beam plasma interactions, the return current will be carried by high-velocity electrons. This reduces the impact of collisions on the beam/return-current system and helps stabilize the system. In a recent simulation, Karlický et al. (2008) have found that for current drift velocities exceeding the electron thermal speed, the return current is carried by both the primary (drifting thermal) current and an extended tail of high-velocity electrons.

The evolution of the electron-beam/return-current system when the return-current drift speed exceeds the electron thermal speed has also been simulated by Lee et al. (2008). They find that double layers form in the return current, regions of enhanced electric field that further increase the energy losses of the electron beam. This, in turn, increases the highest electron energy to which these losses significantly flatten the electron distribution and corresponding hard X-ray spectrum.

The beam/return-current system has been simulated by Karlický (2009), with a focus on the role of the Weibel instability. The Weibel instability tends to isotropize the electron distribution. Karlický and Kašparová (2009) have computed the thin-target X-ray emission from the evolved electron distributions for a model with a weak magnetic field and another model with a strong magnetic field (ratio of the electron gyrofrequency to the plasma frequency $\sim$0 and $\sim$1, respectively). They demonstrate that in both cases the electron distribution is more isotropic and the directivity of the X-ray emission is lower than when the instability of the system is not taken into account, with the greatest isotropization occurring in the weak field limit.

Although we expect plasma instabilities to affect the evolution of the electron beam, observationally identifying them is difficult. The bump-on-tail instability and return current losses both lead to a flat low-energy cutoff. So far we have not established the ability to observationally distinguish a flat low-energy cutoff from a sharp low-energy cutoff. The bump-on-tail instability may be distinguishable from return-current losses by its short time scale and, therefore, the short distance from the acceleration region over which it effectively removes the unstable positive slope from the electron energy distribution. The instabilities that isotropize the electron pitch-angle distribution may be responsible for evidence from albedo measurements that flare electron distributions are isotropic or nearly isotropic (Kontar and Brown 2006b).

7 Height Dependence and Size of X-ray Sources with Energy and Time

7.1 Footpoint Sources

Hard X-ray footpoint sources result from collisional bremsstrahlung radiated by precipitating electrons, which produce most of the emission in the chromosphere according to the collisional thick-target model. Depending on the density structure in the legs of the coronal magnetic loop, mildly energetic electrons lose their energy in the lower corona or transition region, while the more energetic electrons penetrate deeper into the chromosphere (see (2.2)).

The altitude of these hard X-ray footpoint sources could never be measured accurately before *RHESSI*, because of a lack of spatial and spectral resolution. With *RHESSI*, we can measure the centroid of the footpoint location with an accuracy of order an arcsecond for every photon energy in steps as small as 1 keV. For a flare near the limb (Fig. 9), the centroid location translates directly into an altitude.

Aschwanden et al. (2002) studied such a flare, SOL2002-02-20T11:07 (C7.5). The heights of the footpoint sources were fitted with a power-law function of the photon energy. This yielded altitudes $h \approx 1000$–5000 km in the energy range $\varepsilon = 10$–60 keV, progressively lower with higher energy, as expected from the thick-target model (Fig. 9, right frame).

Since the stopping depth of the precipitating electrons is a function of column density, the integrated density along their path in the chromosphere (2.4), the measured height dependence of the hard X-ray centroids can be inverted to yield a density model of the chromosphere (Brown et al. 2002). Assuming the decrease in density with height had a power-law dependence and the plasma is fully ionized, the inversion of the *RHESSI* data in the example shown in Fig. 9 yielded a chromospheric density model that has a significantly higher electron density in the $h = 2000$–5000 km range than the standard chromospheric models based on UV spectroscopy and hydrostatic equilibrium (VAL and FAL models). The *RHESSI*-based chromospheric density model was therefore found to be more consistent with the "spicular extended chromosphere," similar to the results from sub-mm radio observations during solar eclipses carried out at Caltech (Ewell et al. 1993).

Forward fitting *RHESSI* X-ray visibilities to an assumed circular Gaussian source shape, Kontar et al. (2008b) found for a limb flare the full width at half maximum (FWHM) size and centroid positions of hard X-ray sources as a function of photon energy with a claimed resolution of $\sim 0''.2$. They show that the height variation of the chromospheric density and of the magnetic flux density can be found with a vertical resolution of ~ 150 km by mapping the 18–250 keV X-ray emission of energetic electrons propagating in the loop at chromospheric heights of 400–1500 km. Assuming collisional losses in neutral hydrogen with an

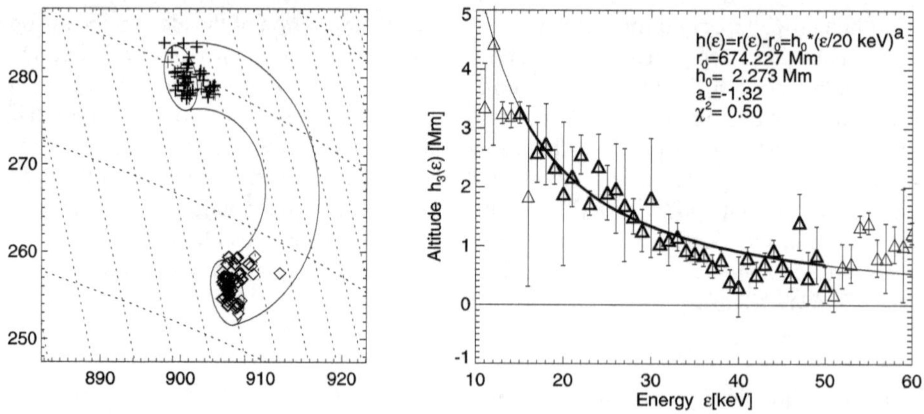

Fig. 9 The centroids of footpoint hard X-ray emission are marked for different photon energies between 10 keV and 60 keV for SOL2002-02-20T11:07 (C7.5), which occurred near the solar west limb and was imaged with *RHESSI* (*left panel*). The altitude $h(\varepsilon)$ as a function of energy ε shows a systematic height decrease with increasing energy (*right panel*) (from Aschwanden et al. 2002)

exponential decrease in density with height, their observations of SOL2004-01-06T06:29 (M5.8) suggest that the density of the neutral gas is in good agreement with hydrostatic models with a scale height of around 140 ± 30 km. FWHM sizes of the X-ray sources decrease with energy, suggesting the expansion (fanning out) of magnetic flux tubes in the chromosphere with height. The magnetic scale height $B(z)(dB/dz)^{-1}$ is found to be on the order of 300 km and a strong horizontal magnetic field is associated with noticeable flux tube expansion at a height of ~900 km. A subsequent analysis with an assumed elliptical Gaussian source shape (Kontar et al. 2010) confirms these results and shows that the vertical extent of the X-ray source decrease with increasing X-ray energy. The authors find the vertical source sizes to be larger than expected from the thick-target model and suggest that a multi-threaded density structure in the chromosphere is required. The thick-target model to which the results were compared, however, did not account for partial occultation of the X-ray sources by the solar limb.

The flare SOL2002-02-20T11:07 (C7.5) has been reanalyzed by Prato et al. (2009) using both photon maps over a range of photon energies and mean electron flux maps deduced from *RHESSI* visibilities over a range of electron energies. Using source centroids computed from the maps and assuming an exponential decrease in density with height, they found the density scale height to be an order of magnitude larger than the expected chromospheric scale height on the quiet Sun, but consistent with the scale height in a non-static, flaring atmosphere. This is also consistent with the enhanced plasma densities found at ~1000–5000 km altitudes by Aschwanden et al. (2002).

If the results for the 400–1500 km height range (Kontar et al. 2008b) and for the ~1000–5000 km height range (Aschwanden et al. 2002; Prato et al. 2009) are typical of flare loops, they imply that the upper chromosphere and transition region respond with a non-hydrostatic, expanded atmosphere while the low chromosphere does not respond to the flare energy release. These results could, of course, depend on the magnitude of the flare. More studies of this kind are clearly desirable, especially in coordination with observations of spectral lines from the chromosphere and transition region.

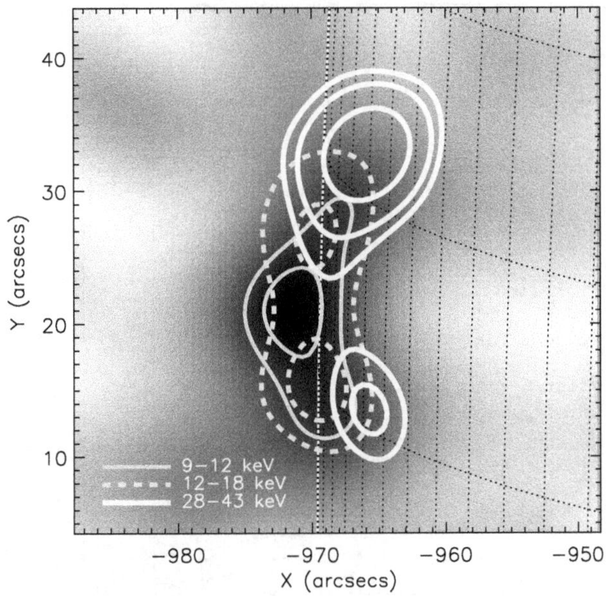

Fig. 10 CLEAN images at 04:58:22-04:58:26 UT during the impulsive phase of SOL2003-11-13T05:01 (M1.6). The background shows the image at 9–12 keV. The contour levels are 75% and 90% of the peak flux at 9–12 keV (*looptop*), 70% and 90% at 12–18 keV (*legs*), and 50%, 60%, & 80% at 28–43 keV (*footpoints*) (from Liu et al. 2006). Reproduced by permission of the AAS

7.2 Loop Sources and Their Evolution

As discussed above, footpoint sources are produced by bremsstrahlung emission in the thick-target chromosphere. The compactness of such sources results from the rapid increase of the density from the tenuous corona to the much denser chromosphere. This also gives rise to the compact height distribution of emission centroids at different energies as shown in Fig. 9. However, if the density distribution has a somewhat gradual variation, one would expect a more diffuse height distribution. Specifically, at some intermediate energies, we expect that HXR emission would appear in the legs of the loop, rather than the commonly observed looptop sources at low energies and footpoint sources at high energies. This has been observed by *RHESSI* in SOL2003-11-13T05:01 (M1.6) (Liu et al. 2006) (Fig. 10) and in SOL2002-11-28T04:37 (C1.0) (Sui et al. 2006b).

To reveal more details of the energy-dependent structure of SOL2003-11-13T05:01 (M1.6), Figs. 11a–c show the X-ray emission profile along the flare loop at different energies for three time intervals in sequence. The high energy emission is dominated by the footpoints, but there is a decrease of the separation of the footpoints with decreasing energy and with time. At later times the profile becomes a single source, peaking at the looptop. The general trend suggests an increase of the plasma density in the loop with time (Liu et al. 2006), which can be produced by chromospheric evaporation and can give rise to progressively shorter stopping distances for electrons at a given energy. Such a density increase also smooths out to some extent the sharp density jump at the transition region. This results in the non-thermal bremsstrahlung HXRs at intermediate energies appearing in the legs of the loop, at higher altitudes than the footpoints, as shown in Fig. 10.

From the emission profiles in the non-thermal regimes of the photon spectra, Liu et al. (2006) derived the density distribution along the loop, using the empirical formula for non-thermal bremsstrahlung emission profiles given by Leach and Petrosian (1983, Eq. 11). Leach and Petrosian found that this formula closely approximates their numerical results for a steady-state, power-law injected electron distribution with a uniform pitch-angle distribution, no return-current losses, and a loop with no magnetic field convergence. Since this

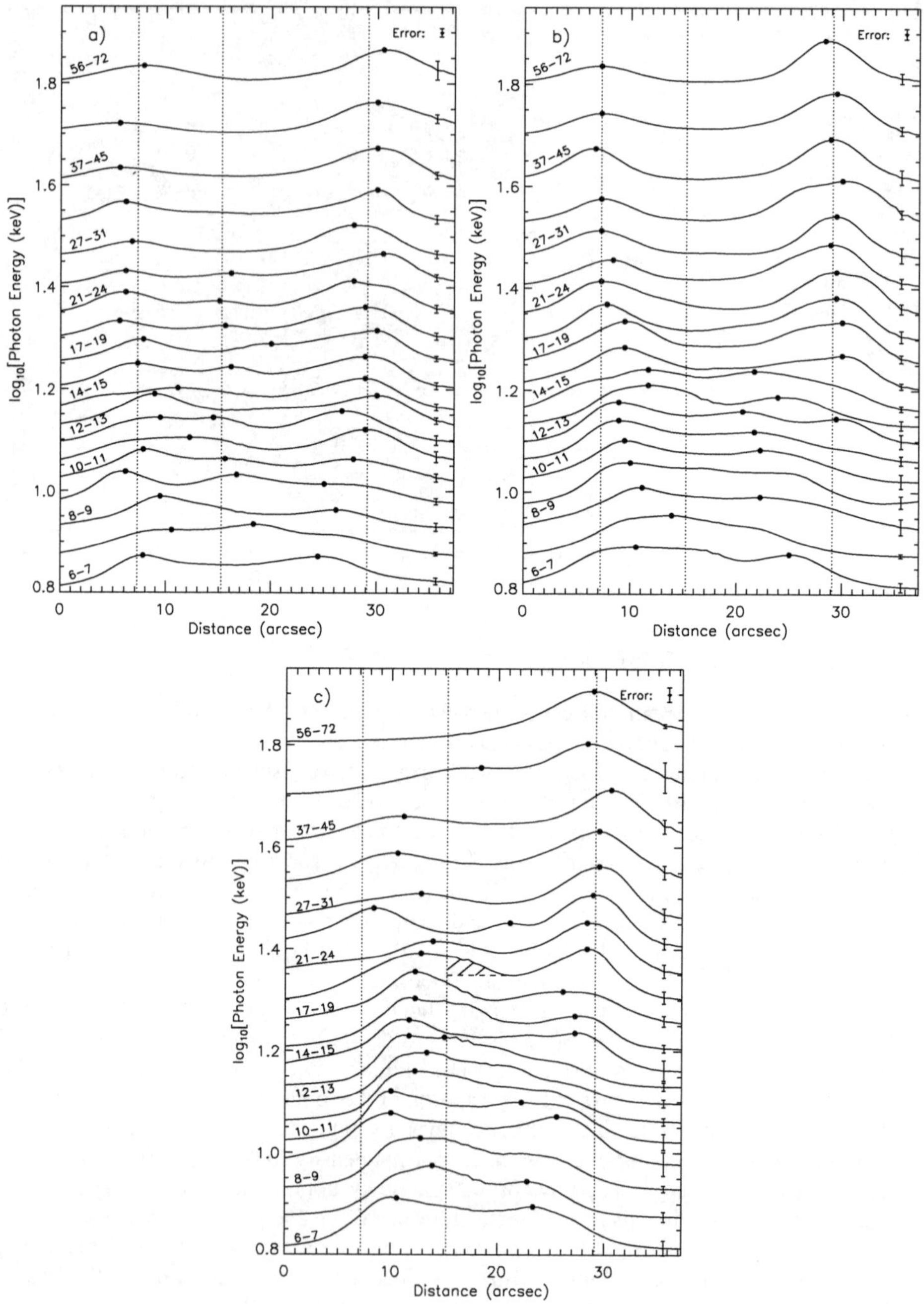

Fig. 11 (**a**) Brightness profiles in different energy bands measured along a semi-circular path fit to the flaring loop for the time interval 04:58:00–04:58:24 UT of SOL2003-11-13T05:01. The *vertical axis* indicates the average photon energy (logarithmic scale) of the energy band for the profile. Representative energy bands (in units of keV) are labeled above the corresponding profiles. The *filled circles* mark the local maxima, and the *vertical dotted lines* are the average positions of the centroids of the looptop and footpoint sources. (**b**, **c**) Same as (**a**), but for 04:58:24–04:58:48 and 04:58:48–04:59:12 UT, respectively. The *error bars* show the uncertainty of the corresponding profile (from Liu et al. 2006). Reproduced by permission of the AAS

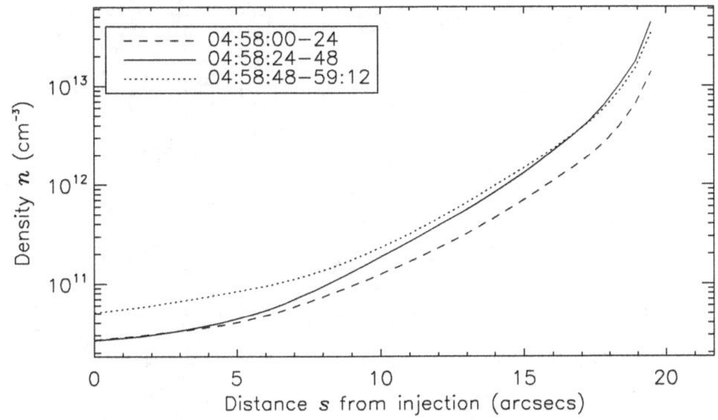

Fig. 12 Averaged density profiles along one leg of the loop inferred from the HXR brightness profiles during the three time intervals in Fig. 11. The distance is measured along the leg extending from the centroid of the thermal looptop source at about 15 arcsec in Fig. 11 to the end of the fitted semi-circle at about 37 arcsec (from Liu et al. 2006). Reproduced by permission of the AAS

formula is a function of the column density, one does not need to assume any model form of the density distribution (cf. Aschwanden et al. 2002). Figure 12 shows the density profiles derived from the emission profiles in the three time intervals shown in Fig. 11. Between the first and second intervals, the density increases dramatically in the lower part of the loop, while the density near the looptop remains essentially unchanged. The density enhancement then shifts to the looptop from the second to the third interval. This indicates a mass flow from the chromosphere to the looptop, most likely caused by chromospheric evaporation. For papers studying chromospheric evaporation using coordinated *RHESSI* HXR and EUV Doppler-shift observations, see Milligan et al. (2006a, 2006b) and Brosius and Holman (2007).

The flare SOL2002-11-28T04:37 (C1.0) was an early impulsive flare, meaning that there was minimal pre-heating of plasma to X-ray-emitting temperatures prior to the appearance of impulsive hard X-ray emission (see Sect. 3.5). *RHESSI* observations of this flare showed coronal X-ray sources that first moved downward and then upward along the legs of the flare loop (Sui et al. 2006b). The bottom panel of Fig. 13 shows the motion of the sources observed in the 3–6 keV band. *RHESSI* and *GOES* light curves are shown in the top panel. The sources originated at the top of the flare loop and then moved downward along both legs of the loop until the time of peak emission at energies above 12 keV. Afterward the source in the northern leg of the loop was no longer observable, but the source in the southern leg moved back to the top of the loop. Its centroid location at the looptop was slightly but significantly lower than the centroid position at the beginning of the flare. Higher-energy sources showed a similar evolution, but they had lower centroid positions than their lower energy counterparts, again in agreement with the predictions of the thick-target model.

The early downward source motion along the legs of the loop is a previously unobserved phenomenon. At this time we do not know if the occurrence is rare, or if it is simply rarely observed because of masking by the radiation from the thermal plasma. Sui et al. (2006b) argue that the motion results from the hardening of the X-ray spectrum, and possibly an increase in the low-energy cutoff, as the flare hard X-ray emission rises to its peak intensity. A flatter spectrum results in a higher mean energy of the electrons contributing to the radiation at a given X-ray energy. In a loop with a plasma density that increases significantly from the top to the footpoints, these higher energy electrons will propagate to a lower altitude in the loop as the spectrum hardens. The softening of the spectrum after peak emission would also contribute to the upward motion of the source after the peak. However, at that time chromospheric evaporation would likely be increasing the density in the loop, as discussed above for SOL2003-11-13T05:01 (M1.6), and thermal emission would be more important. All of these can contribute to an increase in the height of the centroid of the X-ray source. The

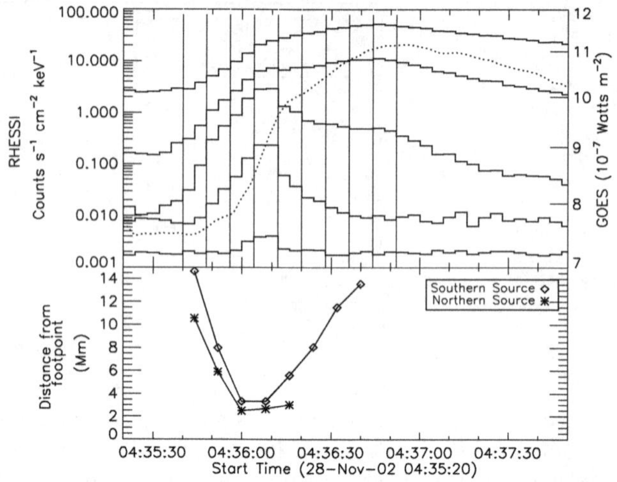

Fig. 13 *RHESSI* (*solid lines*) and *GOES* 1–8 Å (*dotted line*) *light curves* are shown in the *top panel*. The *RHESSI* energy bands (*from top to bottom*) are 3–6, 6–12, 12–25, and 50–100 keV, with scaling factors of 5, 1, 4, 3, and 0.5, respectively. The *RHESSI* and *GOES* integration times are 4 and 3 s, respectively. The *bottom panel* shows the distance between the 3–6 keV moving source centroids and their corresponding footpoint centroids located in the 25–50 keV image of the flare at the time of peak emission. The distances are plane-of-sky values with no correction for motions away from or toward the observer (from Sui et al. 2006b). Reproduced by permission of the AAS

downward motion may only occur in initially cool flare loops, i.e., early impulsive flares, because these loops are most likely to contain the density gradients that are required.

In an attempt to differentiate between thermal and non-thermal X-ray emission, Xu et al. (2008) modeled the size dependence with photon energy of coronal X-ray sources observed by *RHESSI* in ten M-class limb flares. They determined the one-sigma Gaussian width of the sources along the length of the flare loops by obtaining forward fits to the source visibilities. The integration times ranged from one to ten minutes and the source sizes were determined in up to eight energy bins ranging in energy from as low as 7 keV to as high as 30 keV. They found the source sizes to increase slowly with photon energy, on average as $\varepsilon^{1/2}$. The results were compared with several models for the variation of the source size with energy. The source size was expected to vary as $\varepsilon^{-1/2}$ for a thermal model with a constant loop density and a temperature that decreased with a Gaussian profile along the legs of the loop from a maximum temperature at the top of the loop. For the injection of a power-law electron flux distribution into a high-density loop so that the loop is a collisional thick target, the source size was expected to increase as ε^2. Neither of these models are consistent with the observed $\varepsilon^{1/2}$ dependence. A hybrid thermal/non-thermal model and a non-thermal model with an extended acceleration region at the top of the loop were found to be consistent with the deduced scaling, however. The extended acceleration region was deduced to have a half-length in the range $10''$–$18''$ and density in the range (1–5) $\times$ 10^{11} cm^{-3}. We note that the extended acceleration region model implies a column density in the range 0.73–6.5 $\times$ 10^{20} cm^{-2} along the half length and, from (2.4), all electrons with energies less than somewhere in the range of 23–68 keV that traverse this half length will lose all of their energy to collisions. The acceleration process would therefore need to be efficient enough to overcome these losses. On the other hand, the 7–30 keV energy range is the range in which fits to spatially integrated X-ray spectra typically show a combination of both thermal and non-thermal bremsstrahlung emission.

Studies of flare hard X-ray source positions and sizes as a function of photon energy and time hold great promise for determining the height structure of flare plasma and its evolution, as well as providing information about the magnetic structure of the flare loop. Such studies are currently in their early stages, in that they usually assume an over simplified power-law or exponential height distribution for the plasma and do not take into account the variation of the plasma ionization state with height. They also assume the simple, one-dimensional collisional thick-target model, without consideration of the pitch-angle distribution of the beam electrons or the possibility of additional energy losses to the beam (such as return-current losses). Given the potential for obtaining a better understanding of flare evolution, we look forward to the application of more sophisticated models to the flare hard X-ray data.

8 Hard X-ray Timing

The analysis of energy-dependent time delays allows us to test theoretical models of physical time scales and their scaling laws with energy. In the wavelength domain of hard X-rays, there are at least three physical processes known in the observation of solar flares that lead to measurable time delays as a function of energy (for a review, see Aschwanden 2004): (1) time-of-flight dispersion of free-streaming electrons, (2) magnetic trapping with the collisional precipitation of electrons, and (3) cooling of the thermal plasma.

8.1 Time-of-Flight Delays

The first type, the *time-of-flight* (*TOF*) delay, has a scaling of $\Delta t(\varepsilon) \propto \varepsilon^{-1/2}$ and is caused by velocity differences of electrons that propagate from the coronal acceleration site to the chromospheric energy-loss region. The time differences are of order $\Delta t \approx$ 10–100 ms for non-thermal electrons at energies $E \approx$ 20–100 keV (e.g., Aschwanden et al. 1995, 1996). The measurement of such tiny time delays requires high photon statistics and high time resolution. Such data were provided by *CGRO*/BATSE, which had 8 detectors, each with an effective collecting area of $\sim$2000 cm^2 and oriented at different angles to the Sun so that detector saturation at high count rates was not a problem. (For comparison, the total effective collecting area of *RHESSI*'s detectors is less than 100 cm^2.)

These studies of TOF delays have provided important evidence that electrons are accelerated in the corona, above the top of the hot flare loops observed in soft X-rays. The fine structure in the light curves of most, but not all, of the studied flare bursts showed energy-dependent time delays consistent with the free streaming of electrons to the footpoints of the flare loops from an origin somewhat more distant than the half-length of the loops (Aschwanden et al. 1995, 1996; Aschwanden and Schwartz 1995).

8.2 Trapping Delays

The second type, the *trapping delay*, is caused by magnetic mirroring of coronal electrons which precipitate toward the chromosphere after a collisional time scale $\Delta t(\varepsilon) \propto \varepsilon^{3/2}$. This is observable for time differences of $\Delta t \approx$ 1–10 s for non-thermal electrons at $E \approx$ 20–100 keV (e.g., Vilmer et al. 1982; Aschwanden et al. 1997). For trapping delays the higher energy X-rays lag the lower energy X-rays, as opposed to time-of-flight delays where the higher energy X-rays precede the lower energy X-rays.

Aschwanden et al. (1997) filtered variations on time scales $\sim$1 s or less out of *CGRO* BATSE flare HXR light curves. They found time delays in the remaining gradually varying component to be consistent with magnetic trapping and collisional precipitation of the particles. Trap plasma densities $\sim 10^{11}$ cm^{-3} were deduced. No evidence was found for a discontinuity in the delay time as a function of energy and, therefore, for second-step (two-stage) acceleration of electrons at energies ≤ 200 keV.

8.3 Thermal Delays

The third type, the *thermal delay*, can be caused by the temperature dependence of cooling processes, such as by thermal conduction, $\tau_c(T) \propto T^{-5/2}$ (e.g., Antiochos and Sturrock 1978; Culhane et al. 1994), or by radiative cooling, $\tau_r(T) \propto T^{5/3}$ (e.g., Fisher and Hawley 1990; Cargill et al. 1995). The observed physical parameters suggest that thermal conduction dominates in flare loops at high temperatures as observed in soft X-ray wavelengths, while radiative cooling dominates in the later phase in postflare loops as observed in EUV wavelengths (Antiochos and Sturrock 1978; Culhane et al. 1994; Aschwanden and Alexander 2001). When the temperature drops in the decay phase of flares, the heating rate can justifiably be neglected and the conductive or radiative cooling rate dominate the temperature evolution. Before *RHESSI*, the cooling curve $T(t)$ in flare plasmas had been studied in only a few flares (e.g., McTiernan et al. 1993; Culhane et al. 1994; Aschwanden and Alexander 2001).

The high spectral resolution of *RHESSI* data is particularly suitable for any type of thermal modeling, because we can probe the thermal plasma from $\sim$3 keV up to $\sim$30 keV with a FWHM resolution of $\sim$1 keV thanks to the cooled germanium detectors (Lin et al. 2002; Smith et al. 2002). This allows us to measure flare temperatures with more confidence. A statistical study of flare temperatures measured in the range of $T \approx 7$–20 MK indeed demonstrates some agreement between the values obtained from spectral fitting of *RHESSI* data with those obtained from *GOES* flux ratios (Battaglia et al. 2005), although *RHESSI* has a bias for the high-temperature tail of the differential emission measure (DEM) distribution (Aschwanden et al. 2008; Väänänen and Pohjolainen 2007). Of course, we expect an agreement between the deduced emission-measure-weighted temperatures only when both instruments are sensitive to a temperature range that covers the flare DEM peak.

A close relationship between the non-thermal and thermal time profiles was found early on, in the sense that the thermal emission often closely resembles the integral of the non-thermal emission, a relationship that is now known as the *Neupert effect* (Neupert 1968; Hudson 1991; Dennis and Zarro 1993). This relationship is, however, strictly only expected for the asymptotic limit of very long cooling times, while a physically more accurate model would quantify this effect by a convolution of the non-thermal heating with a finite cooling time. The deconvolution of the e-folding cooling time in such a model has never been attempted statistically and as a function of energy or temperature. Theoretical discussions of the Neupert effect, including multiple energy release events, chromospheric evaporation, and cooling, can be found in Warren and Antiochos (2004), Liu et al. (2010), and Reeves and Moats (2010).

The cooling time at a given energy can be estimated from the decay time of a flare time profile. For instance, the decay times measured with *GOES* in soft X-rays were found to have a median of $\tau_{decay} \approx 6$ min (Veronig et al. 2002a, 2002b). The observed cooling times have typically been found to be much longer than predicted from classical conduction, but shorter than the radiative cooling time (e.g., McTiernan et al. 1993; Jiang et al. 2006; Raymond et al. 2007). This discrepancy could result from either continuous heating or suppression of conduction during the decay phase, or a combination of both.

The Neupert effect was tested by correlating the soft X-ray peak flux with the (time-integrated) hard X-ray flux. A high correlation and time coincidence between the soft X-ray peak and hard X-ray end time was generally found, but a significant fraction of events also had a different timing (Veronig et al. 2002c). A delay of 12 s was found in the soft X-ray flux time derivative with respect to the hard X-ray flux in SOL2003-11-13T05:01 (M1.6) (Liu et al. 2006, also see Sect. 7.2). Time delays such as this could be related to the hydrodynamic flow time during chromospheric evaporation. Tests of the "theoretical Neupert effect," i.e., comparisons of the beam power supply of hard X-ray-emitting electrons and the thermal energy of evaporated plasma observed in soft X-rays, found it to strongly depend on the low-energy cutoff to the non-thermal electron distribution (Veronig et al. 2005). This provides another approach to deducing the energy at which the low-energy cutoff in the electron distribution occurs in individual flares. The Neupert effect has also been studied in several flares by Ning (2008, 2009), who finds a high correlation between the hard X-ray flux and the time derivative of the thermal energy deduced from X-ray spectral fits (Ning 2008) and an anti-correlation between the hard X-ray spectral index and the time rate of change of the UV flare area observed by *TRACE* (Ning 2009).

8.4 Multi-thermal Delay Modeling with *RHESSI*

Since major solar flares generally produce a large number of individual postflare loops, giving the familiar appearance of loop arcades lined up along the flare ribbons, it is unavoidable that each loop is heated up and cools down at different times, so that a spatially integrated spectrum always contains a multi-thermal differential emission measure distribution (cf. Warren 2006). The resulting multi-thermal bremsstrahlung spectrum (for photon energies ε) observed in soft X-rays (neglecting the Gaunt factor of order unity),

$$I(\varepsilon) = I_0 \int \frac{\exp(-\varepsilon/k_B T)}{T^{1/2}} \frac{dEM(T)}{dT} dT, \tag{8.1}$$

is then a function of a multi-thermal *differential emission measure* distribution $dEM(T) = n^2(T)\,dV$. An example of a multi-thermal spectrum from a differential emission measure proportional to T^{-4} up to a maximum temperature of 50 MK is shown in Fig. 14.

As discussed above, the initial cooling of the hot flare plasma (say at $T \gtrsim 10$ MK) is generally dominated by conductive cooling (rather than by radiative cooling, which can dominate later after the plasma cools to EUV-emitting temperatures of $T \lesssim 2$ MK). The thermal conduction time has the following temperature dependence:

$$\tau_{cond}(T) = \frac{\varepsilon_{th}}{dE/dt_{cond}} = \frac{3n_e k_B T}{\frac{d}{ds}\kappa T^{5/2}\frac{dT}{ds}} \approx \frac{21}{2}\frac{n_e L^2 k_B}{\kappa} T^{-5/2} = \tau_{c0}\left(\frac{T}{T_0}\right)^{-5/2}; \tag{8.2}$$

see Aschwanden (2007) for parameter definitions. Since the thermal bremsstrahlung at decreasing photon energies is dominated by radiation from lower temperature flare plasma, the conductive cooling time is expected to become longer at lower temperatures ($\tau_{cond} \propto T^{-5/2}$). Thus, the soft X-ray peak is always delayed with respect to the harder X-ray peaks, reflecting the conductive cooling of the flare loops.

Aschwanden (2007) has measured and modeled this conductive cooling delay $\tau_{cond}(\varepsilon)$ for a comprehensive set of short-duration (≤ 10 min) flares observed by *RHESSI*. One example is shown in Fig. 15. He finds that the cooling delay Δt expressed as a function of the photon energy ε and photon spectral index γ can be approximated by

$$\Delta t(\varepsilon, \gamma) \approx \tau_g \frac{7}{4}\left[\log\left(1 + \frac{\tau_{c0}}{\tau_g}\left(\frac{\varepsilon}{(\gamma-1)\varepsilon_0}\right)^{-\beta}\right)\right]^{3/4} \tag{8.3}$$

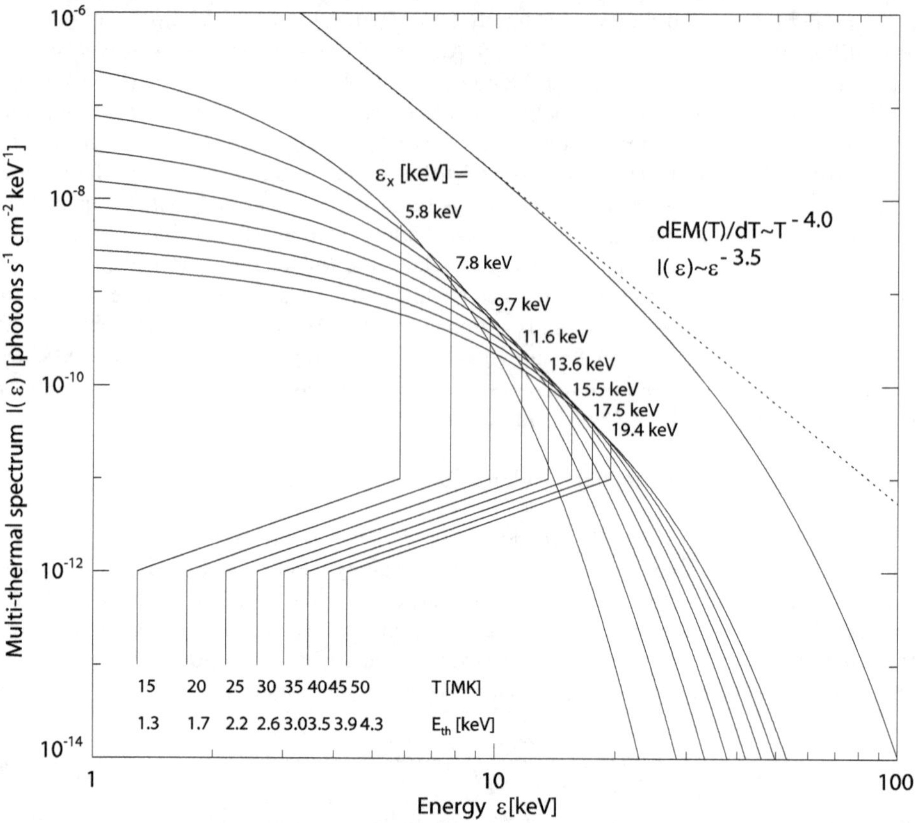

Fig. 14 Example of a multi-thermal spectrum with contributions from plasmas with temperatures of $T = 15, 20, \ldots, 50$ MK and a DEM distribution of $dEM(T)/dT \propto T^{-4}$. The individual thermal spectra and their sum are shown with thin linestyle, where the sum represents the observed spectrum. Note that the photons in the energy range $\varepsilon = 5.8$–19.4 keV are dominated by temperatures of $T = 15$–50 MK, which have a corresponding thermal energy that is about a factor of $(4 + 1/2) = 4.5$ lower than the corresponding photon energy ($\varepsilon_{th} = 1.3$–4.3 keV). The summed photon spectrum without the high-temperature cutoff approaches the power-law function $I(\varepsilon) \propto \varepsilon^{-3.5}$ (*dotted line*) (from Aschwanden 2007). Reproduced by permission of the AAS

(where τ_g is the Gaussian width of the time profile peak) and yields a new diagnostic of the process of conductive cooling in multi-thermal flare plasmas. In a statistical study of 65 flares (Aschwanden 2007), 44 (68%) were well fit by the multi-thermal model with a best fit value for the exponent of $\beta = 2.7 \pm 1.2$, which is consistent with the theoretically expected value of $\beta = 2.5$ according to (8.2). The conductive cooling time at $T_0 = 11.6$ MK ($\varepsilon_0 = 1$ keV) was found to range from 2 to 750 s, with a mean value of $\tau_{c0} = 40$ s.

We note that these timing data, as well as thick-target fits to the non-thermal part of spectra that reveal the evolution of the energy content in accelerated electrons, provide additional constraints on models such as the multithread flare model of Warren (2006).

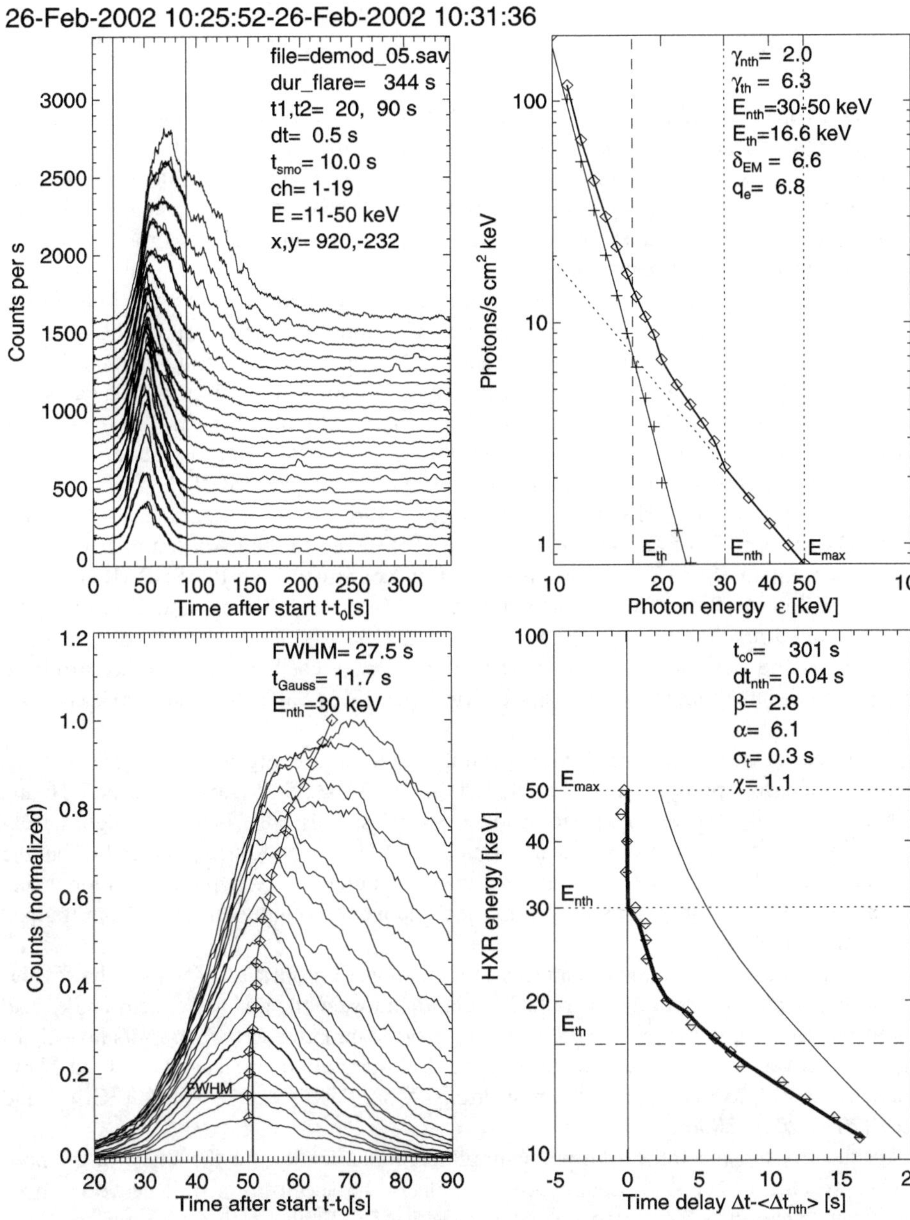

Fig. 15 X-ray light curves are shown for SOL2002-02-26T10:27 (C9.6), for energies of 10 keV to 30 keV in intervals of 1 keV, observed with *RHESSI* (*left panels*). The spectrum is decomposed into thermal and non-thermal components (*top right panel*) and the delay of the peaks at different energies is fitted with a thermal conduction cooling time model that has a scaling of $\tau_{cond}(T) \approx T^{-\beta}$ (*right bottom panel*). The best fit shows a power index of $\beta = 2.8$, which is close to the theoretically expected value of $\beta = 5/2$ (8.2). The full delay of the thermal component is indicated with a thin curve (*bottom left panel*), while the weighted (thermal+non-thermal) fit is indicated with a thick curve (from Aschwanden 2007). Reproduced by permission of the AAS

9 Hard X-ray Spectral Evolution in Flares

9.1 Observations of Spectral Evolution

The non-thermal hard X-ray emission from solar flares, best observed in the 20 to 100 keV range, is highly variable. Often several emission spikes with durations ranging from seconds to minutes are observed. In larger events, sometimes a more slowly variable, long duration emission can be observed in the later phase of the flare. Hence, most flares start out with an *impulsive* phase, while some events, mostly large ones, show the presence of a late *gradual* phase in the hard X-ray time profile.

While these two different behaviors can already be spotted by looking at lightcurves, they also are distinct in their spectral evolution. The impulsive spikes tend to be harder at the peak time, and softer both in the rise and decay phase. The spectrum starts soft, gets harder as the flux rises and softens again after the maximum of the emission. This pattern of the spectral evolution is thus called *soft-hard-soft* (SHS). On the other hand, in the gradual phase, the flux often slowly decreases, while the spectrum stays hard or gets even harder. This different kind of spectral evolution is called *soft-hard-harder* (SHH).

Historically, both the SHS (Parks and Winckler 1969; Kane and Anderson 1970) and the SHH behavior (Frost and Dennis 1971) were observed in the early era of hard X-ray observations of the Sun. Subsequent investigation confirmed both the SHS (Benz 1977; Brown and Loran 1985; Lin and Schwartz 1987; Gan 1998; Fletcher and Hudson 2002; Hudson and Fárník 2002) and the SHH (Cliver et al. 1986; Kiplinger 1995; Saldanha et al. 2008; Grigis and Benz 2008) behavior. The SHH behavior has been found to be correlated with proton events in interplanetary space (Kiplinger 1995; Saldanha et al. 2008; Grayson et al. 2009).

Evidence for hard-soft-hard (HSH) spectral evolution at energies above ~50 keV has been reported for multiple spikes in SOL2004-11-03T03:35 (M1.6) (Shao and Huang 2009b). SHS behavior was observed at lower energies. This HSH behavior might be explained by albedo, which typically peaks around 30–40 keV (see Kontar et al. 2011), but the authors corrected for albedo from isotropically emitted photons. A likely explanation is that the spikes overlie a harder, gradually varying component, possibly emission from trapped electrons (Sect. 8.2).

While all these observations established the qualitative properties of the spectral evolution, a statistical analysis of the quantitative relation between the flux and spectral index had not been performed in the pre-*RHESSI* era. Here, we summarize *RHESSI* results investigating quantitatively the spectral evolution of the non-thermal component of the hard X-ray emission, as well as the theoretical implications. More details can be found in Grigis and Benz (2004, 2005, 2006).

To quantify the spectral evolution, a simple parameterization for the shape of the non-thermal spectrum is needed. Luckily, in solar flares the spectrum is well described by a power law in energy, which often steepens above 50 keV. Such a softening of the spectrum can be modeled by a broken power-law model. However, it is difficult to observe such a downward bending at times of weak flux, because the high-energy region of the spectrum is lost in the background. As a compromise, Grigis and Benz (2004) fitted the data to a single power-law function at all times. Although the single power law does not always provide a good fit to the spectra, it provides a characteristic spectral slope and ensures a uniform treatment of the spectra at different times.

The two free parameters of the power-law model are the spectral index γ and the power-law normalization I_{ε_0} at the reference energy ε_0. The reference energy ε_0 is arbitrary, but

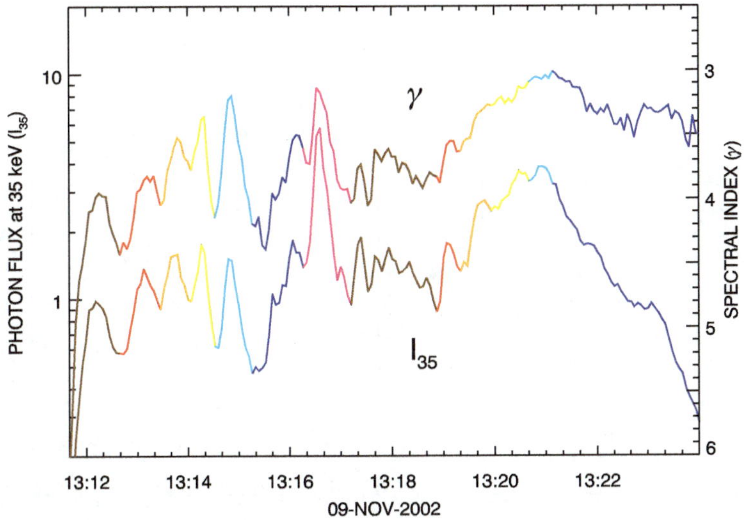

Fig. 16 Time evolution of the spectral index γ (*upper curve, linear scale on right*) and the flux normalization I_{35} (*lower curve, logarithmic scale on left*) of the non-thermal component in SOL2002-11-09T13:23 (M4.9). Different emission spikes are shown in different colors (after Grigis and Benz 2004). Reproduced with permission ©ESO

fixed, usually near the logarithmic mean of the covered energy range. In the *RHESSI* spectral analysis software, OSPEX, $\varepsilon_0 = 50$ keV by default. The time dependent spectrum is given by

$$I(\varepsilon, t) = I_{\varepsilon_0}(t)\left(\frac{\varepsilon}{\varepsilon_0}\right)^{-\gamma(t)}. \tag{9.1}$$

A representative sample of 24 solar flares of *GOES* magnitudes between M1 and X1 was selected by Grigis and Benz (2004). The spectral model (9.1), with the addition of an isothermal emission component at low energies, was fitted with a cadence of one *RHESSI* spin period (about 4 s). This delivered a sequence of measurements of the quantities $I_{\varepsilon_0}(t)$ and $\gamma(t)$ for each of the 24 events, covering a total time of about 62 minutes of non-thermal hard X-ray emission. For these events, $\varepsilon_0 = 35$ keV was chosen, a meaningful energy which lies about in the middle of the range where the non-thermal emission is best observed in these M-class flares.

An example of the measured time evolution of the spectral index γ and the flux normalization I_{35} for the longer-lasting event of the set is shown in Fig. 16. A correlation in time between the two curves can be readily seen. Single emission spikes are plotted in different colors, so that the soft-hard-soft evolution can be observed during each spike (with the exception of the late, more gradual phase, where the emission stays hard as the flux decays).

As there is an anti-correlation in time between $\log I_{35}(t)$ and $\gamma(t)$, a plot of one parameter as a function of the other, eliminating the time dependence, shows the relationship between them. Figure 17 shows plots of γ vs. I_{35} for 3 events where there are only one or two emission peaks. The points in the longer uninterrupted rise or decay phase during each event are marked by plus symbols. A linear relationship between $\log I_{35}$ and γ can be seen during each phase, although it can be different during rise and decay.

On the other hand, a plot of all the 911 fitted model parameters for all the events show a large scatter, as shown in Fig. 18. The large scatter can be understood as originating from

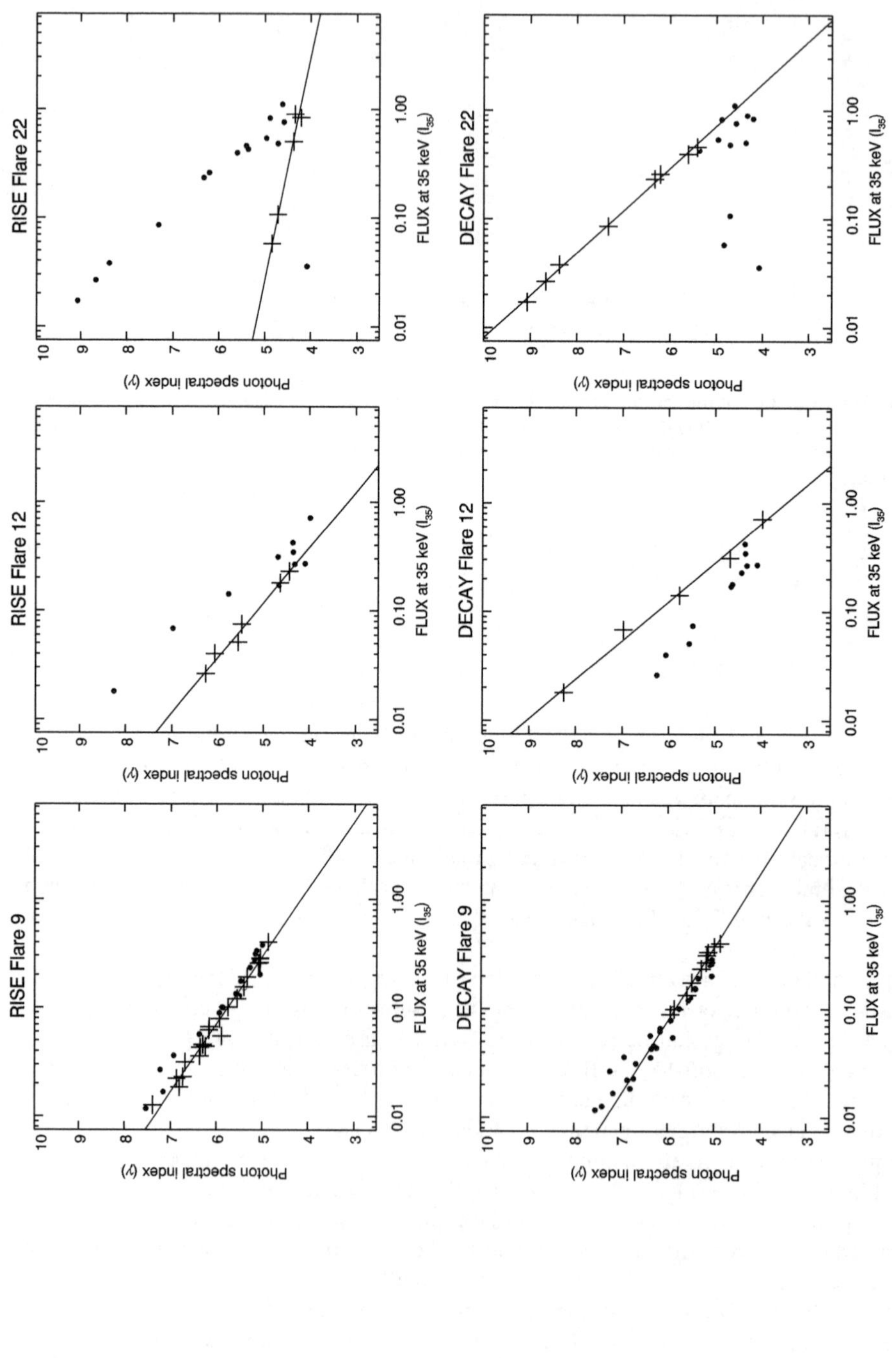

Fig. 17 Spectral index γ vs. flux normalization I_{35} for three events, showing the linear dependence of single rise and decay phases of emission spikes on a log-linear scale. *Dots* mark results from individual spikes, while *pluses* mark the longer rise or decay phase (from Grigis and Benz 2004). Reproduced with permission ©ESO

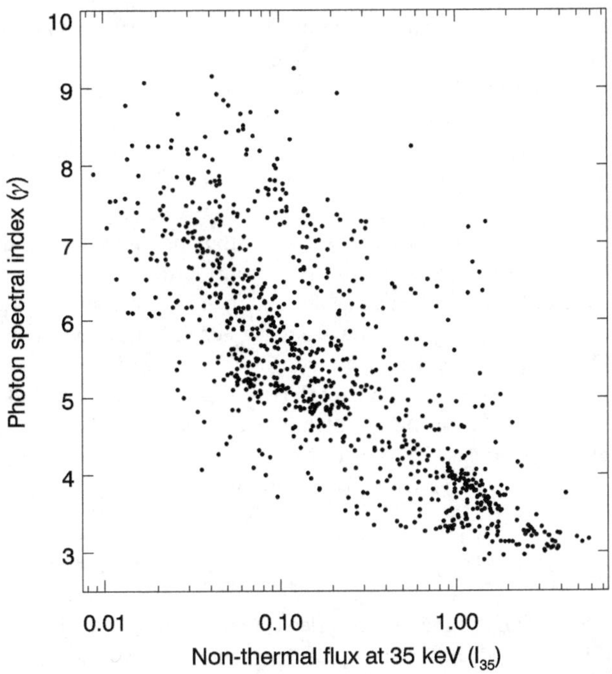

Fig. 18 Plot of the spectral index γ versus the fitted non-thermal flux at 35 keV (given in photons s^{-1} cm^{-2} keV^{-1}). All 911 data points from the 24 events are shown (from Grigis and Benz 2004). Reproduced with permission ©ESO

the superposition of data from a large numbers of different emission spikes, each featuring linear trends with different parameters. This plot does demonstrate, however, the tendency for flatter spectra to be associated with more intense flares.

RHESSI observations of the gradual phase of large solar flares (Grigis and Benz 2008) and its relation with proton events (Saldanha et al. 2008; Grayson et al. 2009) have shown that the hardening behavior is complex and cannot be characterized by a continuously increasing hardness during the event. Therefore the soft-hard-harder (SHH) denomination does not accurately reflect the observed spectral evolution. Rather, phases of hardening (or even approximatively constant hardness) are often seen in larger events as the flux decays (Kiplinger 1995). The start of the hardening phase can happen near the main peak of the flare, or later. The end of hardening can even be followed by new impulsive SHS peaks. The most recent statistical study of the correlation of SHH behavior with proton events (Grayson et al. 2009) found that in a sample of 37 flares that were magnetically well-connected to Earth, 18 showed SHH behavior and 12 of these produced solar energetic particle (SEP) events. None of the remaining 19 flares that did not show SHH behavior produced SEP events.

9.2 Interpretation of Spectral Evolution

Can we explain the soft-hard-soft spectral behavior theoretically? The problem here is that many effects contribute to the properties of the high-energy electron distribution whose bremsstrahlung hard X-rays are observed by *RHESSI* and similar instruments. We can identify three main, closely related classes of physical processes that affect the distribution of the electrons and the spectrum of the X-ray photons they generate: (1) the *acceleration* of part of the thermal ambient plasma, (2) the *escape* from the acceleration region, and (3) the *transport* to the emitting region. The photon spectrum also depends on the properties of the bremsstrahlung emission mechanism.

Miller et al. (1996) proposed a stochastic acceleration mechanism where electrons are energized by small-amplitude turbulent fast-mode waves, called the transit-time damping model. They showed that their model could successfully account for the observed number and energy of electrons accelerated above 20 keV in subsecond spikes or energy release fragments in impulsive solar flares. However, they made no attempt to explain the observed hard X-ray spectra (which are softer than predicted by the transit-time damping model) and did not consider spectral evolution. Furthermore, this approach does not account for particle escape. Grigis and Benz (2006) extended the model with the addition of a term describing the escape of the particles from the acceleration region, as in the model of Petrosian and Donaghy (1999). To ensure conservation of particles, they also add a source term of cold particles coming into the accelerator (such as can be provided by a return current).

The stochastic nature of this acceleration model implies that the electrons undergo a diffusion process in energy space. Mathematically, the acceleration is described by the following convective-diffusive equation:

$$\frac{\partial f}{\partial t} = \frac{1}{2}\frac{\partial^2}{\partial E^2}[(D_{\mathrm{COLL}} + D_{\mathrm{T}})f] - \frac{\partial}{\partial E}[(A_{\mathrm{COLL}} + A_{\mathrm{T}})f] - S(E)\cdot f + Q(E), \qquad (9.2)$$

where $f(E)$ is the electron density distribution function, D_{T} and A_{T} are, respectively, the diffusion and convection coefficients due to the interactions of the electrons with the accelerating turbulent waves, D_{COLL} and A_{COLL} are, respectively, the diffusion and convection coefficients due to collisions with the ambient plasma, $S(E)$ is the sink (escape) term, and $Q(E)$ is the source (return current) term. The escape term is proportional to $v(E)/\tau$, where $v(E)$ is the electron speed, and τ is the escape time. The escape time can be energy-dependent, but for simplicity it is initially kept constant. The longer the escape time, the better the particles are trapped in the accelerator. The source term is in the form of a Maxwellian distribution of electrons with the same temperature as the ambient plasma.

The coefficients D_{T} and A_{T} are proportional to the dimensionless acceleration parameter

$$I_{\mathrm{ACC}} = \frac{U_{\mathrm{T}}}{U_{\mathrm{B}}}\cdot\frac{c\langle k\rangle}{\Omega_{\mathrm{H}}}, \qquad (9.3)$$

where U_{T} and U_{B} are, respectively, the energy densities of the turbulent waves and of the ambient magnetic field, $\langle k\rangle$ is the average wave vector, and Ω_{H} is the proton gyrofrequency.

Equation (9.2) can be solved numerically until an equilibrium state ($\partial f/\partial t = 0$) is reached. The equilibrium electron spectra from the model are controlled by two parameters: the acceleration parameter I_{ACC} described above and the escape time τ. Above 10–20 keV, the collision and source terms in (9.2) can be neglected, since they apply to the ambient Maxwellian, and thus the equilibrium spectra depend to a first approximation only on the product $I_\tau = I_{\mathrm{ACC}}\cdot\tau$.

Figure 19 shows the equilibrium electron spectra for different values of $I_\tau = I_{\mathrm{ACC}}\cdot\tau$. As I_τ increases, the spectrum gets harder and harder. To explain the soft-hard-soft effect, either the acceleration or the trapping efficiency (or both) must increase until the peak time, and then decrease again. We note that this model does not include magnetic trapping (other than in the magnetic turbulence itself), which can alter the computed electron spectra and their time evolution (e.g., Metcalf and Alexander 1999).

To see whether this produces the linear relation between the spectral index and the log of the flux normalization, Grigis and Benz (2006) computed the hard X-ray emission from

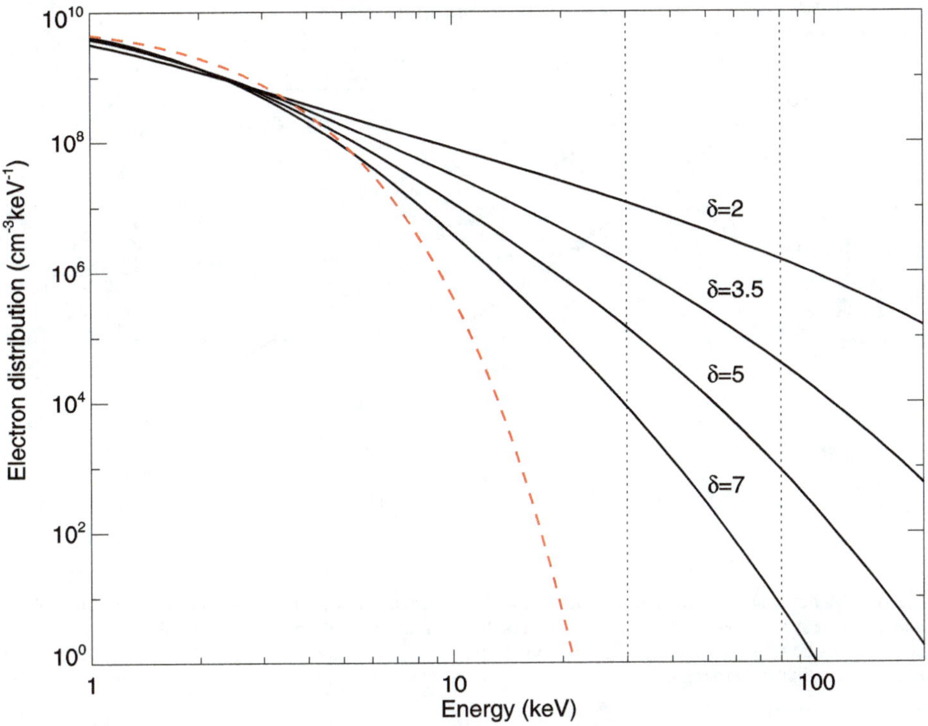

Fig. 19 Accelerated electron density distributions with different values of the power-law index resulting from changes in $I_\tau = I_{\mathrm{ACC}} \cdot \tau$. The *dashed curve* represents the ambient Maxwellian distribution. The two *dotted lines* indicate the energy range used for the computation of the power-law index δ shown above each spectrum. Harder spectra have a larger I_τ value (from Grigis and Benz 2006). Reproduced with permission ©ESO

these model electron spectra. Since these are equilibrium spectra, thin-target emission was computed. They then plotted the spectral index vs. the flux normalization of the resulting photon spectra. Since the spectra are not power-law, but bend down, they fit a power-law model to the model photon spectrum in a similar range as the one used for the observations.

Figure 20 shows the computed values for the spectral indices and flux normalizations for both the electron and the photon spectrum from the model. The results show that there is indeed a linear relation between the spectral index and the log of the flux normalization.

An alternative mechanism that could be responsible for soft-hard-soft spectral evolution is return current losses as the electrons propagate to and within the thick-target footpoints of the flare loop (Zharkova and Gordovskyy 2006). The highest electron energy to which return current losses are significant is proportional to the return current electric field strength, which is in turn proportional to the electron beam flux density (see Sect. 5). Therefore, as the electron flux density increases and then decreases, the low-energy part of the X-ray spectrum flattens to higher and then lower energies as the return current electric field strength increases and then decreases. The net effect is SHS spectral evolution below the maximum energy for which return current losses are significant during the flare. The observation of SHS behavior in coronal X-ray sources, however, indicates that this spectral evolution is a property of the acceleration mechanism rather than a consequence of energy losses during electron propagation (Battaglia and Benz 2006, see Sect. 10.3).

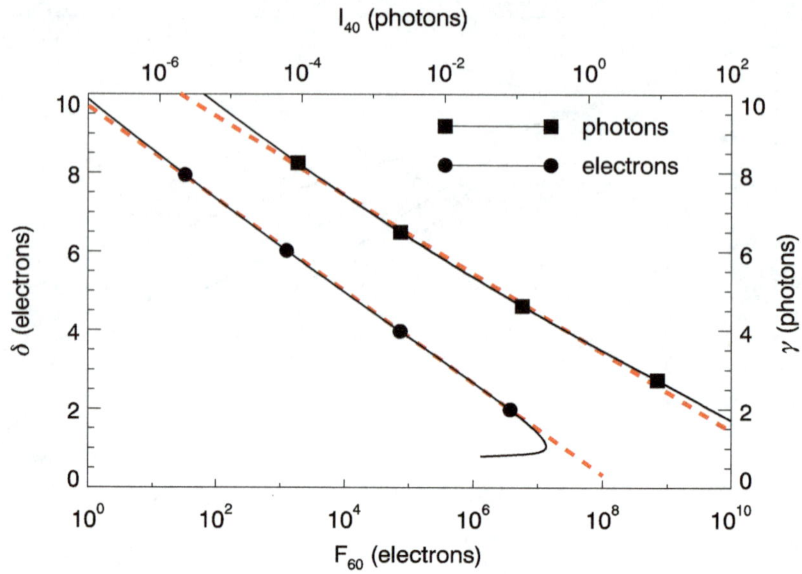

Fig. 20 Model results for the spectral index and flux normalization for electrons and photons. The *dashed line* is the best straight-line fit to the model results (in the range of spectral indices from 2 to 8 for the electrons, and 3 to 9 for the photons), corresponding to a pivot-point behavior (from Grigis and Benz 2006). Reproduced with permission ©ESO

Are there two stages of electron acceleration, one responsible for the impulsive phase and one for the gradual phase? *RHESSI* spectroscopy and imaging of a set of 5 flares with hardening phases (Grigis and Benz 2008) showed that there is no discontinuity in the motion of footpoints at the onset of hardening and no clear separation between the impulsive and the gradual phase: the former seems to smoothly merge into the latter. This supports the view that the same acceleration mechanism changes gradually in the later phase of the flare, rather than a two stage acceleration theory. The hardening phase may in fact be caused by an increase in the efficiency of trapping of the electrons above 100 keV.

The underlying cause of the SHS spectral evolution has been addressed in terms of the stochastic acceleration model by Bykov and Fleishman (2009) and Liu and Fletcher (2009). Bykov and Fleishman consider acceleration in strong, long-wavelength MHD turbulence, taking into account the effect of the accelerated particles on the turbulence. They argue that the electron spectrum flattens during the linear acceleration phase, while the spectrum steepens during the nonlinear phase when damping of the turbulence because of the particle acceleration is important, giving SHS spectral evolution. They argue that SHH evolution will be observed when the injection of particles into the acceleration region is strong. Liu & Fletcher also argue that the SHS evolution results from dependence of the electron distribution power-law index on the level of turbulence as it increases and subsequently decreases. They attribute changes in the SHS correlation during a flare to changes in the background plasma, likely due to chromospheric evaporation.

We note that simple direct-current (DC) electric field acceleration of electrons out of the thermal plasma can produce the SHS spectral evolution. The flux of accelerated electrons and the maximum energy to which electrons are accelerated and, therefore, the high-energy cutoff to the electron distribution, increase and decrease together as the electric field strength increases and decreases (Holman 1985). The X-ray spectrum is steeper at energies within one to two orders of magnitude below the high-energy cutoff (Holman 2003). In large flares, however, where the X-ray spectrum is observed to continue to MeV energies or higher, there is no evidence for a high-energy cutoff in the appropriate energy range. Therefore, at least for large flares with spectra extending to high energies, a simple DC electric field acceleration model does not appear to be appropriate.

10 The Connection Between Footpoint and Coronal Hard X-ray Sources

Hard X-ray (HXR) sources at both footpoints of a coronal loop structure have been observed since Hoyng et al. (1981). As reviewed in Sects. 1 and 2, they are understood to be thick-target bremsstrahlung emission produced by precipitating electrons, accelerated somewhere in or above the loop. A third HXR source situated above the looptop (see Krucker et al. 2008a, for a review) was first noted by Masuda et al. (1994) in *Yohkoh* observations. The nature of this coronal HXR source has remained uncertain, but in simple solar flare models with reconnection and particle acceleration in the corona, we expect some relation between coronal HXR sources and footpoints. *RHESSI* has enabled us to study events featuring coronal HXR sources and footpoints simultaneously. By studying the behavior of the sources in time and the relations between them, we can address questions such as: Are both coronal and footpoint emissions caused by the same electron population? How is such an electron beam modified in the loop (collisions, return currents, trapping, etc.)? Is SHS behavior (Sect. 9.1) a transport effect produced by collisions or return currents, or is it a feature imposed by the acceleration mechanism?

10.1 *RHESSI* Imaging Spectroscopy

RHESSI has provided the possibility of obtaining simultaneous, high-resolution imaged spectra at different locations on the Sun. One can therefore study each source separately in events with several contemporaneous HXR sources. The high spectral resolution has allowed a reliable differentiation between thermal and non-thermal emission to be made in many flares. Furthermore, *RHESSI*'s imaging spectroscopy has allowed differences in individual flare source spectra and their evolution to be studied in considerable detail.

Imaged spectra and the relative timing of sources in three flares, including the limb flare SOL2002-02-20T11:07, were studied by Krucker and Lin (2002). Sui et al. (2002) analyzed and modeled the two footpoint sources and a high, above-the-looptop hard X-ray source observed in this flare. Emslie et al. (2003) analyzed SOL2002-07-23T00:35 (X4.8) flare with four HXR sources observed by *RHESSI*. They found a coronal source with a strong thermal component, but the non-thermal component could not be studied due to severe pulse pile-up. Battaglia and Benz (2006) studied five M-class events. Due to the smaller pile-up amount in those events, studying the non-thermal coronal emission was possible. The results of these studies are summarized below.

10.2 Relation Between Coronal and Footpoint Sources

The quantitative relations between the footpoints and the coronal source and between the two footpoints can give information about the physical mechanisms at work in a solar flare. Simple models envision a beam of accelerated electrons encountering a low-density region in the corona, leading to thin-target bremsstrahlung. When the same electron beam reaches the chromosphere, the particles are fully stopped in the dense material, producing thick-target emission. Assuming an electron power-law distribution for the electron energy E of the form

$$\mathscr{F}(E) = AE^{-\delta} \tag{10.1}$$

producing thin-target bremsstrahlung in the coronal source, the observed photon spectrum has spectral index $\gamma_{thin} = \delta + 1$ (2.10). Reaching the chromosphere, the accelerated electrons will be fully stopped, producing thick-target bremsstrahlung with a photon spectral index $\gamma_{thick} = \delta - 1$ (2.12). In such a simple scenario one would therefore expect a difference in the photon spectral index $\gamma_{thin} - \gamma_{thick} = 2$ between the coronal source and the footpoints. Further, the two footpoints should be of equal hardness and intensity if one assumes a symmetric loop and symmetric injection of particles into the legs of the loop.

10.2.1 Observed Difference Between Coronal and Footpoint Spectral Indices

A sample of flares observed with *Yohkoh* to have coronal HXR sources was studied by Petrosian et al. (2002). They found that the spectral index of the coronal sources was, on the average, steeper by 1 than the spectral indices of the footpoint sources. Sui et al. (2002) also found a spectral index difference of 1 for SOL2002-02-20T11:07 (C7.5) observed with *RHESSI*.

Battaglia and Benz (2006) found that the coronal source was softer than both footpoints for all of their five events in nearly all analyzed time bins. Figure 21 (top left) shows an image of SOL2005-07-13T14:49 (M5.0) in the 34–38 keV energy band. The two footpoints are visible, as well as the 50 and 80% contours of the coronal source taken from a 10–12 keV image. Spectra and spectral fits are shown for the two footpoints and the coronal source. The steepness of the coronal source spectrum (number 3 in the figure) relative to the spectra from the footpoints is apparent. However, the quantitative difference between the values of the spectral index obtained for the coronal source and the footpoints often differed significantly from 2. For the five flares analyzed, the smallest mean difference in the spectral indices, averaged over time, was 0.59±0.24. The maximum mean difference, averaged over time, was 3.68±0.14. These clearly contradict the theoretical expectation summarized above. Simple thin-thick target scenarios do not seem to work in most cases and additional effects need to be considered.

Evidence for two populations of coronal source non-thermal spectra was found by Shao and Huang (2009a). They compare coronal and footpoint spectral indices at 28 hard X-ray peaks from 13 single-loop flares observed by *RHESSI*. The spectral index in the coronal sources was determined from an isothermal plus power-law fit below 30 keV, while the footpoint spectral indices were determined from a power-law fit at 30–60 keV photon energies. They argue that the coronal spectra can be divided into two groups. One, for which the coronal spectral index is greater than 5, is well correlated with the footpoint spectral index, and the difference in the indices ranges from 2–4. For the other, where the spectral indices are anticorrelated, the coronal spectral index is less than 5, and the difference in the indices ranges from 0–2. For the group of anticorrelated spectral indices, the coronal spectral index is correlated with the photon flux, while the footpoint spectral index is anticorrelated with the photon flux for both groups. These are intriguing results if confirmed by future studies.

10.2.2 Differences Between Footpoints

No significant difference was found in the spectral indices for the two footpoints in SOL2002-02-20T11:49 (C7.5) by Krucker and Lin (2002) and Sui et al. (2002). Piana et

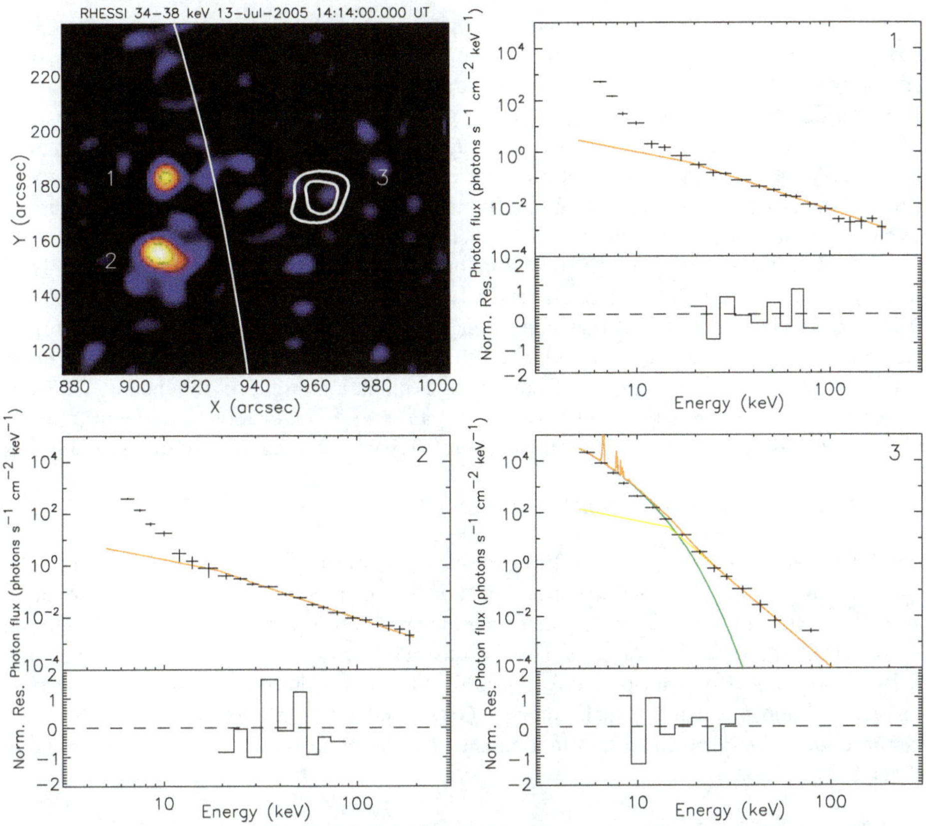

Fig. 21 *Top left*: Composite CLEAN image of a *RHESSI* event with three hard X-ray sources. The footpoints (labeled 1 & 2) are visible on the solar disc in an image made at 34–38 keV. The position of the coronal source (labeled 3) high above the limb is indicated by the 50 and 80% white contours taken from a 10–12 keV image. *Plots 1–3* show spectra and normalized residuals over the fitted energy range for the north footpoint (1), south footpoint (2), and coronal source (3) (after Battaglia and Benz 2006). Reproduced with permission ©ESO

al. (2007) inverted count visibility spectra for this flare to obtain mean electron flux distributions for the footpoints. They found the mean electron flux distribution function at the northern footpoint to be somewhat steeper ($\Delta\delta \approx 0.8$) than that derived for the southern footpoint. They also found the distribution function for the region between the footpoints (not the coronal source studied by Sui et al.) to be steeper than the footpoint distribution functions ($\Delta\delta \approx 1.6$ relative to the southern footpoint) and to substantially steepen at energies above ~60 keV.

Krucker and Lin (2002) found that, when a connection between footpoints could be determined, the footpoints brightened simultaneously (to within the ~1 s time resolution of the observations) and had similar spectra.

Differences of 0.3–0.4 between the spectral indices of two footpoints in SOL2002-07-23T00:35 (X4.8) were reported by Emslie et al. (2003).

For the flares analyzed by Battaglia and Benz (2006), a significant difference was found in only one out of five events. For all other events, the mean difference in $\gamma_{\rm fp}$ was zero within the statistical uncertainty.

Different spectra at the two footpoints imply an asymmetric loop. Such an asymmetry can result, for example, from different column densities or different beam fluxes and corresponding return current energy losses in the legs of the loop. It could also result from asymmetric magnetic trapping within the loop (e.g., Alexander and Metcalf 2002). In a study of 53 flares showing two HXR footpoints, Saint-Hilaire et al. (2008) found that footpoint asymmetry was greatest at the time of peak HXR flux and the difference in the footpoint spectral indices $\Delta\gamma$ rarely exceeded 0.6. In most cases they found the footpoint asymmetry to be inconsistent with different column densities in the two legs of the loops.

In SOL2003-10-29T20:49 (X10.0) Liu et al. (2009a) found that the brighter HXR footpoint was marginally, but consistently harder than the dimmer footpoint by $\Delta\gamma = 0.15 \pm 0.13$. They concluded that neither asymmetric magnetic mirroring nor asymmetric column density *alone* can explain the full time evolution of the footpoint HXR fluxes and spectral indices. However, a self-consistent explanation might be obtained by considering these two effects together and/or in combination with one or more additional transport effects, such as nonuniform target ionization, relativistic beaming, and return current losses.

10.3 Spectral Evolution in Coronal Sources

Previous observations of SHS spectral evolution (see Sect. 9.1) were made with full-Sun spectra which, except for over-the-limb events, are typically dominated by footpoint emission. Battaglia and Benz (2006), in their imaging spectroscopy study, found that the coronal source itself shows SHS evolution. This is illustrated in Fig. 22. This finding implies that SHS is not caused by transport effects within the flare loop, but is rather a property of the acceleration mechanism itself. Indeed, Grigis and Benz (2006) showed that SHS can be reproduced for electron spectra in a transit-time-damping, stochastic-acceleration model (Sect. 9.2).

10.4 Interpretation of the Connection Between Footpoints and The Coronal Source

In the above account, emphasis was given to the difference in the spectral index between the coronal source and footpoints. Assuming a thin target in the corona and a thick target at the footpoints, one would expect a difference of two. However, whether the coronal source acts as thin- or thick-target depends on the energy of the accelerated electrons and the column density in the corona. Veronig and Brown (2004), for example, found coronal sources with column densities high enough to act as thick targets for electrons with energies up to 60 keV.

As early as 1976, Melrose and Brown (1976) showed that magnetic trapping with collisional scattering of electrons out of the trap can lead to a thick-target coronal source. The coronal source transitions through a thin-thick period, with the time scale for this transition depending on the electron energy and the plasma density in the trap. The trapping essentially increases the effective column density in the corona. Metcalf and Alexander (1999) analyzed six flares with coronal sources observed by *Yohkoh* and found that three of the six flares showed properties consistent with trapping.

A simple 1-D model that described the coronal emission as intermediate thin-thick, depending on electron energy, was developed by Wheatland and Melrose (1995). In this model a high-density region ($\gtrsim 10^{12}$ cm^{-3}) is hypothesized to be present at or above the top of the flare loop. The model makes predictions for the shape of the coronal and footpoint spectra and the relations between them. Fletcher (1995) obtained Monte Carlo solutions to the Fokker-Planck equation to show that, with the inclusion of high electron pitch angles and collisional scattering, a compact coronal X-ray source is produced at the top of a loop with a constant coronal density $\sim 3 \times 10^{10}$ cm^{-3}. Holman (1996)

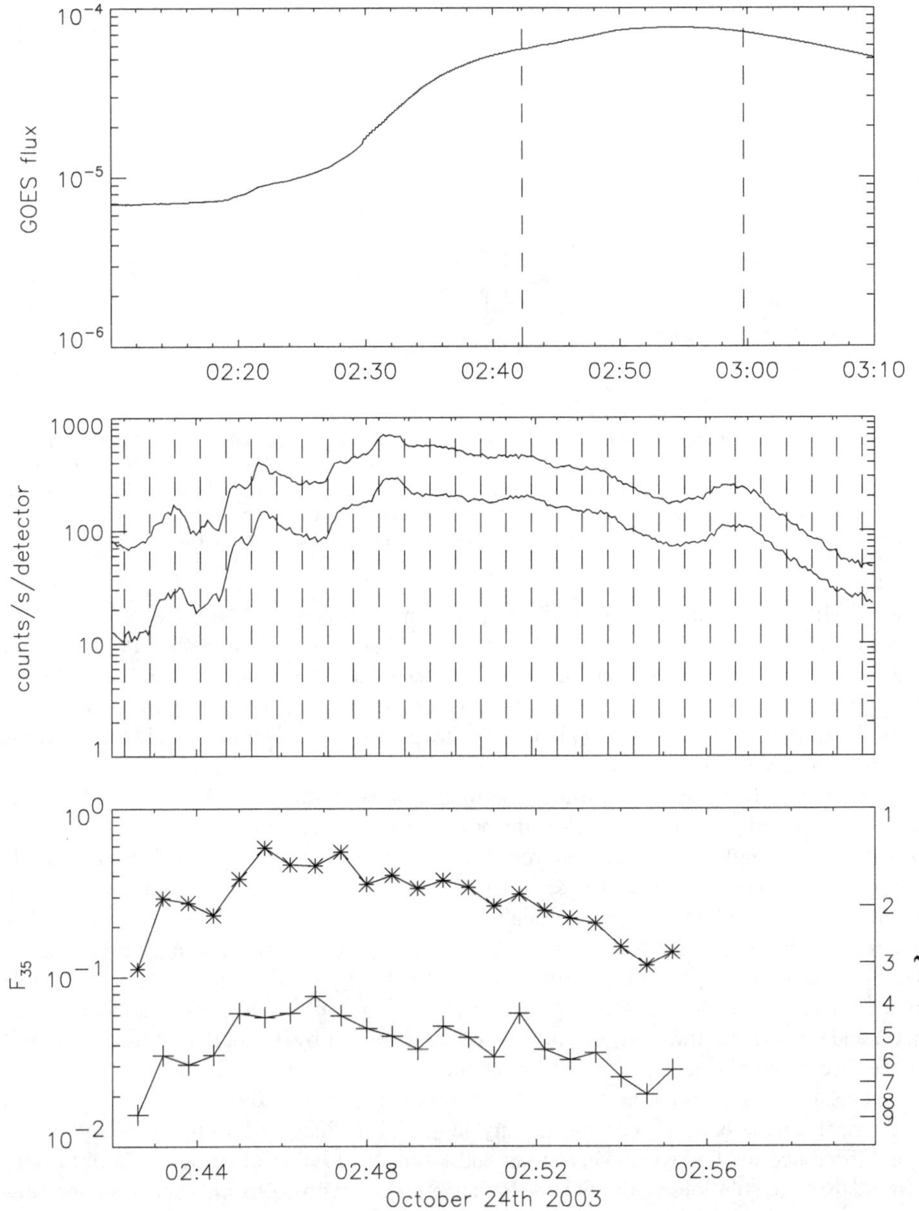

Fig. 22 *Top: GOES* 1–8 Å light curve of SOL2003-10-24T02:54 (M7.6). *Middle: RHESSI* 25–50 and 50–100 keV light curves near the peak of the *GOES* flare. *Bottom:* time evolution of fitted coronal source flux at 35 keV (F_{35}, * symbols, left log scale) and spectral index (γ, + symbols, right log scale) displaying SHS evolution (after Battaglia and Benz 2006). Reproduced with permission ©ESO

showed that, even in the simple 1-D model, a compact coronal source is produced when electrons are injected into a loop with a constant coronal density $\sim 2 \times 10^{11}$ cm^{-3} (see hesperia.gsfc.nasa.gov/sftheory/loop.htm). A compact coronal HXR source can also be produced if there is a compact magnetic trap at or above the top of the loop. Fletcher and

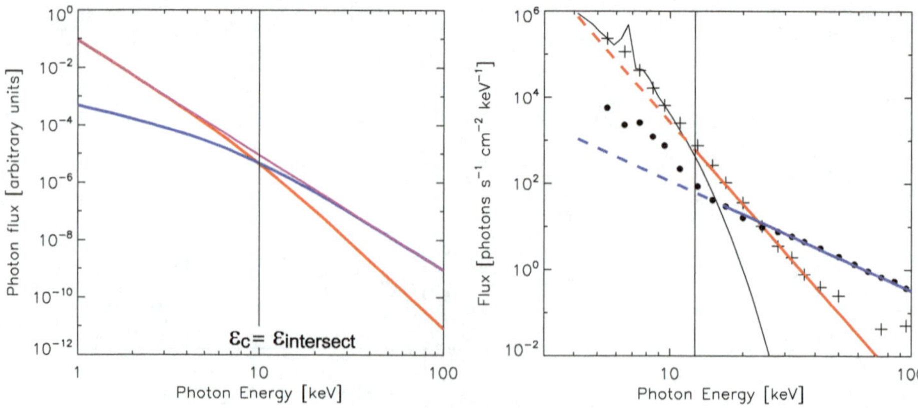

Fig. 23 *Left:* spectra for coronal source (*red*) and footpoints (*blue*) according to the model of Wheatland and Melrose (1995). The spatially integrated spectrum is shown in *violet*. *Right:* observed *RHESSI* spectra for the event SOL2003-10-24T02:54 (M7.6). Isothermal and power-law fits to the coronal (*crosses*) and footpoint (*dots*) spectra are shown. The *vertical line* indicates the predicted critical energy for the transition between thin and thick target (after Battaglia and Benz 2007). Reproduced with permission ©ESO

Martens (1998) showed that, with such a trap, a significant coronal X-ray source can be produced at plasma densities as low as $\sim 4 \times 10^9$ cm^{-3}. Petrosian and Donaghy (1999) showed that the coronal HXR source can be a consequence of acceleration and trapping by turbulence or plasma waves. In their stochastic acceleration model, the difference between the coronal and footpoint spectra is explained by the energy-dependent time scale for electrons to escape the acceleration region.

The left panel of Fig. 23 illustrates the model of Wheatland and Melrose (1995). The spatially integrated spectrum (violet) is the power-law spectrum (thick-target, $\gamma_{thick} = \delta - 1$) expected for a single-power-law electron distribution with no low- or high-energy cutoffs and no thermal component. For $\varepsilon \ll \varepsilon_c = \sqrt{2KN}$ (see (2.4)), the spectrum is dominated by thick-target radiation from the coronal source (red). There is a low-energy cutoff in the electron distribution at the footpoints at $E \approx \sqrt{2KN}$ because of the energy losses in the coronal source. The spectrum is dominated by thick-target radiation from the footpoints (blue) where $\varepsilon \gg \varepsilon_c$. It is in this regime that the radiation from the coronal source is thin-target and the spectral index of the coronal source is steeper by 2 than that of the footpoints. These spectra are characteristic of all the models reviewed above.

Sui et al. (2002) compared the *RHESSI* observations of SOL2002-02-20T11:07 (C7.5) to a model with a constant-coronal-density loop and no magnetic trapping. They used a finite difference method (e.g., McTiernan and Petrosian 1990; Holman et al. 2002) to obtain steady-state solutions to the Fokker-Planck equation with collisional scattering and energy losses. Model images were convolved with the *RHESSI* response to produce simulated *RHESSI* observations for direct comparison with the SOL2002-02-20T11:07 flare images and imaged spectra. They found that, after obtaining a power-law model spectrum with an index of $\gamma = 3$ that agreed with the observed footpoint spectra, the effective spectral index of the coronal source from the model ($\gamma = 4.7$) was significantly steeper than that obtained for the flare ($\gamma = 4$).

Battaglia and Benz (2007) compared the model of Wheatland and Melrose (1995) to the results of their study of five flares observed by *RHESSI*. The right panel of Fig. 23 shows observed spectra and spectral fits for one particular event. The observed spectra were dominated by thermal coronal emission at low energies. Therefore, not all of the model

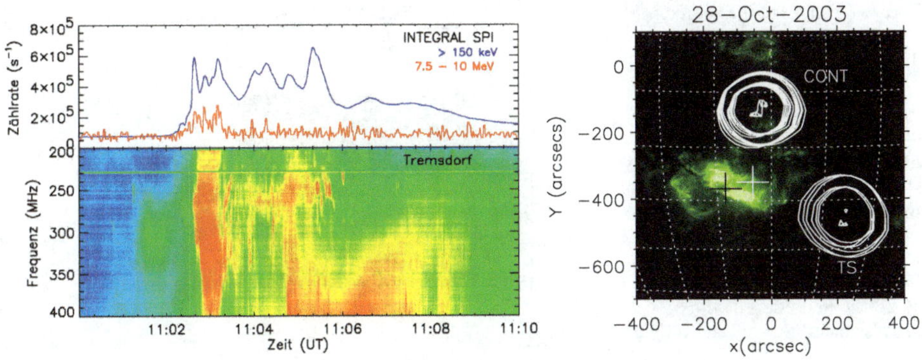

Fig. 24 SOL2003-10-28T11:10 (X17.2). *Left, bottom*: (see Warmuth et al. 2007) 200–400 MHz radio spectrum (Astrophysical Institute Potsdam) showing the signature of the outflow termination shock (TS, starting at 11:02:47 UT). *Left, top*: *INTEGRAL* count rates at 150 keV and 7.5–10 MeV. *Right*: (after Aurass et al. 2007, reproduced with permission ©ESO.) radio source positions (Nançay Radio Heliograph, 327 MHz) overlaid on a *SOHO*-EIT image (11:47 UT 195 Å). The bright areas are EUV flare ribbons in AR10486. *RHESSI* HXR centroids are shown as "+". The integration time intervals are: for the TS source SW of AR10486 11:02:45–11:03:15 UT, for the continuum source CONT N of AR10486 11:13–11:17 UT, respectively (see also Fig. 25). The radio contours are at 50, 70, and 99.5% of the peak flux value

predictions could be tested. However, the observed relations between the spectra did not agree with the predictions of the model. For the flare in Fig. 23, for example, the difference between the coronal source and footpoint spectral indices at the higher photon energies is 3.8 ± 0.1, not 2. Also, an estimate of the column density in the coronal source gives $\sqrt{2KN} \approx$10–15 keV, while the intersection of the coronal and footpoint spectra is found to be at $\varepsilon \approx 23$ keV. Battaglia and Benz (2008) have found that this large difference in the spectral indices is consistent with spectral hardening caused by return current losses (see Sect. 5).

11 Identification of Electron Acceleration Sites from Radio Observations

While energetic electrons excite hard X-ray emission during their precipitation into the dense layers of the solar atmosphere, they can also excite decimeter and meter wave radio emission during propagation and trapping in magnetic field structures in the dilute solar corona. The radio emission pattern in dynamic spectrograms can give information about the electron acceleration process, the locations of injection of electrons in the corona, and the properties of the coronal magnetoplasma structures.

Here we take as an example SOL2003-10-28T11:10 (X17.2) (shown in Fig. 24). Different acceleration sites can be discriminated during the impulsive and the gradual flare phases. Radio spectral data from the Astrophysical Institute Potsdam (AIP; Mann et al. 1992), imaging data from the Nançay Radio Heliograph (NRH, Kerdraon and Delouis 1997), and hard X-ray (*RHESSI*, *INTEGRAL*) data were combined in the analysis of this event. The conclusion was reached that a nondrifting, high-frequency type II radio burst signature in the radio spectrum coincided with a powerful electron acceleration stage. Simultaneously with the nondrifting type II signature, highly relativistic ($\geq$10 MeV) electrons were observed in the impulsive phase of the flare (Fig. 24, upper left). The radio spectrum suggests that this can be due to acceleration at the reconnection outflow termination shock (Aurass and Mann 2004), as predicted for a classical two-ribbon flare (Forbes 1986; Tsuneta and Naito 1998;

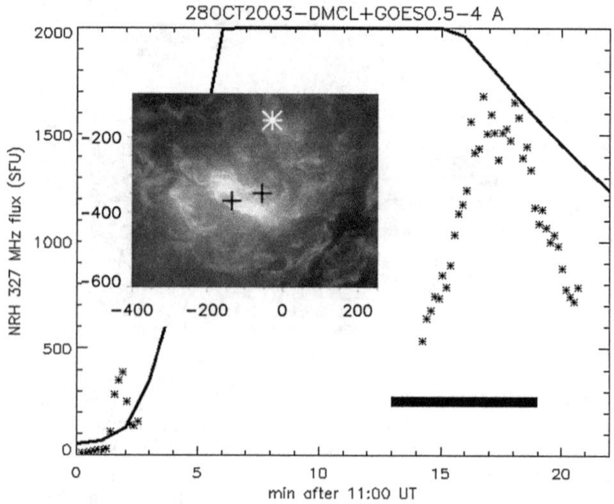

Fig. 25 Timing of the source CONT in Fig. 24: the NRH 327 MHz flux curve (*asterisks*) versus the *GOES* 0.5–4 Å flux curve (*solid line*, partly off-scale). *Inset*: *SOHO*/EIT image showing the radio source centroid (*white asterisk*) and *RHESSI* HXR centroids as in Fig. 24. *Thick bar*: the start time of GeV-energy proton injection in space (after Aurass et al. 2007). Reproduced with permission ©ESO

Aurass et al. 2002). The radio source site is observed about 210 Mm to the SW of the flaring active region (TS in Fig. 24, right). In this direction, *TRACE* and *SOHO*/LASCO[1] C2 images reveal dynamically evolving magnetoplasma structures in an erupting arcade (Aurass et al. 2006). For realistic parameters derived from these observations (the geometry, density, temperature, and low magnetic field values of ∼5 Gauss), Mann et al. (2006) have found that a fully relativistic treatment of acceleration at the fast-mode outflow shock can explain the observed fluxes of energetic particles (see Zharkova et al. 2011).

In the main flare phase of the same event, an additional radio source (CONT in Fig. 24) was found, lasting for ∼10 min, indicating the presence of another acceleration site. No X-ray, EUV, or Hα emission was observed at the location of this radio source. Figure 25 gives the timing and the source position with respect to the flaring active region. CONT is a m-dm-continuum source with fiber burst fine structure. Fiber bursts are excited by whistler waves propagating along field lines of the coronal magnetic field. As marked by a bold bar in the figure, the time of the CONT emission is also the start time of GeV proton injection in space. Aurass et al. (2006) have shown that this source site is not far from an open field (particle escape) region in the potential coronal magnetic field. The source briefly flashes up already in the early impulsive phase. Based on a new method of fiber burst analysis (Aurass et al. 2005; Rausche et al. 2007), Aurass et al. (2007) argue that this source most likely indicates acceleration at a contact between separatrix surfaces of different magnetic flux systems.

Radio observations of flares and their implications are further addressed in White et al. (2011).

12 Discussion and Conclusions

12.1 Implications of X-ray Observations for the Collisional Thick-Target Model

As discussed in Sect. 2, the core assumption of the collisional thick-target model is that the spatially integrated hard X-ray emission from non-thermal electrons is bremsstrahlung

[1] Large Angle and Spectrometric COronagraph.

(free-free radiation) from electrons that lose all their suprathermal energy through collisional losses in the ambient plasma as they simultaneously radiate the hard X-rays. "Simultaneously" means within the observational integration time. This implies that all electrons that contribute significantly to the observed radiation reach a plasma dense enough or, more precisely, traverse a high enough column density for all of their suprathermal energy above the observed photon energies to be collisionally lost to the ambient plasma within the integration time. For typical $\gtrsim 1$ s integration times, these conditions are met when the electrons stream downward from the corona into the increasingly dense plasma of the solar transition region and chromosphere.

Since the thick-target model is often implicit in our interpretation of the hard X-ray emission from flares, it is important to keep the underlying assumptions in mind and test the model while at the same time applying it to flare observations. We have discussed above several physical processes that, if significant, change the conclusions of the simple collisional thick-target model regarding the electron distribution produced in the acceleration region. These processes occur in either the thick-target region itself, or during the propagation of the electrons from the acceleration region to the thick-target region. Only with the high spectral resolution and imaging of *RHESSI* has it become possible to observationally address these processes. Even with the *RHESSI* observations, however, it is difficult to conclusively determine the importance of each process.

A physical process that distorts the emitted X-ray spectrum is albedo (Sect. 3.4 and Kontar et al. 2011). Fortunately, the albedo contribution to the X-ray spectrum can be easily corrected on the assumption that the X-ray photons are isotropically emitted. This correction is available in the *RHESSI* spectral analysis software. If the photons are significantly beamed downward, however, the distortion of the spectrum can be substantially greater than that from isotropically emitted photons. An anisotropic photon distribution results from emitting electrons with an anisotropic pitch-angle distribution. The degree of anisotropy of the electron pitch-angle distribution also quantitatively affects conclusions from the thick-target model concerning the acceleration process. Therefore, it is important to better determine the pitch-angle distribution of the emitting electrons and the contribution of albedo to the hard X-ray spectrum (see Kontar et al. 2011).

The simple collisional thick-target model assumes that the target plasma is fully ionized. We have seen, however, that a nonuniformly ionized target region can produce an upward kink, or "chicane," in an otherwise power-law X-ray spectrum (Sect. 4). This spectral shift can provide a valuable diagnostic of the ionization state of the target plasma and its evolution. It is likely, however, that the power-law spectrum below the chicane is hidden by thermal radiation. The chicane is then observed only as a downward break in the spectrum at energies above those dominated by the thermal emission. The upper limit on the magnitude of the break provides a method for ruling out nonuniform ionization as the sole cause of large spectral breaks. To further distinguish this break from spectral breaks with other causes, it is important to better determine the degree of ionization as a function of column density at the thick-target footpoints.

Return-current energy losses can also produce a downward break in the X-ray spectrum (Sect. 5). The break energy depends on both the thermal structure of the plasma in the flare loop and on the non-thermal electron flux density distribution. These spectral modifications and their evolution throughout flares provide an important test for the presence of initially un-neutralized electron beams and the return currents they must drive to neutralize them. Although *RHESSI* observations provide substantial information about the structure and evolution of flare spectra, only a lower limit on the electron flux density can usually be determined. Observations and analysis sufficiently accurate and comprehensive to verify the presence of return current energy losses as the cause of a spectral break are yet to be obtained. On the other hand, significant evidence exists (Sect. 5.3) indicating that return current losses do have an impact on flare hard X-ray emission.

A thorough comparison of measured flare spectra with theoretical spectra computed from models incorporating collisional and return current energy losses (including their effect on the angular distribution of the non-thermal electrons), as well as nonuniform target ionization and albedo, is still needed. Spectral fitting alone, however, is not likely to distinguish the importance of these different mechanisms. Comparison of the time evolution of the spectra, as well as of the spatial structure of the X-ray emission, with expectations would certainly enhance the possibility of success for such an endeavor.

The analysis of the evolution of X-ray source positions and sizes with photon energy and time provides another important test of the collisional thick-target model (Sect. 7). For these flares that show non-thermal source evolution in the corona and upper transition region, the source position and size are sensitive to the energy losses experienced by the non-thermal electrons. They are, in fact, sensitive to the very assumption that the sources are produced by electrons as they stream downward from an acceleration region higher in the corona. Further studies of the evolution of these coronal X-ray sources should substantially clarify the applicability of the collisional thick-target model.

For completeness, we note that under some circumstances other radiation mechanisms may significantly contribute to the X-ray emission from non-thermal electrons. The possibility that recombination (free-bound) radiation from the non-thermal electrons is sometimes important is discussed in Brown and Mallik (2008, 2009) (also see Kontar et al. 2011). However, the contribution of non-thermal free-bound radiation has recently been found to be less significant than originally estimated (Brown et al. 2010). MacKinnon and Mallik (2010) have concluded that inverse Compton radiation may significantly contribute to the X-ray/γ-ray emission from low-density coronal sources.

Another testable aspect of the collisional thick-target model is the heating of the flare plasma by the non-thermal electrons. If the flare plasma is primarily heated by these electrons and the thick-target region is primarily in the chromosphere and lower transition region, heating originating in the footpoints and expanding into the rest of the flare loop through "chromospheric evaporation" should be observed. On the other hand, if the loop is dense enough for the thick-target region to extend into the corona or if return-current heating is important, localized coronal heating and different ion abundances should be observed.

It has generally been difficult to establish a clear connection between the location and evolution of X-ray sources produced by non-thermal electrons and by thermal plasma at different temperatures. This is largely because of a lack of high-cadence EUV images and spectra covering a broad range of coronal and transition region temperatures prior to the launch of *SDO*. Future studies of the coevolution of non-thermal X-ray sources and thermal sources in flares will be important in determining the extent to which heating mechanisms other than collisional heating by non-thermal electrons is significant.

Predicting the expected evolution of the heated plasma is hampered by insufficient knowledge of the dominant heat transport mechanisms. We have seen evidence that many flares cool by classical thermal conduction or radiation once the heating has subsided (Sect. 8.4), but this is not likely to be the dominant transport mechanism during rapid heating. Nevertheless, the spatial evolution of flare X-ray sources has so far been found to be consistent with chromospheric evaporation (Sect. 7.2). Also, the Neupert effect, observed in most flares, and Doppler-shift measurements qualitatively support the thick-target model (Sect. 8.3), but these do not rule out the possibility of other heating mechanisms temporally correlated with the electron beam collisional heating. As discussed in Sect. 3, substantial progress has been made in deducing the energy flux (total power) carried by non-thermal electrons, but we usually can deduce only a lower limit to this energy flux. Continuing studies of flares similar to SOL2002-04-15T03:55 (M1.2) and the initially cooler, early-impulsive flares (Sect. 3.5) may provide a better handle on this energy flux for comparison with thermal evolution. The thermal properties, energetics, and evolution of flares are discussed further in Fletcher et al. (2011).

12.2 Implications of X-ray Observations for Electron Acceleration Mechanisms and Flare Models

In Sect. 10 we addressed the X-ray spectra of hard X-ray sources sometimes observed above the top of the hot loops or arcades of loops observed in flares. We reviewed results indicating that the spectra are qualitatively, but not quantitatively consistent with expectations for electrons passing through a thin-target or quasi-thick-target region on their way to the thick-target footpoints of the flare loops. The apparent failure of these relatively simple models is probably a manifestation of the more complex above-the-looptop X-ray source structure revealed by *RHESSI* observations.

Before *RHESSI*, time-of-flight delays in hard X-ray timing indicated that electrons were accelerated in a region somewhat above the looptops of the hot flare loops in most flares (Sect. 8.1). Also, cusps were observed at the top of flare loops by *Yohkoh* (e.g., Sect. 10), indicating a magnetic connection to the region above the hot loops.

RHESSI images have revealed flares with double coronal sources, one at or just above the top of the hot loops and the other at a higher altitude above the lower source. The centroid of the lower source is higher in altitude at higher X-ray energies, while the centroid of the upper source is lower in altitude at higher X-ray energies, indicating that energy release occurred between these coronal sources (Sui and Holman 2003; Sui et al. 2004; Liu et al. 2008, 2009b). In one flare, the upper source accelerated outward to the speed of a subsequent coronal mass ejection. The white-light coronagraph on the *Solar Maximum Mission* (Webb et al. 2003), the Large Angle and Spectrometric Coronagraph (LASCO) and the Ultraviolet Coronagraph Spectrometer (UVCS) on *SOHO* (Ko et al. 2003; Lin et al. 2005), *RHESSI* (Sui et al. 2005b), and *TRACE* (Sui et al. 2006a) have all provided direct evidence for the presence of an extended, vertical current sheet above the hot flare loops and below the coronal mass ejection associated with eruptive flares. These and related observations are discussed further in Fletcher et al. (2011).

These recent observations strongly support the "standard" model of eruptive solar flares, in which the hot flare loops build up below a vertical current sheet where inflowing magnetic fields reconnect and a magnetic flux rope forms above the current sheet to become a coronal mass ejection (see Fletcher et al. 2011; Zharkova et al. 2011). The rate of electron acceleration has been observed to be correlated with the rate at which magnetic flux is swept up by the expanding footpoints of flare loops and with the rate of looptop expansion (Qiu et al. 2004; Sui et al. 2004; Holman 2005), indicating that the electron acceleration rate is correlated with the rate of magnetic reconnection. On the other hand, the observations also indicate that the rate of electron acceleration in the impulsive phase of flares is greatest *before* a large-scale current sheet or soft X-ray cusp is observed (Sui et al. 2008).

Initially, when the electron acceleration rate is highest, the current sheet may be short and associated with slow-mode shock waves, as in Petschek reconnection. Fast reconnection jets (e.g., Wang et al. 2007) can stream upward and downward from the current sheet, possibly ending in fast-mode shock waves where they collide with slower magnetized plasma at the flare loop tops and the lower boundary of the magnetic flux rope (termination shocks). The pair of above-the-looptop X-ray sources may be associated with these fast-mode shock waves. We have described possible evidence for these shock waves from radio observations in Sect. 11.

The most difficult task is determining the dominant acceleration mechanism or mechanisms responsible for the energetic particles. The region above the flare loops contains or can contain quasi-DC electric fields, plasma turbulence, slow- and fast-mode shock waves, and collapsing magnetic traps, allowing for almost any acceleration mechanism imaginable. The problem is as much one of ruling out mechanisms as of finding mechanisms that work (cf. Miller et al. 1997). Acceleration mechanisms are addressed in Zharkova et al. (2011).

In Sect. 9 we addressed the soft-hard-soft evolution of flare X-ray spectra. This spectral evolution could occur during the propagation of the electrons from the acceleration region to the thick-target footpoints. Return current losses, with their dependence on the electron beam flux (Sect. 5), for example, could be responsible for this evolution. However, the observation that above-the-looptop sources also show this spectral evolution (Sect. 10.3) indicates that it is a property of the acceleration process rather than electron beam propagation. We saw in Sect. 9.2 that the soft-hard-soft behavior can be reproduced in the acceleration region if the acceleration or trapping efficiency first increases and then decreases.

Flares displaying soft-hard-harder spectral evolution are of special interest, because they have been shown to be associated with high-energy proton events in space (Kiplinger 1995; Saldanha et al. 2008; Grayson et al. 2009). What is the connection between the appearance of energetic protons in space and X-ray spectral hardening late in flares? The answer to this question is important to both space weather prediction and understanding particle acceleration in flares.

12.3 Implications of Current Results for Future Flare Studies in Hard X-rays

What characteristics should a next-generation hard X-ray telescope have to make substantial progress in understanding electron propagation and acceleration in flares? The advances made with *RHESSI* have depended on its high-resolution count spectra that could generally be convolved with the detector response to obtain reliable photon flux spectra. These have been the first observations to allow detailed information about the evolution of accelerated electrons and associated hot flare plasma to be deduced for many flares. Equally important has been the ability to produce hard X-ray images in energy bands determined by the user during the data analysis process. This imaging capability has been critical to determining the origin of the X-ray emission at a given photon energy and in obtaining spectra for individual imaged source regions. These high-resolution imaging spectroscopy capabilities will remain important for continued progress.

RHESSI's X-ray imaging capability has allowed a clear spatial separation to be made for many flares between footpoint sources with non-thermal spectra at higher energies and looptop sources with thermal spectra at lower energies. However, in the energy range of overlap between ~10 keV and 50 keV, where both types of sources may coexist, it is often difficult to distinguish weaker coronal sources (both thermal and non-thermal) in the presence of the stronger footpoint sources. This is because of the limited dynamic range of <100:1 (and significantly less for weaker events) that is possible in any one image made from *RHESSI* data. This is a consequence of the particular form of the Fourier-transform imaging technique that is used. Thus, in most flares the usually intense footpoints mask the much weaker coronal hard X-ray sources that can sometimes be seen in over-the-limb flares when the footpoints are occulted (e.g., Krucker and Lin 2008). In fact, these coronal hard X-ray sources can extend to high energies (up to ~800 keV, Krucker et al. 2008b) and seem to be non-thermal in origin, thus making them of great interest in locating and understanding the particle acceleration process. It is important to study these non-thermal coronal sources in comparison with the footpoint sources, something that is currently not possible with *RHESSI*'s limited dynamic range except in the few cases with exceptionally strong coronal emission (see Sect. 10). In addition, again because of the *RHESSI* dynamic range and sensitivity limitations, it has not generally been possible to observe the thin-target bremsstrahlung emission from the corona that must be present from the electrons streaming down the legs of magnetic loops and also from electrons streaming out from the Sun and producing type-III bursts (see, however, Krucker et al. 2008c; Saint-Hilaire et al. 2009). For all of these reasons, a significantly greater dynamic range will be an important goal for future advanced solar hard X-ray instruments.

Flares at the solar limb for which the hard X-ray footpoints are occulted by the disk provide an important way of observing coronal hard X-ray sources, but these flares do not allow a comparison to be made between the coronal emission and the thick-target footpoint emission. A possible substitute for a high-dynamic-range instrument is hard X-ray observations from two or more spacecraft. Under the right conditions, one spacecraft can observe all the flare emission while the other observes only the coronal emission, with the footpoint emission occulted by the solar disk. Multi-spacecraft observations would also be important for deducing the directivity of the flare emission (especially in conjunction with X-ray polarization measurements—see Kontar et al. 2011) and 3-D source structure. This multi-spacecraft approach, however, limits the number of flares for which the coronal and footpoint emissions can be compared.

Hard X-ray timing studies have provided valuable information about electron propagation and the location of the acceleration region (Sect. 8). Since the time of flight of energetic electrons from a coronal acceleration region to the thick-target loop footpoint is typically $\sim$10–100 milliseconds, the photon count rate must be high enough to distinguish differences in flux on these time scales. Time-of-flight studies have not been successful with *RHESSI*, because of its relatively low collecting area and, therefore, count rate. An instrument with the collecting area and pulse-pileup avoidance of *CGRO*/BATSE, and the imaging and spectral resolution of *RHESSI*, would provide a new generation of studies on the characteristic time scales of propagation for the hard-X-ray-emitting electrons accelerated in flares. Alternatively, smaller instruments sent closer to the Sun on, for example, *Solar Orbiter* or *Solar Probe* could achieve the required sensitivity. Flare studies on these time scales would provide important insights into the physical processes that impact the acceleration and propagation of energetic electrons in flares.

Acknowledgements We thank the chapter editor, Brian Dennis, and the two reviewers for comments that led to many improvements to the text. G.D.H. acknowledges support from the *RHESSI* Project and NASA's Heliophysics Guest Investigator Program. M.J.A. acknowledges support from NASA contract NAS5-98033 of the *RHESSI* mission through University of California, Berkeley (subcontract SA2241-26308PG), and NASA contract NAS5-38099 for the *TRACE* mission. H.A. acknowledges support by the German Space Agency *Deutsches Zentrum für Luft- und Raumfahrt* (DLR), under grant No. 50 QL 0001. M.B. acknowledges support by the Leverhulme Trust. P.C.G. acknowledges support from NASA contract NNM07AB07C. EPK acknowledges support from a Science and Technology Facilities Council Advanced Fellowship. NASA's Astrophysics Data System Bibliographic Services have been an invaluable tool in the writing of this article.

References

N. Agueda, D. Lario, R. Vainio, B. Sanahuja, E. Kilpua, S. Pohjolainen, Astron. Astrophys. **507**, 981 (2009). doi:10.1051/0004-6361/200912224

D. Alexander, A.G. Daou, Astrophys. J. **666**, 1268 (2007). doi:10.1086/520331
D. Alexander, T.R. Metcalf, Sol. Phys. **210**, 323 (2002). doi:10.1023/A:1022457413628
S.K. Antiochos, P.A. Sturrock, Astrophys. J. **220**, 1137 (1978). doi:10.1086/155999
M.J. Aschwanden, Astron. Astrophys. Suppl. Ser. **85**, 1141 (1990)
M.J. Aschwanden, *Physics of the Solar Corona. An Introduction* (Springer, Berlin, 2004)
M.J. Aschwanden, Astrophys. J. **661**, 1242 (2007). doi:10.1086/516814
M.J. Aschwanden, D. Alexander, Sol. Phys. **204**, 91 (2001). doi:10.1023/A:1014257826116
M.J. Aschwanden, J.C. Brown, E.P. Kontar, Sol. Phys. **210**, 383 (2002). doi:10.1023/A:1022472319619
M.J. Aschwanden, R.M. Bynum, T. Kosugi, H.S. Hudson, R.A. Schwartz, Astrophys. J. **487**, 936 (1997). doi:10.1086/304633
M.J. Aschwanden, T. Kosugi, H.S. Hudson, M.J. Wills, R.A. Schwartz, Astrophys. J. **470**, 1198 (1996). doi:10.1086/177943
M.J. Aschwanden, R.A. Schwartz, Astrophys. J. **455**, 699 (1995). doi:10.1086/176617
M.J. Aschwanden, R.A. Schwartz, D.M. Alt, Astrophys. J. **447**, 923 (1995). doi:10.1086/175930
M.J. Aschwanden, R.A. Stern, M. Güdel, Astrophys. J. **672**, 659 (2008). arXiv:0710.2563, doi:10.1086/523926
H. Aurass, G. Mann, Astrophys. J. **615**, 526 (2004). doi:10.1086/424374
H. Aurass, G. Mann, G. Rausche, A. Warmuth, Astron. Astrophys. **457**, 681 (2006). doi:10.1051/0004-6361:20065238
H. Aurass, G. Rausche, G. Mann, Astron. Astrophys. **471**, L37 (2007). doi:10.1051/0004-6361:20077544
H. Aurass, G. Rausche, G. Mann, A. Hofmann, Astron. Astrophys. **435**, 1137 (2005). doi:10.1051/0004-6361:20042199
H. Aurass, B. Vršnak, G. Mann, Astron. Astrophys. **384**, 273 (2002). doi:10.1051/0004-6361:20011735
T.S. Bastian, A.O. Benz, D.E. Gary, Annu. Rev. Astron. Astrophys. **36**, 131 (1998). doi:10.1146/annurev.astro.36.1.131
M. Battaglia, A.O. Benz, Astron. Astrophys. **456**, 751 (2006). arXiv:astro-ph/0606353, doi:10.1051/0004-6361:20065233
M. Battaglia, A.O. Benz, Astron. Astrophys. **466**, 713 (2007). arXiv:astro-ph/0702309, doi:10.1051/0004-6361:20077144
M. Battaglia, A.O. Benz, Astron. Astrophys. **487**, 337 (2008). arXiv:0806.1701, doi:10.1051/0004-6361:200809418
M. Battaglia, P.C. Grigis, A.O. Benz, Astron. Astrophys. **439**, 737 (2005). arXiv:astro-ph/0505154, doi:10.1051/0004-6361:20053027
S.G. Benka, G.D. Holman, Astrophys. J. **435**, 469 (1994). doi:10.1086/174829
A. Benz (ed.), *Plasma Astrophysics*, 2nd edn. Astrophysics and Space Science Library, vol. 279 (2002)
A.O. Benz, Astrophys. J. **211**, 270 (1977)
J.W. Brosius, G.D. Holman, Astrophys. J. Lett. **659**, L73 (2007). doi:10.1086/516629
J.C. Brown, Sol. Phys. **18**, 489 (1971)
J.C. Brown, Sol. Phys. **28**, 151 (1973)
J.C. Brown, M.J. Aschwanden, E.P. Kontar, Sol. Phys. **210**, 373 (2002). doi:10.1023/A:1022469402781
J.C. Brown, A.G. Emslie, G.D. Holman, C.M. Johns-Krull, E.P. Kontar, R.P. Lin, A.M. Massone, M. Piana, Astrophys. J. **643**, 523 (2006). doi:10.1086/501497
J.C. Brown, A.G. Emslie, E.P. Kontar, Astrophys. J. Lett. **595**, L115 (2003). doi:10.1086/378169
J.C. Brown, J. Kašparová, A.M. Massone, M. Piana, Astron. Astrophys. **486**, 1023 (2008). arXiv:0802.0621, doi:10.1051/0004-6361:200809496
J.C. Brown, J.M. Loran, Mon. Not. R. Astron. Soc. **212**, 245 (1985)
J.C. Brown, P.C.V. Mallik, Astron. Astrophys. **481**, 507 (2008). arXiv:0706.2823, doi:10.1051/0004-6361:20078103
J.C. Brown, P.C.V. Mallik, Astrophys. J. Lett. **697**, L6 (2009). doi:10.1088/0004-637X/697/1/L6
J.C. Brown, P.C.V. Mallik, N.R. Badnell, Astron. Astrophys. **515**, C1+ (2010). doi:10.1051/0004-6361:20078103e
J.C. Brown, D.B. Melrose, D.S. Spicer, Astrophys. J. **228**, 592 (1979). doi:10.1086/156883
A.M. Bykov, G.D. Fleishman, Astrophys. J. Lett. **692**, L45 (2009). arXiv:0901.2677, doi:10.1088/0004-637X/692/1/L45
P.J. Cargill, J.T. Mariska, S.K. Antiochos, Astrophys. J. **439**, 1034 (1995). doi:10.1086/175240
A. Caspi, R.P. Lin, Astrophys. J. Lett. **725**, L161 (2010). doi:10.1088/2041-8205/725/2/L161
E.W. Cliver, B.R. Dennis, A.L. Kiplinger, S.R. Kane, D.F. Neidig, N.R. Sheeley Jr., M.J. Koomen, Astrophys. J. **305**, 920 (1986). doi:10.1086/164306
J.L. Culhane, A.T. Phillips, M. Inda-Koide, T. Kosugi, A. Fludra, H. Kurokawa, K. Makishima, C.D. Pike, T. Sakao, T. Sakurai, Sol. Phys. **153**, 307 (1994)
B.R. Dennis, Sol. Phys. **100**, 465 (1985)

B.R. Dennis, D.M. Zarro, Sol. Phys. **146**, 177 (1993)

S.V. Diakonov, B.V. Somov, Sol. Phys. **116**, 119 (1988)

E. Dzifčáková, M. Karlický, Sol. Phys. **250**, 329 (2008). doi:10.1007/s11207-008-9219-3

A.G. Emslie, Astrophys. J. **224**, 241 (1978). doi:10.1086/156371

A.G. Emslie, Astrophys. J. **235**, 1055 (1980). doi:10.1086/157709

A.G. Emslie, Astrophys. J. Lett. **595**, L119 (2003). doi:10.1086/378168

A.G. Emslie, E.P. Kontar, S. Krucker, R.P. Lin, Astrophys. J. Lett. **595**, L107 (2003). doi:10.1086/378931

A.G. Emslie, H. Kucharek, B.R. Dennis, N. Gopalswamy, G.D. Holman, G.H. Share, A. Vourlidas, T.G. Forbes, P.T. Gallagher, G.M. Mason, T.R. Metcalf, R.A. Mewaldt, R.J. Murphy, R.A. Schwartz, T.H. Zurbuchen, Journal of Geophysical Research (Space). Physics **109**, 10104 (2004). doi:10.1029/2004JA010571

M.W. Ewell Jr., H. Zirin, J.B. Jensen, T.S. Bastian, Astrophys. J. **403**, 426 (1993). doi:10.1086/172213

G.H. Fisher, S.L. Hawley, Astrophys. J. **357**, 243 (1990). doi:10.1086/168911

L. Fletcher, Astron. Astrophys. **303**, L9 (1995)

L. Fletcher, H.S. Hudson, Sol. Phys. **210**, 307 (2002). doi:10.1023/A:1022479610710

L. Fletcher, P.C.H. Martens, Astrophys. J. **505**, 418 (1998). doi:10.1086/306137

L. Fletcher et al., Space Sci. Rev. (2011, this issue). doi:10.1007/s11214-010-9701-8

T.G. Forbes, Astrophys. J. **305**, 553 (1986). doi:10.1086/164268

K.J. Frost, B.R. Dennis, Astrophys. J. **165**, 655 (1971)

W.Q. Gan, Astrophys. Space Sci. **260**, 515 (1998)

W.Q. Gan, Y.P. Li, J. Chang, Astrophys. J. **552**, 858 (2001). doi:10.1086/320574

J.A. Grayson, S. Krucker, R.P. Lin, Astrophys. J. **707**, 1588 (2009). doi:10.1088/0004-637X/707/2/1588

P.C. Grigis, A.O. Benz, Astron. Astrophys. **426**, 1093 (2004). arXiv:astro-ph/0407431, doi:10.1051/0004-6361:20041367

P.C. Grigis, A.O. Benz, Astron. Astrophys. **434**, 1173 (2005). arXiv:astro-ph/0501431, doi:10.1051/0004-6361:20042265

P.C. Grigis, A.O. Benz, Astron. Astrophys. **458**, 641 (2006). arXiv:astro-ph/0606339, doi:10.1051/0004-6361:20065809

P.C. Grigis, A.O. Benz, Astrophys. J. **683**, 1180 (2008). arXiv:0708.2472, doi:10.1086/589826

I.G. Hannah, E.P. Kontar, O.K. Sirenko, Astrophys. J. Lett. **707**, L45 (2009). arXiv:0911.0314, doi:10.1088/0004-637X/707/1/L45

E. Haug, Astron. Astrophys. **326**, 417 (1997)

S. Hayakawa, K. Kitao, Prog. Theor. Phys. **16**, 139 (1956)

J. Hénoux, M. Karlický, Astron. Astrophys. **407**, 1103 (2003). doi:10.1051/0004-6361:20030616

G.D. Holman, Astrophys. J. **293**, 584 (1985). doi:10.1086/163263

G.D. Holman, in *Bulletin of the American Astronomical Society*. Bulletin of the American Astronomical Society, vol. 28 (1996), p. 939

G.D. Holman, Astrophys. J. **586**, 606 (2003). doi:10.1086/367554

G.D. Holman, Adv. Space Res. **35**, 1669 (2005). doi:10.1016/j.asr.2004.11.022

G.D. Holman, D. Eichler, M.R. Kundu, in *Radio Physics of the Sun*, ed. by M.R. Kundu, T.E. Gergely. IAU Symposium, vol. 86 (1980), pp. 457–459

G.D. Holman, M.R. Kundu, K. Papadopoulos, Astrophys. J. **257**, 354 (1982). doi:10.1086/159994

G.D. Holman, L. Sui, J. McTiernan, V. Petrosian, in *Multi-wavelength Observations of Coronal Structure and Dynamics*, ed. by P.C.H. Martens, D. Cauffman (2002), p. 405

G.D. Holman, L. Sui, R.A. Schwartz, A.G. Emslie, Astrophys. J. Lett. **595**, L97 (2003). doi:10.1086/378488

P. Hoyng, A. Duijveman, M.E. Machado, D.M. Rust, Z. Svestka, A. Boelee, C. de Jager, K.T. Frost, H. Lafleur, G.M. Simnett, H.F. van Beek, B.E. Woodgate, Astrophys. J. Lett. **246**, L155 (1981). doi:10.1086/183574

P. Hoyng, H.F. van Beek, J.C. Brown, Sol. Phys. **48**, 197 (1976)

H. Hudson, J. Ryan, Annu. Rev. Astron. Astrophys. **33**, 239 (1995). doi:10.1146/annurev.aa.33.090195.001323

H.S. Hudson, in *Bulletin of the American Astronomical Society*. Bulletin of the American Astronomical Society, vol. 23 (1991), p. 1064

H.S. Hudson, F. Fárník, in *ESA Special Publication*, ed. by J. Kuijpers. ESA Special Publication, vol. 506 (2002), pp. 261–264

Y.W. Jiang, S. Liu, W. Liu, V. Petrosian, Astrophys. J. **638**, 1140 (2006). arXiv:astro-ph/0508532, doi:10.1086/498863

C.M. Johns, R.P. Lin, Sol. Phys. **137**, 121 (1992)

S.W. Kahler, Space Sci. Rev. **129**, 359 (2007). doi:10.1007/s11214-007-9143-0

S.R. Kane, K.A. Anderson, Astrophys. J. **162**, 1003 (1970)

M. Karlický, Astrophys. J. **690**, 189 (2009). doi:10.1088/0004-637X/690/1/189

M. Karlický, J. Kašparová, Astron. Astrophys. **506**, 1437 (2009). arXiv:0909.0146, doi:10.1051/0004-6361/200912616
M. Karlický, D.H. Nickeler, M. Bárta, Astron. Astrophys. **486**, 325 (2008). doi:10.1051/0004-6361:200809455
J. Kašparová, M. Karlický, E.P. Kontar, R.A. Schwartz, B.R. Dennis, Sol. Phys. **232**, 63 (2005). arXiv:astro-ph/0508636, doi:10.1007/s11207-005-1581-9
J. Kašparová, E.P. Kontar, J.C. Brown, Astron. Astrophys. **466**, 705 (2007). arXiv:astro-ph/0701871, doi:10.1051/0004-6361:20066689
A. Kerdraon, J.-M. Delouis, in *Coronal Physics from Radio and Space Observations*, ed. by G. Trottet. Lecture Notes in Physics, vol. 483 (Springer, Berlin, 1997), p. 192
A.L. Kiplinger, Astrophys. J. **453**, 973 (1995). doi:10.1086/176457
J.W. Knight, P.A. Sturrock, Astrophys. J. **218**, 306 (1977)
Y.-K. Ko, J.C. Raymond, J. Lin, G. Lawrence, J. Li, A. Fludra, Astrophys. J. **594**, 1068 (2003). doi:10.1086/376982
H.W. Koch, J.W. Motz, Rev. Mod. Phys. **31**, 920 (1959). doi:10.1103/RevModPhys.31.920
E.P. Kontar, J.C. Brown, Adv. Space Res. **38**, 945 (2006a). arXiv:astro-ph/0508418, doi:10.1016/j.asr.2005.09.029
E.P. Kontar, J.C. Brown, Astrophys. J. Lett. **653**, L149 (2006b). arXiv:astro-ph/0611170, doi:10.1086/510586
E.P. Kontar, J.C. Brown, A.G. Emslie, R.A. Schwartz, D.M. Smith, R.C. Alexander, Astrophys. J. Lett. **595**, L123 (2003). doi:10.1086/378172
E.P. Kontar, J.C. Brown, G.K. McArthur, Sol. Phys. **210**, 419 (2002). doi:10.1023/A:1022494318540
E.P. Kontar, E. Dickson, J. Kašparová, Solar Phys., 149 (2008a). arXiv:0805.1470, doi:10.1007/s11207-008-9249-x
E.P. Kontar, I.G. Hannah, N.L.S. Jeffrey, M. Battaglia, Astrophys. J. **717**, 250 (2010). arXiv:1005.0599, doi:10.1088/0004-637X/717/1/250
E.P. Kontar, I.G. Hannah, A.L. MacKinnon, Astron. Astrophys. **489**, L57 (2008b). arXiv:0808.3334, doi:10.1051/0004-6361:200810719
E.P. Kontar et al., Space Sci. Rev. (2011, this issue). doi:10.1007/s11214-011-9804-x
N.A. Krall, A.W. Trivelpiece, Principles of plasma physics, in *International Student Edition—International Series in Pure and Applied Physics* (McGraw-Hill Kogakusha, Tokyo, 1973)
S. Krucker, M. Battaglia, P.J. Cargill, L. Fletcher, H.S. Hudson, A.L. MacKinnon, S. Masuda, L. Sui, M. Tomczak, A.L. Veronig, L. Vlahos, S.M. White, Astron. Astrophys. Rev. **16**, 155 (2008a). doi:10.1007/s00159-008-0014-9
S. Krucker, G.J. Hurford, A.L. MacKinnon, A.Y. Shih, R.P. Lin, Astrophys. J. Lett. **678**, L63 (2008b). doi:10.1086/588381
S. Krucker, R.P. Lin, Sol. Phys. **210**, 229 (2002). doi:10.1023/A:1022469902940
S. Krucker, R.P. Lin, Astrophys. J. **673**, 1181 (2008). doi:10.1086/524010
S. Krucker, P. Saint-Hilaire, S. Christe, S.M. White, A.D. Chavier, S.D. Bale, R.P. Lin, Astrophys. J. **681**, 644 (2008c). doi:10.1086/588549
J. Leach, V. Petrosian, Astrophys. J. **269**, 715 (1983). doi:10.1086/161081
K.W. Lee, J. Büchner, N. Elkina, Astron. Astrophys. **478**, 889 (2008). doi:10.1051/0004-6361:20078419
J. Lin, Y.-K. Ko, L. Sui, J.C. Raymond, G.A. Stenborg, Y. Jiang, S. Zhao, S. Mancuso, Astrophys. J. **622**, 1251 (2005). doi:10.1086/428110
R.P. Lin, B.R. Dennis, G.J. Hurford, D.M. Smith, A. Zehnder, P.R. Harvey, D.W. Curtis, D. Pankow, P. Turin, M. Bester, A. Csillaghy, M. Lewis, N. Madden, H.F. van Beek, M. Appleby, T. Raudorf, J. McTiernan, R. Ramaty, E. Schmahl, R. Schwartz, S. Krucker, R. Abiad, T. Quinn, P. Berg, M. Hashii, R. Sterling, R. Jackson, R. Pratt, R.D. Campbell, D. Malone, D. Landis, C.P. Barrington-Leigh, S. Slassi-Sennou, C. Cork, D. Clark, D. Amato, L. Orwig, R. Boyle, I.S. Banks, K. Shirey, A.K. Tolbert, D. Zarro, F. Snow, K. Thomsen, R. Henneck, A. Mchedlishvili, P. Ming, M. Fivian, J. Jordan, R. Wanner, J. Crubb, J. Preble, M. Matranga, A. Benz, H. Hudson, R.C. Canfield, G.D. Holman, C. Crannell, T. Kosugi, A.G. Emslie, N. Vilmer, J.C. Brown, C. Johns-Krull, M. Aschwanden, T. Metcalf, A. Conway, Sol. Phys. **210**, 3 (2002). doi:10.1023/A:1022428818870
R.P. Lin, R.A. Schwartz, Astrophys. J. **312**, 462 (1987). doi:10.1086/164891
R.P. Lin, R.A. Schwartz, R.M. Pelling, K.C. Hurley, Astrophys. J. Lett. **251**, L109 (1981). doi:10.1086/183704
S. Liu, L. Fletcher, Astrophys. J. Lett. **701**, L34 (2009). arXiv:0907.1874, doi:10.1088/0004-637X/701/1/L34
S. Liu, F. Han, L. Fletcher, Astrophys. J. **709**, 58 (2010). arXiv:0912.0402, doi:10.1088/0004-637X/709/1/58
W. Liu, S. Liu, Y.W. Jiang, V. Petrosian, Astrophys. J. **649**, 1124 (2006). arXiv:astro-ph/0603510, doi:10.1086/506268
W. Liu, V. Petrosian, B.R. Dennis, G.D. Holman, Astrophys. J. **693**, 847 (2009a). arXiv:0805.1055, doi:10.1088/0004-637X/693/1/847

W. Liu, V. Petrosian, B.R. Dennis, Y.W. Jiang, Astrophys. J. **676**, 704 (2008). arXiv:0709.1963, doi:10.1086/527538
W. Liu, T. Wang, B.R. Dennis, G.D. Holman, Astrophys. J. **698**, 632 (2009b). arXiv:0902.1805, doi:10.1088/0004-637X/698/1/632
A.L. MacKinnon, P.C.V. Mallik, Astron. Astrophys. **510**, A29 (2010). arXiv:0908.3903, doi:10.1051/0004-6361/200913190
G. Mann, H. Aurass, W. Voigt, J. Paschke, in *Coronal Streamers, Coronal Loops, and Coronal and Solar Wind Composition*, ed. by C. Mattok. ESA Special Publication, vol. 348 (1992), pp. 129–132
G. Mann, H. Aurass, A. Warmuth, Astron. Astrophys. **454**, 969 (2006). doi:10.1051/0004-6361:20064990
A.M. Massone, A.G. Emslie, E.P. Kontar, M. Piana, M. Prato, J.C. Brown, Astrophys. J. **613**, 1233 (2004). doi:10.1086/423127
S. Masuda, T. Kosugi, H. Hara, S. Tsuneta, Y. Ogawara, Nature **371**, 495 (1994). doi:10.1038/371495a0
J.M. McTiernan, S.R. Kane, J.M. Loran, J.R. Lemen, L.W. Acton, H. Hara, S. Tsuneta, T. Kosugi, Astrophys. J. Lett. **416**, L91 (1993). doi:10.1086/187078
J.M. McTiernan, V. Petrosian, Astrophys. J. **359**, 524 (1990). doi:10.1086/169084
D.B. Melrose, Sol. Phys. **130**, 3 (1990)
D.B. Melrose, J.C. Brown, Mon. Not. R. Astron. Soc. **176**, 15 (1976)
D.B. Melrose, G.A. Dulk, Astrophys. J. **259**, 844 (1982). doi:10.1086/160219
T.R. Metcalf, D. Alexander, Astrophys. J. **522**, 1108 (1999). doi:10.1086/307679
J.A. Miller, P.J. Cargill, A.G. Emslie, G.D. Holman, B.R. Dennis, T.N. LaRosa, R.M. Winglee, S.G. Benka, S. Tsuneta, J. Geophys. Res. **102**, 14631 (1997). doi:10.1029/97JA00976
J.A. Miller, T.N. Larosa, R.L. Moore, Astrophys. J. **461**, 445 (1996). doi:10.1086/177072
R.O. Milligan, P.T. Gallagher, M. Mathioudakis, D.S. Bloomfield, F.P. Keenan, R.A. Schwartz, Astrophys. J. Lett. **638**, L117 (2006a). arXiv:astro-ph/0509664, doi:10.1086/500555
R.O. Milligan, P.T. Gallagher, M. Mathioudakis, F.P. Keenan, Astrophys. J. Lett. **642**, L169 (2006b). arXiv:astro-ph/0603652, doi:10.1086/504592
V.M. Nakariakov, C. Foullon, E. Verwichte, N.P. Young, Astron. Astrophys. **452**, 343 (2006). doi:10.1051/0004-6361:20054608
W.M. Neupert, Astrophys. J. Lett. **153**, L59 (1968). doi:10.1086/180220
Z. Ning, Sol. Phys. **248**, 99 (2008). doi:10.1007/s11207-008-9124-9
Z. Ning, Sci. China Ser. G, Phys. Astron. **52**, 1686 (2009). doi:10.1007/s11433-009-0219-6
N. Nitta, B.R. Dennis, A.L. Kiplinger, Astrophys. J. **353**, 313 (1990). doi:10.1086/168618
L. Ofman, L. Sui, Astrophys. J. Lett. **644**, L149 (2006). doi:10.1086/505622
G.K. Parks, J.R. Winckler, Astrophys. J. Lett. **155**, L117 (1969)
L. Peterson, J.R. Winckler, Phys. Rev. Lett. **1**, 205 (1958). doi:10.1103/PhysRevLett.1.205
L.E. Peterson, J.R. Winckler, J. Geophys. Res. **64**, 697 (1959). doi:10.1029/JZ064i007p00697
V. Petrosian, T.Q. Donaghy, Astrophys. J. **527**, 945 (1999). arXiv:astro-ph/9907181, doi:10.1086/308133
V. Petrosian, T.Q. Donaghy, J.M. McTiernan, Astrophys. J. **569**, 459 (2002). arXiv:astro-ph/0112363, doi:10.1086/339240
M. Piana, A.M. Massone, G.J. Hurford, M. Prato, A.G. Emslie, E.P. Kontar, R.A. Schwartz, Astrophys. J. **665**, 846 (2007). doi:10.1086/519518
M. Piana, A.M. Massone, E.P. Kontar, A.G. Emslie, J.C. Brown, R.A. Schwartz, Astrophys. J. Lett. **595**, L127 (2003). doi:10.1086/378171
M. Prato, A.G. Emslie, E.P. Kontar, A.M. Massone, M. Piana, Astrophys. J. **706**, 917 (2009). doi:10.1088/0004-637X/706/1/917
J. Qiu, H. Wang, C.Z. Cheng, D.E. Gary, Astrophys. J. **604**, 900 (2004). doi:10.1086/382122
G. Rausche, H. Aurass, G. Mann, M. Karlický, C. Vocks, Sol. Phys. **245**, 327 (2007). doi:10.1007/s11207-007-9036-0
J.C. Raymond, G. Holman, A. Ciaravella, A. Panasyuk, Y.-K. Ko, J. Kohl, Astrophys. J. **659**, 750 (2007). arXiv:astro-ph/0701359, doi:10.1086/512604
K.K. Reeves, S.J. Moats, Astrophys. J. **712**, 429 (2010). doi:10.1088/0004-637X/712/1/429
H.L. Rowland, L. Vlahos, Astron. Astrophys. **142**, 219 (1985)
P. Saint-Hilaire, *Energy release in solar flares*, Ph.D. thesis, 2005. 157 pages; Switzerland: Eidgenoessische Technische Hochschule Zuerich (Switzerland); 2005. Publication Number: AAT C819508. DAI-C 66/02, p. 431, Summer 2005 (2005)
P. Saint-Hilaire, A.O. Benz, Astron. Astrophys. **435**, 743 (2005). doi:10.1051/0004-6361:20041918
P. Saint-Hilaire, S. Krucker, S. Christe, R.P. Lin, Astrophys. J. **696**, 941 (2009). doi:10.1088/0004-637X/696/1/941
P. Saint-Hilaire, S. Krucker, R.P. Lin, Sol. Phys. **250**, 53 (2008). doi:10.1007/s11207-008-9193-9
R. Saldanha, S. Krucker, R.P. Lin, Astrophys. J. **673**, 1169 (2008). doi:10.1086/524929
C. Shao, G. Huang, Astrophys. J. **691**, 299 (2009a). doi:10.1088/0004-637X/691/1/299

C. Shao, G. Huang, Astrophys. J. Lett. **694**, L162 (2009b). doi:10.1088/0004-637X/694/2/L162
T.V. Siversky, V.V. Zharkova, Astron. Astrophys. **504**, 1057 (2009). arXiv:0907.1911, doi:10.1051/0004-6361/200912341
D.M. Smith, R.P. Lin, P. Turin, D.W. Curtis, J.H. Primbsch, R.D. Campbell, R. Abiad, P. Schroeder, C.P. Cork, E.L. Hull, D.A. Landis, N.W. Madden, D. Malone, R.H. Pehl, T. Raudorf, P. Sangsingkeow, R. Boyle, I.S. Banks, K. Shirey, R. Schwartz, Sol. Phys. **210**, 33 (2002). doi:10.1023/A:1022400716414
Y. Su, G.D. Holman, B.R. Dennis, A.K. Tolbert, R.A. Schwartz, Astrophys. J. **705**, 1584 (2009). doi:10.1088/0004-637X/705/2/1584
L. Sui, G.D. Holman, Astrophys. J. Lett. **596**, L251 (2003). doi:10.1086/379343
L. Sui, G.D. Holman, B.R. Dennis, Astrophys. J. **612**, 546 (2004). doi:10.1086/422515
L. Sui, G.D. Holman, B.R. Dennis, Astrophys. J. **626**, 1102 (2005a). doi:10.1086/430086
L. Sui, G.D. Holman, B.R. Dennis, Astrophys. J. **646**, 605 (2006a). doi:10.1086/504885
L. Sui, G.D. Holman, B.R. Dennis, Astrophys. J. Lett. **645**, L157 (2006b). doi:10.1086/506325
L. Sui, G.D. Holman, B.R. Dennis, Astrophys. J. **670**, 862 (2007). doi:10.1086/522198
L. Sui, G.D. Holman, B.R. Dennis, Adv. Space Res. **41**, 976 (2008). doi:10.1016/j.asr.2007.03.031
L. Sui, G.D. Holman, B.R. Dennis, S. Krucker, R.A. Schwartz, K. Tolbert, Sol. Phys. **210**, 245 (2002). doi:10.1023/A:1022417832283
L. Sui, G.D. Holman, S.M. White, J. Zhang, Astrophys. J. **633**, 1175 (2005b). doi:10.1086/462410
E. Tandberg-Hanssen, A.G. Emslie, *The Physics of Solar Flares* (Cambridge University Press, Cambridge, 1988), p. 286
S. Tsuneta, T. Naito, Astrophys. J. Lett. **495**, L67 (1998). arXiv:astro-ph/9801109, doi:10.1086/311207
M. Väänänen, S. Pohjolainen, Sol. Phys. **241**, 279 (2007). doi:10.1007/s11207-007-0235-5
G.H.J. van den Oord, Astron. Astrophys. **234**, 496 (1990)
A. Veronig, M. Temmer, A. Hanslmeier, W. Otruba, M. Messerotti, Astron. Astrophys. **382**, 1070 (2002a). arXiv:astro-ph/0207234, doi:10.1051/0004-6361:20011694
A. Veronig, B. Vršnak, B.R. Dennis, M. Temmer, A. Hanslmeier, J. Magdalenić, Astron. Astrophys. **392**, 699 (2002b). arXiv:astro-ph/0207217, doi:10.1051/0004-6361:20020947
A. Veronig, B. Vršnak, M. Temmer, A. Hanslmeier, Sol. Phys. **208**, 297 (2002c). arXiv:astro-ph/0208088
A.M. Veronig, J.C. Brown, Astrophys. J. Lett. **603**, L117 (2004). doi:10.1086/383199
A.M. Veronig, J.C. Brown, B.R. Dennis, R.A. Schwartz, L. Sui, A.K. Tolbert, Astrophys. J. **621**, 482 (2005). doi:10.1086/427274
N. Vilmer, G. Trotter, S.R. Kane, Astron. Astrophys. **108**, 306 (1982)
N. Vilmer et al., Space Sci. Rev. (2011, this issue). doi:10.1007/s11214-010-9728-x
L. Vlahos, H.L. Rowland, Astron. Astrophys. **139**, 263 (1984)
T. Wang, L. Sui, J. Qiu, Astrophys. J. Lett. **661**, L207 (2007). arXiv:0709.2329, doi:10.1086/519004
A. Warmuth, G.D. Holman, B.R. Dennis, G. Mann, H. Aurass, R.O. Milligan, Astrophys. J. **699**, 917 (2009). doi:10.1088/0004-637X/699/1/917
A. Warmuth, G. Mann, H. Aurass, Cent. Eur. Astrophys. Bull. **31**, 135 (2007)
H.P. Warren, Astrophys. J. **637**, 522 (2006). arXiv:astro-ph/0507328, doi:10.1086/497904
H.P. Warren, S.K. Antiochos, Astrophys. J. Lett. **611**, L49 (2004). doi:10.1086/423840
D.F. Webb, J. Burkepile, T.G. Forbes, P. Riley, Journal of Geophysical Research (Space). Physics **108**, 1440 (2003). doi:10.1029/2003JA009923
M.S. Wheatland, D.B. Melrose, Sol. Phys. **158**, 283 (1995)
S.M. White et al., Space Sci. Rev. (2011, this issue). doi:10.1007/s11214-010-9708-1
R.M. Winglee, A.L. Kiplinger, D.M. Zarro, G.A. Dulk, J.R. Lemen, Astrophys. J. **375**, 366 (1991). doi:10.1086/170196
Y. Xu, A.G. Emslie, G.J. Hurford, Astrophys. J. **673**, 576 (2008). doi:10.1086/524184
V.V. Zharkova, M. Gordovskyy, Astron. Astrophys. **432**, 1033 (2005). doi:10.1051/0004-6361:20041102
V.V. Zharkova, M. Gordovskyy, Astrophys. J. **651**, 553 (2006). doi:10.1086/506423
V.V. Zharkova et al., Space Sci. Rev. (2011, this issue). doi:10.1007/s11214-011-9803-y

Space Sci Rev (2011) 159:167–224
DOI 10.1007/s11214-010-9728-x

Properties of Energetic Ions in the Solar Atmosphere from γ-Ray and Neutron Observations

N. Vilmer · A.L. MacKinnon · G.J. Hurford

Received: 30 April 2010 / Accepted: 2 November 2010 / Published online: 11 August 2011

Abstract Gamma-rays and neutrons are the only sources of information on energetic ions present during solar flares and on properties of these ions when they interact in the solar atmosphere. The production of γ-rays and neutrons results from convolution of the nuclear cross-sections with the ion distribution functions in the atmosphere. The observed γ-ray and neutron fluxes thus provide useful diagnostics for the properties of energetic ions, yielding strong constraints on acceleration mechanisms as well as properties of the interaction sites. The problem of ion transport between the accelerating and interaction sites must also be addressed to infer as much information as possible on the properties of the primary ion accelerator. In the last couple of decades, both theoretical and observational developments have led to substantial progress in understanding the origin of solar γ-rays and neutrons. This chapter reviews recent developments in the study of solar γ-rays and of solar neutrons at the time of the *RHESSI* era. The unprecedented quality of the *RHESSI* data reveals γ-ray line shapes for the first time and provides γ-ray images. Our previous understanding of the properties of energetic ions based on measurements from the former solar cycles is also summarized. The new results—obtained owing both to the gain in spectral resolution (both with *RHESSI* and with the non solar-dedicated *INTEGRAL*/SPI instrument) and to the pioneering imaging technique in the γ-ray domain—are presented in the context of this previous knowledge. Still open questions are emphasized in the last section of the chapter and future perspectives on this field are briefly discussed.

Keywords Sun: flares · Sun: gamma-rays · Sun: energetic particles · Sun: acceleration mechanisms

N. Vilmer (✉)
LESIA-Observatoire de Paris, CNRS, UPMC Univ Paris 06, Univ. Paris-Diderot, Paris, France
e-mail: nicole.vilmer@obspm.fr

A.L. MacKinnon
DACE/Physics and Astronomy, University of Glasgow, Glasgow, G12 8QQ, UK
e-mail: a.mackinnon@educ.gla.ac.uk

G.J. Hurford
Space Sciences Laboratory, University of California, Berkeley, CA 94720-7450, USA
e-mail: ghurford@ssl.berkeley.edu

1 Introduction

The first evidence of acceleration of charged particles (electrons and ions) to relativistic energies in association with solar flares was found in 1942 with the discovery of large increases in the count rates of several ground level cosmic-ray intensity monitors. These observations told us, more than fifty years ago, that solar flares are capable of accelerating protons to GeV energies. While Biermann et al. (1951) had predicted that relativistic protons accelerated during a solar flare could produce a flux of high-energy neutrons observable at the Earth (as well as some γ-ray emission), Morrison (1958) published the first prediction of γ-ray fluxes expected from several celestial objects including solar flares. These first estimates were at the base of former detailed theoretical studies of high-energy neutral emissions which were initiated by Lingenfelter and Ramaty (1967a). They provided a detailed treatment of the production of nuclear reactions in the solar atmosphere and calculated the yield of neutrons, positrons and γ-ray lines using normalized rigidity spectra for the accelerated ions. The first observational evidence for γ-rays came from the observations of the flare SOL1972-08-04 with the *OSO-7*[1] γ-ray spectrometer (Chupp et al. 1973, see Fig. 1), which confirmed the prediction by Morrison that nuclear reactions in the solar atmosphere could produce γ-ray lines. Pioneering observations of solar γ-ray lines were also obtained for later flares of August 1972 by spectrometers on *Prognoz-6* (Talon et al. 1975) and for the flares SOL1978-07-11T10:54 by *HEAO 1* (Hudson et al. 1980) and SOL1979-11-09T03:20 by *HEAO 3* (Prince et al. 1982). Since then solar γ-ray astronomy has provided a new window for the study of ion acceleration in solar flares.

Energetic ions interacting with the solar atmosphere produce a wealth of γ-ray emissions. A complete γ-ray line spectrum is produced through interactions of ions in the range $\sim$1–100 MeV/nucleon and consists of several nuclear de-excitation lines, neutron capture and positron annihilation lines. (See, for example, Chupp (1984), Ramaty (1986), Chupp (1996), Trottet and Vilmer (1997), Share and Murphy (2000, 2006), Vilmer and MacKinnon (2003)). The temporal and spectral characteristics of all these radiations provide strong constraints on acceleration timescales, ion energy spectra and numbers as well as energetic ion abundances. If more energetic ions are present, over a few hundred MeV/nucleon, nuclear interactions with the ambient medium produce secondary pions whose decay products lead to a broad-band continuum at photon energies above 10 MeV (with a broad peak around 70 MeV from neutral pion radiation) (Stecker 1970; Murphy et al. 1987) and also secondary energetic neutrons which, if energetic enough, may escape from the Sun and be directly detected in interplanetary space ($\geq$ 10 MeV neutrons at 1 AU) or at ground level ($\geq$ 200 MeV neutrons) (see, e.g., Chupp and Dunphy 2000; Chupp and Ryan 2009). The importance of observing neutrons was pointed out by Biermann et al. (1951) and by Lingenfelter et al. (1965a, 1965b). Lower-energy neutrons (kinetic energies below 100 MeV) can only be observed in space because they are strongly attenuated in the Earth's atmosphere and cannot reach the ground. Neutrons with kinetic energies higher than 100 MeV can be observed on the ground and hence with simultaneous observations in space and on the ground, it is possible to obtain the energy spectrum of solar neutrons and of accelerated particles, in a wide energy range.

The first detection of solar high-energy neutrons was provided by the observations of the *SMM*/GRS[2] spectrometer for the flare SOL1980-06-21T02:00 (X2.6) which showed

[1] Orbiting Solar Observatory-7.

[2] *Solar Maximum Mission*/Gamma Ray Spectrometer.

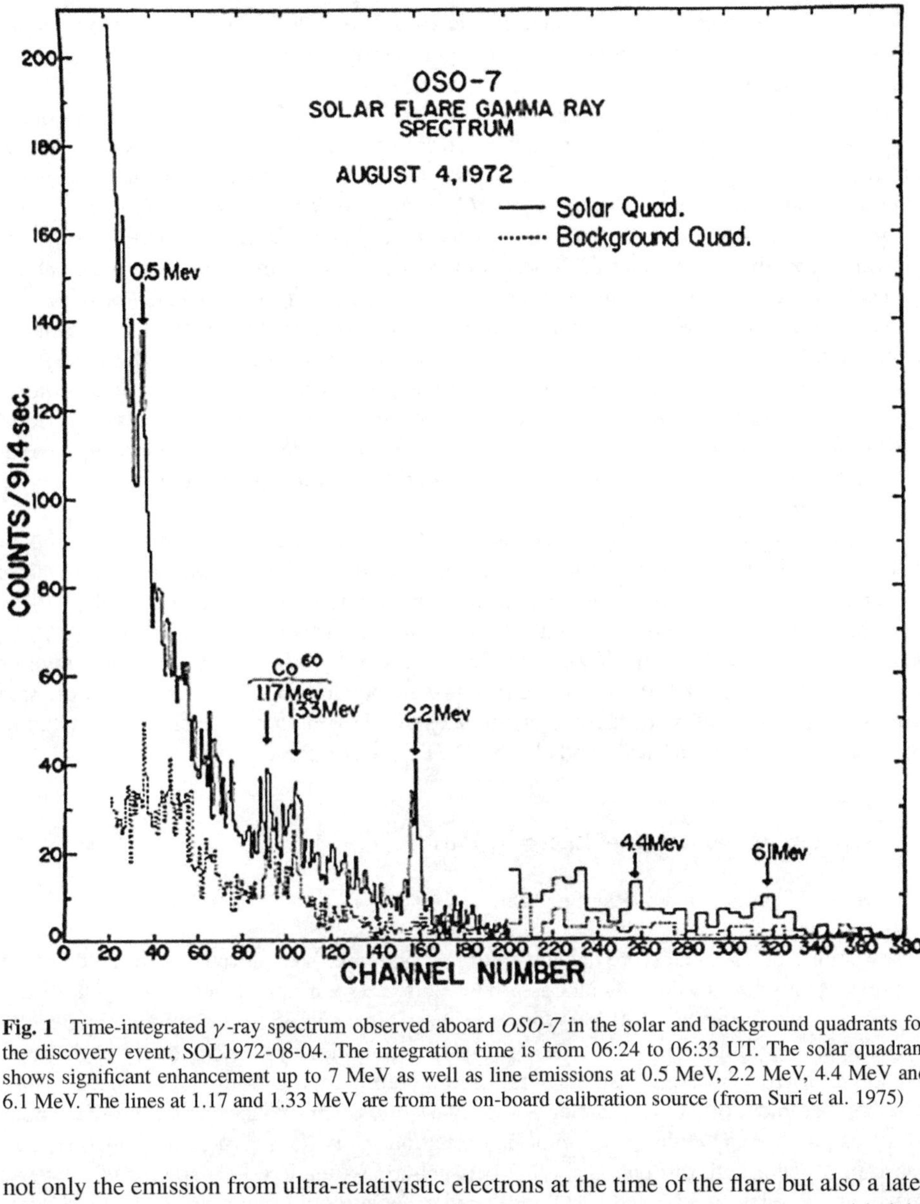

Fig. 1 Time-integrated γ-ray spectrum observed aboard *OSO-7* in the solar and background quadrants for the discovery event, SOL1972-08-04. The integration time is from 06:24 to 06:33 UT. The solar quadrant shows significant enhancement up to 7 MeV as well as line emissions at 0.5 MeV, 2.2 MeV, 4.4 MeV and 6.1 MeV. The lines at 1.17 and 1.33 MeV are from the on-board calibration source (from Suri et al. 1975)

not only the emission from ultra-relativistic electrons at the time of the flare but also a later signature from high-energy neutrons (Chupp et al. 1982). This same experiment provided the first observations of pion-decay radiation in solar flares (high energy radiation produced by ultra-relativistic ions) (Forrest et al. 1985) for the event SOL1982-06-03T13:26 (X8.0). For this event, solar neutrons were also detected for the first time at ground-level by the IGY-type neutron monitor installed at Jungfraujoch, Switzerland, as well as by *SMM*/GRS (Chupp et al. 1987). With these observations, all the components predicted by Lingenfelter and Ramaty (1967a) had been detected (see, e.g., Chupp 1996, for a detailed review of early γ-ray observations of solar flares). Since these first observations, solar energetic neutrons have been detected with ground-based detectors on a handful of occasions. 10–120 MeV neutrons were also detected in space with COMPTEL on the *Compton Gamma-Ray Observatory*

(CGRO) mission, whose mode of operation enabled direct determination of their energies (Ryan et al. 1993; Kocharov et al. 1998). We consider these various neutron observations in more detail below (Sect. 4).

Before the launch of *RHESSI* and *INTEGRAL* in 2002, high energy observations of solar flares had been obtained for two solar cycles respectively with *SMM*/GRS as well as on *Hinotori* for cycle 21 and by the PHEBUS and SIGMA experiments aboard *Granat*, the wide-band spectrometer on *Yohkoh*, *GAMMA-1* and *CGRO* for cycle 22 (see, e.g., Chupp 1984, 1996; Ramaty 1986; Trottet and Vilmer 1997; Vilmer and MacKinnon 2003, for reviews). Before the advent of *RHESSI* observations, the solar electromagnetic radiation above 100 keV (including the γ-ray line region) was one of the last spectral domains with no spatially resolved observations, so that no information was available on the location of the interaction sites of the energetic ions (nor of electrons above 100 keV) on the Sun. γ-ray line emission from ions had been observed in many flares but the quantitative constraints from γ-ray line spectroscopy, which provides information on the bulk of the flare fast ions, had been deduced from observations with limited spectral resolution. Higher energy radiations and neutrons have been observed in fewer events, but they provide crucial information on the extremes of particle acceleration.

Since the launch of the solar-dedicated *RHESSI* mission (Lin et al. 2002) and of *INTEGRAL* in 2002 (Vedrenne et al. 2003), both equipped with high-resolution germanium detectors, several γ-ray line flares produced strong enough fluxes so that a detailed γ-ray line shape analysis could be achieved for a few events. Furthermore, since February 2002, *RHESSI* has observed a total of 29 events above 300 keV with 18 clearly showing γ-ray line emission (Shih et al. 2009b). Pioneering results were obtained by *RHESSI* on a few of these events on the localization of the interaction sites of energetic ions. We shall review below the results provided by *RHESSI* and also by *INTEGRAL* on these issues.

2 γ-Ray Line Observations of Energetic Ions

2.1 General Features of a γ-Ray Spectrum

Many studies performed before the launch of *RHESSI* have been devoted to the study of the complete γ-ray line spectrum which is produced in the solar atmosphere through the interactions of ions in the ~1–100 MeV/nucleon range. In contrast to the monotonic electron bremsstrahlung continuum, the nuclear γ-ray spectrum displays a rich variety of features: broad and narrow lines and various sorts of continuum arising from several mechanisms. The bombardment of ambient nuclei of the solar atmosphere by accelerated protons and α particles produces a whole spectrum of de-excitation γ-ray lines (prompt narrow lines), almost all of them with rest energies lying between a few hundred keV and 8 MeV. Inverse reactions in which accelerated C, O and heavier nuclei collide on ambient H and He are the origin of prompt broad lines centered at the same line energy. Since the emitting nuclei retain significant momentum after the exciting collisions, these lines are substantially Doppler-broadened and/or shifted, to the extent that they are not individually resolvable. These broad lines merge with many other weaker lines from nuclei heavier than oxygen to form an unresolved continuum (see Sect. 2.11) (Murphy et al. 2009). Figure 2 from Share and Murphy (2000), based on the *CGRO*/OSSE[3] observations of the large flare SOL1991-06-04T03:37 (X12.0), shows the major features of the γ-ray line spectrum: strong narrow

[3]Compton Gamma Ray Observatory/Oriented Scintillation Spectrometer Experiment.

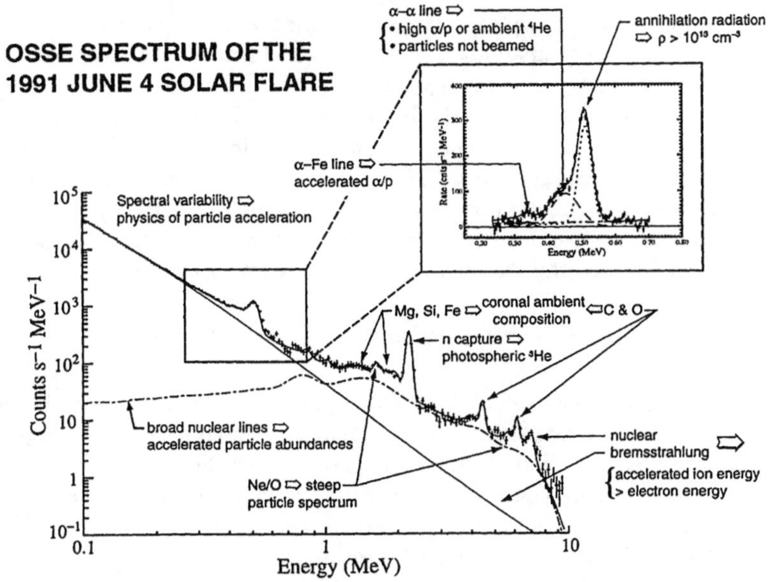

Fig. 2 OSSE spectrum of the flare SOL1991-06-04T03:37 (X12.0) summarizing the different components and the physics from γ-ray line spectroscopy (from Share and Murphy 2000). Reproduced by permission of the ASP

de-excitation lines are found at 6.129 MeV from ^{16}O, 4.438 MeV from ^{12}C, 1.779 MeV from ^{28}Si, 1.634 MeV from ^{20}Ne, 1.369 MeV from ^{24}Mg and 0.847 MeV from ^{56}Fe. The broad lines merge into a quasi continuum dominating over the bremsstrahlung emission from electrons in the $\sim$1–8 MeV range. The single strongest line is the extremely narrow 2.223 MeV deuterium formation line. Also present are photons of flare origin at 0.511 MeV from positron annihilation and the broad feature at 0.4–0.5 MeV formed from the two lines, at 0.429 and 0.47 MeV, that accompany formation of ^{7}Li and ^{7}Be in α-^{4}He fusion reactions.

2.2 Magnetic Loop Transport and Interaction Model

2.2.1 Gamma-Ray Line Production Including Transport

Information derived from γ-ray line emissions on the characteristics of the accelerated ions requires a comprehensive understanding of the transport of these energetic ions between the acceleration site, presumably located in the low corona, and the interaction and emitting sites in the denser layers of the solar atmosphere. This requires a model for ion acceleration, transport in flare loops and interaction. Several models have been developed since the 1980s on this topic, a couple of which we describe. The recent model described by Murphy et al. (2007) addresses particle transport and interaction and includes energy losses due to Coulomb collisions, removal by nuclear reactions, magnetic mirroring in the converging magnetic flux and MHD pitch-angle scattering in the corona, treated via the quasilinear formalism. The loop formalism is the one previously developed by Hua et al. (1989). The loop consists of a semicircular coronal portion of length L and two straight portions extending vertically from the transition region through the chromosphere into the photosphere. Below the transition region, the magnetic field strength is assumed proportional to a power δ of the pressure (Zweibel and Haber 1983) taken here as $\delta \simeq 0.2$. Such a converging magnetic field results in mirroring of accelerated particles. Pitch-angle scattering is characterized by Λ, the mean free path required for an arbitrary initial angular distribution to relax to an isotropic distribution. The dependence of Λ on particle energy is expected to be weak (see discussion by Hua et al. 1989) and is assumed to be independent of particle energy. In the model, the level of pitch-angle scattering of the energetic particles is simply characterized by λ, the ratio of Λ to the loop half-length $L_c = L/2$. Magnetic convergence and pitch-angle scattering determine the angular distribution of the accelerated particles when they interact with the ambient medium, which is crucial to several aspects of the observable emissions. Several height-density profiles $n(h)$ for the solar atmosphere are also assumed.

2.2.2 *Time History of the Nuclear Interaction Rate*

Murphy et al. (2007) studied the effects of transport on temporal behavior by assuming instantaneous release of all particles at $t = 0$. Actual observed time profiles represent the convolution of this behavior with the time profile of the particle accelerator. In the absence of magnetic convergence, ions do not mirror and the nuclear interaction time history depends only on the energy loss rate in the lower chromosphere and upper photosphere where the density is greatest and most of the interactions occur.

In the presence of magnetic convergence, particles behave differently depending on pitch-angle α. Increase of the magnetic field and of the density towards the deeper atmosphere leads to partial particle trapping and to the formation of a loss cone of angular half-width (α_0). Ions with $\alpha < \alpha_0$ precipitate in a loop transit time, behaving as described above. Ions outside the loss cone mirror and lose energy much more slowly as they traverse the low-density corona. These interactions thus occur on longer time scales, with the time scale increasing with increasing convergence. Pitch-angle scattering causes the loss cone to be continuously repopulated, and in fact is essential if observed, impulsive time profiles are to be understood (Zweibel and Haber 1983; Hua et al. 1989). As a result, with increasing pitch-angle scattering rate (i.e., decreasing λ) the nuclear interaction rate increases at early times and correspondingly decreases at later times. However, the time history is no longer affected by increasing pitch-angle scattering when the rate of loss-cone replenishment exceeds the rate of nuclear reactions in the loss cone. Increasing the loop length increases the time scale of the interaction rate since mirroring particles spend more time at lower coronal densities where nuclear reactions are less likely. All these effects are illustrated in Fig. 3, from Murphy et al. (2007). On short timescales (<1 s) the effects of velocity dispersion may be seen, as higher energy particles transit the loop and encounter the deeper regions more rapidly. On longer timescales, trapping and precipitation, via pitch-angle scattering, dominate the temporal behavior.

The results of these detailed calculations are summarized and put to work in the "trap-plus-precipitation" model, in which the flaring atmosphere is divided into a coronal, magnetic trap characterized by a single, uniform density n_0, and a high density, chromosphere-photosphere region. The escape of particles from the trap to the deep atmosphere is summarized in an escape rate. The saturation of escape rate with increasing scattering rate, described above, may be recognized as the "strong pitch-angle scattering" limit identified by Kennel and Petschek (1966), in which case the coronal pitch-angle distribution is always more or less isotropic and the escape time from the trap is given by (see, e.g., Hulot et al. 1989)

$$\mu_{prep} = \frac{\alpha_0^2}{2} \frac{V}{L}, \tag{2.1}$$

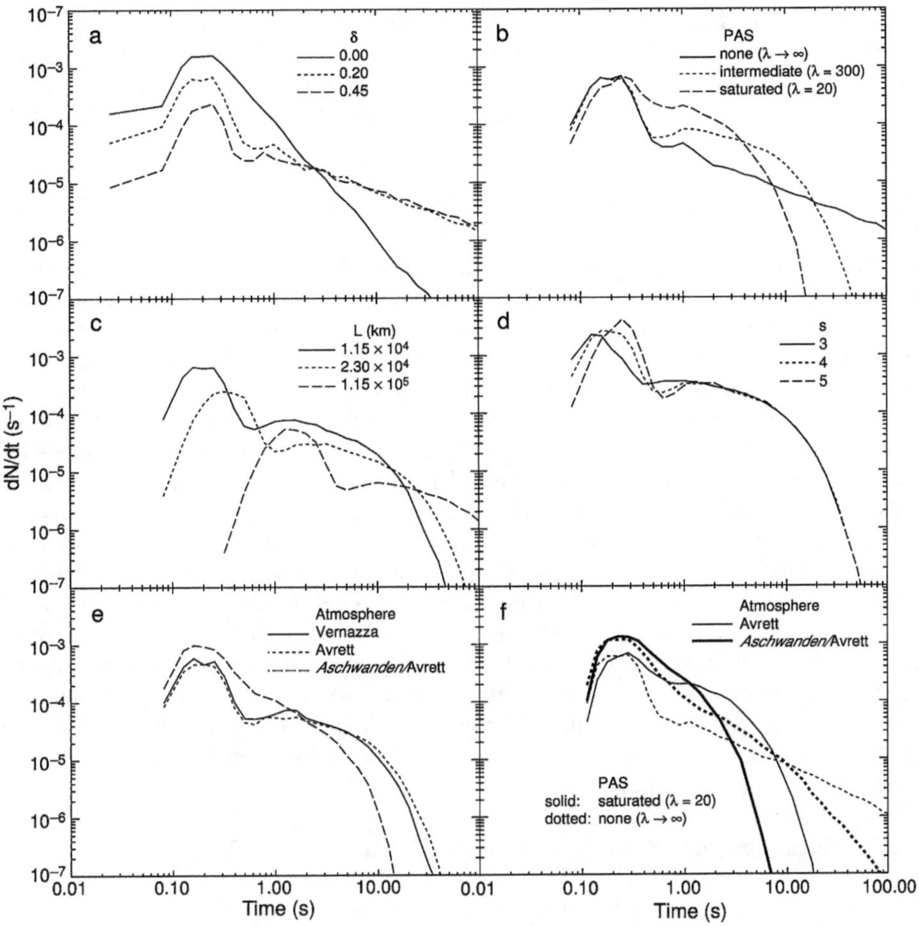

Fig. 3 Nuclear interaction rate as a function of the different parameters: (**a**) magnetic convergence, (**b**) pitch-angle scattering, (**c**) loop length, (**d**) spectral index, (**e**) atmospheric model, (**f**) combined effect of the pitch-angle scattering and of the atmospheric model. The yields are normalized to one proton of energy greater than 30 MeV (from Murphy et al. 2007). Reproduced by permission of the AAS

where α_0 and L are as above and V is the particle velocity. For scattering rates below the strong limit, the resulting long coronal particle lifetimes become inconsistent with the short duration usually observed in solar γ-ray bursts (Zweibel and Haber 1983). The trap-plus-precipitation model has been applied to observations, with results described below (Sect. 2.2.6).

2.2.3 Angular Distribution of the Interacting Ions

In the presence of magnetic convergence, particles encounter the greatest densities at their mirror points when they also have the greatest pitch angles. Observations (e.g., of line shapes) reveal the density-weighted, source-integrated pitch-angle distributions and these may consequently take a "fan-beam" form (e.g., Hua and Lingenfelter 1987b). As pitch-angle scattering approaches the strong limit ($\lambda \approx 20$) (see Sect. 2.2.2 for the definition of λ), however, more and more particles precipitate in the loss cone, at small pitch angles, and the density-weighted distribution becomes more strongly beamed downward. So the overall pitch-angle distribution reflects a combination of the initial pitch-angle distribution, the degree of magnetic convergence in the loop, and the pitch-angle scattering rate (see Murphy et al. 2007 for details).

2.2.4 *Depth Distribution of the Interaction Site*

In the absence of magnetic convergence, there is no mirroring and the depth distribution directly reflects the grammage required for the accelerated ions to interact as they move downward through the solar atmosphere. In the presence of magnetic convergence, mirroring results in interactions occurring at higher elevations (and therefore lower densities) as more and more particles are prevented from penetrating the lower atmosphere. Even with minimal convergence and in case of no pitch-angle scattering, a significant fraction of the interactions occurs at low densities (almost 20% of the interactions occur at densities less than 10^{10} cm^{-3}). As the scattering is increased, more particles are able to precipitate and the bulk of the interactions move deeper, to higher densities. Because higher-energy ions tend to interact farther along their paths, harder particle spectra result in interactions occurring at higher densities since more higher-energy ions are producing the interactions. Similarly, interactions whose cross-sections have higher threshold energies tend to occur preferentially at higher densities. This dependence of the depth distribution on the interaction cross-section also explains why the depth distribution is affected by the accelerated α/proton ratio. Alpha-particle interactions generally have lower threshold energies and, as just discussed, such interactions occur at lower densities. When the α/proton ratio is high, a larger fraction of the line yield is due to such reactions and the depth distribution shifts to lower densities (see Murphy et al. 2007, for details).

2.2.5 *Effective Energies of the Accelerated Ions Producing γ-Ray Lines*

The main factor determining the effective energy range for a nuclear reaction is the cross-section, but the shape of the accelerated-ion spectrum can be important. For thin-target interactions, only ions with initial energies where the cross-section is significant contribute. For thick target interactions, however, even ions with higher initial energies also contribute, since they can lose energy and then interact at energies where the cross-section is significant.

Murphy et al. (2007) determined the effective ion energies to produce a γ-ray line for a particular ion spectrum by weighting yields from monoenergetic ions with the spectrum. The ambient abundances are assumed to be coronal (Ramaty et al. 1996). As an example, Fig. 4 from Murphy et al. (2007) shows the differential yield of the 1.634 MeV ^{20}Ne line as a function of accelerated ion energy for power-law spectra with indices of 3 and 5 and accelerated α/proton $= 0.5$. These energies are the energies of the ions leaving the acceleration region. Separate contributions to the line from accelerated protons (dashed curves) and α particles (dotted curves) are plotted. For soft spectra, the α-particle contributions dominate due to their lower threshold energies, and the most effective ion energies are around a few MeV/nucleon. For hard spectra, the proton interactions become important due to their higher threshold energies and because the effective energies are higher. An effective ion energy range for producing the line is defined as follows: the ion energy where the yield is maximum is determined and the effective range is the one for which the yield has fallen to 50% on each side of the maximum. We note that if the effective energy distribution is very broad, the arbitrary value of 50% could be misleading.

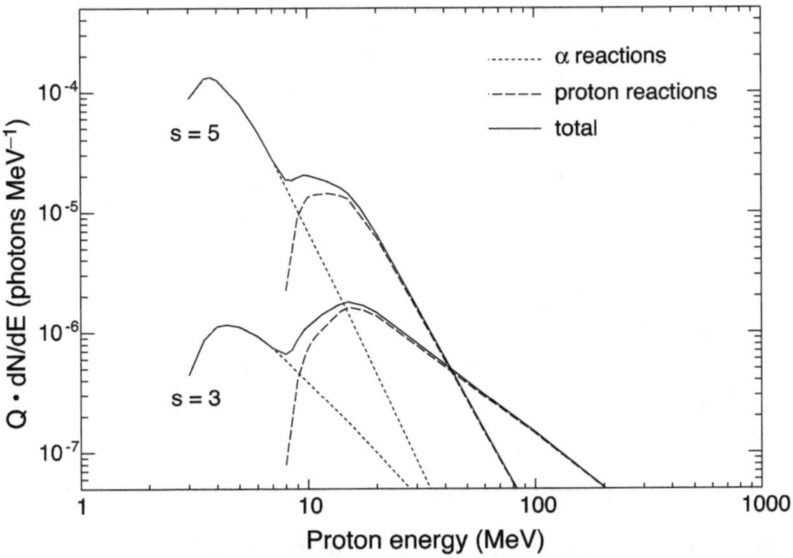

Fig. 4 Yield of the 6.129 MeV ^{16}O line weighted by accelerated ion power-law spectra with indices $s = 3$ and 5. The contributions to the line from accelerated protons and α particles are separately shown (from Murphy et al. 2007). Reproduced by permission of the AAS

The fluence ratio of the 6.129 MeV ^{16}O and 1.634 MeV ^{20}Ne lines has been frequently used as a measure of the accelerated-ion spectral index. Figure 5 shows the effective accelerated-ion energy ranges for producing the 1.634 MeV ^{20}Ne line (white boxes) and the 6.129 MeV ^{16}O line (grey boxes) for several power-law spectral indices and for accelerated α/proton $= 0.1$ (panel a) and 0.5 (panel b). The horizontal line within each box corresponds to the peak of the distribution. The figure shows that the effective energy range for producing the ^{16}O line is usually shifted to energies higher than for the ^{20}Ne line. The extent of the separation of the effective energy ranges for the two lines determines the sensitivity of the ratio to the spectral index. Because the separation for these two lines is not large, the ratio is not very sensitive. For very hard spectra, the upper range of the effective ion energies can be ~100 MeV/nucleon, but for most (softer) spectra it is less than 10 MeV/nucleon. The lower range is typically a few MeV/nucleon but for very soft spectra (particularly when the α/p ratio has the commonly assumed value of 0.5) it can be less than 2 MeV/nucleon. As the spectrum hardens, the higher-energy proton reactions begin to contribute and the effective ion energy range shifts to higher energies and becomes broader as interactions of both α particles and protons contribute. The interpretation of this particular energy-distribution diagnostic is discussed further below (Sect. 2.3), where the roles of ambient chemical abundances and the relative contributions of protons and α-particles are discussed.

The fluence ratio of the 4.438 MeV ^{12}C de-excitation line and the 2.223 MeV line has also been used as a measure of the accelerated-ion spectral index. Figure 6 shows the effective accelerated ion energy ranges for the 4.438 MeV line (white boxes) and the 2.223 MeV line (grey boxes) for several power-law spectral indices and for accelerated α/proton $= 0.1$ (panel a) and 0.5 (panel b). For hard spectra (spectral index $s < 4$), the effective ion energies for the neutron capture line extend to very high energies, up to ~100 MeV/nucleon, and are generally much higher than those producing the ^{12}C line. However, for soft spectra ($s > 4$), the effective ion energies for the neutron capture line can be very low (<7 MeV/nucleon)

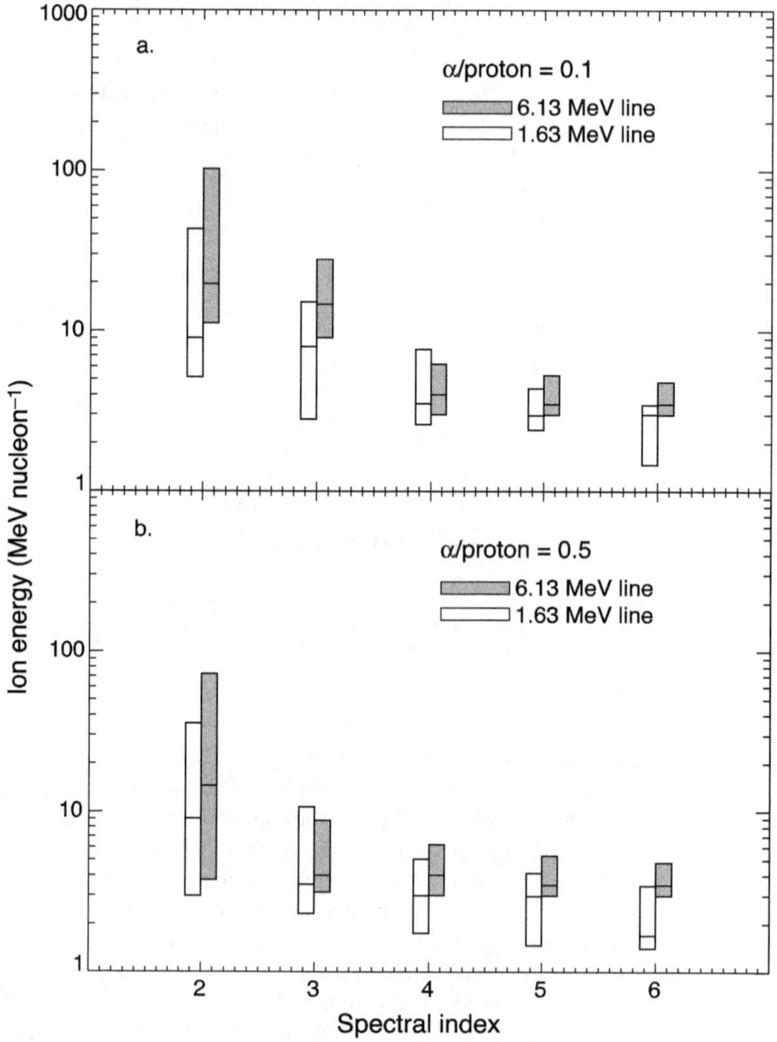

Fig. 5 Effective accelerated ion energies for production of the 1.634 MeV ^{20}Ne line (*white boxes*) and 6.129 MeV ^{16}O line (*grey boxes*) as a function of spectral index. The effective energy ranges are the 50% yield range as defined in the text. Panel (**a**) is for accelerated α/proton = 0.1 and panel (**b**) is for 0.5 (from Murphy et al. 2007). Reproduced by permission of the AAS

due to neutron production by the exothermic (α, n) reactions on heavy elements. For such soft spectra the effective ion energies for the neutron capture line are even less than those for the ^{12}C line. As the flare location moves from disk center to limb, the effective energies shift to slightly lower energies since the neutrons from higher energy reactions are generally produced deeper and are subsequently more attenuated by Compton scattering. When a spectral index is derived using the ratio of these two lines, the relevant ion energies cover a broad range of energies. For hard spectra, the relevant ion energy range extends from a few MeV/nucleon up to and greater than 100 MeV/nucleon. For soft spectra ($s > 4$), the relevant ion energy range is much narrower, from around 1 to a few MeV/nucleon. For very soft

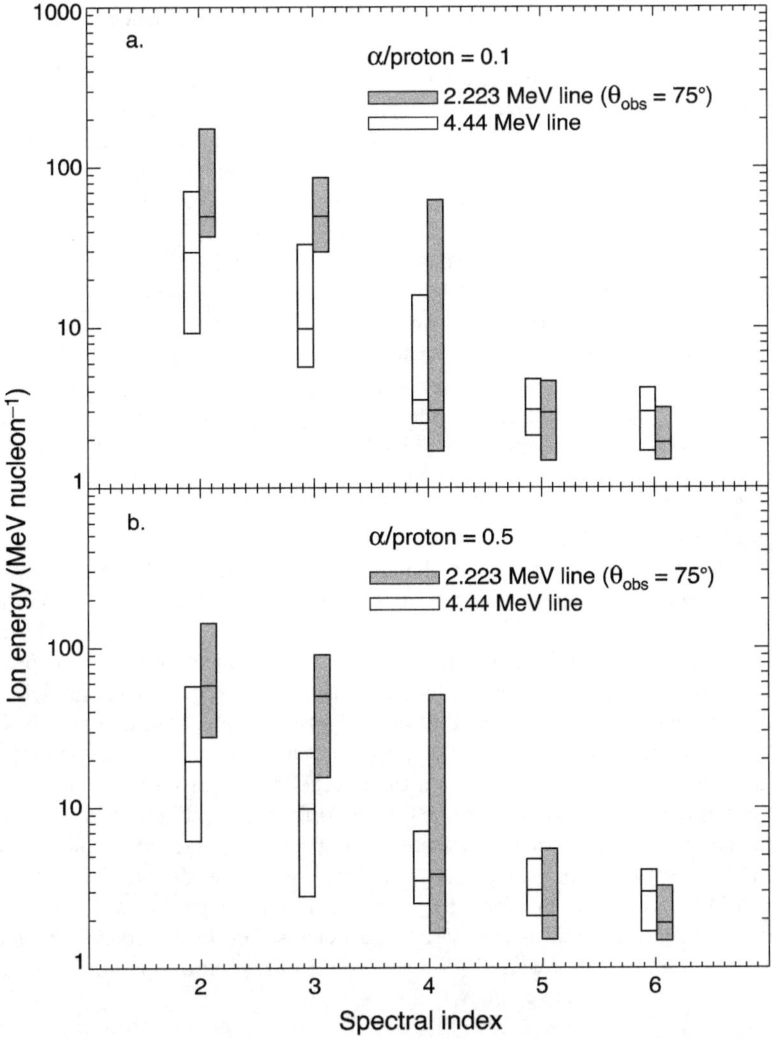

Fig. 6 Effective accelerated ion energies for production of the 4.438 MeV ^{12}C line (*white boxes*) and 2.223 MeV neutron capture line (*grey boxes*) as a function of spectral index. The effective energy ranges are the 50% yield range as defined in the text. Panel (**a**) is for accelerated α/proton $= 0.1$ and panel (**b**) is for 0.5. For the neutron capture line, $\lambda = 300$, $\delta = 0.2$, $L = 11{,}500$ km, and $\theta_{\mathrm{obs}} = 75°$ (from Murphy et al. 2007). Reproduced by permission of the AAS

spectra, the lack of separation of the effective ion energies producing the two lines reduces the sensitivity of the ratio to the spectral index.

2.2.6 Loop Parameters and Time Development of γ-Radiation

Murphy et al. (2007) discuss an integrated approach to the interpretation of γ-ray measurements, distinguishing in particular between the deduction of quantities describing the accelerated particles ("acceleration parameters") and quantities describing the atmospheric environment in which the particles evolve and radiate ("physical parameters"—loop lengths, target abundances, parameters describing the state of plasma turbulence in the corona). γ-ray measurements in turn yield various quantities, each of which contains information on some combination of acceleration and/or physical parameters. De-excitation line fluences, for example, are determined by acceleration parameters and otherwise only by source region chemical abundances. Line widths and centroid energies, on the other hand, are determined by the energy distribution and composition of energetic ions, but also by the amount of magnetic convergence in the loop and the degree of pitch-angle scattering, i.e., in principle, the energy distribution over wavenumbers of coronal MHD turbulence, but typically parametrized in terms of the ion mean free path λ (normalized to the loop half-length). Below we concentrate on deduction of loop parameters from timing measurements.

Peak time delays, i.e., differences between the times of peak flux measured at different photon energies, have been reported in a few flares between hard X-rays and γ-ray line emissions (e.g., Hulot et al. 1989; Yoshimori 1989). Their interpretation was focused on the deduction of loop parameters. The relative timing of prompt γ-ray line emissions and of hard X-rays was studied in this context by Hulot and collaborators. These models were again used to interpret the delays of the hard X-ray and γ-ray line emissions observed for two flares by *ISEE-3* and *SMM*/GRS (Hulot et al. 1992) and also applied to one of the γ-ray line flares observed with *RHESSI* (Dauphin and Vilmer 2007).

For the study of temporal behavior, the basic characteristics of the hard X-ray and γ-ray trap-plus-precipitation models are summarized above, and described in Vilmer et al. (1982) and Hulot et al. (1989). The principle of the model is the following: electrons and ions are supposed to be continuously injected over a finite injection time in a coronal loop where they experience trapping and pitch-angle scattering as above. While particles trapped in the loops produce X-ray and γ-ray thin-target emissions, electrons and protons escaping from the trap propagate downwards to the denser chromospheric layers where they produce hard X-ray and γ-ray line emissions in a thick-target approximation. In these works, the precipitation rate is supposed to be in the "strong diffusion limit" given in (2.1). The free, "physical" parameters of the model describing the loop thus reduce to the precipitation rate (Sect. 2.1) of the trapped particles into the chromosphere and the density of the ambient medium. Additional parameters describe the production of energetic particles: the spectral index and the time evolution of the injected particles which in this model are not instantaneously, but continuously released. Energetic electrons, protons and alphas are assumed injected in the corona at a rate given by

$$q(E,t) = q_0 f(E) f(t), \tag{2.2}$$

where q_0 is the amplitude of the injection and $f(t)$ is given by a simple parametric form with rise-and-fall behavior:

$$\begin{aligned} &f(t) = (t - t_0)(2t_{max} - (t - t_0)) \quad \text{for } t_0 \le t \le 2t_{max} + t_0; \\ &f(t) = 0 \quad \text{elsewhere,} \end{aligned} \tag{2.3}$$

where t_{max} and t_0 are, respectively, the peak time and the starting time of the injection. Here $f(E)$ represents the energy spectrum of injected electrons and ions; $f(E)$ is taken as a power law in energy with a spectral parameter δ defined by

$$f(E) = E^{-\delta}. \tag{2.4}$$

These models compute electron bremsstrahlung radiation and the strong γ-ray lines at 6.129 MeV from ^{16}O, 4.438 MeV from the first excited state of ^{12}C populated directly

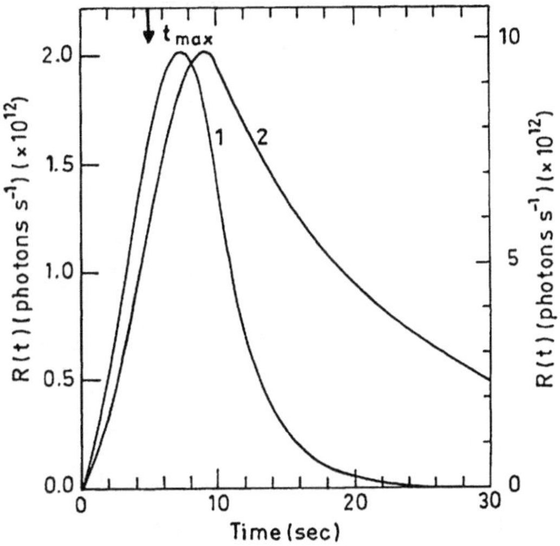

Fig. 7 Time profile of the production of the line at 4.4 MeV in the case of the trap-plus-precipitation model developed by Hulot et al. (1989) in the case of a short injection duration ($2t_{max} = 10$ s) t_{max} and for two values of the precipitation rate $\alpha_0^2/L = 10^{-5}$ km^{-1} and 10^{-6} km^{-1} (from Hulot et al. 1989)

and by the spallation reactions on ^{16}O, and 1.779 MeV from ^{28}Si, 1.634 MeV from ^{20}Ne, 1.369 MeV from ^{24}Mg (with the photospheric target abundance for γ-ray lines). The free parameters are finally the parameter α_0^2/L linked to the precipitation rate, the density, the starting time t_0 and the peak time t_{max} of the injection, the amplitude of the injection and the spectral index of the energy spectrum of injected particles. Figure 7 shows the effect of the precipitation rate on the delay with respect to the injection time and on the time decay of the production of the prompt γ-ray line. Decreasing the value of the precipitation rate has a similar effect on the decay time as decreasing the pitch-angle scattering in the previous model.

Hulot et al. (1989), by using a trap-plus-precipitation model, were the first to calculate the relative time delay of γ-ray flux with respect to the hard X-ray flux. They showed that such delays strongly depend on both the trapping time of the accelerated particles and the density of the loop. In particular, they showed that the time delays of the hard X-ray and γ-ray line emissions observed by *ISEE-3*[4] and *SMM*/GRS during the SOL1980-06-07T03:22 and SOL1981-04-27T09:45 events could be explained by this model and that parameters of the particles and of the ambient model could be obtained. However, these studies were performed using the time profile of the 4–8 MeV range as a proxy for the time profiles of the nuclear lines, and no images at X-rays or γ-rays were used as additional constraints on the model.

A similar analysis was performed by Dauphin and Vilmer (2007) on the first γ-ray line event observed by *RHESSI*, SOL2002-07-23T00:35 (X4.8). Indeed, the analysis of the hard X-ray and γ-ray flux time evolution (Lin et al. 2003; Share et al. 2003a) showed that there is a slight delay between hard X-ray emission at 150 keV and γ-ray line emission. Figure 8 shows the temporal evolution of the HXR emissions at 150 keV and 1 MeV and of the γ-ray line time profile. These time profiles are obtained from the spectroscopic analysis presented in Share et al. (2003a). The spectral analysis is performed for successive time intervals of 20 s, and results in the fitting to observed count spectra of a model photon spectrum including a double power law for the bremsstrahlung continuum, a nuclear γ-ray

[4]International Sun-Earth Explorer-3.

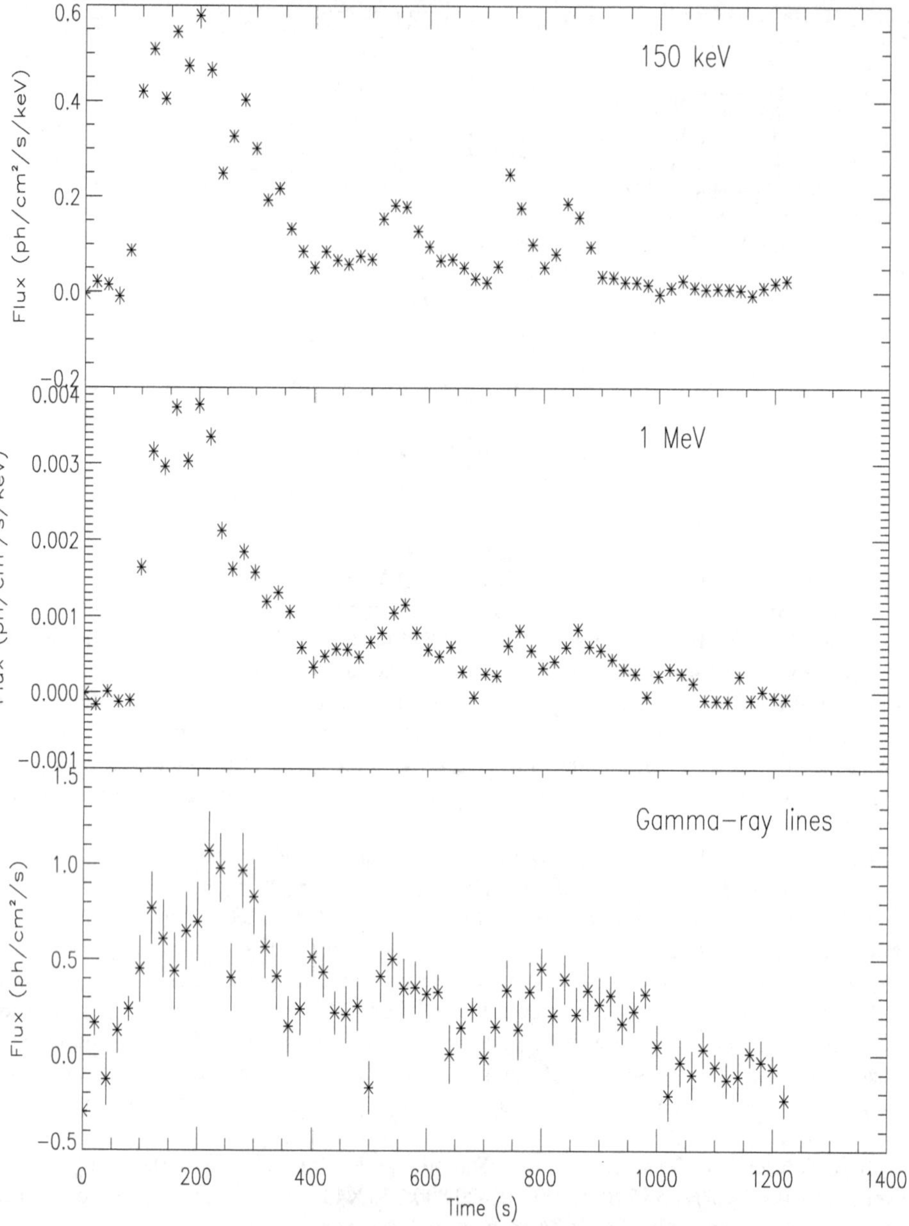

Fig. 8 Time evolution of the bremsstrahlung flux at 150 keV and 1 MeV, and of the prompt de-excitation lines, observed with a time integration of 20 s with *RHESSI* for SOL2002-07-23T00:35 (X4.8). The starting time of the figure is 00:26 UT (from Dauphin and Vilmer (2007), adapted from Share et al. (2003a))

line function made of 15 narrow and broad Gaussians (Smith et al. 2003), a neutron-capture line (Murphy et al. 2003), an α-^{4}He fusion line complex between ∼400 and 500 keV (Share et al. 2003a), and the solar annihilation line at 511 keV and its positronium continuum. Figure 8—adapted from Share et al. (2003a)—shows the results of this fit as a function of time: the bremsstrahlung flux at 150 keV and 1 MeV and the nuclear de-excitation line flux. This figure shows that the hard X-ray and γ-ray time evolutions are roughly similar, indicating a common origin of the accelerated electrons and ions, but that for the first main peak (from 0 s to 400 s) a time delay of around 12 s is observed between the time profile at 150 keV and the γ-ray line time profile. There is, however, no significant delay between the time profile at 150 keV and at 1 MeV.

Dauphin and Vilmer (2007) investigated whether this delay could be reproduced by the trap-plus-precipitation model, given the additional constraint provided by the X-ray and γ-ray images. They found that the time profiles of both HXR and prompt γ-ray line fluxes could be reproduced (given that electrons and ions are injected and partially trapped in different coronal loop systems with slightly different characteristics discussed below) but they also found that the detailed X-ray and γ-ray time profiles could be well reproduced if the ratio of energetic electrons with respect to ions slightly varies from peak to peak during the flare. Such a variability of electron-to-ion production from peak to peak had been observed previously in flares based on spectral analysis (e.g., Chupp et al. 1993) and Dauphin and Vilmer (2007) showed here that a similar result can be found from time-delay analysis. They finally found, given this variability in the electron-to-ion ratio, that a good reproduction of the evolution of both X-ray and γ-ray line time profiles (see Fig. 9) was obtained if compared to electrons—ions were injected in a system of coronal loops of lower density and with a slightly larger mirror ratio leading to a slightly more efficient trapping. If we consider a similar value for the loss cone for the two coronal loops, we find that ions should propagate in slightly longer loop lengths, which is consistent with the imaging observations of this flare (see Sect. 5).

The emphasis above has been on inter-comparison of X-ray and γ-ray time profiles. As discussed in detail by Murphy et al. (2007), analysis of γ-ray measurements on their own can also provide constraints on the loop parameters, e.g., from the flare γ-ray line shapes and the fluence and time evolution of the 2.223 MeV neutron-capture line.

2.3 Characteristics of the Solar Atmosphere During Flares from γ-Ray Line Measurements

2.3.1 Atmospheric Abundances: Deduction from and Consequences for γ-Ray Lines

The fluence in any particular de-excitation line depends on the number and energy distribution of fast particles, and on the abundance of the species that can contribute to it (either directly or via spallation reactions). A set of measured fluences in these lines thus includes information on both source chemical abundances, and the numbers and energy distributions of the exciting particles. If the energy distributions of the fast particles were known, we could deduce a set of source relative abundances from the measured fluences. Solar photospheric abundance determinations are fundamental to an enormously wide range of astrophysical topics. Determinations in the γ-ray source region would complement these, extending knowledge of the outer solar atmosphere. This could be particularly important at a time when 3-D radiative-transfer modeling has produced several significant revisions of solar heavy element abundances (Asplund et al. 2005), with the caveat that exceptional conditions might occur in the flaring atmosphere.

If, on the other hand, the source chemical abundances were known, we could deduce the energy distributions of protons and α particles in principle. This depends on the extent to which the various cross-sections have different dependences on fast particle energy (Toner and MacKinnon 2004). In practice neither of these sets of variables is known with much

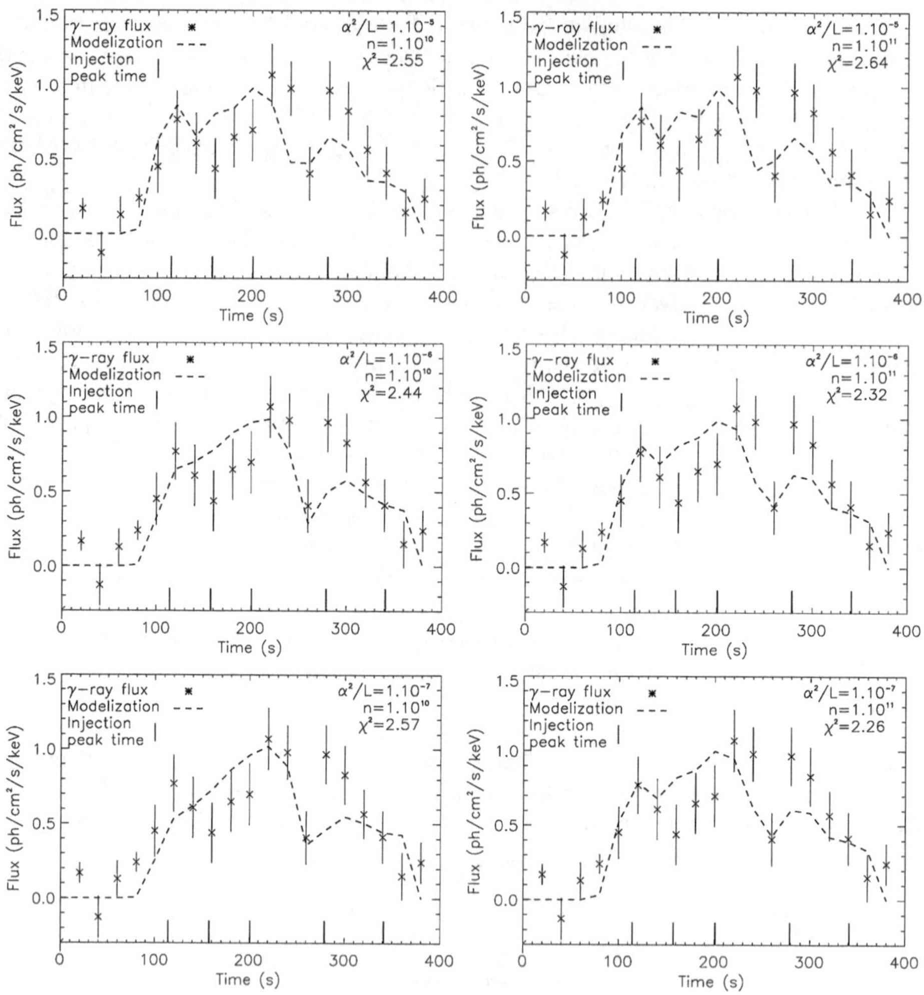

Fig. 9 Time evolution of the observed and computed γ-ray flux derived for the RHESSI SOL2002-07-23T00:35 (X4.8) flare for different parameters of the trap plus precipitation model (from Dauphin and Vilmer 2007)

reliability, and both the abundances and the parameters of the energy distributions should be deduced together. To date, the *SMM*/GRS work of Murphy et al. (1991) remains the only published study where source abundances and particle energy distributions are determined together. Other studies have determined particle distributions using, e.g., coronal or photospheric abundances (Ramaty et al. 1995), further deciding between these possibilities using goodness-of-fit tests. Trends in line fluence ratios across many events have been used to make at least qualitative statements (Share and Murphy 1995); with many events one can attempt to disentangle the separate influences of particle distributions and abundances. Target abundances may be determined non-parametrically or as a choice among several alternative sets. Ion energy distributions are always represented by assumed analytical forms, however (e.g., power-law or Bessel function K_2 in momentum) whose parameters are determined as best fits to data. From these various studies, certain general statements may be made (see below)—always noting that these might be subject to change via future revisions of nuclear cross-sections or refinements of data analysis techniques. We now survey some results of these studies.

Abundances of several species are enhanced compared to standard photospheric values. As is found using other measures of coronal abundances (e.g., Meyer 1993), enhancements may be ordered by first ionization potential (FIP) of the species. Low-FIP elements (Mg, Si, Fe) generally show enhancements by an amount that varies from one event to the next but may be as great as 3–4. There is a definite "FIP effect," with low-FIP elements showing more variability than high. At least in *SMM*/GRS data, the 1.63 MeV line from ^{20}Ne needs both a ^{20}Ne enhancement (Ne/O $= 0.25$, rather than the photospheric value of 0.15) and a steeply falling ion energy distribution that can exploit its low threshold for excitation by protons (Share and Murphy 1995). Although an enhanced Ne abundance in the solar interior might have helped with certain problems in understanding helioseismology data (Drake and Testa 2005), there is no convincing evidence for such an enhancement at other wavelengths (Schmelz et al. 2005). It now seems likely that we are identifying some peculiarity of the flaring atmosphere rather than the normal state of affairs in the non-flaring Sun.

Rather than a complete fit of the whole spectrum, one often tries to make progress using ratios of pairs of line fluences, for instance the ratio mentioned in Sect. 2.2.5 of fluences in the 1.63 and 6.13 MeV lines. Let Φ_ϵ stand for the measured fluence (cm^{-2}) in a line at photon energy ϵ (keV). Then the assumed abundances of the emitting species play a crucial role. With a photospheric or a coronal abundance ratio for neon and oxygen, one can only accommodate the observed values of $\Phi_{6.13}$, $\Phi_{1.63}$ and $\Phi_{0.4-0.5}$ by allowing protons and αs to have significantly different energy distributions (Toner 2004; Toner and MacKinnon 2004). Because the true degree of enhancement of ^{20}Ne is uncertain, the proton and α energy distributions are correspondingly uncertain. This uncertainty adds to that introduced by the different energy-dependence of the cross-sections for excitation of the lines by protons and αs. For example, assuming $\alpha/\mathrm{p} = 0.1$ among the fast particles, $\Phi_{6.13}/\Phi_{1.63}$ for SOL1981-04-27T09:45 (X5.5) implies a power-law energy spectral index δ roughly in the range 4.3–4.8 if Ne/O $= 0.15$, but 3.6–4.1 if Ne/O $= 0.25$ (Ramaty et al. 1995), the former value believed appropriate to the photosphere (Asplund et al. 2005). Ratios involving the 2.223 MeV line give a valuable further constraint in the face of this uncertainty. All target species can contribute to free neutron production so some uncertainty in the abundance of any one has much less impact on deductions. On the other hand, the intensity of the 2.223 MeV line depends on other quantities that must be constrained, if at all, by other means: pitch-angle distribution and magnetic geometry in which the fast ions move.

Usually one needs to integrate over the whole of a flare to get a γ-ray spectrum of adequate statistical quality. There have, however, been a few attempts to study the time development of the γ-ray spectrum. Murphy et al. (1997) found evidence for an increase with time of the abundances of low-FIP elements relative to high-FIP ones, from the *CGRO*/OSSE data of SOL1991-06-04T03:37 (X12.0). Evidence for a similar trend was found in *RHESSI* data for the SOL2002-07-23T00:35 (X4.8) by Shih et al. (2003). Share and Murphy (2006) additionally show such a trend in the flare of SOL2003-11-03T09:55 (X3.9). It is interesting that a similar trend is found in data from three different flares, involving two completely different instruments, although the ubiquity or otherwise of this effect remains to be shown. Assuming, unjustifiably, that this effect proves commonplace we may make some preliminary comments. We can imagine at least three reasons why γ-ray source region abundances might change (see also MacKinnon 2006). First, abundances might vary with position in an active region, and the regions of fast ion precipitation change as the flare progresses. It is not at first easy to see why this would produce a systematic change in the ratio of low-FIP to high-FIP elements, however. Second, abundances may well vary with height in the atmosphere. Even if the atmosphere remains static as the flare progresses, an ion energy distribution evolving in time (e.g., hardening) would sample an evolving range of atmospheric heights and thus relative abundances. Third, relative abundances might indeed evolve in the source region during the flare (e.g., Winglee 1989).

Kiener et al. (2006) studied the *INTEGRAL* data from the SOL2003-10-28T11:10 (X17.2) flare, noting a change in time of the relative intensities of the 4.44 MeV ^{12}C and 6.13 MeV ^{16}O lines. With line shape data providing additional constraints, their preferred explanation involves a change in time of the α/p fast ion abundance ratio, rather than the abundances of the target species. This example shows the importance of the extra information obtainable from line shapes.

The time development of the 2.223 MeV line gives a quite different sort of constraint on the abundance of ^{3}He. Indeed, after thermalization (typical duration around 100 s), the free neutrons produced in nuclear reactions are captured in the solar photosphere in two competitive processes: radiative capture on protons to form deuterium (and the 2.2 MeV line) and non-radiative capture on ^{3}He (^{3}He(n, p)^{3}H). This method was first described and employed by Wang and Ramaty (1974). They showed that the greater the ^{3}He abundance, the more rapid the decay of the 2.2 MeV line. The determination of ^{3}He/H was thereafter done on a few flares observed by *HEAO-1* (Hudson et al. 1980), *SMM/GRS* (Chupp et al. 1981; Hua and Lingenfelter 1987a), *GRANAT*/PHEBUS (Trottet et al. 1993), *CGRO*/OSSE (Share and Murphy 1998), *CGRO*/EGRET (Dunphy et al. 1999; Dunphy 1999), *CGRO*/COMPTEL (Rank et al. 2001) and *Yohkoh*/GRS (Yoshimori 1999a); see, e.g., Mandzhavidze and Ramaty (2000) for a review of observations before the *RHESSI* era. An upper limit on the photospheric ^{3}He/H of $1.6 \pm 0.1 \times 10^{-5}$ was found in the review of Mandzhavidze and Ramaty (2000). The analysis of the time profile of the 2.223 MeV line was also achieved for SOL2003-07-23T00:35 (X4.8) observed with *RHESSI* (Murphy et al. 2003) to get a measurement of ^{3}He/H. It is one of the factors included in the analysis of Naiman et al. (2008).

2.3.2 Conditions in the Positron Annihilation Region

Positron production by flare fast ions has been studied in detail by Kozlovsky et al. (1987), Murphy et al. (1987) and Kozlovsky et al. (2004). Positrons may be produced via the decay of radioactive daughter nuclei, beta decay of excited states of target nuclei, or via π^+ production, e.g., in p(p,nπ^+X)p$'$. In the first two cases cross-sections have thresholds mostly in the 1–10 MeV energy range, while π^+ production starts at 200–300 MeV/nucleon. Thus positron production, slowing down and annihilation will occur over a wide range of heights and the resulting γ-rays potentially carry information on deep layers of the atmosphere.

The very broad energy range of ions contributing to positron production is illustrated in Fig. 10 from Murphy et al. (2007). Effective accelerated ion energies are shown for the production of positrons for accelerated ^{3}He/^{4}He $= 0$ (white boxes) and 1 (grey boxes) and for accelerated α/proton $= 0.1$ (panel a) and 0.5 (panel b). For very hard spectra dominated by protons, the effective ion energies range from 0.1 to $>$1 GeV/nucleon because the positrons result mainly from very high-energy reactions producing charged pions. At the opposite extreme, positron production from soft energy distributions rich in αs and ^{3}He is dominated by ions of 1–15 MeV/nucleon and even includes a significant contribution from ^{3}He above $\sim$1 MeV/nucleon. Note that the effective energies shown in Fig. 10 are for the production of *positrons*. The escaping 0.511 MeV annihilation photons can be significantly attenuated if the positrons are produced deep in the solar atmosphere. Pion production can be very deep

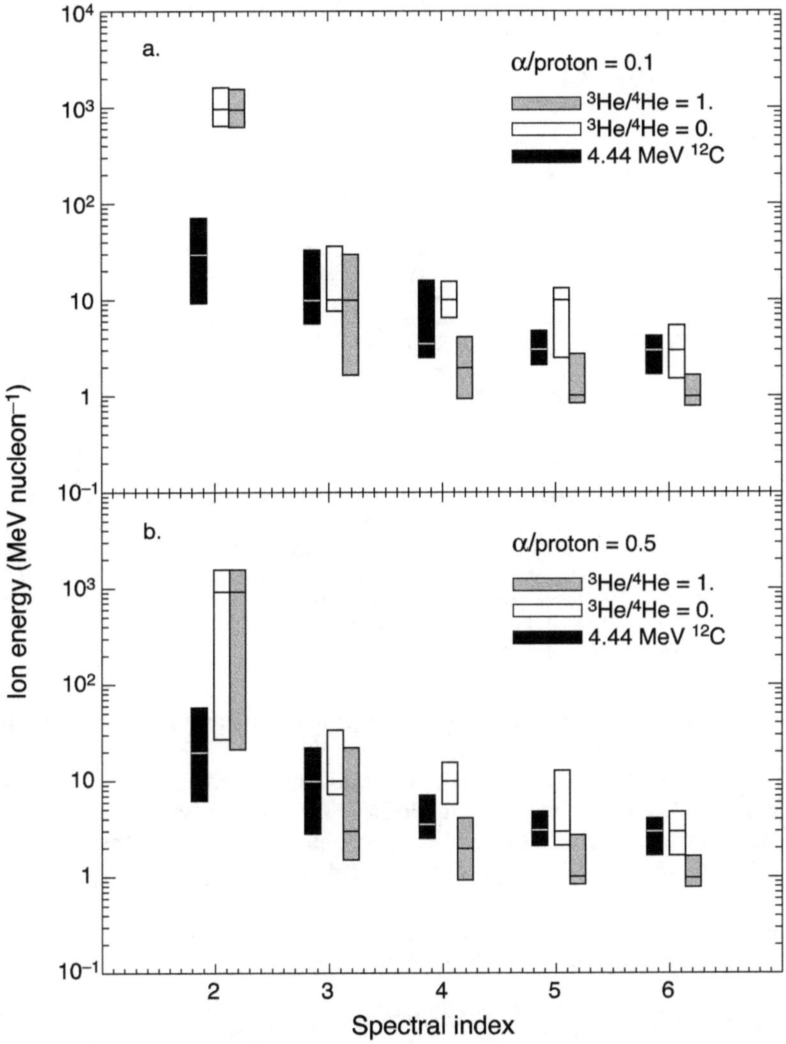

Fig. 10 Effective accelerated ion energies for production of positrons as a function of spectral index for accelerated ^{3}He/^{4}He = 0 (*white boxes*) and 1 (*grey boxes*). The effective energy ranges are the 50% yield range as defined in Sect. 2.2.5. Panel (**a**) is for accelerated α/proton = 0.1 and panel (**b**) for 0.5. Also shown are the effective energies for the 4.438 MeV ^{12}C line (*black boxes*) from Fig. 6 (from Murphy et al. 2007). Reproduced by permission of the AAS

since the high energy ions responsible have very long ranges. As a result, even for very hard spectra, the annihilation photons that escape can be mainly from decay of radioactive nuclei rather than pions. Such nuclei are produced by ions with energies that are much lower than those indicated in the figure.

Positron annihilation and consequent radiation are dealt with in great detail by Murphy et al. (2005). Positrons must slow down and thermalize to contribute to 0.511 MeV radiation. They may annihilate directly on free electrons, or via positronium formation. Positronium formation can involve both free or atomic electrons, in the latter case via charge exchange re-

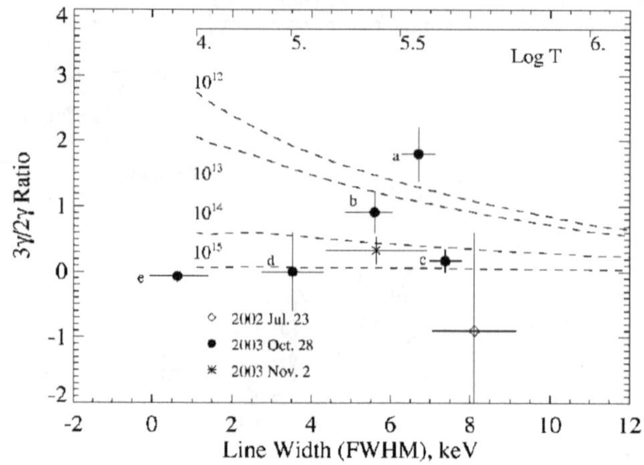

Fig. 11 *RHESSI* measurements of the $3\gamma/2\gamma$ ratio vs. 511 keV line width for three flares. Time intervals for SOL2003-10-28T11:10 (X17.2) are: (*a*) 11:06:20–11:08:20 UT, (*b*) 11:08:20–11:10:20 UT, (*c*) 11:10:20–11:16:20 UT, (*d*) 11:16:20–11:18:20 UT, and (*e*) 11:18:20-11:30:20 UT. The SOL2003-11-02T17:25 (X8.3) measurement was made between 17:16 and 17:26 UT, when the line was broad. The data point when the line was narrow is consistent with point (*e*). The SOL2002-07-23T00:35 (X4.8) measurement was integrated over the entire flare. The temperature scale and curves showing the calculated $3\gamma/2\gamma$ ratio vs. 511 keV line width for different densities are for a fully ionized medium (from Share et al. 2004). Reproduced by permission of the AAS

actions. Positronium in turn may exist as para-positronium (net spin 0) or ortho-positronium (net spin 1). Para-positronium annihilates to give two photons in the 0.511 keV line (2γ mode). To conserve spin, ortho-positronium must annihilate producing three photons, at energies $\leq$ 0.511 MeV (3γ mode). The lifetimes of the two spin states are different, so the proportions that annihilate (i.e., the ratio of the two-photon (line) and three-photon (continuum) spectral components) depend on the rate of collisional destruction and thus on ambient density.

Positrons in an ionized medium of temperature T (K) produce a line of FWHM (full width at half maximum) width σ (keV) (Murphy et al. 2005)

$$\sigma = 0.011T^{1/2}. \tag{2.5}$$

For low temperatures an additional, broader component may be discriminated, from positronium formed via charge exchange in flight. Equation (2.5) is valid for 0.511 MeV line emission formed via both direct annihilation and para-positronium formation on free electrons, but not for positronium formation via charge exchange with atoms or ions. Thus the ionization state of the medium also influences the shape and width of the line.

In summary, then, the shape of the line and the relative intensities of the 2γ and 3γ components of the annihilation spectrum include information on the temperature, density and ionization state of the positron-annihilation region

RHESSI has observed 0.511 MeV line emission from the flares SOL2002-07-23T00:35 (X4.8), SOL2003-10-28T11:10 (X17.2), SOL2003-11-02T17:25 (X8.3) and SOL2005-01-20T07:01 (X7.1) (Share et al. 2003a, 2004; Murphy et al. 2005). During the greater part of these flares, until quite late times, the width of the line was determined to be 4–8 keV (see Fig. 11).

The SOL2003-10-28T11:10 flare provided the best-observed 0.511 MeV line. For the first 10 minutes of the flare its width is between ~5 and 8 keV. Equation (2.5) then im-

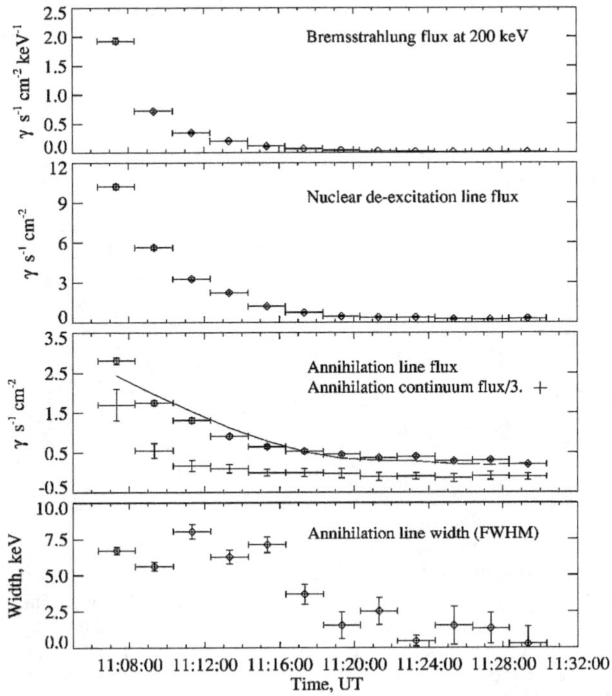

Fig. 12 Time histories of the bremsstrahlung, total nuclear de-excitation line, annihilation line and positronium continuum fluxes, and the width of the annihilation line during SOL2003-10-28T11:10 (X17.2) (from Share et al. 2004). Reproduced by permission of the AAS

plies an ambient temperature of $\sim 5 \times 10^5$ K for a width $\sim$8 keV (see Fig. 12). The measurements of the positronium continuum to line ratio ($3\gamma/2\gamma$ ratio) implies an ambient density in the range $\sim 10^{13-15}$ cm^{-3} (see above) (Murphy et al. 2005). In this initial phase the positron annihilation region appears characterized by an unexpected combination of chromospheric density and transition-region temperature (Share et al. 2004; Murphy et al. 2005). (Note that the similarly broad line found in the SOL2002-07-23T00:35 (X4.8) carries similar implications (Share et al. 2004)). The line then narrows in $\sim$2 min, settling down to a width of $\sim$1 keV for the remaining 12 min of the flare (Fig. 12). The best explanation of the late-time narrow line is in terms of a medium at a temperature of $\sim$5000 K, but with a relatively high degree of ionization ($\sim$20%), to suppress the broader feature that results from positronium formation in flight (Fig. 11).

The line in SOL2003-11-02T17:25 (X8.3) is more consistent with a "normal" solar atmosphere, specifically with the conditions in the layer of the Vernazza et al. (1981) model C atmosphere at which the temperature is 5000 K. So far the 0.511 MeV line shapes have been calculated for single regions with homogeneous conditions. As we see, these yield significant insight into at least average annihilation region conditions. There remains a need for such calculations to take account of the broad range of depths over which positrons will be produced, and the consequent wide range of conditions in which they will annihilate.

The conditions found for SOL2003-10-28T11:10 (X17.2) are particularly surprising, unlike those found at any layer of existing models for quiet-Sun or flaring atmospheres. Raymond et al. (2007) study lines of O VI UV lines and X-ray emission from SOL2002-07-23T00:35 (X4.8). The density and temperature implied by the positron annihilation line would have to apply only in a very narrow layer, $\sim$100 m, to be consistent with the measured O VI line strengths. Even if such a narrow layer could be produced, bearing in mind the range of heights over which fast electrons would deposit energy, we would also have

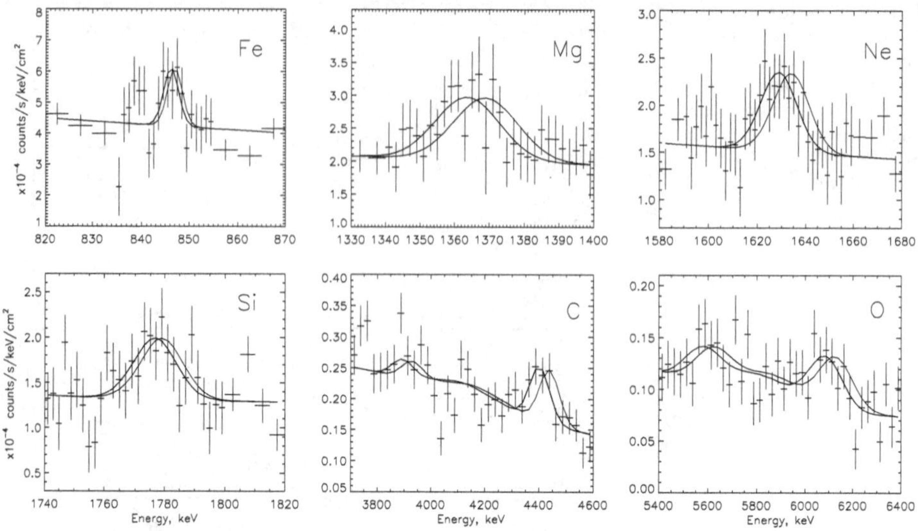

Fig. 13 *RHESSI* background-subtracted count spectra for SOL2002-07-23T00:35 (X4.8). Each panel is labeled with the element primarily responsible for the line shown. The carbon and oxygen lines also show the secondary peak from the escape of a 511 keV positron-annihilation photon, which also contains information on the line shape. The *thick curve* shown in each panel is the Gaussian fit with a redshift of respectively 0.11% for the Fe line, 0.40% for the Mg line, 0.32% for the Ne line, 0.12% for the Si line, 0.79% for the C line and 0.58% for the O line plus the underlying bremsstrahlung continuum and broad lines, convolved with the instrument response. The *thinner line* is the same fit forced to zero redshift for comparison. The *error bars* are one σ from Poisson statistics (from Smith et al. 2003). Reproduced by permission of the AAS

to find some mechanism which would localize most positron annihilations there. A multi-wavelength study of this flare (Schrijver et al. 2006) further underlines the difficulty of accounting for the positron annihilation line width.

2.3.3 Inclination of Magnetic Field Lines

With the unprecedented spectral resolution in the γ-ray domain provided by *RHESSI*, Smith et al. (2003) were able to resolve for the first time the shapes of several of the strongest flare γ-ray lines in SOL2002-07-23T00:35 (X4.8), determining line widths and Doppler shifts (Fig. 13) (for a complete discussion of the interpretation of line shapes, see Sect. 2.6). All lines were found to be redshifted by amounts that decrease with increasing target species mass, as expected for kinematic recoil. The absolute values of these redshifts were larger than expected for a flare at S13E72, however, on the assumption of field lines normal to the solar surface. Smith et al. (2003) thus suggested that field lines might be inclined to the vertical, in this flare at least by $\sim 40°$. The alternative explanation, that fast ions are highly beamed, appears to be inconsistent with deductions from the α–α line complex in the 0.4–0.5 MeV range (Share et al. 2003b), as well as potentially raising difficult questions of velocity space stability (Tamres et al. 1989). Systematic field line tilts to the vertical have been suspected in the past, on the basis of Zeeman split lines formed at different heights (Wiehr 1978). MacKinnon and Brown (1990) had also suggested such a systematic tilt because of an apparent asymmetry in the azimuthal distribution of events visible at photon energies >10 MeV. Field lines will fan out with altitude as a result of pressure balance (Gabriel 1976) but this would produce only a broadening, not a systematic red or blue shift.

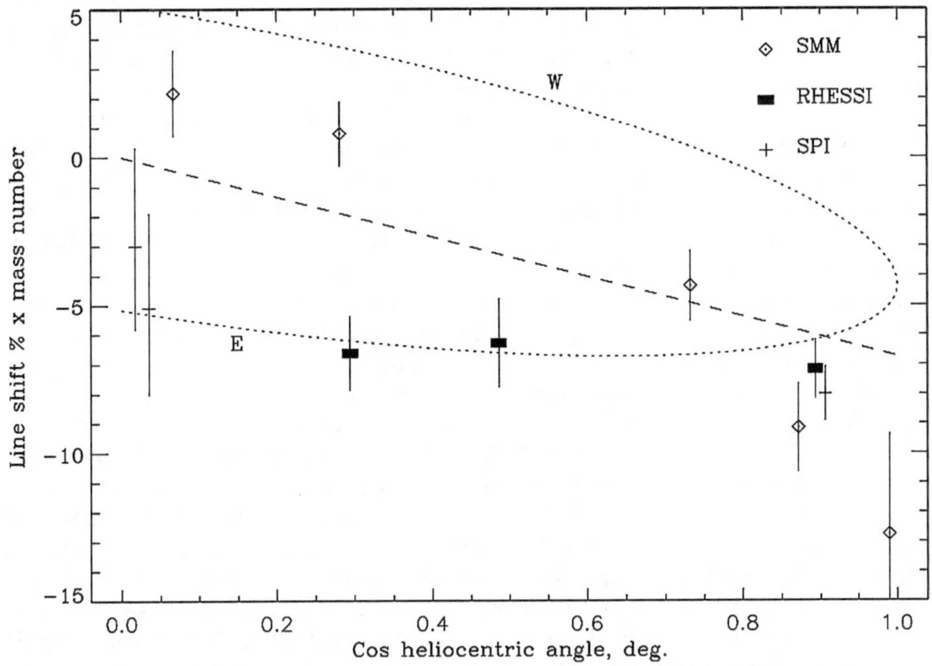

Fig. 14 Doppler shifts of ^{12}C and ^{16}O line energies from rest, multiplied by the atomic mass of the recoiling nucleus. *Dashed line*: expected Doppler shift from a simple model of 10 MeV protons incident on ambient solar atmosphere nuclei in a downward isotropic beam at 90°. *Dotted lines*: expected shifts if the protons are incident along magnetic field lines tilted at 50° to the horizontal. If this is the explanation for the *RHESSI* point at $\cos\theta = 0.29$ towards the east limb, then the Doppler shifts should follow the branches labeled "E" and "W" to the east and west of the Sun's center respectively. Sources of data points are from *SMM* (19 measurements in bins of heliocentric angle; Share et al. 2002), *RHESSI* measurements for the flares SOL2002-07-23T00:35, SOL2003-10-28 and SOL2003-11-02 (Share and Murphy 2006), and *INTEGRAL*/SPI measurements for the flares SOL2003-10-28 (Kiener et al. 2006), SOL2003-11-04T19:53 and SOL2005-09-07T11:10 (Harris et al. 2007); from Harris et al. (2007)

Harris et al. (2007) studied *INTEGRAL* spectra of SOL2003-11-04T19:53 (X17.4) and SOL2005-09-07T09:52 (X17.0). These took place on the west and east limbs respectively. Systematically tilted loops would produce a systematic difference in Doppler shifts: the 40° tilt suggested by Smith et al. (2003) would imply blueshifted lines from the west limb. The two flares studied both display (statistically weak) evidence for redshifts, but none for blueshifts. Thus it appears that both flares plus SOL2002-07-23T00:35 (X4.8) involve field lines oriented predominantly away from the observer, but that there is no evidence so far for a systematic tilt in any preferred sense (Fig. 14). The small number of flares renders this study preliminary. At the heart of the interpretation is a picture of unidirectional field lines, either normal to the solar surface or all tilted at a single angle to the vertical, instructive but very simple. Major γ-ray flares originate from complex active regions and will undoubtedly involve a broad distribution of magnetic field directions.

2.4 Narrow Lines and Ion Energy Distribution

Since the first detection of solar γ-ray lines in 1972, many solar γ-ray line flares have been observed with detectors aboard *SMM*, *Hinotori*, *GRANAT*, *CGRO* and *Yohkoh*. The analysis of these observations have led to quantitative analysis for more than 20 γ-ray line events. Apart from information on the solar atmospheric elemental abundances which have been deduced from the analysis of narrow γ-ray lines (see previous sections), energy spectra of protons have been deduced for 19 *SMM*/GRS flares (Share and Murphy 1995; Ramaty et al. 1995) and for one *CGRO*/OSSE flare (Murphy et al. 1997). Each line is characterized by energy-dependent cross-sections for excitation by protons and by α particles. Thus the measured fluence in each line provides a sum of energy-weighted measures of the numbers of protons and αs (e.g., Ramaty 1986), above the lesser of the threshold energies for proton and α particle excitation. The measured fluences in more than one line contain information on the energy distributions of protons and αs. The 1.63 MeV cross-section of ^{20}Ne has a lower cross-section than the other de-excitation lines (Share and Murphy 1995), so the ratio of the fluences in the lines at 1.63 and 6.13 MeV, for instance, gives a measure of fast ion spectral hardness, representing a variable range of ion energies of around 1–20 MeV (see Sect. 2.2.5). Higher ion energies, on average, are needed to liberate neutrons in nuclear collisions so the ratio of the 2.223 MeV line to one or more of the de-excitation lines similarly gives a measure of spectral shape. The authors listed above assumed the power-law form for ion energy distributions and deduced the values of energy power-law spectral index from either or both of these measures, for the 20 flares observed by *SMM* or *CGRO*. The measured ratios imply that the accelerated ion proton spectra should extend as unbroken power laws down to at least about 2 MeV/nucleon if a reasonable ambient Ne/O abundance ratio is used, i.e., in agreement with measurements of the Ne/O abundance ratio in the corona (Ramaty et al. 1995). The spectrum deduced is thereafter used to estimate the energy contained in accelerated protons above 1 MeV. This last determination is, however, largely dependent on the value of the ambient Ne/O abundance ratio and also on the composition of accelerated particles, in particular the ratio of energetic He nuclei to energetic protons (α/p) (see Ramaty et al. 1995).

Four γ-ray line events so far observed with *RHESSI* provide new opportunities for such analyses: SOL2002-07-23T00:35 (X4.8) (see Fig. 15) (see Lin et al. 2003; Smith et al. 2003; Share et al. 2003a), SOL2003-10-28T11:10 (X17.2) (Share et al. 2004); SOL2003-11-02T17:25 (X8.3) and SOL2005-01-20T07:01 (X7.1) (see Share and Murphy 2006, for a review). *RHESSI* has for the first time the energy resolution necessary to resolve all the γ-ray lines, except the intrinsically narrow 2.223 MeV line, and to determine the detailed line shapes expected from Doppler shifts, thus allowing deduction of velocity distributions of the interacting energetic ions (Smith et al. 2003). The comparison of fluxes in the ^{20}Ne line at 1.63 MeV and of the ^{12}C and ^{16}O lines at 4.44 and 6.13 MeV was performed for these four flares to provide information on the proton spectrum. Apart from SOL2002-07-23T00:35, for which the proton spectral slope was found to be $\sim$3.5 (Lin et al. 2003), the other events have been reported to have much harder slopes ($\sim$2.5) (Share and Murphy 2006), harder than the average ($\sim$4.3) measured previously for the 19 *SMM* flares. New analyses of the most recent events are currently being undertaken using the improved nuclear continuum modeling developed in Murphy et al. (2009), which could lead to softer proton spectra than was estimated in Share and Murphy (2006).

There are also well-resolved *INTEGRAL* spectra for three flares: SOL2003-10-28T11:10 (X17.2 (Kiener et al. 2006), SOL2003-11-04T19:53 (X17.4) and SOL2005-09-07T09:52 (X17.0) (Harris et al. 2007). The proton spectral index was deduced for these events from the ratio of the 2.2 MeV line to the sum of the 4.44 and 6.13 MeV lines. Values of the power-law index in the range [3.5, 4] were found. For SOL2003-10-28T11:10 (X17.2), observed both by *RHESSI* and *INTEGRAL*, the difference in the deduced power-law indices may result from the different methods used to infer the spectral index from observations. New analysis

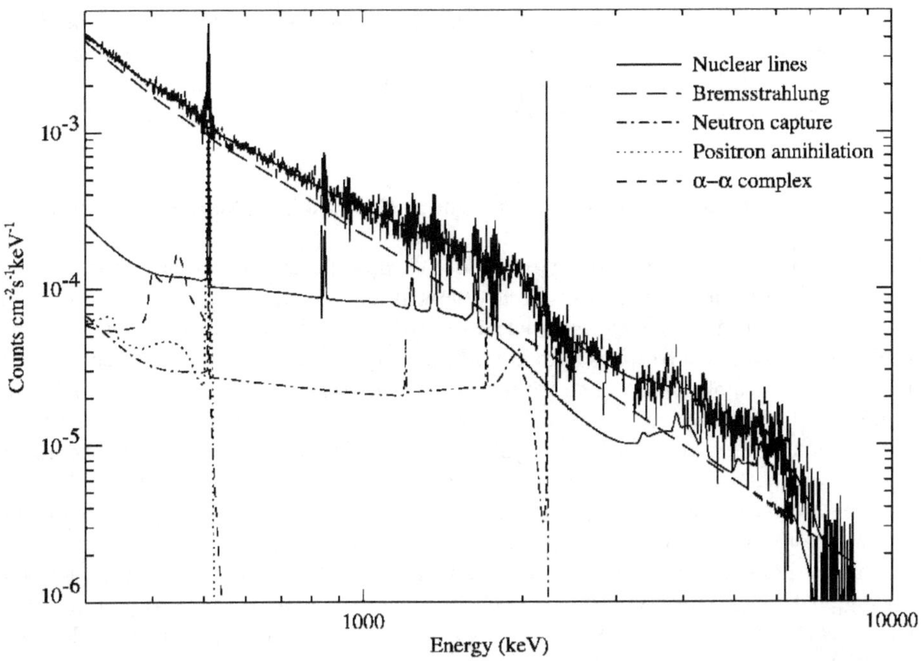

Fig. 15 *RHESSI* γ-ray count spectrum for SOL2002-07-23T00:35 (X4.8). The *lines* show the different components of the model used to fit the spectrum (from Lin et al. 2003). Reproduced by permission of the AAS

using improved nuclear line modeling (Murphy et al. 2009) could help to understand the apparent existing discrepancy.

2.5 Alpha Particles

Both protons and αs excite most of the strong de-excitation lines. Definitive information on the relative numbers of protons and αs must be obtained by determining fluences of lines to which only one or the other species contribute. Information on the α/p ratio had been deduced for five flares observed with *SMM* (Share and Murphy 1998; Mandzhavidze et al. 1999) from the fluence ratio of two lines: the prompt Fe line at 0.847 MeV which is produced by the interaction of accelerated protons and αs on ambient iron, and a pure α line at 0.339 MeV which only results from the interaction of energetic α particles on iron, producing an excited state of nickel. It was found by Mandzhavidze et al. (1999) that for these five flares α/p exceeds the standard value of 0.1 and can even reach 0.5. Nuclear interactions of accelerated ^{3}He with ambient ^{16}O result in three γ-ray lines at 0.937, 1.04 and 1.08 MeV. While the fluence of the line at 0.937 MeV could be determined for a few flares observed by *SMM*/GRS and *CGRO*/OSSE, the other two lines cannot be separated from α lines at 1.05 and 1.00 MeV (Share and Murphy 1998). These lines combine to yield to an unresolved feature centered at 1.02 MeV. Using the information from this unresolved feature and from the line at 0.937 MeV, an estimate could be obtained of the ratio of accelerated ^{3}He with respect to accelerated ^{4}He. In 7 flares this led to an enhancement of this ratio (0.1 to 1) with respect to coronal values (Mandzhavidze et al. 1999). In these analyses it was assumed that all accelerated species have the same energy distributions. As will be discussed in Sect. 2.7, Toner and MacKinnon (2004) attempted to show how line fluences might be analyzed without this assumption. Finally, observations with good energy resolution help to reach stronger conclusions on these issues, and the additional information from γ-ray line shapes also contributes, as discussed in the next section, to the determination of the α/p ratio.

2.6 Angular Distributions of Ions and α/p Ratio as Deduced from Narrow Line Shapes

Profiles of narrow γ-ray lines can be used to derive the angular distributions of the emitting ions (in the interaction site) and of the accelerated ions, if the emission mechanisms are coupled with transport models. The nuclear excited states produced by collisions between accelerated and ambient particles have indeed very short lifetimes so that the decay γ-ray is emitted with a Doppler shift due to the recoil from the original collision. Furthermore, the collisions with the heavier α particles will produce greater recoil and should produce a broad "red" tail to the line shape. Fitting this tail, i.e., looking at the line shape, can constrain the α/p ratio independently of line fluxes and ratios. Accelerated nuclei heavier than He also produce flare γ-ray lines but they are largely broadened and merge into a quasi-continuum γ-radiation see Sect. 2.11). The shape of the narrow γ-ray lines ultimately depends on the angular distribution and spectrum in the interaction site, the assumed α/p ratio, and the viewing angle of the observer (see, e.g., Ramaty and Crannell 1976; Ramaty et al. 1979; Murphy et al. 1988; Werntz et al. 1990; Lang and Werntz 1991; Kiener et al. 2001; Murphy et al. 2007).

Before the launch of *RHESSI* and *INTEGRAL*, line redshifts and widths had been studied for 19 flares using data from *SMM* (Share and Murphy 1995; Share et al. 2002). Measurements of energies and widths of the strong flare γ-ray lines of C, O and Ne were performed as a function of heliocentric angle using data with moderate spectral resolution provided by *SMM*. Flares were grouped in five bins in heliocentric angle, with analysis performed on these sums of flares located in the same heliocentric interval (see Fig. 14). Redshifts of the Ne, C and O lines of the order of 1% are found for flares close to the disk center but no redshifts are found for flares close to the limb. The mean widths of the de-excitation lines were of the order of 3% and did not exhibit any significant variation with heliocentric longitude. It should be noted that the line shapes result from the γ-ray spectral analysis and are thus not completely independent of the way the analysis is performed. The observed redshifts are compared with predictions of line shapes in the case of angular distributions in the interacting sites being represented by a downward beam, a fan beam, and a downward or upward isotropic distribution. Assuming a mean spectral index for ions of ~4.2, $\alpha/\mathrm{p} = 0.3$ and Ne/O = 0.25, Share et al. (2002) concluded that the measured redshifts as a function of heliocentric angle are not consistent with beams in the interacting site but are rather consistent with a broad angular distribution in the downward direction. These angular distributions are of course descriptive of the interaction sites and should be related to the angular distributions in the acceleration region itself by developing transport models (see Murphy et al. 2007).

Line widths were clearly observed for the first time in the flare SOL2002-07-23T00:35 (X4.8) observed with *RHESSI* (Smith et al. 2003; Share et al. 2003a) (see Fig. 13). It was found that redshifts varied as expected with the mass of the target species, but were larger than expected for line production by vertically precipitating ions in a flare at this location. The large shifts obtained rather suggest that the magnetic loops along which ions propagate are tilted with respect to the vertical (see Sect. 2.3). Two other γ-ray line flares have been observed with *RHESSI* for which line shapes have been analyzed

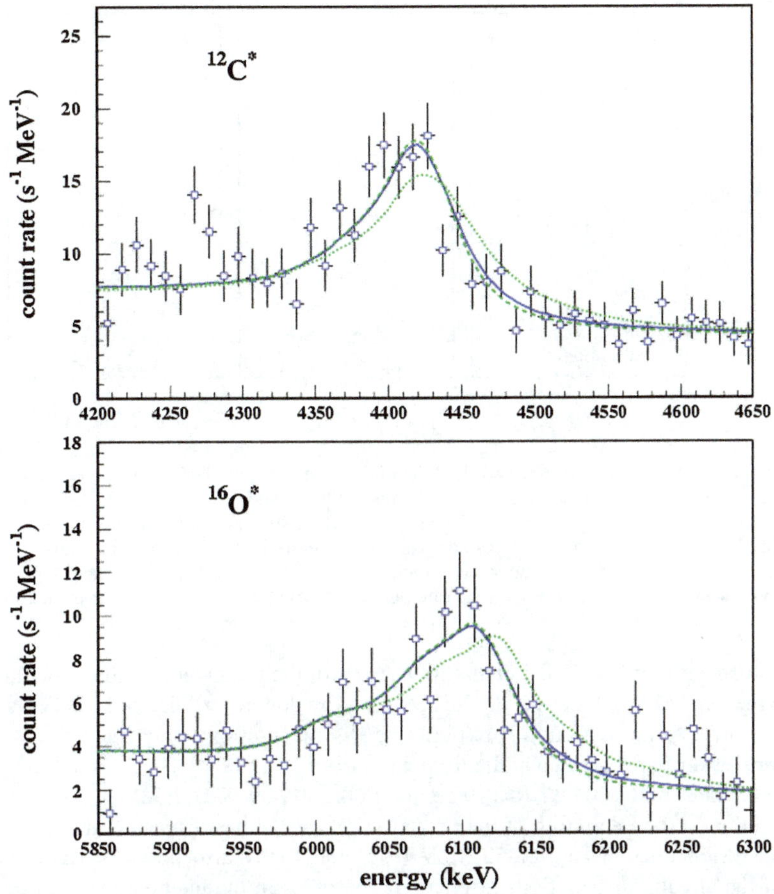

Fig. 16 Observed and calculated line shapes of the 4.44 and 6.13 MeV ambient ^{12}C and ^{16}O de-excitation lines observed by *INTEGRAL*/SPI for SOL2003-10-28T11:10 (X17.2). The *dashed line* represents the calculated line shape for a downward isotropic distribution. The *full* and *dotted lines* represent the calculated line shapes for the distribution computed with pitch-angle scattering as in Murphy et al. (2007) for $\lambda = 30$ (*full line*), $\lambda = 300$ (*dotted line*) (from Kiener et al. 2006)

(SOL2003-10-28T11:10 (X17.2) and SOL2003-11-02T17:25 (X8.3)) (see, e.g., Gan 2005; Share and Murphy 2006). The SOL2003-10-28T11:10 (X17.2) event was also observed at high spectral resolution with the *INTEGRAL*/SPI spectrometer (Gros et al. 2004; Kiener et al. 2006), which provided high-resolution data for the 4.44 MeV ^{12}C and 6.13 MeV ^{16}O lines (see Fig. 16). The comparison of the results of detailed calculations of line shapes to the observations places strong constraints on the α/p ratio (Kiener et al. 2006). In particular, the observed line width is inconsistent with an interpretation where too many α particles relative to the number of protons would be produced or otherwise the lines would be too broad when compared with the observations (Fig. 16). However, the line shapes depend on many parameters: the angular distribution of the interacting ions, the spectral index of the energetic ions, and the α/p ratio; these cannot be determined independently so the analysis results in extended regions of allowed parameter space.

Using a spectral index between 3 and 4 deduced from the ratio of the 2.2 MeV to the fluence at 4.4 and 6.1 MeV (Gros et al. 2004), simultaneous fits of the C and O line shapes

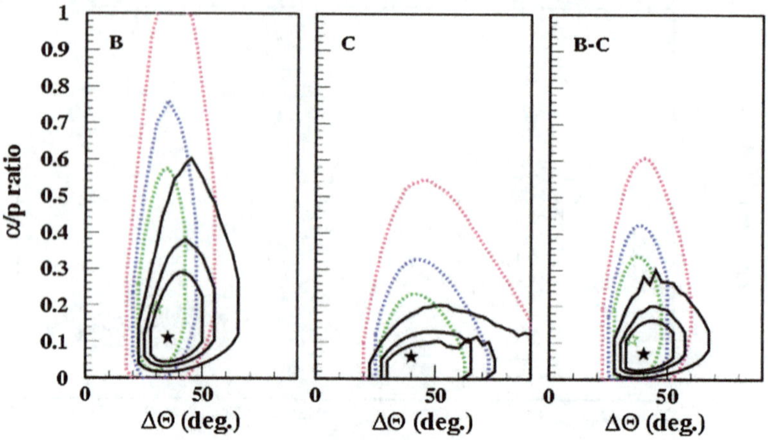

Fig. 17 *Contour lines* in the α/p–$\Delta\Theta$ (width of the injected particle distribution) parameter space of downward directed particle distributions for simultaneous fits to the 4.4 and 6.1 MeV lines for different phases of SOL2003-10-28T11:10 (X17.2). Confidence levels at 50%, 70%, 90% are plotted for fits using an abundance of C reduced by 50% with respect to O as compared to the chromospheric abundance and for power-law particle spectra of 3.0 (*dotted lines*) and 4.0 (*continuous lines*). The best values of α/p and $\Delta\Theta$ for the two spectral indices are indicated by *open* ($s = 3.0$) and *filled symbols* ($s = 4.0$) (from Kiener et al. 2006)

and line fluences have led Kiener et al. (2006) to favor the production of the γ-ray line emissions in the SOL2003-10-28T11:10 (X17.2) flare, by downward directed ion beams with a relatively low α/p ratio, around 0.1 (Fig. 17). High-resolution γ-ray line spectroscopy of two other weaker flares located at the limb was also performed by Harris et al. (2007) using *INTEGRAL* data (SOL2003-11-04T19:53 and SOL2005-09-07T09:52; see also Sect. 2.3).

In addition to the line width of the narrow flare γ-ray lines discussed above, the profile of $\alpha/^4\mathrm{He}$ fusion lines around 0.452 MeV (α/α lines) may also provide a potential diagnostic of the angular distributions of energetic particles in the interaction site. From *SMM* observations of two flares, it had been shown that the line shape of the $\alpha/^4\mathrm{He}$ fusion line was inconsistent with its production by a downward beam and was rather consistent with an isotropic or a fan-beam distribution (i.e., as obtained at a magnetic mirror point) in the interacting site (Share and Murphy 1995, 1998). This is consistent with what was found by Share et al. (2002) by analyzing line shapes for *SMM* flares.

Since the launch of *RHESSI*, high-resolution spectra of the α/α lines have been obtained for the SOL2002-07-23T00:35 (X4.8) (Fig. 18) and SOL2003-10-28T11:10 (X17.2) events (Share et al. 2003a, 2004). For these two flares the shape of the line complex is also found to be consistent with a downward isotropic distribution.

2.7 Energy Content in Ions

De-excitation γ-ray lines provide information about ions above ~1 MeV/nucleon. The energetic ion distribution below this energy is essentially unknown but is of particular interest for the total ion energy content (see Sect. 2.10). Since the first detection of γ-ray lines in 1972, quantitative results from γ-ray spectroscopy have been obtained for more than 20 events, yielding information on the energy content in ions. Several ratios of pairs of nuclear lines have been used to infer the ion energy spectrum and then the ion energy content: $\Phi_{2.223}/\Phi_{4-7}$ and $\Phi_{1.63}/\Phi_{6.13}$ (Ramaty et al. 1996). As an example, the threshold energies for excitation of Ne and O lines by fast protons are significantly different, >2 and

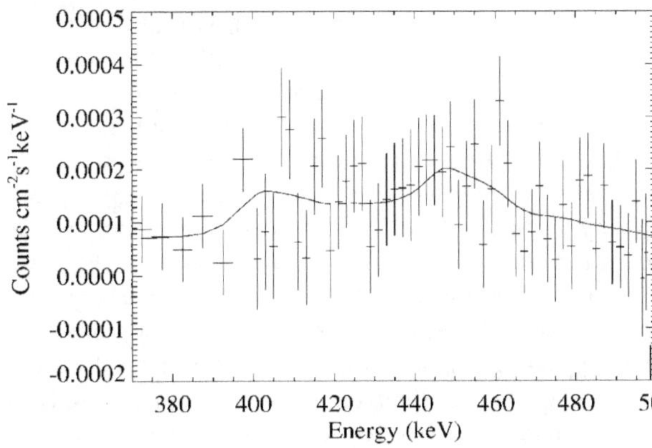

Fig. 18 Count spectrum of the α-^{4}He line measured by *RHESSI* for the SOL2002-07-23T00:35 (X4.8) event after subtraction of the other spectral components. The *curve* shows the *RHESSI* response to the line shape in the case of an α particle distribution with saturated pitch-angle scattering (see Sect. 2.2.1) propagating in magnetic loops perpendicular to the solar surface at a heliocentric angle of 73° (from Share et al. 2003b). Reproduced by permission of the AAS

>8 MeV, respectively, which allows the use of the ratio of their measured fluxes to infer the ion energy distribution (Share and Murphy 1995). This determination is not independent of the assumed target abundances but combining results from several such methods reduces some of this uncertainty. The comparison of the steepness of the ion spectra deduced from the different ratios, $\Phi_{2.223}/\Phi_{4-7}$ and $\Phi_{1.63}/\Phi_{6.13}$ for different flares, shows indeed some consistency only when an enhanced Ne abundance is assumed, Ne/O = 0.25 instead of the more standard value of 0.14 (Ramaty et al. 1996) (see also Sect. 2.3.1). As most of the narrow lines can be produced by energetic protons and αs, the usual hypothesis is that the energy distributions of accelerated protons and αs expressed in energy/nucleon are identical. The relaxation of this assumption will be discussed below. The determination of the proton spectrum from the ratio of the excitation lines from Ne and O depends on the assumption made on the ratio of Ne/O in the flaring region and on the ratio α/p of fast α particles to energetic protons. It is indeed found in Ramaty et al. (1995) that steeper ion spectra are needed to reproduce the same ratio of line fluences for smaller values of Ne/O and larger values of the α/p ratio. The analysis based on the 19 events observed by *SMM*/GRS led to the conclusion that the energy contained in >1 MeV ions lies in the range 10^{29-32} erg for γ-ray line flares (see Ramaty et al. 1995, 1996, and the summary by Miller et al. 1997). These estimations were done under the following assumptions: the slope of the ion power-law spectrum is deduced from the ratio of the Ne/O line assuming thick-target emission and values of Ne/O = 0.25, α/p = 0.1 and "impulsive" composition for other species (i.e. increase of Ne/O, Mg/O, Si/O, and S/O by a factor of ~3 and by a factor of ~10 for Fe/O with respect to coronal composition for the accelerated particles). The power-law spectrum is extended down to 1 MeV with a flat extension to lower energies. For the giant flares SOL1991-06-01T16:14 (X12) and SOL1991-06-04T03:37 (X12.0), observed either by *GRANAT*/PHEBUS and *CGRO*/OSSE, the energy contained in 1 MeV/nucleon ions has also been deduced and is found to be in the 10^{32}–10^{33} ergs range (Ramaty et al. 1997; Murphy et al. 1997). For SOL1991-06-04T03:37, the total energy contained in accelerated protons is 1.7×10^{32} erg and the total energy contained in accelerated ions is 1.0×10^{33} erg under the following assumptions: impulsive flare abundances for accelerated particles, identical energy dependence for all ion spectra, and α/p = 0.5.

The relaxation of the assumption of similar proton and α energy spectra in MeV/nucleon has been studied by Toner and MacKinnon (2004) and applied to the estimates of the ion energy content in flares. As some of the narrow lines (in particular the α–α lines at 0.429 and 0.478 MeV) result only from the interaction of energetic αs with the solar atmosphere, the measured fluences of these lines can be used to infer the number of α particles

>1 MeV/nucleon once the steepness (δ_α) of the α spectrum has been chosen. This allows a computation of the flux of the narrow lines at 1.63 MeV and 6.13 MeV produced by the αs. The remaining fluences in the narrow lines should be produced by fast protons. The ratio of the remaining fluences is then used to determine the steepness (δ_p) of the proton spectrum and then the number of fast protons required to produce the remaining fluences. Note that some values of the steepness of the α spectra (excessively hard ones) can be excluded directly by the fact that the fluences of the produced narrow lines at 1.63 and 6.13 MeV would exceed the observed ones. The reexamination of the spectra and numbers of particles deduced from *SMM*/GRS spectra of four flares chosen in the list of Share and Murphy (1995) shows that harder spectra for α particles than for protons are generally deduced, that most of the energy still goes to protons (by a factor of 100 to 1000) and that as a whole less energy goes to ions. As an example, for the well studied SOL1981-04-27T09:45 (X5.5) event, a total ion energy of 1.6×10^{31} erg is found with $\delta_\alpha = 2.5$, $\delta_p = 4.4$, $\alpha/\mathrm{p} = 0.004$ from the method of Toner and MacKinnon (2004), while the usual analysis (using the ratio of the fluences of the 1.63 MeV and 6.13 MeV lines) would lead, for the usually chosen value of $\alpha/\mathrm{p} = 0.5$, to a total ion energy of 5.5×10^{31} erg with an α and proton spectral index of 5.2. This discussion clearly shows the present limits and the uncertainties in deducing ion energy content from γ-ray line spectroscopy. It must finally be noted that in the analysis developed in Toner and MacKinnon (2004) the meaning of α/p is different from the standard one when both ion spectra are similar. In the case of different ion spectra, the value of α/p which is indicated above is the total number of α particles $\geq$1 MeV/nucleon to the total number of protons above 1 MeV.

Other effects can also alter the amount of energy estimated for accelerated ions as deduced from γ-ray line spectroscopy. As stated above, the production of the γ-ray line emission is usually estimated under the assumption of thick-target production. This means that the energy loss rate of the radiating ions is also a parameter (even if well hidden) of the γ-ray line production. Most of the models of γ-ray line production in flares assume that the target in which the fast ions interact and produce lines is a "cold" neutral target (i.e., energetic particle speeds much larger than the target particle speeds). This may no longer be true if ions with energies close to the threshold energies for γ-ray line production interact in a hot flaring plasma (temperature of a few 10^7 K), i.e., if the γ-ray line emission site is partly coronal (see Sect. 5 for some discussion on possible coronal sites of γ-ray line emission). Emslie et al. (1997), and more recently MacKinnon and Toner (2003), have examined the effects of introducing a "warm target" for the production of γ-ray lines. In such a warm medium, fast ions with energies just above the threshold for nuclear line production will lose less energy to collisions than in a cold medium. This has the effect of increasing the efficiency of line production by ions just above the threshold. This increase is counterbalanced by the fact that the energy losses in an ionized (because hot) medium are larger than in a neutral medium. MacKinnon and Toner (2003) estimated the strength of the Ne 1.63 MeV line relative to the strength of the O 6.13 MeV line, used to estimate ion spectra, as a function of the temperature of the target and as function of the α/p ratio, i.e. assuming as in most works a similar spectral shape for energetic protons and αs in MeV/nucleon. They show that in a warm target, there is an increase of the relative production of the Ne 1.63 MeV line to the O 6.13 MeV line. This increase is more important for steeper spectra and for lower values of α/p (this simply illustrates the fact that the Ne line cross-sections have much lower thresholds for proton production than for α production). Flatter ion spectra and thus lower ion energy contents are then required to reproduce the observed line fluences ratio than usually deduced (a reduction by more than a factor of $\sim$10 of the total energy contained in protons and αs can be achieved if the emission site has a temperature of 6×10^7 K and α/p is 0.1; see

MacKinnon and Toner 2003). Taking into account the effects of a warm target could finally reduce the discrepancies which were found in the ion spectral indices deduced from the different line ratio $\Phi_{2.223}/\Phi_{4-7}$ and $\Phi_{1.63}/\Phi_{6.13}$ for a standard value of Ne/O. Indeed, even with a standard value of the Ne/O abundance ratio, the inclusion of a warmer emission site leads to harder ion spectra to produce the same $\Phi_{1.63}/\Phi_{6.13}$ (MacKinnon and Toner 2003).

Observations performed by *RHESSI* brought information on the ion energy content for three more γ-ray line events: SOL2002-07-23T00:35 (X4.8) (Lin et al. 2003; Emslie et al. 2004a; Dauphin and Vilmer 2007), SOL2003-10-28T11:10 (X17.2) (Gan 2005; Share and Murphy 2006), and SOL2003-11-02T17:25 (X8.3) (Gan 2005; Share and Murphy 2006). Using a procedure similar to the one which has been used previously for *SMM*/GRS and *CGRO* events by Murphy et al. (1997) and Ramaty et al. (1997), Lin et al. (2003) found a minimum total energy in accelerated protons above 2.5 MeV of 1.4×10^{30} ergs (10^{31} erg for all ions) for SOL2002-07-23T00:35 (X4.8) (assuming $\alpha/\mathrm{p} = 0.5$). Consistent values of the total energy in protons and α particles above 2.5 MeV/nucleon were found by Dauphin and Vilmer (2007) from the modeling of the γ-ray line time profiles. Values of the proton energy content derived for SOL2003-10-28T11:10 (X17.2) are consistently found to be around 1.5×10^{31} ergs by Gan (2005) and by Share and Murphy (2006). For SOL2003-11-02T17:25 (X8.3) the proton energy content ranges from 0.5×10^{31} ergs to 4.2×10^{31} erg depending on the value of the proton spectral index used (3–3.6). This clearly shows the still-large uncertainties in the determination of these quantities.

2.8 Ion and Electron Acceleration in Flares and Variability

Using the accelerated ion and electron spectra derived from γ-ray and X-ray spectroscopy, it is possible to compare the energies contained in flare-accelerated ions and electrons. Such a study performed by Ramaty and Mandzhavidze (2000) for the 19 γ-ray line flares observed by *SMM* showed that the energy contained in >1 MeV ions may be comparable to the energy contained in subrelativistic electrons (>20 keV). Emslie et al. (2004a) obtained a similar result for the first γ-ray line flare observed by *RHESSI*. At present, such comparisons of subrelativistic electron and ion energy contents have not been performed for more flares. The published comparisons show that there is a large dispersion of the relative electron (>20 keV) and ion (>1 MeV/nucleon) energy contents from one flare to the other, the energy contained in the ions sometimes exceeding the energy in the electrons. However, it must be recalled that all these estimations strongly depend on the low energy cut-off of both energetic electrons and ions which is for electrons not always easy to determine (see the discussion in Holman et al. 2011, this issue).

Another way of dealing with the question of the relative ion and electron acceleration in solar flares is to investigate the relationship between ion acceleration diagnosed by γ-ray line fluences and relativistic electron acceleration diagnosed by the >300 keV bremsstrahlung radiation. A first comparison was carried out on *SMM*/GRS data between the >300 keV bremsstrahlung radiation and the excess rate in the 4–8 MeV range (i.e., the rate above the bremsstrahlung continuum which is attributed to the production of γ-ray lines by ions (see Fig. 2)) showing that there is a good correlation between ion production (normalized above 30 MeV/nucleon) and electron production above 300 keV (see e.g. Forrest 1983; Chupp 1984). The relationship between energetic ion and relativistic electron acceleration can also be studied by comparing the 2.223 MeV line fluence to the >300 keV X-ray emission (see, e.g., Murphy et al. 1993). Since the launch of *RHESSI*, 29 flares have been observed with significant >300 keV continuum fluence. The fluences of the 2.223 MeV lines have been estimated for these flares from spectral analysis and corrected for attenuation for

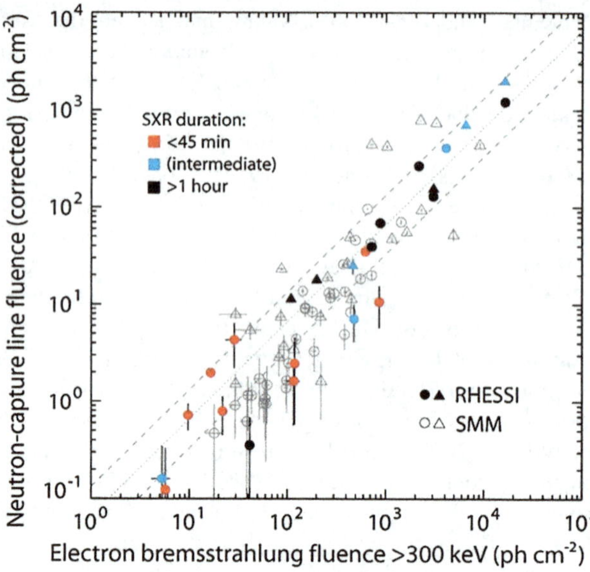

Fig. 19 Neutron-capture line fluence as a function of the electron bremsstrahlung fluence above 300 keV for flares with heliocentric angle <80° (*RHESSI* in *solid symbols*, *SMM* in *open symbols*). *Circles* (*triangles*) represent flares with complete (incomplete) coverage. *Colors* represent the *GOES* soft X-ray (SXR) duration in three bins. The *dotted line* indicates the best fit in linear space that passes through the origin with a ratio of 0.066, and the *dashed lines* have slopes that differ by factors of ∼2 from the best-fit line (from Shih et al. 2009b). Reprinted by permission of AAS

flares close to the limb. The neutron-capture line fluences are found to be highly correlated to the >300 keV bremsstrahlung fluences for all the flares located at heliocentric angles less than 80° and over a range of more than three orders of magnitude in fluence. A similar correlation was found for the flares previously observed by *SMM*. The ion energy range which produces the 2.223 MeV line depends in fact on the ion spectral index and on the accelerated and ambient abundances (see Sect. 2.2.5 and Murphy et al. 2007). As a first approximation, this line can be considered as a good indicator of the number of >30 MeV protons and its fluence is taken as being proportional to the number of protons above this energy. When both *SMM* and *RHESSI* data sets are taken into account, the strong correlation obtained between the neutron-capture line fluence (corrected for attenuation) and the bremsstrahlung fluence (>300 keV) (see Fig. 19) clearly shows a link between the total energy content in protons above 30 MeV and the total energy content in electrons above 300 keV (Shih et al. 2009b). This suggests that high-energy electrons and ions are directly linked by acceleration processes. This may be a more direct link than for subrelativistic electrons and ions, given the larger variation between electron energy contents above 20 keV and ion energy contents above 1 MeV/nucleon discussed above. This is a constraint which should be borne in mind when discussing acceleration mechanisms in solar flares.

The correlation found by Shih et al. (2009b) between ion and relativistic electron acceleration seems to indicate that energetic ions are produced as soon as there is a significant production of energetic electrons above, e.g., 300 keV. From this correlation, there does not appear to be a distinct class of electron-dominated flares that produce much more high-energy bremsstrahlung radiation relative to nuclear emission (e.g., Rieger et al. 1998) although some episodes of the flares reported by Shih et al. (2009b) may exhibit higher relative numbers of electrons with respect to ions than the mean value over several flares as well as over the whole event. Extreme cases of the variability of the electron bremsstrahlung component to the nuclear line component have been reported in the *SMM* era during short duration (a few seconds to a few tens of seconds) transient bursts observed above 10 MeV at any time during a flare. They are referred to as electron-dominated events (Rieger 1994) and are characterized by weak or no detectable γ-ray line emission and by hard $\geq$1 MeV electron bremsstrahlung spectra. They were first reported from *SMM*/GRS observations (e.g., Rieger and Marschhäuser 1991), and were afterwards observed by *GAMMA-1* (Akimov et al. 1992; Leikov et al. 1993), PHEBUS and SIGMA experiments aboard *GRANAT* (Pelaez et al. 1992; Vilmer et al. 1994), *CGRO* (e.g., Dingus et al. 1994) and *Yohkoh* (Yoshimori 1999b). The spectral analysis of 12 electron-dominated events observed by *SMM*/GRS (Rieger et al. 1998) confirmed the hardness of the bremsstrahlung spectra above 1 MeV (mean value of the power-slope around 1.84). The mean value of the spectral index between 0.3 and 1 MeV ($\sim$2.7) does not differ significantly from that of other flares. However, the apparent lack of γ-ray line emission does not rule out a simultaneous production of relativistic electrons and ions (Trottet et al. 1998; Vilmer et al. 1999) for the electron-dominated events observed with *GRANAT*/PHEBUS. Indeed, if one assumes that the energy content in ions above 1 MeV is similar to that contained in electrons (i.e., around a few 10^{29} ergs for the cases studied by Trottet et al. (1998) and Vilmer et al. (1999)), as appears to be the case for γ-ray line flares, one finds that no detectable γ-ray line fluence could be observed, given the hardness of the electron bremsstrahlung component and the limited spectral resolution (and thus line sensitivity) of the PHEBUS experiment. The only remaining question would then be to understand why the spectrum of accelerated electrons is so much harder in these events than at most other times.

Variability of the accelerated electron-to-proton ratio not only occurs from flare to flare as discussed above, but also on time scales of tens of seconds within individual events. Chupp et al. (1993) reported, e.g., for an *SMM* event a variation of the relative number of electrons above 500 keV and of ions above 30 MeV on time scales of tens of seconds. Dauphin and Vilmer (2007) also reported variation of the relative number of electrons above 150 keV and of the number of energetic protons and α particles above 1 MeV/nucleon in the evolution of SOL2002-07-23T00:35 (X4.8) observed by *RHESSI*. These variations were deduced from the analysis and modeling of time delays between γ-ray line emissions and hard X-rays. A question of obvious interest would be to determine whether these changes are connected to spatial source evolution. Such variations of the ratio of accelerated electrons to accelerated protons should also be taken into account when discussing acceleration processes.

Another interesting result follows from the statistical analysis of Shih et al. (2009b). For eight of the flares with the largest 2.223 MeV line fluence, a correlation is found between the 2.223 MeV line fluence (i.e., the number of $>$30 MeV protons) and the fluence above 50 keV as well as with the *GOES* class. However, this correlation is found only when a threshold in the production of ions is reached. A similar relationship had been discussed at the time of *SMM* by Cliver et al. (1994). This ultimately suggests that while the acceleration of protons above 30 MeV is closely related to the acceleration of relativistic electrons, the acceleration of subrelativistic electrons is only proportional to the acceleration of relativistic electrons and ions when a given threshold of high energy particles is reached. This may suggest two acceleration processes, one producing proportional quantities of relativistic electrons and ions and the other one producing mostly subrelativistic electrons.

2.9 Long-Lived Radioactivity from Solar Flares

Interactions of flare-accelerated ions with the solar atmosphere can synthesize radioactive nuclei, whose decay can produce observable, delayed γ-ray lines in the aftermath of large flares. The detection of these delayed lines would provide new insights into the spectrum and composition of the flare fast ions. Many of these radioisotopes would be produced predominantly by interactions of fast heavy ions with ambient hydrogen and helium, especially since accelerated heavy nuclei are believed to be significantly enhanced compared to the ambient medium composition (e.g., Murphy et al. 1991).

One of the most promising of such lines is at 0.847 MeV resulting from the first excited state of ^{56}Fe which in turn has been produced via the decay of ^{56}Co (half-life $T_{1/2} = 77.2$ days) (Ramaty and Mandzhavidze 2000; Kozlovsky et al. 2002; Tatischeff et al. 2006). Another line produced in this decay is at 1.238 MeV (Shih et al. 2007). For a few days after a large flare however, the most intense delayed line is predicted to be the 511 keV positron-electron annihilation line resulting from the decay of several long-lived β^+ radioisotopes. Tatischeff et al. (2006) have pointed out other delayed γ-ray lines that appear to be promising for detection, e.g., at 1.434 MeV from the radioactivity of both the isomer ^{52}Mn^m ($T_{1/2} = 21.1$ min) and the ground state ^{52}Mn^g ($T_{1/2} = 5.59$ days), 1.332 and 1.792 MeV from ^{60}Cu ($T_{1/2} = 23.7$ min), 1.369 and 2.754 MeV from ^{24}Na ($T_{1/2} = 15.0$ hours), and 0.931 MeV from ^{55}Co ($T_{1/2} = 17.5$ hours). The delayed lines should be very narrow, because the radioactive nuclei are stopped by energy losses in the solar atmosphere before they decay. Unless the flare is very close to the solar limb, the γ-ray line attenuation in the solar atmosphere should not be significant (see Hua et al. 1989), as long as the radioactive nuclei do not plunge deep into the solar convection zone. Searches for upper limits of the radioactivity line at 1.238 MeV were performed with *RHESSI* (Shih et al. 2007) for periods after the large flares of July 2002, October-November 2003 and January 2005. No strong conclusion has been drawn so far.

In addition to γ-ray lines emitted from de-excitation of daughter nuclei, radioactive X-ray lines can be produced from the decay of proton-rich isotopes by orbital electron capture or the decay of isomeric nuclear levels by emission of a conversion electron. The strongest delayed X-ray line is predicted to be the Co Kα at 6.92 keV (Tatischeff et al. 2006), which is produced from both the decay of the isomer ^{58}Co^m ($T_{1/2} = 9.04$ hours) by the conversion of a K-shell electron and the decay of ^{57}Ni ($T_{1/2} = 35.6$ hours) by orbital electron capture. The attenuation of this line by photoelectric absorption in the solar atmosphere should be $<10\%$ for flares occurring at low heliocentric angles, as long as the radioisotopes produced in the chromosphere and upper photosphere are not transported to greater depths. Distinguishing this atomic line from the thermal X-ray emission can be challenging until the flare plasma has significantly cooled down. However, a few hours after the flare end time the thermal emission will be gone, or significantly reduced, and the delayed Co Kα line will be more easily detected.

To illustrate the potential of these lines for revealing details of the heavy ion energy distribution, we show in Fig. 20 calculated fluences of the 0.847 MeV and 1.434 MeV lines as a function of the accelerated α/p abundance ratio. The calculations were normalized to a total fluence of the summed 4.44 and 6.13 MeV prompt narrow lines of 300 photons cm^{-2}, which is the approximate fluence observed in the SOL2003-10-28T11:10 (X17.2) flare. We assumed for the abundances of fast C and heavier elements relative to α-particles the average composition of solar energetic particles measured in impulsive events (Reames 1999). Thus, the predicted fluence variations with accelerated α/p actually show the relative contributions of reactions induced by fast protons. The fluences decrease for decreasing α/p ratio (i.e., increasing proton abundance), because, for $\alpha/\mathrm{p} > 0.05$, the radioisotopes are predominantly produced by spallation of accelerated Fe, whereas the ambient ^{12}C and ^{16}O lines largely result from fast proton interactions. This effect is less pronounced for $s = 5$, because for this very soft spectrum, the contribution of α-particle reactions to the prompt line emission is more important. Thus, a concomitant detection of the 0.847 MeV and 1.434 MeV lines would allow measurement of not only the abundance of accelerated ions, but also of their energy spectrum.

As well as serving as diagnostics of the flare fast ion population, the radioisotopes can serve as tracers to study solar atmospheric mixing (Ramaty and Mandzhavidze 2000).

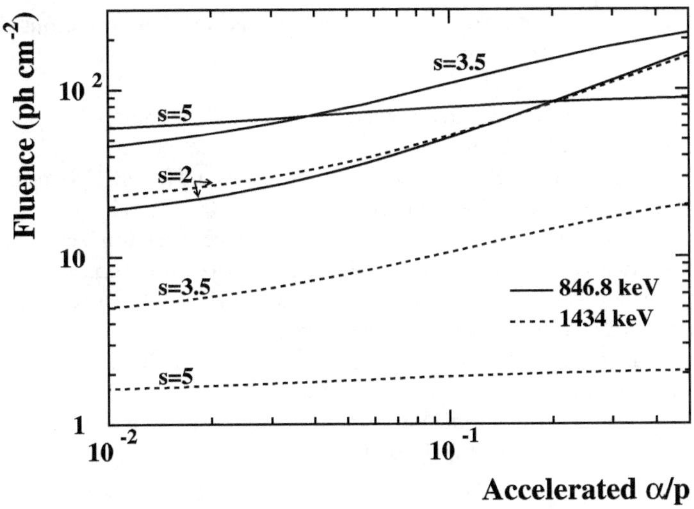

Fig. 20 Total fluences of the delayed lines at 0.847 MeV (*solid curves*) and 1.434 MeV (*dashed curves*) as a function of the accelerated α/p abundance ratio (see text), for spectral indices of the accelerated ion power-law spectrum $s = 2$, 3.5 and 5. The calculations are normalized to a total fluence of 300 photons cm^{-2} emitted during the γ-ray flare in the sum of the 4.44 and 6.13 MeV ambient ^{12}C and ^{16}O de-excitation lines (from Tatischeff et al. 2006). Reproduced by permission of the AAS

A measurement of the decay profile with time of several delayed lines would place very useful constraints on the extent and timescale of mixing processes in the outer convection zone. A future observation of the size and development of the radioactive patch on the solar surface would furthermore provide unique information on both the transport of flare-accelerated particles and dynamics of solar active regions. Note that solar radioactivity can be the only way to study flares that had recently occurred over the east limb.

2.10 Low-Energy Protons

The total energy content of a power-law particle energy distribution, $F(E) \propto E^{-s}$, is dominated by the lowest energy E for which this form holds (as long as $s > 2$). The low-energy form of the distribution is thus important for assessing the ions' overall importance in flare energetics and energy transport. Power-law ion distributions measured in space appear unbroken to 0.02 MeV/nucleon (Reames et al. 1997). If this were to be true also at the flare site, extrapolation to these low energies of the ion distributions deduced above ~2 MeV (Murphy et al. 1997) would yield estimates of ion energy content well in excess of any other measure of flare total energy. Unfortunately, there are few diagnostics bearing on ion distributions below 1 MeV.

Proton-capture cross-sections typically have resonances in the 0.1–1 MeV energy range. Since these resonances result from formation of the compound nucleus in excited states, they also give rise to γ-ray lines as these states decay to lower-lying energy levels. The cross-sections for these lines are small compared to those for de-excitation, but useful constraints on total ion energy could result from upper limits at the 10^{-5} cm^{-2} level (MacKinnon 1989). The two strongest lines are at 2.37 MeV, from ^{12}C(p, γ)^{13}N, and at 8.07 MeV, from ^{13}C(p,γ)^{14}N, constraining the proton energy distribution above 0.46 and 0.555 MeV, respectively. Attempts to constrain the fluxes in these lines observationally have proven inconclusive, however, whether in flares (Share et al. 2001) or the quiet Sun (McConnell et al. 1997). The line at 8.07 MeV would suffer less competition from the strong de-excitation lines, but attempts to constrain it would still require fine energy resolution and/or low instrumental background.

Other possible diagnostics for lower energy ions include Doppler-shifted Lyman-α line emission (Canfield and Chang 1985) and line impact polarization of Hα, Hβ and other lines (Henoux et al. 1990; Vogt and Hénoux 1996). Observations of Hα and Hβ linear polarization in flares seem increasingly well-established (Xu et al. 2005, 2006; Firstova et al. 2008) although there remains an ambiguity in its interpretation between ion beams and streaming electron distributions (e.g., return currents). While there is a claimed observation of Doppler-shifted Lyman-α emission in a stellar flare (Woodgate et al. 1992); Brosius (2001) finds no evidence for α-particle beams in He II Lyman-α from a C3.8 flare.

Their importance for the flare energy budget notwithstanding, flare-site ions below 1 MeV/nucleon remain more or less unconstrained by observations. A promising new technique could be to detect suprathermal ions through charge exchange with the ambient hydrogen (see the first detection of energetic neutral hydrogen atoms in a solar flare by Mewaldt et al. 2009).

2.11 Heavy Energetic Ions as Diagnosed from Broad Lines

The relative abundances of accelerated ions heavier than He can be determined from the relative fluences of the broad de-excitation lines. These lines are produced when heavy energetic ions interact with ambient H and He. Because the heavy ions lose little of their kinetic energy in the reactions, the excited nucleus produced has a large velocity and the de-excitation γ-ray line can be strongly Doppler-broadened by up to $\sim$20% FWHM. Because the lines are so broad, many of them overlap, making the heavy-ion abundance determination much more difficult than the corresponding determination of ambient abundances using the narrow de-excitation lines. In addition, such broad lines are difficult to distinguish from the underlying continua formed by the electron bremsstrahlung and the nuclear continuum. The nuclear continuum is composed of hundreds of relatively weak, closely spaced de-excitation lines that also result from the nuclear interactions along with the strong lines clearly visible in flare spectra. These lines are so numerous and closely spaced that they merge and appear as a continuum to γ-ray detectors. This continuum, however, does have structure, on the order of several hundred keV, and knowledge of this structure is critical to reliably separate it from the broad lines. The early attempts to derive heavy-ion abundances from flare γ-ray data that we discuss here used relatively crude modeling of this nuclear continuum. On-going analyses using improved nuclear-continuum modeling recently obtained from modern nuclear reaction codes will improve the reliability of such determinations (see Murphy et al. 2009).

Murphy et al. (1991) used the γ-ray de-excitation line code from Ramaty et al. (1979) (see the revised version in Kozlovsky et al. 2002) to calculate both narrow and broad line spectra from the most abundant elements in the solar atmosphere, and the most abundant accelerated ions. The abundances were varied to obtain the best fit to the *SMM* data for SOL1981-04-27T09:45 (X5.5). The resulting accelerated-ion abundances relative to accelerated carbon are shown in Fig. 21. Also shown for comparison are similar element abundances measured in space from large proton flares (LPF) and impulsive, ^{3}He-rich events. The uncertainties for most of the elements are quite large, but Mg and Fe are seen to be significantly ($>$3–4σ) enhanced relative to C and O, similar to the enhancements seen in the impulsive ^{3}He-rich events.

Observations of the behind-the-limb flare SOL1991-06-01T16:14 (X12) with *GRANAT*/ PHEBUS (Barat et al. 1994) have moreover shown that the enhancements in heavy ions

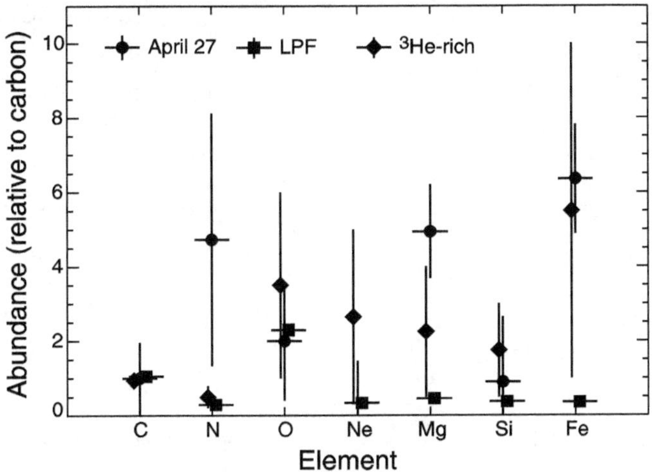

Fig. 21 Derived accelerated-ion abundances relative to accelerated carbon (*circles*) for SOL1981-04-27T09:45 (X5.5) (adapted from Murphy et al. 1991). Also shown are large proton flare (LPF, *squares*) and impulsive ^{3}He-rich event (*diamonds*) abundances (Murphy, private communication)

may increase with time in the course of the flare, reaching towards its end the highest values observed for solar energetic particles in space (Trottet et al. 1996; Ramaty et al. 1997). As a thin-target production of the emission is required to account for the very high observed ratio of 1.1–1.8 MeV flux to 4.1–7.6 MeV flux, the temporal evolution of the abundances of accelerated ions is to be related to the evolution of the accelerated particles themselves. It must finally be noticed that, although a behind-the-limb flare, this event is associated with one of the largest γ-ray line fluences observed so far. This flare is also surprisingly associated with the observation of a strong flux of neutrons by the OSSE experiment (Murphy et al. 1999). As the neutrons are also expected to be produced as a thin target, a hard ion spectral index (power-law index around 2) is deduced by Murphy et al. (1999) up to at least 50 MeV.

3 Pion-Decay Radiation in Solar Flares

Observations of pion-decay radiation from solar flare γ-rays and of neutrons (see next section) combined with observations of flare γ-ray lines give a complete picture of the accelerated ion distribution above a few MeV/nucleon to several GeV/nucleon (see, e.g., Ramaty 1986; Chupp 1984, 1996; Vilmer and MacKinnon 2003; Chupp and Ryan 2009, for reviews). Studies of high energy emissions from the flare site started with *SMM* (Murphy et al. 1987), continued with events observed with *GAMMA-1* (Akimov et al. 1992), *CGRO* (e.g., Mandzhavidze and Ramaty 1992; Kanbach et al. 1993; Dunphy et al. 1999), and now with *CORONAS-F* (Kudela et al. 2003). Pion-decay radiation together with observations of neutrons provide information on the highest ion energies produced during flares. Charged pions decay to yield electrons and positrons, which in turn produce γ-rays by bremsstrahlung. Positrons also contribute to the continuum by annihilating in flight. Neutral-pion (π^0) decay results in two photons, with one emitted at a high energy. This results in a very flat, broad "bump" feature which has a maximum at 67 MeV. Synchrotron losses of electrons and positrons may be important and shorten their lifetimes, therefore reducing their contribution with respect to the radiation from neutral-pion decay (Murphy et al. 1987). Energetic

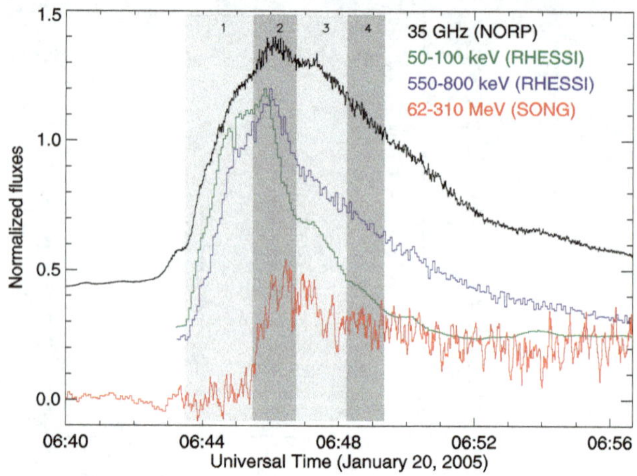

Fig. 22 Time profiles of the normalized flux density at 35 GHz (*top*) and normalized count rates of hard X-rays and γ-rays (*RHESSI*, *CORONAS-F*/SONG) at different energies. The high energy γ-rays observed by SONG are pion-decay photons from primary protons at energies above 300 MeV (from Masson et al. 2009)

electrons above 10 MeV are also produced in solar flares and produce bremsstrahlung continuum, potentially masking pion-decay radiation.

Pion decay radiation was first observed with the *SMM*/GRS instrument for the flare SOL1980-06-21T02:00 (X2.6) (upper limit) and then for the flare SOL1982-06-03T13:26 (X8.0) (Forrest et al. 1985). The latter event provided the first convincing observations of π^0-decay radiation in a solar flare. This event showed that the production of pions can occur early in the impulsive phase defined by the production of hard X-rays near 100 keV. However, a significant portion of the high-energy emission was also observed well after the impulsive phase in an extended phase mostly observed at high energies. High-energy neutrons were also detected for that event (see next section). Together with the development of models for ion acceleration and γ-ray production, these observations led to the first determination of high-energy ion distributions in flares (Murphy et al. 1987). Given the assumed spectral shape of energetic protons (either a Bessel function or a power-law spectrum), numbers and spectra of energetic protons were estimated for both the impulsive and extended phases from the ratio of the differential pion decay radiation at 100 MeV to the 4.1–6.4 MeV nuclear emission. It was found that the first phase was characterized by a significantly steeper proton spectrum than the extended phase. Pion-decay radiation was subsequently reported for other flares observed with *SMM*/GRS (Dunphy and Chupp 1991, 1992; see also Chupp and Dunphy 2000, for a review).

After the end of the *SMM* mission in November 1989 and before the launch of *CORONAS-F* at the end of July 2001, there was no specifically solar-dedicated mission for high-energy radiation, but several satellites still provided γ-ray data (*GRANAT*/PHEBUS, *GAMMA-1*. and *CGRO*).

Until now, around 20 events have been observed with significant pion production (see Lockwood et al. 1997; Myagkova et al. 2007; Chupp and Ryan 2009). Some of the most recent events have been observed in a wide energy range by both *RHESSI* and *CORONAS-F* (e.g., Kuznetsov et al. 2008; Trottet et al. 2008; Masson et al. 2009) (see, e.g., Fig. 22). For some of the events, high-energy emission has been observed for hours after the impulsive phase of the flare, revealing that high energy ions must be continuously accelerated on long timescales in some flares (e.g., Kanbach et al. 1993; Ryan et al. 1994; Ryan 2000; Rank et al. 2001). Quantitative analysis of a few of the events with significant pion production has been performed providing information on the ion energy spectrum at energies greater than 300 MeV and allowing a comparison of this spectrum with the one deduced at lower

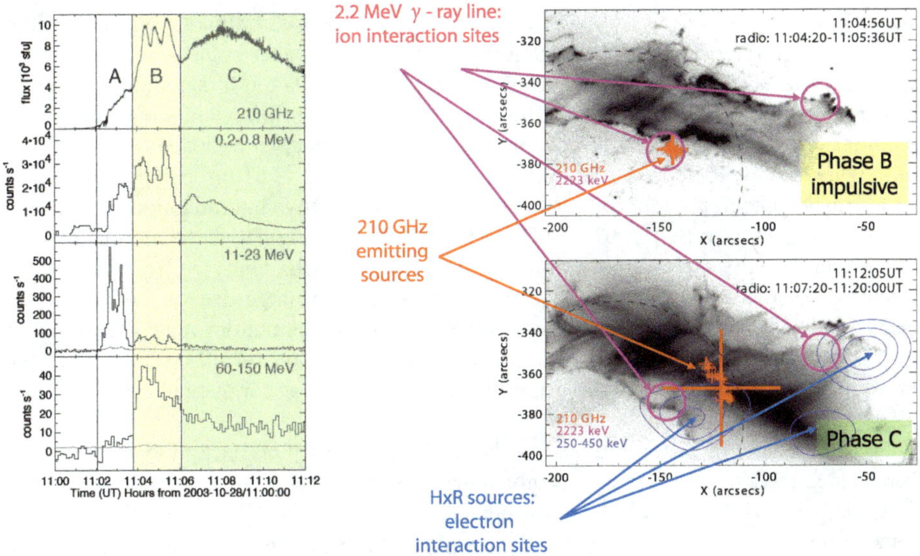

Fig. 23 *Left:* 210 GHz submillimeter radio flux from KOSMA (Kölner Observatorium für SubMillimeter Astronomie) and X-ray/γ-ray fluxes from *CORONAS-F*/SONG; *right:* imaging in radio at 210 GHz (*red crosses*), UV (background image from *TRACE*), X-rays and γ-rays from *RHESSI* (*blue* and *pink contours* from Hurford et al. 2006) in phase B and C showing, respectively, electron and ion interaction sites (adapted from Trottet et al. 2008). Reproduced by permission of the AAS

energies from γ-ray line spectroscopy (see e.g., Alexander et al. 1994; Dunphy et al. 1999; Kocharov et al. 1994, 1998; Vilmer et al. 2003). These comparisons generally show that the ion energy distribution does not have a simple power-law form from the γ-ray line emitting energy domain (1–10 MeV) to the pion-producing energy domain (>300 MeV).

In recent years, some of the events with significant pion-decay radiation have been found to be associated with submillimeter (above 200 GHz) emissions (Kaufmann et al. 2004; Lüthi et al. 2004). These observations have revealed the existence of a new emission component with intensities well above the extrapolation of the radio synchrotron spectrum seen at lower frequencies. Moreover, the radio spectrum increases with increasing frequency contrary to what is expected from an optically thin synchrotron spectrum. The origin of this new component is still under discussion. One possibility could be that this emission be produced by synchrotron emission from pion-decay positrons. This last process was described a long time ago (Lingenfelter and Ramaty 1967b) and could be considered as a good candidate for the high frequency observations, given the association between events exhibiting pion-decay radiation and events showing a spectral increase above 200 GHz (Myagkova et al. 2004; Silva et al. 2007). This is illustrated in Fig. 23 from Trottet et al. (2008) which shows the observations of SOL2003-10-28T11:10 (X17.2) with KOSMA[5] at 210 GHz and with the SONG experiment on *CORONAS-F* at energies above 0.2 MeV. The onset of the impulsive component of the submillimeter component (phase B on the figure) is clearly associated with the start of the radiation above 60 MeV, i.e., the start of >300 MeV/nucleon ion acceleration. The 210 GHz source size is compact ($<10''$) in this phase and its location is co-spatial with the site of interacting ions (revealed by the 2.2 MeV line emission site) but not to the

[5] Kölner Observatorium für SubMillimeter Astronomie.

site of interacting electrons. In phase C, when no impulsive submillimeter emission and no strong emission above 60 MeV are observed, the submillimeter sources are quite different. The close correlation in time and space of the impulsive submillimeter emission and of the strong production of neutral pions thus suggests that synchrotron emission from charged pion-decay positrons could be responsible for the submillimeter emission. However the submillimeter flux predicted from the number of positrons derived from charged pion decay seems inadequate to account for observed fluxes. An alternative interpretation was given by Silva et al. (2007) for SOL2003-11-02T09:55 (X3.9). It was concluded there that the most likely source of the submillimeter emission is gyrosynchrotron emission from non-thermal electrons but this requires high local magnetic fields, very large densities of accelerated electrons and very small source sizes. Although many interpretations have been proposed in the literature (see above and Kaufmann and Raulin 2006; Sakai and Nagasugi 2007; Fleishman and Kontar 2010), the origin of this submillimeter component remains a challenging topic of discussion.

Another challenging topic linked to the observations of pion decay radiation in a few flares is related to the origin of coronal hard X-ray sources reported in, particularly, SOL2005-01-20T07:01 (X7.1). MacKinnon and Mallik (2010) have indeed shown that some coronal hard X-ray sources might be interpreted as photospheric, optical photons inverse Compton scattered to deka-keV energies by electrons, or indeed positrons in the 10–100 MeV energy range. In particular, the coronal hard X-ray source reported in SOL2005-01-20T07:01 (Krucker et al. 2008b) could be explained in this way if just a few percent of the secondary positrons implied by estimated fast ion numbers (Masson et al. 2009) were present in the corona. Coronal hard X-ray sources in locations that seem too tenuous for a bremsstrahlung explanation (see Krucker et al. 2008a, for a review) may thus offer a novel window on ~GeV-energy ions.

4 Neutron Observations in Space and on Earth

Many reactions between energetic flare ions and ambient nuclei produce energetic neutrons as secondaries. These secondary neutrons undergo multiple, elastic scattering and either thermalize in the solar atmosphere or escape into interplanetary space. Measurements of the escaping neutrons augment knowledge of flare ion distributions gained from γ-ray lines. In particular they fill a "diagnostic gap" between the de-excitation lines, dominated by ions in the 1–100 MeV energy range, and the pion-decay radiation which needs ions of $\geq$300 MeV (Lockwood et al. 1997). Neutrons in space give valuable information on the most energetic flare ions, especially when measurements of pion decay γ-rays are not available. They may also play a role in constraining the numbers of heavier fast ions, prolific producers of free neutrons via evaporation reactions.

Having no electric charge, they arrive at the detector undeviated by active region magnetic fields or interplanetary medium plasma turbulence. Free neutrons are unstable to β-decay, with a rest-frame mean lifetime of 886 s, so many decay before they can be detected at Earth.

4.1 Neutron Detections

Energetic neutrons from flares have been detected both in space and on the ground. They give a signal in scintillators that may be distinguished by various means from photons or charged particles (pulse shape, combination of signals in multiple elements, etc.) (e.g.,

Chupp et al. 1987; Murphy et al. 1999). On the ground they have been detected by neutron monitors, and by specially designed neutron telescopes (e.g., Watanabe et al. 2006).

The global neutron-monitor network has great value for studying extremes of solar particle acceleration, offering detectors of a stopping power and effective area that would be impractical in space. They are located inside Earth's magnetic field in a way that aids the discrimination of neutrons from other particle signals (Usoskin et al. 1997; Lockwood and Debrunner 1999). Neutron-monitor count-rate enhancements following solar flares are generally due to solar energetic particles interacting with Earth's atmosphere (e.g., Lopate 2006) but neutrons genuinely of solar origin may be distinguished on grounds of timing and geographical distribution. Neutron telescopes provide neutron spectral and angular resolution in the $>$50 MeV energy range and can discriminate neutrons from charged particles (Muraki et al. 1992; Tsuchiya et al. 2001). They have been installed in seven high-altitude locations around the world.

A neutron of 150 MeV travels a distance of 1 AU in a neutron lifetime (MacKinnon 2007). Lesser energies are more likely to decay *en route* from the Sun while time dilatation extends the lifetime of neutrons at energies above 1 GeV. The neutron survival probability to Earth thus plummets below $\sim$100 MeV. Nonetheless, solar flare neutrons have arrived at Earth in detectable numbers in the energy ranges 10–100 MeV (Ryan et al. 1993) and 50–360 MeV (Muraki et al. 1992). Minimum neutron energies for detection on the ground are 50 MeV (for neutron telescopes, see e.g., Muraki et al. 1992) or 200–500 MeV (for neutron monitors; see Debrunner et al. 1989; Shibata 1994).

The first detection of flare neutrons was by the *SMM*/GRS instrument, from the flare SOL1980-06-21T02:00 (X2.6) (Chupp et al. 1982). The flare SOL1982-06-03T13:26 (X8.0) produced energetic neutrons that were detected in space by *SMM*/GRS and on the ground using the Jungfraujoch and other neutron monitors (Chupp et al. 1987). Neutron-decay protons were also detected in space from this flare (Evenson et al. 1983).

Ground-based detections of solar neutrons have been recently reviewed by Watanabe et al. (2005) and Valdés-Galicia et al. (2009). Since SOL1982-06-0313:26 (X8.0), ten more solar neutron events have been clearly recognized in neutron monitor data: SOL1990-05-24T21:45 (X9.3) (Shea et al. 1991; Smart et al. 1995; Debrunner et al. 1993, 1997), SOL1991-03-22T22:45 (X9.4) (Pyle and Simpson 1991), SOL1991-06-04T03:37 (X12.0) and SOL1991-06-06T03:37 (X12.0) (Muraki et al. 1992; Struminsky et al. 1994) and several occasions during Cycle 23 (Watanabe et al. 2003, 2006; Bieber et al. 2005; Sako et al. 2006). The second such event was observed by the IGY-type neutron monitor located at Climax and several other stations in North America: SOL1990-05-24T21:45 (X9.3) (Shea et al. 1991; Smart et al. 1995; Debrunner et al. 1993, 1997). In the SOL1991-06-04T03:37 (X12.0) event, solar neutron signals were recorded by three different detectors (neutron monitor, solar neutron telescope, and muon telescope) located at Mt. Norikura.

In space, apart from the *SMM*/GRS instrument, energetic flare neutrons have been detected with the OSSE and COMPTEL instruments on the *CGRO* mission (Murphy et al. 1999; Dunphy et al. 1999) as well as with the PHEBUS instrument aboard *GRANAT* (Debrunner et al. 1997) (see, e.g., Chupp and Ryan 2009, for a review). Pulse-shape analysis in OSSE's NaI scintillators discriminated neutrons from photons (Murphy et al. 1999). In COMPTEL, signals from two arrays of detectors were combined to deduce arrival directions and energies for individual neutrons, via elastic scattering kinematics. Such imaging spectroscopy of 10–100 MeV neutrons was possible for several flares in June 1991 (e.g., Ryan et al. 1993).

4.2 Key Features of Neutron Observations

Solar neutron events are too few in number for statistical studies. Nonetheless detections so far share some common features summarized in Watanabe et al. (2005). Detectable neutrons have come from large flares, SOL2000-11-24T15:13 (X2.3) being the smallest of these (Watanabe et al. 2003). Of course, this could be observational bias rather than a physical feature of flares. No correlation is evident between the flare soft X-ray flux and the neutron emissivity. Detectable neutrons at Earth do not appear to come preferentially from limb or disk centre flares.

Using, e.g., the γ-ray de-excitation line flux time profile as a proxy for the rate of ion production at the Sun, neutron energy distributions have been deduced from neutron-monitor count-rate time profiles. Neutrons at the Sun are generally found to have $E^{-\delta}$ distributions, with δ in the range 3–4 (Watanabe et al. 2003, 2006). Via model calculations using the code of Hua et al. (2002), the ion energy distribution at the Sun is then found to be roughly one power of E softer. Total emissivity at the Sun is in the range 4×10^{28}–8×10^{30} sr^{-1} for neutrons in the range 0.05–1.5 GeV.

Lockwood et al. (1997) and Debrunner et al. (1997) emphasized that the de-excitation line flux does not always give the best proxy for the time development of neutron production, urging use instead of the pion-decay flux when this is measured. Neutron and pion production by energetic ions (>300 MeV) are closely related (see Sect. 4.3) so this is not surprising. The flare SOL2005-09-07T09:52 (X17.0), for instance, was detected on the ground by both neutron telescopes and neutron monitors and found to produce neutrons over a more extended period than implied by the flare impulsive phase, as defined by hard X-ray emission. The class of extended γ-ray flares (Ryan 2000) often show pion-decay γ-radiation over tens of minutes. Similarly time-extended neutron emission would be a corollary, as was found using COMPTEL for SOL1991-06-09T04:24 (X10.0) and SOL1991-06-15T11:17 (X12.0) (Ryan et al. 1993).

4.3 Production of Solar Neutrons

Ions are believed to be accelerated in the corona, losing their energy mostly in the chromosphere and photosphere and producing the bulk of their secondaries (photons, neutrons) there. The eventual escaping neutron velocity distribution is governed by the initial velocity distribution of neutrons, which in turn depends on the primary ion distribution throughout the atmosphere, and the consequences of multiple neutron elastic scattering. Thus a complete modeling of neutron production needs to include ion transport, neutron production and scattering. Such a code was constructed by Hua and Lingenfelter (1987b) and improved and updated by Hua et al. (2002), and the consequences of varying loop and acceleration parameters was described in detail by Murphy et al. (2007). The code of Hua et al. (2002) is freely available and may be used to calculate 2.223 MeV line flux and neutron distributions expected at 1 AU under a wide variety of assumptions for both physical and acceleration parameters.

Neutron-producing reactions of fast ions have been enumerated by Hua and Lingenfelter (1987b), Murphy et al. (1987), Hua et al. (2002). The reaction $\mathrm{p(p, n\pi^+)p'}$ has a threshold of 270 MeV. It is a major contributor to total neutron yields for harder ion distributions (e.g., energy power law E^{-s} with $s > 3$; Murphy et al. 2007). Thus neutron emission from flares will often be accompanied by pion-decay γ-rays.

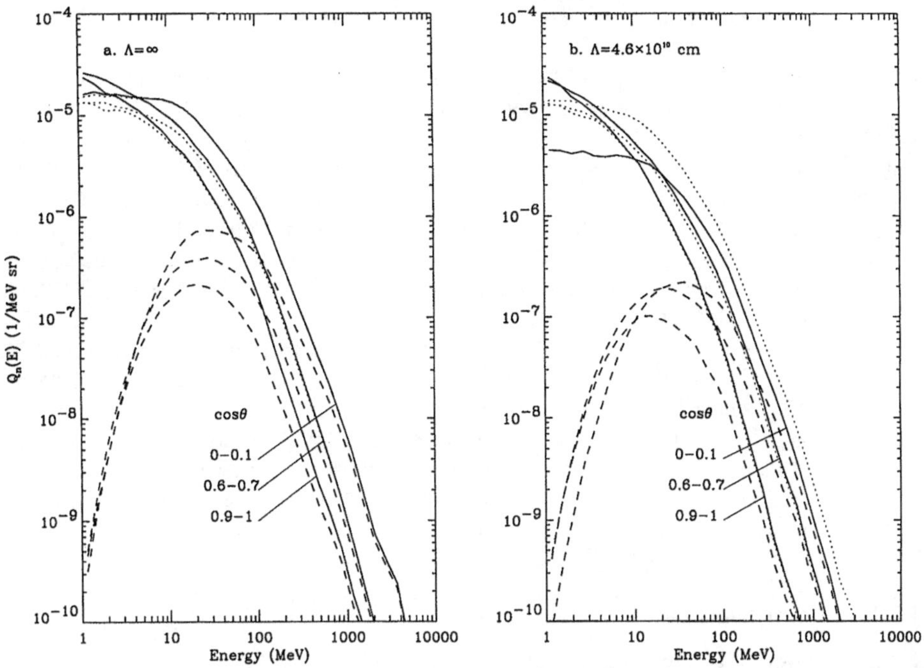

Fig. 24 Angular-dependent energy spectra of the escaping neutrons at the Sun (*solid curves*) and of those surviving at 1 AU (*dashed curves*) at different cos θ (*zenith angle*) compared to their production spectra at the Sun (*dotted curves*) in the case with no pitch-angle scattering of the ions (*left*) and of nearly saturated pitch-angle scattering (*right*). See Sect. 2.2.6 for a discussion of scattering. (From Hua et al. 2002). Reproduced by permission of the AAS

In contrast are the cross-sections for collisions involving heavier nuclei. Neutrons in these cases can be produced via evaporation processes, without the creation of pions, so cross-sections often have much lower threshold energies, below 1 MeV/nucleon in some cases. Neutron measurements thus have some potential for constraining heavy ion acceleration in flares. Especially with softer assumed ion energy distributions, neutron yields will depend sensitively on assumptions about fast ion abundances (Murphy et al. 2007).

Figure 24 shows neutron distributions produced by a population of fast ions, calculated using the code of Hua et al. (2002). A power-law ion energy distribution has been assumed, $\propto E^{-s}$ with $s = 2$, and fast ion abundances representative of "gradual" particle events in space. No magnetic field convergence or MHD pitch-angle scattering have been assumed in the corona so ions precipitate freely to the dense atmosphere. The two figures show the energy distribution of neutrons produced in the solar atmosphere, and the energy distribution of neutrons that actually escape from the Sun. Above ∼100 MeV the neutron energy distribution reflects the primary ion energy distribution; for assumed power laws the neutron distribution is about one power of energy harder (Watanabe et al. 2005). At lower energies the neutron energy distribution is dominated by the relevant nuclear physics, particularly since many of the reactions are near threshold, and by neutron scattering in the atmosphere. Any factor that increases neutron production at greater depth (e.g., pitch-angle scattering in the strong limit which maximizes precipitation—Sect. 2.2.6) decreases the escaping neutron flux (Murphy et al. 2007).

4.4 Neutron Decay Products

Energetic protons from decaying flare neutrons were detected in space from SOL1980-06-21T02:00 (X2.6) (Evenson et al. 1990), SOL1982-06-03T13:26 (X8.0) (Evenson et al. 1983) and SOL1984-04-25T01:40 (X13) (Evenson et al. 1990). When the detector is not magnetically well connected to the flare site, neutron decay products may be clearly distinguished from flare-associated fast particles. An energetic proton "precursor" in SOL1990-05-24T21:45 (X9.3), arriving at the *GOES* spacecraft before the bulk of the flare-associated fast ions, has also been interpreted as a signature of decaying neutrons (Kocharov et al. 1996). Energetic neutron-decay electrons were observed from SOL1980-06-21T02:00 (X2.6) (Droege et al. 1996).

Neutron-decay protons carry off most of the energy of their parent particles and are straightforwardly detected, given an appropriately placed spacecraft (Evenson et al. 1990). These advantages have to be weighed against the complication of accounting correctly for proton propagation in the turbulent interplanetary medium (Ruffolo 1991). Clearly they could offer a useful window on flare ion acceleration but we are aware of only one such claimed detection in Cycles 22 and 23 (Kocharov et al. 1996), and the argument in this case rested on fast proton arrival time rather than a lack of magnetic connection to the flare location.

4.5 Low-Energy Neutrons

Neutrons <100 MeV mostly decay long before they reach 1 AU. A neutron detector placed inside the orbit of Mercury would detect four to five orders of magnitude more neutrons in the 1–20 MeV energy range than a detector at 1 AU (in contrast to a detector of photons, which would gain roughly a single order of magnitude). There are several good scientific reasons to attempt neutron measurements in the inner heliosphere (e.g., Vilmer et al. 2001; Share et al. 2011). Much greater neutron fluxes will be measured from large flares by detectors on future inner-heliosphere space missions. As we mentioned above, heavy-ion populations may be usefully constrained, even though these do not produce narrow de-excitation lines. Constraints on ion acceleration in small flares and even the quiet Sun will be greatly strengthened. It should be emphasized that very recently, the *MESSENGER*[6] neutron spectrometer made the first claimed detection of 1–8 MeV solar neutrons continuously produced for several hours, in good consistency with the previously observed extended production of γ-ray line emissions or pion-decay radiation (see previous sections) (Feldman et al. 2010). However, local neutron production by solar energetic protons and α particles cannot be completely excluded in these observations (see Share et al. 2011). These considerations motivate ongoing development of multiple scatter neutron detectors (Pirard et al. 2009), whose angular resolution in particular is crucial to discriminate genuine solar neutrons from those produced locally by interactions of solar energetic particles in the spacecraft.

5 Imaging Solar γ-Rays

5.1 Introduction

The spatial distribution of γ-ray emission provides a channel of information on the acceleration and transport of energetic ions that is independent of spectroscopic analyses. As with

[6]MErcury Surface, Space ENvironment, GEochemistry, and Ranging.

electron bremsstrahlung, nuclear γ-rays disclose not where particle acceleration occurs, but rather where accelerated particles interact with the ambient atmosphere. Nevertheless it provides the only currently available method for placing accelerated ions in a solar magnetic context.

Prior to the launch of *RHESSI*, the only information on the spatial distribution of solar flare γ-ray emission originated from non-imaging observations of γ-ray line emissions from flares believed to be behind the limb. Vestrand and Forrest (1993), for example, reported a strong neutron-capture line emission from a behind-the-limb flare. Such an observation would imply that in addition to a compact source predicated in many models (e.g., Hua and Lingenfelter 1987a), the γ-ray line emission also originated from a large scale source, probably diffuse, that was far from the flare itself. Another event observed with strong prompt γ-ray line emission by Barat et al. (1994) was also believed to be associated with a behind-the-limb flare. In this case, the absence of neutron capture line emission as well as the smooth time profile of the prompt γ-ray line emission led to the conclusion that this emission was produced in the low corona at densities below 1–5 $\times$ 10^{11} cm^{-3}. On the other hand, measurements of γ-ray line emissions (time profiles with respect to HXR time profiles) have been interpreted as generally consistent with their production in a compact region located in the chromosphere at densities $\geq 10^{12}$ cm^{-3} (e.g., Chupp 1996).

RHESSI has provided the first opportunity to locate the γ-ray emission from solar flares directly. *RHESSI* γ-ray imaging to date is based mainly on the narrow neutron-capture line at 2.223 MeV. This line is generated when fast neutrons, created by the interaction of accelerated protons and ions with the ambient atmosphere, are thermalized and subsequently captured by protons, producing deuterium and a narrow γ-ray line at 2.223 MeV. Narrow because of the absence of Doppler broadening, the line is an optimal choice for imaging for four reasons: (i) it is relatively intense; (ii) since locally-generated fast neutrons cannot thermalize in the limited mass of the spacecraft, there is no locally-generated background line emission; (iii) the line is narrow so that the continuum background is minimized; (iv) the narrowness of the line also limits the contribution of the solar electron-bremsstrahlung continuum to the resulting images. The multiple elastic collisions required for neutron thermalization require an average of $\sim$100 s (e.g., Murphy et al. 2007), which results in a characteristic time delay in this line compared to the prompt lines. Nevertheless, the corresponding average spatial offset between the primary interaction site and the capture site is only $\sim$500 km (R. Murphy, private communication). As a result, despite its two-step formation mechanism, a neutron-capture line image can faithfully indicate the site of the original nuclear interaction to within an arcsecond.

RHESSI imaging of nuclear γ-ray emission is based on time- and energy-tagged counts from the thick rear-segments of *RHESSI* detectors 6 and 9 (35″ and 183″ resolution, respectively). These are the only detectors for which the grids are sufficiently thick (2 and 3 cm, respectively) to modulate the high energy photon flux. Rear segments of the detectors are shielded from the intense flux of low energy X-rays, and so potential concerns about instrumental live time or pulse pileup effects (Smith et al. 2002) can be discounted. On the other hand, since only two of the nine collimators (RMCs) are relevant to imaging and their thick grids block some of the flux, the γ-ray throughput for such imaging is only about 14% of that for spectroscopy, and so the sensitivity for imaging is necessarily lower than that for spectroscopy. The count data provide the starting point for algorithms such as back-projection to reconstruct the image of the source (Hurford et al. 2002). The count statistics and the relatively few spatial frequencies measured do not permit *RHESSI* to reproduce complex source morphologies. However, the data *are* well-suited to the quantitative characterization of simple sources and γ-ray images for five flares have been obtained.

Table 1 *RHESSI* Events with Neutron Capture Line Imaging

Parameter	2002-07-23	2003-10-28	2003-10-29	2003-11-02	2005-01-20
Time range	00:27:20–00:43:20	11:06:20–11:29:40	20:43:00–21:00:00	17:16:00–17:29:00	06:44:00–06:56:00
GOES class	X4.8	X17	X10	X8	X7
Energy range (keV)	2223 ± 5	2223 ± 5	2223 ± 5	2223 ± 5	2223 ± 8
Total counts in RMC 6, 9	336,240	2280,1643	463,313	781,557	961,518
Relative visibility RMC 6	1.61 ± 0.45	0.91 ± 0.20	...	0.76 ± 0.26	0.98 ± 0.27
Relative visibility RMC 9	1.28 ± 0.26	0.82 ± 0.10	1.28 ± 0.28	0.84 ± 0.15	0.93 ± 0.21
Source size FWHM (arcsec) (2σ upper limit)	<22	<30	<94	<44	<20

5.2 Neutron-Capture Line Imaging

The complete set of images obtained to date is shown in Fig. 25 with quantitative details summarized in Table 1 (Hurford et al. 2003, 2006a, 2006b). Rather than deal with each flare in turn, we summarize these observations from three perspectives: the number of distinct spatial components; their size; and their location.

Considering first the number of distinct components, four of the five events presented a single, unresolved source whereas the fifth event, SOL2003-10-28T11:10 (X17.2), showed a double source that straddled an arcade of loops. This pattern does not necessarily imply, however, that double footpoint sources are an atypical feature of the γ-ray emission. Two conditions must be met to image a double footpoint source: the separation of the footpoints must be sufficiently large that they can be resolved with the available 35 arcsec resolution; and there must be sufficient signal-to-noise that two sources (rather than one) can be confidently detected. Referring to the count statistics in Table 1 and using the separation of the electron-bremsstrahlung components as a guide, SOL2003-10-28T11:10 (X17.2) was the only one of the five that satisfied both criteria and that event *did* display a double source. Had γ-ray emission from the other events also been in the form of double sources, this could not have been seen. Therefore, based on the number of observed components, there is no evidence to suggest that the true neutron-capture line sources are predominantly single sources, a result that would be in contrast to typical hard X-ray double sources.

A second perspective is provided by the estimates of the size of the unresolved components. Such estimates can be obtained from comparisons of the total observed line flux to that imaged in the unresolved source. The ratio of imaged flux to total flux, called the relative visibility, can be used to put upper limits on the size of an unresolved source. Ignoring the modulation, the total number of observed counts in the line provides a direct indication of the total, spatially-integrated flux in the narrow energy range associated with the line. Such counts are due to a combination of real solar neutron-capture line events, continuum

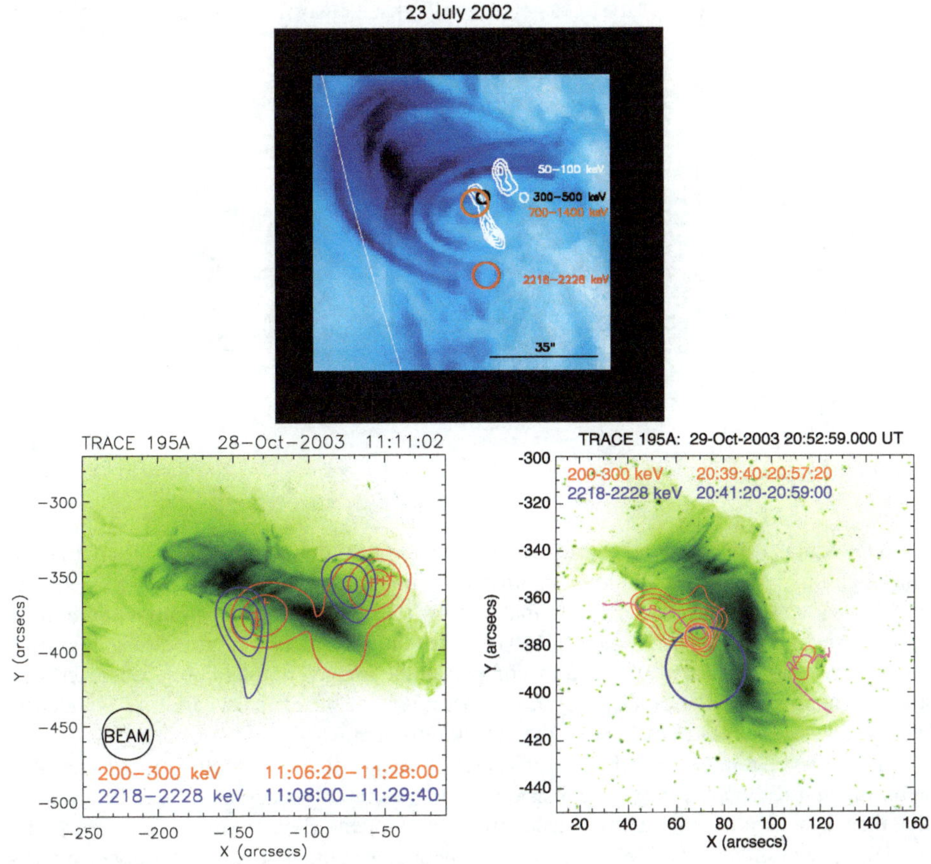

Fig. 25 Respective locations of γ-ray sources at 2.223 MeV and of hard X-ray sources observed with *RHESSI* for different flares. The SOL2002-07-23T00:35 (X4.8) event; the *circles* represent the 1σ errors for the 300–500 keV (*black*), 700–1400 keV (*red*) and 2218–2228 keV (*pink*) maps made with identical parameters. The *white contours* show the high resolution 50–100 keV map with 3 arcsec resolution. The background is a *TRACE* image showing the post-flare loops (adapted from Hurford et al. 2003). The SOL2003-10-28T11:10 (X17.2) event; overlay of contours at 50%, 70%, and 90% of 35″, γ-ray images made with detectors 6 and 9 on a *TRACE* image. The *red plus signs* indicate 200–300 keV footpoint locations at different times. The SOL2003-10-29T20:49 (X10.0) event; *fine red contours*: (50–90%) of the 200–300 keV high resolution maps superposed on a *TRACE* image. The *purple line* indicates the motions of the 200–300 keV footpoints. *Thick red* and *blue circles*: centroid locations with 1 σ error for detector 9 imaging at 200–300 and 2218–2228 keV, respectively. The SOL2003-11-02T17:25 (X8.3) event; overlay of 200–300 keV (*red contours*) and 2218-2228 keV (*blue contours*) sources on a postflare *SOHO*/EIT image. The *dashed* and *thin solid red contours* (50%) show the result of 200–300 keV mapping with 7″ and 35″ respectively. The *thin blue contour* (50%) is the corresponding map at 2218–2228 keV made with detectors 6 and 9. The *thicker* 1σ error *circles* show the detector 6 source centroid location (adapted from Hurford et al. 2006b). The SOL2005-01-20T07:01 (X7.1) event; contours of the 250–500 keV emission overlaid on a *TRACE* image and centroids (in *red* and *blue*) of the map performed in the 250–500 keV and 2215–2231 keV range

solar emission and background. Since the number of such counts that are due to solar continuum or background continuum can be reliably estimated from energy ranges just above 2.223 MeV, an accurate determination of the spatially integrated solar neutron-capture line flux can be made, independent of background models. The imaged flux does not contain

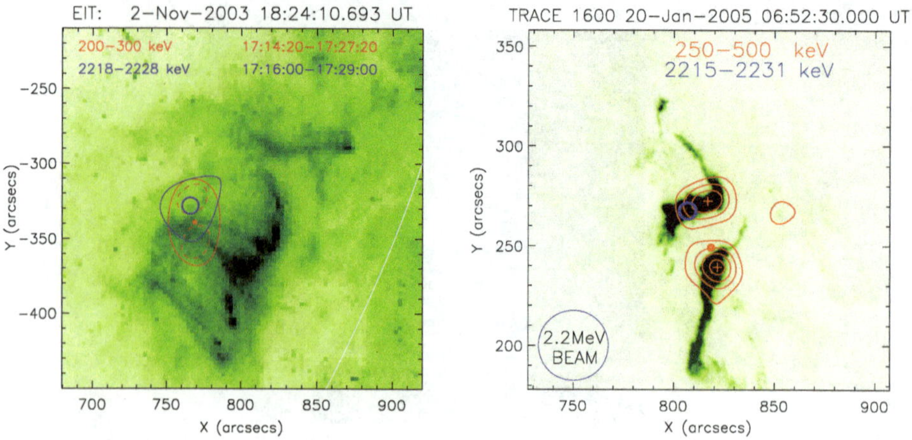

Fig. 25 (*Continued*)

any unmodulated background, but could include a contribution from the solar continuum. Although this contribution could be estimated by imaging in a narrow energy range above 2.223 MeV, it is easy to demonstrate that, thanks to the narrow energy range required for imaging the neutron-capture line (10–16 keV), its contribution is negligible. Note that since the relative visibilities are estimated by comparing the imaged to total counts observed with a given rotating modulation collimator, the measurement of relative visibilities is independent of potential errors in estimating the detector efficiency.

The observed relative visibilities (Table 1) show that in all cases the flux in the unresolved neutron-capture line source(s) is consistent with the spatially-integrated flux. This implies that there is no evidence for a large-scale diffuse component. By assuming a Gaussian source model, upper limits to the FWHM source diameters are obtained that correspond to a few tens of arcseconds.

A third perspective on the neutron-capture line sources is provided by their location. An unanticipated result was that in some, if not all, cases, the neutron-capture line sources were not co-spatial with the corresponding electron-bremsstrahlung sources. Since the character and statistical significance of the displacements differed from event to event, we first consider them individually. For SOL2003-10-28T11:10 (X17.2), double footpoint γ-ray sources straddled the same loop arcade as the electron sources. However, they were displaced along the loop arcade by $14''$ and $17''\pm5''$ respectively (Fig. 25). For the other events, the location of the single unresolved sources are expressed as a centroid location with appropriate statistical errors. For SOL2003-07-23T00:35 (X4.8), the neutron-capture line source was displaced by $25''\pm5''$ from the corresponding electron source. Although it was not located near any obvious EUV or Hα line emission, its location was consistent with the footpoint of a post-flare loop. For SOL2005-01-20T07:01 (X7.1) the neutron source was clearly associated with one of the two electron-bremsstrahlung footpoints. For SOL2003-11-02T17:25 (X8.3) the location also appeared to be preferentially associated with one of the two electron-bremsstrahlung footpoints, although in this case the displacement of the 2.2 MeV emission compared to the centroid of the electron-bremsstrahlung source was significant only at the 2σ level. For SOL2003-10-29T20:49 (X10.0), the measured separation of the corresponding centroids was $13''\pm17''$, so a separation, if any, could not be established. In summary, statistically significant displacements were observed in three of the five events, marginally in a fourth but not seen in the fifth and statistically least significant event.

5.3 Other γ-Ray Imaging

In addition to the neutron-capture line, imaging of the γ-ray continuum in the 200–800 keV range dominated by electron bremsstrahlung has also been reported (Hurford et al. 2005, 2006a; Krucker et al. 2008b). At higher energies, where nuclear emission is more important, only *RHESSI* detector 9 (183″) was able to modulate effectively with the events studied. The tentative explanation for the reduced efficiency of detector 6 modulation for imaging in a broad energy band at high energies is that photons passing through front grid slits into rear grid slats can scatter into the detector and be counted as part of the modulated flux. The modulation of such events, however, is 180° out of phase with that of the valid (front slit-to-rear slit) events and so tends to reduce the effective modulation efficiency. This issue is a factor for detector 6 (but not detector 9) since the detector 6 grids are not thick enough to suppress the scattered photons. This issue is not relevant for neutron-capture line imaging since photons that are scattered by the front grid scatter away from the narrow window used for spectral analysis.

Nevertheless, the centroid locations obtained for the SOL2005-01-20T07:01 (X7.1) γ-ray continuum with detector 9 in four bands between 600 keV and 8 MeV were of interest particularly because of their energy dependence. At the lowest energies in this band, their centroid locations matched that of the centroid of the X-ray electron-bremsstrahlung source. However, at higher energies where the nuclear contribution became more important, the γ-ray continuum location trended toward the neutron-capture line centroid (Hurford et al. 2006a). Such a pattern is consistent with the displacement between the sites of electron-bremsstrahlung collisions and nuclear interactions.

Imaging in a narrow energy range surrounding the 511 keV line was also attempted for SOL2005-01-20. Images obtained independently from the front and rear segments of detector 9 were found to be self-consistent in terms of strength and intensity. However, modeling suggests that although a narrow 500–520 keV energy range was used, the imaged fluence was dominated by the electron-bremsstrahlung continuum.

5.4 Discussion

One well-established result of the neutron-capture line mapping was that the flux for all five events was consistent with coming from compact source component(s) in the flaring active region. This would suggest that the flare-associated γ-ray emission was associated with flare-accelerated ions, and not with ions that are shock-accelerated at high altitudes. If that were the case, the ions might be expected to interact with ambient material in a more diffuse pattern, perhaps at some distance from the flare. This shows that the γ-ray emission does not result from the interaction of solar energetic particles with the solar wind as alternatively proposed by Kahler and Ragot (2008).

Given the resolution and sensitivity of the imaging and the limited sample of events, there is no evidence that the morphologies of nuclear γ-ray sources (e.g., double vs. single footpoints) differ systematically from that of hard X-ray sources.

The most striking result, of course, is the differences in the locations of the electron-bremsstrahlung and neutron-capture line sources. Hypotheses that attempt to account for such differences generally fall into three categories: (i) the displacement is an instrumental artifact; (ii) the displacement has a solar, but relatively mundane origin; or (iii) the displacement reflects differences in acceleration and/or transport processes for flare-accelerated ions and electrons.

We first consider the possibility that the displacements are an artifact of the imaging process. There are three independent arguments indicating that this is highly unlikely. First, the measurements of the locations of the sources at two different energy regimes were based on analyses that differed only in the energy of the counts upon which the imaging is based. The same time interval, detector grid optics, aspect solution, imaging algorithm and imaging parameters were used for both so that the potential for systematic co-location errors can be discounted. Second, if the relative displacements had a systematic instrumental origin they would likely be with respect to the only axis of symmetry associated with the imaging process, namely the line joining the direction of the rotation axis to that of the flare. However, the displacements were neither systematically parallel to or perpendicular to this direction. Third, in all but SOL2002-07-23T00:35, the direction of the displacement, if any, was in a magnetically significant solar direction, as opposed to some randomly-chosen direction. As result of these three considerations, it is considered highly unlikely that the displacements have an instrumental origin.

A second class of potential explanations for the displacements is that they have a solar, but relatively mundane explanation. One such explanation might be based on the average distance between the locations of the original nuclear interaction and that of the neutron-capture process. However, as mentioned, such a displacement is expected to be less than $\sim 1''$. Another possibility is that the displacement is related to the time-delay associated with thermalization. This scenario notes that the neutron-capture line image reflects the location of nuclear interactions that occurred ~ 100 seconds earlier. To investigate this hypothesis, for each event the neutron-capture line images were compared both to cotemporaneous electron-bremsstrahlung images and to electron-bremsstrahlung images made with integration times shifted by 100 s. No significant differences were observed.

These considerations therefore support a solar-physics-based interpretation of the displacement, based on a difference in acceleration and/or propagation between accelerated ions and electrons. Several such interpretations have been put forward. Emslie et al. (2004b) discussed a stochastic acceleration mechanism whereby ions are preferentially accelerated in larger loops than electrons. Imaging for SOL2002-07-23 (Hurford et al. 2003) would support this interpretation. The interpretation of time delays between γ-ray lines and hard X-rays by trap-plus-precipitation models for the same flare (Dauphin and Vilmer 2007) also confirm the fact that ions are injected in longer loops than electrons.

However, such an explanation would not appear to be appropriate for the SOL2003-10-28T11:10 (X17.2) where the footpoint separations for the electrons and ions were similar. Gradient and curvature drifts were also considered by Hurford et al. (2006b), but the predicted displacements between the ions and electrons were two orders of magnitude too small. Another possibility, raised by Litvinenko and Somov (1993), was that electric-field acceleration would lead to separation of particles with opposite charge (Zharkova and Gordovskyy 2004). Its potential applicability in these cases has not yet been pursued.

In two of the five events (SOL2005-01-20T17:25 and SOL2003-11-02T17:25) the ions were preferentially associated with one of the two electron bremsstrahlung footpoints. This raises the possibility that at least for these two events, the different footpoint weighting between the two species might be attributed to differences between electron and ion propagation in asymmetrical fields. This would be a result of the significant differences in the scattering properties and ionization-loss stopping distances between the accelerated electrons and ions. Modeling to support such a hypothesis has not been carried out, however.

With only five imaged events, it is too early to establish the systematics of the displacements. It is also apparent that *RHESSI*'s $35''$ resolution is insufficient to resolve the individual γ-ray sources. Therefore, it is to be hoped that *RHESSI* will observe additional events in the coming maximum so that their systematic properties can be further explored and that new instrumentation will become available with improved angular resolution and sensitivity in the γ-ray regime.

6 Conclusions and Future Perspectives

Two particular observational advances characterize the *RHESSI* era: spectral resolution adequate to reveal line shapes and to better constrain the line fluences; and γ-ray imaging. The former provides in principle previously inaccessible information on proton and α-particle energy and angular distributions. The latter has revealed the sites of ion interaction and neutron production and most dramatically shown that they do not coincide with the HXR-identified locations of electron interaction. Also, it has not confirmed the *SMM*-era suggestion that 2.223 MeV line emission might be produced over an extended region (Vestrand and Forrest 1993). The observations are also not consistent with a production of γ-rays through the interaction of solar energetic protons with the solar wind as examined by Kahler and Ragot (2008). We may look forward to further advances, both observational (e.g., with GRIPS; Shih et al. 2009a), and theoretical, over the next few years.

At the same time as *RHESSI* as well as *INTEGRAL*/SPI observations have brought pioneering observations of γ-ray spectra at high resolution, it was also realized that the determination of all the parameters which characterize the γ-ray spectrum (target abundances and ionization states, energy and angular distributions of energetic ions, relative abundances of accelerated α particles and heavier ions, transport and energy losses of energetic ions between the acceleration region and interaction region if not co-spatial, etc.) is still very challenging. In particular, the different parameters cannot be determined independently from the assumptions made on others: both ambient abundances and parameters of energetic particles must be deduced together even in the "simple" analysis of a γ-ray spectrum. Combining the information from narrow line shapes and line fluence ratios can, however, help to disentangle the effects of the different parameters. The use of lines purely produced by α particles (in particular the line at 0.339 MeV formed by the interaction of α particles with iron) would also considerably improve the analysis of γ-ray spectra by providing cleaner constraints on the distribution of α particles. A few attempts had been made at the time of *SMM*, but no further studies have been published since then based on spectra obtained at higher resolution. This is also the case for the lines produced by energetic ^{3}He on ambient ^{16}O around 1 MeV. Line shapes and widths of the 511 keV annihilation line and positronium continuum have been measured at high spectral resolution leading for a few flares to unexpected combination of chromospheric densities and transition region temperatures for the annihilation region. Taking this into account in the future, the production and annihilation of positrons over a broad range of heights and thus of conditions in the medium is probably the next step to undertake to better understand the observations of shapes and widths of the annihilation line and continuum.

Heavier ($Z > 2$) accelerated ion species are harder to diagnose but they may nonetheless be significant in overall flare energetics (Murphy et al. 1997; Emslie et al. 2004a). The detailed treatment now available of the unresolved component of nuclear γ-rays should bear significantly on this question (Murphy et al. 2009). Heavier species are prolific producers of free neutrons via evaporation-type reactions so instruments planned to measure, on future inner-heliosphere missions, neutrons in the 1–10 MeV range not yet explored systematically (e.g., Woolf et al. 2009) will yield new constraints.

Do all flares accelerate ions? It seems unlikely that they do not, but all flares do not produce detectable γ-ray lines. *RHESSI*'s eight years of operation have accumulated a data set bearing on this question (Shih et al. 2009b). A good correlation is found between the energy content of electrons above 300 keV and protons above 30 MeV suggesting a strong correlation between the acceleration of high-energy electrons and γ-ray line producing ions. A larger variation is found between the energy content of electrons above 20 keV and the ion energy content. Some episodes in flares or some specific flares exhibit indeed higher relative numbers of electrons with respect to ions. Based on the *RHESSI* data set, it was also found that there is a strong link between the production of energetic ions and sub-relativistic electrons, but only when a threshold in the production of ions is attained. In conclusion, while the acceleration of protons above 30 MeV is related to the acceleration of relativistic electrons, the acceleration of sub-relativistic electrons is linked to the acceleration of these energetic protons and relativistic electrons, but only when a threshold of energetic ions and electrons is reached. These conclusions must be kept in mind when addressing acceleration mechanisms in flares. The content in low energy ions from flares is mostly unknown and no strong constraints have been derived on the production of low energy protons in solar flares. Attempts to constrain fluxes from low energy proton capture lines could be a way to address this issue. Another promising way could be to detect suprathermal ions through charge exchange with the ambient hydrogen (see the first detection of energetic neutral hydrogen atoms in a solar flare by Mewaldt et al. 2009).

Apart from *CORONAS-F*/SONG (e.g., Kuznetsov et al. 2006), solar observations at the highest γ-ray energies have been lacking in recent years. This is one of the only ways (with the detection of neutrons) to address the issue of the highest energetic particles accelerated in solar flares and to determine the shape of the energetic ion spectrum over several decades of energy range. The global network of neutron detectors thus plays at the present time a particularly important role, yielding otherwise unobtainable information on ion energy distributions in the GeV energy range (e.g., Masson et al. 2009). As the solar cycle picks up, *Fermi* observations of the high-energy continuum should become critical to probing the highest ion energies attained in flares. Further observations of flares at submillimeter (and even shorter) wavelengths (see, e.g., Klein et al. 2006) as well as the search for X-ray inverse Compton radiation in coronal sources (MacKinnon and Mallik 2010) could additionally provide information on electrons and positrons produced in the 10–100 MeV energy range in flares.

Recent years have seen a re-examination of the thick-target interpretation of flare HXRs (Fletcher and Hudson 2008; Brown et al. 2009). Since HXR-emitting electrons are believed to embody a large fraction of the total energy released in the flare, this is a question that goes beyond interpretation of radiation signatures to the heart of the flare energy release process. The existing thick-target models assume that electrons radiate negligibly in the (coronal) acceleration region, and that they slow down collisionally in the (thick-target, chromospheric) interaction region. If acceleration and interaction regions were one and the same, fewer electrons would be needed to account for HXRs. A similar reconsideration of γ-ray source regions may prove necessary.

Acknowledgements This chapter is dedicated to Reuven Ramaty who was one of the initiators of the field of high energy physics in solar flares, inducting two of us (NV and ALM) into the field and inspiring many other high energy solar physicists. The authors are very grateful to the organizers of the series of *RHESSI* workshops and to the team leaders of the working group on energetic ions and particle acceleration who stimulated the many discussions and collaborations that led to many recent results summarized in this chapter. We would like to express our special thanks to G. Share, R. Murphy, V. Tatischeff and K. Watanabe for useful suggestions for this chapter and discussions. Nicole Vilmer acknowledges support from the Centre National d'Etudes Spatiales (CNES) and from the French program on Solar-Terrestrial Physics (PNST) of INSU/CNRS for the participation to the *RHESSI* project. Collaborative work on the chapter in Meudon was supported by a British Council Franco-British Alliance grant. Solar physics research in Glasgow is supported by an STFC Rolling Grant. We are grateful to several institutions for granting permission for the use of previously published figures. In addition to those noted in figure captions, we also acknowledge the following figures from A&A, reproduced with permission (©) ESO: Fig. 14 (Harris et al., **461**, 723, 2007); Figs. 16 and 17 (Kiener et al., **445**, 725, 2006). The following figures are reproduced with kind permission from Springer Science+Business Media: Fig. 1 (Suri et al., Solar Physics **43**, 415, 1975); Fig. 22 (Masson et al., Solar Physics **257**, 305, 2009).

References

V.V. Akimov, V.G. Afanasev, A.S. Belousov, I.D. Blokhintsev, V.A. Volzhenskaya, L.F. Kalinkin, N.G. Leikov, V.E. Nesterov, A.M. Galper, S.A. Voronov, V.M. Zemskov, V.G. Kirillov-Ugryumov, B.I. Luchkov, Y.V. Ozerov, A.V. Popov, V.A. Rudko, M.F. Runtso, V.Y. Chesnokov, L.V. Kurnosova, M.A. Rusakovich, N.P. Topchiev, M.I. Fradkin, E.I. Chuikin, V.Y. Tugaenko, T.N. Tian, V.I. Ishkov, M. Gros, I. Grenier, E. Barouch, P. Wallin, A.R. Baser-Bachi, J.M. Lavigne, J.F. Olive, J. Juchniewicz, Sov. Astron. Lett. **18**, 69 (1992)

D. Alexander, P.P. Dunphy, A.L. MacKinnon, Sol. Phys. **151**, 147 (1994)

M. Asplund, N. Grevesse, A.J. Sauval, in *Cosmic Abundances as Records of Stellar Evolution and Nucleosynthesis*, ed. by T.G. Barnes III, F.N. Bash. Astronomical Society of the Pacific Conference Series, vol. 336 (2005), p. 25

C. Barat, G. Trottet, N. Vilmer, J. Dezalay, R. Talon, R. Sunyaev, O. Terekhov, A. Kuznetsov, Astrophys. J. **425**, L109 (1994)

J.W. Bieber, J. Clem, P. Evenson, R. Pyle, D. Ruffolo, A. Sáiz, Geophys. Res. Lett. **32**, 3 (2005)

L. Biermann, O. Haxel, A. Schlüter, Z. Naturforsch. Teil A **6**, 47 (1951)

J.W. Brosius, Astrophys. J. **555**, 435 (2001)

J.C. Brown, R. Turkmani, E.P. Kontar, A.L. MacKinnon, L. Vlahos, Astron. Astrophys. **508**, 993 (2009)

R.C. Canfield, C.-R. Chang, Astrophys. J. **295**, 275 (1985)

E.L. Chupp, Annu. Rev. Astron. Astrophys. **22**, 359 (1984)

E.L. Chupp, in *American Institute of Physics Conference Series*, vol. 374, ed. by R. Ramaty, N. Mandzhavidze, X.-M. Hua (1996), p. 3

E.L. Chupp, H. Debrunner, E. Flueckiger, D.J. Forrest, F. Golliez, G. Kanbach, W.T. Vestrand, J. Cooper, G. Share, Astrophys. J. **318**, 913 (1987)

E.L. Chupp, P.P. Dunphy, in *High Energy Solar Physics Workshop - Anticipating HESSI*, ed. by R. Ramaty, N. Mandzhavidze. Astronomical Society of the Pacific Conference Series, vol. 206 (2000), p. 400

E.L. Chupp, J.M. Ryan, Res. Astron. Astrophys. **9**, 11 (2009)

E.L. Chupp, D.J. Forrest, P.R. Higbie, A.N. Suri, C. Tsai, P.P. Dunphy, Nature **241**, 333 (1973)

E.L. Chupp, D.J. Forrest, J.M. Ryan, M.L. Cherry, C. Reppin, G. Kanbach, E. Rieger, K. Pinkau, G.H. Share, R.L. Kinzer, Astrophys. J. **244**, L171 (1981)

E.L. Chupp, D.J. Forrest, J.M. Ryan, J. Heslin, C. Reppin, K. Pinkau, G. Kanbach, E. Rieger, G.H. Share, Astrophys. J. **263**, L95 (1982)

E.L. Chupp, G. Trottet, H. Marschhauser, M. Pick, L. Soru-Escaut, E. Rieger, P.P. Dunphy, Astron. Astrophys. **275**, 602 (1993)

E.W. Cliver, N.B. Crosby, B.R. Dennis, Astrophys. J. **426**, 767 (1994)

C. Dauphin, N. Vilmer, Astron. Astrophys. **468**, 289 (2007)

H. Debrunner, E.O. Flückiger, P. Stein, Nucl. Instrum. Methods Phys. Res., Sect. A, Accel. Spectrom. Detect. Assoc. Equip. **278**, 573 (1989)

H. Debrunner, J.A. Lockwood, C. Barat, R. Buetikofer, J.P. Dezalay, E. Flueckiger, A. Kuznetsov, J.M. Ryan, R. Sunyaev, O.V. Terekhov, G. Trottet, N. Vilmer, Astrophys. J. **479**, 997 (1997)

H. Debrunner, J.A. Lockwood, J.M. Ryan, Astrophys. J. **409**, 822 (1993)

B.L. Dingus, P. Sreekumar, D.L. Bertsch, C.E. Fichtel, R.C. Hartman, S.D. Hunter, D.J. Thompson, E.J. Schneid, K.T.S. Brazier, G. Kanbach, C. von Montigny, H.A. Mayer-Hasselwander, Y.C. Lin, P.F. Michelson, P.L. Nolan, D.A. Kniffen, J.R. Mattox, in *High-Energy Solar Phenomena - a New Era of Spacecraft Measurements*, ed. by J. Ryan, W.T. Vestrand. American Institute of Physics Conference Series, vol. 294 (1994), p. 177

J.J. Drake, P. Testa, Nature **436**, 525 (2005)

W. Droege, D. Ruffolo, B. Klecker, Astrophys. J. **464**, L87+ (1996)

P.P. Dunphy, in *International Cosmic Ray Conference*, vol. 6 (1999), p. 9

P.P. Dunphy, E.L. Chupp, in *International Cosmic Ray Conference*, vol. 3 (1991), p. 65

P.P. Dunphy, E.L. Chupp, in *Particle Acceleration in Cosmic Plasmas*, ed. by G.P. Zank, T.K. Gaisser. American Institute of Physics Conference Series, vol. 264 (1992), p. 253

P.P. Dunphy, E.L. Chupp, D.L. Bertsch, E.J. Schneid, S.R. Gottesman, G. Kanbach, Sol. Phys. **187**, 45 (1999)

A.G. Emslie, J.C. Brown, A.L. MacKinnon, Astrophys. J. **485**, 430 (1997)

A.G. Emslie, H. Kucharek, B.R. Dennis, N. Gopalswamy, G.D. Holman, G.H. Share, A. Vourlidas, T.G. Forbes, P.T. Gallagher, G.M. Mason, T.R. Metcalf, R.A. Mewaldt, R.J. Murphy, R.A. Schwartz, T.H. Zurbuchen, Journal of Geophysical Research (Space). Physics **109**, 10104 (2004a)

A.G. Emslie, J.A. Miller, J.C. Brown, Astrophys. J. **602**, L69 (2004b)

P. Evenson, R. Kroeger, P. Meyer, D. Reames, Astrophys. J. Suppl. Ser. **73**, 273 (1990)

P. Evenson, P. Meyer, K.R. Pyle, Astrophys. J. **274**, 875 (1983)

W.C. Feldman, D.J. Lawrence, J.O. Goldsten, R.E. Gold, D.N. Baker, D.K. Haggerty, G.C. Ho, S. Krucker, R.P. Lin, R.A. Mewaldt, R.J. Murphy, L.R. Nittler, E.A. Rhodes, J.A. Slavin, S.C. Solomon, R.D. Starr, F. Vilas, A. Vourlidas, J. Geophys. Res. **115**, 1102 (2010)
N.M. Firstova, V.I. Polyakov, A.V. Firstova, Sol. Phys. **249**, 53 (2008)
G.D. Fleishman, E.P. Kontar, Astrophys. J. **709**, L127 (2010)
L. Fletcher, H.S. Hudson, Astrophys. J. **675**, 1645 (2008)
D.J. Forrest, in *Positron-Electron Pairs in Astrophysics*, ed. by R. Ramaty. American Institute of Physics Conference Series, vol. 101 (1983), p. 3
D.J. Forrest, W.T. Vestrand, E.L. Chupp, E. Rieger, J.F. Cooper, G.H. Share, in *International Cosmic Ray Conference*, vol. 4, ed. by F.C. Jones, (1985), p. 146
A.H. Gabriel, Philos. Trans. R. Soc. Lond. Ser. A, Math. Phys. Sci. **281**, 339 (1976)
W.Q. Gan, Adv. Space Res. **35**, 1833 (2005)
M. Gros, V. Tatischeff, J. Kiener, B. Cordier, C. Chapuis, G. Weidenspointner, G. Vedrenne, A. von Kienlin, R. Diehl, A. Bykov, M.N. Méndez, in *5th INTEGRAL Workshop on the INTEGRAL Universe*, ed. by V. Schoenfelder, G. Lichti, C. Winkler. ESA Special Publication, vol. 552 (2004), p. 669
M.J. Harris, V. Tatischeff, J. Kiener, M. Gros, G. Weidenspointner, Astron. Astrophys. **461**, 723 (2007)
J.C. Henoux, G. Chambe, D. Smith, D. Tamres, N. Feautrier, M. Rovira, S. Sahal-Brechot, Astrophys. J. Suppl. Ser. **73**, 303 (1990)
Holman et al., Space Sci. Rev. (2011, this issue). doi:10.1007/s11214-010-9680-9
X. Hua, R.E. Lingenfelter, Sol. Phys. **113**, 229 (1987a)
X.-M. Hua, R.E. Lingenfelter, Sol. Phys. **107**, 351 (1987b)
X. Hua, R. Ramaty, R.E. Lingenfelter, Astrophys. J. **341**, 516 (1989)
X. Hua, B. Kozlovsky, R.E. Lingenfelter, R. Ramaty, A. Stupp, Astrophys. J. Suppl. **140**, 563 (2002)
H.S. Hudson, T. Bai, D.E. Gruber, J.L. Matteson, P.L. Nolan, L.E. Peterson, Astrophys. J. **236**, L91 (1980)
E. Hulot, N. Vilmer, E.L. Chupp, B.R. Dennis, S.R. Kane, Astron. Astrophys. **256**, 273 (1992)
E. Hulot, N. Vilmer, G. Trottet, Astron. Astrophys. **213**, 383 (1989)
G.J. Hurford, E.J. Schmahl, R.A. Schwartz, A.J. Conway, M.J. Aschwanden, A. Csillaghy, B.R. Dennis, C. Johns-Krull, S. Krucker, R.P. Lin, J. McTiernan, T.R. Metcalf, J. Sato, D.M. Smith, Sol. Phys. **210**, 61 (2002)
G.J. Hurford, R.A. Schwartz, S. Krucker, R.P. Lin, D.M. Smith, N. Vilmer, Astrophys. J. **595**, L77 (2003)
G.J. Hurford, S. Krucker, R.P. Lin, R.A. Schwartz, D.M. Smith, AGU Fall Meeting AbstractsA313+ (2005)
G.J. Hurford, S. Krucker, R.P. Lin, R.A. Schwartz, G.H. Share, D.M. Smith, Bull. Am. Astron. Soc. **38**, 255 (2006a)
G.J. Hurford, S. Krucker, R.P. Lin, R.A. Schwartz, G.H. Share, D.M. Smith, Astrophys. J. **644**, L93 (2006b)
S.W. Kahler, B.R. Ragot, Astrophys. J. **675**, 846 (2008)
G. Kanbach, D.L. Bertsch, C.E. Fichtel, R.C. Hartman, S.D. Hunter, D.A. Kniffen, P.W. Kwok, Y.C. Lin, J.R. Mattox, H.A. Mayer-Hasselwander, Astron. Astrophys. Suppl. **97**, 349 (1993)
P. Kaufmann, J. Raulin, Phys. Plasmas **13**(7), 070701 (2006)
P. Kaufmann, J. Raulin, C.G.G. de Castro, H. Levato, D.E. Gary, J.E.R. Costa, A. Marun, P. Pereyra, A.V.R. Silva, E. Correia, Astrophys. J. **603**, L121 (2004)
C.F. Kennel, H.E. Petschek, J. Geophys. Res. **71**, 1 (1966)
J. Kiener, N. de Séréville, V. Tatischeff, Phys. Rev. C **64**(2), 025803 (2001)
J. Kiener, M. Gros, V. Tatischeff, G. Weidenspointner, Astron. Astrophys. **445**, 725 (2006)
K. Klein, G. Trottet, G. Molodij, A. Sémery, in *Visions for Infrared Astronomy, Instrumentation, Measure, Métrologie*, ed. by V. Coudé du Foresto, D. Rouan, G. Rousset (2006), p. 49
L.G. Kocharov, G.A. Kovaltsov, G.E. Kocharov, E.I. Chuikin, I.G. Usoskin, M.A. Shea, D.F. Smart, V.F. Melnikov, T.S. Podstrigach, T.P. Armstrong, Sol. Phys. **150**, 267 (1994)
L.G. Kocharov, J. Torsti, R. Vainio, G.A. Kovaltsov, I.G. Usoskin, Sol. Phys. **169**, 181 (1996)
L. Kocharov, H. Debrunner, G. Kovaltsov, J. Lockwood, M. McConnell, P. Nieminen, G. Rank, J. Ryan, V. Schoenfelder, Astron. Astrophys. **340**, 257 (1998)
B. Kozlovsky, R.E. Lingenfelter, R. Ramaty, Astrophys. J. **316**, 801 (1987)
B. Kozlovsky, R.J. Murphy, R. Ramaty, Astrophys. J. Suppl. Ser. **141**, 523 (2002)
B. Kozlovsky, R.J. Murphy, G.H. Share, Astrophys. J. **604**, 892 (2004)
S. Krucker, M. Battaglia, P.J. Cargill, L. Fletcher, H.S. Hudson, A.L. MacKinnon, S. Masuda, L. Sui, M. Tomczak, A.L. Veronig, L. Vlahos, S.M. White, Astron. Astrophys. Rev. **16**, 155 (2008a)
S. Krucker, G.J. Hurford, A.L. MacKinnon, A.Y. Shih, R.P. Lin, Astrophys. J. **678**, L63 (2008b)
K. Kudela, S.N. Kuznetsov, N. Myagkova, B.Y. Yushkov, in *International Cosmic Ray Conference*, vol. 6 (2003), p. 3183
S.N. Kuznetsov, V.G. Kurt, I.N. Myagkova, B.Y. Yushkov, K. Kudela, Sol. Syst. Res. **40**, 104 (2006)
S.N. Kuznetsov, V.G. Kurt, B.Y. Yushkov et al., in *International Cosmic Ray Conference*, vol. 1 (2008), p. 121

F.L. Lang, C.W. Werntz, in *Gamma-Ray Line Astrophysics*, ed. by P. Durouchoux, N. Prantzos. American Institute of Physics Conference Series, vol. 232 (1991), p. 445
N.G. Leikov, V.V. Akimov, V.A. Volzhenskaia, L.F. Kalinkin, V.E. Nesterov, A.M. Gal'Per, V.M. Zemskov, I.V. Ozerov, N.P. Topchiev, M.I. Fradkin, Astron. Astrophys. Suppl. **97**, 345 (1993)
R.P. Lin, B.R. Dennis, G.J. Hurford, D.M. Smith, A. Zehnder, P.R. Harvey, D.W. Curtis, D. Pankow, P. Turin, M. Bester, A. Csillaghy, M. Lewis, N. Madden, H.F. van Beek, M. Appleby, T. Raudorf, J. McTiernan, R. Ramaty, E. Schmahl, R. Schwartz, S. Krucker, R. Abiad, T. Quinn, P. Berg, M. Hashii, R. Sterling, R. Jackson, R. Pratt, R.D. Campbell, D. Malone, D. Landis, C.P. Barrington-Leigh, S. Slassi-Sennou, C. Cork, D. Clark, D. Amato, L. Orwig, R. Boyle, I.S. Banks, K. Shirey, A.K. Tolbert, D. Zarro, F. Snow, K. Thomsen, R. Henneck, A. Mchedlishvili, P. Ming, M. Fivian, J. Jordan, R. Wanner, J. Crubb, J. Preble, M. Matranga, A. Benz, H. Hudson, R.C. Canfield, G.D. Holman, C. Crannell, T. Kosugi, A.G. Emslie, N. Vilmer, J.C. Brown, C. Johns-Krull, M. Aschwanden, T. Metcalf, A. Conway, Sol. Phys. **210**, 3 (2002)
R.P. Lin, S. Krucker, G.J. Hurford, D.M. Smith, H.S. Hudson, G.D. Holman, R.A. Schwartz, B.R. Dennis, G.H. Share, R.J. Murphy, A.G. Emslie, C. Johns-Krull, N. Vilmer, Astrophys. J. **595**, L69 (2003)
R.E. Lingenfelter, E.J. Flamm, E.H. Canfield, S. Kellman, J. Geophys. Res. **70**, 4077 (1965a)
R.E. Lingenfelter, E.J. Flamm, E.H. Canfield, S. Kellman, J. Geophys. Res. **70**, 4087 (1965b)
R.E. Lingenfelter, R. Ramaty, *High Energy Nuclear Reactions in Solar Flares* (1967a)
R.E. Lingenfelter, R. Ramaty, Planet. Space Sci. **15**, 1303 (1967b)
Y.E. Litvinenko, B.V. Somov, Sol. Phys. **146**, 127 (1993)
J.A. Lockwood, H. Debrunner, Space Sci. Rev. **88**, 483 (1999)
J.A. Lockwood, H. Debrunner, J.M. Ryan, Sol. Phys. **173**, 151 (1997)
C. Lopate, Geophys. Monogr. **165**, 283 (2006)
T. Lüthi, A. Magun, M. Miller, Astron. Astrophys. **415**, 1123 (2004)
A.L. MacKinnon, Astron. Astrophys. **226**, 284 (1989)
A.L. MacKinnon, Geophys. Monogr. **165**, 157 (2006)
A.L. MacKinnon, Astron. Astrophys. **462**, 763 (2007)
A.L. MacKinnon, J.C. Brown, Astron. Astrophys. **232**, 544 (1990)
A.L. MacKinnon, P.C.V. Mallik, Astron. Astrophys. **510**, A29+ (2010)
A.L. MacKinnon, M.P. Toner, Astron. Astrophys. **409**, 745 (2003)
N. Mandzhavidze, R. Ramaty, Astrophys. J. **389**, 739 (1992)
N. Mandzhavidze, R. Ramaty, in *High Energy Solar Physics Workshop - Anticipating HESSI*, ed. by R. Ramaty, N. Mandzhavidze. Astronomical Society of the Pacific Conference Series, vol. 206 (2000), p. 64
N. Mandzhavidze, R. Ramaty, B. Kozlovsky, Astrophys. J. **518**, 918 (1999)
S. Masson, K. Klein, R. Bütikofer, E. Flückiger, V. Kurt, B. Yushkov, S. Krucker, Sol. Phys. **257**, 305 (2009)
M.L. McConnell, K. Bennett, A.L. MacKinnon, R. Miller, G. Rank, J.M. Ryan, V. Schönfelder, in *Proc. 25th ICRC, Durban (South Africa)*, vol. 113 (1997)
R.A. Mewaldt, R.A. Leske, E.C. Stone, A.F. Barghouty, A.W. Labrador, C.M.S. Cohen, A.C. Cummings, A.J. Davis, T.T. von Rosenvinge, M.E. Wiedenbeck, Astrophys. J. **693**, L11 (2009)
J. Meyer, Adv. Space Res. **13**, 377 (1993)
J.A. Miller, P.J. Cargill, A.G. Emslie, G.D. Holman, B.R. Dennis, T.N. LaRosa, R.M. Winglee, S.G. Benka, S. Tsuneta, J. Geophys. Res. **102**, 14631 (1997)
P. Morrison, in *Fifth International Congress on Cosmic Radiation*, ed. by M.S. Vallarta (1958), p. 305
Y. Muraki, K. Murakami, M. Miyazaki, K. Mitsui, S. Shibata, S. Sakakibara, T. Sakai, T. Takahashi, T. Yamada, K. Yamaguchi, Astrophys. J. **400**, L75 (1992)
R.J. Murphy, C.D. Dermer, R. Ramaty, Astrophys. J. Suppl. Ser. **63**, 721 (1987)
R.J. Murphy, B. Kozlovsky, R. Ramaty, Astrophys. J. **331**, 1029 (1988)
R.J. Murphy, R. Ramaty, D.V. Reames, B. Kozlovsky, Astrophys. J. **371**, 793 (1991)
R.J. Murphy, G.H. Share, J.E. Grove, W.N. Johnson, R.L. Kinzer, R.A. Kroeger, J.D. Kurfess, M.S. Strickman, S.M. Matz, D.A. Grabelsky, W.R. Purcell, M.P. Ulmer, R.A. Cameron, G.V. Jung, C.M. Jensen, W.T. Vestrand, D.J. Forrest, in *American Institute of Physics Conference Series*, vol. 280, ed. by M. Friedlander, N. Gehrels, D.J. Macomb. (1993), p. 619
R.J. Murphy, G.H. Share, J.E. Grove, W.N. Johnson, R.L. Kinzer, J.D. Kurfess, M.S. Strickman, G.V. Jung, Astrophys. J. **490**, 883 (1997)
R.J. Murphy, G.H. Share, K.W. Delsignore, X.-M. Hua, Astrophys. J. **510**, 1011 (1999)
R.J. Murphy, G.H. Share, X.-M. Hua, R.P. Lin, D.M. Smith, R.A. Schwartz, Astrophys. J. **595**, L93 (2003)
R.J. Murphy, G.H. Share, J.G. Skibo, B. Kozlovsky, Astrophys. J. Suppl. Ser. **161**, 495 (2005)
R.J. Murphy, B. Kozlovsky, G.H. Share, X.-M. Hua, R.E. Lingenfelter, Astrophys. J. Suppl. Ser. **168**, 167 (2007)
R.J. Murphy, B. Kozlovsky, J. Kiener, G.H. Share, Astrophys. J. Suppl. Ser. **183**, 142 (2009)

I.N. Myagkova, S.N. Kuznetsov, B.Y. Yushkov, K. Kudela, in *35th COSPAR Scientific Assembly*. COSPAR, Plenary Meeting, vol. 35 (2004), p. 1511
I.N. Myagkova, S.N. Kuznetsov, V.G. Kurt, B.Y. Yuskov, V.I. Galkin, E.A. Muravieva, K. Kudela, Adv. Space Res. **40**, 1929 (2007)
J.P. Naiman, D.M. Smith, R.J. Murphy, G.H. Share, A.Y. Shih, J. Kiener, AGU Fall Meeting Abstracts B 1670+ (2008)
F. Pelaez, P. Mandrou, M. Niel, B. Mena, N. Vilmer, G. Trottet, F. Lebrun, J. Paul, O. Terekhov, R. Siuniaev, Sol. Phys. **140**, 121 (1992)
B. Pirard, R.S. Woolf, U. Bravar, P.J. Bruillard, E.O. Flückiger, J.S. Legere, A.L. MacKinnon, J.R. Macri, P.C.V. Mallik, M.R. Moser, J.M. Ryan, Nucl. Instrum. Methods Phys. Res., Sect. A, Accel. Spectrom. Detect. Assoc. Equip. **603**, 406 (2009)
T.A. Prince, J.C. Ling, W.A. Mahoney, G.R. Riegler, A.S. Jacobson, Astrophys. J. **255**, L81 (1982)
K.R. Pyle, J.A. Simpson, in *International Cosmic Ray Conference*, vol. 3 (1991), p. 53
R. Ramaty, in *Physics of the Sun*, vol. 2 (1986), p. 2291
R. Ramaty, C.J. Crannell, Astrophys. J. **203**, 766 (1976)
R. Ramaty, N. Mandzhavidze, in *Highly Energetic Physical Processes and Mechanisms for Emission from Astrophysical Plasmas*, ed. by P.C.H. Martens, S. Tsuruta, M.A. Weber. IAU Symposium, vol. 195 (2000), p. 123
R. Ramaty, B. Kozlovsky, R.E. Lingenfelter, Astrophys. J. Suppl. Ser. **40**, 487 (1979)
R. Ramaty, N. Mandzhavidze, B. Kozlovsky, R.J. Murphy, Astrophys. J. **455**, L193+ (1995)
R. Ramaty, N. Mandzhavidze, B. Kozlovsky, in *American Institute of Physics Conference Series*, vol. 374, ed. by R. Ramaty, N. Mandzhavidze, X.-M. Hua, (1996), p. 172
R. Ramaty, N. Mandzhavidze, C. Barat, G. Trottet, Astrophys. J. **479**, 458 (1997)
G. Rank, J. Ryan, H. Debrunner, M. McConnell, V. Schönfelder, Astron. Astrophys. **378**, 1046 (2001)
J.C. Raymond, G. Holman, A. Ciaravella, A. Panasyuk, Y.-K. Ko, J. Kohl, Astrophys. J. **659**, 750 (2007)
D.V. Reames, Space Sci. Rev. **90**, 413 (1999)
D.V. Reames, L.M. Barbier, T.T. von Rosenvinge, G.M. Mason, J.E. Mazur, J.R. Dwyer, Astrophys. J. **483**, 515 (1997)
E. Rieger, Astrophys. J. **90**, 645 (1994)
E. Rieger, H. Marschhäuser, in *Max '91/SMM Solar Flares: Observations and Theory*, ed. by R.M. Winglee, A.L. Kiplinger (1991), p. 68
E. Rieger, W.Q. Gan, H. Marschhäuser, Sol. Phys. **183**, 123 (1998)
D. Ruffolo, Astrophys. J. **382**, 688 (1991)
J.M. Ryan, Space Sci. Rev. **93**, 581 (2000)
J. Ryan, K. Bennett, H. Debrunner, D. Forrest, J. Lockwood, M. Loomis, M. McConnell, D. Morris, V. Schönfelder, B.N. Swanenburg, W. Webber, Adv. Space Res. **13**, 255 (1993)
J. Ryan, D. Forrest, J. Lockwood, M. Loomis, M. McConnell, D. Morris, W. Webber, K. Bennett, L. Hanlon, C. Winkler, H. Debrunner, G. Rank, V. Schönfelder, B.N. Swanenburg, in *High-Energy Solar Phenomena - a New Era of Spacecraft Measurements*, ed. by J. Ryan, W.T. Vestrand. American Institute of Physics Conference Series, vol. 294 (1994), p. 89
J.I. Sakai, Y. Nagasugi, Astron. Astrophys. **474**, L33 (2007)
T. Sako, K. Watanabe, Y. Muraki, Y. Matsubara, H. Tsujihara, M. Yamashita, T. Sakai, S. Shibata, J.F. Valdés-Galicia, L.X. González, A. Hurtado, O. Musalem, P. Miranda, N. Martinic, R. Ticona, A. Velarde, F. Kakimoto, S. Ogio, Y. Tsunesada, H. Tokuno, Y.T. Tanaka, I. Yoshikawa, T. Terasawa, Y. Saito, T. Mukai, M. Gros, Astrophys. J. **651**, L69 (2006)
J.T. Schmelz, K. Nasraoui, J.K. Roames, L.A. Lippner, J.W. Garst, Astrophys. J. **634**, L197 (2005)
C.J. Schrijver, H.S. Hudson, R.J. Murphy, G.H. Share, T.D. Tarbell, Astrophys. J. **650**, 1184 (2006)
G.H. Share, R.J. Murphy, Astrophys. J. **452**, 933 (1995)
G.H. Share, R.J. Murphy, Astrophys. J. **508**, 876 (1998)
G.H. Share, R.J. Murphy, in *High Energy Solar Physics Workshop - Anticipating HESSI*, ed. by R. Ramaty, N. Mandzhavidze. Astronomical Society of the Pacific Conference Series, vol. 206 (2000), p. 377
G.H. Share, R.J. Murphy, Geophys. Monogr. **165**, 177 (2006)
G.H. Share, R.J. Murphy, J. Kiener, N. de Séréville, Astrophys. J. **573**, 464 (2002)
G.H. Share, R.J. Murphy, E.K. Newton, Sol. Phys. **201**, 191 (2001)
G.H. Share, R.J. Murphy, J.G. Skibo, D.M. Smith, H.S. Hudson, R.P. Lin, A.Y. Shih, B.R. Dennis, R.A. Schwartz, B. Kozlovsky, Astrophys. J. **595**, L85 (2003a)
G.H. Share, R.J. Murphy, D.M. Smith, R.P. Lin, B.R. Dennis, R.A. Schwartz, Astrophys. J. **595**, L89 (2003b)
G.H. Share, R.J. Murphy, D.M. Smith, R.A. Schwartz, R.P. Lin, Astrophys. J. **615**, L169 (2004)
G.H. Share, R.J. Murphy, A.J. Tylka, B. Kozlovsky, J.M. Ryan, C. Gwon, J. Geophys. Res. **116**(A15), A03102 (2011)
M.A. Shea, D.F. Smart, K.R. Pyle, Geophys. Res. Lett. **18**, 1655 (1991)

S. Shibata, J. Geophys. Res. **99**, 6651 (1994)

A.Y. Shih, R.P. Lin, G.J. Hurford, S.E. Boggs, A.C. Zoglauer, C.B. Wunderer, J.G. Sample, P. Turin, S. McBride, D.M. Smith, H. Tajima, P.N. Luke, M.S. Amman, in *AAS/Solar Physics Division Meeting*, vol. 40 (2009a). #18.10

A.Y. Shih, R.P. Lin, D.M. Smith, Astrophys. J. **698**, L152 (2009b)

A.Y. Shih, D.M. Smith, R.P. Lin, G.J. Hurford, in *Bulletin of the American Astronomical Society*, vol. 38 (2007), p. 175

A.Y. Shih, D.M. Smith, R.P. Lin, G.J. Hurford, S. Krucker, R.A. Schwartz, G.H. Share, R.J. Murphy, AGU Fall Meeting Abstracts D 1132+ (2003)

A.V.R. Silva, G.H. Share, R.J. Murphy, J.E.R. Costa, C.G.G. de Castro, J. Raulin, P. Kaufmann, Sol. Phys. **245**, 311 (2007)

D.F. Smart, M.A. Shea, K. O'Brien, in *International Cosmic Ray Conference*, vol. 4 (1995), p. 171

D.M. Smith, R.P. Lin, P. Turin, D.W. Curtis, J.H. Primbsch, R.D. Campbell, R. Abiad, P. Schroeder, C.P. Cork, E.L. Hull, D.A. Landis, N.W. Madden, D. Malone, R.H. Pehl, T. Raudorf, P. Sangsingkeow, R. Boyle, I.S. Banks, K. Shirey, R. Schwartz, Sol. Phys. **210**, 33 (2002)

D.M. Smith, G.H. Share, R.J. Murphy, R.A. Schwartz, A.Y. Shih, R.P. Lin, Astrophys. J. **595**, L81 (2003)

F.W. Stecker, Astrophys. Space Sci. **6**, 377 (1970)

A. Struminsky, M. Matsuoka, K. Takahashi, Astrophys. J. **429**, 400 (1994)

A.N. Suri, E.L. Chupp, D.J. Forrest, C. Reppin, Sol. Phys. **43**, 415 (1975)

R. Talon, G. Vedrenne, A.S. Melioranskii, N.F. Pissarenko, V.M. Shamolin, O.B. Likin, in *Solar Gamma-, X-, and EUV Radiation*, ed. by S.R. Kane. IAU Symposium, vol. 68 (1975), p. 315

D.H. Tamres, D.B. Melrose, R.C. Canfield, Astrophys. J. **342**, 576 (1989)

V. Tatischeff, B. Kozlovsky, J. Kiener, R.J. Murphy, Astrophys. J. Suppl. Ser. **165**, 606 (2006)

M.P. Toner, High-energy neutral emissions from solar flares. Ph.D. thesis, University of Glasgow (2004)

M.P. Toner, A.L. MacKinnon, Sol. Phys. **223**, 155 (2004)

G. Trottet, C. Barat, R. Ramaty, N. Vilmer, J.P. Dezalay, A. Kuznetsov, N. Mandzhavidze, R. Sunyaev, R. Talon, O. Terekhov, in *American Institute of Physics Conference Series*, vol. 374, ed. by R. Ramaty, N. Mandzhavidze, X.-M. Hua (1996), p. 153

G. Trottet, S. Krucker, T. Lüthi, A. Magun, Astrophys. J. **678**, 509 (2008)

G. Trottet, N. Vilmer, in *European Meeting on Solar Physics*, ed. by G.M. Simnett, C.E. Alissandrakis, L. Vlahos. Lecture Notes in Physics, vol. 489 (Springer, Berlin, 1997), p. 219

G. Trottet, N. Vilmer, C. Barat, A. Benz, A. Magun, A. Kuznetsov, R. Sunyaev, O. Terekhov, Astron. Astrophys. **334**, 1099 (1998)

G. Trottet, N. Vilmer, C. Barat, J.P. Dezalay, R. Talon, R.A. Siuniaev, A. Kuznetsov, O. Terekhov, Astron. Astrophys. Suppl. **97**, 337 (1993)

H. Tsuchiya, Y. Muraki, K. Masuda, Y. Matsubara, T. Koi, T. Sako, S. Ohno, T. Hoshida, S. Shibata, Y. Munakata, K. Hatanaka, T. Wakasa, H. Sakai, Nucl. Instrum. Methods Phys. Res., Sect. A, Accel. Spectrom. Detect. Assoc. Equip. **463**, 183 (2001)

I.G. Usoskin, G.A. Kovaltsov, H. Kananen, P. Tanskanen, Ann. Geophys. **15**, 375 (1997)

J.F. Valdés-Galicia, Y. Muraki, K. Watanabe, Y. Matsubara, T. Sako, L.X. Gonzalez, O. Musalem, A. Hurtado, Adv. Space Res. **43**, 565 (2009)

G. Vedrenne, J. Roques, V. Schönfelder, P. Mandrou, G.G. Lichti, A. von Kienlin, B. Cordier, S. Schanne, J. Knödlseder, G. Skinner, P. Jean, F. Sanchez, P. Caraveo, B. Teegarden, P. von Ballmoos, L. Bouchet, P. Paul, J. Matteson, S. Boggs, C. Wunderer, P. Leleux, G. Weidenspointner, P. Durouchoux, R. Diehl, A. Strong, M. Cassé, M.A. Clair, Y. André, Astron. Astrophys. **411**, L63 (2003)

J.E. Vernazza, E.H. Avrett, R. Loeser, Astrophys. J. Suppl. Ser. **45**, 635 (1981)

W.T. Vestrand, D.J. Forrest, Astrophys. J. **409**, L69 (1993)

N. Vilmer, S.R. Kane, G. Trottet, Astron. Astrophys. **108**, 306 (1982)

N. Vilmer, A.L. MacKinnon, in *Energy Conversion and Particle Acceleration in the Solar Corona*, ed. by L. Klein. Lecture Notes in Physics, vol. 612 (Springer, Berlin, 2003), p. 127

N. Vilmer, A.L. MacKinnon, G. Trottet, C. Barat, Astron. Astrophys. **412**, 865 (2003)

N. Vilmer, M. Maksimovic, R.P. Lin, G. Trottet, in *Solar encounter. Proceedings of the First Solar Orbiter Workshop*, ed. by B. Battrick, H. Sawaya-Lacoste, E. Marsch, V. Martinez Pillet, B. Fleck, R. Marsden. ESA Special Publication, vol. 493 (2001), p. 405

N. Vilmer, G. Trottet, C. Barat, J.P. Dezalay, R. Talon, R. Sunyaev, O. Terekhov, A. Kuznetsov, in *Advances in Solar Physics*, ed. by G. Belvedere, M. Rodono, G.M. Simnett. Lecture Notes in Physics, vol. 432 (Springer, Berlin, 1994), p. 197

N. Vilmer, G. Trottet, C. Barat, R.A. Schwartz, S. Enome, A. Kuznetsov, R. Sunyaev, O. Terekhov, Astron. Astrophys. **342**, 575 (1999)

E. Vogt, J.C. Hénoux, Sol. Phys. **164**, 345 (1996)

H.T. Wang, R. Ramaty, Sol. Phys. **36**, 129 (1974)

K. Watanabe, M. Gros, P.H. Stoker, K. Kudela, C. Lopate, J.F. Valdés-Galicia, A. Hurtado, O. Musalem, R. Ogasawara, Y. Mizumoto, M. Nakagiri, A. Miyashita, Y. Matsubara, T. Sako, Y. Muraki, T. Sakai, S. Shibata, Astrophys. J. **636**, 1135 (2006)

K. Watanabe, Y. Muraki, Y. Matsubara, K. Murakami, T. Sako, H. Tsuchiya, S. Masuda, M. Yoshimori, N. Ohmori, P. Miranda, N. Martinic, R. Ticona, A. Velarde, F. Kakimoto, S. Ogio, Y. Tsunesada, H. Tokuno, Y. Shirasaki, Astrophys. J. **592**, 590 (2003)

K. Watanabe, Y. Muraki, Y. Matsubara, T. Sako, T. Sakai, H. Tsuchiya et al., in *International Cosmic Ray Conference*, vol. 1 (2005), p. 37

C. Werntz, Y.E. Kim, F.L. Lang, Astrophys. J. Suppl. Ser. **73**, 349 (1990)

E. Wiehr, Astron. Astrophys. **69**, 279 (1978)

R.M. Winglee, Astrophys. J. **343**, 511 (1989)

B.E. Woodgate, R.D. Robinson, K.G. Carpenter, S.P. Maran, S.N. Shore, Astrophys. J. **397**, L95 (1992)

R.S. Woolf, J.M. Ryan, P.F. Bloser, U. Bravar, E.O. Flückiger, J.S. Legere, A. MacKinnon, P.C. Mallik, M.L. McConnell, B. Pirard, in *Society of Photo-Optical Instrumentation Engineers (SPIE) Conference Series*, vol. 7438 (2009)

Z. Xu, J.-C. Hénoux, G. Chambe, M. Karlický, C. Fang, Astrophys. J. **631**, 618 (2005)

Z. Xu, J.C. Henoux, G. Chambe, A.G. Petrashen, C. Fang, Astrophys. J. **650**, 1193 (2006)

M. Yoshimori, Space Sci. Rev. **51**, 85 (1989)

Y. Yoshimori, in *International Cosmic Ray Conference*, vol. 6 (1999a), p. 5

Y. Yoshimori, in *International Cosmic Ray Conference*, vol. 6 (1999b), p. 1

V.V. Zharkova, M. Gordovskyy, Astrophys. J. **604**, 884 (2004)

E.G. Zweibel, D.A. Haber, Astrophys. J. **264**, 648 (1983)

Space Sci Rev (2011) 159:225–261
DOI 10.1007/s11214-010-9708-1

The Relationship Between Solar Radio and Hard X-ray Emission

S.M. White · A.O. Benz · S. Christe · F. Fárník · M.R. Kundu · G. Mann · Z. Ning · J.-P. Raulin · A.V.R. Silva-Válio · P. Saint-Hilaire · N. Vilmer · A. Warmuth

Received: 22 September 2010 / Accepted: 3 November 2010 / Published online: 23 July 2011

Abstract This review discusses the complementary relationship between radio and hard X-ray observations of the Sun using primarily results from the era of the *Reuven Ramaty High Energy Solar Spectroscopic Imager* satellite. A primary focus of joint radio and hard X-ray studies of solar flares uses observations of nonthermal gyrosynchrotron emission at radio

S.M. White (✉)
Space Vehicles Directorate, AFRL, Kirtland AFB, Albuquerque, NM 87123, USA
e-mail: Stephen.White@kirtland.af.mil

A.O. Benz
Institute of Astronomy, ETH Zürich, 8093 Zürich, Switzerland

S. Christe
NASA/Goddard Space Flight Center, Mail Code 671, Greenbelt, MD 20771, USA

F. Fárník
Astronomical Institute, Academy of Sciences, 251 65 Ondřejov, Czech Republic

S.M. White · M.R. Kundu
Dept. of Astronomy, University of Maryland, College Park, MD 20742, USA

G. Mann · A. Warmuth
Astrophysikalisches Institut Potsdam, An der Sternwarte 16, 14482 Potsdam, Germany

Z. Ning
Purple Mountain Observatory, Nanjing 210008, China

J.-P. Raulin · A.V.R. Silva-Válio
Centro de Radio Astronomia e Astrofisica Mackenzie, R. da Consolação 896, 01302-907 São Paulo, SP, Brazil

P. Saint-Hilaire
Space Sciences Laboratory, University of California, Berkeley, CA 94720, USA

N. Vilmer
Observatoire de Paris, LESIA—CNRS UMR 8109, 5 Place Jules Janssen, 92190 Meudon, France

wavelengths and bremsstrahlung hard X-rays to study the properties of electrons accelerated in the main flare site, since it is well established that these two emissions show very similar temporal behavior. A quantitative prescription is given for comparing the electron energy distributions derived separately from the two wavelength ranges: this is an important application with the potential for measuring the magnetic field strength in the flaring region, and reveals significant differences between the electrons in different energy ranges. Examples of the use of simultaneous data from the two wavelength ranges to derive physical conditions are then discussed, including the case of microflares, and the comparison of images at radio and hard X-ray wavelengths is presented. There have been puzzling results obtained from observations of solar flares at millimeter and submillimeter wavelengths, and the comparison of these results with corresponding hard X-ray data is presented. Finally, the review discusses the association of hard X-ray releases with radio emission at decimeter and meter wavelengths, which is dominated by plasma emission (at lower frequencies) and electron cyclotron maser emission (at higher frequencies), both coherent emission mechanisms that require small numbers of energetic electrons. These comparisons show broad general associations but detailed correspondence remains more elusive.

Keywords Sun: radio radiation · Sun: X-rays · Sun: flares · Sun: energetic particles

1 Introduction

From its earliest days, the study of hard X-ray emission from the Sun has had a natural ally in solar radio emission (Kundu 1961). The reason is straightforward: the electrons that produce hard X-ray emission have energies of order 10 keV or more, and such energetic electrons are also very efficient emitters of radio emission in the solar corona. That they can produce radio emission by a number of different physical mechanisms, in contrast to the bremsstrahlung-dominated hard X-rays, means that the radio data provide a range of diagnostics that complement the hard X-ray measurements. Between these two wavelength ranges we should expect to be sensitive to most sources of energetic electrons, both thermal and nonthermal, in the solar corona.

In this article we will review developments in our understanding of the joint use of radio and hard X-ray emission since the launch of the *Reuven Ramaty High Energy Solar Spectroscopic Imager* (*RHESSI*) satellite in 2002. In all cases discussed here, the hard X-ray emission most likely arises from bremsstrahlung emitted when energetic electrons are accelerated by Coulomb forces in collisions with ambient ions, either in the chromosphere or in the corona. Bremsstrahlung hard X-ray emission is proportional to the product of the nonthermal electron density and the ambient ion density. Inverse Compton emission is also a possible emission mechanism for the hard X-rays, but more extreme conditions are required and are not believed to be relevant for our discussion (e.g., see Krucker et al. 2008a). There are two general areas in which comparison of radio and hard X-ray (HXR) emission is relevant: phenomena (usually at higher radio frequencies, i.e., above 5 GHz) in which the radio emission is incoherent gyrosynchrotron emission[1], and those (usually at lower frequencies, i.e., below 2 GHz) where the radio emission is due to a coherent emission mechanism such as plasma emission or cyclotron maser.

[1] A note on terminology: it is conventional to reserve the term "synchrotron" for emission by highly relativistic electrons (Lorentz factor $\gamma \gg 1$) spiraling along magnetic field lines, while emission by mildly relativistic electrons is referred to as "gyrosynchrotron" in solar radiophysics.

2 Quantitative Comparison of Hard X-ray and Microwave Spectra

The study of radio emission from solar flares predates the detection of hard X-rays from the Sun, and the presence of activity at microwave frequencies in the impulsive phase of solar flares had been well established by 1958 when the first observation of a flare at hard X-ray/γ-ray wavelengths was carried out with a balloon-borne telescope over Cuba (Peterson and Winckler 1958). The authors of that paper noted the coincidence of the HXRs with microwave emission, and speculated that the "radio spectrum is emitted as betatron radiation from [electrons at energies of about 1 MeV] as they spiral in the intense magnetic fields associated with the sunspot group. These same electrons finally stop in the solar photosphere, where a small fraction of their energy is lost as bremsstrahlung γ-rays" (see also Kundu 1961). This remains our overall picture of the relationship between hard X-ray and microwave emission in the impulsive phase of flares: electrons with energies above 30 keV produce gyrosynchrotron radio emission as they spiral in ubiquitous magnetic fields in the corona, and bremsstrahlung HXRs when they precipitate into the dense chromosphere and lose their energy through collisions. The magnetic field strength B in the corona is typically of order 500 G over active regions, leading to electron gyrofrequencies ($2.8 \times 10^6 B$ Hz) of order 1–2 GHz in the low corona, and electrons with energies of order 100 keV and above radiate gyrosynchrotron emission at harmonics from the third upwards, usually dominating radio emission from flares at frequencies above 3 GHz. Typical flare radio spectra peak at 10 GHz (Guidice and Castelli 1975), although in larger flares the spectral peak can often occur at much higher frequencies.

In this picture we expect images of the radio emission at microwave frequencies to outline the field lines in the corona occupied by nonthermal electrons, while the HXRs show the same electron population as they strike the chromosphere at the footpoints of those field lines. With sufficiently good imaging, one could establish field-line connectivity by such a comparison. This picture can be distorted in several ways: the strength of the radio emission is proportional to a high power of B (e.g., Dulk and Marsh 1982) and thus electrons trapped in regions of strong magnetic fields will appear very bright, whereas strong magnetic fields near the surface tend to prevent precipitation of electrons with non-zero pitch angles due to magnetic mirroring and this effect could actually reduce gyrosynchrotron emission close to the footpoints compared to the rest of the loop.

Such comparisons of gyrosynchrotron and HXR emission remain a staple of high energy solar physics. One aspect to be borne in mind is that although radio telescopes are capable of excellent imaging and are generally more sensitive to nonthermal electrons than are HXR detectors, solar-dedicated radio telescopes capable of excellent imaging are few in number and it is difficult to arrange observations during flaring periods on non-solar-dedicated major facilities such as the Very Large Array (VLA). Therefore most of the radio work at microwave frequencies utilizes data from solar-dedicated arrays such as the Nobeyama Radio Heliograph (NoRH), the Owens Valley Solar Array (OVSA) or the Siberian Solar Radio Telescope.

We discuss the formalism for quantitative comparison of gyrosynchrotron radio data and bremsstrahlung hard X-ray measurements in the current section. Examples of scientific topics addressed specifically by comparison of *RHESSI* and microwave observations are given in the following sections.

2.1 Hard X-ray Bremsstrahlung Spectra

In the following discussion of HXR emission we follow the prescription of Hudson et al. (1978), based in turn on Brown (1971) and the Bethe-Heitler form for the bremsstrahlung cross-section (see the discussion of cross-sections in Kontar et al. 2011). Essentially the same calculation was used by Nitta et al. (1991) and White et al. (2003) for quantitative comparison of radio- and HXR-emitting electrons. Assume that the hard X-ray photon spectrum is a power law of the form

$$\Phi(\epsilon) = A_0 \left(\frac{\epsilon}{E_0} \right)^{-\gamma} \quad \text{photons}\,\text{cm}^{-2}\,\text{s}^{-1}\,\text{keV}^{-1}, \tag{2.1}$$

where ϵ is the photon energy, γ is the power-law slope of the photon spectrum, and A_0 is the normalization constant at a fiducial photon energy E_0 keV. In the case of the widely-used OSPEX package of Richard Schwartz, $E_0 = 50$ keV is the default for the broken power-law fit to the spectrum and the normalization obtained from the fit refers to this energy. Given this photon spectrum, the corresponding electron flux energy spectrum into the target is given by

$$\frac{d^2\mathcal{N}(E)}{dE\,dt} = 3.28 \times 10^{33} \frac{A_0\, b(\gamma)}{E_{0,\text{keV}}} \left(\frac{E}{E_0} \right)^{-(\gamma+1)} \quad \text{electrons}\,\text{keV}^{-1}\,\text{s}^{-1} \tag{2.2}$$

(Hudson et al. 1978), where $b(\gamma) = \gamma^2\ (\gamma - 1)^2\ B(\gamma - 0.5, 1.5)$ and $B(x, y)$ is the beta function. The parameter $b(\gamma)$ is of order 10 to 60 for typical values of γ (3 to 6). The factor of E_0 appearing in the denominator here is in units of keV (per Brown 1971, p. 498, equation 15). We use $\mathcal{N}(E)$ to denote the total number of electrons at energy E in some volume: derivatives of this quantity with respect to volume and energy yield the volume number density and the energy distribution, respectively.

2.2 Thick-Target Hard X-rays

The electron flux into the HXR target cannot be compared directly with the microwave parameters since the relevant quantity for the radio emission is the total number of radiating electrons in the coronal volume, rather than a flux. Comparison of the two depends on the nature of the hard X-ray source. In this section we consider the so-called "thick-target" case, in which the radiating electrons immediately lose all their energy in a high-density source. The density of the source then does not appear in the formulae. To derive the nonthermal electron volume density from the energy distribution of the electron flux into the thick target, we assume the classic expression relating a density to a flux,

$$\frac{d^2\mathcal{N}(E)}{dE\,dt} = A_X\, v\, \frac{d^2\mathcal{N}(E)}{dE\,dV} \tag{2.3}$$

where A_X is the area of the X-ray source and v is the velocity of the electrons into the target. A_X must be obtained from observations, and its determination relies on high-quality imaging data.

The velocity v is treated as follows. For ultrarelativistic electrons it is assumed to be of order c with no energy dependence. For nonrelativistic electrons we assume, as an approximation, that the component of velocity into the target (i.e., the downwards component for thick-target emission from the chromosphere) carries one-third of the electron energy (equipartition between directions of motion). Thus we set

$$\frac{1}{2} m v^2 = \frac{1}{3} E \tag{2.4}$$

which leads to

$$v = c\sqrt{\frac{2}{3}\frac{E}{mc^2}} = 0.0361 E_{\mathrm{keV}}^{0.5} c = 1.08 \times 10^9 E_{\mathrm{keV}}^{0.5}\ \mathrm{cm\,s^{-1}}. \tag{2.5}$$

Substituting Eqs. 2.2 and 2.5 into Eq. 2.3, we find the following expression for the electron volume number density energy distribution in the nonrelativistic limit:

$$\frac{d^2\mathcal{N}(E)}{dE\,dV} = 3.04 \times 10^{24} \frac{A_0 b(\gamma)}{E_{0,\mathrm{keV}}^{1.5} A_X} \left(\frac{E}{E_0}\right)^{-(\gamma+1.5)} \mathrm{electrons\,cm^{-3}\,keV^{-1}}. \tag{2.6}$$

This formula indicates that for a given photon power-law index γ, the electron energy power-law index $\delta = \gamma + 1.5$ at nonrelativistic energies in a thick-target model, i.e., the energy spectrum of the electrons entering the target is actually steeper than the resulting thick-target bremsstrahlung photon spectrum. This occurs because the electron energy spectrum in the target is flatter than the injected energy spectrum, due to the fact that electrons with low energy lose their energy faster by collisions with ambient electrons, and thus are depleted in the target faster, than high-energy electrons.

The ultrarelativistic limit is somewhat complicated and simple expressions are not available. In terms of the spectrum, setting $v = c$ in Eq. 2.3 means that no additional power of E is added when the electron flux is converted to a number density, i.e., if this were the only change in the ultrarelativistic limit we would find $\frac{d^2\mathcal{N}(E)}{dE\,dV} \propto E^{-(\gamma+1)}$. In terms of the photon spectrum, this would actually imply that for a pure electron energy power law δ the bremsstrahlung photon spectrum would steepen in the ultrarelativistic limit compared to the nonrelativistic limit by 0.5 in the spectral index, i.e., $\gamma = \delta - 1$ instead of $\gamma = \delta - 1.5$. However, the electron-ion cross-section changes form in the ultrarelativistic limit, and electron-electron collisions play a more important role in producing bremsstrahlung at energies above the electron rest mass. Electron-electron bremsstrahlung results in a photon spectrum shallower by 1 in the spectral index than electron-ion bremsstrahlung (Vestrand 1988; Kontar et al. 2011). (The main energy loss of the incident electrons occurs in *e–e* collisions, but at lower energies HXR photons are predominantly produced by the large deflections in *e–p* collisions.) The combined result of all these effects seems to be to flatten the photon spectrum above 500 keV by about 0.5, i.e., $\gamma \approx \delta - 2$. McTiernan and Petrosian (1990) state that a rough integration of the product of electron flux and cross-section gives a flattening of the photon spectrum above 500 keV by $\log(E_\gamma / m_e c^2)$, although their calculation refers to fluxes rather than volume number densities. In any case, it does not seem to be readily feasible to derive quasi-analytical approximations in the relativistic limit: a crude approach is to join the nonrelativistic limit smoothly onto a power law 0.5 flatter above 500 keV.

2.3 Thin-Target Limit

In contrast to the thick-target limit, which is generally believed to apply to footpoint HXR sources, in the thin-target limit the electron energy distribution evolves only slowly under the influence of collisions. The rate of bremsstrahlung emission depends on the ambient ion number density n_i ($\mathrm{cm^{-3}}$) in the source (assumed to be much larger than the density of any accelerated ions present). In this case, for a given photon spectrum (Eq. 2.1), we find that the nonthermal electron energy distribution is

$$\frac{d^2\mathcal{N}(E)}{dE\,dV} = 7.9 \times 10^{41} \frac{A_0\, C(\gamma)}{n_i V_X E_{0,\mathrm{keV}}^{0.5}} \left(\frac{E}{E_0}\right)^{-(\gamma-0.5)} \mathrm{electrons\,cm^{-3}\,keV^{-1}}, \tag{2.7}$$

where $C(\gamma) = (\gamma - 1)/B(\gamma - 1, 0.5)$ and V_X is the physical volume of the hard X-ray source. In the thin-target limit, $\delta = \gamma - 0.5$, i.e., the electron energy distribution is flatter than the photon energy spectrum by about 0.5 in the index, whereas in the thick-target case the electron energy distribution is steeper than the observed photon energy spectrum.

2.4 Simplified Gyrosynchrotron Formulae

Exact calculations of the gyromagnetic emission by an energetic electron distribution are somewhat complicated and do not lend themselves to simple analytic formulae (e.g., Ramaty 1969; Takakura 1972). Faster numerical approaches have also been developed (e.g., Petrosian 1981; Klein 1987; Fleishman and Kuznetsov 2010). In order to have rough analytic estimates of the number density of the radio-emitting electrons, we use the Dulk and Marsh (1982) approximation formulae (see Dulk 1985, for corrected versions) for gyrosynchrotron emission. These describe a power-law distribution of mildly relativistic electrons, and are valid for isotropic electron pitch-angle distributions, harmonics in the range 20–100 and viewing angles (between the line of sight and the magnetic field direction) in the range 30°–80°. The physical property of the radio source that is easiest to measure and least dependent on a specific geometrical model is the brightness temperature at a point in the image, obtained at the highest frequency and spatial resolution possible. Other quantities, such as fluxes, require source areas to be known, and these are much more model-dependent than a localized brightness temperature. On the other hand, measuring the true brightness temperatures requires excellent high-spatial-resolution imaging. It is assumed that the radio spectral index α_r can be measured from two optically-thin frequencies (i.e., two frequencies above the flux peak in the radio spectrum). The expression for radio brightness temperature at a frequency f produced by an electron number density energy distribution

$$\frac{d^2\mathcal{N}(E)}{dE\,dV} = \mathcal{N}_r \frac{(\delta - 1)}{E_r}\left(\frac{E}{E_r}\right)^{-\delta} \text{electrons cm}^{-3}\,\text{keV}^{-1} \tag{2.8}$$

(here $\mathcal{N}_r$ is the number of electrons per unit volume in the distribution above the fiducial electron energy E_r, taken to be 10 keV by Dulk and Marsh 1982) in a magnetic field of strength B Gauss is

$$T_B = \kappa\, T_{\mathit{eff}} = 3.1 \times 10^{-0.53\delta} (\sin\theta)^{-0.45+0.66\delta} \left(\frac{f}{f_B}\right)^{-0.80-0.90\delta} \frac{\mathcal{N}_r L}{B}, \tag{2.9}$$

where θ is the angle between the magnetic field and the line of sight, L is the line-of-sight depth through the source at the point where T_B is measured, and $f_B = 2.8 \times 10^6\, B$ Hz is the electron gyrofrequency. Throughout this review we will use cgs units except for photon and electron energies, where in solar physics it is traditional to use keV rather than ergs as the energy unit.

Equation 2.9 results from multiplying together the Dulk (1985) expressions for opacity κ and effective temperature T_{eff}. We proceed to use Eq. 2.9 for T_B to determine the parameter $\mathcal{N}_r$ that appears in Eq. 2.8; note that $\mathcal{N}$ can also be derived directly from the X-ray Eq. 2.6 by equating it to Eq. 2.8, as presented below.

For convenience we write

$$\frac{f}{f_B} = \frac{10^9 f_{\mathrm{GHz}}}{2.8 \times 10^6\, B} = 357 \frac{f_{\mathrm{GHz}}}{B} \tag{2.10}$$

(where B is everywhere in units of Gauss) and find

$$T_B = 2.81 \times 10^{-2.00-2.83\delta} (\sin\theta)^{-0.45+0.66\delta} f_{\rm GHz}^{-0.80-0.90\delta} \mathcal{N}_r \, L \, B^{-0.20+0.90\delta}. \tag{2.11}$$

Inverting this expression to obtain $\mathcal{N}_r$ and substituting into Eq. 2.8 we find

$$\frac{d^2\mathcal{N}(E)}{dE\,dV} = 35.3\,(\delta-1)\,10^{3.83\delta-1} \frac{T_B}{L\,B^{-0.20+0.90\delta}} (\sin\theta)^{0.45-0.66\delta} f_{\rm GHz}^{0.80+0.90\delta} E_{\rm keV}^{-\delta}, \tag{2.12}$$

in units of electrons cm^{-3} keV^{-1}. The quantity T_B can be measured directly if imaging observations of sufficient resolution are available; θ is usually assumed to be close to 90° since this maximizes the gyrosynchrotron emissivity; L can be estimated from the radio images; and since $\alpha_r = 1.20 - 0.90\delta$ is the radio flux spectral index, the value of δ obtained from the hard X-ray spectrum can be independently checked from the high frequency radio spectrum. The physical quantity in this equation that cannot be derived from observations is B, which appears as a very high power. In principle this too can be determined using the location of the radio peak frequency, which determines B via the following gyrosynchrotron approximation:

$$f_{peak} = 2.72 \times 10^{3.00+0.27\delta} (\sin\theta)^{0.41+0.03\delta} (\mathcal{N}_r\,L)^{0.32-0.03\delta} B^{0.68+0.03\delta}. \tag{2.13}$$

A peak frequency of 10 GHz typically corresponds to $B \approx 600$ G (roughly the 6th harmonic of the gyrofrequency), with only a weak dependence on other physical parameters in the source (e.g., Dulk and Marsh 1982).

Equations 2.6 and 2.12 are two separate expressions for a volume density energy distribution and can be compared. When the spectral indices disagree by a large amount, as is often the case (with the energy distribution derived from radio data usually being flatter than that derived from HXR data), then it is usually assumed that they refer to different energy ranges: the optically-thin microwaves are more sensitive to electrons above 300 keV, while hard X-rays are usually dominated by electrons below 300 keV. This is illustrated in Fig. 1, which

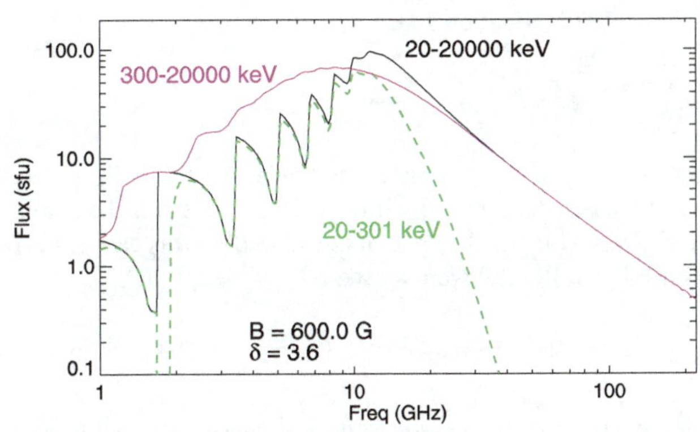

Fig. 1 A comparison of gyrosynchrotron radio spectra from electrons with different energy ranges. The *black line* is the spectrum produced by an electron energy power law from 20 to 20000 keV with a spectral index of 3.6 in a constant magnetic field of 600 G. The *green dashed line* is the spectrum produced by the lower-energy electrons from 20 to 300 keV, and the *purple dash-dot line* is the spectrum produced by the 300 to 20000 keV electrons. The calculation assumes a homogeneous loop of length 10″, width 3″ and depth 1″, and an isotropic pitch-angle distribution for the radiating electrons. Radiative transfer is taken into account in the calculations

compares the gyrosynchrotron spectra of high- and low-energy electrons: electrons below 300 keV produce little radio emission above the spectral peak near 10 GHz (in the case of Fig. 1, at 30 GHz, the 20–300 keV emission is already two orders of magnitude weaker than the emission from the full energy range). When the electrons below 100 keV are present, they are sufficient to make the source optically thick at frequencies below the spectral peak, and the harmonic structure is clear (but will only be observed if the magnetic field is constant throughout the source, which is unlikely to be the case); when only electrons above 300 keV are present, they produce a smoother spectrum below the spectral peak because their harmonics are intrinsically broader. The spectrum shows some flattening above 40 GHz in the regime where relativistic particles dominate the emission. Note that the Dulk and Marsh formulae were derived from a full relativistic calculation for the gyrosynchrotron emission, so there is no contradiction in comparing the nonrelativistic thick-target bremsstrahlung results dominated by <500 keV electrons with the relativistic gyrosynchrotron emission dominated by >500 keV electrons, as long as the underlying electron energy spectrum has an unbroken power law (Eq. 2.2) from nonrelativistic to relativistic energies.

The discussion in this section relies on the paradigm that the nonthermal electrons are accelerated in the corona and then propagate down to the footpoints of the loops. Another complication for the comparison between radio and HXR results that needs to be borne in mind is that the number density appearing in Eq. 2.6 refers to the density entering the thick target at the footpoints, i.e., at some depth in the chromosphere depending on the electron energy (e.g., Brown et al. 2002), while the radio emission comes from the corona and the number density therefore refers to a different location in the loop. Implicit is also the assumption that the pitch-angle distribution is everywhere isotropic, corresponding to the fast pitch-angle scattering limit; an anisotropy in the pitch-angle distribution will affect the comparison (discussed further below).

2.5 Radio Flux from a Thick-Target Hard X-ray Spectrum

We may now derive an expression for the radio flux from an integrated hard X-ray spectrum under the thick-target assumption. Given the radio brightness temperature, in this case the Dulk and Marsh approximation (Eq. 2.9), we obtain the total radio flux from the standard expression (e.g., Dulk 1985, equation 14):

$$S = \frac{k_B f^2}{c^2} \int T_B \, d\Omega, \tag{2.14}$$

where the integral is over the solid angle Ω subtended by the radio source on the sky. We may replace the solid angle element $d\Omega$ with dA/D^2, where A is the actual physical area of the source on the sky and D is the distance of the source (in this case, D is one astronomical unit, 1.5×10^{13} cm). From Eq. 2.9 we may write

$$S = \frac{k_B f^2}{c^2} \int 2.81 \times 10^{-2.00-2.83\delta} (\sin\theta)^{-0.45+0.66\delta} f_{\rm GHz}^{-0.80-0.90\delta} B^{-0.20+0.90\delta} \mathcal{N}_r \frac{L\,dA}{D^2}. \tag{2.15}$$

Now we note that $\int \mathcal{N}_r L\, dA$ is the integral of the number density per unit volume (above the reference energy E_r) over the volume of the source, i.e., it is the total number of electrons in the emitting volume, $\mathcal{N}_{tot}$. Combining the physical constants (note that one solar flux unit, or "sfu," is 10^{-19} ergs cm^{-2} s^{-1} Hz^{-1} in cgs units) and setting the angular factor to unity as an approximation (e.g., for $\theta = 70°$ and $\delta = 3.0$, $(\sin\theta)^{-0.45+0.66\delta} = 0.9$), we find

$$S_{sfu} = 1.9 \times 10^{-28.0-2.83\delta} \mathcal{N}_{tot} f_{\rm GHz}^{1.20-0.90\delta} B^{-0.20+0.90\delta}. \tag{2.16}$$

2.6 The Effect of Anisotropy

The discussion presented above assumes an isotropic pitch-angle distribution for the radiating electrons. This is appropriate, e.g., when strong pitch-angle scattering in the flare loops, by collisions or by wave-particle interactions, isotropizes the particle distribution. This assumption is necessary to derive simple analytic formulae that can be applied (weak anisotropy is included in the papers by Petrosian 1982; Robinson 1985; Klein 1987), but may not be appropriate in many cases. Fleishman and Melnikov (2003a) discuss the effect of pitch-angle anisotropies on gyrosynchrotron emission in some detail. A single relativistic electron emits most of its radiation in a narrow cone around the direction of motion, but particles with small pitch angles barely radiate at all, so there can be large differences in the emission of electrons with large and small pitch angles. Tied to this question is the evolution of the pitch-angle distribution of a population of magnetically-trapped electrons both in time and in space. In a loop with significant variation in magnetic field strength between the footpoints and the loop top, conservation of the first adiabatic invariant means that the pitch angle of an electron will increase as it propagates towards stronger magnetic fields, and the electron will reflect at the height where its pitch angle reaches 90° ("magnetic mirroring"). Electrons with pitch angles sufficiently small for them to reach the solar chromosphere will "precipitate" (i.e., lose their energy immediately in collisions and stop) there and be lost from the corona. When gradual pitch-angle scattering takes place, without renewed injection of electrons, one ends up with a population of electrons with large pitch angles trapped near the loop top.

Fleishman and Melnikov (2003a) (with corrections in Fleishman and Melnikov 2003b) investigate changes to the gyrosynchrotron spectrum due to various forms of pitch-angle anisotropy at different observer viewing angles. Changes are more pronounced for viewing angles close to the magnetic field direction rather than perpendicular to the magnetic field direction in the source. The degree of polarization tends to increase as the anisotropy of the pitch-angle distribution strengthens, and the spectral index in the optically-thin limit is larger (steeper spectra) than for the isotropic pitch-angle distribution. In particular, loss-cone distributions viewed along the magnetic field show much weaker emission than isotropic distributions, and steeper high-frequency (i.e., optically thin) spectra: this can explain some observed differences between loop ends and loop tops for disk flares where the legs of the loop are viewed along the magnetic field and the loop top is viewed orthogonal to the magnetic field.

2.7 Transport Effects

Particle transport effects can also influence the appearance of microwave and hard X-ray sources, and delays between the two types of emission. Simultaneous imaging observations of radio and HXR sources can place physical constraints (loop lengths, source separations) on the flare site. We will not address these issues in detail here. We note that Melnikov et al. (2006) discuss the ways in which both pitch-angle distributions and particle transport effects can affect the microwave appearance of flaring loops. Minoshima et al. (2008) carry out detailed calculations of a trap-plus-precipitation model for a flare observed by *RHESSI* and NoRH. This flare exhibited HXR emission from two footpoints and a loop-top microwave source, and the authors argue that their model can explain the observation that the radio-emitting electrons appear to have a spectrum harder than that of the HXR-emitting electrons. As time proceeds, higher energy electrons will tend to concentrate in the trapped population of coronal loops, and this can explain observations such as the gradual hardening of the microwave spectrum even when the HXR spectrum is softening (e.g., Ning 2007, 2008a).

3 Thermal Bremsstrahlung Radio Emission from Flares

For completeness we include a brief discussion of thermal radio emission. The opacity due to bremsstrahlung of a uniform plasma of typical solar composition at temperature T_e and density n can be approximated by (e.g., Dulk 1985)

$$\kappa = 0.2 \frac{n^2}{T_e^{1.5} f^2} \text{ cm}^{-1}, \tag{3.1}$$

where n is measured in cm^{-3}, T_e in Kelvin and frequency f in Hz. From Eq. 2.14, the corresponding radio flux is given by

$$S = k_B/c^2 \Omega \begin{cases} T_e f^2 & \text{optically thick: } \kappa L \gg 1, \\ 0.2 n^2 L / T_e^{0.5} & \text{optically thin: } \kappa L \ll 1, \end{cases} \tag{3.2}$$

where L is the thickness of the source along the line of sight (in units of cm) and Ω is the projected solid angle on the sky occupied by the source. As above, the radio flux is traditionally quoted in units of sfu. As a crude rule of thumb, we may write Eq. 3.2 as

$$S = 2 \left(\frac{T_B}{10^6 \text{ K}} \right) \left(\frac{f}{10 \text{ GHz}} \right)^2 \left(\frac{r}{10''} \right)^2 \text{ sfu}, \tag{3.3}$$

where r is the radius of the area in which the brightness temperature is T_B (with $T_B = T_e$ in the optically-thick limit and $T_B = \kappa L T_e$ in the optically-thin limit).

Generally in solar flares, we find that the heated coronal plasma is optically thin to thermal bremsstrahlung above about 10 GHz, and this can be verified because, by Eq. 3.2, the flux spectrum should be almost independent of frequency (or equivalently, the brightness temperature spectrum, obtained from spatially-resolved images, varies as f^{-2}). A *GOES* class M1 flare typically produces a peak thermal flux of 1–5 sfu in the optically-thin flat-spectrum limit from the 10^7 K plasma to which *GOES* is sensitive, while an X1 flare produces of order 10–50 sfu of thermal emission. At low, optically thick, frequencies the spectrum rises as f^2. The $n^2/T^{0.5}$ dependence in the optically thin limit means that thermal bremsstrahlung is strongest in cool dense plasmas. In fact, if we take a given volume of optically thin plasma and heat it up without changing the density, the radio flux from the plasma will actually decrease due to the $T^{-1.5}$ dependence of the opacity. The *GOES* soft X-ray data are most sensitive to plasma at a temperature of order 10 MK, and if any cooler plasma is also present then the thermal radio emission will be larger than the flux level inferred from the *GOES* data alone. It should also be noted that the radio emission is produced by electrons whereas the soft X-ray emission has a large contribution from lines of highly-ionized atoms of elements heavier than hydrogen, so quantitative comparison of the two also depends on the abundance distribution of the radiating plasma (see the discussion in White et al. 2005).

4 Physical Properties of Flare-Accelerated Particles from Hard X-rays and Microwaves

As discussed in Sect. 2, if electrons with the same energy distribution produce both the microwave and hard X-ray emission from a solar flare, there should be straightforward relationships between the spectral indices of the two emissions. However, there has been a long record of comparisons finding that in fact the radio and hard X-ray spectral indices are not compatible with the same electron energy distribution (e.g., Kundu et al. 1994; Silva et al. 2000). Since the typical (initial) energy of an electron emitting a HXR photon of energy ϵ is $1.5\epsilon - 3\epsilon$, with higher electron energies for flatter energy spectra (e.g., Kosugi et al. 1988; Aschwanden and Schwartz 1996), and microwave emission at higher frequencies is typically dominated by electrons with energies above 300 keV, it is desirable to have the hard X-ray spectral index at energies up to several hundred keV when making this comparison. Previously, such comparisons between microwave and HXR/γ-ray photon spectra up to several hundreds of keV have been possible with *GRANAT*/PHEBUS data (Trottet et al. 1998; Vilmer et al. 1999). Those studies found consistency between the spectral indices of the radio-emitting electrons and the HXR/γ-ray-emitting electrons at energies above several hundred keV.

4.1 SOL2002-07-23T00:35 (X4.8)

The first *RHESSI* flare for which this was possible was SOL2002-07-23T00:35 (X4.8). Before the main phase of this event, a remarkable distended precursor radio loop is seen, shown in Fig. 2 (Asai et al. 2006). The elongated loop structure is visible in both the 17 and 34 GHz images, extending over 80″ to the north-east from the subsequent site of the flare (located at the base of this loop, marked in the middle panel of Fig. 2). In the 75 minutes for which NoRH images are available prior to the flare, the loop does not seem to change shape, nor are there significant changes in its brightness. On the other hand, the compact source at the base of the loop, which is not visible at 34 GHz and therefore may be nonthermal, declines steadily over this period until 00:18 UT, when pre-flare activity commences. EUV images of this region show no feature corresponding to the radio loop: it is difficult to measure its spectrum, but it may be thermal, suggesting that it is material much hotter than the $\sim 10^6$ K range to which the EUV images are sensitive.

Once the flare starts, intense nonthermal gyrosynchrotron emission from the main flare loops takes up the dynamic range of the images and the pre-flare loop is no longer visible. White et al. (2003) carried out a careful study of the main flare using radio imaging and spectral data from NoRH together with *RHESSI* data. In common with other large flares, this event had a high turnover frequency in its radio spectrum, necessitating the use of radio

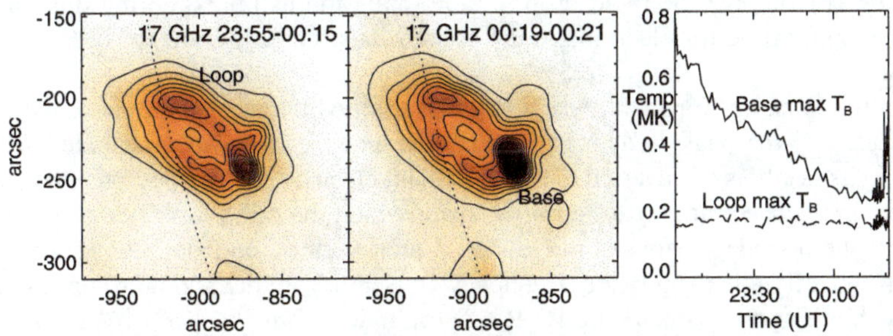

Fig. 2 The evolution of pre-flare emission at 17 GHz in SOL2002-07-23T00:35 (X4.8). The *left panel* shows an image averaged over the period 20 minutes prior to the impulsive phase, showing a distended loop with a compact source at its base. The *central panel* shows the image at 00:20 UT, with the source at the base brightening significantly. The contour interval is 2×10^4 K in both images. The *right panel* shows the light curves of the maximum 17 GHz brightness temperature in the compact source at the base (*solid curve*) and in the rest of the loop (*dashed line*)

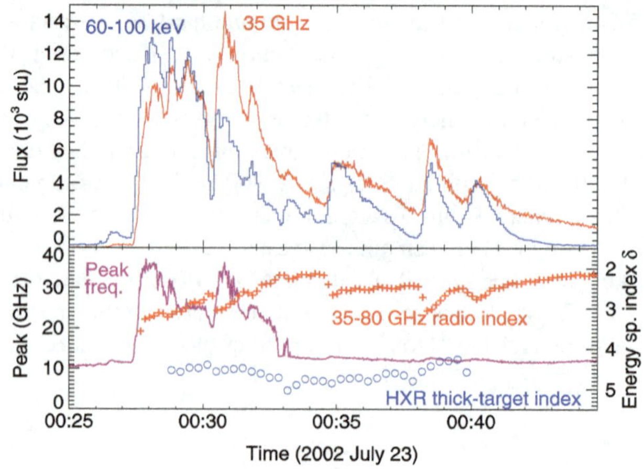

Fig. 3 Comparison (*upper panel*) of the *RHESSI* 60–100 keV hard X-ray light curve (*blue histogram*) and the NoRP 35 GHz light curve (*red solid curve*) for the well-observed flare SOL2002-07-23T00:35 (X4.8), as well as (*lower panel*) the time evolution of the radio spectral peak frequency (*purple solid line*) and the radio spectral index from 35 to 80 GHz converted to an electron energy spectral index assuming gyrosynchrotron emission from an optically-thin source (*red plus symbols*, uncertainty ± 0.3). For comparison, the thick-target electron energy index obtained from the *RHESSI* 100–400 keV spectrum is also shown (*blue open circles*; formal uncertainty ± 0.2). From White et al. (2003)

fluxes above 30 GHz in order to estimate the radio spectral index on the optically-thin high-frequency side. The imaging data were consistent with the picture in which the high-energy HXRs were emitted from the footpoints of coronal loops visible in the radio images.

Radio and HXR light curves for this event and the results of fits to the radio and HXR spectra are shown in Fig. 3. The structure of the 35 GHz light curve is representative of all radio frequencies above 4 GHz, and 60–100 keV is representative of the light curves for all hard X-rays above 30 keV; below 30 keV, the X-rays show less temporal fine structure in their light curves (Holman et al. 2003). The impulsive phase consists of a number of spikes and dips that are all seen in both the radio and HXR light curves. Different peaks have relatively different heights at the two wavelengths, but over the $\sim$20 minutes of the impulsive phase the similarity in time profiles is striking, including the brief sharp dip at 00:30:20 UT. Note that the rapid drop in radio emission exhibited during this dip implies that the trapping time for electrons in the corona must be very short (of order seconds) in this event.

The lower panel in Fig. 3 represents the spectral evolution of the radio emission using two parameters: the peak frequency (solid line), and the radio spectral index from 35 to 80 GHz (plus symbols), converted to the equivalent electron energy spectral index. The peak frequency in the spectrum, i.e., the frequency at which the radio flux is largest at any time, represents the boundary between the lower frequency emission from the regime where the source is optically thick and higher frequency emission from the regime where the source is optically thin. For comparison, the HXR spectral index from 100–400 keV, converted to an electron energy spectral index by assuming thick-target emission, is also plotted (as circles) in the lower panel, and it is clear that the two energy spectral indices are inconsistent with one another: as usual (e.g., Kundu et al. 1994; Silva et al. 2000), the radio data indicate a much flatter energy spectrum than do the HXR data.

Applying the quantitative analysis above to the hard X-ray data, we can estimate electron number densities in this event. The 40–400 keV photon energy spectrum at

00:35:00 UT, when the radio spectral peak is well below 34 GHz, has the form $9(E_\gamma/50\ \text{keV})^{-3.2}$ photons cm^{-2} keV^{-1} s^{-1}. The corresponding rate of electrons striking the chromosphere is $2 \times 10^{35} (E/20\ \text{keV})^{-4.2}$ electrons keV^{-1} s^{-1}. To convert this to a number density we need to divide by the electron speed v and the footpoint area A. The *RHESSI* images do not resolve the footpoints at a resolution of 7″ and images made using grid 1 suggest a size as small as 2″. If we adopt $A = 10^{16}$ cm^2, corresponding to a diameter of 2″, using the formalism of the previous section we derive an energy distribution for the electron number density at the footpoints of $5 \times 10^9\ (E/20\ \text{keV})^{-4.7}\ (A/10^{16}\ \text{cm}^2)$ electrons cm^{-3} keV^{-1}.

If we take the electron energy spectral index of 4.7 derived from HXR and apply it to the radio brightness temperature as discussed in the previous section, then the radio data also require very high densities.[2] Based on gyrosynchrotron theory, the brightness temperatures of over 10^9 K achieved at 17 GHz at the peak of the flare can only be produced by an electron energy spectrum as steep as 4.7 if the harmonic number is in excess of 30 (see figures in Dulk and Marsh 1982, confirmed by accurate numerical calculations). Such high harmonic numbers imply a relatively small magnetic field strength, no more than 200 G, and to achieve the same radio brightness temperature with the lower value of B, nonthermal densities over 10^{10} cm^{-3} above 20 keV are required, similar to those found above from the HXR data. The low value for B is consistent with the location of the brightest radio emission at the top of the arcade of loops. Caspi and Lin (2010) found that the superhot component in this flare (temperature of order 30–40 MK) required a magnetic field strength of at least 500 G to confine it, consistent with a picture in which the bulk of the nonthermal particles are located at a greater height than the superhot component. Even larger values of nonthermal electron density, with much stronger magnetic fields, were deduced by Raulin et al. (2004) in the case of SOL2001-08-25T16:45 (X5.3).

So, unfortunately, this extremely well observed event only adds to the problem of reconciling energy spectra derived from radio and HXR data. Since the radio and HXR profiles are so similar, we cannot argue that the radio emission comes from a long-lived population of trapped electrons while the hard X-rays come from directly precipitating electrons, as occurs in other events where the time profiles clearly differ (e.g., Raulin et al. 1999): the radio-emitting electrons have the same time behavior as the hard X-ray emitting electrons and should have a common origin and common evolution. There is a flattening in the *RHESSI* γ-ray spectra by about 0.5 in the index, but no sign in the data up to 8 MeV of the flattening by 2 implied by the radio data (Smith et al. 2003), so we argue that the radio spectral indices in Fig. 3 are not compatible with the *RHESSI* observations. We are forced to assume that the high-frequency radio emission is dominated by a region with a high turnover frequency so that the 35–80 GHz spectrum does not represent optically-thin emission, even in the later stages of the flare, and the true optically-thin radio spectral index is steeper than derived above. Nitta et al. (1991) reached a similar conclusion for their event. Both radio and HXR data require extreme number densities of nonthermal electrons to be accelerated in the energy release: over 10^{10} cm^{-3} above 20 keV. These numbers imply an acceleration mechanism of very high efficiency.

4.2 SOL2003-06-17T22:55 (M6.8)

Another well-observed flare is SOL2003-06-17T22:55 (M6.8) (Kundu et al. 2009). This flare is one of the few in which emission up to energies exceeding 200 keV can be imaged in

[2] Note that the fact that the energy spectral index derived from HXR is steeper than that derived from radio spectra implies that there are fewer high-energy electrons in the spectrum, and since these are the electrons that dominate the microwave emission, a larger total number of electrons is then required.

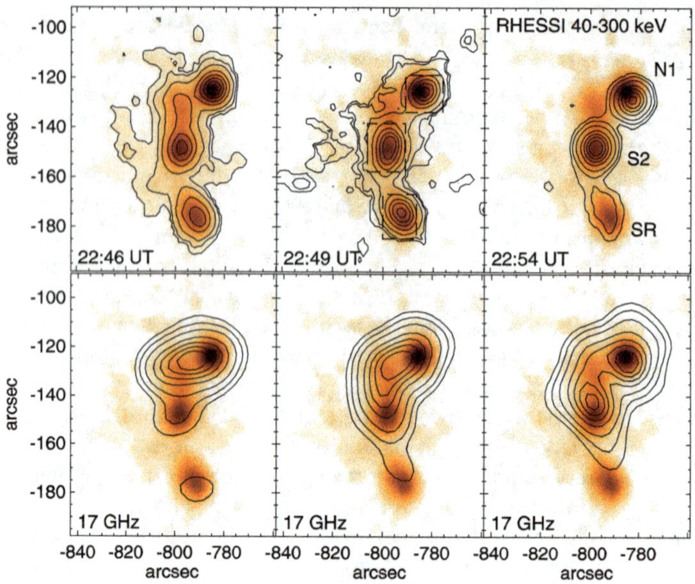

Fig. 4 *Upper panels*: hard X-ray images at each of the three main peaks in the HXR light curve (*contours*) of SOL2003-06-17T22:55 (M6.8). The images result from summing over the 40–300 keV range. The background image in each panel is the HXR image for 22:50 UT so that changes in morphology from one peak to the next can be seen. The resolution of the HXR images is 9″. The *middle panel* shows the regions used for the HXR spectra of sources S2, N1, and SR as *dotted boxes*. *Lower panels*: contours of simultaneous 17 GHz emission plotted over the HXR image of the flare from the *upper left panel*. Contours are at 4, 8, 16, 32, 48, 64, and 80% of the peak in each image. From Kundu et al. (2009)

hard X-rays, and we can investigate the HXR spectra of individual sources up to this energy. The morphology of the event is shown in Fig. 4. It takes the form of a filament eruption followed by a large double-ribbon flare with nonthermal emission stretching along a north-south axis over a neutral line. Three peaks each separated by about 5 minutes (times labeled in Fig. 4) are seen in the HXR light curve, and are matched in the radio data. In this case the radio and HXR morphologies are very similar, with both showing sources seeming to straddle the neutral line: the southernmost source is brightest during the first peak and then fades relative to the other peaks. Note that HXR and radio emission are both seen over a distance of more than 60 Mm along the neutral line in this one event: nonthermal energy release must have taken place throughout a large volume.

The distinctive feature of this event is that we can clearly separate three spatially distinct HXR sources during three different temporal peaks, and compare the energy spectra in each case. Figure 5 shows the *RHESSI* spectra for the three sources outlined by boxes in the upper middle panel of Fig. 4 during each of the three peaks, and Table 1 reports the spectral indices determined from power-law fits to these spectra (Kundu et al. 2009).

The fits to the spatially-resolved spectra assume a single power law over the 40–240 keV range, while the fits to the integrated spectra assume a broken power law over the range 50–400 keV. For peaks 1 and 2 the spectral break in the power law is above 200 keV and the fitted spectral index below 200 keV generally matches the fits to the spatially resolved spectra, which are dominated by photons below 120 keV, while the fit above the break gives a flatter spectrum. The uncertainties in the fits above the break are large due to the small number of high-energy photons detected, so one cannot conclude that there is flattening of

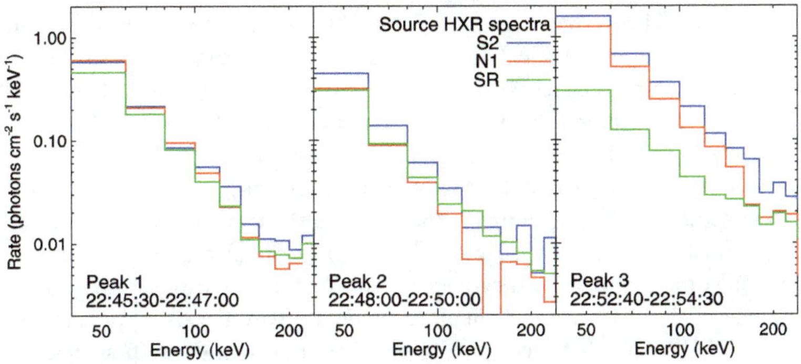

Fig. 5 Hard X-ray spectra at each of the three main peaks in the light curve for each of the three sources in the HXR images of SOL2003-06-17T22:55 (M6.8). These are derived from image cubes made in 20 keV channels from 40 to 300 keV. From Kundu et al. (2009)

Table 1 Power-law fits to the photon spectral index γ of individual sources in each of the three main peaks in the SOL2003-06-17T22:55 (M6.8) light curve derived from images in different energy bins, together with the fit to background-subtracted 50–400 keV spectra from the *RHESSI* front detectors. For the spatially-integrated spectra, the numbers are the results of a broken power-law fit: the spectral index at energies below the break, the break energy (keV) in parentheses, and the spectral index above the break. Uncertainties in the fits to the break energies are typically large (tens of keV)

Peak	S2 (middle)	N1 (north)	SR (south)	Spatially integrated
22:45:30–22:47:00 UT	3.1 ± 0.3	3.3 ± 0.3	3.0 ± 0.4	3.3 (280) 2.5 ± 0.2
22:48:00–22:50:00 UT	3.3 ± 0.1	3.5 ± 0.1	2.8 ± 0.5	3.4 (210) 2.0 ± 0.2
22:52:40–22:54:30 UT	2.7 ± 0.1	2.9 ± 0.2	2.1 ± 0.6	2.5 (120) 3.4 ± 0.2

the electron energy spectrum at high energies during the first two peaks (although note that fits to the *RHESSI* rear-detector spectra, more sensitive than the front detectors at high photon energies, agree with the results in Table 1 to within the uncertainties). We can also look at radio spectra for the individual peaks. NoRH images can be used to measure the spectral index for the three individual sources from 17 to 34 GHz, and again it is found that the inferred energy spectra are flatter than the energy spectra inferred from the HXR data with a difference in spectral index of order 1–2. Non-imaging data from the Nobeyama polarimeters (NoRP) up to 80 GHz suggest steeper radio spectra above 34 GHz, but there is a large uncertainty in the calibration of the NoRP 80 GHz data during this period and this leads to large uncertainties in the high-frequency spectral index. We do see significant anomalies between the radio and HXR properties of the sources: source SR is relatively bright at 34 GHz until late in the event, suggesting a very flat energy spectrum; the radio spectra do not show the same trend that all the spectra are flatter during peak 3. However, again we have major uncertainties in the location of the peak in the radio spectra of the individual sources since we only have images at two frequencies, and therefore we cannot determine whether the spectral index from 17 to 34 GHz actually represents the true optically-thin spectral index: if the 17 GHz source is partially optically thick then this frequency range will show a spectrum flatter than the true optically-thin spectral index. These problems again emphasize the need to have routine data at more than one frequency above 40 GHz if we are to be confident in measuring the true optically-thin radio spectral index in large flares.

The general conclusion of Fig. 5 and Table 1 is that a given peak shows the same energy spectrum (from HXRs) in all three spatial locations, but it may differ from one peak to the next: peaks 1 and 2 clearly have steeper spectra than peak 3 (below 120 keV) in all three sources and in the integrated spectra. This suggests either that the electron acceleration mechanism has the same physical characteristics over a large spatial scale (5×10^4 km) or a more localized accelerator distributes electrons over the full volume. The challenge for the first interpretation is the fact that all sources show a flattening of their spectra in peak 3 after being steeper in peaks 1 and 2: how can sources so far apart have their characteristics change in the same way? For example, if acceleration is due to stochastic acceleration by wave turbulence, how is turbulence generated with identical properties over such a large volume? On the other hand, a localized accelerator that can distribute nonthermal electrons over a distance of 5×10^4 km is difficult to reconcile with the usual picture of post-flare loops in two-ribbon flares that are typically much shorter than the ribbons and straddle the neutral line rather than parallel the ribbons.

4.3 Coronal Hard X-ray Sources

The traditional picture of HXR arising from bremsstrahlung when nonthermal electrons strike the dense chromosphere is consistent with most observations. HXR sources in the corona are much less often observed, at least in part due to instrumental issues: imaging HXR telescopes typically have a limited dynamic range, and this limits their ability to see fainter thin-target emission from the low-density corona if it is competing with very bright thick-target footpoint emission from the high-density chromosphere. Coronal HXR emission is almost certainly present in most flares but difficult to observe, and a recent comprehensive review is devoted to this subject (see Krucker et al. 2008a).

There is one circumstance in which coronal emission does not have to compete with bright footpoint emission, and that is when the footpoints of flaring loops are obscured from our view by the solar limb. Such occulted flares have long been our main source of knowledge of coronal emission (e.g., Frost and Dennis 1971; McKenzie 1975; Kane et al. 1979), and this continues to be true in the *RHESSI* era. Krucker et al. (2010) observed a flare on 2007 December 31 that occurred 12° behind the limb as seen from Earth, but happily the event was within the field of view of telescopes on the *STEREO-B* spacecraft trailing the Earth in its orbit. *RHESSI* observed remarkably strong emission from this flare, up to 80 keV, despite the fact that the footpoints were occulted from *RHESSI*'s view. The 50 keV photon flux was comparable to what we see from the footpoints in a *GOES* class M flare.

In this flare (Fig. 6), the nonthermal HXR come from a source seen to lie at a height in the corona that is above the lower-energy thermal soft X-rays. This offset was seen previously in the famous "Masuda" flare (SOL1992-01-13T17:25, M2.0) (Masuda et al. 1994). NoRH images of the 2007 event show that the 17 GHz emission from the flare is co-spatial with the >30 keV source, but the larger occultation height at radio wavelengths (9 Mm above the optical limb) only allows us to observe the top part of the nonthermal HXR source at 17 GHz. The main flare loop seen during the soft X-ray peak time is occulted in the 17 GHz observations and emission from the initial source continues through the soft X-ray peak time. The microwave spectrum during the hard X-ray peak decreases with frequency as expected for gyrosynchroton emission and has a low turnover frequency, consistent with a coronal origin high above the solar surface. Based on the low turnover frequency of 2.5 GHz in the NoRP radio spectrum, the magnetic field strength in the 17 GHz source is estimated to be 30–50 G at a height 25 Mm above the solar photosphere. Since the electron gyrofrequency for this field strength is only 0.1 GHz, emission at 17 GHz is at harmonics in excess of 100 and requires relativistic electrons.

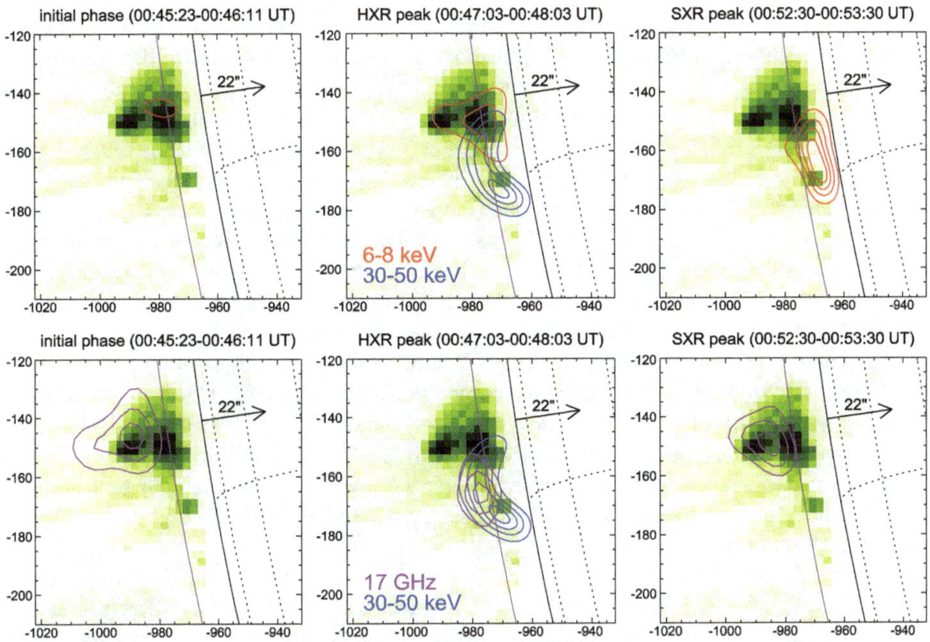

Fig. 6 X-ray and microwave imaging during the initial phase of SOL2007-12-31T01:11 (C8.3) (*left*), the hard X-ray peak (*middle*), and the soft X-ray peak (*right*). *Top row*: Contours in the thermal (6–8 keV, *red*) and the nonthermal (30–50 keV, *blue*) range are shown on an EIT 195 Å image taken towards the end of the main hard X-ray peak. The 6–8 keV contour levels are the same for all time intervals at 20, 40, 60, 80% of the maximum emission during the soft X-ray peak time, while the nonthermal emission is shown at 30, 50, 70, 90% levels. *Bottom row*: Microwave contours (*magenta*) at 17 GHz are shown at the 50, 70, 90% levels on the same EIT image. For comparison, the 30–50 keV source seen during the impulsive phase is shown as well (*blue*). The *magenta curve* gives the location of the limb at 17 GHz, about 12$''$ above the photosphere; for this reason emission can be seen to lower heights in the HXR images than in the radio images. The *black arrows* indicate the occultation height for the flare location. From Krucker et al. (2010)

In this event the energy power-law spectral index of 3.4 derived from the radio data is not very different from the value of 3.7 derived from the *RHESSI* spectrum under the thin-target assumptions normally made for such coronal HXR sources. Because the occultation height for this event is so large and the ambient density in the corona is correspondingly low, extreme number densities of accelerated nonthermal electrons are required in order to explain the large HXR fluxes. No source of any kind is visible in the corona at the heights of the flare HXR emission prior to the event. From the absence of thermal HXR emission at the site of the harder photons during the flare, an upper limit of 8×10^9 cm^{-3} can be placed on the ambient density. The observed HXR flux at 50 keV then requires a density for the nonthermal electrons above 16 keV in excess of 10^9 cm^{-3}; for a more plausible ambient density of 2×10^9 cm^{-3} at a height of 16 Mm above the photosphere, the nonthermal electrons would have to outnumber the ambient ions in order to explain the observed HXR fluxes.

Thus in this event the standard electron-beam scenario where suprathermal electrons move through an ambient plasma producing hard X-ray emission by bremsstrahlung does not work. The main problems are that the number of accelerated electrons is comparable to the total number of electrons in the pre-flare source and that collisional heating would increase the ambient plasma temperature to superhot temperatures within seconds, producing a bright thermal HXR source. Since no such source is seen, this picture cannot be correct. Rather, the nonthermal HXR source seems to be produced by a mechanism that accelerates all electrons and produces an entirely nonthermal electron distribution. This suggests that the above-the-looptop source is the acceleration region itself. In a purely nonthermal electron distribution, collisional losses to electrons are much reduced, making it an efficient hard X-ray source. Once a significant fraction of the electrons is accelerated, collisional losses of an accelerated electron are reduced due to a smaller number of ambient electrons. This simple picture suggests that once a significant fraction of electrons is accelerated, it might not take that much more to accelerate all of the electrons. Observations of a similar event in which the pre-flare density in the region of the nonthermal HXR source can be measured are needed to confirm this picture.

Asai et al. (2007) also report joint *RHESSI* and NoRH observations of an over-the limb M4 flare (SOL2005-07-27T05:02) whose footpoints were occulted in HXR. In that event, a filament eruption preceded the flare and was clearly visible in the NoRH 34 GHz images; again, nonthermal coronal sources are seen in both 25–40 keV HXR and in the microwave data, but in this event the energy spectra inferred from the two wavelength ranges differ greatly. The radio spectrum in this event has a spectral peak near 10 GHz, i.e., much higher than in the Krucker et al. (2010) event, suggesting that effectively none of the radio emission is obscured from us.

In the Krucker et al. (2010) event, the radio-deduced and HXR-deduced electron energy spectra are compatible with one another. It is evident that the discrepancy between the energy spectral indices derived from the microwave and HXR data in events such as SOL2002-07-23T00:35 (X4.8) and SOL2003-06-17T22:55 (M6.8) would be largely removed if the appropriate comparison were with thin-target bremsstrahlung rather than thick-target: as noted earlier, the actual energy spectrum of electrons in a thick-target HXR source is flatter than the injected spectrum due to the faster energy loss rate of lower-energy electrons. For the thick-target spectrum to be the appropriate comparison, the radio emission from the flare would have to be dominated by the electrons in the HXR source. In active regions where flares occur, the filling factor of strong magnetic fields at the surface is high and there is little if any difference between the field strength in the upper chromosphere and that in the lower corona. Therefore, since the volume of the footpoint thick-target HXR sources is so much smaller than the volume of the coronal loop that connects the footpoints (the footpoint sources have a depth typically less than $1''$ at the top of the chromosphere, whereas a coronal loop can be $20''$ long), and the incident nonthermal density at the footpoints will be smaller than the average nonthermal density in the loop if there is any magnetic mirroring, the only way that the electrons in the thick target can dominate the radio emission is if there are no nonthermal electrons in the coronal loop, i.e., the nonthermal electrons are entirely confined to the chromosphere. Images of disk flares such as Fig. 4 may seem to suggest that the locations of the radio emission are consistent with the HXR footpoints, but such images can be misleading due to their limited spatial resolution: loops smaller than the resolution of the radio images cannot be identified as loops in such images, and the distinction between footpoint sources and loop emission cannot be made. In general (see Sect. 5), radio images of flares are consistent with the paradigm that the radio emission is dominated by coronal sources. In any model in which the electrons are accelerated in the corona, there must always be many more electrons in the coronal volume than in the small volume occupied by the bremsstrahlung-emitting electrons in the narrow layer of the chromosphere at the footpoints of the coronal loop. In this case it remains appropriate to use the thick-target determination of electron energy spectra described in Sect. 2 for comparison with the spectral index determined from gyrosynchrotron observations.

4.4 Oscillations

Any periodic or quasi-periodic behavior observed in flares is valuable because it can reveal the characteristic frequency of a physical process or environmental property and thus has diagnostic value. There have been a number of events in which such periodicity has been seen simultaneously in microwaves and hard X-rays, confirming a common origin for the electrons radiating in the two wavelength domains and providing strong constraints on the physical processes involved (e.g., Nakajima et al. 1983; Asai et al. 2001; Grechnev et al. 2003). Fleishman et al. (2008) discuss another event in which oscillations are seen both in microwave data from the Owens Valley Radio Observatory and the Nobeyama Radioheliograph, as well as in the *RHESSI* hard X-ray light curves. Periodicities in the range 15–20 s are found in both wavelength ranges, with degrees of modulation typically around 10% or lower; at radio wavelengths, the degree of modulation rises with frequency, suggesting that it is larger in the optically-thin regime and thus that it is the number density of nonthermal electrons that is being modulated. They also find that the modulations at lower radio frequencies lead the higher frequencies in time by a small amount, and that modulation of the circular polarization is in the opposite sense to the modulation of the intensity: circular polarization is weakest when the intensity is strongest.

Another event exhibiting modulations simultaneous in HXRs and radio is SOL2003-11-03T09:55 (X3.9) (Dauphin et al. 2005). In this event, the later phase of HXR emission shows a quasi-periodic train of peaks with a period of order 35 s. These peaks are clear both in the 150–300 keV HXR light curve and in the microwave light curves. Surprisingly, the decimetric emission at 432 and 327 MHz shows the same modulations. Images from the Nançay Radio Heliograph at 432 MHz during the modulations show three different sources: one close to the limb, just above the HXR source locations in *RHESSI* images, and two other sources several hundred arcsec higher in the corona. The higher 432 MHz sources show more pronounced modulations than the lower source, despite being further from the modulated HXR footpoint sources. The correlation of the decimetric emission with microwaves and HXRs suggests that the emission mechanism is still gyrosynchrotron emission at decimetric wavelengths. Since the Razin effect is usually thought to suppress gyrosynchrotron emission at such low frequencies, this observation suggests that Razin suppression is not effective in this event, at least for the higher sources, and further that the gyrosynchrotron-emitting nonthermal electrons have access to large loops at the height of the 432 MHz sources. (Note that this event was also remarkable for the fact that it exhibited a Type II radio burst from a shock apparently driven by soft X-ray loops rising rapidly into the corona; see Dauphin et al. (2006).)

Possible interpretations of such oscillations have been reviewed by Nakariakov and Verwichte (2005) and Nakariakov and Melnikov (2006). Fleishman et al. (2008) test two models with their event: oscillations driven by MHD loop oscillations, or quasi-periodic injections of fresh nonthermal electrons into a stable source. An important constraint is that the oscillations are observed to be essentially in phase across the whole microwave spectrum, whereas a number of MHD modes (sausage, torsional) would predict that the optically thick (lower) frequencies and optically thin (higher) frequencies should show opposite behavior. Thus, the authors conclude that quasi-periodic injections of fresh nonthermal electrons provide a better fit to the observations.

4.5 Microflares

This topic is discussed at greater length by Hannah et al. (2011) and here we briefly mention radio aspects of HXR microflares. Kundu et al. (2006) compared the radio and HXR emission in a sample of microflares, using *RHESSI* and NoRH data. Microflares, X-ray bright-point flares, coronal X-ray jets and active region transient brightenings are all consequences of energy releases in the corona that are smaller than traditional flares, but are nonetheless clearly identifiable phenomena. The sample of 30 microflares studied by Kundu et al. (2006) showed relatively soft (i.e., steep) HXR spectra, which would lead one to expect that they would not possess sufficient mildly relativistic nonthermal electrons radiate strongly at microwave frequencies. Yet most of the microflares exhibit nonthermal microwave emission that requires electrons with energies exceeding 100 keV. This suggests that the radio-emitting electrons must have a harder energy spectrum than the HXR-emitting electrons: the imaging results show that typically the radio and HXR emission are not exactly coincident, supporting the possibility that the two electron populations are different. On the other hand, Ning (2008b) investigates a sample of apparently radio-quiet microflares and finds that their HXR properties are normal for microflares, as in the sample of Kundu et al. (2006). Ning (2008b) used less sensitive microwave patrol data for the microwave fluxes, and it is possible that more sensitive NoRH observations would have detected nonthermal radio emission.

A prior study by Qiu et al. (2004) investigated somewhat larger events (up to *GOES* B2 in soft X-ray class) using *RHESSI* and OVSA data. The advantage of OVSA data is that complete radio spectral information is available, but their detection level of around 1 sfu is not capable of addressing events weaker in the radio domain. About 40% of their sample of about 200 events are detected in microwaves, with emission up to about 10 GHz and spectral peaks in the 5–10 GHz range: thus their microwave spectra behave like standard nonthermal gyrosynchrotron spectra and strongly resemble those of normal flares. Where measured, the radio spectra typically suggest harder electron energy distributions than the HXR spectra exhibit, and some microflares with negligible responses in the *GOES* soft X-ray range nonetheless can have strong nonthermal microwave emission, emphasizing again the possible differences between the lower-energy nonthermal electrons to which HXRs below 25 keV are sensitive, and the higher-energy electrons responsible for the microwave emission.

5 Morphological Comparison of Microwaves and Hard X-rays

As noted earlier, we have the expectation that radio images will illuminate the paths of nonthermal electrons along magnetic loops in the corona whose footpoints are outlined in HXR by precipitating nonthermal electrons. We have to bear in mind the differences in HXR and radio imaging data: the best *RHESSI* images have a dynamic range of order 100, whereas radio images (e.g., from NoRH and the VLA) can achieve a dynamic range of order 10^4, and thus the radio images can show secondary sources in locations where the *RHESSI* images do not have enough dynamic range for sources to be visible above the noise. On the other hand, the brightness of the radio emission is strongly weighted by magnetic field strength, and this can lead to different emphases in the images at the two wavelengths.

5.1 "Loop" Flares

We expect radio images of flares to outline magnetic field lines carrying trapped energetic particles: any such particles on open field lines rapidly move out into the solar wind (it takes 5 seconds for an electron at speed $0.5c$ to propagate to 1 $R_\odot$) and are not present in the regions of strong magnetic fields low in the Sun's atmosphere long enough to contribute significantly to the microwave emission (but they may be seen via their plasma emission as

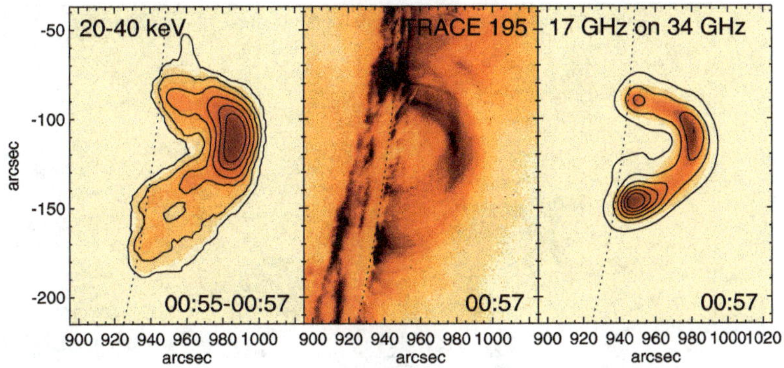

Fig. 7 Images of the early phase of the loop flare SOL2002-08-24T01:12 (X3.1) on the west limb. The *left panel* shows the *RHESSI* image of nonthermal 20–40 keV HXR photons covering the period 00:55–00:57 UT. The *middle panel* shows a *TRACE* 195 Å image at 00:57 UT, while the *right panel* shows contours of the emission at 17 GHz overlaid on a color image of the 34 GHz emission, also at 00:57 UT. Contours in the HXR and radio images are at 5, 20, 35, 50, 65 & 80% of the maximum in each image. The placement of the *TRACE* image relative to the other wavelengths is uncertain

Type III bursts instead). In fact, flare radio images that clearly look like magnetic loops are the exception, not the rule. This may be partly due to limited spatial resolution: since NoRH has a typical resolution of order 10″, any loop needs to be quite large to appear as such in the radio images. Another factor is that the magnetic field variation along a loop can emphasize emission from strong-field regions or regions with $\theta \approx 90^\circ$ so much that a loop morphology is not evident. Accordingly, events that do have a loop-like morphology tend to be heavily analyzed.

By far the favorite loop event for radio studies during the *RHESSI* period is SOL2002-08-24T01:12 (X3.1) on the west limb (Karlický 2004; Li and Gan 2005; Melnikov 2006; Tzatzakis et al. 2008; Romano et al. 2009; Reznikova et al. 2009; Ning and Cao 2009), which also produced a fast (1900 km s^{-1}) CME and an energetic particle event at the Earth. Figure 7 shows images in HXRs, microwaves and the EUV 195 Å band (Fe XII and Fe XXIV) during the early part of the impulsive phase, for which *RHESSI* data are available. The apparent loop has a footpoint separation of order 70″ and height of order 50″. The images provide an interesting contrast: in HXRs at tens of keV, presumably from nonthermal electrons, we see a pronounced brightening at the top of the "loop" in this early stage of the flare (here we use quotation marks to emphasize to the reader that while the source appears from our line of sight to be a "single loop"; in fact it could equally well be an arcade of loops extended significantly east-west, but co-located in projection along our line of sight). Line-of-sight effects can also play an important role in the appearance of the radio source, as discussed in Sect. 2. The flare apparently occurred on the visible disk and hence footpoint HXR sources should be visible, but they are not pronounced in these images; there is some evidence for multiple sources in the region of the southern footpoint at times before 00:55 UT. The radio images also show a bright peak at the top of the loop, but at 17 GHz the brightest source is at the southern footpoint, in a region that appears extended along the limb in the HXR images. The EUV image shows dense cool (Fe XII) and hot (Fe XXIV) material, the latter also detectable by *RHESSI* via its observations of the Fe and Fe–Ni X-ray emission-line complexes (Caspi et al. 2010).

The radio light curve for this event shows a number of intensive peaks over a period of 10 minutes starting at 01:00 UT, requiring repeated injections of energetic particles onto

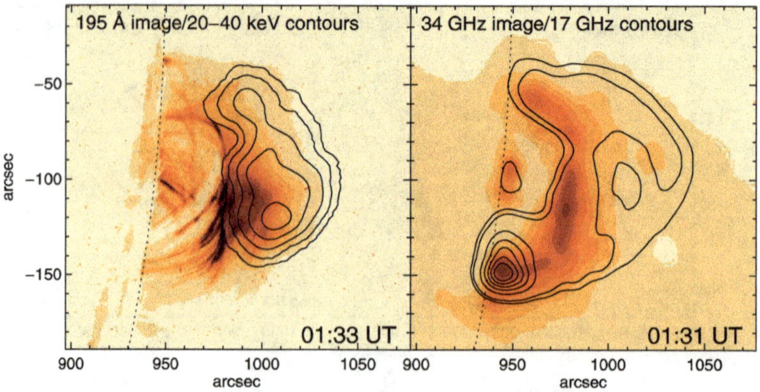

Fig. 8 Images of SOL2002-08-24T01:12 later in the event during the extended gradual phase (near 01:30 UT). At this time the 17 GHz image (contours at 4, 6.5, 20, 35, ..., 80% of the maximum, *right hand panel*) shows two loops as well as a bright source at the southern footpoint, while the 34 GHz image (underlying image in *right hand panel*) only shows the lower loop and a (relatively) much weaker southern footpoint source. The *RHESSI* 20–40 keV image of nonthermal HXR (contours at 10, 30, 50, 70, 90% of maximum, *left panel*) only shows a high coronal source, roughly filling the region between the two loops in the 17 GHz image, and lying above the cooler (2 MK) loops visible in the TRACE 195 Å Fe XII image (*left panel*, image). The 2 MK loops appear to occupy a smaller volume than the 17 GHz radio source

the coronal loops. Reznikova et al. (2009) analyzed the evolution of the radio morphology and found that each peak showed the same behavior: during the rise of a peak the southern footpoint is brightest at 17 GHz, but as each peak decays the loop top starts to dominate the images. Reznikova et al. (2009) interpret this as requiring that the highest density of nonthermal electrons always be located at the loop-top where the magnetic field is weakest, while the nonthermal density in the footpoints is initially high but decreases with time. An implication of this interpretation is that pitch-angle scattering must always be weak in the radio-bright region at the loop top. The question arises as to whether these injections occur onto increasingly higher field lines, as in the helmet-streamer reconnection scenario, or onto essentially the same field lines, as expected for a confined flare. Li and Gan (2005) measured the radius of a circular fit to the radio "loop" at 34 GHz and found that the radius of the loop shrank steadily until 01:02 UT, and increased thereafter.

Later in the flare (01:30 UT; see Fig. 8) the 17 GHz radio images show two loops well separated in height, while the 34 GHz images only seem to show the lower loop. The brightness temperature of the lower loop is 2–3 MK at 17 GHz and around 0.6 MK at 34 GHz, while the upper loop has a 17 GHz brightness temperature of 1–2 MK but is less than 0.05 MK at 34 GHz. The radio spectrum of the lower-altitude loop is consistent with thermal emission from a post-flare arcade of loops. Karlický (2004) argues that the higher loop, with its nonthermal radio spectrum, is emitting by gyrosynchrotron emission and its appearance requires ongoing acceleration of electrons to nonthermal energies in the extended phase of the flare. The presence of a very bright source at the southern footpoint, with brightness temperatures (over 20 MK at 17 GHz, less than 0.6 MK at 34 GHz) indicating nonthermal emission, supports this argument, as does the presence of significant emission in the 20–40 keV range in the same volume as the nonthermal radio emission (Fig. 8, left panel).

Loop morphologies can also show up in thermal emission. Bone et al. (2007) discuss the development of a bright compact loop which has a flat radio flux spectrum and thus appears to be radiating by thermal bremsstrahlung. Densities of close to 10^{11} cm^{-3} are inferred. Another flare occurs in the same set of loops 30 minutes later, and Veronig and Brown

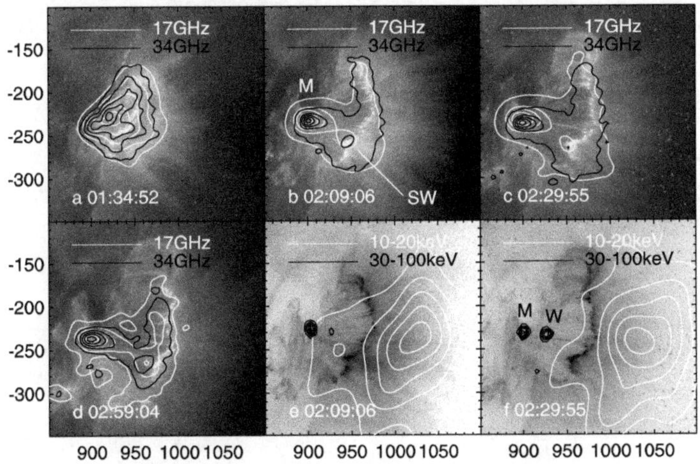

Fig. 9 The evolution of the radio and hard X-ray sources later in SOL2002-08-24T01:12 (X3.1). The first four images show a sequence of overlays of 17 GHz (*white contours*) and 34 GHz (*black contours*) images on *TRACE* 195 Å FeXII/FeXXIV images during the extended phase of the flare when the loop system is expanding above the west limb. The upper four contour levels are at 30, 50, 70 & 90% of the maximum in each image, while the lowest contours are at 10, 4, 2 & 1%, respectively, to match the changes in the peak intensity. *RHESSI* observes the early impulsive phase of the event and then goes into eclipse until 02:08 UT, and the last two images show contours of the *RHESSI* 10–20 keV (*white*) and 30–100 keV (*black*) emission overlaid on the *TRACE* 195 Å FeXII/FeXXIV images at 02:09:56 UT and 02:29:55 UT (displayed with inverted color table). The prominent nonthermal radio source is visible at the eastern base of the radio emission from 02:00 UT onwards. The radio emission from the loop tops at the northern end of the arcade has a flat spectrum and hence is probably thermal free-free emission. The 10–20 keV hard X-rays originate above the EUV loops, in higher soft X-ray emitting loops, while the harder 30–100 keV X-rays originate at the same location as the nonthermal radio source. From Kundu et al. (2004)

(2005), from analysis of the *RHESSI* hard X-ray spectra, inferred that the flare must have occurred in a very dense environment. The subsequent analysis of the radio data by Bone et al. (2007) confirmed this conclusion. Further discussion of this topic may be found in the chapter by Fletcher et al. (2011).

5.2 SOL2002-04-21T01:51 (X1.5)

The well-studied limb flare SOL2002-04-21T01:51 (X1.5) exhibited a spectacular arcade of post-flare loops in EUV images. Kundu et al. (2004) studied the radio emission from this flare and found, much as in SOL2002-08-24T01:12 (X3.1), that bright nonthermal sources were seen at a number of locations low in the corona, while the top of the post-flare arcade clearly shows thermal bremsstrahlung emission from the dense plasma there. Distinct regions of radio emission with very different time behavior can be identified in the radio images, and in particular a peculiar nonthermal source seen in radio and hard X-rays low in the corona at the base of the arcade is seen to turn on at 01:45 UT, some 30 minutes after the start of the impulsive phase (Fig. 9).

This event is striking for the wide range of types and locations of both radio and hard X-ray emission it displays. The energy distribution inferred for the radio-emitting electrons during the impulsive phase is quite similar to that inferred for the hard X-ray-emitting electrons, and the radio and hard X-ray light curves show similar time structure. In the radio images we can identify at least four spatially-distinct regions that show quite different temporal behaviors (Fig. 10). The brightest radio emission in the flare comes from a location to

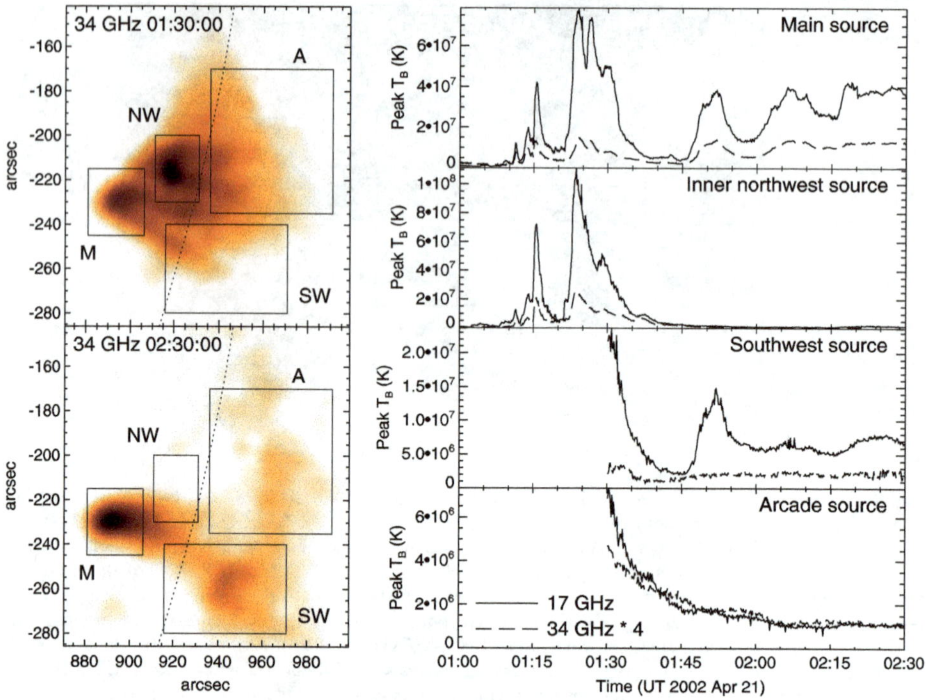

Fig. 10 Light curves for individual locations in SOL2002-04-21T01:51 (X1.5). The 34 GHz NoRH *images in the left hand panels* show the locations of the four regions chosen for investigation relative to the morphology of the source at two different times, while the *four panels on the right* show the evolution of the maximum brightness temperature in each of the regions. The light curves show both 17 and 34 GHz curves: the 34 GHz brightness temperatures are multiplied by 4, so that they should be the same as the 17 GHz values if the emission is optically-thin thermal emission (as is clearly the case for the arcade source)

the north-west of the site where the flare starts, under the middle of the loop arcade, but the radio emission from this location fades after the impulsive phase and is unimpressive during the extended decay phase. On the other hand, the main source and the thermal and non-thermal sources in the arcade (distinguishable by their brightness temperature spectra) all participate in the extended phase. This indicates that energy release is ongoing throughout this phase and apparently distributed throughout the coronal volume above the flare site.

6 Millimeter- and Submillimeter-Wavelength Emission from Flares

Radio emission from solar flares at millimeter and shorter wavelengths is expected to have two main characteristics: since optically-thin thermal emission has a flat flux spectrum, it should become increasingly important relative to the falling spectrum of nonthermal gyrosynchrotron emission as one goes to higher frequencies. However, higher-frequency non-thermal gyrosynchrotron emission requires emission at ever higher harmonics of the electron gyrofrequency, and this makes it sensitive to electrons with much higher energies than are required for microwave emission (MeV and higher energies rather than tens to hundreds of keV: Ramaty 1969; White and Kundu 1992; Ramaty et al. 1994, and Fig. 1). Bremsstrahlung is increasingly inefficient as electron energy increases because faster electrons suffer less

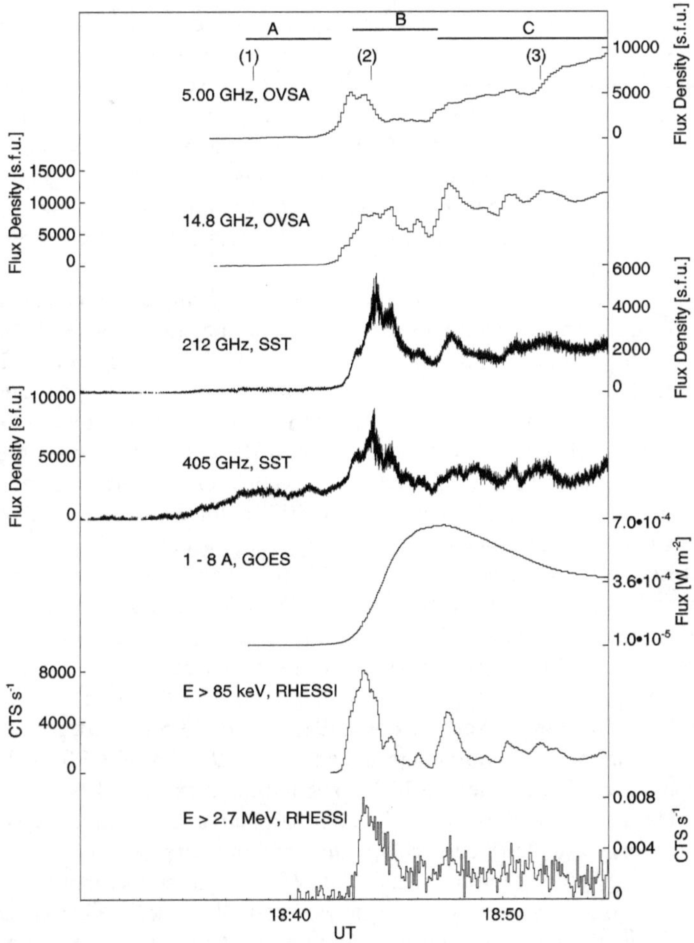

Fig. 11 Light curves for microwave, millimeter, submillimeter, hard X-ray and γ-rays from SOL2006-12-06T18:47 (X6.5) (from Kaufmann et al. 2009)

deflection by a nucleus. However, synchrotron emission increases in efficiency as electron energy increases. Thus millimeter-wavelength observations are a very sensitive diagnostic of γ-ray-emitting electrons, more sensitive, in fact, than current γ-ray detectors in the sense that nonthermal emission from smaller flares is more easily detected at millimeter wavelengths (e.g., Kundu et al. 1994; White 1994, 1999; Silva et al. 1996; Raulin et al. 1999).

During the period of *RHESSI* observations, the Solar Submillimeter Telescope (SST; Kaufmann et al. 2008) has routinely monitored the Sun at 212 and 405 GHz from the Argentinian time zone. Typically only major flares can be detected at these wavelengths, where the atmosphere has a major influence on detectability (the sky opacity at 405 GHz is generally in excess of 1).

An example of SST data is shown in Fig. 11, from SOL2006-12-06T18:47 (X6.5). This event is typical in that there is in general a good correlation between the radio emission and the HXR/γ-ray emission, consistent with the picture of the radio emission resulting from gyrosynchrotron emission by the same nonthermal electrons that produce the HXR/γ-ray emission by bremsstrahlung. However, this event, in common with a number of others, shows a major inconsistency with this picture: the radio spectrum shows a normal nonthermal behavior with a spectral peak in the 10–20 GHz range, with a falling spectrum at high microwave frequencies, but the SST data show an increasing spectrum from 212 to 405 GHz. Such behavior is not understood. Other events detected by SST show radio spectra that are consistent with a single nonthermal electron population, as in the case described by Giménez de Castro et al. (2009): in that flare no flux was detected above 250 keV, yet the 212 GHz emission was clearly detected.

As discussed earlier, it is well known that there is generally a discrepancy between the energy distribution of the nonthermal electrons that produce the HXRs and those that produce the radio emission at high frequencies: the radio data generally show a harder spectrum (e.g., Kundu et al. 1994; Raulin et al. 1999; Silva et al. 2000). This is attributed to the fact that the energies of electrons emitting the HXRs (many tens of keV) are smaller than those of the radio-emitting electrons (hundreds of keV): if the typical electron energy spectrum breaks up at energies above a few hundred keV, this could explain the observations. In principle, γ-ray spectra should show a break up at higher photon energies if the electron energy spectrum has such a break, but in the relevant energy range (photon energies above 500 keV) it is difficult to separate the electron bremsstrahlung spectrum from the nuclear line spectrum that usually dominates γ-rays above 1 MeV. There is one class of events for which the electron bremsstrahlung spectrum above 0.5 MeV is more easily determined, the so-called "electron-dominated events" (Marschhauser et al. 1991). These events result from extremely hard electron energy spectra whose bremsstrahlung dominates the usual nuclear line region. The prototype of this class of event was SOL1989-03-06T13:56 (X15), in which the photon spectrum during the main emission peak had a spectral index of 2.5 from 0.3–0.8 MeV, flattening to about 1.4 from 1 MeV to 10 MeV (Marschhäuser et al. 1994). Petrosian et al. (1994) argued that the entire 0.3–10 MeV photon spectrum in this event can be explained by bremsstrahlung from a single electron energy distribution with power-law index 2.2. In their model, the steeper photon spectrum below 1 MeV is due to the fact that the lower-energy photons are emitted nearly isotropically whereas the higher-energy photons are preferentially beamed along the direction of electron motion into the Sun (e.g., Vestrand et al. 1987). As noted in Sect. 2.2, relativistic and electron-electron effects in a single power-law energy distribution can produce a flattening of order 0.5 in the photon spectral index above 500 keV. Trottet et al. (1998) discuss the electron-dominated flare SOL1990-06-11T09:43 (M4.5), which exhibited a flattening of the bremsstrahlung photon spectrum at high energies but also major variations in the spectrum with time. During the brightest γ-ray peak the spectrum flattened by about 1 at 700 keV, whereas during other emission peaks the lower-energy photons had a spectrum steeper by 2–2.5 than the spectrum above 400 keV. The radio spectrum from 35 to 50 GHz during the brightest peak corresponds to an electron energy distribution with index of order 3, as does the bremsstrahlung photon spectrum above the break, so in this event they appear to be compatible with a single electron energy distribution. Vilmer et al. (1999) study another electron-dominated event in which the bremsstrahlung photon spectrum flattens above 500 keV by about 1 in the spectral index.

But even a break upwards in the electron energy spectrum, as described above, cannot produce a radio spectrum that rises in the submillimeter range. One can envisage a source of ultra-relativistic leptons in a very high magnetic field region: Trottet et al. (2008) suggest that ultra-relativistic positrons resulting from pion decay are a possible source of the impulsive submillimeter component in SOL2003-10-28T11:10 (X17.2) seen in conjunction with high-energy protons (>200 MeV), although it seems unlikely that sufficient positrons can be produced by this means and energetic electrons are still a more plausible interpretation. An optically-thick thermal source could produce a rising spectrum, but it would have to be either exceptionally large or exceptionally hot (e.g., Silva et al. 2007, who prefer an interpretation in terms of gyrosynchrotron emission from electrons in a very strong magnetic field). In the case of SOL2006-12-06T18:47 (X6.5), the SST observations place an upper limit of 15″ on the submillimeter source size, and for a source that small to produce the observed flux via thermal emission requires such high temperatures that the source would be intense in the *GOES* soft X-ray range, and would have been seen at other wavelengths (e.g., Kaufmann et al. 2009). At the time of writing, there is no completely accepted explanation for this spectral feature; a more detailed discussion of this topic may be found in Krucker et al. (2011).

7 Decimeter and Low-Frequency Radio Emission in Association with Hard X-rays

Most decimeter (300–3000 MHz) and low-frequency (here referring to frequencies below 300 MHz) emission from the Sun is dominated in flares by very bright plasma emission, i.e., conversion of electrostatic Langmuir waves at the electron plasma frequency ($f_p = 9000 n_e^{1/2}$) into electromagnetic radiation at the fundamental f_p and the second harmonic $2f_p$. The Langmuir waves are generated by coherent processes, such as a bump-on-tail instability driven by an electron beam, and so this mechanism can reach very high brightness temperatures even though relatively few electrons are involved. Furthermore, by definition, low-frequency plasma emission comes from regions of low density, whereas hard X-ray emission by bremsstrahlung from collisions is preferentially seen from regions of high density. For this reason, we might not expect a close relationship between low-frequency radio bursts and hard X-ray emission: the electrons producing both have similar energies (plasma emission can be very strong from electrons with energies of tens of keV), but they may be located in very different environments. As an example, Vilmer et al. (2003) discuss the radio and HXR data for an M8 flare in which there are multiple HXR sources and multiple spatially-distinct radio sources in the 100–500 MHz range, but little spatial correlation between the two wavelengths: in particular the early HXR emission in the event seems to come from compact closed magnetic loops with little escape of energetic electrons onto the higher field lines where we believe that the low-frequency radio sources are located. In a later phase of this event a new HXR source low in the atmosphere is associated with the appearance of new radio sources at much greater heights, suggesting simultaneous injection of electrons over a wide range of spatial scales.

7.1 Radio Bursts and Energy Release in Flares

Benz et al. (2005) and Arzner and Benz (2005) have carried out surveys of the types of radio bursts in the range 0.1–4 GHz that occur in flares that exhibit hard X-rays, using data from the *Phoenix-2* spectrometer operated by ETH-Zürich. This frequency range is dominated by plasma emission and other forms of coherent processes such as electron cyclotron maser emission.

Benz et al. (2005) investigated 201 flares and found that about 20% of such flares did not exhibit radio emission in their range (although many of the events with no radio emission were close to the limb and that fact may play a role in their non-detection; see Benz et al. 2007), but over half of the HXR-productive flares followed a similar pattern: at lower frequencies Type III bursts (indicating electron beams) are seen propagating outwards, and at the highest frequencies the bottom of the optically-thick gyrosynchrotron continuum from nonthermal electrons is seen, while in between (typically in the range 500 MHz to 2 GHz) pulsations and/or narrowband spikes (often attributed to electron cyclotron maser emission) may occur. Note that a plasma frequency of 1 GHz corresponds to an electron density of 1.2×10^{10} cm^{-3} and an electron cyclotron frequency of 1 GHz corresponds to a magnetic field strength of 360 G, both of which are values characteristic of the low corona and heights where we might expect energy release to be taking place, For this reason, there has long been interest in understanding the relationship between energy releases and coherent radio bursts in the decimetric range (e.g., Aschwanden et al. 1985, 1990; Güdel et al. 1991).

Arzner and Benz (2005) looked in more detail at the relationship between the timing of radio and hard X-ray emission in this frequency range. They found that only about 20% of type III bursts show a close temporal correlation with HXR peaks, generally trailing the HXR by about 0.5 seconds. Some reverse-drift and patchy bursts also show such delays relative to the HXRs. Such delays can be explained by scattering of the radio emission in the inhomogeneous medium around the source.

Narrowband decimetric spikes show the strongest association with HXRs, in the sense that 95% of such spikes occur during HXR emission (but only 2% of HXR flares show spikes; Guedel et al. 1991). The spikes have timescales shorter (<0.1 s) than we can easily measure in hard X-rays, so timing comparisons are difficult. In addition, spikes may be occurring across a range of frequencies at any given instant, so choices have to be made when making comparisons with HXR light curves. Dabrowski and Benz (2009) carry out a careful comparison of decimetric spikes and HXRs in the time domain. They cross-correlated the radio spike and *RHESSI* HXR light curves to measure delays. The mean correlation coefficient was around 0.7, which is statistically significant for the size of the sample. The delays showed a broad distribution with a standard deviation of order 4 seconds but a mean consistent with zero delay.

Battaglia and Benz (2009) investigated the relationship between spike and HXR sources by comparing the spatial locations of the two emissions for a flare at the limb. The spike locations were obtained from Nançay Radio Heliograph observations between 300 and 432 MHz, while *RHESSI* provided HXR images. The radio spikes were found to be significantly displaced from the HXR sources and at a greater height. Low-frequency radio emission is typically scattered and refracted in the corona, but any refraction will make the apparent height lower than the true height (e.g., Melrose and Dulk 1988). It is possible that there is HXR emission from the spike locations but that it is much weaker than the lower-altitude sources and is lost in the limited dynamic range of the HXR images. An occulted event would be needed to test this possibility. Ning et al. (2009) investigate some features in decimetric dynamic spectra that they argue, based on simultaneous *RHESSI* data, represent upflows due to heating of the chromosphere by nonthermal electrons.

7.2 Hard X-rays from Reverse-Drift Bursts

Electron beams are a common phenomenon in the solar corona (even in the absence of flares), visible through both their plasma emission at radio wavelengths and their bremsstrahlung hard X-ray emission if they reach the chromosphere The exact mechanism by which these electron beams are generated is still unknown. For example, X-point reconnection is believed primarily to produce oppositely directed bulk ion flows, but not collisionless electron beams, and so the production of such beams in this mechanism would require a secondary process. Detection of both radio and HXR emission from the same feature would be an important step in understanding their origin, since the HXR data reveal the energy

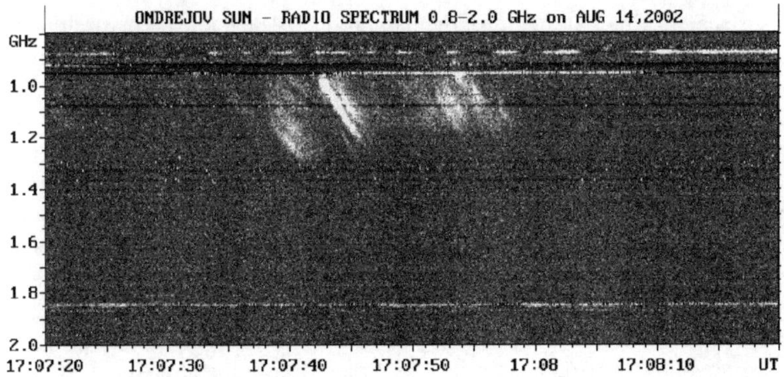

Fig. 12 A dynamic spectrum of reverse-drift radio bursts observed with the Ondrejov spectrograph between 1.0 and 1.2 GHz (Fárník and Karlický 2007)

distribution of accelerated electrons and the radio data have the potential of revealing the location of the acceleration site and the physical conditions therein.

While the most common manifestation of electron beams is the Type III radio burst (discussed next), generated by electrons moving outwards through the corona on open field lines, symmetry suggests that there should be roughly equal numbers of electron beams going downwards. If such beams generate plasma emission, they will be seen to drift from lower to higher frequencies as time proceeds, the reverse of Type III bursts, and hence they are referred to as "reverse-drift" bursts. Since the electrons in such events are moving downwards towards the chromosphere, they seem more likely to be associated with hard X-rays. At frequencies above 1 GHz, such bursts, if due to plasma emission, must arise low in the corona and propagation times down to the chromosphere are typically less than 0.5 seconds.[3]

The relationship between reverse-drift bursts and hard X-rays has been studied by Karlický et al. (2004) and Fárník and Karlický (2007). An example is shown in Fig. 12. From comparison with hard X-ray data in over 20 groups of reverse-drift bursts, these studies find that bursts are mostly observed during the rise phase of the hard X-ray emission, but in the range above 1 GHz there was no one-to-one relationship between individual HXR peaks and individual reverse-drift bursts at a time scale of order 1 second. By contrast, in the frequency range below 1.4 GHz Aschwanden et al. (1995) found correspondence between individual X-ray peaks and fast-drift radio bursts (both Type IIIs and reverse-drift bursts) at a level of 26%. Thus the higher-frequency (and presumably, higher-density environment) reverse-drift bursts above 1 GHz currently do not support the idea that such downgoing electron beams visible at radio wavelengths are responsible for the bulk of the HXRs emitted by flares.

7.3 Hard X-rays from Type III Radio Bursts

Depending on the height at which they originate, Type III radio bursts may seem to be unlikely candidates for correlation with hard X-rays. Those originating at densities below 10^9 cm^{-3} ($f_p = 300$ MHz) do not experience much column density as they travel away from

[3]Historical note: an early paper on flare HXR bursts by Anderson and Winckler (1962) noted the similar timing of Type III bursts and suggested that they may be due to the same electrons that produced the hard X-rays. The more likely connection to downwards-moving reverse-drift bursts, and the absence of such bursts in that event, was pointed out by Kundu (1963), with further discussion presented in Kundu (1965).

the solar surface, making it difficult for them to produce detectable bremsstrahlung HXRs as they propagate outwards. But if they have downgoing counterparts, not necessarily detected as radio sources, then the downgoing electrons will encounter a greater column density and thus could produce observable HXRs.

Type III bursts have a high degree of correlation with the onset of HXR emission in impulsive flares: in about 30% of flares (and a larger proportion of impulsive flares) Type III emission occurs as the impulsive phase begins (e.g., Cane et al. 1986; Cane and Reames 1988; Aschwanden et al. 1990), but typically these impulsive-phase radio bursts do not last for the entire period of HXR emission so a one-to-one association with HXR features cannot be established.

Saint-Hilaire et al. (2009) carry out a detailed study of the quantitative conditions needed for outward-traveling electron beams such as Type-III-emitting streams to be detectable as thin-target nonthermal HXR sources. They find that radially-extended HXR sources can indeed be produced by electron beams propagating through standard coronal density models, but only if the electron beams are intense: they find that 10^{35} electrons above 10 keV are needed for detection of the source, but that 10^{36} electrons above 10 keV are needed if *RHESSI* is to be capable of imaging the HXR source. Such strong beams are more characteristic of the downward flux at an HXR-emitting footpoint in a flare than of the outward-propagating Type-III-burst electron beams as inferred from measurements in the solar wind (Krucker et al. 2007). These calculations assume that the electron beam is generated in a region of density 3×10^9 cm^{-3} (plasma frequency of 500 MHz) in a density model with scale height $\sim 10^{10}$ cm; beams generated at lower starting densities will experience less column density and therefore will be difficult to detect unless the beams are much stronger.

Vilmer et al. (2002) compared Type III bursts with hard X-rays in one of the first flares observed by *RHESSI*, SOL2002-02-20T11:07 (C7.5). In that event the radio emission from outgoing electron beams seems to be very well correlated with hard X-ray emission. Figure 13 shows the relative locations of the HXR sources and 410 MHz emission observed with the Nançay Radio Heliograph. The *RHESSI* images show three separate sources in locations consistent with thick-target footpoint emission, each with a different temporal behavior (only two are visible in the figure). At one point there is a transition in which the source initially brightest in HXRs fades and is overtaken by a different source. This transition occurs at a time when the 410 MHz radio emission shows a distinct brightening, suggesting a causal connection between the radio and HXR emission. Nançay Radio Heliograph radio images of this event also show a transition. Initially the HXR footpoint sources occupy a spatial scale of order 20″ while the 410 MHz radio emission shows two sources

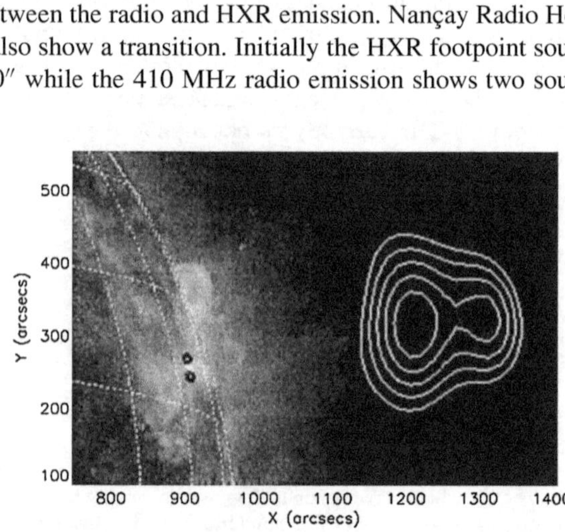

Fig. 13 *RHESSI* iso-contours (*black*) (40, 60, 80% of the maximum) at 25–40 keV and NRH contours at 410 MHz (*white*) (50, 60, 70, 80, 90%) for the SOL2002-02-20T11:07 (C7.5) (Vilmer et al. 2002). The *RHESSI* and NRH contours are superposed on an EIT image obtained at 11:12 UT

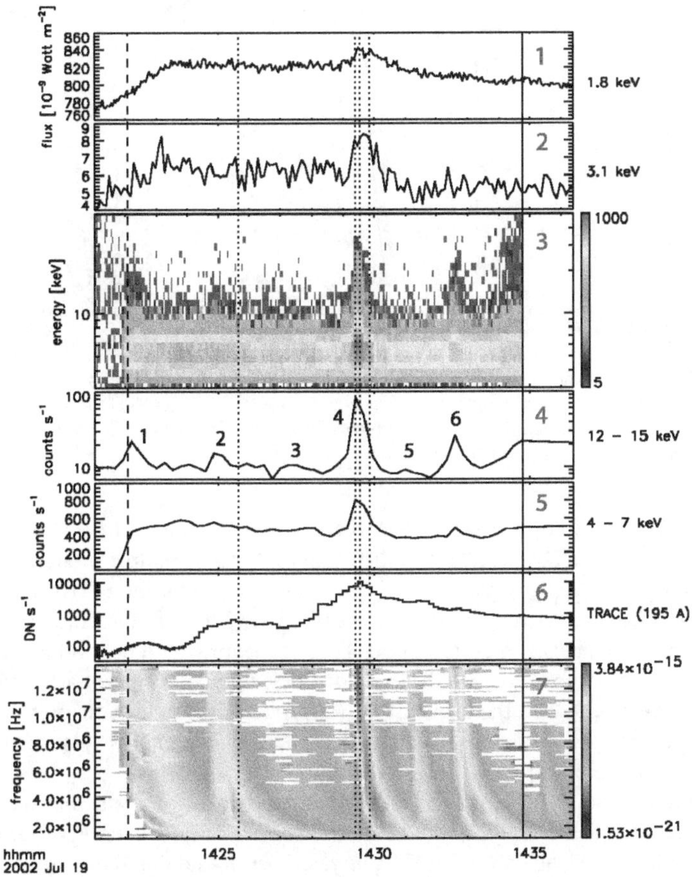

Fig. 14 Plot of a group of Type III bursts (SOL2002-07-19T14:30). *Panels 1* and *2*: *GOES* 1–8 Å (1.6 keV) and 0.5–4 Å (3.1 keV) light curves in a linear scale. *Panel 3*: *RHESSI* spectrogram plot, nighttime background-subtracted. *Panel 4*: *RHESSI* 12–15 keV light curve (nonthermal). *Panel 5*: *RHESSI* 4–7 keV light curve (thermal). *Panel 6*: *TRACE* light curve integrated over the active region from which the HXRs originated. *Panel 7*: Radio spectrogram from the *WIND*/WAVES instrument. The *dashed line* delimits the end of eclipse for *RHESSI*. The *solid line* indicates passage through the South Atlantic Anomaly during which *RHESSI* data is unavailable. From Christe et al. (2008)

aligned almost radially above the flare site at altitudes of order 300″ & 400″ above the photosphere, with the lower source being the brighter: after the transition in the location of the brightest HXR emission, the higher 410 MHz source becomes brighter. Vilmer et al. (2002) interpret the good correlation between the HXR and radio emission, which extends down to timescales as short as a few seconds, as evidence for a common acceleration site with simultaneous injection of electrons both in low-lying magnetic features where they produce hard X-rays, and in larger scale and higher magnetic structures where the radio emission is produced. Yan et al. (2006) discuss another flare in which Type III bursts are seen for a long period coincident with HXR emission, but the locations of the Type III sources are found to be a large distance from the flare site, at the edge of an associated CME, and the connection to the HXR emission is less clear.

Christe et al. (2008) investigated a group of 6 interplanetary Type III bursts that showed associated HXR emission at *RHESSI*. Figure 14 shows *RHESSI* data during a group of

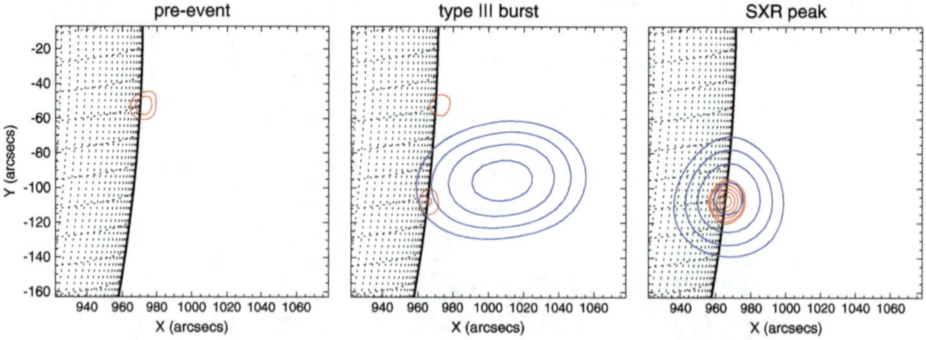

Fig. 15 *RHESSI* X-ray imaging before (*left*), during (*middle*), and after (*right*) radio Type III bursts. Thermal emission (4–8 keV) is shown in red contours (contours at 5, 10, 30, 50, 70, 90% of the level in the *final panel*) with 10″ FWHM resolution, and the 14–30 keV emission is given in blue (levels are 65, 75, 85, and 95%) with a resolution of 60″. From Krucker et al. (2008b)

Type III bursts observed by the *WIND*/WAVES instrument from 2–14 MHz; there is a clear correlation between the HXR peaks and the onset of Type III radio emission for this group. Christe et al. (2008) discuss (quantitatively) the possibility that the HXRs are produced by the Type III-emitting electrons via thin-target bremsstrahlung in the corona, and conclude that the number of electrons required in the Type III beam would be too large to be consistent with the observations. Instead, these events look like HXR microflares, with the HXR spectra being very soft, and the Type III bursts produced during the energy release in the same way that they are at the onset of impulsive flares. The energy releases in the events shown in Fig. 14 are less than 10^{27} erg.

On the other hand, Krucker et al. (2008b) discuss a larger event that occurred at the solar limb such that the main footpoint HXR sources are occulted and *RHESSI* could easily determine the height of the coronal HXR sources. They found that the event did show HXR sources at the solar surface, but when Type III radio emission was present there was a coronal HXR source elongated in the radial direction (see Fig. 15), consistent with expectations for thin-target HXRs from the Type III-emitting electrons. The *WIND* spacecraft was able to measure the number of nonthermal electrons in the Type III burst reaching the Earth and again it appeared to be an order of magnitude too low to explain the observed HXR flux. These quantitative comparisons are fraught with assumptions that need to be made but cannot easily be tested: for example, we must assume that *WIND* measures the bulk of the Type III electrons in the solar wind when it might be merely sampling the edge of the electron beam, and we make assumptions about the pitch-angle distribution of the nonthermal electrons in the HXR source, how long the electrons spend in the source, and how open magnetic field lines connected to the surface diverge with distance in the solar wind. These assumptions introduce uncertainties in the result, but at this point we can say that there are no cases where we can quantitatively attribute the HXR to the Type III-emitting electrons. On the other hand, based on this event we can expect that there are more cases of partially-occulted flares exhibiting radially-elongated HXR emission in conjunction with Type III bursts waiting to be identified in the *RHESSI* archive.

7.4 Acceleration Sites in the Corona

As noted above the combination of radio and HXR data is a valuable tool for understanding the location and nature of energy release sites: the radio data, in particular, identify the density in the accelerator if radio emission at the local plasma frequency is visible. This principle has been applied by Aurass et al. (2006) in identifying a particular feature in radio dynamic spectra with the termination shock of a reconnection outflow. The argument is that reconnection sites produce bulk outflows at the Alfvén speed and these outflows are likely to terminate at a shock where the bulk flow energy in the ions can be converted into heat. Such a shock can also accelerate electrons and thus lead to observable plasma emission (Aurass et al. 2002; Aurass and Mann 2004). Figure 16 shows a slow-drift feature in the dynamic radio spectrum of SOL2003-10-28T11:10 (X17.2) that Aurass et al. (2006) interpret as such a termination shock. In this event, Nançay Radio Heliograph observations above 300 MHz (at frequencies interpreted as harmonic plasma emission) show a source at the time of the termination-shock feature that is located 0.3 $R_\odot$ (in projection onto the sky) away from the flare site (the physical distance may be larger since the flare is close to disk center; Pick et al. 2005). The radio termination-shock feature is seen at the same time that intense γ-ray emission is seen, and Aurass et al. (2006) argue that the termination shock is the acceleration site for the relativistic γ-ray-emitting electrons. Warmuth et al. (2009) and Mann et al. (2009) investigate shock-drift acceleration as a means by which the energy in such outflows may be converted into electron energy at a termination shock, utilizing the relativistic theory for shock-drift acceleration developed by Mann et al. (2006). As yet there is no independent confirmation of the presence either of reconnection outflows or a "wall" in the low-density corona suitable to terminate such a flow, so the true nature of spectral features such as that in Fig. 16 continues to be a topic for research. Additional discussion of this topic may be found in Holman et al. (2011).

This model is not alone in placing the acceleration site for very energetic electrons that produce observable HXRs and γ-rays at a considerable height such that the ambient density (in this case, with the fundamental plasma frequency being around 150 MHz in the termination shock region, the electron density is less than 3×10^8 cm^{-3}) and the magnetic field in the acceleration site are both relatively low. In such models, acceleration takes place in a relatively low-energy-density region, and then the energetic electrons must propagate a large distance back to a region of higher energy density at the flare site low in the atmosphere to produce observable bremsstrahlung. If the γ-ray sources consist of footpoints on either side of a neutral line, then presumably the oppositely-directed field lines from the active region must also thread the termination shock despite the large distance between the acceleration site and the HXR/γ-ray sources (on the other hand, non-footpoint locations for the γ-ray sources, as for the 2.223 MeV neutron-capture line, may point to an acceleration site that is not threaded by loops from the active region; Hurford et al. 2006). These complications apply to any model in which acceleration does not take place low in the corona over an active

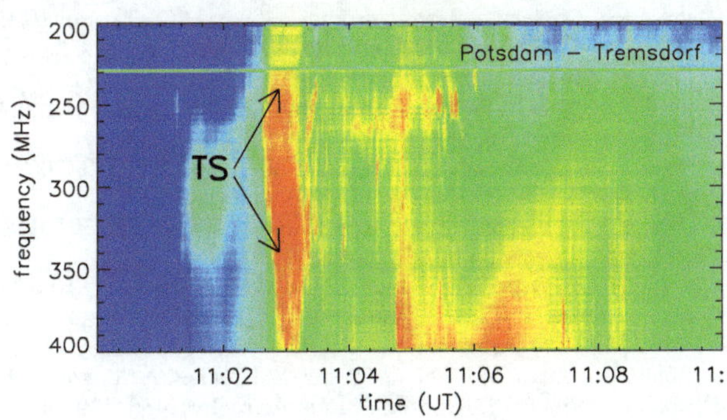

Fig. 16 The dynamic spectrum of SOL2003-10-28T11:10 (X17.2), from the Tremsdorf spectrograph operated by the Astrophysikaliches Institut Potsdam. The bright chevron-shaped feature centered at 11:03 UT (labelled "TS") is identified as a termination-shock feature

region, and distinguishing between these two classes of models remains an important step in identifying the energy release and acceleration mechanisms operating in solar flares.

8 Summary

Progress during the *RHESSI* era has confirmed the value of utilizing radio and HXR data simultaneously to study accelerated electrons in the solar corona. The complementary nature of the diagnostics in the two wavelength regimes allows one to study phenomena from differing viewpoints. However, we have not yet solved several very important problems, notably the discrepancy between the nonthermal electron energy spectral index derived from radio and HXR observations of electrons which, since they show the same temporal behavior, must be related; and the anomalous rising-spectrum submillimeter component. Both of these problems need high-frequency radio data (>30 GHz) for proper study.

Acknowledgements This paper is dedicated to the memory of Mukul Kundu, who passed away during revision. Mukul was a pioneer of solar radiophysics and one of the first to realize the close association of radio and hard X-ray emission and the importance of this relationship for advancing our understanding of flares. We thank the organizers of the *RHESSI* Workshop series (Gordon Emslie and Brian Dennis in particular) and the local organizing committees for each of the workshops for their hospitality and great efforts to make the workshop series such a success. Two referees and Bob Lin provided thoughtful comments that helped to improve the manuscript. Research using *RHESSI* and radio observations was supported at the University of Maryland by NSF grant ATM 02-33907 and NASA contracts NAG 5-12860, NNG-05-GI91G, and NNX-06-AC18G.

References

K.A. Anderson, J.R. Winckler, J. Geophys. Res. **67**, 4103 (1962). doi:10.1029/JZ067i011p04103
K. Arzner, A.O. Benz, Sol. Phys. **231**, 117 (2005)
A. Asai, M. Shimojo, H. Isobe, T. Morimoto, T. Yokoyama, K. Shibasaki, H. Nakajima, Astrophys. J. Lett. **562**, L103 (2001)
A. Asai, H. Nakajima, M. Shimojo, S.M. White, H.S. Hudson, R.P. Lin, Publ. Astron. Soc. Jpn. **58**, L1 (2006)
A. Asai, H. Nakajima, M. Oka, K. Nishida, Y.T. Tanaka, Adv. Space Res. **39**, 1398 (2007). doi:10.1016/j.asr.2007.03.077
M. Aschwanden, A.O. Benz, S.R. Kane, Astron. Astrophys. **229**, 206 (1990)
M. Aschwanden, H.J. Wiehl, A.O. Benz, S.R. Kane, Sol. Phys. **97**, 159 (1985)
M.J. Aschwanden, A.O. Benz, B.R. Dennis, R.A. Schwartz, Astrophys. J. **455**, 347 (1995)
M.J. Aschwanden, R.A. Schwartz, Astrophys. J. **464**, 974 (1996). doi:10.1086/177385
H. Aurass, G. Mann, Astrophys. J. **615**, 526 (2004). doi:10.1086/424374
H. Aurass, K. Shibasaki, M. Reiner, M. Karlický, Astrophys. J. **567**, 610 (2002)
H. Aurass, G. Mann, G. Rausche, A. Warmuth, Astron. Astrophys. **457**, 681 (2006). doi:10.1051/0004-6361:20065238
M. Battaglia, A.O. Benz, Astron. Astrophys. **499**, L33 (2009). doi:10.1051/0004-6361/200912143
A.O. Benz, P.C. Grigis, A. Csillaghy, P. Saint-Hilaire, Sol. Phys. **226**, 121 (2005)
A.O. Benz, R. Brajša, J. Magdalenić, Sol. Phys. **240**, 240 (2007)
L. Bone, J.C. Brown, L. Fletcher, A. Veronig, S. White, Astron. Astrophys. **466**, 339 (2007)
J.C. Brown, Sol. Phys. **18**, 489 (1971)
J.C. Brown, M.J. Aschwanden, E.P. Kontar, Sol. Phys. **210**, 373 (2002). doi:10.1023/A:1022469402781
H.V. Cane, D.V. Reames, Astrophys. J. **325**, 895 (1988)
H.V. Cane, R.E. McGuire, T.T. von Rosenvinge, Astrophys. J. **301**, 448 (1986). doi:10.1086/163913
A. Caspi, Ph.D. Dissertation, U.C. Berkeley, 2010
A. Caspi, R.P. Lin, Astrophys. J. Lett. **725**, L161 (2010). doi:10.1088/2041-8205/725/2/L161
S. Christe, S. Krucker, R.P. Lin, Astrophys. J. Lett. **680**, L149 (2008). doi:10.1086/589971
B.P. Dabrowski, A.O. Benz, Astron. Astrophys. **504**, 565 (2009). doi:10.1051/0004-6361/200811108
C. Dauphin, N. Vilmer, T. Lüthi, G. Trottet, S. Krucker, A. Magun, Adv. Space Res. **35**, 1805 (2005)
C. Dauphin, N. Vilmer, S. Krucker, Astron. Astrophys. **455**, 339 (2006)

G.A. Dulk, Annu. Rev. Astron. Astrophys. **23**, 169 (1985)
G.A. Dulk, K.A. Marsh, Astrophys. J. **259**, 350 (1982)
F. Fárník, M. Karlický, Sol. Phys. **240**, 121 (2007). doi:10.1007/s11207-006-0238-7
G.D. Fleishman, A.A. Kuznetsov, Astrophys. J. **721**, 1127 (2010). doi:10.1088/0004-637X/721/2/1127
G.D. Fleishman, V.F. Melnikov, Astrophys. J. **587**, 823 (2003a)
G.D. Fleishman, V.F. Melnikov, Astrophys. J. **593**, 598 (2003b)
G.D. Fleishman, T.S. Bastian, D.E. Gary, Astrophys. J. **684**, 1433 (2008). doi:10.1086/589821
L. Fletcher, B.R. Dennis, H.S. Hudson, S. Krucker, A. Veronig, M. Battaglia, L. Bone, Q. Chen, P. Gallagher, P.T. Grigis, H. Ji, W. Liu, R.O. Milligan, M. Temmer, Space Sci. Rev. (2011, this issue)
K.J. Frost, B.R. Dennis, Astrophys. J. **165**, 655 (1971)
C.G. Giménez de Castro, G. Trottet, A. Silva-Valio, S. Krucker, J.E.R. Costa, P. Kaufmann, E. Correia, H. Levato, Astron. Astrophys. **507**, 433 (2009). doi:10.1051/0004-6361/200912028
V.V. Grechnev, S.M. White, M.R. Kundu, Astrophys. J. **588**, 1163 (2003)
M. Güdel, M. Aschwanden, A.O. Benz, Astron. Astrophys. **251**, 285 (1991)
M. Guedel, A.O. Benz, M.J. Aschwanden, Astron. Astrophys. **251**, 285 (1991)
D.A. Guidice, J.P. Castelli, Sol. Phys. **44**, 155 (1975)
I.G. Hannah, H.S. Hudson, M. Battaglia, S. Christe, J. Kašparová, M. Kundu, S. Krucker, A. Veronig, Space Sci. Rev. (2011)
G.D. Holman, L. Sui, R.A. Schwartz, A.G. Emslie, Astrophys. J. Lett. **595**, L97 (2003)
G.D. Holman, M.J. Aschwanden, H. Aurass, M. Battaglia, P. Grigis, E.P. Kontar, W. Liu, P. Saint-Hilaire, V.V. Zharkova, Space Sci. Rev. (2011, this issue)
H.S. Hudson, R.C. Canfield, S.R. Kane, Sol. Phys. **60**, 137 (1978)
G.J. Hurford, S. Krucker, R.P. Lin, R.A. Schwartz, G.H. Share, D.M. Smith, Astrophys. J. Lett. **644**, L93 (2006). doi:10.1086/505329
S.R. Kane, K.A. Anderson, W.D. Evans, R.W. Klebesadel, J. Laros, Astrophys. J. Lett. **233**, L151 (1979). doi:10.1086/183095
M. Karlický, New Astron. **9**, 383 (2004). doi:10.1016/j.newast.2004.01.002
M. Karlický, F. Fárník, S. Krucker, Astron. Astrophys. **419**, 365 (2004). doi:10.1051/0004-6361:20035817
P. Kaufmann, H. Levato, M.M. Cassiano, E. Correia, J.E.R. Costa, C.G. Giménez de Castro, R. Godoy, R.K. Kingsley, J.S. Kingsley, A.S. Kudaka, R. Marcon, R. Martin, A. Marun, A.M. Melo, P. Pereyra, J. Raulin, T. Rose, A. Silva Valio, A. Walber, P. Wallace, A. Yakubovich, M.B. Zakia, in *Society of Photo-Optical Instrumentation Engineers (SPIE) Conference Series*, vol. 7012 (2008). doi:10.1117/12.788889
P. Kaufmann, G. Trottet, C.G. Giménez de Castro, J.-P. Raulin, S. Krucker, A.Y. Shih, H. Levato, Sol. Phys. **255**, 131 (2009). doi:10.1007/s11207-008-9312-7
K.-L. Klein, Astron. Astrophys. **183**, 341 (1987)
E.P. Kontar, J.C. Brown, A.G. Emslie, W. Hajdas, G.D. Holman, G.J. Hurford, J. Kašparová, P.C.V. Mallik, A.M. Massone, M.L. McConnell, M. Piana, M. Prato, E.J. Schmahl, E. Suarez-Garcia, Space Sci. Rev. (2011, this issue)
T. Kosugi, B.R. Dennis, K. Kai, Astrophys. J. **324**, 1118 (1988)
S. Krucker, E.P. Kontar, S. Christe, R.P. Lin, Astrophys. J. Lett. **663**, L109 (2007). doi:10.1086/519373
S. Krucker, M. Battaglia, P.J. Cargill, L. Fletcher, H.S. Hudson, A.L. MacKinnon, S. Masuda, L. Sui, M. Tomczak, A.L. Veronig, L. Vlahos, S.M. White, Astron. Astrophys. Rev. **16**, 155 (2008a)
S. Krucker, P. Saint-Hilaire, S. Christe, S.M. White, A.D. Chavier, S.D. Bale, R.P. Lin, Astrophys. J. **681**, 644 (2008b). doi:10.1086/588549
S. Krucker, H.S. Hudson, L. Glesener, S.M. White, S. Masuda, J.-P. Wuelser, R.P. Lin, Astrophys. J. **714**, 1108 (2010). doi:10.1088/0004-637X/714/2/1108
S. Krucker, T.S. Bastian, C.G. Gimenénez de Castro, A.S. Hales, H.S. Hudson, J. Kašparová, K.-L. Klein, M. Kretzschmar, T. Lüthi, A.L. MacKinnon, S. Pohjolainen, G. Trottet, S.M. White, Space Sci. Rev. (2011, in preparation)
M.R. Kundu, J. Geophys. Res. **66**, 4308 (1961)
M.R. Kundu, Space Sci. Rev. **2**, 438 (1963). doi:10.1007/BF00173952
M.R. Kundu, *Solar Radio Astronomy* (Interscience, New York, 1965)
M.R. Kundu, S.M. White, N. Gopalswamy, J. Lim, Astrophys. J. Suppl. Ser. **90**, 599 (1994)
M.R. Kundu, V.I. Garaimov, S.M. White, S. Krucker, Astrophys. J. **600**, 1052 (2004). doi:10.1086/379876
M.R. Kundu, E.J. Schmahl, P.C. Grigis, V.I. Garaimov, K. Shibasaki, Astron. Astrophys. **451**, 691 (2006). doi:10.1051/0004-6361:20053987
M.R. Kundu, V.V. Grechnev, S.M. White, E.J. Schmahl, N.S. Meshalkina, L.K. Kashapova, Sol. Phys. **260**, 135 (2009). doi:10.1007/s11207-009-9437-3
Y.P. Li, W.Q. Gan, Astrophys. J. Lett. **629**, L137 (2005). doi:10.1086/444617
G. Mann, H. Aurass, A. Warmuth, Astron. Astrophys. **454**, 969 (2006). doi:10.1051/0004-6361:20064990
G. Mann, A. Warmuth, H. Aurass, Astron. Astrophys. **494**, 669 (2009). doi:10.1051/0004-6361:200810099

H. Marschhauser, E. Rieger, G. Kanbach, in *International Cosmic Ray Conference*, vol. 3 (1991), p. 61
H. Marschhäuser, E. Rieger, G. Kanbach, in *High-Energy Solar Phenomena—A New Era of Spacecraft Measurements*, ed. by J. Ryan, W.T. Vestrand. American Institute of Physics Conference Series, vol. 294 (1994), pp. 171–176. doi:10.1063/1.45186
S. Masuda, T. Kosugi, H. Hara, S. Tsuneta, Y. Ogawara, Nature **371**, 495 (1994)
D.L. McKenzie, Sol. Phys. **40**, 183 (1975)
J.M. McTiernan, V. Petrosian, Astrophys. J. **359**, 541 (1990). doi:10.1086/169085
V.F. Melnikov, in *Solar Physics with the Nobeyama Radioheliograph* (2006), pp. 11–22
V.F. Melnikov, V.E. Reznikova, S.P. Gorbikov, K. Shibasaki, in *Recent Advances in Astronomy and Astrophysics*, ed. by N. Solomos. American Institute of Physics Conference Series, vol. 848 (2006), pp. 123–132. doi:10.1063/1.2347968
D.B. Melrose, G.A. Dulk, Sol. Phys. **116**, 141 (1988)
T. Minoshima, T. Yokoyama, N. Mitani, Astrophys. J. **673**, 598 (2008). doi:10.1086/523884
H. Nakajima, T. Kosugi, K. Kai, S. Enome, Nature **305**, 292 (1983)
V.M. Nakariakov, V.F. Melnikov, Astron. Astrophys. **446**, 1151 (2006). doi:10.1051/0004-6361:20053944
V.M. Nakariakov, E. Verwichte, Living Rev. Sol. Phys. **2**, 3 (2005)
Z. Ning, Astrophys. J. Lett. **671**, L197 (2007). doi:10.1086/525249
Z. Ning, Sol. Phys. **247**, 53 (2008a). doi:10.1007/s11207-007-9101-8
Z. Ning, Astrophys. J. **686**, 674 (2008b). doi:10.1086/590652
Z. Ning, W. Cao, Sol. Phys. **257**, 335 (2009). doi:10.1007/s11207-009-9388-8
Z. Ning, W. Cao, J. Huang, G. Huang, Y. Yan, H. Feng, Astrophys. J. **699**, 15 (2009). doi:10.1088/0004-637X/699/1/15
N. Nitta, S.M. White, E.J. Schmahl, M.R. Kundu, Sol. Phys. **132**, 125 (1991)
L.E. Peterson, J.R. Winckler, Phys. Rev. Lett. **1**, 205 (1958)
V. Petrosian, Astrophys. J. **251**, 727 (1981). doi:10.1086/159517
V. Petrosian, Astrophys. J. Lett. **255**, L85 (1982)
V. Petrosian, J.M. McTiernan, H. Marschhauser, Astrophys. J. **434**, 747 (1994)
M. Pick, J.-M. Malherbe, A. Kerdraon, D.J.F. Maia, Astrophys. J. **631**, L97 (2005)
J. Qiu, C. Liu, D.E. Gary, G.M. Nita, H. Wang, Astrophys. J. **612**, 530 (2004). doi:10.1086/422401
R. Ramaty, Astrophys. J. **158**, 753 (1969)
R. Ramaty, R.A. Schwartz, S. Enome, H. Nakajima, Astrophys. J. **436**, 941 (1994)
J. Raulin, S.M. White, M.R. Kundu, A.R. Silva, K. Shibasaki, Astrophys. J. **522**, 547 (1999)
J.-P. Raulin, V.S. Makhmutov, P. Kaufmann, A.A. Pacini, T. Lüthi, H.S. Hudson, D.E. Gary, Sol. Phys. **223**, 181 (2004)
V.E. Reznikova, V.F. Melnikov, K. Shibasaki, S.P. Gorbikov, N.P. Pyatakov, I.N. Myagkova, H. Ji, Astrophys. J. **697**, 735 (2009). doi:10.1088/0004-637X/697/1/735
R.D. Robinson, in *Solar Radiophysics* (Cambridge University Press, Cambridge, 1985), p. 385
P. Romano, F. Zuccarello, L. Fletcher, F. Rubio da Costa, H.M. Bain, L. Contarino, Astron. Astrophys. **498**, 901 (2009). doi:10.1051/0004-6361/200811309
P. Saint-Hilaire, S. Krucker, S. Christe, R.P. Lin, Astrophys. J. **696**, 941 (2009). doi:10.1088/0004-637X/696/1/941
A.V.L. Silva, S.M. White, R.P. Lin, I. de Pater, K. Shibasaki, H.S. Hudson, M.R. Kundu, Astrophys. J. Lett. **458**, L49 (1996)
A.V.R. Silva, H. Wang, D.E. Gary, Astrophys. J. **545**, 1116 (2000)
A.V.R. Silva, G.H. Share, R.J. Murphy, J.E.R. Costa, C.G. Giménez de Castro, J.-P. Raulin, P. Kaufmann, Sol. Phys. **245**, 311 (2007). doi:10.1007/s11207-007-9044-0
D.M. Smith, G.H. Share, R.J. Murphy, R.A. Schwartz, A.Y. Shih, R.P. Lin, Astrophys. J. Lett. **595**, L81 (2003)
T. Takakura, Sol. Phys. **26**, 151 (1972)
G. Trottet, N. Vilmer, C. Barat, A. Benz, A. Magun, A. Kuznetsov, R. Sunyaev, O. Terekhov, Astron. Astrophys. **334**, 1099 (1998)
G. Trottet, S. Krucker, T. Lüthi, A. Magun, Astrophys. J. **678**, 509 (2008). doi:10.1086/528787
V. Tzatzakis, A. Nindos, C.E. Alissandrakis, Sol. Phys. **253**, 79 (2008). doi:10.1007/s11207-008-9263-z
A.M. Veronig, J.C. Brown, Astrophys. J. Lett. **603**, L117 (2005)
W.T. Vestrand, Sol. Phys. **118**, 95 (1988)
W.T. Vestrand, D.J. Forrest, E.L. Chupp, E. Rieger, G.H. Share, Astrophys. J. **322**, 1010 (1987)
N. Vilmer, G. Trottet, C. Barat, R.A. Schwartz, S. Enome, A. Kuznetsov, R. Sunyaev, O. Terekhov, Astron. Astrophys. **342**, 575 (1999)
N. Vilmer, S. Krucker, R.P. Lin, the *RHESSI* Team. Sol. Phys. **210**, 261 (2002)
N. Vilmer, S. Krucker, G. Trottet, R.P. Lin, Adv. Space Res. **32**, 2509 (2003). doi:10.1016/S0273-1177(03)90423-6

A. Warmuth, G. Mann, H. Aurass, Astron. Astrophys. **494**, 677 (2009). doi:10.1051/0004-6361:200810101
S.M. White, in *High Energy Solar Phenomena: A New Era of Spacecraft Measurements*, vol. 294, ed. by J. Ryan, W.T. Vestrand (American Institute of Physics, New York, 1994), p. 199
S.M. White, in *Nobeyama Symposium on Solar Physics with Radio Observations*, ed. by T.S. Bastian, N. Gopalswamy, K. Shibasaki (Nobeyama Radio Observatory, Nobeyama, 1999), p. 223
S.M. White, M.R. Kundu, Sol. Phys. **141**, 347 (1992)
S.M. White, S. Krucker, K. Shibasaki, T. Yokoyama, M. Shimojo, M.R. Kundu, Astrophys. J. Lett. **595**, 111 (2003)
S.M. White, R.J. Thomas, R.A. Schwartz, Sol. Phys. **227**, 231 (2005)
Y. Yan, M. Pick, M. Wang, S. Krucker, A. Vourlidas, Sol. Phys. **239**, 277 (2006). doi:10.1007/s11207-006-0202-6

Space Sci Rev (2011) 159:263–300
DOI 10.1007/s11214-010-9705-4

Microflares and the Statistics of X-ray Flares

I.G. Hannah · H.S. Hudson · M. Battaglia · S. Christe · J. Kašparová · S. Krucker · M.R. Kundu · A. Veronig

Received: 16 May 2009 / Accepted: 28 October 2010 / Published online: 23 July 2011

Abstract This review surveys the statistics of solar X-ray flares, emphasising the new views that *RHESSI* has given us of the weaker events (the microflares). The new data reveal that these microflares strongly resemble more energetic events in most respects; they occur solely within active regions and exhibit high-temperature/nonthermal emissions in approximately the same proportion as major events. We discuss the distributions of flare parameters (e.g., peak flux) and how these parameters correlate, for instance via the Neupert effect. We also highlight the systematic biases involved in intercomparing data representing many decades of event magnitude. The intermittency of the flare/microflare occurrence, both in space and in time, argues that these discrete events do not explain general coronal heating, either in active regions or in the quiet Sun.

Keywords Sun · Flares · X-rays

I.G. Hannah (✉) · M. Battaglia
School of Physics & Astronomy, University of Glasgow, Glasgow G12 8QQ, UK
e-mail: iain@astro.gla.ac.uk

H.S. Hudson · S. Krucker
Space Sciences Laboratory, University of California at Berkeley, Berkeley, CA 94720-7450, USA

S. Christe
NASA Goddard Space Flight Center, Greenbelt, MD 20771, USA

J. Kašparová
Astronomický ústav AV ČR, v.v.i., Fričova 298, Ondřejov 251 65, Czech Republic

M.R. Kundu
Dept. of Astronomy, Univ. of Maryland, College Park, MD 20740, USA

A. Veronig
Institute of Physics/IGAM, University of Graz, Universitätsplatz 5, 8010 Graz, Austria

1 Introduction

A solar flare is a rapid and transient release of energy in the solar corona, associated with electromagnetic radiation from radio waves to γ-rays, local plasma heating to tens of MK, particles accelerated to GeV, violent mass plasma motions, and shock waves. In the largest events, a spectacular variety of phenomena can be studied in detail across many wavelength ranges to try to understand the processes involved (see examples in Fletcher et al. 2011). The X-ray emission observed is of particular interest as it shows the accelerated particles and intense heating, and therefore provides the most direct insights into the physics of the basic energy release. Hard X-rays (HXRs), from about 10 keV to hundreds of keV, are primarily produced via thick-target bremsstrahlung (Brown 1971; Kontar et al. 2010, this issue) in which the coronal accelerated electrons are stopped instantaneously through Coulomb collisions with denser material in the lower solar atmosphere. Soft X-rays (SXRs), typically the component below tens of keV, are thermal emission (lines and continua) by plasma of a few to tens of MK. This emission results to some extent from heating at the site of energy release, but mainly from new coronal material evaporated from the chromosphere during the impulsive phase (see Fletcher et al. 2011). Given the wealth of X-ray flare data, it is possible to study large samples of these events, allowing the statistics of the events to provide clues to the underlying processes behind the emission. This article reviews such flare studies, the results, interpretation and limitations, from a predominantly X-ray viewpoint. In particular, we discuss the advances that have been made in extending this analysis to weaker HXR events (*microflares*) using the *Reuven Ramaty High Energy Solar Spectroscopic Imager (RHESSI)* (Lin et al. 2002).

One of the most remarkable features of flare surveys is the range of magnitudes these transient phenomena span. The observed SXR flux extends over five orders of magnitude (see Fig. 1) and over nine orders of magnitude in energy (see Fig. 2; we return to discuss these Figures in detail in Sect. 3). The observations come from many spacecraft: e.g., *Orbiting Solar Observatory (OSO-3)* and *(OSO-7)*, the *Solar Maximum Mission (SMM)*, the *International Cometary Explorer (ICE, a.k.a. ISEE-3)*, the *Compton Gamma-Ray Observatory (CGRO)*, *Yohkoh*, the *Solar Heliophysical Observatory (SOHO)*, *GRANAT*/WATCH, and the *Transition Region and Coronal Dynamics Explorer (TRACE)*. The distributions of these and other flare parameters, such as emission at other wavelengths and for other durations, have the further striking property that they all can be well represented by a power law of the form

$$f(x;\alpha) = Cx^{-\alpha}, \tag{1.1}$$

where f is the probability density function (PDF; this is often called the flare frequency distribution) of the flare parameter x, $\alpha > 0$ is the power-law index and C is a scaling constant. In the case of the flare's energy U, the quantity $f(U)dU$ is the fraction of events per unit time releasing energy between U and $U + dU$. This power-law nature of the frequency distribution has long been observed, first noticed in solar radio bursts by Akabane (1956). The fact that so many of the flare characteristics have such a distribution is thought to arise from the corona being in a self-organized critical state (Lu and Hamilton 1991), discussed further in Sect. 3.3. Here, the underlying energy release from the largest to the smallest flares occurs as an "avalanche," similar to the theory for earthquake occurrence or stress relief in general (Katz 1986). These properties have brought forth interesting interpretations one of which is that small flares are more numerous.

Could these weaker events at the currently unobservable "nanoflare" magnitude level (about 10^{-9} times the energy of a large flares) directly constitute the hot corona, thus explaining its heating? This is often called Parker's nanoflare mechanism (Parker 1988). The

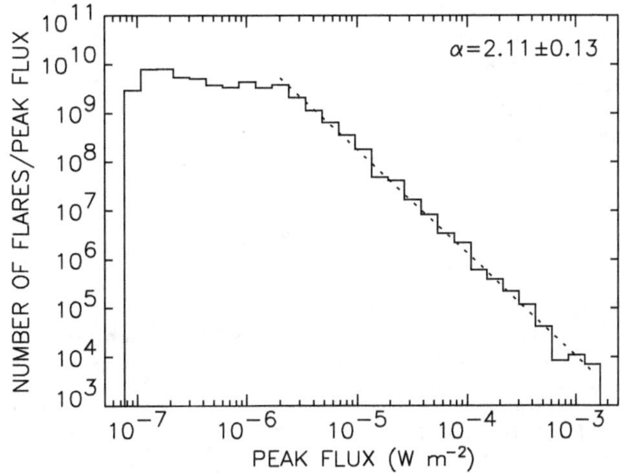

Fig. 1 Frequency distribution of the peak 1–8 Å SXR *GOES* flux of 49,409 flares between 1976 and 2000 (Veronig et al. 2002a). Note that this study did not include the A-class events ($<10^{-7}$ W m^{-2}). In addition, there was no attempt at background subtraction, so the resulting value of α should be regarded as an upper limit (background predominantly affects smaller flares, artificially steepening the distribution)

answer to this question requires knowledge of the total power that is contained in the flare distribution, i.e.

$$P = \int_{U_{\min}}^{U_{\max}} f(U;\alpha)U dU = \frac{C}{2-\alpha}\left[{U_{\max}}^{2-\alpha} - {U_{\min}}^{2-\alpha}\right]. \tag{1.2}$$

Since $U_{\min} \ll U_{\max}$, we can see that in the case $\alpha > 2$ in (1.2) the low-energy half of the distribution would contain the most energy (Hudson 1991b). In this situation the smallest flares fit the requirement for heating the corona as they have a high occurrence rate and release more net energy than the large flares. Therefore, to understand the role of flares in coronal heating, the power-law index α has to be accurately estimated; this is problematic as the energy is inferred from observations and thus subject to potentially large errors and biases. In addition, (1.2) assumes that the distributions continue into the unobservable low-energy range. Even if the deduced α were large enough (i.e., larger than two), that would not automatically mean that small flares heat the corona. For further proof we would require observations showing the power-law index to maintain its value down to the distribution's physical limit, rather than its observational limit.

Between the largest flares and these theoretical nanoflares lie the microflares (nominally with energies of order 10^{-6} times those of large flares), and it is with these events that *RHESSI*'s HXR observations have made major advances. *RHESSI*'s continuous spectral coverage, down to 3 keV, uniquely allows the study of the transition between the thermal and nonthermal emissions with the same instrument. In particular, *RHESSI*'s view of their nonthermal characteristics has allowed us to study smaller-scale active region events with simpler structures than the major events that often attract the most attention. Perhaps these can help us to isolate the essential physics, which could then be applied in more complex situations.

We begin this article with a general overview of flares and flare-like brightenings across the range of magnitudes (Sect. 2). This discloses similar properties between major and minor events, but systematic differences do occur. In Sect. 3, we learn what we can from these differences about the physics, taking advantage of the very large numbers of events in major surveys, including *RHESSI*'s new HXR views. We also discuss in this section the biases that arise in the statistical surveys and possible methods for obtaining the unbiased intrinsic

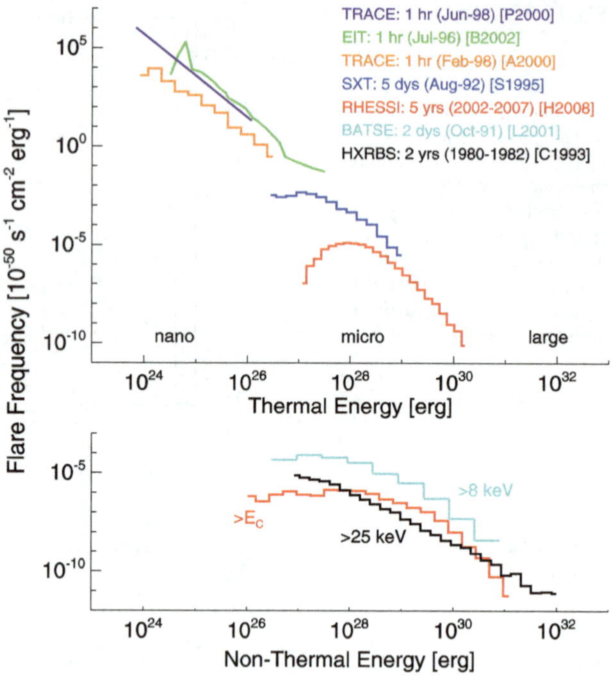

Fig. 2 The energy distributions for solar flares. The nonthermal energy distribution is shown for large flares >25 keV observed with *SMM*/HXRBS (Crosby et al. 1993), microflares >8 keV from *CGRO*/BATSE (Lin et al. 2001a) and microflares > E_C (above the low energy cutoff) from *RHESSI* (Hannah et al. 2008a). The thermal energy distribution is shown for microflares with *RHESSI* (Hannah et al. 2008a) and *Yohkoh*/SXT (Shimizu 1995) and EUV nanoflares with *TRACE* (Parnell and Jupp 2000; Aschwanden et al. 2000) and *SOHO*/EIT (Benz and Krucker 2002). This figure is deceptive as it is comparing energy distributions of different flare energy components, each involving different instrument and selection effects, and were obtained over different periods of different solar cycles

distributions. In Sect. 3.3 we briefly discuss how the power-law nature of the flare parameters arises. Conclusions and discussion are given in Sect. 4.

2 From Major to Minor Flares

2.1 Flare Classification & General Properties

The most powerful ordinary flares have energies estimated at above 10^{33} ergs and present a spectacular range of phenomena, easily observed across the wavelengths. The first flare observed was a powerful event in 1859, detectable through its small, intense white-light emission patches as described by Carrington (1859) and corroborated by Hodgson (1859). Remarkable terrestrial effects accompanied this flare and also followed it after an interval of half a day. This event anticipated much of the complexity of flares as we know them today, but it was not until the 1940s that "flare" was accepted as the term to describe these transient phenomena (Newton 1943; Richardson 1944). Events with total energy about a millionth smaller than large flares (about 10^{27} erg), became known as "microflares" (Schadee et al. 1983; Lin et al. 1984). Parker hypothesized that even smaller flares, "nanoflares," with energies of order one billionth of large flares or about 10^{24} erg, could be the basic unit of a localized impulsive energy release (Parker 1988).

Quantitative flare classification is based on the 1–8 Å SXR flux observed by *GOES*. Large flares have Xn-class, indicating a peak flux of $n \times 10^{-4}$ W m^{-2}, the largest events being above X10. This classification decreases through the decades of M, C and B-class flares down to the smallest An-class events with $n \times 10^{-8}$ W m^{-2} and the sensitivity limit of the detector. The classification of flares and the associated range of *GOES* fluxes is shown in Table 1. The largest *GOES* flare was SOL2003-11-04T19:53 (X17.4), which saturated the

Table 1 Different flare classifications and the associated ranges of *GOES* flux (SXRs) and HXRs

Flare size	Description	Energy [erg]	*GOES* class	*GOES* flux [W m^{-2}]	HXR emission
Large\normal	Active region phenomena	$\leq 10^{33}$	C, M, X	10^{-6}–10^{-3}	>25 keV
Micro\ARTB[1]	Active region phenomena	$\sim 10^{27}$	A, B	10^{-8}–10^{-6}	>10 keV
Nano	Unobserved basic unit of localized impulsive energy release **or** very small (EUV) brightening	$\sim 10^{24}$	? $\ll$ A	? $\ll 10^{-8}$	?

[1]"Active Region Transient Brightening" observed in SXRs (Shimizu 1995)

detectors at 18×10^{-4} W m^{-2} (i.e., class X18). It is estimated that this flare was an X28 event with range X25 to X31 (Kiplinger and Garcia 2004). In terms of *GOES* classification, the Carrington flare appears to have been a large soft X-ray flare of magnitude >X10 (Tsurutani et al. 2003) and like many large flares resulted in a major geomagnetic storm (Cliver and Svalgaard 2004). This region also produced one of the largest solar energetic proton fluences at the Earth in the last 500 years (McCracken et al. 2001). These most energetic flares occur in active regions, often when new flux emerges into an already-complex magnetic structure. From the start of the *GOES* observations in 1975, to the start of cycle 24 in 2009, 22 flares >X10 had been observed. In total, 359 *GOES* X-class flares, 4708 M-class, 32784 C-class and 11558 B-class flares occurred between 1976 and 2000 (Veronig et al. 2002a). Although more small flares are expected due to the inherent power-law nature of the flare distribution, they are hidden by the higher background during active times. The majority of large flares seem to occur during the peak of the solar cycle heading into the decay phase, though this behavior can vary dramatically from cycle to cycle (e.g., Hudson 2007). These two features can be seen in the *GOES* flaring rate shown in Fig. 5, discussed in Sect. 2.2.1.

Flare phenomena encompass all of the accessible wavelength ranges of electromagnetic radiation, as well as the emission of neutral particles, gaseous ejecta, and large-scale shock waves (e.g. Fletcher et al. 2011). In the most energetic flares, all of these phenomena appear, considerably more intensely than in smaller flares and regardless of the detailed physics; this is the so-called *big-flare syndrome* (e.g., Kahler 1982). This makes these events helpful diagnostically. This property suggests that if a particular phenomenon (e.g., white-light continuum) is not observed in a weaker flare (as not distinguishable from the background), the process may still function in the same way. Non-scalable properties might exist; one that is often cited is the maximum energy attained by accelerated particles. An emission such as π^0-decay γ-ray emission might thus have a non-linear threshold dependence; see Vilmer et al. (2010). Of course, given that there weak flares are considerably more numerous, exceptional cases can often be found, such as microflares with nonthermal emission to remarkably high energies (e.g., Hannah et al. 2008b).

2.1.1 Flare Relationship to CMEs and SEPs

There has been a long debate whether ejection of material from the corona, namely a Coronal Mass Ejection (CME), triggers a flare, or vice-versa, but it is now generally accepted that they are both consequences of coronal energy release through the reconfiguration of the magnetic field. The issue of whether a flare or CME is produced will depend on the magnetic field configuration (Wang and Zhang 2007). Using a large survey of nearly 7000 CMEs,

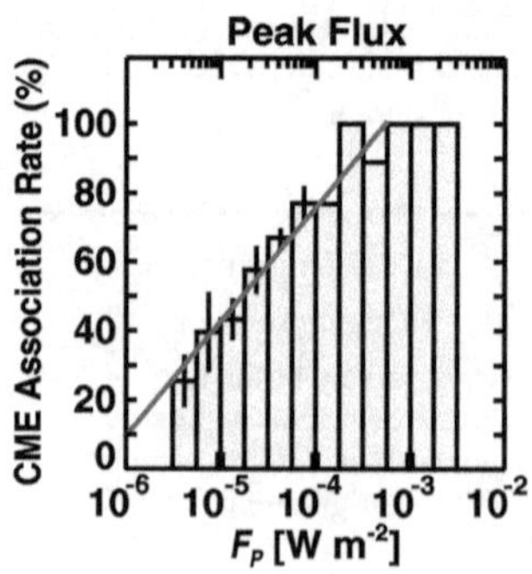

Fig. 3 The association rate of CMEs with flares as a function of the flare peak *GOES* flux (Yashiro et al. 2006). Only flares from C-class (10^{-6} W m^{-2}) to X-class ($>10^{-4}$ W m^{-2}) are included in this survey. Reproduced by permission of the AAS

Yashiro et al. (2004) found that all the most energetic (>X2) flares have CMEs; see Fig. 3 (Yashiro et al. 2006). The largest and fastest CMEs are generally associated with such large flares. The rate for C-class flares is about 20% and for M-class about 40%. Of course, there may be a sensitivity issue which limits the ability to observe the small CMEs associated with microflares. Studies of flares that do not have associated CMEs have found these events to have a correlation between peak intensity and flare duration (Kay et al. 2003). No such correlation was found for CME flare events, suggesting a physical difference in the time development of a flare when a CME is involved. CMEs appear to have broadly the same range of characteristics whether or not they are associated with a flare, but the fastest and broadest CMEs are always associated with the most energetic flares (Vršnak et al. 2005). In terms of the energy partition between flares and CMEs, the CME kinetic energy ranges may be comparable to the total flare energy (Emslie et al. 2005).

The ejection of highly accelerated (up to several MeV) electrons, protons and heavy ions from the Sun into interplanetary space, called Solar Energetic Particles (SEPs), are sometimes associated with flares (e.g. Reames 1999). Impulsive SEP events demonstrate a rapid burst of electron enhancement often related to the impulsive HXR phase of flares, suggesting that the same coronal energy release is responsible for the flare and SEPs. These energetic electrons therefore provide an important diagnostic of flare particle acceleration, especially when combined with HXR flare observations. Type III radio bursts are also produced by energetic electrons escaping from the Sun into interplanetary space, see Sect. 2.2.3. Gradual SEP events show a slower enhancement of energetic protons, thought to be accelerated by a CME-associated shock in interplanetary space rather than in the coronal energy release. Although these proton events are not directly related to flares, Kiplinger (1995) showed that flares that were associated with 10 MeV proton events predominantly demonstrated progressive hardening of the HXR photon spectra into the decay phase of the HXR emission. In contrast, the other population of HXR events not associated with SEPs exhibit softening spectra in the decay phase. A similar result was found for X-class flares during the January 2005 solar storm event (Saldanha et al. 2008). In these flares, four out of five showed spectral hardening with *RHESSI* and were associated with interplanetary proton events. A study of 37 events with *RHESSI* found that the majority of events that demonstrated HXR spectral flattening (12 out of 18) produced SEPs (detected with *GOES* >10 MeV proton data and *WIND*/3DP 0.1–1 MeV proton and 30–500 keV electron data) and all without flattening (19 flares) did not produce SEPs (Grayson et al. 2009).

2.2 Microflares

2.2.1 *Association with Active Regions*

"Subflares" had always been known to the Hα observers (e.g., Smith and Smith 1963) as events that were small in area but not necessarily faint. The term "microflare" was introduced in the 1980s by Schadee et al. (1983) for SXRs, and by Lin et al. (1984) for HXRs. The presence of X-rays, the occurrence of subflares with "brilliant" Hα classification (high intensity), and the discovery of faint microwave bursts (Gopalswamy et al. 1994; Gary et al. 1997) in association with SXR microflares made it clear that the basic flare physics extended over a wide magnitude range. Flares generally (including subflares and microflares) show both thermal and nonthermal emission, indicating the presence of particle acceleration and plasma heating. In terms of *GOES* classification these events are typically A- and B-class events, down to and beyond the sensitivity limit of the *GOES* detectors.

The HXR observations prior to *RHESSI* normally suffered from poor sensitivity due to lack of access to lower hard X-ray energies (<15 keV), hence these observations missed most of the microflares. The problem was the use of thick fixed attenuators to reduce the excessive low energy counts in large flares. One exception to this was the HXIS[1] imager on the *Solar Maximum Mission (SMM)* (Schadee et al. 1983), which covered the 3–30 keV range with a restricted field of view and very small detectors, thus avoiding saturation. Because of this, the main HXIS results were in a spectral region comparable with the *GOES* data. At lower energies (0.25–4 keV), the Soft X-ray Telescope (SXT) on *Yohkoh* found microflares to occur in active regions (Shimizu 1995).

The first observation of HXR microflares was with balloon-borne detectors that could observe only down to about 15 keV due to atmospheric absorption, and which had high sensitivity from the use of large-area, low-background detectors (Lin et al. 1984). Subsequently with *Yohkoh*/HXT, emission at 14–23 keV was found to be associated with SXR brightenings in active regions (Nitta 1997). Microflares were also detected at 8–13 keV with *CGRO*/BATSE (Lin et al. 2001b). However, it was not until *RHESSI* that the detailed associations of these HXR microflares could be readily investigated.

RHESSI's introduction of a movable attenuating shutter system (Lin et al. 2002) allowed the attenuators only to be deployed when the detectors were saturated. With no shutters, the detectors become saturated once solar emission (either background emission from active regions or flares) is approximately greater than *GOES* C1 level. *RHESSI* microflares are therefore flares that occur when the attenuating shutters are out; they are typically sub-C-class events, $<10^{-6}$ W m^{-2} in the 1–8 Å *GOES* band. The full view and sensitivity of *RHESSI* is available to such events, providing the first comprehensive HXR view of microflares, with both imaging and spectroscopy. In addition to this, *RHESSI* introduced Ge detectors with resolution superior to the scintillation counters previously used for HXR observations, allowing the properties of steep spectra to be accurately determined.

RHESSI microflares appear to be exclusively localized to active regions. Figure 4 (left panel) shows this explicitly with the positions of 24,097 events, all of which are identified as coming from active regions (Christe et al. 2008a). These events were identified from analyzing all count-rate spikes in the *RHESSI* 6–12 keV energy range between March 2002 and March 2007, at times when the attenuating shutters were out. A total of 25,705 events were found, though only 24,097 events could be imaged to give positions on the solar disk.

[1] Hard X-ray Imaging Spectrometer.

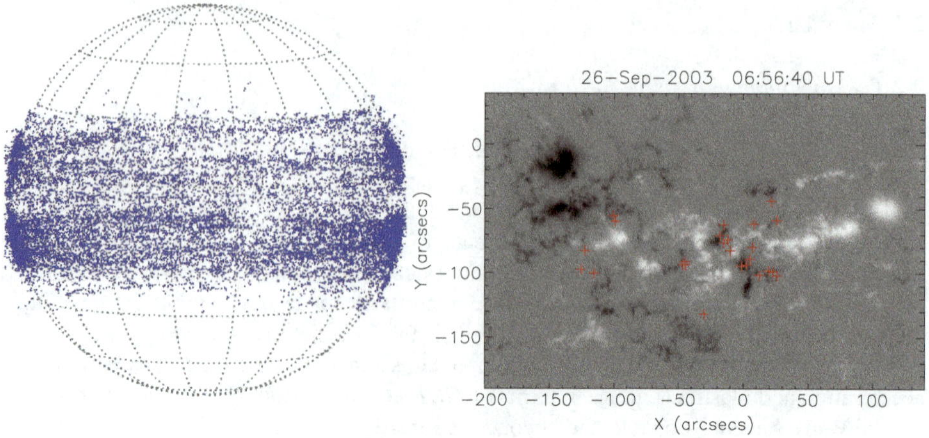

Fig. 4 (*Left*) The locations of 24,097 microflares observed with *RHESSI* between March 2002 and March 2007 (Christe et al. 2008a). All of the confirmed events could be associated with active regions. Reproduced by permission of the AAS. (*Right*) The locations of one day of *RHESSI* microflares from active region 10456 (Stoiser et al. 2007). Of 53 events found, 24 could be imaged. The centroid positions (3–6 keV) are indicated by crosses and overplotted on the MDI (Michelson Doppler Imager of *SOHO*) magnetogram

The events not counted include some of the faintest true microflares, but are predominantly misidentified particle events (which gave a non-solar disk position) or events close to the rotation axis of *RHESSI* (events close to where the spacecraft rotation axis is pointing will have no modulation in their detected signal and hence no spatial information can be recovered). There is repeated microflaring from active regions, thus tracing out the location of a region as it moves across the disk. The microflaring rate varies greatly between active regions. Detailed *RHESSI* imaging of all the microflares from one active region compared to MDI magnetograms shows that the microflare emission occurred at multiple locations throughout the active region (Stoiser et al. 2007), though many originated in one localized area of the region (Fig. 4, right panel). Another sample of microflares showed that they mostly occur near magnetic neutral lines in active regions (Liu et al. 2004).

As expected, then, Fig. 5 (left panel) shows that these events tend to follow the solar cycle in their occurrence frequency. We note that it is difficult to follow the microflare rate over a whole solar cycle because the solar X-ray background level also follows the solar cycle. Fewer microflares are observed during periods of high solar activity as they are hidden by the high background from active regions. This can be clearly seen in the number of *GOES* B-class flares (right panel of Fig. 5). Their rate appears unexpectedly to be almost anti-correlated with the solar cycle. This observational bias means that the microflaring rate is only accurately known during times of lower activity/solar background.

As the HXR microflares observed by *RHESSI* are inherently an active-region phenomenon, it seems entirely reasonable to associate them with microflares (Schadee et al. 1983) and the active-region transient brightenings (ARTBs) seen in soft X-rays (Shimizu 1995). This had been previously postulated (Tsuneta and Lemen 1993) and early evidence suggested it to be correct (Nitta 1997), but the confirmation was only possible with *RHESSI's* hard X-ray imaging and sensitivity.

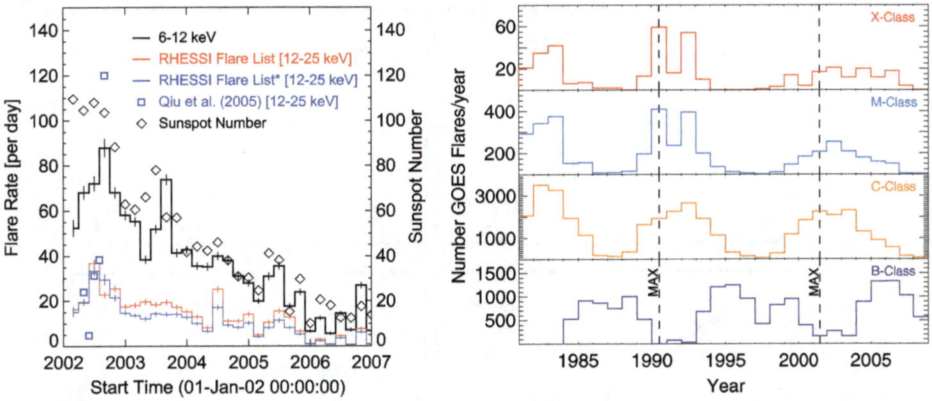

Fig. 5 (*Left*) Occurrence rate of *RHESSI* microflares (<C1 events) (Christe et al. 2008a), showing a clear decline over the waning years of Cycle 23. Reproduced by permission of the AAS. (*Right*) The number of *GOES* flares per year per flare class for X-class (*top*) through M, C to B-class (*bottom*). The microflares seem to be anti-correlated with the larger flares, which would be expected from the increased background during times of high activity. Note that B-class flare data was only available from 1984 (data from Milligan 2009)

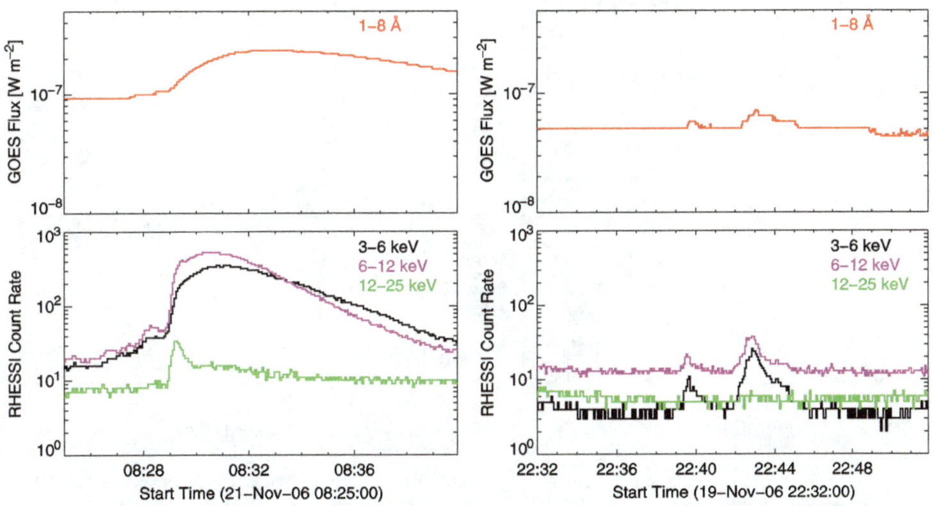

Fig. 6 The *GOES* 1–8 Å lightcurve (*top panels*) and *RHESSI* lightcurves for some example microflares. (*Left*) A single B-class microflare (B1 with pre-flare background subtracted) which shows the classic flare time profile: impulsive at higher energies with a slower rise and more gradual fall at lower energies. This event also shows some pre-flare emission. (*Right*) Two A-class events (A1 and A3, background-subtracted) both of which have no emission above 12 keV observed by *RHESSI*; their nonthermal emission hidden by the background. Note that the background level in these microflares is predominantly solar for *GOES*, but terrestrial and instrumental in origin for *RHESSI*

2.2.2 Physical Properties

The X-ray time profiles of typical microflares (examples given in Fig. 6) demonstrate a similar structure to those in typical large flares: a short impulsive burst at higher energies (HXRs) followed by a more gradual thermal emission at lower energies (SXRs). Microflares

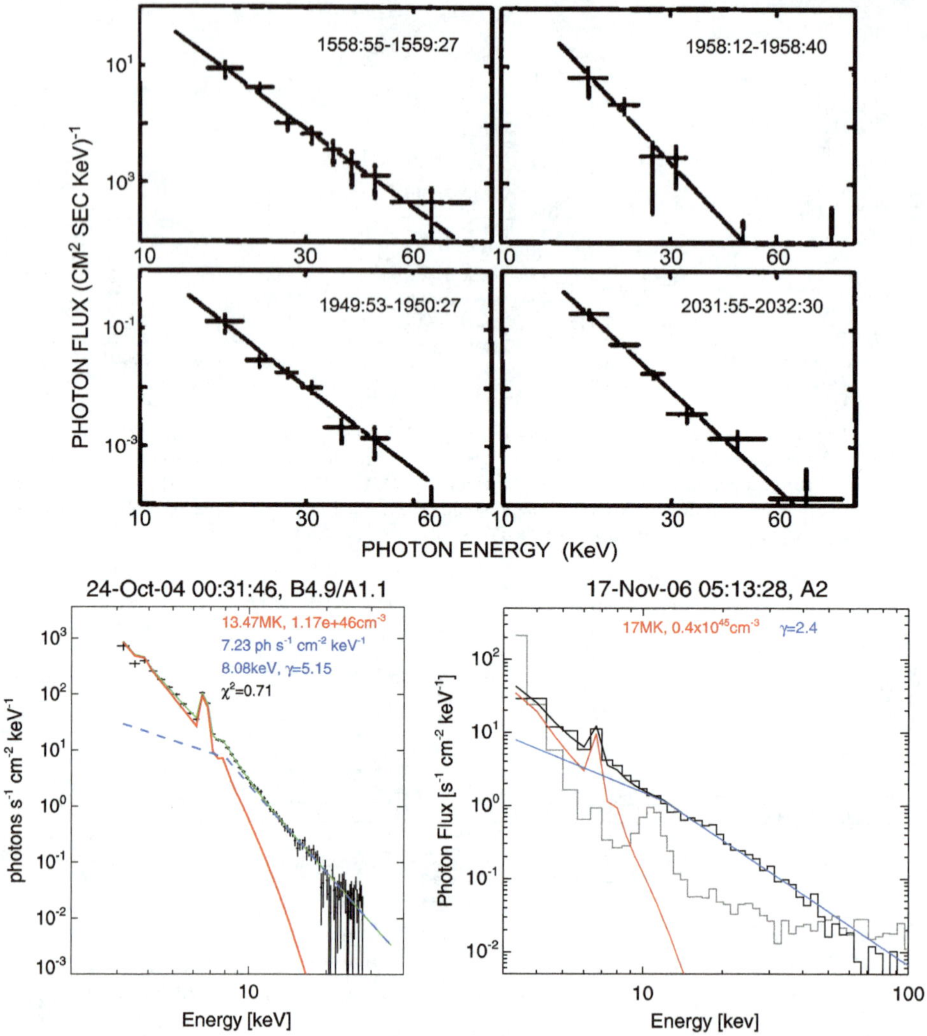

Fig. 7 (*Top four panels*) the first HXR microflare spectra from a balloon-borne instrument (Lin et al. 1984), and (*bottom two panels*) spectra from *RHESSI* showing a typical microflare (Hannah et al. 2008a) (*left*) and one with a particularly hard spectrum (*right*) (Hannah et al. 2008b). The *red* and *blue lines* are the thermal and nonthermal components of the model, respectively. In the *RHESSI* spectra, the upper limit of the nonthermal component is hidden by the instrumental background (the *large errors bars* in the *bottom-left spectrum* and the *grey histogram* in the *bottom-right spectrum*). Reproduced by permission of the AAS

also demonstrate other temporal phenomena seen in large flares like more gradual emission or pre-impulsive behavior though the timescales are generally shorter (seconds to minutes instead of several to tens of minutes).

The first HXR spectra of microflares (top panel of Fig. 7) showed only a power-law component, characteristic of nonthermal emission, due to the detector being sensitive only above about 13 keV (Lin et al. 1984). *RHESSI* spectra (bottom row of Fig. 7) can be obtained with moderate energy resolution down to a few keV. This allows the thermal component, and the transition from it to the nonthermal component, to be seen. These spectra (from the time of peak emission just after the impulsive phase) are similar to those seen in more energetic flares, except that (a) the thermal emission tends to be at lower temperatures and (b) the nonthermal component tends to have a steeper spectrum (fewer electrons accelerated to high energies). In the bottom left panel of Fig. 7, we have a typical *RHESSI* microflare with a fitted temperature of about 13 MK. The Fe feature (e.g., Phillips 2004) is clearly in evidence at about 6–7 keV and a nonthermal spectrum above 8 keV with an $E^{-5.15}$ power law. The other *RHESSI* spectrum shown (bottom right panel of Fig. 7) is of an anomalously hard microflare, which despite being a *GOES* A2 event (near the limit of the *GOES* sensitivity) demonstrates strong nonthermal emission to relatively high energies, just as in much more powerful events (Hannah et al. 2008b). One should note, however, that the early microflare HXR spectra of Lin et al. (1984) were more sensitive above about 30 keV since their shielded detectors were able to reduce the background clearly evident in the *RHESSI* spectra. These Lin et al. (1984) microflares also demonstrate steep nonthermal emission of $I(\epsilon) \propto \epsilon^{-\gamma}$ with $\gamma = 4$ to 6 that continue to energies normally obscured by background in *RHESSI* observations.

RHESSI microflares are not spatially small and typically show an elongated loop-like structure, with higher-energy HXR emission from the footpoints at the ends of the loop (e.g., Krucker et al. 2002; Liu et al. 2004; Stoiser et al. 2007; Hannah et al. 2008b). Figure 8 shows an example of this overlaid on an EUV image from *TRACE*. We interpret the EUV loop as the cooling post-flare loop which was first seen at higher temperatures in the *RHESSI* image. This hot material could be driven (evaporated) from the chromosphere by the energy deposition of the nonthermal electrons penetrating to the loop footpoints (Fisher et al. 1985; Abbett and Hawley 1999; Brosius and Phillips 2004) and producing HXR sources there. Several *RHESSI* microflares have also been observed with the HXR loop connecting two bright Hα kernels (Liu et al. 2004), another possible signature of the energy deposition of nonthermal electrons in the lower atmosphere. Previous work had shown the association between microflare SXR and Hα emissions (Shimizu et al. 2002), consistent with the relationship known from ordinary flares (Thomas and Teske 1971).

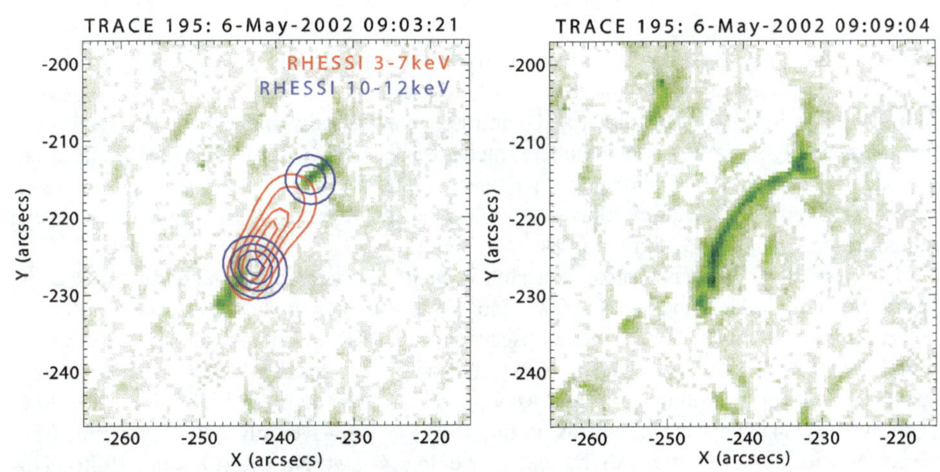

Fig. 8 Overlay of *RHESSI* thermal (*red*) and nonthermal (*blue*) contours on *TRACE*, showing the typical flare and microflare scenario of (*left*) footpoints depositing their energy to heat chromospheric material that expands upwards, filling the loops connecting them. (*Right*) his material then cools, emitting brightly in EUV. Updated version of figure from Krucker et al. (2002)

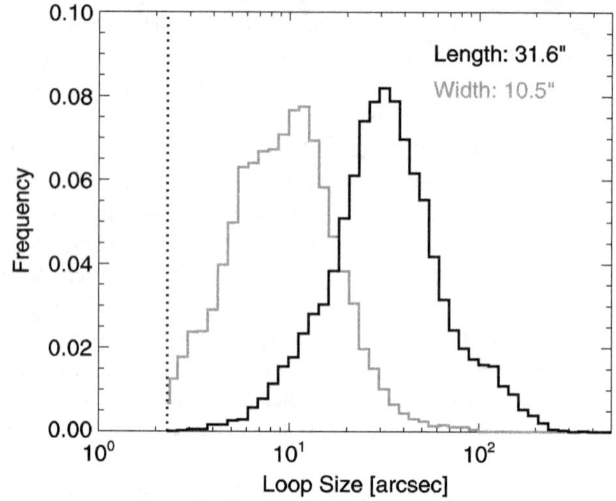

Fig. 9 Loop arc length and width distributions for 18,656 microflares imaged by *RHESSI* indicating that they are predominantly elongated structures (Hannah et al. 2008a). Thus microflares are not necessarily spatially small. Reproduced by permission of the AAS

Figure 9 shows the fitted lengths and widths of the thermal emission (4–8 keV) for 16 seconds about the time of peak emission in 6–12 keV for almost 19,000 *RHESSI* microflares (Hannah et al. 2008a). These parameters are from simple forward-fitted image models and show that the microflares have a clear tendency to be elongated, as we would expect from their identification with the loops that characterize SXR microflares (e.g., Shimizu 1995). The distribution of the loop length has a peak well above *RHESSI*'s angular resolution, so we conclude that these microflares may not be inherently small-scale. Many ordinary flares, though more energetic, have dimensions in this range. In addition, these lengths are not correlated with the microflare peak *GOES* flux or peak *RHESSI* emission (Hannah et al. 2008a).

2.2.3 Associated Ejecta and Radio Emission

Radio signatures generally guide us to the corona, since the opacity (and the minimum height of formation of the emission) increases in height almost monotonically with wavelength (e.g., White et al. 2010). Millimeter and centimeter wavelengths (microwave GHz emission) originate mainly in the lower atmosphere, meter-waves (100s MHz to MHz radio emission) in the lower corona, and the longest wavelengths right out into the solar wind. The longer-wavelength radio observations thus often show ejecta, especially via nonthermal plasma signatures (e.g., Pick and Vilmer 2008).

The microflare gyrosynchrotron emission is closely correlated in time to the HXR emission (Qiu et al. 2004; Kundu et al. 2006), indicating that they are both signatures of the accelerated flare electrons. The previous observations of SXR microflares also demonstrated a correlation with microwave emission (Gary et al. 1997; Nindos et al. 1999). The microwave spectral index for microflares is found to be flatter (harder) than the *RHESSI* HXR spectra (Qiu et al. 2004). This is thought to be due to the microwave emission originating from electrons with higher energies than those producing X-rays (Nitta and Kosugi 1986). This discrepancy in spectral indices is also observed in larger flares (Silva et al. 2000) though it is possibly greater in microflares. This may be due to the HXR spectra being typically steeper in microflares than large flares. A B-class microflare was observed to have nonthermal microwave emission from MeV-energy electrons but showed no HXRs at lower energies

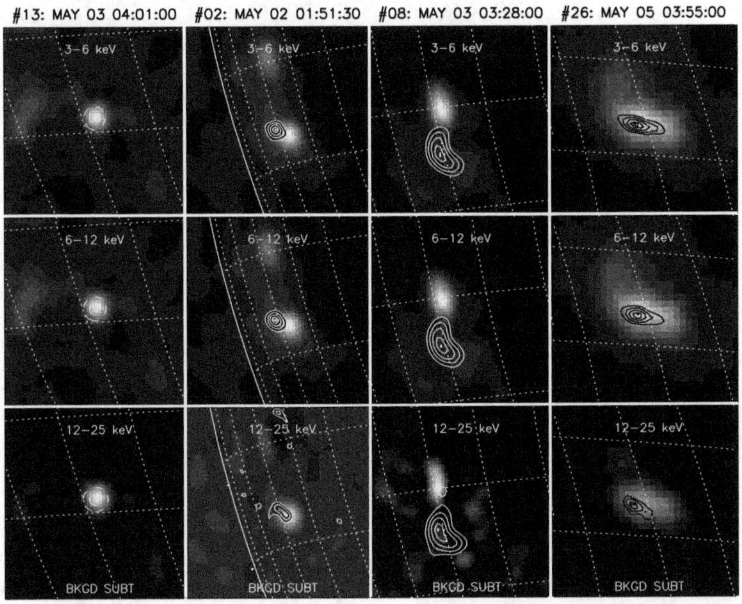

Fig. 10 Four microflares (one per column) imaged in different HXR energy bands (increasing from top row to bottom) with *RHESSI* (contours), overplotted on NoRH 17 GHz images (Kundu et al. 2006). The bottom microwave image is background-subtracted. Similar-sized footpoints and loop structures are seen in both HXRs and microwaves

(Raulin et al. 1999). This behavior could result from trapping of the lower-energy electrons during the peak of the microwave emission from the higher-energy electrons.

Imaged microwave emission using the Nobeyama radioheliograph (NoRH) for *RHESSI* microflares indicated similarly sized and separated footpoints in both HXR and microwave emission (Kundu et al. 2006); see Fig. 10. The higher-frequency microwaves come from the footpoints, and the lower-frequency from the connecting loop (Kundu et al. 2005, 2006).

RHESSI microflares are often associated with Type III radio bursts, produced by electron beams escaping from the corona and often subsequently detected as SEPs. These are the "fast drift" bursts identified with weakly relativistic electron beams (e.g., Wild et al. 1963). The HXR emission in these cases still seems to be a signature of accelerated electrons reaching the lower atmosphere at loop footpoints, although the radio emission comes from accelerated electrons of similar energies moving outwards from a postulated coronal acceleration site. An example of six microflares identified with *RHESSI* and their associated MHz radio emission observed with *WIND*/WAVES is shown in Fig. 11. Here we can see the time correlation of the HXR emission and the high-frequency radio emission. The microflare with the largest HXR flux produced the brightest radio burst. The association with Type III bursts establishes that electrons accelerated in the microflares have access to open field lines (or field lines that extend into the upper corona), not requiring an eruption prior to the event. An association between Type III bursts and small SXR flares had previously been found by Fuerst et al. (1982).

The presence of open field lines in microflares has also been seen with the association of jets of material flowing out of the flare region (Shimojo et al. 1996). For *RHESSI* HXR microflares, such jets have been observed in EUV with *TRACE* (Christe et al. 2008b) and in

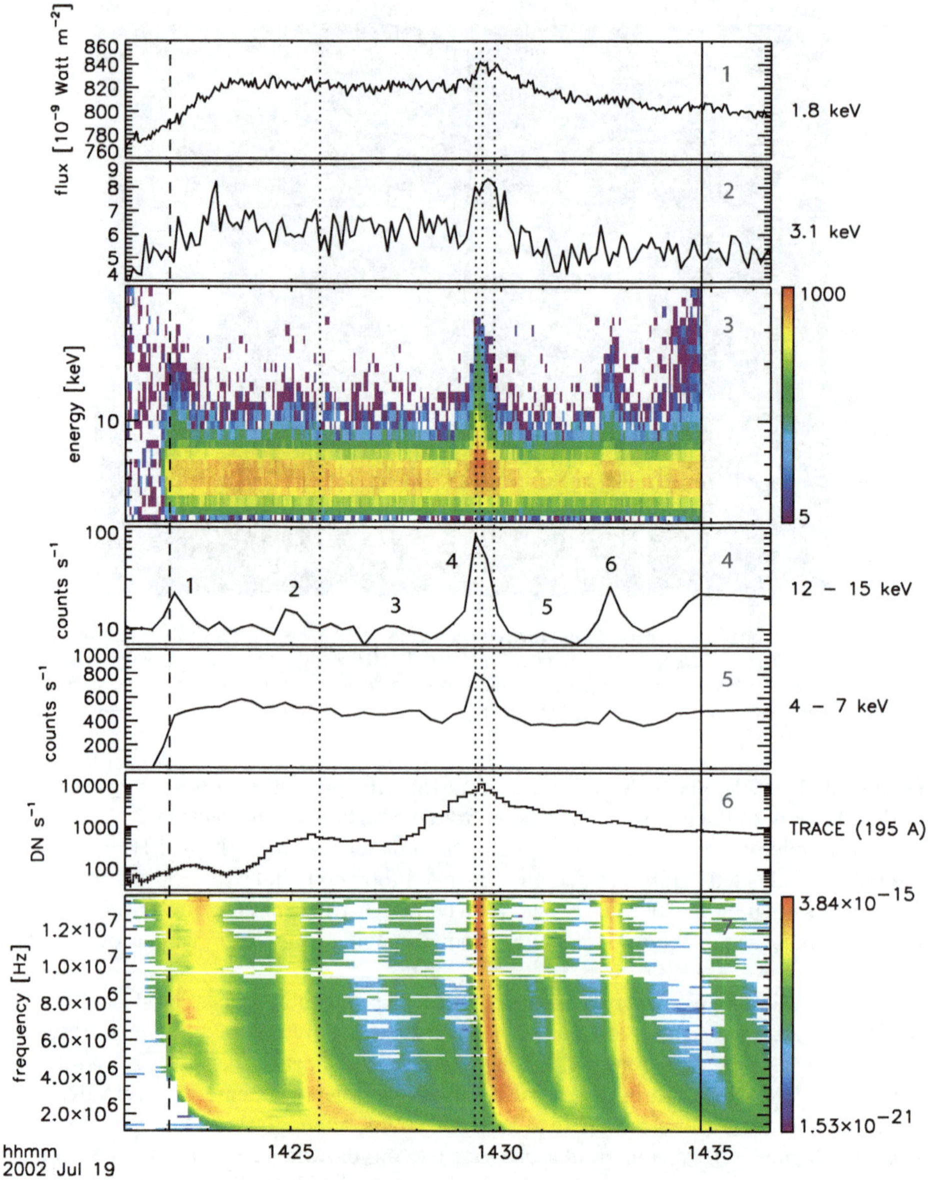

Fig. 11 The lightcurves and radio spectrograms of six *RHESSI* microflares from July 19, 2002 (Christe et al. 2008b). The *top two panels* show the *GOES* lightcurves in 1–8 Å (with a B8 background level) and 0.5–4 Å. The *next three panels* show the *RHESSI* spectrogram with night-time background removed and lightcurves in 12–15 keV (nonthermal) and 4–7 keV (thermal). The EUV lightcurve from *TRACE* and a radio spectrogram from *WIND*/WAVES are shown in the *bottom two panels*. The radio emission starts at around 14 MHz and drifts down to 2 MHz. It is well correlated with the *RHESSI* HXR lightcurve. Reproduced by permission of the AAS

EUV and SXRs with *Hinode* (Chifor et al. 2008). In the former case, multiple microflares and type III bursts were observed, occurring every few minutes, with EUV jets associated with the largest microflares (Christe et al. 2008b). In the latter case, the recurring jets occurred on a timescale of hours and were attributed to chromospheric evaporation flows due to recurring coronal magnetic reconnection (Chifor et al. 2008).

This association with jets is also well observed in soft X-rays (Strong et al. 1992; Shibata et al. 1992). Indeed all such jets, including those in the quiet Sun, appear to have loop brightenings at their base. A jet appears as a collimated flow, implying the prior existence of large-scale or even open magnetic fields. Type III radio bursts, known to be produced by electron beams escaping from the corona into the solar wind, have a strong association with the soft X-ray jets as well (Aurass et al. 1994; Kundu et al. 1995). Thus the microflare/jet events also have thermal and nonthermal attributes just as major flares do, but the jet is relatively more prominent in the microflare domain.

2.3 Nanoflares and Non-active-region Phenomena

We have seen above that the *RHESSI* microflares occur only in active regions and have a strong kinship with "ordinary" flares. How do these relate to X-ray bright points (Golub et al. 1974), network flares (Krucker et al. 1997), plumes (Ahmad and Webb 1978), or any of many more types of weak flare-like brightenings seen both inside and outside the active regions? Note that EUV observations such as those of *SOHO*/EIT or *TRACE* may be much more sensitive than SXR (or HXR) observations of transient features at ordinary coronal temperatures of order 1 MK (Porter et al. 1987; Krucker and Benz 1998). Many of these phenomena have been related to Parker's nanoflares and to the problem of coronal heating. The nanoflare description of these EUV events is justified, in the sense that an EUV brightening may have a much smaller event energy than a microflare observed in SXRs, HXRs, or microwaves. However it is often difficult to place these different sorts of observations on a uniform energy scale. We deal with this at length in Sect. 3.1 below.

Outside of active regions, the most prevalent X-ray features of SXR images are the X-ray bright points (XBPs), originally discovered with rocket-borne X-ray imagers in the late 1960s and studied statistically with data from the X-ray telescope on *Skylab* (Golub et al. 1974). Subsequent observations had sufficient resolution to detect hot (about 2 MK) loop-like (about $10''$ in length) properties and footpoints in the XBPs (Moses et al. 1994; Strong et al. 1992; Kotoku et al. 2007), pointing to a possible relationship with flare physics. Typically hundreds of XBPs are visible, uniformly spread across the solar disk (Golub et al. 1974), their number changing little once observational bias, due to the dominant active region emission, is removed (Hara and Nakakubo-Morimoto 2003). This spatially-uniform and time-invariant distribution across the solar surface is expected as the XBPs are associated with the magnetic fields related to surface convective flows (the granulation and supergranulation of the quiet photosphere) rather than with the active-region fields responsible for sunspot fields. X-ray bright points are associated with connectivity changes in the network magnetic field. The underlying fundamental physical processes could be the same as for active-region microflares. Often, XBPs may persist for hours to days, unlike the transient active region flares, and thus their behavior may resemble that of quiescent active-region loops (e.g., Yoshida and Tsuneta 1996). In the absence of solar active regions, the XBPs and the corona of the quiet Sun provide a basal X-ray emission level. Jets are sometimes observed from XBPs, particularly in polar regions (Shibata et al. 1992; Shimojo et al. 2007; Cirtain et al. 2007; Savcheva et al. 2007) where there is easy access to open field lines.

There have been no HXR observations of XBPs. Such emission therefore must be below the sensitivity limits of current spacecraft, which have been optimized to observe brighter

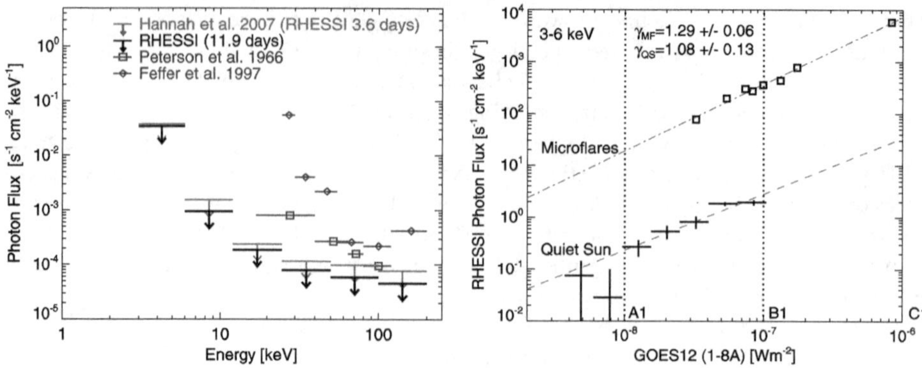

Fig. 12 Observations and upper limits of the quiet-Sun hard X-ray flux found with *RHESSI*. (*Left*) The *RHESSI* upper limits to the quiet Sun emission at 3–200 keV in comparison to previous observations (Hannah et al. 2010), including the first *RHESSI* quiet Sun limits (Hannah et al. 2007b). (*Right*) comparison of the 3–6 keV flux observed with *RHESSI* to the 1–8 Å *GOES* emission for the quiet Sun and offpointed microflares. The γ_{MF} and γ_{QS} indicate the index of power-law fits to the data, the *dashed* and *dashed-dotted lines* (Hannah et al. 2007b). Reproduced by permission of the AAS

active-region flares. Radio emission associated with XBPs has been observed, but in many instances it is consistent with thermal instead of nonthermal emission (Nitta et al. 1992). Network flares, associated with the magnetic network boundaries, are about an order of magnitude fainter than XBPs and are more transient, lasting for only about 10 minutes (Krucker et al. 1997). Radio emission has also been observed in these events, but again in the majority of the cases it is consistent with thermal emission, and in only a few cases could it be associated with nonthermal gyrosynchrotron emission from accelerated electrons.

Determining whether there is nonthermal emission from a population of accelerated electrons outside of active regions and in the quiet Sun would provide important insights into the nature of possible small-scale steady-state energization processes in the solar corona. Although *RHESSI* has unprecedented sensitivity over 3–25 keV, which has greatly aided the study of small active-region flares (see Sect. 2.2), emission outside active regions still remains elusive.

Investigating the non-flaring properties of active regions (McTiernan 2009) or the quiet Sun with *RHESSI* is a non-trivial task. *RHESSI*'s imaging method is designed for bright compact flares and ill-suited for the spatially-diffuse emission expected from coronal heating. Instead, upper limits to the quiet Sun HXR flux can be obtain by offpointing *RHESSI* from the Sun, so that the quiet solar signal is "chopped" as *RHESSI* rotates (Hannah et al. 2007a). This process has obtained limits that are smaller and cover a wider energy range than previously found (Hannah et al. 2007b, 2010), as shown in Fig. 12. A wide range of possible thermal and nonthermal emission is still consistent with these limits with only a hint as to the relationship between active region and non-active region flares. However, the nanoflare model for coronal heating, operating in a manner similar to active-region flares, does seem unlikely (Hannah et al. 2010). Future instruments with higher sensitivity and dynamic range could possibly find such HXR emission, which would be expected from a nanoflare heating model even in the absence of flare-like brightenings (e.g., Cargill and Klimchuk 2004).

3 Flare Distributions & Parameter Scaling

The occurrence distribution function provides one good way to characterize a large number of observations. Many flare distribution functions have been published, and we can separate them into two classes: distributions of raw observable parameters (e.g., peak flux), and distributions of derived parameters (e.g., energy content). A derived parameter such as the total event energy might seem a more physically correct object, but the data manipulations required to get it will necessarily involve uncertain calibrations and "plausible" assumptions about the unobserved parts of the event. To make this latter point more concrete, consider the two *GOES* energy bands (0.5–4 Å and 1–8 Å). These define an isothermal temperature and emission measure, but have little sensitivity at low coronal temperatures (e.g., 1–2 MK). Thus in principle a large emission measure and energy can be masked from view and go unaccounted for, systematically biasing the energy estimate to be a lower limit. In the following we provide a critical review of efforts to understand these different occurrence distribution patterns.

3.1 X-ray Flux Distributions

We summarize the reported X-ray flux and fluence (the time-integrated flux) distributions in Table 2. Most of the power-law indices α fall below the critical value of 2.0, meaning that the energetic events dominate, rather than the microflares (information on how to determine α accurately is detailed in Appendix A). An exception to this is the huge *GOES* event sample of Veronig et al. (2002a), which we showed in Fig. 1, but in this case the authors deliberately did not attempt to correct for background counting rates. Indeed, this is a fundamentally ambiguous procedure, as noted by Bornmann (1990), and a typical source of systematic error. The SXR observations suffer more from this than the HXR observations because of their longer time scales. Determining the fluence also introduces bias: how does one determine the time interval for the integration, and how does one allow for sensitivity variations (either between instruments or as a function of time)?

Most of the entries in Table 2 (see also the representative plots of several distributions in Fig. 13) refer to directly-observed peak fluxes. This is the simplest way to handle the data and one that suffers the least from the introduction of systematic errors due to unknown or imprecise corrections. Uncertainty due to background rates in the detector still remains, as with the intrinsic biases of the sampling. This will have virtually no effect on the largest events but dominate over the smallest ones. If the instrument measures spectra in detail, it is also possible to take one step away from instrumental bias by fitting a spectrum and then evaluating the spectral flux density at a well-measured photon energy, e.g., at 30 keV.

The representative distributions shown in Fig. 13 have several common features: they match power laws well over a certain magnitude range, and they roll off towards higher and lower magnitudes. The roll-off at the low end is either due entirely to the sensitivity limit of the particular instrument, or is heavily confused by selection effects due to this limit. In particular, faint events may be missed as their emission is obscured by brighter simultaneous events. A fit to the distribution function that attempts to characterize this roll-over (e.g., a log-normal or a Weibull distribution; Parnell 2002) probably has little relevance to the physics of the events in the well-observed part of the distribution. At larger magnitudes the deficit also might have systematic errors (the saturation of a given detector would be an obvious one), but at some point a power-law fit for $\alpha < 2$ will diverge unphysically. This implies the existence of an upper limit of some sort (e.g., Lingenfelter and Hudson 1980). Alexander and Daou (2007) analyzed 10 M- and X-class flares, finding that each was

Table 2 Table of the power-law indices α found from statistical surveys of the fluxes of X-ray flares. The top set are predominantly SXR surveys, the bottom HXR surveys

Index α	Energy Range	Duration	Number	Quantity	Instrument & Reference
1.86	7.7–12.5 keV	1967 (2 weeks)	177	Peak	*OSO-3* (Hudson et al. 1969)
1.75	1–6.2 keV[1]	1966–1968	4028	Peak	*Explorer-33/35* (Drake 1971a)
1.44	1–6.2 keV[1]	1966–1968	4028	Fluence	*Explorer-33/35* (Drake 1971a)
1.79	CA XIX	1980–1989		Peak	BCS (Lee et al. 1995)
1.86	1.5–12.4 keV[2]	1980–1989		Peak	*GOES* (Lee et al. 1995)
1.88	1.5–12.4 keV[2]	1993–1995	1054	Peak	*GOES* (Feldman et al. 1997)
2.11	1.5–12.4 keV[2]	1976–2000	49409	Peak	*GOES* (Veronig et al. 2002a)
2.03	1.5–12.4 keV[2]	1976–2000	49409	Fluence	*GOES* (Veronig et al. 2002a)
1.8	20 keV	1971–1972	123	Peak^{ph}	*OSO-7* (Datlowe et al. 1974)
2.0	20 keV	27-06-1980	25	Peak^{ph}	Balloon (Lin et al. 1984)
1.8	>30 keV	1980–1985	>7000	Peak	HXRBS (Dennis 1985)
1.66–1.75	>25 keV	1980–1989	>7000	Peak	BATSE (Schwartz et al. 1992)
1.61	>25 keV	1991	1262	Peak	HXRBS (Schwartz et al. 1992)
1.75	>26 keV	1978–1986	4356	Peak	*ICE* (Lee et al. 1993)
1.70–1.86	>30 keV	1980–1984	3578	Peak	HXRBS (Bai 1993)
1.73	>25 keV	1980–1989	7045	Peak	HXRBS (Crosby et al. 1993)
1.59	>25 keV	1980–1989	2878	Peak^{ph}	HXRBS (Crosby et al. 1993)
1.60–1.74	>25 keV	1991–1994		Peak	*CGRO* (Biesecker a994)
1.86–2.00	>26 keV	1978–1986	3468	Peak^{ph}	*ICE* (Bromund et al. 1995)
1.74	>60 keV	1980–1989	12327	Peak	HXRBS (Kucera et al. 1997)
1.58	>10 keV	1989–1992	1551	Peak	WATCH (Crosby et al. 1998)
1.56	>25 keV	1991–1994	5430	Peak	BATSE (Aschwanden et al. 1998)
1.46	>50 keV	1991–1994	5430	Peak	BATSE (Aschwanden et al. 1998)
1.50	3–6 keV	2002–2007	24097	Peak	*RHESSI* (Christe et al. 2008a)
1.51	6–12 keV	2002–2007	24097	Peak	*RHESSI* (Christe et al. 2008a)
1.58	12–25 keV	2002–2007	24097	Peak	*RHESSI* (Christe et al. 2008a)
1.71	4–8 keV	2002–2007	18656	Peak^{ph}	*RHESSI* (Hannah et al. 2008a)

[1] 2–12 Å, [2] 1–8 Å

The default units for the hard X-ray results are count rate (observed flux) except those indicated by ^{ph} which are nominally instrument-independent photon fluxes

composed of several bursts that saturated above 20 keV, suggesting a saturation of the HXR emission for electron-beams in large flares.

The sensitivity and bias effects (whether instrumental or observational) result in a clear change in the distributions at the extremes, but they can also have a more subtle effect on the power law in the mid-regions. Although the power-law nature of the distribution is not dramatically changed, the value of α can be different (Aschwanden and Charbonneau 2002; Aschwanden and Parnell 2002). Some attempts have been made to correct for these biases in the context of the derived energy distributions and will be discussed further in Sect. 3.5.

The distribution of event energies is obtained by integrating over the spectrum to estimate an energy flux, and then over time. Such conversions are not model-independent. Therefore, the direct flux measurement may be a better guide to the general conclusion from all such

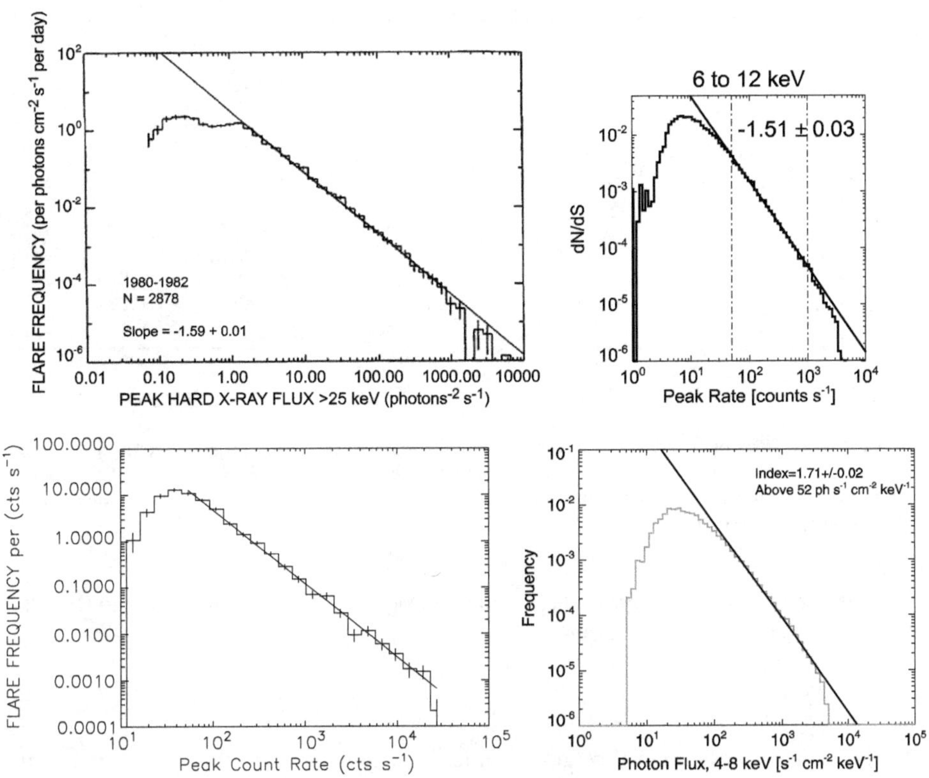

Fig. 13 Several of the flux distributions listed in Table 2 (Crosby et al. 1993, 1998; Christe et al. 2008a; Hannah et al. 2008a) (left to right, top to bottom). Another example (Veronig et al. 2002a) is shown in Fig. 1. Reproduced by permission of the AAS

distributions: *flare occurrence is scale-invariant*. That is, the length scale does not change when multiplied by a common factor, a property of power-law distributions. Solar flares thus have behavior resembling that of earthquakes as described by the Gutenberg-Richter law (Gutenberg and Richter 1956) but how general is it for flares? First, we note that stellar flares, on a variety of stellar types, tend to follow similar distributions (e.g. Shakhovskaia 1989). From region to region, there can be slight variations in the distribution, in particular there being a varying upper energy cutoff (Kucera et al. 1997). This is lower for smaller active regions as there is less free energy, but even in large regions there is a finite amount of energy available (making "super flares" not just infrequent but impossible). An active region from a period of low solar activity has been found to have a frequency distribution which clearly rolls over, this deviation from a power-law being attributed to the low finite energy available (Wheatland 2010).

There can also be variations on intermediate time scales (Bai 1993; Bromund et al. 1995). However, in general, these are not strong deviations from the general pattern. From *RHESSI* we find (Fig. 14, right panel) little evidence for change in the slope of the microflare distribution as a function of phase of the solar cycle, with excellent statistical significance. This is consistent with the results of Veronig et al. (2002a) for *GOES* events (Fig. 14, left panel). The scale-invariant property of the flare occurrence distributions is thought to provide the

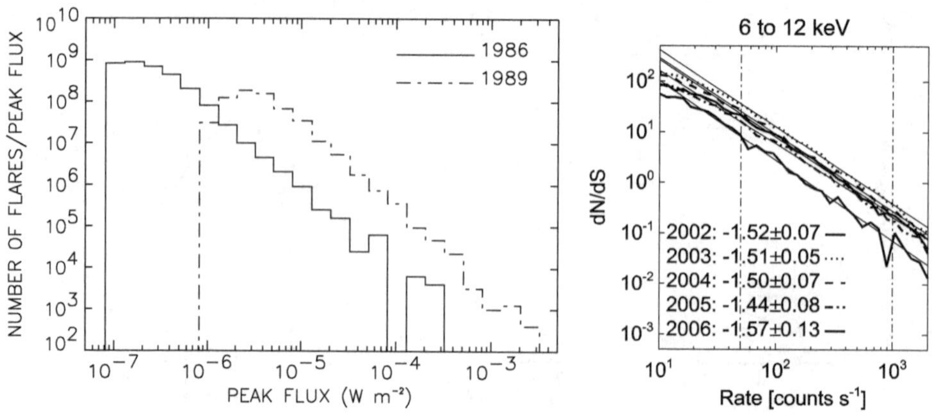

Fig. 14 Comparisons of occurrence distributions with phase in the solar cycle. (*Left*) Veronig et al. (2002a) using *GOES* 1–8 Å data from solar minimum (1986) to maximum (1989); (*right*) Christe et al. (2008a) using *RHESSI* microflare data from leaving solar maximum (2002) through to approaching solar minimum (2006). Reproduced by permission of the AAS

evidence that the coronal magnetic field is in a self-organized critical state (Lu and Hamilton 1991) and this is detailed further in Sect. 3.3.

3.2 Time Distributions

The duration of an event must be known for an estimation of the fluence and therefore also for an estimation of the event energy. Figure 15 shows two examples of duration distributions: from *GOES* (Veronig et al. 2002a) and from *RHESSI* (Christe et al. 2008a). The *GOES* data have a clear power-law falloff to long durations (see also Drake 1971b), whereas the *RHESSI* distributions (shown separately for rise, fall, and total times) have more symmetrical distributions. This is because the *RHESSI* data are considerably more affected by selection bias. Trying to determine the duration of the flare crucially depends on being able to distinguish the flare from the background. The *RHESSI* distribution shown is for microflares (smaller than *GOES* C-class) for which it is difficult to separate their start and end times from the background. The *GOES* distribution is only for B-class events and larger ($>10^{-7}$ W m^{-2}) and the background rate in *GOES* varies less than in *RHESSI*.

Lee et al. (1993) explicitly deal with the simultaneous flux and duration distributions of an HXR data set, that of (*ICE*, formerly *ISEE-3*) above 26 keV (Anderson et al. 1978). This analysis of joint variables requires an explicit consideration of the parameter-space domain as a means of understanding sampling bias. Lee et al. (1993) concluded that a correlation between the duration and peak flux was not present; this would be a requirement if microflares or nanoflares were to outweigh the energy present in the ordinary flares. Thus, the power laws observed in the distributions of peak fluxes most likely provide a guide to the "true" distribution of total event energies.

The other important temporal signature of flares is the time between one flare and the next one, the *waiting-time*. This is generally taken to be the time between the peak emission of one flare and the next. The waiting-time distributions provide information about the probability of a flare occurring. These distributions can be predicted by self-organized criticality (avalanche) models of the coronal magnetic fields, discussed further in Sect. 3.3. These avalanche models have an exponential waiting-time distribution which corresponds

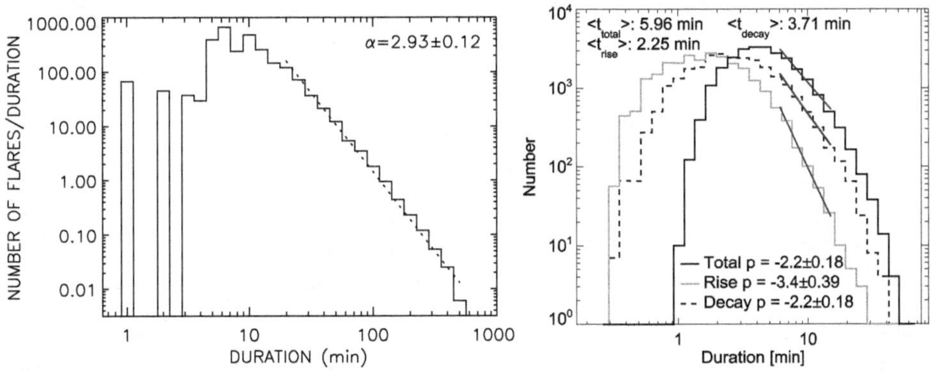

Fig. 15 *Left:* the distribution of the durations of flares *GOES* >B class (Veronig et al. 2002a); *right:* distribution of durations of microflares (Christe et al. 2008a), typically below *GOES* C class. Reproduced by permission of the AAS

to a constant Poisson flare occurrence rate (Wheatland et al. 1998). Such a flare occurrence model produces the desired power-law distribution of energy and duration. Individual active regions demonstrate waiting-time distributions that are either exponentials or the sum of exponentials (Wheatland 2001; Moon et al. 2001). The latter results can be explained by the Poisson occurrence rate varying as the active region crosses the solar disk.

The waiting-time distribution found for large samples of X-ray flares is shown in Fig. 16. The sample of 6,919 flares observed >30 keV with *ICE/ISEE 3* is found to have a waiting-time distribution that is neither power-law nor exponential (Wheatland et al. 1998), shown left in Fig. 16. A larger study of 32,563 *GOES* C-class and above flares demonstrates a power-law tail in its waiting-time distribution over long timescales (Wheatland 2000), shown right in Fig. 16. The index of this power law varies with the solar cycle (Wheatland and Litvinenko 2002), again consistent with a Poisson occurrence with a time-varying rate. An alternative model explains this tail using an occurrence rate with a Lévy distribution (Lepreti et al. 2001) but this also requires a "memory" in the underlying process. This suggests that not only can flare rates be determined but features of the underlying physical processes can be understood, although this is still under debate. The waiting-time distribution of Type III radio bursts from an active region have been found to be consistent with a Poisson process (Eastwood et al. 2010).

An overabundance of short waiting-times compared to simulations has now been found (Wheatland et al. 1998), suggesting that HXR bursts are not independent events. This *sympathetic flaring* behavior has been suggested previously (Fritzova-Svestkova et al. 1976), referring to temporally close flares in different active regions. This may also relate to the misidentification of several peaks within a single flare as multiple events. It is difficult with non-imaging instruments to exclude closely related flares from the same active regions, but statistically significant evidence for sympathetic flaring has been found (Moon et al. 2002). *RHESSI* would be able to provide such information, as well as investigating waiting-times for smaller HXR flares, but the highly discontinuous nature of the data (with gaps due to nighttime, South Atlantic Anomaly passage, etc.) would make such analysis highly subject to selection effects.

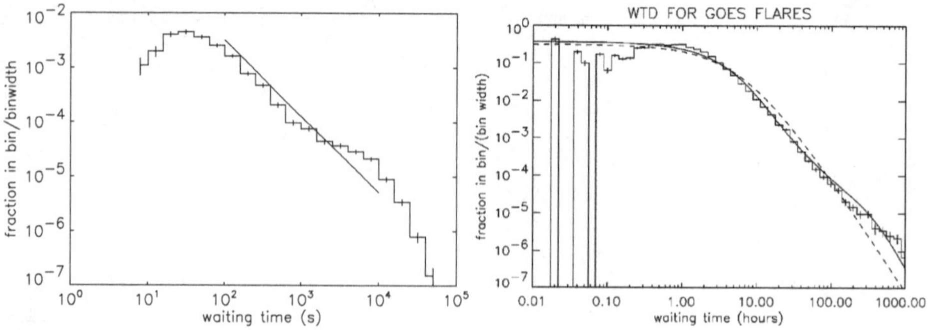

Fig. 16 Waiting-time distributions for different sets of X-ray flares. *Left: ICE/ISEE-3* study of 6919 flares >30 keV (Wheatland et al. 1998), the overplotted *solid line* is a power-law fit. *Right:* 32,563 *GOES* flares above C-class (Wheatland 2000), the overplotted lines indicate a model for a time-dependent Poisson process (*solid*) and independent events with an exponential distribution of rates (*dashed*). Reproduced by permission of the AAS

3.3 Origin of Power-Law Distribution

An early analytic model for the statistics of solar-flare occurrence assumed that the available energy grew exponentially with time and would be released as a flare with a Poisson probability distribution (Rosner and Vaiana 1978). This results in a power-law frequency distribution. This problem was later recast as a steady-state transport equation, allowing the inclusion of an arbitrary energy resupply rate (Litvinenko 1994). These models assume that all of the free energy is released in a flare, as in a relaxation oscillator. Such behavior has never been found in the occurrence patterns of solar flares, although there are other astrophysical contexts in which it has (e.g., Lewin et al. 1976). Wheatland and Glukhov (1998) and Wheatland (2008) generalized these models to take into account that the flares do not release all of the free energy by finding a "master equation" to describe the system. This model again produces a power-law flare frequency distribution as well as a high-energy turnover.

A complementary view is that the scale-invariant behavior of the flare distributions implies that the system dynamics can be described as variations around a self-organized critical state (Lu and Hamilton 1991; Lu et al. 1993). In the standard view, magnetic energy builds up in complicated and stressed magnetic field structures in the corona due to the motion and emergence of magnetic fields through the photosphere. Eventually, the coronal structure loses equilibrium catastrophically, and its restructuring (usually thought to involve magnetic reconnection) suddenly liberates some of the built-up magnetic energy. This newly released energy goes into the various forms we observe in a solar flare. In the scenario of the self-organized critical state, the instabilities are spontaneous, independent of the history of how the energy accumulated, and directly trigger a cascade of energy releases, ending in a temporarily stable state. Therefore, the same process could easily produce events at the nanoflare level as well as the more energetic flares. The avalanche-like behavior of such a system is frequently described using an idealised "cellular automaton" model (Katz 1986; Bak et al. 1987). Such a model does not depend upon the actual physical mechanisms involved at the microscopic level, replacing them with a schematic set of *ad-hoc* rules for the system evolution. The links between the physics of such a model and its statistical description remain unclear, though taking a steady-state energy release in the "master equation" is similar to assuming an underlying avalanche process (Wheatland and Glukhov 1998). Nevertheless an extensive literature applying cellular automata to solar flares has arisen (e.g., Charbonneau et al. 2001).

3.4 Nonthermal and Thermal Spectral Parameters

The *RHESSI* X-ray spectra bridge the thermal and nonthermal spectral domains, at the same time providing better spectral resolution than any earlier HXR imager. The nonthermal emission of the impulsive phase is often characterized by brief spikes of emission with soft-hard-soft (steep to flat to steep) spectral evolution, and its simplest characterization is a power law in energy. The SXR thermal emission roughly follows the time integral of the HXR impulsive-phase emission, following the Neupert effect (Neupert 1968). This empirical result is thought to demonstrate that the total nonthermal energy deposited heats the chromospheric plasma, driving it up into flaring loops, producing bright thermal emission.

3.4.1 Nonthermal

As regards energetics, the HXR parameter of interest is the slope of the power-law fit to the photon spectrum in the impulsive phase, namely γ in $I(\epsilon) \propto \epsilon^{-\gamma}$. This spectrum gives us the best possible guide to the behavior of the energy in the impulsive-phase electrons, which can be highly significant (Brown 1971; Lin and Hudson 1976). Figure 17 shows the distributions of the power-law index γ from several data sets. Each shows clearly distinguishable selection effects. The left panel of Fig. 17 displays the spectral index of large flares above 30 keV as found from *ICE/ISEE-3* (pre-*RHESSI*). *RHESSI* observations of large flares indicate similar spectral indices, with occulted flares (those with their footpoints hidden behind the limb so the HXRs are coronal in origin) having systematically softer spectra (Krucker and Lin 2008) as seen in the middle panel of Fig. 17. *RHESSI* microflares (Fig. 17, right panel) show many much steeper spectra which are often difficult to distinguish from the thermal emission (Hannah et al. 2008a). Kašparová et al. (2007) were able to isolate one important source of systematic error, namely the albedo resulting from X-rays backscattered off the photosphere (Santangelo et al. 1973; Kontar et al. 2010). The albedo correction affects mainly low energies, ~20 keV. The result of the correction is a steeper photon spectrum at those energies, as shown in detail in Fig. 18. *RHESSI* allows imaging spectroscopy of individual HXR footpoints and the difference in spectral index between footpoints in large flares (above *GOES* M-class) with two well-resolved footpoints ranges between 0 to 0.6 (Saint-Hilaire et al. 2008).

The essential problem in the interpretation of the hard X-ray spectrum lies in the behavior of the bremsstrahlung cross-section (e.g., Brown 1971). The photon spectrum is

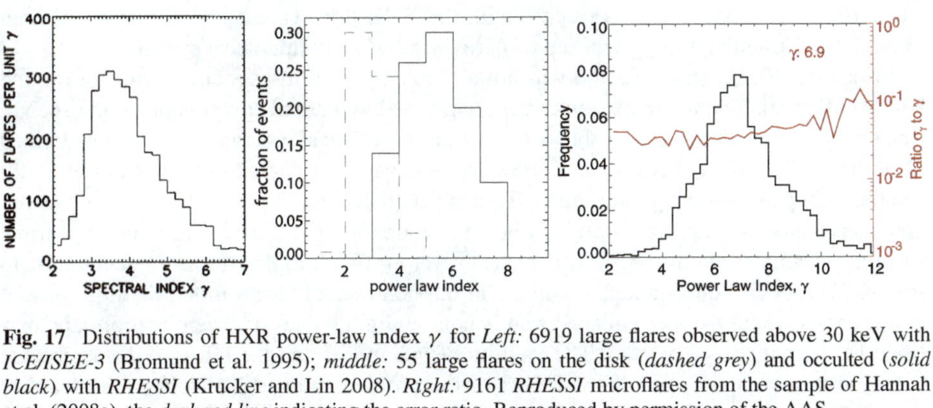

Fig. 17 Distributions of HXR power-law index γ for *Left:* 6919 large flares observed above 30 keV with *ICE/ISEE-3* (Bromund et al. 1995); *middle:* 55 large flares on the disk (*dashed grey*) and occulted (*solid black*) with *RHESSI* (Krucker and Lin 2008). *Right:* 9161 *RHESSI* microflares from the sample of Hannah et al. (2008a), the *dark red line* indicating the error ratio. Reproduced by permission of the AAS

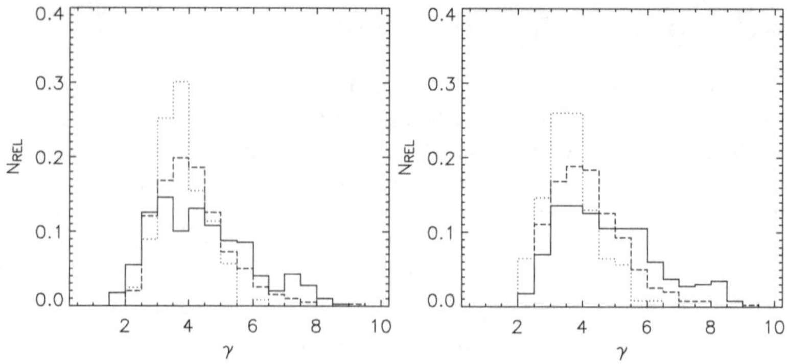

Fig. 18 The effect of albedo correction on the observed HXR power-law index, with histograms of γ (*left*) found from spectrum fitting and (*right*) then corrected for albedo (Kašparová et al. 2007). The different lines show the power-law fits made over different energy ranges: 398 flares with γ_0 (*solid*) 15–20 keV and γ_1 (*dashed*) 20–35 keV, and 123 flares with γ_2 (*dotted*) 35–50 keV

an integral over the electron spectrum at all higher energies. This smearing effect means that, in practice, it is difficult to determine the parent electron spectrum at low energies, and yet these lower-energy electrons contain most of the energy supplied to the flaring plasma via collisions. Analytically, the electron spectrum can be assumed to exist above some "cutoff" energy E_C, with means the accelerated electron population appears abruptly at energies where the thermal spectrum is negligible. This results in a flattening of the expected photon spectra, with a "break" energy ϵ_B occurring at photon energies below the electron cutoff energy (Kosugi et al. 1988). Although the spectral indices of the source electron distributions and observed photon spectra are related in a simple way (Brown 1971; Kontar et al. 2010), this is not the case for the photon break and electron cutoff energies. There are large and ill-defined uncertainties in the inverse or forward deconvolutions of the HXR spectrum for these parameters.

Sui et al. (2007) discuss the behavior of the break energy in a selection of "early impulsive" events, which have less confusion between thermal and nonthermal components. They found that 9 out of 33 early impulsive flares demonstrated spectral flattening in the X-ray spectrum towards low energies, a signature of a low-energy cutoff in the electron spectrum. In three of these events, albedo correction removed the flattening, but in the remaining six, the flattening was found to be consistent with forward-fitting a low-energy cutoff in the range 15–50 keV, which also correlates with the HXR flux. This is at lower energies than found from fitting the X-ray spectrum of flares without early impulsive emission.

Prior to *RHESSI* this confusion was not a problem, as an instrumental cutoff of around 20–30 keV meant that the transition of the nonthermal to thermal components of the spectra was rarely observed. Figure 19 shows two examples of break-energy distributions derived from the *RHESSI* data, the left panel showing large occulted flares (Krucker and Lin 2008) and the right panel showing microflares (Hannah et al. 2008a). For microflares, the observed break tends down to very low energies where there are multiple emission lines in the thermal spectra, as can be seen in Fig. 7. This makes it very difficult to determine ϵ_B accurately. In addition, a model requiring a sharp cutoff in the nonthermal electron distribution down at low energies where the thermal spectrum is non-negligible does not seem appropriate, as a smoother transition is expected. The validity of this model and the effort to determine the energy in a microflare's nonthermal electrons is problematic and is discussed in detail in Sect. 3.5.

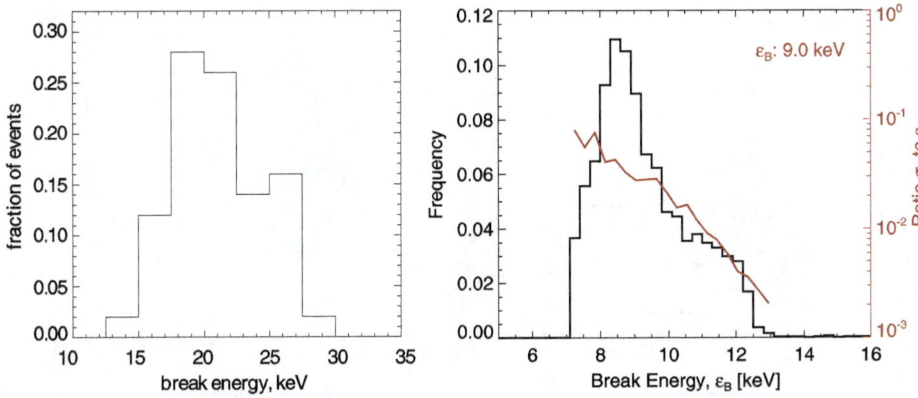

Fig. 19 Two examples of low-energy break energy distributions from fitting *RHESSI* photon flare spectra, (*left*) for 55 large occulted flares (Krucker and Lin 2008) and (*right*) 4236 microflares (Hannah et al. 2008a), the *dark red line* indicating the error ratio. Reproduced by permission of the AAS

Fig. 20 Histogram of the measured local minimum of the *dip* in the mean electron flux spectrum for a sample of 18 flares (Kontar et al. 2008). These dips can be removed by correcting for albedo and do not represent a low-energy cutoff

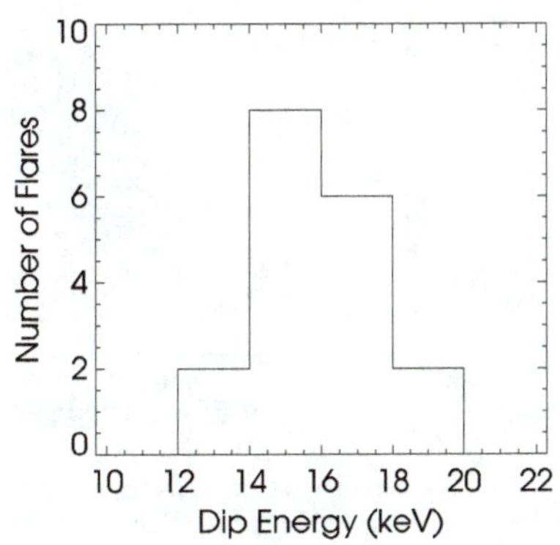

Instead of trying to fit models to the photon spectrum based upon the expected analytical derivation from the electron distribution, Kontar et al. (2008) inverted the photon spectrum directly to obtain the mean electron spectrum for several flares. They found that there was no sharp cutoff in the electron distribution, but instead a dip between 12–20 keV. A histogram of these dip energies is shown in Fig. 20. Making albedo corrections to these spectra completely removed the dip, suggesting that there is a smooth transition between the thermal and nonthermal emissions. Their study also suggests that if low-energy cutoffs exist in the mean electron spectra of these flares, they should be located at energies less than $\sim$12 keV. This implies that applying a cutoff model is only appropriate when considering an electron population well away from the thermal distribution.

3.4.2 Thermal

For the basic parameters of the thermal distribution, the present standard approach is to make an isothermal fit to a model spectrum derived from the Chianti atomic-physics database (Landi and Phillips 2006), using standard assumptions about abundances and ionization states. The isothermal fit determines an effective temperature T and emission measure n^2V. Figure 21 shows comparable regions of this parameter space for microflares observed with

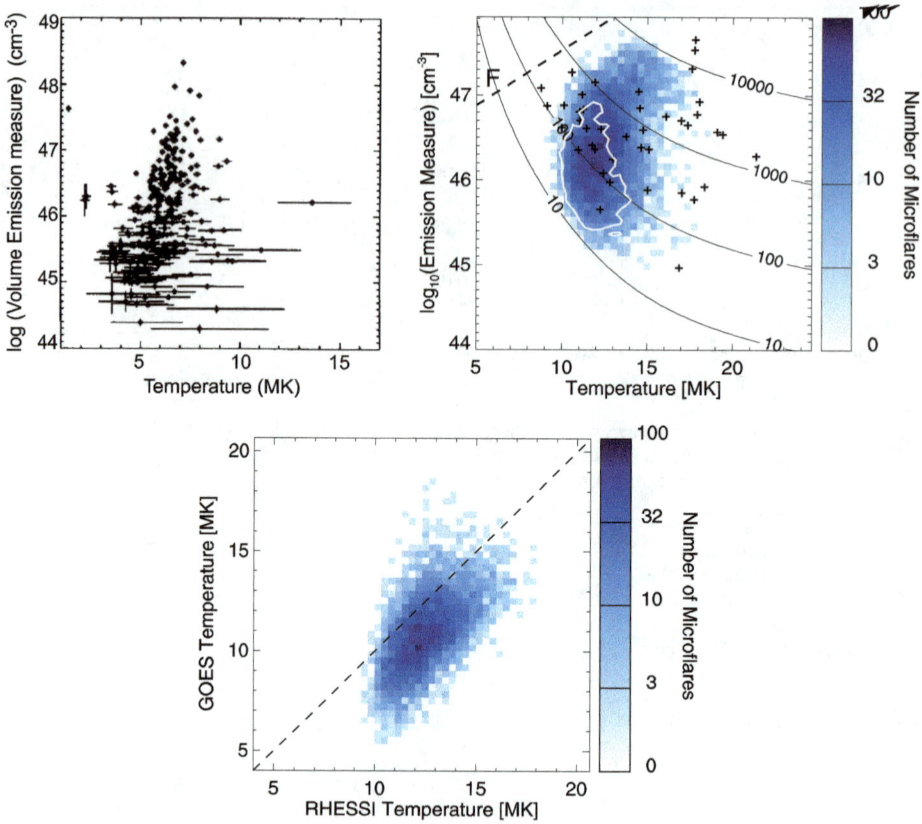

Fig. 21 Thermal parameter surveys of microflares. (*Top left*) 291 *Yohkoh*/SXT microflares (Shimizu 1995) (his "active region transient brightenings"). Reproduced by permission of the PASJ. (*Top right*) 9161 *RHESSI* microflares (Hannah et al. 2008a). The numbered solid lines show expected 4–8 keV *RHESSI* counting-rate levels as a function of model T and EM, the white contour shows the events with *GOES* T < 10 MK from the bottom panel and the dashed line (F) the relationship between emission measure and temperature found by (Feldman et al. 1996); see Fig. 22. Note that neither of these studies individually confirms this correlation, and that the two data sets are almost disjoint. (*Bottom*) The microflare temperature derived using both *RHESSI* and *GOES* for 6740 microflares (Hannah et al. 2008a). Reproduced by permission of the AAS

the *Yohkoh*/SXT grazing-incidence telescope, at energies below about 2 keV (Shimizu 1995), top left, and by *RHESSI* at energies around 6–12 keV (Hannah et al. 2008a), top right. The two samples, though taken at different times, arguably represent the same class of events, and yet the sets of points are almost disjoint. This illustrates the effects of experimental bias, in that the isothermal approximation made for each instrument will produce different weightings of the full DEM (differential emission measure) distribution. This can be further seen in the bottom panel of Fig. 21 where the temperature has been derived for the same microflares at the same time using *RHESSI* and *GOES* separately. Clearly each instrument is responding to different parts of a DEM distribution.

Extending this to large flares, the thermal parameters continue to correlate in the sense that brighter flares have higher temperatures, but with a relatively slow growth of temperature with emission measure. Feldman et al. (1996) found it to extend over three decades of flare magnitude and a factor of five in temperature (5 to 25 MK) when using *GOES* emission

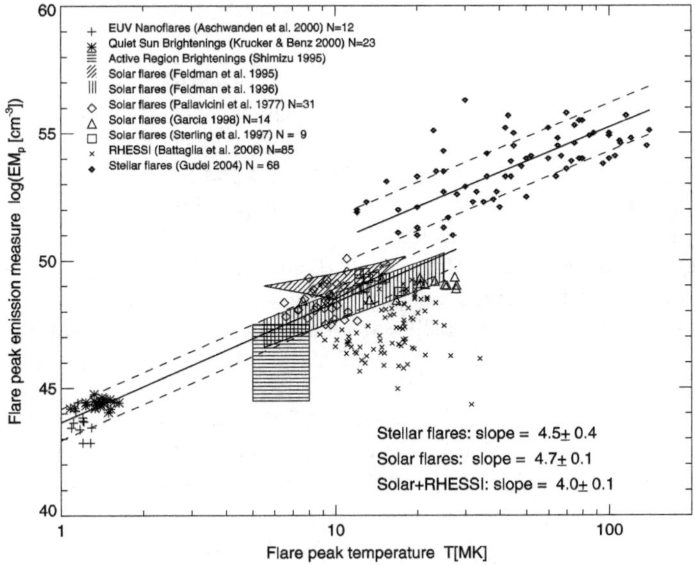

Fig. 22 Thermal fits from various surveys showing the correlation of emission measure (log scale) with temperature of solar flare observations compared with observations of stellar flares and with quiet-Sun brightenings, from the sources shown (Aschwanden et al. 2008). Reproduced by permission of the AAS

measure and *Yohkoh*/BCS[2] peak temperature estimates of 868 A- to X-class flares. But this correlation cleanly misses the *RHESSI* microflare points, as shown by the dashed line in Fig. 21; these tend to be much hotter than predicted. The fit misses the SXT parameter space in a similar way. We presume these two samples represent the same physical objects, but with different observational bias. Kay et al. (2003) found a similar correlation to Feldman et al. (1996) using *GOES* emission measure and temperature of 89 B- to X-class events. Analysis of super-hot flares (>30 MK) with *RHESSI*, 37 M- and X-class flares, is able to extend parameter space up to 50 MK. The *GOES* emission of these events still correlates in a similar manner to temperature as in the Feldman et al. (1996) survey, but it produces a somewhat flatter correlation in linear-log space (Caspi and Lin 2010).

For individual studies over narrow ranges, there appears to be a poor correlation between temperature and emission measure, shown in Fig. 22. A possible correlation appears if many studies over a wider range are considered (Aschwanden et al. 2008) incorporating even stellar flares and quiet-Sun brightenings. Taken together one seemingly obtains a definite correlation from sample to sample, if not within a given data set. However, this may be force-fitting a single concept to different things; a single power-law fit to all of the points would be described approximately by $EM \propto T^7$, but if one ignores the quiet-Sun events one might prefer a much steeper relationship such as $EM \propto T^{15}$. Given the strong systematic biases among the different kinds of observation represented, perhaps linked only by the word "flare," it is no doubt premature to draw any strong conclusions and caution is required when making such comparisons.

[2] Bent (or Bragg) Crystal Spectrometer.

3.4.3 Thermal-Nonthermal Relationships

For major flares we note that the impulsive-phase nonthermal signatures often have a simple physical relationship with the gradual thermal signatures, namely the Neupert effect where the time integral of the HXR (and microwave) emission is empirically seen to match the SXR time profile (Neupert 1968; Hudson 1991a; Dennis and Zarro 1993). Does this relationship extend into the microflare domain as well? Figure 23 shows how well the thermal and nonthermal signatures relate. The *GOES* peak flux and HXRBS (Hard X-Ray Burst Spectrometer on the *Solar Maximum Mission, SMM*) fluences show a large scatter for weaker events, but for the better-observed large energetic events approach a proportional relationship with little scatter (Veronig et al. 2002b). One would expect greater scatter in the fainter events purely due to selection effects, instrumental sensitivity and analysis procedure (e.g., background subtraction performed).

For *RHESSI*, a similar effect can be observed (Battaglia et al. 2005; Stoiser et al. 2007), as shown in the lower panel of Fig. 23, although here the fluxes are directly compared, rather than the SXR flux and HXR fluence. Also shown is one of the most energetic stellar

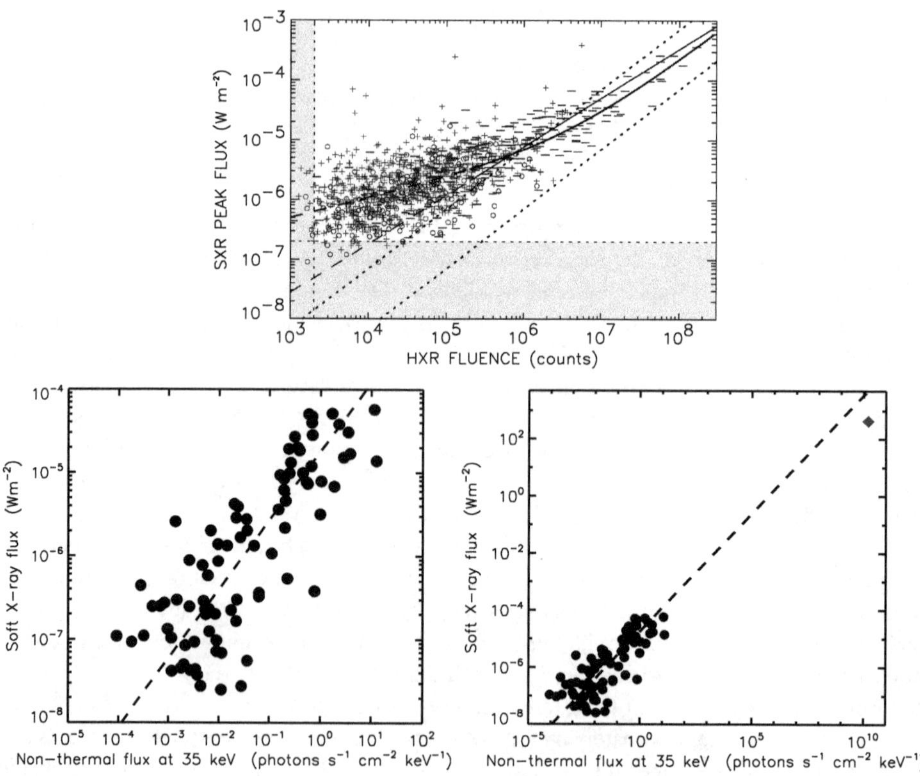

Fig. 23 Comparisons of thermal and nonthermal fluxes for ordinary flares. *Upper:* background-subtracted SXR peak flux from *GOES* vs. HXR fluence from HXRBS (Veronig et al. 2002b). The *solid lines* show various functional fits, the *dashed lines* show proportionality, with extension to smaller fluence, and the *grey areas* indicate regions below instrumental threshold. The different plot symbols indicate the time difference between SXR and HXR peaks. *Lower:* a representative sample of *RHESSI* flares (Battaglia et al. 2005), plus the same sample but including the enormous stellar flare observed by Osten et al. (2007) (the *red point*). Again, the *dashed line* shows proportionality

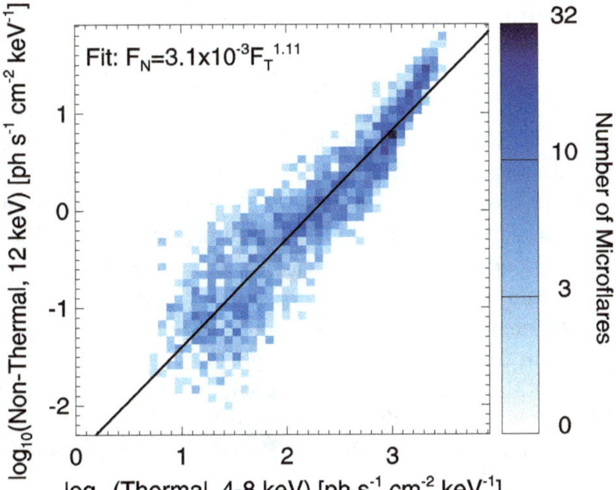

Fig. 24 Thermal/nothermal correlation for 4236 *RHESSI* microflares (Hannah et al. 2008a). The nonthermal flux is the estimated 12 keV emission using the fitted nonthermal model. The thermal flux is found from fitting the X-ray visibilities over 4–8 keV. Note that the axes are inverted relative to those of Fig. 23, but that the line of proportionality is the same. Reproduced by permission of the AAS

flares observed (Osten et al. 2007), over six orders of magnitude brighter in SXRs than the largest solar flare. Here again the SXR and HXR fluxes scaled together as in the "big flare syndrome." This relationship appears to extend into the microflare domain. Figure 24 shows that it does, using the large sample of *RHESSI* microflares studied by Hannah et al. (2008a) and Christe et al. (2008a). Note that the deviation from the trend line for the largest events is instrumental and due to increased detector deadtime before *RHESSI*'s thin shutter deploys. In general, Fig. 24 provides additional strong evidence that the microflares simply represent an extension, to lower energies, of the same physical processes at work in flares. Aschwanden (2007) has pointed out that such a relationship depends upon the scaling of the event environment; at some point we would expect the weakest and the most powerful events to show some kind of deviation from this behavior—for example, low-altitude loops might have shorter coronal cooling times and therefore depart systematically from the observed thermal/nonthermal correlation.

Detailed quantitative analysis of flare thermal and nonthermal emission suggests that the underlying mechanisms are more complicated than empirically shown via the Neupert effect. The differential emission measures (DEMs) for 80 flares were studied with the *Yohkoh*/SXT and BCS instruments with the finding that the high-temperature plasma ($>$16.5 MK) is more likely to demonstrate the Neupert effect than lower temperature plasma (McTiernan et al. 1999). Given that the Neupert effect is thought to show that the accelerated electrons are responsible for heating the chromospheric plasma, Veronig et al. (2005) investigated the similarity of the power in the electron beam compared to the power required to produce the observed SXRs. They expected the powers to be better correlated than the HXR fluence and SXR flux but found a similar correlation. One possibility for this could be that the heating of SXR-emitting plasma is not solely due to the electron beam that produces the HXR emission.

3.5 Energy Distributions

As previously introduced, the thermal and nonthermal energies of flares cannot be determined directly from observations but have to be inferred using model assumptions. The ambiguities of these models and the errors and bias on the observational parameters result in large uncertainties in the energy, and hence in the energy distribution.

The thermal energy U_{th} can be estimated via

$$U_{\mathrm{th}} = 3n_{\mathrm{e}}k_{\mathrm{B}}TV = 3k_{\mathrm{B}}T\sqrt{EM \cdot V_{\mathrm{obs}}f}, \tag{3.1}$$

where V is the volume of the emitting plasma and f is the filling factor (the ratio of the actual volume to the measured value, such that $V = fV_{\mathrm{obs}}$, where n_{e} is the plasma density, T is the temperature, and $EM = n_{\mathrm{e}}^2V$ is the emission measure). The temperature and emission measure can be found through fitting an isothermal model spectrum to either the flare spectrum or the ratio of images in multiple wavelengths. The volume is estimated from the 2-D area of emission from the images, with some model assumption to convert to a volume. The thermal energy (given by (3.1)) is an upper limit estimate over the time of the observed spectrum and images, since a filling factor of $f = 1$ is normally assumed (see, e.g., Fletcher et al. 2011), though in reality it could be considerably smaller, 10^{-3} to 10^{-2} (Cargill and Klimchuk 1997; Takahashi and Watanabe 2000).

The power in the nonthermal electrons of energies above E_{C} (in keV) is estimated via

$$P_{\mathrm{N}}(> E_{\mathrm{C}}) = 9.5 \times 10^{24}\gamma^2(\gamma - 1)\beta(\gamma - 0.5, 1.5)\, E_{\mathrm{C}}^{1-\gamma} I_0 \text{ erg s}^{-1}, \tag{3.2}$$

where γ and I_0 are the index and normalization of the power law in the observed photon spectrum (in units of photon spectral flux) and $\beta(m, n)$ is the beta function (Brown 1971; Lin 1974). To estimate the energy in the nonthermal electrons U_{n} the total duration of the HXR emission needs to be known. The standard "cold thick-target model" model assumes a chromospheric thick target in which a beam of electrons stops, such that the thermal energy of the ambient electrons is negligible compared to the energies of the electrons emitting the HXR.

RHESSI crucially allows the thermal and nonthermal energies to be determined in a flare using a single instrument, due to the energy range it covers and to its imaging and spectroscopic capabilities. Studies of several large flares (C-, M- and X-class) have found that the thermal and nonthermal energies are the same, to within an order of magnitude (e.g., Holman et al. 2003; Saint-Hilaire and Benz 2005). This suggests that the conversion of the energy in the accelerated particles into thermal plasma energy is highly efficient. For C- and M-class events this is typically about 10^{30} erg (Saint-Hilaire and Benz 2005), and at minimum about 10^{31} erg for large X-class events (Holman et al. 2003). Another study found that the power required for the white-light emission was comparable to the electron beam power required to produce the HXR emission, but only if the cutoff was less than 25 keV (Fletcher et al. 2007).

For microflares, the situation is more complicated, given the difficulty in determining the properties of the photon spectrum and the uncertainty this leads to in estimating the break energy. The first detailed analysis on individual *RHESSI* microflares (A- and B-class flares) found that the steepness of their spectra results in considerable power in the these low-energy electrons, but only estimated the energy above 25 keV at 10^{26}–10^{27} erg (Krucker et al. 2002). In a larger study of microflares, where the observed break found in the photon spectrum was used as the cutoff energy, the nonthermal energy ranged over 10^{28-31} erg (Stoiser et al. 2007). This gives an overestimate in the energy since $\epsilon_B < E_{\mathrm{C}}$, and using slightly larger values for E_{C} produced typical nonthermal energies of about 10^{28} erg, similar to the thermal energies found. In the largest *RHESSI* microflare study (Hannah et al. 2008a) the thermal and nonthermal energies at about the time of peak emission over 6–12 keV were found to be similar, to within an order of magnitude, once an empirical correction factor had been used to convert the measured ϵ_B to E_{C}. This factor was found by fitting a broken power laws to model thick-target spectra with a range of indices and low energy cutoffs. Further discussion on the calculation of flare energy and the associated issues is given in Holman et al. (2010).

The distribution of these *RHESSI* microflare values is shown in Fig. 2, in comparison to previous HXR nonthermal energy distributions (Crosby et al. 1993; Lin et al. 2001a), microflares from the SXR thermal distribution (Shimizu 1995) and the thermal distribution of EUV nanoflares (Parnell and Jupp 2000; Aschwanden et al. 2000; Benz and Krucker 2002). Although such a figure nicely shows the scaling of nanoflares through microflares to large flares, and is crucial for determining coronal heating, it is dangerously deceptive. Each of these distributions was found for a different type of event, using various instruments and for different periods during the solar cycle, and so each will be affected by different types of selection effect and biases. For instance, the SXT energies (Shimizu 1995) are from 291 brightenings in one active region over five days in 1992 August, whereas the *RHESSI* microflare energies (Hannah et al. 2008a) are from 9161 events taken over five years, March 2002 to March 2007. This is a comparison of probably the same type of event but even then they are still observed at different wavelengths and over different time periods. Extending this comparison to EUV nanoflares is very difficult and may not even be appropriate given that we do not know if these are similar events or completely distinct physical processes. When looking at the actual values of α that these distributions provide, we see the striking observational result that flare peak flux distributions are power laws flatter than $\alpha = 2$, a result that goes back at least to Akabane (1956). This appears to exclude nanoflares as the source of coronal heating (Hudson 1991b). On the other hand, distributions that attempt to reflect the event energies themselves have often found steeper spectra that are consistent with Parker's idea. Parnell and Jupp (2000) found that α ranged from 2.1 to 2.6 for different model assumptions but was consistently in the >2 range. Krucker and Benz (1998) found that α was between 2.3 and 2.6 but later showed that these values and those from other EUV nanoflare studies were highly dependent on the flare selection criteria (Benz and Krucker 2002).

So, could selection effects be responsible for this discrepancy between nanoflare and larger flare distributions? Attempts have been made to recover the intrinsic distributions from the observationally derived ones (Aschwanden et al. 2000; Aschwanden and Parnell 2002) using a technique similar to that applied to Malmquist bias, a threshold selection effect that biases galaxy number counts in cosmology (Hendry and Simmons 1990; Willick 1994). In this situation, the counting statistics are biased because the brighter, more distant galaxies are more likely to be counted than the fainter ones. This Malmquist-bias procedure has been used on the thermal energies derived for EUV nanoflares; it takes the means and covariances of the observed data parameters and iterates them back to the intrinsic unbiased values. To do this requires model assumptions about how the parameters required to calculate the thermal energy, in (3.1), relate to each other, so that the probability distribution of the intrinsic distribution can be analytically described. However, determining how these parameters scale is, in itself, subject to biases and selection effects, as discussed in Sect. 3.4.2. The analysis procedure may then only adjust the scalings and not test whether the parameters actually relate in such a manner. Attempting to recover the intrinsic distributions, free of instrumental and selection effect biases, should certainly be the priority of any study deriving flare distributions, but in practice this does not readily appear to be practicable given that the parameter of greatest interest, the energy, is not directly obtainable from the existing observations. An alternative approach of determining flare energy, which is much less prone to these effects, is to measure the luminous component directly via the resulting change in the total solar irradiance (TSI). This has been done for the X17 event SOL2003-10-28T11:10 using the TIM (Total Irradiance Monitor) instrument on the *Solar*

Irradiance and Climate Experiment (SORCE) (Kopp et al. 2005), which is sensitive to the solar emission from X-rays to far infrared. The estimated total luminous energy for this flare was approximately 5×10^{32} erg (Woods et al. 2004). This technique is only appropriate at present for the largest flares, because of competition from other sources of solar variation.

Given the evidence so far, it does not appear that flare-like events can heat the quiescent corona. Moreover it seems that making the comparison of active-region and quiet-corona flare events is unwise given that these transient events have different populations in appear in different physical environments (Benz and Krucker 2002).

4 Conclusions & Discussion

The *RHESSI* data have allowed us to study microflares effectively while using the same instrumentation for major events. The statistical study of solar flares has great importance in understanding the underlying processes involved in the energy release and subsequent emission we observe in the solar corona. Often, emphasis is placed on detailed multi-wavelength studies of individual events. Such studies have their merits in revealing insights to the processes in these events. However without being able to place the flare in context of other events, it is very difficult to determine whether it is typical or unusual behavior that is being studied. Only by studying events in large numbers with a minimum of selection bias can one really approach an understanding of the general physics.

The discovery that "ordinary" flare physics extends down to the tiniest events observable by *RHESSI* (or by *GOES*) allows us to conclude that such events do not explain coronal heating. Even the smallest events *RHESSI* observes are in the active regions and are flare-like, distinguishable as individual events. The nanoflare hypothesis instead requires an apparently continuous flare population that is many orders of magnitude smaller. They also continue to show the flat peak-flux scaling that puts most of the energy in the most powerful events, rather than the weakest ones. In a sense, this conclusion simply confirms the appearance of the *GOES* data—sometimes flare-dominated, and sometimes showing steady emission. This is inconsistent with the universality of the flat flare power-law distribution. Of course, numerous individually unobservable nanoflares could create the apparently steady emissions.

To make further progress in this field, three crucial things have to happen. Firstly, we need spacecraft that have a higher sensitivity, lower background and wider dynamic range than *RHESSI* while maintaining the energy range covered across both imaging and spectroscopy. This would allow the faintest events in active regions and the quiet Sun, as well as the fainter emission components of large flares, to be confidently analyzed. One such suitable implementation would be a HXR focusing-optics telescope dedicated to solar observations. Second, we must understand better how biased our observations are, and how we can obtain the intrinsic unbiased physics from our observations. This will require a third advance, more sophisticated modeling of these faint events to match improvements in the data and their implications.

Acknowledgements The authors thank the editor (B.R. Dennis) and the two referees for providing many constructive comments that have greatly improved this review. IGH would also like to thank the Glasgow SSH (Solar Self Help) group for their comments. This work is financially supported through NASA contract NAS 5-98033, an STFC rolling grant, and by the European Commission through the SOLAIRE Network (MTRN-CT-2006-035484), and all are gratefully acknowledged. JK acknowledges support from Grant 205/06/P135 of the GA CR and the research plan AV0Z10030501.

Appendix A: Determining Distribution Parameters

For a power-law distribution (1.1), the standard way to determine the index α is to perform a linear fit to the log-log histogram of the data. However this is a highly subjective approach as there is considerable choice as to the "best" bin width and fitting method. An alternative and more objective approach is to estimate the power-law index using the maximum-likelihood method (Crawford et al. 1970; Bai 1993). This approach leads to a remarkably simple calculation on the sample to determine the index above some chosen threshold:

$$\alpha_{\mathrm{m}} = \frac{N}{\sum_{i=0}^{N} \ln(U_i/U_0)} + 1. \tag{A.1}$$

In this example, the energies U are used, where N is the total number of events and U_i is the energy of the ith event normalized by a threshold energy of U_0. The error in this most likely value of α_{m} can be estimated (Wheatland 2004) as

$$\sigma_\alpha = (\alpha_{\mathrm{m}} - 1)N^{-1/2}. \tag{A.2}$$

The observed distribution, however, is often affected by instrumental and selection effects, resulting in a deviation from a power law, so fitting a power law alone would be unwise. A common problem is that the smallest events are hard to detect and analyze successfully, resulting in a flattening of the power-law for these missing smallest events. To fit this observed biased distribution a skew-Laplace distribution can be used instead of a single power law (Parnell and Jupp 2000). This fits the desired power-law distribution for the larger events, but below some critical value, a power law with a different index is fitted to the under-reported smallest events. The determination of the parameters of this skew-Laplace distribution again can be found using a maximum likelihood estimation method (Parnell and Jupp 2000).

Another possible distribution to describe the observed sample of data is the Weibull distribution (Parnell 2002). This distribution has the form

$$f(x;\kappa,\xi) = C\,(x/\xi)^{(\kappa-1)} \exp[-(x/\xi)^{\kappa}], \tag{A.3}$$

where κ is the shape parameter and ξ is the scale parameter. For a shape parameter $\kappa < 1$, the resulting distribution is similar to a power law, but it turns over at the smallest and largest events. This can then represent deficiencies in the smallest and largest events. Again, the under-reporting of the smallest events is likely to arise from instrumental sensitivity and selection effect bias. The largest events may be missing due to the limited dynamic range of the instrument, if these events saturate the detector; or, there might actually be fewer of these events if a critical physical upper limit is being reached (for instance maximum energy available in an active region (Kucera et al. 1997). Such a feature is consistent with the predictions of avalanche models (Lu et al. 1993), as discussed in Sect. 3.3. The parameters of this Weibull distribution can be determined again using the maximum likelihood method (Parnell 2002).

As there are several distributions that could successfully fit the data, a statistical test is needed to determine which is best. Such a test is the Kolmogorov-Smirnov statistic (KS), which is the maximum difference between the cumulative distribution function (CDF) and the empirical distribution function (EDF) (Press et al. 1992). For a data set of, energies U, the CDF for the ith energy U_i is the integral of the PDF, using the fitted parameters, to U_i. The EDF is derived from the observed/calculated parameters and in this example is the number of events with energy less than or equal to U_i, which turns out to be $(i - 1/2)/N$. The

KS statistic then provides a measure of the significance level of each distribution (Press et al. 1992). Plotting CDF versus EDF provides a graphical way of determining how consistently the data belong to the chosen distribution. A graph similar to the familiar histogram can be obtained by plotting 1− CDF and EDF against the event parameters (i.e., energy).

A detailed example of using the maximum likelihood method to determine the parameters, and testing the goodness of the fit using the KS statistic for power-law and Weibull distributions in the solar context, is given in Parnell (2002).

References

W.P. Abbett, S.L. Hawley, Astrophys. J. **521**, 906 (1999). doi:10.1086/307576

I.A. Ahmad, D.F. Webb, Sol. Phys. **58**, 323 (1978). doi:10.1007/BF00157278

K. Akabane, Publ. Astron. Soc. Jpn. **8**, 173 (1956).

D. Alexander, A.G. Daou, Astrophys. J. **666**, 1268 (2007). doi:10.1086/520331

K.A. Anderson, S.R. Kane, J.H. Primbsch, R.H. Weitzmann, W.D. Evans, R.W. Klebesadel, W.P. Aiello, IEEE Trans. Geosci. Electron. **16**, 157 (1978)

M.J. Aschwanden, Advances in Space Research **39**, 1867 (2007). doi:10.1016/j.asr.2007.03.050

M.J. Aschwanden, P. Charbonneau, Astrophys. J. Lett. **566**, L59 (2002). doi:10.1086/339451

M.J. Aschwanden, C.E. Parnell, Astrophys. J. **572**, 1048 (2002). doi:10.1086/340385

M.J. Aschwanden, B.R. Dennis, A.O. Benz, Astrophys. J. **497**, 972 (1998). doi:10.1086/305484

M.J. Aschwanden, T.D. Tarbell, R.W. Nightingale, C.J. Schrijver, A. Title, C.C. Kankelborg, P. Martens, H.P. Warren, Astrophys. J. **535**, 1047 (2000)

M.J. Aschwanden, R.A. Stern, M. Güdel, Astrophys. J. **672**, 659 (2008). arXiv:0710.2563, doi:10.1086/523926

H. Aurass, K.-L. Klein, P.C.H. Martens, Sol. Phys. **155**, 203 (1994). doi:10.1007/BF00670741

T. Bai, Astrophys. J. **404**, 805 (1993). doi:10.1086/172335

P. Bak, C. Tang, K. Wiesenfeld, Phys. Rev. Lett. **59**, 381 (1987). doi:10.1103/PhysRevLett.59.381

M. Battaglia, P.C. Grigis, A.O. Benz, Astron. Astrophys. **439**, 737 (2005). arXiv:astro-ph/0505154, doi:10.1051/0004-6361:20053027

A.O. Benz, S. Krucker, Astrophys. J. **568**, 413 (2002)

D.A. Biesecker, Ph.D. thesis, University of New Hampshire, 1994

P.L. Bornmann, Astrophys. J. **356**, 733 (1990). doi:10.1086/168880

K.R. Bromund, J.M. McTiernan, S.R. Kane, Astrophys. J. **455**, 733 (1995). doi:10.1086/176619

J.W. Brosius, K.J.H. Phillips, Astrophys. J. **613**, 580 (2004). doi:10.1086/422873

J.C. Brown, Sol. Phys. **18**, 489 (1971)

P.J. Cargill, J.A. Klimchuk, Astrophys. J. **478**, 799 (1997). doi:10.1086/303816

P.J. Cargill, J.A. Klimchuk, Astrophys. J. **605**, 911 (2004). doi:10.1086/382526

R.C. Carrington, Mon. Not. R. Astron. Soc. **20**, 13 (1859)

A. Caspi, R.P. Lin, Astrophys. J. Lett. **725**, L161–L166 (2010). doi:10.1088/2041-8205/725/2/L161. arXiv:1105.2839

P. Charbonneau, S.W. McIntosh, H.-L. Liu, T.J. Bogdan, Sol. Phys. **203**, 321 (2001). doi:10.1023/A:1013301521745

C. Chifor, H. Isobe, H.E. Mason, I.G. Hannah, P.R. Young, G. Del Zanna, S. Krucker, K. Ichimoto, Y. Katsukawa, T. Yokoyama, Astron. Astrophys. **491**, 279 (2008). doi:10.1051/0004-6361:200810265

S. Christe, I.G. Hannah, S. Krucker, J. McTiernan, R.P. Lin, Astrophys. J. **677**, 1385 (2008a). doi:10.1086/529011

S. Christe, S. Krucker, R.P. Lin, Astrophys. J. Lett. **680**, L149 (2008b). doi:10.1086/589971

J.W. Cirtain, L. Golub, L. Lundquist, A. van Ballegooijen, A. Savcheva, M. Shimojo, E. DeLuca, S. Tsuneta, T. Sakao, K. Reeves, M. Weber, R. Kano, N. Narukage, K. Shibasaki, Science **318**, 1580 (2007). doi:10.1126/science.1147050

E.W. Cliver, L. Svalgaard, Sol. Phys. **224**, 407 (2004). doi:10.1007/s11207-005-4980-z

D.F. Crawford, D.L. Jauncey, H.S. Murdoch, Astrophys. J. **162**, 405 (1970). doi:10.1086/150672

N. Crosby, N. Vilmer, N. Lund, R. Sunyaev, Astron. Astrophys. **334**, 299 (1998)

N.B. Crosby, M.J. Aschwanden, B.R. Dennis, Sol. Phys. **143**, 275 (1993)

D.W. Datlowe, M.J. Elcan, H.S. Hudson, Sol. Phys. **39**, 155 (1974)

B.R. Dennis, Sol. Phys. **100**, 465 (1985)

B.R. Dennis, D.M. Zarro, Sol. Phys. **146**, 177 (1993)

J.F. Drake, Sol. Phys. **16**, 152 (1971a). doi:10.1007/BF00154510
J.F. Drake, Sol. Phys. **16**, 152 (1971b)
J.P. Eastwood, M.S. Wheatland, H.S. Hudson, S. Krucker, S.D. Bale, M. Maksimovic, K. Goetz, J. Bougeret, Astrophys. J. Lett. **708**, L95 (2010). arXiv:0911.4131, doi:10.1088/2041-8205/708/2/L95
A.G. Emslie, B.R. Dennis, G.D. Holman, H.S. Hudson, J. Geophys. Res. **110**(A9), 11103 (2005). doi:10.1029/2005JA011305
U. Feldman, G.A. Doschek, W.E. Behring, K.J.H. Phillips, Astrophys. J. **460**, 1034 (1996). doi:10.1086/177030
U. Feldman, G.A. Doschek, J.A. Klimchuk, Astrophys. J. **474**, 511 (1997). doi:10.1086/303428
G.H. Fisher, R.C. Canfield, A.N. McClymont, Astrophys. J. **289**, 434 (1985). doi:10.1086/162903
L. Fletcher, I.G. Hannah, H.S. Hudson, T.R. Metcalf, Astrophys. J. **656**, 1187 (2007). doi:10.1086/510446
L. Fletcher, B.R. Dennis, H.S. Hudson, S. Krucker, K. Phillips, A. Veronig, M. Battaglia, L. Bone, Q. Chen, P. Gallagher, P. Grigis, H. Ji, H. Liu, R.O. Milligan, M. Temmer, Space Sci. Rev. (2011)
L. Fritzova-Svestkova, R.C. Chase, Z. Svestka, Sol. Phys. **48**, 275 (1976). doi:10.1007/BF00151996
E. Fuerst, A.O. Benz, W. Hirth, Astron. Astrophys. **107**, 178 (1982)
D.E. Gary, M.D. Hartl, T. Shimizu, Astrophys. J. **477**, 958 (1997). doi:10.1086/303748
L. Golub, A.S. Krieger, J.K. Silk, A.F. Timothy, G.S. Vaiana, Astrophys. J. Lett. **189**, L93 (1974)
N. Gopalswamy, T.E.W. Payne, E.J. Schmahl, M.R. Kundu, J.R. Lemen, K.T. Strong, R.C. Canfield, J. de La Beaujardiere, Astrophys. J. **437**, 522 (1994). doi:10.1086/175015
J.A. Grayson, S. Krucker, R.P. Lin, Astrophys. J. **707**, 1588 (2009). doi:10.1088/0004-637X/707/2/1588
B. Gutenberg, C.F. Richter, *Seismicity of the Earth and Associated Phenomena*, 2nd edn. (Princeton University Press, Princeton, 1956)
I.G. Hannah, G.J. Hurford, H.S. Hudson, R.P. Lin, Rev. Sci. Instrum. **78**, 4501 (2007a). doi:10.1063/1.2437120
I.G. Hannah, G.J. Hurford, H.S. Hudson, R.P. Lin, K. van Bibber, Astrophys. J. Lett. **659**, L77 (2007b). arXiv:astro-ph/0702726, doi:10.1086/516750
I.G. Hannah, S. Christe, S. Krucker, G.J. Hurford, H.S. Hudson, R.P. Lin, Astrophys. J. **677**, 704 (2008a). arXiv:0712.2544, doi:10.1086/529012
I.G. Hannah, S. Krucker, H.S. Hudson, S. Christe, R.P. Lin, Astron. Astrophys. **481**, L45 (2008b). arXiv:0712.0369, doi:10.1051/0004-6361:20079019
I.G. Hannah, H.S. Hudson, G.J. Hurford, R.P. Lin, Astrophys. J. (2010)
H. Hara, K. Nakakubo-Morimoto, Astrophys. J. **589**, 1062 (2003). doi:10.1086/374778
M.A. Hendry, J.F.L. Simmons, Astron. Astrophys. **237**, 275 (1990)
R. Hodgson, Mon. Not. R. Astron. Soc. **20**, 15 (1859)
G.D. Holman, L. Sui, R.A. Schwartz, A.G. Emslie, Astrophys. J. Lett. **595**, L97 (2003). doi:10.1086/378488
G.D. Holman, M.J. Aschwanden, H. Aurass, M. Battaglia, P.C. Grigis, E.P. Kontar, W. Liu, P. Saint-Hilaire, V.V. Zharkov, Space Sci. Rev. (2010)
H.S. Hudson, in *Bulletin of the American Astronomical Society*, vol. 23 (1991a), p. 1064
H.S. Hudson, Sol. Phys. **133**, 357 (1991b)
H.S. Hudson, Astrophys. J. Lett. **663**, L45 (2007) arXiv:0707.1118, doi:10.1086/519923
H.S. Hudson, L.E. Peterson, D.A. Schwartz, Astrophys. J. **157**, 389 (1969)
S.W. Kahler, J. Geophys. Res. **87**, 3439 (1982). doi:10.1029/JA087iA05p03439
J.I. Katz, J. Geophys. Res. **91**, 10412 (1986). doi:10.1029/JB091iB10p10412
J. Kašparová, E.P. Kontar, J.C. Brown, Astron. Astrophys. **466**, 705 (2007). arXiv:astro-ph/0701871, doi:10.1051/0004-6361:20066689
H.R.M. Kay, J.L. Culhane, L.K. Harra, S.A. Matthews, Advances in Space Research **32**, 1051 (2003). doi:10.1016/S0273-1177(03)00308-9
A.L. Kiplinger, Astrophys. J. **453**, 973 (1995). doi:10.1086/176457
A.L. Kiplinger, H.A. Garcia, in *Bulletin of the American Astronomical Society*, vol. 36 (2004), p. 739
E. Kontar, J. Brown, A. Emslie, W. Hajdas, G. Holman, G. Hurford, J. Kasparova, P. Mallik, A. Massone, M. McConnell, M. Piana, M. Prato, E. Schmahl, E. Suarez-Garcia, Space Sci. Rev. (2010)
E.P. Kontar, E. Dickson, J. Kašparová, Sol. Phys. **252**, 139 (2008). arXiv:0805.1470, doi:10.1007/s11207-008-9249-x
G. Kopp, G. Lawrence, G. Rottman, Sol. Phys. **230**, 129 (2005). doi:10.1007/s11207-005-7433-9
T. Kosugi, B.R. Dennis, K. Kai, Astrophys. J. **324**, 1118 (1988). doi:10.1086/165967
J. Kotoku, R. Kano, S. Tsuneta, Y. Katsukawa, T. Shimizu, T. Sakao, K. Shibasaki, E.E. Deluca, K.E. Korreck, L. Golub, A. Bobra, Publ. Astron. Soc. Jpn. **59**, 735 (2007)
S. Krucker, A.O. Benz, Astrophys. J. Lett. **501**, L213 (1998). doi:10.1086/311474
S. Krucker, R.P. Lin, Astrophys. J. **673**, 1181 (2008). doi:10.1086/524010
S. Krucker, A.O. Benz, T.S. Bastian, L.W. Acton, Astrophys. J. **488**, 499 (1997). doi:10.1086/304686

S. Krucker, S. Christe, R.P. Lin, G.J. Hurford, R.A. Schwartz, Sol. Phys. **210**, 445 (2002). doi:10.1023/A:1022404512780
T.A. Kucera, B.R. Dennis, R.A. Schwartz, D. Shaw, Astrophys. J. **475**, 338 (1997). doi:10.1086/303532
M.R. Kundu, J.P. Raulin, N. Nitta, H.S. Hudson, M. Shimojo, K. Shibata, A. Raoult, Astrophys. J. Lett. **447**, L135 (1995). doi:10.1086/309567
M.R. Kundu, G. Trottet, V.I. Garaimov, P.C. Grigis, E.J. Schmahl, Advances in Space Research **35**, 1778 (2005). doi:10.1016/j.asr.2005.04.052
M.R. Kundu, E.J. Schmahl, P.C. Grigis, V.I. Garaimov, K. Shibasaki, Astron. Astrophys. **451**, 691 (2006). doi:10.1051/0004-6361:20053987
E. Landi, K.J.H. Phillips, Astrophys. J. Suppl. Ser. **166**, 421 (2006). doi:10.1086/506180
T.T. Lee, V. Petrosian, J.M. McTiernan, Astrophys. J. **412**, 401 (1993). doi:10.1086/172929
T.T. Lee, V. Petrosian, J.M. McTiernan, Astrophys. J. **448**, 915 (1995). doi:10.1086/176020
F. Lepreti, V. Carbone, P. Veltri, Astrophys. J. Lett. **555**, L133 (2001). doi:10.1086/323178
W.H.G. Lewin, J. Doty, G.W. Clark, S.A. Rappaport, H.V.D. Bradt, R. Doxsey, D.R. Hearn, J.A. Hoffman, J.G. Jernigan, F.K. Li, W. Mayer, J. McClintock, F. Primini, J. Richardson, Astrophys. J. Lett. **207**, L95 (1976). doi:10.1086/182188
R.P. Lin, Space Sci. Rev. **16**, 189 (1974). doi:10.1007/BF00240886
R.P. Lin, H.S. Hudson, Sol. Phys. **50**, 153 (1976). doi:10.1007/BF00206199
R.P. Lin, R.A. Schwartz, S.R. Kane, R.M. Pelling, K.C. Hurley, Astrophys. J. **283**, 421 (1984). doi:10.1086/162321
R.P. Lin, P.T. Feffer, R.A. Schwartz, Astrophys. J. Lett. **557**, L125 (2001a)
R.P. Lin, P.T. Feffer, R.A. Schwartz, Astrophys. J. Lett. **557**, L125 (2001b). doi:10.1086/323270
R.P. Lin, B.R. Dennis, G.J. Hurford, D.M. Smith, A. Zehnder, P.R. Harvey, D.W. Curtis, D. Pankow, P. Turin, M. Bester, A. Csillaghy, M. Lewis, N. Madden, H.F. van Beek, M. Appleby, T. Raudorf, J. McTiernan, R. Ramaty, E. Schmahl, R. Schwartz, S. Krucker, R. Abiad, T. Quinn, P. Berg, M. Hashii, R. Sterling, R. Jackson, R. Pratt, R.D. Campbell, D. Malone, D. Landis, C.P. Barrington-Leigh, S. Slassi-Sennou, C. Cork, D. Clark, D. Amato, L. Orwig, R. Boyle, I.S. Banks, K. Shirey, A.K. Tolbert, D. Zarro, F. Snow, K. Thomsen, R. Henneck, A. McHedlishvili, P. Ming, M. Fivian, J. Jordan, R. Wanner, J. Crubb, J. Preble, M. Matranga, A. Benz, H. Hudson, R.C. Canfield, G.D. Holman, C. Crannell, T. Kosugi, A.G. Emslie, N. Vilmer, J.C. Brown, C. Johns-Krull, M. Aschwanden, T. Metcalf, A. Conway, Sol. Phys. **210**, 3 (2002). doi:10.1023/A:1022428818870
R.E. Lingenfelter, H.S. Hudson, in *The Ancient Sun: Fossil Record in the Earth, Moon and Meteorites*, ed. by R.O. Repin, J.A. Eddy, R.B. Merrill (1980), pp. 69–79
Y.E. Litvinenko, Sol. Phys. **151**, 195 (1994). doi:10.1007/BF00654092
C. Liu, J. Qiu, D.E. Gary, S. Krucker, H. Wang, Astrophys. J. **604**, 442 (2004). doi:10.1086/381799
E.T. Lu, R.J. Hamilton, Astrophys. J. Lett. **380**, L89 (1991). doi:10.1086/186180
E.T. Lu, R.J. Hamilton, J.M. McTiernan, K.R. Bromund, Astrophys. J. **412**, 841 (1993). doi:10.1086/172966
K.G. McCracken, D.F. Smart, M.A. Shea, G.A.M. Dreschhoff, in *International Cosmic Ray Conference*, vol. 8 (2001), p. 3209
J.M. McTiernan, Astrophys. J. **697**, 94 (2009). doi:10.1088/0004-637X/697/1/94
J.M. McTiernan, G.H. Fisher, P. Li, Astrophys. J. **514**, 472 (1999). doi:10.1086/306924
R.O. Milligan, Private communication, 2009
Y.-J. Moon, G.S. Choe, H.S. Yun, Y.D. Park, J. Geophys. Res. **106**, 29951 (2001). doi:10.1029/2000JA000224
Y.-J. Moon, G.S. Choe, Y.D. Park, H. Wang, P.T. Gallagher, J. Chae, H.S. Yun, P.R. Goode, Astrophys. J. **574**, 434 (2002). doi:10.1086/340945
D. Moses, J.W. Cook, J.-D.F. Bartoe, G.E. Brueckner, K.P. Dere, D.F. Webb, J.M. Davis, J.W. Harvey, F. Recely, S.F. Martin, H. Zirin, Astrophys. J. **430**, 913 (1994). doi:10.1086/174461
W.M. Neupert, Astrophys. J. Lett. **153**, L59 (1968)
H.W. Newton, Mon. Not. R. Astron. Soc. **103**, 244 (1943)
A. Nindos, M.R. Kundu, S.M. White, Astrophys. J. **513**, 983 (1999). doi:10.1086/306886
N. Nitta, Astrophys. J. **491**, 402 (1997). doi:10.1086/304938
N. Nitta, T. Kosugi, Sol. Phys. **105**, 73 (1986). doi:10.1007/BF00156378
N. Nitta, T.S. Bastian, M.J. Aschwanden, K.L. Harvey, K.T. Strong, Publ. Astron. Soc. Jpn. **44**, L167 (1992)
R.A. Osten, S. Drake, J. Tueller, J. Cummings, M. Perri, A. Moretti, S. Covino, Astrophys. J. **654**, 1052 (2007). arXiv:astro-ph/0609205, doi:10.1086/509252
E.N. Parker, Astrophys. J. **330**, 474 (1988). doi:10.1086/166485
C.E. Parnell, Mon. Not. R. Astron. Soc. **335**, 389 (2002). doi:10.1046/j.1365-8711.2002.05618.x
C.E. Parnell, P.E. Jupp, Astrophys. J. **529**, 554 (2000). doi:10.1086/308271
K.J.H. Phillips, Astrophys. J. **605**, 921 (2004). doi:10.1086/382523
M. Pick, N. Vilmer, Astron. Astrophys. Rev. 6 (2008). doi:10.1007/s00159-008-0013-x

J.G. Porter, R.L. Moore, E.J. Reichmann, O. Engvold, K.L. Harvey, Astrophys. J. **323**, 380 (1987). doi:10.1086/165835

W.H. Press, S.A. Teukolsky, W.T. Vetterling, B.P. Flannery, *Numerical Recipes in C. The Art of Scientific Computing*, 2nd edn. (Cambridge University Press, Cambridge, 1992)

J. Qiu, C. Liu, D.E. Gary, G.M. Nita, H. Wang, Astrophys. J. **612**, 530 (2004). doi:10.1086/422401

J.-P. Raulin, S.M. White, M.R. Kundu, A.V.R. Silva, K. Shibasaki, Astrophys. J. **522**, 547 (1999). doi:10.1086/322974

D.V. Reames, Space Sci. Rev. **90**, 413 (1999). doi:10.1023/A:1005105831781

R.S. Richardson, Publ. Astron. Soc. Pac. **56**, 156 (1944)

R. Rosner, G.S. Vaiana, Astrophys. J. **222**, 1104 (1978). doi:10.1086/156227

P. Saint-Hilaire, A.O. Benz, Astron. Astrophys. **435**, 743 (2005). doi:10.1051/0004-6361:20041918

P. Saint-Hilaire, S. Krucker, R.P. Lin, Sol. Phys. **250**, 53 (2008). doi:10.1007/s11207-008-9193-9

R. Saldanha, S. Krucker, R.P. Lin, Astrophys. J. **673**, 1169 (2008). doi:10.1086/524929

N. Santangelo, H. Horstman, E. Horstman-Moretti, Sol. Phys. **29**, 143 (1973). doi:10.1007/BF00153445

A. Savcheva, J. Cirtain, E.E. Deluca, L.L. Lundquist, L. Golub, M. Weber, M. Shimojo, K. Shibasaki, T. Sakao, N. Narukage, S. Tsuneta, R. Kano, Publ. Astron. Soc. Jpn. **59**, 771 (2007)

A. Schadee, C. de Jager, Z. Svestka, Sol. Phys. **89**, 287 (1983). doi:10.1007/BF00217252

R.A. Schwartz, B.R. Dennis, G.J. Fishman, C.A. Meegan, R.B. Wilson, W.S. Paciesas, in *The Compton Observatory Science Workshop*, ed. by C.R. Shrader, N. Gehrels, B. Dennis (1992), pp. 457–468

N.I. Shakhovskaia, Sol. Phys. **121**, 375 (1989). doi:10.1007/BF00161707

K. Shibata, Y. Ishido, L.W. Acton, K.T. Strong, T. Hirayama, Y. Uchida, A.H. McAllister, R. Matsumoto, S. Tsuneta, T. Shimizu, H. Hara, T. Sakurai, K. Ichimoto, Y. Nishino, Y. Ogawara, Publ. Astron. Soc. Jpn. **44**, L173 (1992)

T. Shimizu, Publ. Astron. Soc. Jpn. **47**, 251 (1995)

T. Shimizu, R.A. Shine, A.M. Title, T.D. Tarbell, Z. Frank, Astrophys. J. **574**, 1074 (2002). doi:10.1086/340998

M. Shimojo, S. Hashimoto, K. Shibata, T. Hirayama, H.S. Hudson, L.W. Acton, Publ. Astron. Soc. Jpn. **48**, 123 (1996)

M. Shimojo, N. Narukage, R. Kano, T. Sakao, S. Tsuneta, K. Shibasaki, J.W. Cirtain, L.L. Lundquist, K. Reeves, A. Savcheva, Publ. Astron. Soc. Jpn. **59**, 745 (2007)

A.V.R. Silva, H. Wang, D.E. Gary, Astrophys. J. **545**, 1116 (2000). doi:10.1086/317822

H.J. Smith, E.V.P. Smith, *Solar Flares* (Macmillan, New York, 1963)

S. Stoiser, A.M. Veronig, H. Aurass, A. Hanslmeier, Sol. Phys. **246**, 339 (2007). doi:10.1007/s11207-007-9066-7

K.T. Strong, K. Harvey, T. Hirayama, N. Nitta, T. Shimizu, S. Tsuneta, Publ. Astron. Soc. Jpn. **44**, L161 (1992)

L. Sui, G.D. Holman, B.R. Dennis, Astrophys. J. **670**, 862 (2007). doi:10.1086/522198

M. Takahashi, T. Watanabe, Adv. Space Res. **25**, 1833 (2000). doi:10.1016/S0273-1177(99)00628-6

R.J. Thomas, R.G. Teske, Sol. Phys. **16**, 431 (1971). doi:10.1007/BF00162486

S. Tsuneta, J.R. Lemen, in *Physics of Solar and Stellar Coronae: G.S. Vaiana Memorial Symposium*, Palermo, Italy, ed. by J.L. Linsky, S. Serio (1993), p. 113

B.T. Tsurutani, W.D. Gonzalez, G.S. Lakhina, S. Alex, J. Geophys. Res. (Space Phys.) **108**, 1268 (2003). doi:10.1029/2002JA009504

A. Veronig, M. Temmer, A. Hanslmeier, W. Otruba, M. Messerotti, Astron. Astrophys. **382**, 1070 (2002a). arXiv:astro-ph/0207234, doi:10.1051/0004-6361:20011694

A. Veronig, B. Vršnak, B.R. Dennis, M. Temmer, A. Hanslmeier, J. Magdalenić, Astron. Astrophys. **392**, 699 (2002b). arXiv:astro-ph/0207217, doi:10.1051/0004-6361:20020947

A.M. Veronig, J.C. Brown, B.R. Dennis, R.A. Schwartz, L. Sui, A.K. Tolbert, Astrophys. J. **621**, 482 (2005). doi:10.1086/427274

N. Vilmer, A.L. MacKinnon, G.J. Hurford, Space Sci. Rev. (2010)

B. Vršnak, D. Sudar, D. Ruždjak, Astron. Astrophys. **435**, 1149 (2005). doi:10.1051/0004-6361:20042166

Y. Wang, J. Zhang, Astrophys. J. **665**, 1428 (2007). arXiv:0808.2976, doi:10.1086/519765

M.S. Wheatland, Astrophys. J. Lett. **536**, L109 (2000). arXiv:astro-ph/0005121, doi:10.1086/312739

M.S. Wheatland, Sol. Phys. **203**, 87 (2001)

M.S. Wheatland, Astrophys. J. **609**, 1134 (2004). arXiv:astro-ph/0403613, doi:10.1086/421261

M.S. Wheatland, Astrophys. J. **679**, 1621 (2008). arXiv:0802.3931, doi:10.1086/587871

M.S. Wheatland, Astrophys. J. **710**, 1324 (2010). arXiv:1001.1464, doi:10.1088/0004-637X/710/2/1324

M.S. Wheatland, S. Glukhov, Astrophys. J. **494**, 858 (1998). doi:10.1086/305245

M.S. Wheatland, Y.E. Litvinenko, Sol. Phys. **211**, 255 (2002)

M.S. Wheatland, P.A. Sturrock, J.M. McTiernan, Astrophys. J. **509**, 448 (1998). doi:10.1086/306492

S.M. White, K. Arzner, S. Christe, F. Farnik, K.-L. Klein, M.R. Kundu, Z. Ning, J.-P. Raulin, A.V.R. Silva VA'alio, P. Saint-Hilaire, G. Trottet, A. Warmuth, Space Sci. Rev. (2010)
J.P. Wild, S.F. Smerd, A.A. Weiss, Annu. Rev. Astron. Astrophys. **1**, 291 (1963). doi:10.1146/annurev.aa.01.090163.001451
J.A. Willick, Astrophys. J. Suppl. Ser. **92**, 1 (1994). doi:10.1086/191957
T.N. Woods, F.G. Eparvier, J. Fontenla, J. Harder, G. Kopp, W.E. McClintock, G. Rottman, B. Smiley, M. Snow, Geophys. Res. Lett. **3110802** (2004)
S. Yashiro, S. Akiyama, N. Gopalswamy, R.A. Howard, Astrophys. J. Lett. **650**, L143 (2006). arXiv:astro-ph/0609197, doi:10.1086/508876
S. Yashiro, N. Gopalswamy, G. Michalek, O.C. St. Cyr, S.P. Plunkett, N.B. Rich, R.A. Howard, J. Geophys. Res. **109**(A18), 7105 (2004). doi:10.1029/2003JA010282
T. Yoshida, S. Tsuneta, Astrophys. J. **459**, 342 (1996). doi:10.1086/176897

Space Sci Rev (2011) 159:301–355
DOI 10.1007/s11214-011-9804-x

Deducing Electron Properties from Hard X-ray Observations

E.P. Kontar · J.C. Brown · A.G. Emslie · W. Hajdas · G.D. Holman · G.J. Hurford · J. Kašparová · P.C.V. Mallik · A.M. Massone · M.L. McConnell · M. Piana · M. Prato · E.J. Schmahl · E. Suarez-Garcia

Received: 2 February 2008 / Accepted: 2 June 2008 / Published online: 11 August 2011

Abstract X-radiation from energetic electrons is the prime diagnostic of flare-accelerated electrons. The observed X-ray flux (and polarization state) is fundamentally a convolution of the cross-section for the hard X-ray emission process(es) in question with the electron distribution function, which is in turn a function of energy, direction, spatial location and time. To address the problems of particle propagation and acceleration one needs to infer as much information as possible on this electron distribution function, through a deconvolution of this fundamental relationship. This review presents recent progress toward this goal using

E.P. Kontar (✉) · J.C. Brown · P.C.V. Mallik
Department of Physics and Astronomy, University of Glasgow, Kelvin Building, Glasgow, G12 8QQ, UK
e-mail: eduard@astro.gla.ac.uk

A.G. Emslie
Department of Physics, Oklahoma State University, Stillwater, OK 74078, USA

A.G. Emslie
Western Kentucky University, 1906 College Heights Blvd., Bowling Green, KY 42101, USA

W. Hajdas · E. Suarez-Garcia
Paul Scherrer Institut, Villigen PSI, Switzerland

G.D. Holman
NASA Goddard Space Flight Center, Code 671, Greenbelt, MD 20771, USA

G.J. Hurford
Space Sciences Laboratory, University of California, Berkeley, CA 94720, USA

J. Kašparová
Astronomický ústav AV ČR, v.v.i., Fričova 298, Ondřejov, 251 65, Czech Republic

A.M. Massone · M. Prato
CNR-SPIN, Via Dodecaneso 33, 16146 Genova, Italy

M.L. McConnell
Space Science Center, University of New Hampshire, Durham, NH 03824, USA

spectroscopic, imaging and polarization measurements, primarily from the *Reuven Ramaty High Energy Solar Spectroscopic Imager* (*RHESSI*). Previous conclusions regarding the energy, angular (pitch angle) and spatial distributions of energetic electrons in solar flares are critically reviewed. We discuss the role and the observational evidence of several radiation processes: free-free electron-ion, free-free electron-electron, free-bound electron-ion, photoelectric absorption and Compton backscatter (albedo), using both spectroscopic and imaging techniques. This unprecedented quality of data allows for the first time inference of the angular distributions of the X-ray-emitting electrons and improved model-independent inference of electron energy spectra and emission measures of thermal plasma. Moreover, imaging spectroscopy has revealed hitherto unknown details of solar flare morphology and detailed spectroscopy of coronal, footpoint and extended sources in flaring regions. Additional attempts to measure hard X-ray polarization were not sufficient to put constraints on the degree of anisotropy of electrons, but point to the importance of obtaining good quality polarization data in the future.

Keywords Sun: flares · Sun: X-rays · Sun: acceleration · Sun: energetic particles

1 Introduction

X-ray emission, because it is produced promptly in an optically-thin environment, is one of the most direct methods with which to study energetic electrons in solar flares. Such remote radiation measurements are generally functions of photon energy ϵ, direction $\boldsymbol{\Omega}$, and time t. A common description of incoherent and partially polarized X-ray radiation typical of solar flares is in terms of its *photon intensity* $I(\epsilon, \boldsymbol{\Omega}, t)$ (photons s^{-1} cm^{-2} sr^{-1} keV^{-1}; CGS units are used throughout the paper) and the fractional degree and orientation of linear polarization (P, Ψ). The emergent photon intensity is the number of photons dN that escape from the source in the time interval t to $t + dt$ in the energy range ϵ to $\epsilon + d\epsilon$, from a solar source with a direction and angular size defined by the elementary cone $d\boldsymbol{\Omega}$, oriented with respect to the normal of the detector of area dS:

$$dN = I(\epsilon, \boldsymbol{\Omega}, t)\, d\epsilon\, dS\, d\Omega\, dt. \tag{1.1}$$

For optically-thin radiation, the emergent intensity of the radiation is simply a linear convolution of the cross-section for the pertinent emission process and the electron phase-space distribution function $f_e(\mathbf{v}, \mathbf{r}, t)$ (electrons cm^{-3} [cm s^{-1}]$^{-3}$), or equivalently the particle *flux* $F(E, \boldsymbol{\Omega}', \mathbf{r}, t)$ (electrons cm^{-2} s^{-1} sr^{-1} keV^{-1}), differential in energy E and velocity solid angle $\boldsymbol{\Omega}'$. Since $dE = m_e v\, dv$ in the non-relativistic regime, it follows that $F(E, \boldsymbol{\Omega}', \mathbf{r}, t) = v^2 f_e(\mathbf{v}, \mathbf{r}, t)/m_e$, where m_e (g) is the electron mass.

M. Piana
Dipartimento di Matematica, Universita di Genova, Via Dodecaneso 35, 16146 Genova, Italy

E.J. Schmahl
NWRA/CoRA, Boulder, CO 80301, USA

E. Suarez-Garcia
Department of Nuclear and Particle Physics, University of Geneva, Quai Ernest Ansermet 24, 1212 Geneva, Switzerland

For an elementary bremsstrahlung source of ambient plasma density $n(\mathbf{r})$, located at position $\mathbf{r}$ on the Sun, along the line of sight $\boldsymbol{\Omega}$, which is subjected to an electron flux spectrum $F(E, \boldsymbol{\Omega}', \mathbf{r}, t)$, the emergent photon flux spectrum at distance R is the convolution

$$I(\epsilon, \boldsymbol{\Omega}, t) = \int_\ell \int_{\Omega'} \int_\epsilon^\infty n(\mathbf{r}) F(E, \mathbf{r}, \boldsymbol{\Omega}', t) Q(\boldsymbol{\Omega}, \boldsymbol{\Omega}', \epsilon, E)\, dE\, d\boldsymbol{\Omega}'\, d\ell, \tag{1.2}$$

where ℓ is the distance along the line of sight, $Q(\boldsymbol{\Omega}, \boldsymbol{\Omega}', \epsilon, E)$ (cm^2 keV^{-1} sr^{-1}) is the cross-section for the pertinent hard X-ray emission process(es), differential in ϵ and $\boldsymbol{\Omega}$. The dominant hard X-ray emission process in solar flares is *bremsstrahlung* radiation associated with electron deceleration in the Coulomb field of an ion or other electron. For this process, the angular dependence of Q depends only on the angle $\theta' = \widehat{\boldsymbol{\Omega}'\boldsymbol{\Omega}}$ between the incoming electron $\boldsymbol{\Omega}'$ and the emitted photon $\boldsymbol{\Omega}$ directions, so that $Q = Q(\epsilon, E, \theta')$.

To deduce the physical properties of energetic particles in solar flares from observed hard X-ray quantities, the electron flux spectrum $F(E, \mathbf{r}, \boldsymbol{\Omega}', t)$ must be deconvolved from the emission cross-section $Q(\epsilon, E, \theta')$ in this integral. Although the relation between the observable quantity $I(\epsilon, \boldsymbol{\Omega}, t)$ and the physical electron flux $F(E, \mathbf{r}, \boldsymbol{\Omega}', t)$ is linear, it is still nontrivial and the deconvolution requires some rather insightful techniques. This fundamental problem and the progress toward the solution of the problem using spectroscopic, imaging and polarization measurements from the *Reuven Ramaty High Energy Solar Spectroscopic Imager* (*RHESSI*, Lin et al. 2002) are addressed in this review.

In Sect. 2, we review the physical processes leading to X-ray emission in solar flares, including free-free electron-ion bremsstrahlung (Sect. 2.1), free-bound electron-ion emission (Sect. 2.2), and free-free electron-electron bremsstrahlung (Sect. 2.3). This review, however, does not discuss bound-bound transitions, nor the emissions due to energetic ions such as nuclear gamma-ray lines, ion gamma-ray continuum, and pseudo-continuum.

Since the emission mechanisms are well established, hard X-rays (HXRs) are often viewed as one of the most direct (i.e., least affected by propagation effects) diagnostics of solar flare electrons. However, it must not be forgotten that downward-propagating X-ray photons are effectively scattered toward the observer by electrons in the dense layers of the solar atmosphere, thereby complicating the diagnostic potential of hard X-ray radiation. Section 3 reviews spectroscopic and imaging techniques to infer this Compton backscattered (albedo) component. Progress in the deduction of primary, i.e., directly flare-emitted, and photospherically-reflected X-rays from the observed spectrum, is discussed.

Section 4 presents an overview of the results obtained using purely spectroscopic data (i.e., data integrated over the source volume). Both forward fitting and regularized inversion techniques to deduce the energy dependence of the *mean source electron flux spectrum* $\overline{F}(E)$ (averaged over volume and solid angle) are reviewed. Properties of the electron flux distribution, deviations from power-law forms, low-energy cutoffs and interpretation in terms of a thermal source are discussed.

Section 5 provides methods and results for the case of an anisotropic angular distribution $\overline{F}(E, \boldsymbol{\Omega})$ of electrons. The role of X-ray Compton scattering in the solar atmosphere (solar albedo) in deducing this angular distribution is discussed. Recent *RHESSI* polarization measurements and their implications for electron anisotropy are also reviewed.

Section 6 focuses on the spatial structure $F(E, \mathbf{r})$ of the electron flux, using imaging spectroscopy observations from *RHESSI*. Recently-developed, visibility-based techniques to optimize the inference of electron maps are discussed. Section 7 highlights the major finding from *RHESSI* and discusses the open questions.

The role of these results in the context of multi-wavelength observations of solar flares is discussed by Fletcher et al. (2011, this issue). The implications of these findings for electron transport, and acceleration models are discussed by Holman et al. (2011, this issue), Hannah et al. (2011, this issue), and Zharkova et al. (2011, this issue).

2 X-ray Emission Processes and Energetic Electrons

When energetic electrons are deflected in close encounters with ambient particles (both electrons and ions), a *bremsstrahlung* (literally, "braking radiation") photon is produced. As stated in Sect. 1, the cross-section $Q(\epsilon, E, \theta')$ for electron-ion free-free bremsstrahlung is a function of the emitted photon energy ϵ, the pre-collision electron energy E, and the angle θ' between the direction of the pre-collision electron and the outgoing photon (see, e.g., Koch and Motz 1959). For simplicity we often consider only the solid-angle-integrated form of the cross-section $Q(\epsilon, E)$ (and the corresponding scalar electron flux $F[E, \mathbf{r}, t]$); however, it must be remarked that the effects of the angular dependence of the cross-section can, for highly-beamed electron distributions, be quite significant, leading to substantial differences in the number of electrons required to produce a given hard X-ray flux—see Sect. 5.

The photons of energy from a few keV to a few hundred keV under consideration are mostly produced by collisional electron-ion bremsstrahlung in the solar atmosphere. Bremsstrahlung emission from energetic electrons is more efficient than inverse Compton scattering or synchrotron emission from the same electron population (Korchak 1967). The responsible electrons have kinetic energies E up to a few hundreds of keV, and so will collisionally stop within a column density $N \approx E^2/6\pi e^4\Lambda \sim 10^{17}E^2 \lesssim 10^{21}$ cm^{-2}, where e (esu) is the electronic charge and Λ the Coulomb logarithm (e.g., Emslie 1978). Thus, the column density in a solar coronal loop of density 10^{10} cm^{-2} and length 10^9 cm is 10^{19} cm^{-2}, which stops electrons up to $\sim$10 keV, while the upper chromosphere with typical column densities 10^{20}–10^{22} cm^{-2} can stop electrons with the energies of 10–300 keV. The cross-section for scattering the emitted photons is of the order of the Compton cross-section $\sigma_C \approx \pi r_o^2 \approx 2 \times 10^{-25}$ cm^2, where $r_o = e^2/m_e c^2$ is the classical electron radius, so that the optical depth $\tau = N\sigma_C \approx 10^{-4}$ and there is negligible self-absorption in the source, i.e., the source is optically thin. In Sect. 3, we shall address the issue of so-called "albedo" photons—photons that are emitted downward toward the solar photosphere, which is optically thick, and subsequently backscattered toward the observer. For now, we consider only "primary" photons, i.e., those initially directed toward the observer. Integrating the photon intensity $I(\epsilon, \Omega)$ over the solid angle subtended by the source of area A, $d\Omega = dA/R^2$ and making the volume element substitution $d\boldsymbol{\Omega}\, d\ell = d^3r/R^2$, the fundamental equation (1.2) shows that the observed bremsstrahlung flux (at time t; hereafter understood) at the Earth (photons s^{-1} keV^{-1} cm^{-2} of detector area) is

$$I(\epsilon) = \frac{1}{4\pi R^2}\int_\epsilon^\infty \int_V n(\mathbf{r})\, F(E, \mathbf{r})\, Q(\epsilon, E)\, dE\, d^3\mathbf{r}, \tag{2.1}$$

where the second integral is taken over the source volume V.

2.1 Electron-Ion Bremsstrahlung

For electron-ion bremsstrahlung, the full form of the cross-section $Q(\epsilon, E)$ is given by formula 3BN of Koch and Motz (1959). Numerical computations may be facilitated by use of the simplified form published by Haug (1997). A frequently-used analytic approximation to the cross-section $Q(\epsilon, E)$ is the Kramers form

$$Q(\epsilon, E) = Z^2 \frac{\sigma_o}{\epsilon E}, \tag{2.2}$$

where $\sigma_o = (8\alpha/3)(m_e c^2) r_o^2 = 7.9 \times 10^{-25}$ cm^2 keV. Here $\alpha \simeq 1/137$ is the fine structure constant, m_e is the electron mass, Z is the mean ion charge, c is the speed of light. A more accurate analytic form, valid in the non-relativistic limit, is the Bethe-Heitler form

$$Q(\epsilon, E) = Z^2 \frac{\sigma_o}{\epsilon E} \ln \frac{1 + \sqrt{1 - \epsilon/E}}{1 - \sqrt{1 - \epsilon/E}}. \tag{2.3}$$

For purely spectral observations, a spatially-integrated form of the basic equation (2.1) is appropriate. In this case, we can write (e.g., Brown 1971)

$$I(\epsilon) = \frac{1}{4\pi R^2} \int_\epsilon^\infty [\overline{n} V \overline{F}(E)]\, Q(\epsilon, E)\, dE, \tag{2.4}$$

where $\overline{n} = (1/V) \int_V n(\mathbf{r})\, d^3\mathbf{r}$ and $\overline{F}(E) = (1/\overline{n}V) \int_V n(\mathbf{r}) F(E, \mathbf{r})\, d^3\mathbf{r}$. The quantity $\overline{F}(E)$ (electrons cm^{-2} s^{-1} keV^{-1}) is termed the *mean electron flux spectrum* (Brown et al. 2003); it has also been termed the "X-ray emitting electron spectrum" by Johns and Lin (1992a, 1992b). Since the quantity $\overline{n}\, V$ (the number of target particles in the emitting volume) is dimensionless, the units of the quantity $[\overline{n} V \overline{F}(E)]$ are the same as those for electron flux, viz., electrons cm^{-2} s^{-1} keV^{-1}. For large events, typical values of $\overline{F}(E)$ and $[\overline{n} V \overline{F}(E)]$ at a representative energy $E \simeq 20$ keV are of order 10^{18} and 10^{55}, respectively.

The inference of $[\overline{n} V \overline{F}(E)]$ corresponding to an observed $I(\epsilon)$ may be accomplished in several ways, which are described in detail later in this chapter. A review of the different degrees of effectiveness of these techniques in discerning the overall magnitude of, overall spectral shape of, and form of "local" features in $\overline{F}(E)$ has been presented by Brown et al. (2006).

It is of crucial importance to note that the quantity $[\overline{n} V \overline{F}(E)]$ is the *only* quantity that can be inferred unambiguously (i.e., without additional model assumptions) from the source-integrated bremsstrahlung emission $I(\epsilon)$. Use of (2.4) to obtain $[\overline{n} V \overline{F}(E)]$ for a given $I(\epsilon)$ is therefore a fundamental issue in the interpretation of solar hard X-ray spectra. Once $[\overline{n} V \overline{F}(E)]$ has been determined, the actual magnitude of $\overline{F}(E)$ depends on the values[1] of $\overline{n}$ and V.

As an example of the use of the mean source electron spectrum to determine physical properties of solar flares, let us consider the inference of the *accelerated* electron flux spectrum $\mathcal{F}_0(E_0)$. The bremsstrahlung yield, the number of bremsstrahlung photons emitted between ϵ and $\epsilon + d\epsilon$ from an electron of initial energy E_0 in a plasma of density $n(\mathbf{r})$, may be written

$$\nu(\epsilon, E_0) = \int_\epsilon^{E_0} \frac{n(\mathbf{r}) Q(\epsilon, E) v(E)\, dE}{|dE/dt|}, \tag{2.5}$$

where dE/dt is the energy loss rate (here assumed a function of E only). For energy losses in a cold target due to binary collisions with the background electrons (radiation energy losses are much smaller and can be ignored), $dE/dt = -(K/E) n(\mathbf{r}) v(E)$, where $K = 2\pi e^4 \Lambda = 2.6 \times 10^{-18}$ cm^2 keV^2 (Emslie 1978), and so the total observed flux from an *injected* distribution with energy spectrum $\mathcal{F}_0(E_0)$ (electrons cm^{-2} s^{-1} keV^{-1}) is

[1] The astute reader will note that as the source volume $V \to \infty$, the value of $\overline{F}(E) \to 0$. This formal difficulty may be removed in practice by the truncation of the emission volume V at some reasonable upper limit.

$$I(\epsilon) = \frac{A}{4\pi R^2} \int_{E_0=\epsilon}^{\infty} \mathcal{F}_0(E_0) v(\epsilon, E_0)\, dE_0$$
$$= \frac{A}{4\pi R^2} \frac{1}{K} \int_{E_0=\epsilon}^{\infty} \mathcal{F}_0(E_0)\, dE_0 \int_{E=\epsilon}^{E_0} E\, Q(\epsilon, E)\, dE, \tag{2.6}$$

where A (cm^2) is the area of the flare. Reversing the order of integration in (2.6) gives

$$I(\epsilon) = \frac{A}{4\pi R^2} \frac{1}{K} \int_{E=\epsilon}^{\infty} E\, Q(\epsilon, E)\, dE \int_{E_0=E}^{\infty} \mathcal{F}_0(E_0)\, dE_0, \tag{2.7}$$

and comparing this with the fundamental equation (2.4) yields the result

$$\overline{n} V \overline{F}(E) = \frac{A}{K} E \int_{E_0=E}^{\infty} \mathcal{F}_0(E_0)\, dE_0. \tag{2.8}$$

From this it follows straightforwardly that

$$A\mathcal{F}_0(E_0) = -K \frac{d}{dE} \left(\frac{\overline{n} V \overline{F}(E)}{E} \right)_{E=E_0}, \tag{2.9}$$

permitting the determination of the quantity $A\mathcal{F}_0(E_0)$ (electrons s^{-1} keV^{-1}); this quantity represents the *rate* of injection of electrons per unit energy (e.g., Holman et al. 2003). Clearly $\mathcal{F}_0(E_0)$ is a nonnegative function, and so (2.9) constrains $\overline{F}(E)$ to either be a decreasing function of E or, at worst, a function that increases more slowly than E. As we shall see below, certain recovered forms of $[\overline{n} V \overline{F}(E)]$ (e.g., Piana et al. 2003) can have difficulty satisfying this constraint; however it is also shown that such difficulties may be removed if the effects of photospherically-backscattered (albedo) photons (e.g., Kontar et al. 2006) are taken into account in (2.4).

2.2 Free-Bound Emission

In fitting or inferring mean source electron flux spectra $\overline{F}(E)$, free-bound recombination emission by nonthermal electrons had always been neglected compared with free-free electron-ion bremsstrahlung, as argued by Korchak (1967) and Landini et al. (1973). For *hot* plasma hard X-ray sources (coronal or in soft X-ray footpoints) this is inconsistent with inclusion of recombination as significant for thermal electrons of similar energies in thermal spectrum modeling (e.g., Culhane 1969; Culhane and Acton 1970). Of importance also is the fact that the estimated coronal abundance A_Z for Fe is now much higher than in e.g. Landini et al. (1973). The recombination emission rate $\propto Z^4 A_Z$ for hydrogenic ions of charge Ze and abundance A_Z with $A_Z Z^4 \approx 1$ for H and ≈ 40 for Fe^{25+}. Brown et al. (2010) have therefore re-examined the importance of nonthermal electron recombination.

In the hydrogenic Kramers approximation, the free-bound emission rate from a plasma of proton density n_p and volume V from nonthermal electrons with mean source electron spectrum $\overline{F}(E)$ is given by Brown et al. (2010)

$$J_R(\epsilon) \approx \frac{32\pi}{3\sqrt{3}\alpha} \frac{r_e^2 \chi^2}{\epsilon} n_p V \sum_{Z_{\rm eff}} \sum_{n \ge n_{\min}} \frac{p_n}{n^3} Z_{\rm eff}^4 A_{Z_{\rm eff}} \frac{\overline{F}(\epsilon - Z_{\rm eff}^2 \chi/n^2)}{\epsilon - Z_{\rm eff}^2 \chi/n^2}, \tag{2.10}$$

where $\chi = 13.6$ eV is the H ionization potential, $Z_{\rm eff}$ the effective charge on the ion, r_e the classical electron radius and α the fine-structure constant. n is the principal quantum number

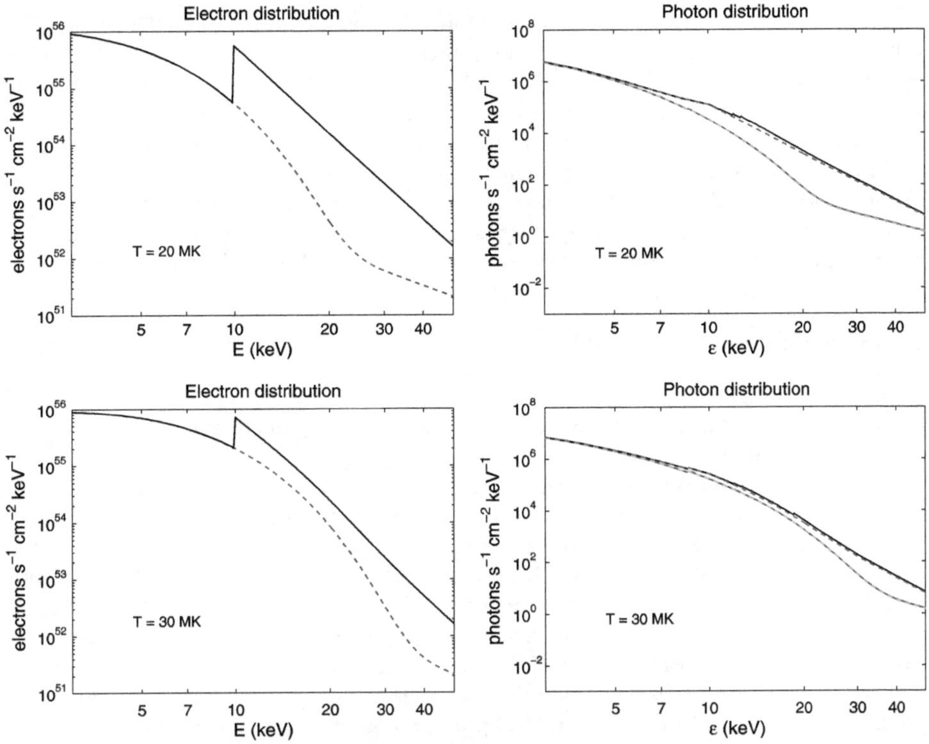

Fig. 1 *Left:* the thin-target (including both thermal and nonthermal) electron and photon (*right*) spectra for two different plasma temperatures, 20 MK (*top*) and 30 MK (*bottom*). Electron nonthermal spectra for various values of low energy cutoff E_c and electron spectral index δ: $E_c = 1$ keV and $\delta = 2$ (*blue and dashed*); $E_c = 10$ keV and $\delta = 5$ (*black and solid*); The corresponding photon spectra: *blue solid line* for $\delta = 2$, $E_c = 1$ keV with free-bound emission; *green dashed line* for $\delta = 2$, $E_c = 1$ keV without free-bound emission; *black solid line* for $\delta = 5$, $E_c = 10$ keV with free-bound; *red dashed line* for $\delta = 5$, $E_c = 10$ keV without free-bound

of the empty shell into which the electron recombines and p_n is a 'vacancy factor' which takes into account the ratio of available to total states in that shell. [Note that (2.10) applies for $\epsilon \geq E_c + Z_{\text{eff}}^2\chi/n^2$; for $\epsilon < E_c + Z_{\text{eff}}^2\chi/n^2$, $J_R(\epsilon) = 0$.] Since the recombination cross-section falls off as $1/n^3$, it is adequate to take $n = n_{\min}$, the value for the first empty n-shell. A key feature of free-bound emission, unlike free-free, is that for any specific shell, each electron energy value E maps to a unique photon energy value ϵ so that $J_R(\epsilon)$ is a much more direct reflection of $\overline{F}(E)$ than is the bremsstrahlung $J_B(\epsilon)$ convolution of $\overline{F}(E)$, the former preserving features in $\overline{F}(E)$.

Comparing (2.10) with the corresponding Kramers free-free expression, Brown and Mallik (2008, 2009) concluded that nonthermal free-bound emission is negligible in cold sources. However, they concluded that in hot plasmas ($T \approx$ 10–30 MK), such as coronal sources and the soft X-ray component of footpoint sources, free-bound emission can dominate for steep spectra and low cut-off energies (Fig. 1). Subsequently, a serious error in Brown and Mallik (2008, 2009), involving the use of incorrect values of $n_{\min}$ in their equation (13) (i.e., (2.10) above) was recognized (Brown et al. 2010). This substantially reduced both the magnitude of the cross-section and the energy shift $\epsilon - E$ involved in recombination. Because of the typically steep electron spectra involved, the latter substantially affects the number of electrons responsible for emitting the photon in question. In the amended results $J_R(\epsilon)$ is never dominant, even for hot sources, but can account for up to ~30% of the flux $J_B(\epsilon) + J_R(\epsilon)$ in the range ~10–30 keV for the non-thermal component dominating thermal as shown in Fig. 1. *For such hot sources*, the free-bound emission could be comparable with such effects as albedo and differences in bremsstrahlung cross-section used, and could be used to diagnose the sharp features in $\overline{F}(E)$ from $J(\epsilon)$. Even more important is the fact that $J_R(\epsilon)$ adds edges to the total $J(\epsilon)$ with the result that, in data with good signal to noise ratio, inversion (essentially differentiation) of data $J(\epsilon)$ to yield $\overline{F}(E)$ based on bremsstrahlung alone (essentially differentiation of J) could result in spurious features in $\overline{F}(E)$ just as happens when albedo is ignored.

2.3 Electron-Electron Bremsstrahlung

Energetic electrons propagating in the solar atmosphere encounter ions and electrons (both free and bound) and hence can produce X-ray emission via both electron-ion and electron-electron bremsstrahlung.

When the maximum electron energy is much larger than the photon energies under consideration, the photon spectrum resulting from a power-law spectrum of electrons $\overline{F}(E) \propto E^{-\delta}$ is also close to the power-law form $I(\epsilon) \propto \epsilon^{-\gamma}$ (Haug 1989). However, while for pure electron-ion bremsstrahlung $\gamma \simeq \delta + 1$, for pure electron-electron bremsstrahlung a significantly shallower photon spectrum, with $\gamma \simeq \delta$, results. Thus, the importance of the electron-electron bremsstrahlung contribution increases with photon energy and the enhanced emission per electron leads to a flattening of the photon spectrum $I(\epsilon)$ above ~300 keV produced by a given $\overline{F}(E)$ (Haug 1975) or, equivalently, a steepening of the $\overline{F}(E)$ form required to produce a given $I(\epsilon)$. Kontar et al. (2007) provide a discussion of the essential differences between electron-electron and electron-ion bremsstrahlung processes. We here provide a succinct summary of that discussion and note that the properties of electrons with the energies above ~400 keV are also crucial for ion diagnostics and radio emission.

As is well known (Koch and Motz 1959), the cross-section for electron-ion bremsstrahlung scales as Z^2. Further, when considering electron-electron bremsstrahlung, the possible binding of target electrons to their host ions in a neutral or partially-ionized medium is not significant. Hence, in a quasi-neutral target of particles with atomic number Z, the bremsstrahlung cross-section per atom for emission of a photon of energy ϵ by an electron of energy E is in general equal to

$$Q(\epsilon, E) = Z^2 Q_{e-p}(\epsilon, E) + Z Q_{e-e}(\epsilon, E), \tag{2.11}$$

where $Q_{e-p}(\epsilon, E)$ and $Q_{e-e}(\epsilon, E)$ are the cross-sections, in the laboratory frame, for electron-proton, and electron-electron bremsstrahlung, as given by Koch and Motz (1959) and Haug (1989), respectively.[2] It is also important to note that while the electron-ion cross-section is finite for all $\epsilon < E$, the cross-section for electron-electron bremsstrahlung vanishes above a maximum photon energy, due to the necessarily finite energy carried by the recoiling target electron (Haug 1975; Kontar et al. 2007). For highly *non*-relativistic electrons ($E \ll m_e c^2$), $\epsilon_{\max} \to E/2$; only for highly *relativistic* electrons ($E \gg m_e c^2$) can $\epsilon_{\max} \to E$ (see Fig. 2). As discussed by Kontar et al. (2007), this result has important implications for the form of the photon spectrum produced by electron-electron bremsstrahlung.

In Sect. 4, we discuss the application to the recovery of the electron spectrum at pertinent (mildly-relativistic) energies.

[2] Note that there is a typographical error in the form of Q_{e-e} in Haug (1989); see Kontar et al. (2007) for details.

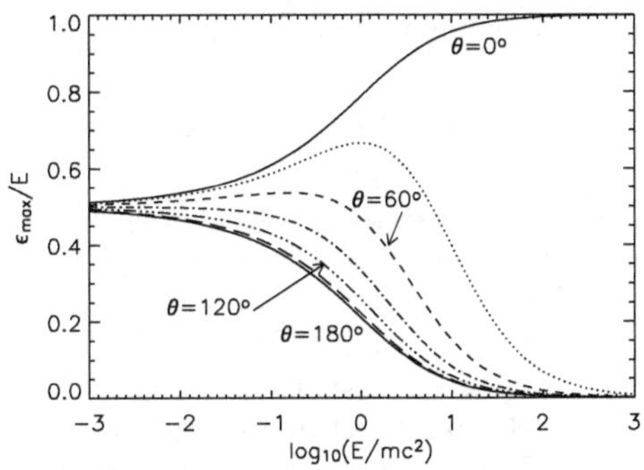

Fig. 2 Maximum photon energy $\epsilon_{\max}$ produced by electron-electron bremsstrahlung, expressed as a fraction of the incident electron energy E (in units of the electron rest mass mc^2), for various values of θ, the angle between the incoming electron and the outgoing photon trajectories. For clarity, only curves for $\theta = 0°$, $60°$, $120°$, and $180°$ are labeled; the curves for $\theta = 30°$, $90°$ and $150°$ lie between these (after Kontar et al. 2007)

3 Primary and Compton Backscattered X-rays

As discussed in Sect. 2, the atmosphere *above* the X-ray bremsstrahlung-producing region is optically thin. However, the very dense lower photospheric layers *below* the primary source are not. Consequently, photons emitted downwards are efficiently Compton backscattered by atomic electrons in the photosphere. The observed hard X-ray (HXR) flux spectrum from solar flares is therefore a combination of primary bremsstrahlung photons with a spectrally modified component from photospheric Compton backscatter of the downward-directed primary emission. This backscattered component can be significant, creating new features and/or distorting or masking features in the primary spectrum, and so substantially modifying key diagnostics such as the electron energy budget.

Photons of energy above ~100 keV penetrate so deeply that they are lost to the observer, while below ~10 keV they are mostly photoelectrically absorbed (e.g., Tomblin 1972). Therefore, the reflectivity of the photosphere has a broad hump in the range 10–100 keV, with a maximum around 30–40 keV. At some energies and view angles the reflectivity approaches 90%, so the observed spectrum may be very substantially affected by backscatter. This effect is well-known in solar physics and more generally in X-ray astronomy (see Magdziarz and Zdziarski 1995), and there have been several discussions of its influence on observed X-ray spectra (e.g., Tomblin 1972; Santangelo et al. 1973; Bai and Ramaty 1978), and on the electron spectra inferred from them (e.g., Johns and Lin 1992b; Alexander and Brown 2002; Kašparová et al. 2005; Kontar et al. 2006).

Understanding and modeling backscatter (albedo) has become even more important with the advent of high quality X-ray spectra from *RHESSI* (Lin et al. 2002) with spectral resolution as high as $\simeq$1 keV, in combination with uncertainties as low as a few percent (for strong flares). Generally, contamination of the observed X-ray spectrum by reflected photons leads to a flattening of the spectrum and hence to an underestimation of the electron spectral index if the contribution of backscattered photons is not taken into account (Bai and Ramaty 1978). Extrapolation to low electron energies using such an underestimated spectral index leads to a substantial underestimation of the total electron energy in a flare. Indeed, a low-energy cutoff in an uncorrected electron spectrum is not even required if a "true" primary (albedo-corrected) electron spectrum can be used (Kontar et al. 2006).

3.1 Spectroscopy of the Photospheric Albedo

Downward-emitted photons are either absorbed or Compton-scattered, with some of the latter returned toward the observer, adding to the total flux detected (X-ray albedo). Scattering takes place on electrons, whether free or atomic. To account for elements heavier than hydrogen, the Compton cross-section is multiplied by an effective mean atomic number $Z = 1.2$ (e.g., Aschwanden 2005, for element abundances). The detailed density structure of the medium is irrelevant (Tomblin 1972).

Absorption, on the other hand, does depend strongly on chemical composition, and the best estimate of photospheric abundances should be included. The heavy elements Fe/Ni play the most important role from 6–8 keV up to ~30 keV, while lighter elements contribute below 6 keV (Morrison and McCammon 1983).

3.1.1 Green's Function Approach

Propagation, absorption, and Compton scattering of primary hard X-rays can be straightforwardly studied using Monte-Carlo simulations (e.g., Bai and Ramaty 1978). This technique is ideal for obtaining the reflected, and hence the total observed, photon spectrum for a given form of the primary X-ray spectrum. However, the primary spectrum is generally unknown and unlikely to be an exact power-law, as is sometimes assumed. Therefore, an approach independent of the primary spectrum is required (Kontar et al. 2006).

For any isotropic primary spectrum $I_P(\epsilon_0)$ (photons cm^{-2} s^{-1} keV^{-1}) we can write the secondary, backscattered spectrum $I_S(\epsilon)$ as

$$I_S(\epsilon,\mu) = \int_{\epsilon}^{\infty} I_P(\epsilon_0) G(\mu,\epsilon,\epsilon_0)\, d\epsilon_0, \tag{3.1}$$

where $G(\mu,\epsilon,\epsilon_0)$ is an angle-dependent Green's function and $\mu = \cos\theta$, where θ is the heliocentric position angle of the primary hard X-ray source, or the angle between the Sun center—observer and Sun center—X-ray source lines. The observed spectrum, at photon energy ϵ and direction μ, is $I_S(\epsilon) + I_P(\epsilon)$. The importance of not averaging over viewing angle may be seen from Bai and Ramaty (1978). Using Green's function analytical fits to Monte Carlo simulations derived by Magdziarz and Zdziarski (1995), Kontar et al. (2006) calculated functions $G(\mu,\epsilon,\epsilon_0)$ shown in Fig. 3 for solar flare parameters. Green's functions, as calculated, account for Compton scattering and bound-free absorption.

The shape of the Green's function depends on the energy of the primary photon. For primary photons with low energies $\epsilon_0 < 30$ keV, the Green's function has a rather simple structure close to a Dirac delta-function (Fig. 3), showing that backscattering is dominated by the first scattering (especially at low energies), with the contributions from higher orders of scattering being generally small.

3.1.2 Reflected X-ray Photon Spectrum

Because the reflectivity is spectrum-dependent (Fig. 4), the albedo spectrum also depends on the shape of the primary spectrum (Kontar et al. 2006; Kotoku et al. 2007). While previous studies considered only the results for prescribed power-law or thermal primary spectra (cf. Bai and Ramaty 1978; Johns and Lin 1992b; Alexander and Brown 2002), the Green's function method allows more general forms of the primary spectrum.

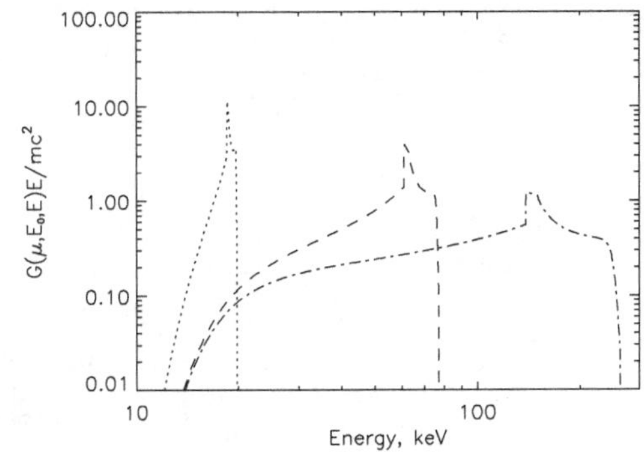

Fig. 3 Green's functions $G(\mu, \epsilon, \epsilon_0)$ [keV^{-1}], including Compton scattering and photoelectric absorption, for three primary photon energies $\epsilon_0 = 20, 80, 300$ keV and $\mu = 0.7$ ($\theta \approx 45°$) calculated using approximations (Magdziarz and Zdziarski 1995) for solar conditions (after Kontar et al. 2006)

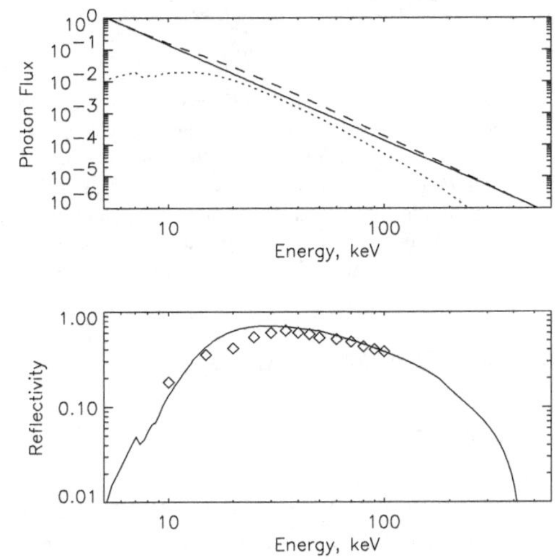

Fig. 4 *Upper panel:* Primary (*solid line*), reflected (*dotted line*) and total (*dashed line*) photon spectra calculated assuming a primary spectrum $I_P(\epsilon) \propto \epsilon^{-3}$, and using the Green's function for an X-ray source at heliocentric angle $\theta = 45°$. *Lower panel:* Reflectivity, defined as the ratio of reflected to primary fluxes $R(\epsilon, \theta = 45°) = I_R(\epsilon)/I_P(\epsilon)$. The reflectivity taken from Bai and Ramaty (1978) is shown with *diamonds*. Two absorption edges of Fe at 7.1 keV and Ni at 8.3 keV are seen in the reflected component (after Kontar et al. 2006)

The total observed spectrum $I(\epsilon)$ is given by

$$I(\epsilon, \mu) = I_P(\epsilon) + \int_\epsilon^\infty I_P(\epsilon_0) G(\mu, \epsilon, \epsilon_0)\, d\epsilon_0. \tag{3.2}$$

For a measured $I(\epsilon, \mu)$, we may obtain the primary spectrum $I_P(\epsilon)$ by solving the integral equation (3.2). In practice measurements yield discrete quantities and the integral equation (3.2) is used in the matrix form

$$I(\epsilon_i, \mu) = I_P(\epsilon_i) + G_{ij}(\mu) I_P(\epsilon_j), \tag{3.3}$$

where we have used the summation convention for repeated indices, and introduced the *Green's matrix*

$$G_{ij}(\mu) = \int_{\epsilon_j}^{\epsilon_{j+1}} G(\mu, \epsilon_i, \epsilon_0)\, d\epsilon_0. \tag{3.4}$$

Due to sharp features in the Green's function (Fig. 3), the integration in (3.4) is best performed via a change of variable to the wavelength (reciprocal energy) domain.

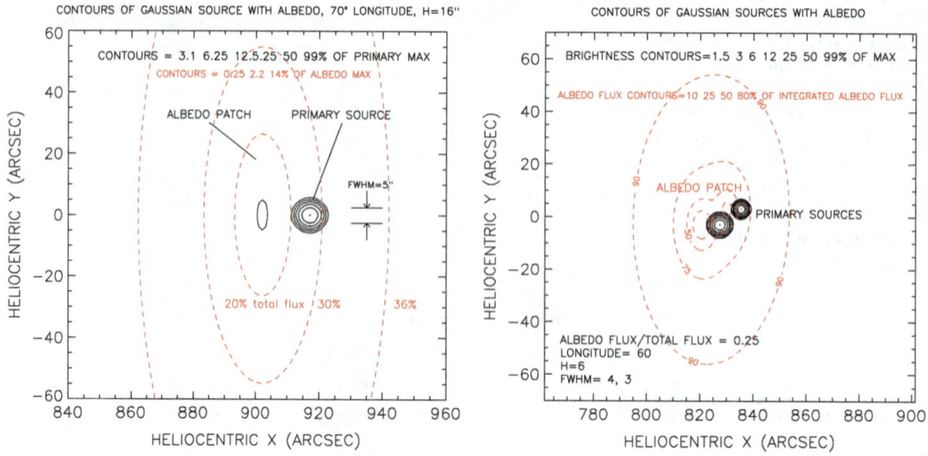

Fig. 5 *Left panel:* Model of a single primary source with albedo contours. The *solid contours* at logarithmically-spaced intervals show the primary source down to 6.25% of the maximum, and the brightest portion of the albedo patch at 3.1% of the peak of the primary. The *dashed red contours* show the integrated flux of the albedo patch. Note that 50% of the albedo flux arises from an area about 10 times larger than the primary source, given its height of 12 Mm. *Right panel:* A model of a double source 10 Mm above the photosphere with the resultant albedo patch shown by flux contours (*red dashes*). There is considerable overlap of the primary source and albedo patch in this case (after Schmahl and Hurford 2002)

3.2 Imaging of Photospheric Albedo

3.2.1 Expected Spatial Signatures of Albedo

Before considering observational approaches to the spatial isolation of albedo, it is instructive to consider an elementary model of the solar backscattering process. To do this, we make four simplifying assumptions (e.g., Brown et al. 1975): first, that the primary bremsstrahlung is generated as a point source at a height h above a planar scattering surface; second, that this surface is perpendicular to the line of sight (e.g., disk center flare); third, that the primary X-ray emission and backscattering processes are isotropic; and fourth, that absorption of the scattered photons can be neglected. With such a scenario, the albedo source would extend out to the horizon as seen from the primary source. The albedo surface brightness would be determined by the distance from the primary source and the scattering location and by a cosine illumination factor and so the surface brightness would fall off as $(1 + [r/h]^2)^{-3/2}$, where r is the radial distance between the scattering location and the sub-source point on the scattering surface. Note that the scale of the radial profile is determined by the height of the primary source.

The implications of easing some of the restrictions in this simple model are illustrated in Fig. 5. For a primary source with finite size or structure, the resulting albedo patch would be a convolution of the primary source with the aforementioned profile. For a primary source located away from disk center, the center of the albedo patch is displaced toward disk center and presents an elliptical shape oriented parallel to the limb.

More rigorous calculation of the spatial properties of an albedo patch are provided by Bai and Ramaty (1978) who show, for example, that limb darkening would make the albedo more easily detected near disk center. In addition, the albedo patch will be energy- and primary-spectrum-dependent (Kontar and Jeffrey 2010). Nevertheless, the simplified model does suggest that the albedo has three potentially observable spatial signatures: first, for sufficiently high primary source altitudes, the albedo would be significantly larger in extent than the primary source, with a size scale that increases with source height; second the albedo source would be displaced toward disk center by a distance $h \sin\theta_{\odot}$, where $\theta_{\odot}$ is the heliocentric angle; third, the albedo source would be elongated parallel to the limb with a minor to major axis ratio of $\cos\theta_{\odot}$. Observationally, however, because of the relative size scales of the primary source and its albedo patch, the albedo surface brightness would be only a small fraction (typically only a few percent) of that of the primary source. This would pose a potential challenge for conventional imaging systems because of scattered light; for reconstructed *RHESSI* hard X-ray images, typically limited in dynamic range to about 10:1, this would seem to make the spatial detection of albedo even more problematic.

3.2.2 The Spatial-Frequency Signature of Albedo

The potential observational difficulties posed by the low surface brightness of the albedo source can be eased if we Fourier transform the source of size $\sim d$ with the distribution $\exp(-x^2/d^2)$ and consider the amplitude of the Fourier components as a function of spatial frequency $\propto \exp(-k^2 d^2)$, where k is the spatial frequency. Compact primary sources have Fourier amplitudes that fall off at high spatial frequencies $kd \gg 1$. For spatial periods large compared to the source dimension $kd \ll 1$, however, the amplitudes are effectively constant. Further, the Fourier transform of the $(1 + [r/h]^2)^{-3/2}$ profile of an albedo source implies an amplitude that varies as e^{-kh}, where k is the spatial frequency. As illustrated in the lower half of Fig. 2 in Schmahl and Hurford (2002), this could be readily distinguished from that of the primary source. In effect. the Fourier transform integrates over the faint, distributed albedo source so that it becomes potentially detectable. The most obvious albedo signature would then be an "excess" in Fourier amplitude at low spatial frequencies over that expected for the compact primary source. The excess would be comparable to the reflected fraction, viz., up to several tenths (Bai and Ramaty 1978; Kontar et al. 2006). This is relevant to *RHESSI*, because with nine logarithmically-distributed spatial frequencies, its rotating modulation collimators (RMCs) directly measure the Fourier components of the source distribution.

3.2.3 Detection of a "Halo" Component

Applying such considerations to flare observations, early analysis of *RHESSI* data (Schmahl and Hurford 2002) directly fit the observed modulated light curves to derive an average modulation amplitude for each RMC. This technique, although simple, assumed circularly symmetric sources and was subject to potential statistical issues since it directly fit the counts which were only sparsely populating the short time bins. Nevertheless, the analysis showed (Fig. 6) that the modulation amplitude continued to increase toward the coarsest subcollimator even though the latter had a 183″ spatial resolution (Hurford et al. 2002). This was taken as evidence of a hard X-ray "halo" component, consistent with expectations from albedo.

The analysis was refined by Schmahl and Hurford (2003), who formed individual back-projection images with the nine RMCs. In this case, the peak calibrated intensity of each back-projection image corresponded to the azimuthally-averaged modulation amplitude for the corresponding spatial frequency. The results, illustrated in Fig. 7, confirmed the presence of a non-Gaussian large-scale component consistent with a halo source. This technique alleviates potential statistical concerns associated with directly fitting the sparsely-populated time bins, but still requires circular sources for unambiguous interpretation.

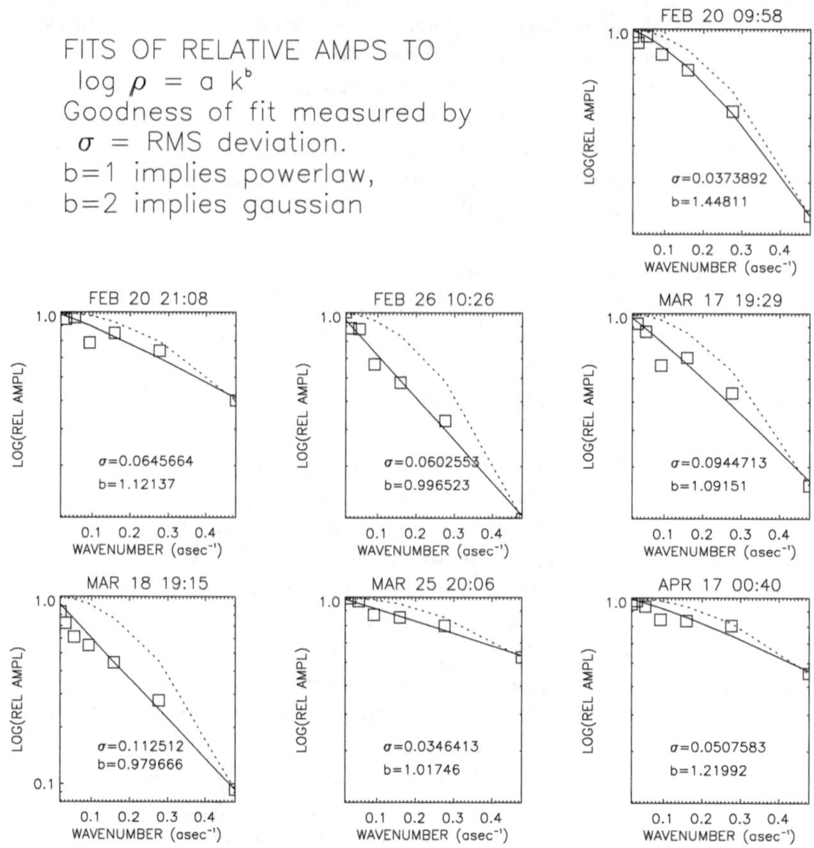

Fig. 6 Relative amplitudes of 7 flares vs. spatial frequency in the energy range 12–25 keV. The departures of the profiles from a Gaussian shape are interpreted as a "halo" component. Since (for computational simplicity) the fits were assumed to be azimuthally symmetric, the profiles must be considered to be polar averages over 360 degrees of rotation (after Schmahl and Hurford 2002)

3.2.4 Direct Use of Visibility Measurements

Since the initial reports of the halo component, analysis techniques have been developed that enable the *RHESSI* data to be directly transformed into calibrated measurements of the visibilities (specific Fourier components) as a function of spatial period and orientation (Schmahl and Hurford 2002; Schmahl et al. 2007). There are several reasons why visibilities are more appropriate for albedo determination: the visibilities are fully calibrated so that instrumental issues can be cleanly separated from solar issues; the expected visibilities can be directly calculated from source models; each visibility measurement is independent and has well-determined statistical errors whose propagation can guide subsequent conclusions; visibilities can be readily calculated using code that is now an integral part of the *RHESSI* object-oriented software package; visibilities are determined from linear transforms of the observed count rates, so that they can be combined in time or energy as desired; visibilities are well-suited to more complex sources since an observed visibility is the sum of the corresponding visibilities of its components.

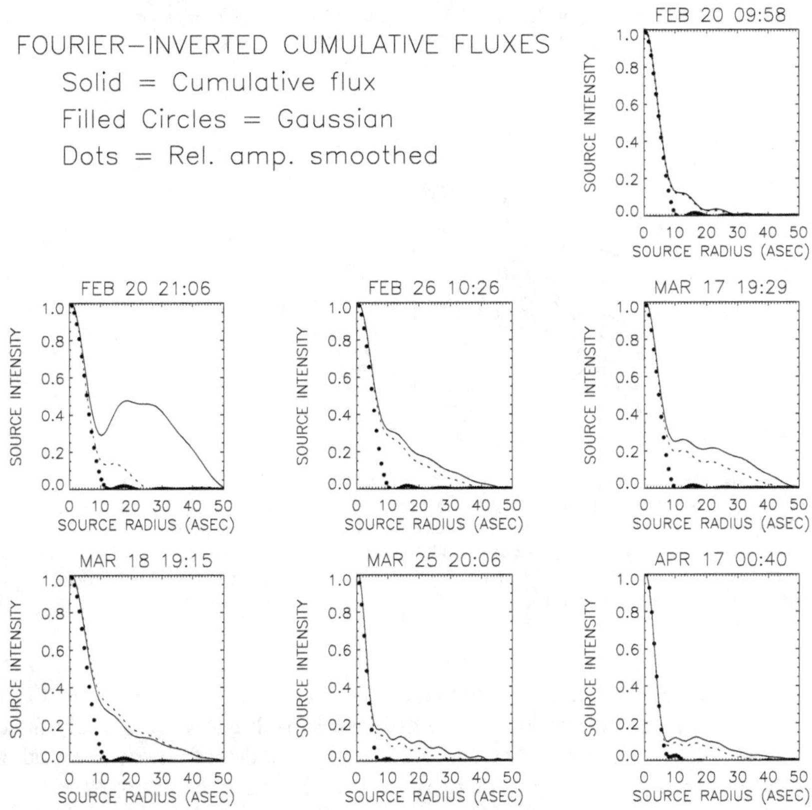

Fig. 7 A Fourier inversion produces the brightness profile in the energy range 12–25 keV. Since the albedo surface brightness is very low, the cumulative integrated flux from r to ∞ is plotted as a function of r. If the primary source profile is Gaussian, the integrated flux profile is Gaussian. The clear departures from a Gaussian shape in these profiles indicates the presence of extended sources inferred to be albedo patches (after Schmahl and Hurford 2003)

Figure 8 illustrates observations expressed in terms of visibilities. This event occurred at about 60° longitude, so the albedo emission is expected to be relatively weak, although there may be a signature of albedo for subcollimators 8 and 9, as indicated by the excess observed flux over the model flux.

3.2.5 Future Prospects for Visibility-Based Albedo Measurements with RHESSI

In the context of albedo, visibilities are used in two ways: first, to generate maps of the source using a direct visibility imaging method [such as back-projection (MEM-NJIT) (Schmahl et al. 2007) to obtain primary source positions and fluxes; and second, to use a visibility-based forward-fit algorithm to parameterize simple models of the primary source and the corresponding parameters of the albedo component (size, intensity, ellipticity, location) or maximum entropy.

The MEM-NJIT mapping algorithm is useful for obtaining the qualitative source configuration, although it is less reliable for quantitative measurement of source sizes. On the other hand the forward-fit algorithm requires a good starting point in parameter space, but can yield reliable source parameters.

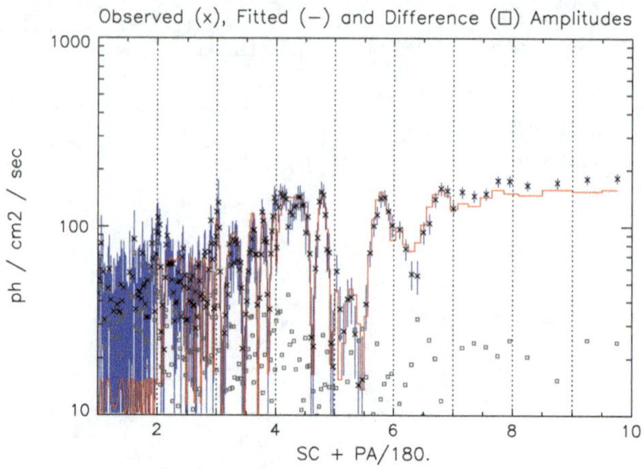

Fig. 8 Observed visibility amplitudes (*crosses* with *blue error bars*) for a flare interval as a function of subcollimator (SC = 1–9) and position angle (PA = 0–180°) of the grids in the energy range 12–25 keV. Each of the 9 *vertical panels* shows the amplitude as a function of PA for one subcollimator (labeled by *digits below the X-axis*). The *red curve* represents a model using two Gaussian sources, and the *squares* show the residuals relative to the model. For a given subcollimator (5 and 6 are good examples), the amplitude rises and falls while the grids rotate from PA = 0 to PA = 180°. Such patterns can correspond to an extended or double source

There are practical difficulties that remain to be overcome. As suggested by Fig. 8, the relative calibration (to within ∼5%) of the different subcollimators (especially the coarse ones) is critical to the isolation of albedo. Regrettably, the detector-to-detector calibration cannot currently support such an objective. However, using relative visibilities (viz., normalizing each RMC's visibilities to that detector's spatially integrated response) eliminates the effects of the detectors' relative efficiencies and so provides a potential approach to bypassing the calibration issue. Confirmation of this approach could be achieved with near-limb flares (presumed to be without significant albedo) or at energies where the albedo component is minimal. Other potential improvements are corrections for azimuthal averaging in each visibility, improvements in the present forward fit algorithm, perhaps with the use of an alternative search algorithm and with the use of third harmonic to add additional spatial frequencies visibilities.

The first flares for which albedo can be parameterized will necessarily be spatially simple (single, compact, and strong) and located within a few arc min of disk center. Eventually, it is expected that spectroscopic tools for albedo will be combined with imaging tools to provide comprehensive albedo information for a large subset of *RHESSI* flares. Since the albedo intensity and location depends on electron directivity, the potential reward of spatially-based albedo diagnostics is well worth the effort required to refine the analysis tools (Kontar and Brown 2006b).

4 The Electron Energy Spectrum

In this section, the angular dependence of the bremsstrahlung cross-section and the angular/spatial/temporal characteristics of the electron distribution are neglected. The primary source photon spectrum $I_P(\epsilon)$ may be therefore treated simply as the convolution in electron energy of the solid-angle-averaged bremsstrahlung cross-section and the mean source electron flux (2.4).

In general, we first correct for instrumental effects (see Smith et al. 2002; Schwartz et al. 2002) such as pulse pileup,[3] then correct the observed hard X-ray spectrum for albedo effects (Sect. 3.1) to obtain the primary source spectrum $I_P(\epsilon)$. As explained in Sect. 2, the functional form of the primary photon energy spectrum $I(\epsilon)$ (subscript 'P' hereafter understood) contains crucial information on the form of the *mean source electron flux spectrum* $\overline{F}(E)$, information that may in turn be used to reveal properties of the electron acceleration and propagation processes.

We first discuss forward fitting (Sect. 4.1) and regularized inversion (Sect. 4.2) methods of extracting the electron energy spectrum $\overline{F}(E)$ from noisy hard X-ray data $I(\epsilon)$.

4.1 Forward Fitting

Forward fitting is the process of quantitatively comparing a parameterized model with observational data. Criteria are established to determine acceptable fits and, if the model is capable of providing acceptable fits, a best fit, that gives the most probable values of the model parameters, is determined by minimizing chi-squared (χ^2), the sum of the squares of the normalized residuals.[4]

Models may be based on the apparent structure of the data, a physical model, or a combination of the two. At the lowest X-ray energies, *RHESSI* flare spectra can contain *thermal bremsstrahlung* and *free-bound* (recombination) continua from the hottest plasma in flares, in addition to spectral lines (e.g., Phillips et al. 2006). The X-rays at higher energies (normally $\gtrsim$10–20 keV) are dominated by electron-ion bremsstrahlung from energetic, nonthermal electrons (Sect. 2.1). Free-bound radiation from nonthermal electrons may also contribute (Sect. 2.2). At higher energies, electron-electron bremsstrahlung can become significant (Sect. 2.3). At γ-ray energies, spectral lines excited by energetic, nonthermal ions and positronium continuum emission can be present. The quality of a fit is also dependent on a careful subtraction of background counts before obtaining the spectral fit. Spectral fits are only reliable over the range of photon energies for which the flare emission is well above the background.

RHESSI's high spectral resolution often makes it easy to distinguish the thermal component of an X-ray spectrum from the nonthermal component, especially in large flares. An example of this is shown in the top panel of Fig. 9, a spatially-integrated spectrum from SOL2002-07-23T00:35 (X4.8); below a photon energy ~40 keV, the thermal component clearly dominates over the flatter nonthermal component. For many spectra, however, the thermal and nonthermal components are not so clearly distinguishable. This was the case during the early rise of SOL2002-07-23T00:35 (X4.8), for example (Holman et al. 2003). For these spectra, the most likely model can sometimes be deduced from the time evolution of the flare spectra and/or flare images (cf. Sui et al. 2005).

The thermal component of the spectrum can typically be well fitted with the form of $I(\epsilon)$ from an isothermal plasma at temperature $T: I(\epsilon) \propto (EM/\epsilon T^{1/2}) \exp(-\epsilon/kT)$, where k is

[3]Pulse pileup is an issue for large solar flare spectra with high count rate in *RHESSI* detectors. Un-physical counts are recorded when pairs (or more) of low-energy photons, arriving nearly simultaneously, are detected as a single energy count at higher energies (for details, see Datlowe 1976b, 1976a). Current pileup corrections for spatially integrated spectra have limited precision that might be inadequate for the events with extremely high count rates. As of April 2011, no standard image pileup corrections yet exist.

[4]A normalized residual is defined as the difference between the measured and model-predicted value, divided by the uncertainty in the measured value. The division by the uncertainty gives greater weight to measured values with smaller relative uncertainties (e.g., Chap. 15 of Press et al. 1992).

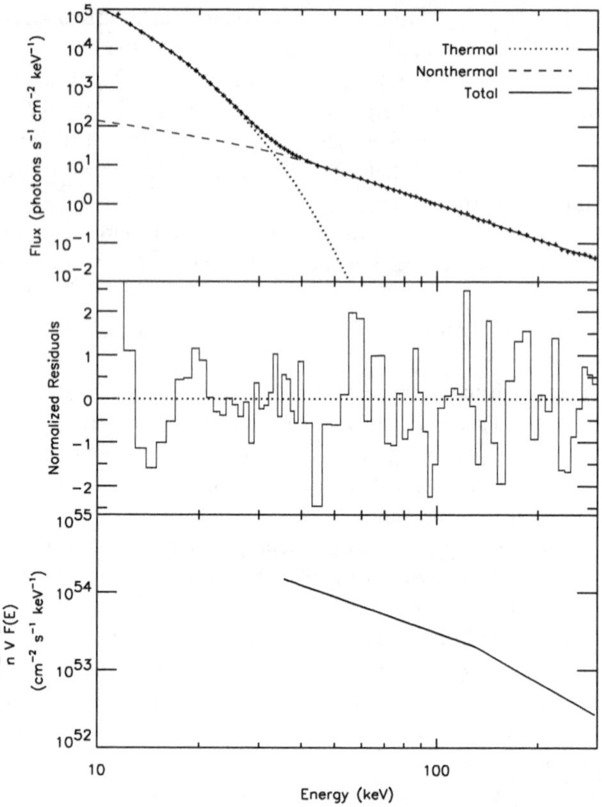

Fig. 9 Fit to a spatially-integrated spectrum from SOL2002-07-23T00:35 (X4.8). *Top panel:* Photon flux spectrum (*plus signs*) integrated over the 20 s time period 00:30:00–00:30:20 UT. A fit to the spectrum (*solid curve*) consisting of the sum of the bremsstrahlung from an isothermal plasma (*dotted curve*) and the bremsstrahlung from a double power-law mean electron flux distribution with a low-energy cutoff. *Middle panel:* Residuals from the fit in the *top panel* (observed flux minus model flux divided by the 1σ uncertainty in the observed flux). *Bottom panel:* Best fit mean electron flux distribution times the mean plasma density and source volume, plotted as a function of electron energy in keV (after Holman et al. 2003). Note that pulse pileup (Smith et al. 2002; Kontar et al. 2003) might be an issue for this flare

Boltzmann's constant and EM is the emission measure ($\int n_e n_i \, dV$). The fit in Fig. 9, for example, gave a temperature of 37 MK and an emission measure of 4.1×10^{49} cm^{-3}. A fit to thermal bremsstrahlung alone is often adequate, but this does not account for the spectral line complexes at $\sim$6.7 keV and $\sim$8 keV or for recombination radiation.[5] The thermal component of *RHESSI* spectra is now routinely fitted (included into standard *RHESSI* software) using the latest version of Chianti (Landi et al. 2006), which incorporates all the emission mechanisms important at low energies. The multithermality of plasma and the corresponding emission measure differential in temperature are addressed in Sect. 4.6.

The nonthermal component of the spectra can usually be fitted adequately with either a single or a double power-law photon flux spectral model. Sometimes a third, flatter power-law component is included at low energies to simulate a low-energy cutoff in the electron distribution. Such fits are useful for examining the evolution of flare spectra with time. They do **not**, however, contain any direct physical information about the electrons responsible for the observed emission. It is therefore more interesting to fit the photon spectra with the radiation from a model *electron* distribution, typically assumed to have the form of a double power law with a possible low- and/or high-energy cutoff. This form allows sharp breaks in

[5] RHESSI spectroscopy in the range of energies below $\sim$10 keV is often complicated by non-diagonal instrument response (Smith et al. 2002; Schwartz et al. 2002) and some instrumental features (Phillips et al. 2006).

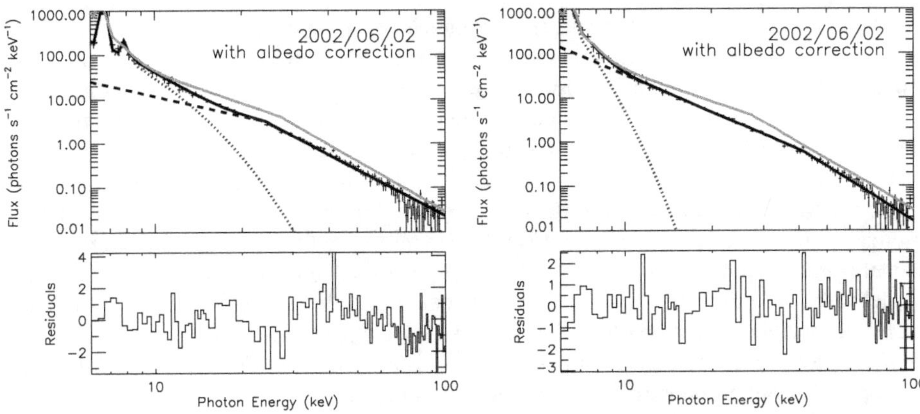

Fig. 10 Albedo-corrected *RHESSI* spectrum (*crosses* with *error bars*) at the hard X-ray peak (11:44:28–11:44:32 UT) of SOL2002-06-02T20:44 (M1.0). The *solid line* shows the combined isothermal (*dotted line*) plus double power-law (*dashed line*) spectral fit. The spectral fit before albedo correction is overlaid (*gray, solid line*). *Left panel:* The nonthermal part of the spectrum is fitted to a double power-law model with a break energy at ∼21 keV. *Right panel:* The same spectrum is fitted to a double power-law model with a break energy at ∼40 keV. The reduced χ^2 values of the fits in the *left and right panels* are 1.4 and 1.0, respectively. The normalized residuals are plotted below each spectrum (after Sui et al. 2007)

the electron distribution (either a mean electron flux $\overline{F}(E)$; Equation (2.4) or, for a thick-target model, an injected electron distribution $\mathcal{F}_0(E_0)$; Equation (2.7)). However, due to the filtering of the bremsstrahlung cross-section $Q(\epsilon, E)$, such breaks are generally smoothed out in the corresponding photon spectrum $I(\epsilon)$.

The nonthermal part of the flare spectrum in Fig. 9 is fit with the bremsstrahlung from a double power-law mean electron flux distribution with a low-energy cutoff

$$\overline{F}(E) = \begin{cases} 0; & E < E_c \\ AE^{-\delta_1}; & E_c < E < E_b \\ AE_b^{\delta_2-\delta_1}E^{-\delta_2}; & E_b < E \end{cases} \tag{4.1}$$

(bottom panel). The highest value of the low-energy cutoff E_c consistent with a good fit to the data was used; the value of E_c is not constrained below this value because of the dominance of thermal radiation. This fit therefore provides a lower limit to the energy in nonthermal electrons. The spectrum could not be acceptably fit with a single power law; note that the location of the break energy E_b is at a higher energy than the apparent location of the break in the photon spectrum; this is because all electrons with energies above a given photon energy contribute to the radiation at that photon energy.

The photon spectrum residuals (using a sum of the isothermal and nonthermal models) are shown in the middle panel of Fig. 9. Besides providing a reduced χ^2 (χ^2 divided by the number of degrees of freedom in the fit) close to 1, the residuals from a good fit should be random and uncorrelated and have a near-normal distribution $N(0, 1)$. For the event in question, the residuals do *not* exhibit this desired behavior, with significant deviation at photon energies between 10 and 15 keV (see Caspi and Lin 2010). Practically, a broad spectral "line" is often included in the model to account for this feature; with the inclusion of this ad hoc feature, it is generally possible above ∼10 keV to obtain good fits to the *RHESSI* spectra without assuming the presence of any systematic uncertainty in the data above the level of Poisson noise.

A comparison of a poor forward fit (left panels) to a good forward fit (right panels) is shown in Fig. 10. Both fits are to the same spectrum and in both the model is bremsstrahlung from an isothermal plasma plus a double power-law photon spectrum. In the fit on the left, the break energy E_b is fixed at a value ~21 keV, so the model is a single power law above this energy. In the fit on the right, the break energy is allowed to adjust to a value that gives the best fit to the data. The reduced χ^2 for the fit on the right is 1.0; in the fit on the left it is 1.4, still consistent with an "acceptable" fit. However, the "long wavelength" oscillation in the residuals for the fit on the left provides the most obvious clue that the model is *not* adequate; the overall shape of the spectrum is not consistent with the assumed model form. Adding more parameters (in this case a variable break energy E_b) results in a more acceptable distribution of residuals, as in the right panel (Fig. 10).

The gray curves in Fig. 10 are fits to the data before albedo, assumed to be from isotropically-emitted photons, is taken into account (see Sect. 3.1). They demonstrate the significant impact that albedo can have on the inferred spectrum of the emitted radiation.

An inadequately resolved issue with forward fitting is determining the uncertainty in the fit parameters and the resulting model function. Since, in general, the fit is not linear, the fit parameters are not independent and the uncertainties are not necessarily even symmetric around the best-fit values. In Fig. 10, for example, notice that the temperature of the isothermal component is adjusted to a higher value to compensate for the low break energy in the double power-law component (Sui et al. 2007). The uncertainty in the value of the low-energy cutoff to the mean electron flux fit function in Fig. 9 is small in the positive (higher-energy) direction, but indefinitely large in the negative direction! As long as a good initial choice is made for the fit parameters, the process of obtaining the best fits is relatively quick. An efficient method for determining the uncertainties in the fit parameters and function is not in place, however. Bayesian Monte Carlo approaches to determining these uncertainties are robust but slow. A practical solution to this important issue is badly needed.

4.2 Regularized Inversion

The unprecedented energy resolution of *RHESSI* hard X-ray spectra has introduced the need, perhaps for the first time in solar hard X-ray spectroscopy, to apply sophisticated mathematical tools for information retrieval in order to fully exploit the physical significance of the recorded spectra. It is well established that the most effective mathematical framework for this problem lies in the theory of linear inverse problems (e.g., Craig and Brown 1986). In this setting, linear integral equations of the first kind relate the photon spectrum ("data") to the electron spectrum ("source function"). Such equations are usually ill-posed in the sense of Hadamard—the effects of the experimental noise can be strongly amplified by the intrinsic numerical instability of the model (for details, see Bertero et al. 1985).

Standard approaches to obtain a solutions of inverse problem in solar X-ray spectroscopy are based on forward-fitting (e.g., Holman et al. 2003) the photon flux spectrum with parametric forms of the electron flux spectrum (see Sect. 4.1). However, in forward-fitting, the number of parameters utilized in the input form is generally small. This imposes severe, possibly artificial, constraints on the allowable form for the source function and is the main reason why inversion techniques (e.g., Bertero et al. 1985; Piana 1994), which find the best model-free non-parametric fit to the data subject to physically sound constraints, are currently a very promising approach to data analysis in solar hard X-ray spectroscopy.

A particularly promising technique is that of regularized inversion (e.g., Bertero et al. 1985). The essence of the regularization technique is to seek a least-squares solution of the pertinent integral equation (e.g., (2.4)) within a subset of the solution space which accounts for some measure of a priori information on the source function. Consider the linear system

$$\mathbf{g} = \mathbf{A}\mathbf{f}, \tag{4.2}$$

which represents a discretized version of the (Volterra) integral equation (2.4), where

$$A_{ij} = \frac{1}{4\pi R^2} Q\left(\frac{\epsilon_i + \epsilon_{i+1}}{2}, \frac{E_j + E_{j+1}}{2}\right)\delta E_j, \quad i = 1, \ldots, N; \ j = 1, \ldots, M; \tag{4.3}$$

$\mathbf{f}$ is the "source vector" $\overline{n}V(\overline{F}(E_1), \ldots, \overline{F}(E_M))$, $\mathbf{g}$ is the "data vector" $(g(\epsilon_1), \ldots, g(\epsilon_N))$ (with $M \geq N$), and the $\delta\epsilon_i$ and δE_j are appropriate weights. The values $g(\epsilon_i)$ correspond to a set of discrete photon counts in energy bands $\epsilon_i \rightarrow \epsilon_i + \delta\epsilon_i$, while the $\overline{F}(E_j)$ are the corresponding values of the mean electron flux in energy bands $E_j \rightarrow E_j + \delta E_j$. Owing to the strong smoothing properties of the integral operators, the matrix $\mathbf{A}$ is quasi-singular and standard inversion routines cannot be effectively applied. However, Tikhonov regularization theory (Tikhonov 1963) obtains sufficiently smooth source functions as the (unique) solution of the minimization problem

$$\|\mathbf{g} - \mathbf{A}\mathbf{f}\|^2 + \lambda\|\mathbf{C}\mathbf{f}\|^2 = \text{minimum}, \tag{4.4}$$

where λ is the *regularization parameter*, which tunes the trade-off between the fitting term $\|\mathbf{g} - \mathbf{A}\mathbf{f}\|^2$ and the penalty term $\|\mathbf{C}\mathbf{f}\|^2$. If $\mathbf{C} = \mathbf{I}$, the method is termed *zero-order regularization* (Piana et al. 2003), while if $\mathbf{C}$ is the matrix corresponding to numerical differentiation, the method is termed *first-order regularization* (Kontar et al. 2004). The optimal choice of the parameter λ can be accomplished by means of some optimization approach or by means of a semi-heuristic technique based on a statistical analysis of the cumulative residuals (Piana et al. 2003).

The main disadvantage of using Tikhonov regularization is that for noisy data, solutions with negative (unphysical) values might result. A possible solution to this is provided by the projected Landweber method (Piana and Bertero 1997)

$$\mathbf{f}_{n+1} = P_+[\mathbf{f}_n + \tau\mathbf{A}^T(\mathbf{g} - \mathbf{A}\mathbf{f}_n)], \qquad \mathbf{f}_0 = 0, \tag{4.5}$$

where P_+ sets to zero all negative components at each iteration and τ is a relaxation parameter. In this framework, the tuning between stability and fitting is realized by applying some optimal stopping rule to the iterative procedure.

4.2.1 Validation of Regularization Techniques

The effectiveness of different inversion algorithms, including a standard forward-fitting technique, have been tested by Brown et al. (2006) using synthetic data. In this test six forms of the mean source electron spectrum $\overline{F}(E)$ in Fig. 11 (bottom), each one characterized by specific features like bumps or cutoffs, were used to generate the corresponding hard X-ray spectra $I(\epsilon)$ in Fig. 11 (top), and a realistic amount of Poisson noise added. In all cases, the photon spectra look smooth and quite similar in their shape, while the corresponding mean electron spectra exhibit very irregular behavior that is filtered out by the smoothing effect of the bremsstrahlung cross-section. These differences epitomize the mathematical concept of ill-posedness.

The comparisons used four different techniques: zero-order (Piana 1994) and first-order (Kontar et al. 2004) Tikhonov regularization, triangular matrix row elimination with variable

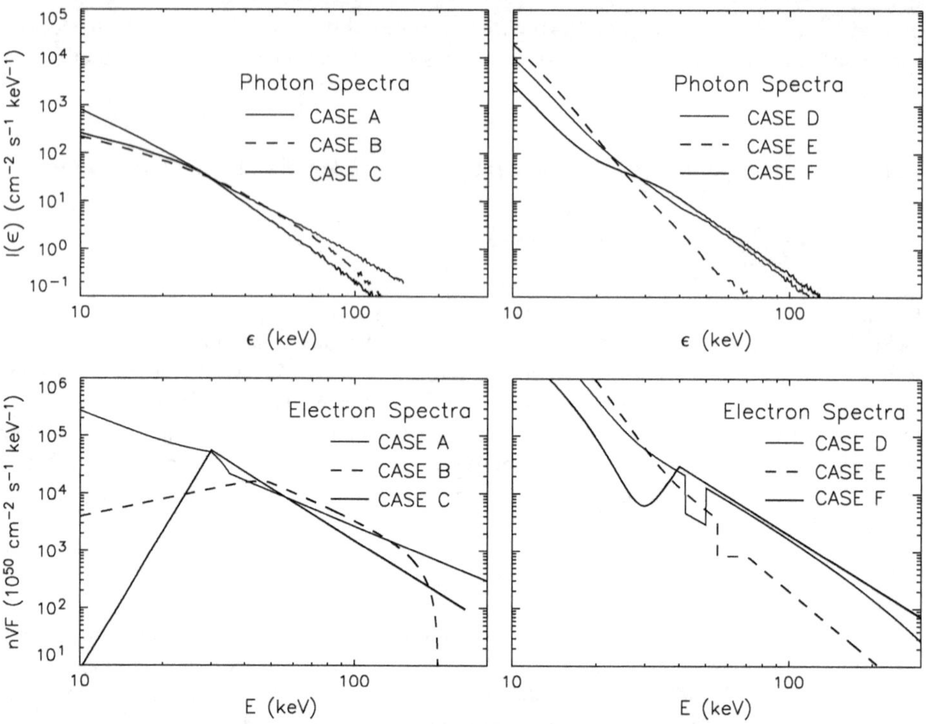

Fig. 11 Synthetic mean source electron spectra (*bottom*) and the corresponding photon spectra (*top*) in a "blind" experiment for assessing the effectiveness of different reconstruction methods in *RHESSI* X-ray spectroscopy (after Brown et al. 2006)

energy binning (Johns and Lin 1992b) and forward-fitting with a parametric form consisting of a double power law with low- and high-energy cutoffs plus an isothermal component (e.g., Holman et al. 2003). These tests were done "in the blind" to recover $\overline{F}(E)$ for later comparison with the true forms (Brown et al. 2006). All of these approaches were able to reconstruct the general magnitude and form of $\overline{F}(E)$, although forward-fitting inevitably fails to recover small features which are not coded within the parameterized form of the model function.

4.2.2 Application to RHESSI Data

The Tikhonov regularization method has been applied to hard X-ray measurements recorded by *RHESSI* for SOL2002-02-20T11:07 (C7.5), SOL2002-03-17T19:31 (M4.0), SOL2002-08-06T12:59 (C7.9) (Massone et al. 2003), SOL2002-02-26T10:27 (C9.6) (Kontar et al. 2005) and SOL2002-07-23T00:35 (X4.8) (Piana et al. 2003). In this last paper a "dip-hump" feature in the recovered mean source electron spectrum $\overline{n}V\overline{F}$ was noted near $E = 55$ keV (Fig. 12); such a feature is (by construction) absent in the superimposed forward-fit spectrum (Holman et al. 2003) using the same (Haug 1997) bremsstrahlung cross-section.

The 3σ error bars plotted in Fig. 12 clearly show that this "dip-hump" feature is statistically significant. Its physical interpretation is still an open issue, but it may reflect the depletion of low-energy nonthermal electrons due to the effect of Coulomb collisions for an injected distribution with a low-energy cutoff (Emslie and Smith 1984). On the other hand,

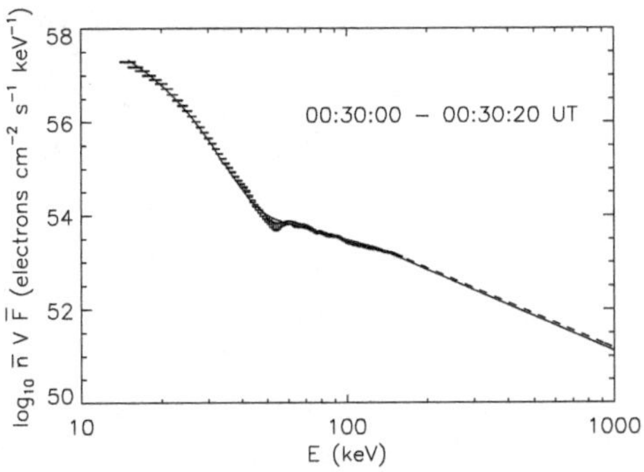

Fig. 12 Regularized spectrum $\overline{n}V\overline{F}$ versus electron energy E for the time interval shown in SOL2002-07-23T00:35 (X4.8). The spectrum has been extended at high energies using a power law of index $\delta = 2.45$ (*dashed lined*). The vertical size of the *error boxes* reflects the 3σ limit caused by statistical noise in the observed $I(\epsilon)$. The spectrum obtained by a forward-fitting procedure using the same bremsstrahlung cross-section is shown as a *solid line* (after Piana et al. 2003)

it may simply reflect an inadequate correction for the pulse pileup effect in the *RHESSI* detectors, with a resultant aliasing of the photon spectrum used in the construction of Fig. 12 (Smith et al. 2002; Kontar et al. 2003).

4.3 High-Energy Cutoffs in the Electron Distribution

It should be noted that the matrix **A** in (4.2) need not be square, so that the energy range corresponding to the electron flux (source) vector **f** may extend over a larger range than the photon flux (data) vector **g**. Physically, this corresponds to the production (through free-free bremsstrahlung only) of photons of energy ϵ by electrons of energy $E > \epsilon_{\max}$, where $\epsilon_{\max}$ is the largest photon energy observed. It should be noted that, because the bremsstrahlung cross-section $Q(\epsilon, E)$ is a non-diagonal matrix of E, the form of the photon spectrum at photon energy ϵ does provide information on the electron spectrum for all energies $E > \epsilon$, including those with $E > \epsilon_{\max}$. Consequently, a distinct advantage of the regularization methodologies (over, say, the more straightforward matrix inversion method of Johns and Lin 1992b) is that some information can be obtained on the form of the electron spectrum $F(E)$ at energies beyond the maximum photon energy observed.

A particular example of this is the possible existence of a high-energy cutoff $E_{\max}$ in the *electron* spectrum $\overline{F}(E)$. Although for free-free emission this would correspond to a maximum photon energy $\epsilon_{\max} = E_{\max}$, $\epsilon_{\max}$ may lie beyond the range of statistically useful photon (or count) data. However, as shown by Kontar et al. (2004), the regularized inversion method can, in principle, detect the presence of the high-energy cutoff at $E = E_{\max}$ through its effect on the photon spectrum at observable energies ϵ that are all significantly less than $E_{\max}$, and this technique was indeed used to discern a high-energy cutoff (or at least a very sharp downward spectral break) in the flare of SOL2002-02-26T10:27 (C9.6) by Kontar et al. (2005). The ability of the regularization technique to detect such high-energy cutoffs in the electron spectrum was dramatically highlighted in the validation study of Brown et al. (2006). In that study, a synthetic electron spectrum $\overline{F}(E)$ with a high-energy cutoff at 200 keV (case B in Fig. 11) was used to generate noisy photon "data" in the range wholly below 100 keV. Analysis of these "data" using both zero-order and higher-order regularization techniques rather faithfully reproduced the high-energy cutoff in the electron spectrum, with an accuracy better than 30%. High-energy cutoffs have

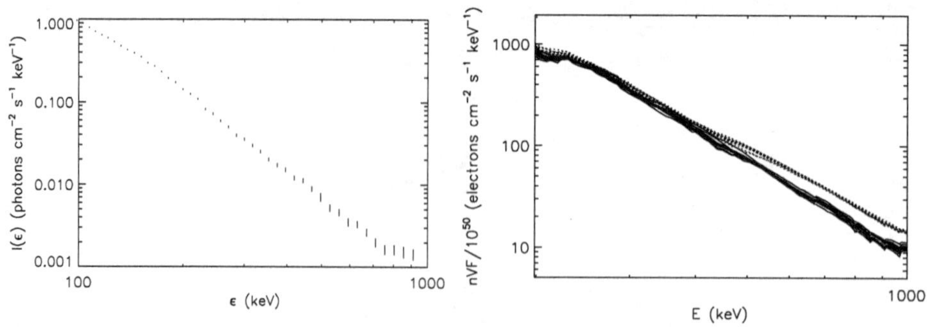

Fig. 13 *Left panel:* Photon spectrum for the time interval 09:43:16–09:44:24 UT in SOL2005-01-17T09:52 (X3.8), with gamma-ray lines removed. *Right panel:* Recovered forms of the quantity $\bar{n}V\overline{F}(E)$ (in units of 10^{50} electrons $\mathrm{cm}^{-2}\,\mathrm{s}^{-1}\,\mathrm{keV}^{-1}$; see (2.4)) using a zero-order regularization technique and presented as a "confidence strip," i.e., a series of solutions, each based on a realization of the data consistent with the size of the uncertainties. The *dashed lines* assume electron-ion emission only; the *solid lines* include the additional electron-electron emission term (after Kontar et al. 2007)

had very limited use to date as a parameter in forward-fitting methods (Sui et al. 2002; Holman et al. 2003), although both forward-fit and direct matrix inversion methods (Johns and Lin 1992b) should be able to say something about the electron spectrum above maximum photon "data" energy, in that if there is a high energy cutoff $E_{\max}$ close to the maximum observed photon energy the spectral shape is much different.

4.4 Spectral Breaks in the Electron Distribution

Figure 13 shows the photon spectrum for the time interval 09:43:16–09:44:24 UT (the time of approximate peak flux) for SOL2005-01-17T09:52 (X3.8) (after Kontar et al. 2007). This event, which produced several strong gamma-ray lines, was previously studied by Kontar and Brown (2006b), who concluded that the pitch-angle distribution for electrons up to $\sim$300 keV was close to isotropic. There is evidence for an upward break in the spectrum at energies $\gtrsim$300 keV. For completeness, we note that the energy range $\gtrsim$300 keV also contains some pseudo-continuum γ-ray contributions from ions (e.g. Smith et al. 2003).

The mean source electron spectrum $\overline{F}(E)$ recovered from the photon data, using two different bremsstrahlung cross-sections—electron-ion only and (electron-ion + electron-electron) is shown in Fig. 13. (The results are presented in the form of two "confidence strips," bundles of solutions using different noisy realizations of the same data.) The $\overline{F}(E)$ recovered using the full cross-section (2.11), including electron-electron bremsstrahlung, is, for $E \gtrsim 300$ keV, steeper (spectral index greater by $\sim$0.4) than the $\overline{F}(E)$ recovered assuming purely electron-ion emission. Kontar et al. (2007) point out that while the upward break at $\epsilon \approx 300$ keV in Fig. 13 is real, the break at $E \simeq 400$ keV in the $\overline{F}(E)$ recovered using the electron-ion bremsstrahlung cross-section alone (Fig. 13) is an artifact of the neglect of electron-electron emission at energies $\gtrsim$300 keV. The true form of $\overline{F}(E)$, obtained using the full cross-section, has a rather straightforward power-law form over the energy range from 200–1000 keV. As pointed out by Kontar et al. (2007), the inclusion of electron-electron bremsstrahlung may remove the need to explain "break energies" in several events (cf. Dennis 1985; Heristchi 1986; Li 1995; Trottet et al. 1998; Vestrand 1988). However electron-electron bremsstrahlung cannot explain spectral breaks at low energies $\lesssim$300 keV often observed in *RHESSI* spectra (e.g., Conway et al. 2003).

4.5 Low-Energy Cutoffs in the Electron Distribution

Given that nonthermal electron spectra often have a form close to a steep power law $E^{-\delta}$, with $\delta > 2$ (Dennis 1985), an accurate value for the low-energy cutoff E_c parameter is required to obtain values of the total energy in nonthermal electrons (see, however, Emslie 2003; Hannah et al. 2009). Thus, the determination of E_c plays a key role in the interpretation of hard X-ray data (see also Holman et al. 2011, this issue).

In most cases, the value of E_c must be somehow disentangled from the combined (thermal + nonthermal) form of $\overline{F}(E)$, and the value of E_c is, in general, only weakly constrained by observations. Some flares require rather high values of E_c to explain observations. The presence of a low-energy cutoff introduces a flattening of the photon spectrum at energies below E_c (see, e.g., Holman 2003; Huang 2009; Han et al. 2009). Nitta et al. (1990) show that the spectrum of the impulsive component flattens toward low energies, suggesting a value of E_c as high as 50 keV. Fárník et al. (1997), using *Yohkoh* data, observed a few flares with flat spectra below 30 keV.

Studies before *RHESSI* mostly assumed an arbitrary, fixed, value of E_c. More recently, using *RHESSI* high-spectral-resolution data (without albedo correction), several studies have reported clear evidence for a low-energy cutoff or even a dip (or gap) (e.g., Kontar and Brown 2006a) in the mean electron flux distribution $\overline{F}(E)$ between its thermal and nonthermal components, leading to a clear identification of a low-energy cutoff for the nonthermal electron distribution (Fig. 14).

Forward-fitting methods (Sect. 4.1) usually assume a strict low-energy cutoff, i.e., a sharp change in the $\overline{F}(E)$. However, as showed by, e.g., Kucera et al. (1996) or Kašparová et al. (2005), it is often not possible to determine a unique best spectral fit; fit parameters can be varied substantially without unacceptably large changes in the photon spectrum, with other free parameters compensating. In particular, because of the dominance of thermal emission at low energies, forward-fit approaches can reliably infer only *upper limits* to the value of E_c; values of E_c well below such upper limits provide equally good fits to an observed X-ray spectrum, with a somewhat different value of the thermal source temperature T. Moreover, as mentioned in Sect. 4.1 determining confidence intervals for model parameters is an extremely time-consuming task.

On the other hand, regularized inversion methods (Sect. 4.2) have proven their ability to detect dips in $\overline{F}(E)$ (see Piana et al. 2003). These methods also provide estimates of uncertainties in the solution, through so-called *confidence strips* in which 99% of data-consistent solutions lie; see Brown et al. (2006). Using this approach, clear low-energy cutoffs or dips in the mean electron flux distribution $\overline{F}(E)$ at an energy around 20–40 keV were reported by Kašparová et al. (2005) for SOL2002-08-20T08:25 (M3.4). Forward-fit methods yielded $E_c = 44 \pm 6$ keV, a somewhat higher value. Kontar and Brown (2006a) found a clear dip around 20 keV for the SOL2002-09-17T05:54 (C2.0) event. For further examples of low-energy cutoffs obtained through both forward-fitting and regularization, see Figs. 14 and 15, respectively.

Some flares, such as SOL2002-04-25T06:02 (C2.5), SOL2002-09-17T05:54 (C2.0), or SOL2002-08-20T08:25 (M3.4), have quite hard (flat) spectra with a relatively weak thermal components. Such a flat form of the photon spectrum can require (Kontar and Brown 2006a) a low-energy cutoff, or local minimum, in the corresponding mean electron distribution $\overline{F}(E)$. These local minima are particularly interesting, since if $\overline{F}(E)$ is sufficiently steep (steeper than E^1; see the remarks after (2.9)), it could have a form inconsistent with the widely-used collision-dominated thick-target model for X-ray production (Kontar and Brown 2006a; Brown et al. 2009).

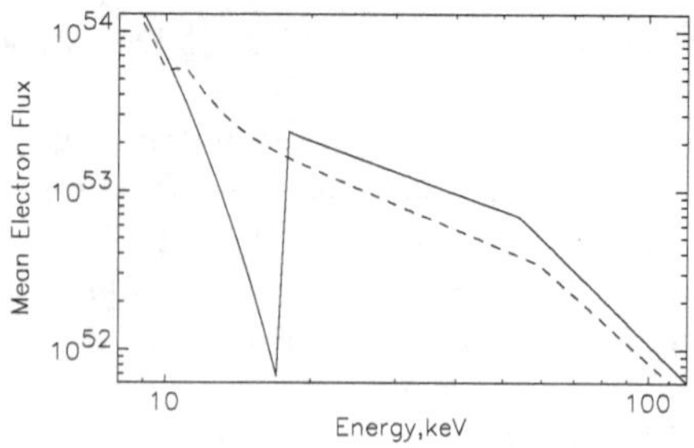

Fig. 14 Mean electron flux spectra $\overline{n}V\overline{F}(E)$ of SOL2002-09-17T05:54 (C2.0) recovered using forward fitting. The *solid/dashed lines* show the spectrum without/with the albedo correction (after Kontar et al. 2006)

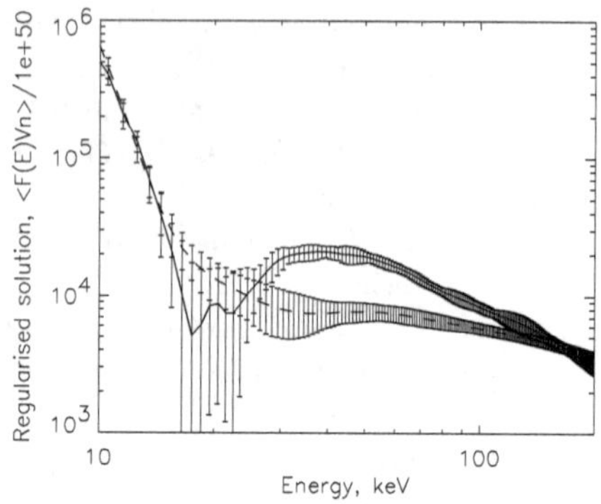

Fig. 15 Mean source electron flux spectrum $\overline{n}V\overline{F}(E)$ for SOL2002-08-20T08:25 (M3.4) for the time interval 08:25:20–08:25:40 UT recovered using regularized inversion. The *solid line* shows the spectrum without albedo correction. The confidence intervals represent the range of solutions found for different statistical realizations of the photon spectrum (after Kontar et al. 2006)

However, the effects of Compton backscattering (Sect. 3.1) on the hard X-ray spectrum are most pronounced for flares with such hard spectra. Kašparová et al. (2007) showed that the flares listed above are all located close to solar disk center ($\mu > 0.5$, $\theta < 60°$; denoted as stars in Fig. 18) and they therefore attributed the flat spectra to the heliospheric-angle-dependent albedo (Sect. 3.1). This result is consistent with earlier observations of Nitta et al. (1990) and Fárník et al. (1997), who observed several flares with flat spectra which were located not far from disk center ($\mu \geq 0.6$) or from near center ($\mu \approx 1.0$), respectively. Interestingly, adding considerations of the albedo in such events (Kontar et al. 2006) removes the spectral hardening, and hence the need for a low-energy cutoff in this photon energy range—see Figs. 14 and 15, Kašparová et al. (2005), and Kontar et al. (2006). Recently, Kontar et al. (2008a) have analyzed a large number of solar flares with weak thermal components and flat photon spectra. It has been shown that if the isotropic albedo correction is applied, all low-energy cutoffs in the mean electron spectrum are removed, and hence the low-energy cutoffs in the mean electron spectrum of solar flares above $\sim$12 keV cannot be viewed as real features. If low-energy cutoffs exist in the mean electron spectra, their energies should be less than $\sim$12 keV. Thus, the apparent low-energy cutoff in the mean electron distribution is most likely to be a feature connected with albedo; it is *not* a true physical property (Hannah et al. 2009). This result can substantially change the total electron energy requirements in a given flare.

In summary, our view on the existence and/or value of low-energy cutoffs has been significantly broadened since high quality *RHESSI* observations have become available. Yet, the determination and even the existence of a low-energy cutoff remains a very complex issue. First, the contribution of the photospheric albedo to the observed photon spectrum must be taken into account and removed before the observed photon spectrum is converted to the electron distribution. Such correction removes apparent energy cutoffs in the mean electron distribution above $\sim$12 keV (Kontar et al. 2008a). Secondly, apparent evidence from hard X-ray emission for a low-energy cutoff in the electron spectrum should be carefully combined with other data sensitive to the low-energy cutoff in nonthermal electron distribution, e.g., microwave spectra (Holman 2003) or plasma radio emission (Mel'nik et al. 1999; Kontar 2001). Finally, effects which may lead to hard X-ray spectral flattening (e.g., albedo and non-uniform ionization; see Kontar et al. 2002, 2003; Liu et al. 2009; Su et al. 2009) should also be carefully assessed before reaching conclusions on the value of the crucial parameter E_c, with its attendant implications for the energy content in nonthermal electrons (e.g., Falewicz et al. 2009).

4.6 Temperature Distribution of Thermal Plasma

The thermal free-free (bremsstrahlung) continuum emission (photons $\mathrm{s}^{-1}\,\mathrm{keV}^{-1}$) at photon energy ϵ from an element of plasma of density n (cm^{-3}), temperature T (K) and volume dV is, neglecting factors of order unity,

$$dI(\epsilon) = a\frac{n^2\,dV}{\epsilon\,T^{1/2}}\exp(-\epsilon/kT), \tag{4.6}$$

where k is Boltzmann's constant and a is a constant. For an extended source, with nonuniform density and temperature, integration of (4.6) over the spatial extent of the source gives

$$I(\epsilon) = \frac{a}{\epsilon}\int_0^\infty \frac{1}{T^{1/2}}\xi(T)\exp(-\epsilon/kT)\,dT, \tag{4.7}$$

where $\xi(T) = n^2\,dV/dT$ ($\mathrm{cm}^{-3}\,\mathrm{K}^{-1}$) is the *differential emission measure* at temperature T (see (10) of Craig and Brown 1976). As first pointed out by Brown (1974), (4.7) may be written as a Laplace transform with respect to the inverse temperature variable $x = 1/kT$:

$$\frac{k^{1/2}}{a}\epsilon I(\epsilon) = \int_0^\infty e^{-\epsilon x} f(x)\,dx \equiv \mathcal{L}[f(x);\epsilon], \tag{4.8}$$

where $f(x) = \xi(1/kx)x^{-3/2}$ or, equivalently, $\xi(T) = (kT)^{-3/2} f(1/kT)$.

A solution of (4.8) for $f(x)$ (and hence $\xi(T)$) is formally possible for a large variety of (but, it should be noted, not *all*; see Brown and Emslie 1988) forms of $I(\epsilon)$. Even when a solution does exist, however, the solution of (4.8) is not a trivial task. This integral equation is of Fredholm-type (Bertero et al. 1985) and is highly ill-posed, with a large class of solutions $\xi(T)$ corresponding to a given $I(\epsilon)$ when (even very small) uncertainties in $I(\epsilon)$ are taken into account. Mathematically, this extreme ill-posedness arises from the very broad form of the Laplace kernel $\exp(-\epsilon x)$. *Physically*, the problem exists because of the broad range of temperatures T that contribute to the emission at a given photon energy ϵ. Unlike, for example, the bremsstrahlung photon-to-electron inversion problem (see Sect. 4.2), in which only electrons with energy $E > \epsilon$ contribute to the emission at photon energy ϵ (and so the corresponding integral equation is of Volterra type, and so less ill-posed), in the electron-energy-to-temperature problem considered here *all* temperatures T contribute to the emission at energy ϵ (and conversely a source at a single temperature T produces emission at all photon energies).

An early study of the inversion of (4.8), using data from a balloon-borne instrument (Lin and Schwartz 1987), was carried out by Piana et al. (1995). They deduced not only forms of $\xi(T)$ but also (in a one-dimensional geometry) the corresponding conductive flux $F_c(T)$

and its derivative dF_c/dx. However, later work by Prato et al. (2006) examined the relation between $\xi(T)$ and the mean source electron spectrum $\overline{F}(E)$, viz.,

$$\overline{F}(E) = b\int_0^{\infty} \frac{1}{T^{1/2}}\xi(T)\exp(-E/kT)\,dT. \tag{4.9}$$

This study utilized the Landweber method, which ensures positivity of $\xi(T)$ everywhere—see (4.5) in Sect. 4.2—and which is effective in recovering narrow, δ-function-like forms of the differential emission measure $\xi(T)$. It also showed that the form of $\xi(T)$ deduced from inversion of this (rigorous) equation is much less well-determined than suggested by the earlier work of Piana et al. (1995), which involved inversion of the (inexact) equation (4.6).

The method was applied to three photon spectra emitted during the flares SOL2002-08-21T01:41 (M1.4), SOL2003-11-03T09:55 (X3.9), and SOL2002-07-23T00:35 (X4.8). The photon spectra were first inverted by applying zero-order Tikhonov regularization to recover the form of $\overline{F}(E)$. For the first two events, the recovered $\xi(T)$ was consistent with a roughly isothermal low-temperature plasma plus a very broad form of $\xi(T)$ at high temperatures. However, for the SOL2002-07-23T00:35 (X4.8) event, the reconstruction method produced unacceptably large residuals at low temperatures, consistent with the fact that this same spectrum fails to satisfy the derivative test (Brown and Emslie 1988) necessary for compatibility with a purely thermal interpretation of the event. Observation of optically-thin lines is often used as an alternative approach to infer the DEM of solar plasma (as a review, see Phillips et al. 2008).

5 The Electron Angular Distribution

Flare studies typically derive the properties of accelerated particles in the target region from observations of the radiation spectrum, but such radiation spectra are strongly dependent on the angular distribution of the energetic particles. Consequently, knowledge of both the angular distribution and the energy distribution of energetic particles, as they interact in the solar material, is necessary for understanding the acceleration and transport of particles in flaring regions.

Attempts to measure the angular distribution of the accelerated electrons rely on the fact that an anisotropic ensemble of bremsstrahlung-producing electrons will generate a radiation field that is both polarized and anisotropic. Efforts to measure electron beaming have therefore concentrated on studies of the hard X-ray continuum emission, by looking at either the directivity or the polarization of the emitted radiation. Theoretical studies have considered the evolution of the electron pitch-angle distribution as the particles are transported along magnetic field lines (e.g., Brown 1972; Leach and Petrosian 1981; McTiernan and Petrosian 1990a, 1990b). It is generally expected that, even if the particles are strongly beamed when injected, the net effect of the particle transport through the atmosphere will be to broaden the angular distribution (Holman et al. 2011, this issue). Ideally, the measurements of the electron angular distribution should therefore be performed as a function of both time and space. No such measurements are available as yet. However, those measurements that are available (as discussed below) already indicate some evidence for electron beaming.

The angular distribution of accelerated ions can be studied by measuring the energies and widths of broad γ-ray lines (for details, see Smith et al. 2003). Studies of γ-ray line data from the *Solar Maximum Mission (SMM)* Gamma Ray Spectrometer (GRS) suggest that protons and α-particles are likely being accelerated in a rather broad angular distribution (Share and Murphy 1997; Share et al. 2002). It is currently still unclear whether electrons and ions are being accelerated (Zharkova et al. 2011, this issue) in a similar fashion or by the same process.

5.1 Early Results

One technique for studying hard X-ray directivity is to look for center-to-limb variations on a statistical basis. Correlations between flare longitude and flare intensity or spectrum reflect the anisotropy of the X-ray emission and hence an associated anisotropy of the energetic electrons. For example, if the radiation is preferably emitted in a direction parallel to the surface of the Sun, then a flare located near the limb will look brighter than the same flare near the disk center. Various statistical studies of X-ray flares at energies below 300 keV reported no significant center-to-limb variation of the observed intensity (e.g., Datlowe et al. 1974; Pizzichini et al. 1974). A statistically significant center-to-limb variation in the shape of the spectra of these events *was* found by Roy and Datlowe (1975), suggesting that perhaps some directivity may be present.

The large sample of flares detected at energies greater than 300 keV by *SMM*/GRS allowed, for the first time, a statistical search for directivity at higher energies. Analysis of these data collected during Solar Cycle 21 provided the first clear evidence for directed emission, with a tendency for the high energy events to be located near the limb (Vestrand et al. 1987; Bai 1988). Bogovalov et al. (1985) and McTiernan and Petrosian (1991) also reported evidence for anisotropies at hard X-ray energies. Observations from *SMM*/GRS during Solar Cycle 22 provided further support for directivity (Vestrand et al. 1991) at energies above 300 keV. However, several high-energy events were also observed near the disk center by a number of different experiments during Cycle 22 (e.g., on *GRANAT* and the *Compton Gamma-ray Observatory, CGRO*). For a summary, see Vilmer 1994; this perhaps suggests a more complex pattern. Li (1995) used data from *SMM* to confirm the general results of Vestrand et al. (1987) and concluded that there was evidence for increasing directivity with increasing energy. Quantifying the magnitude of the directivity from these statistical measurements is difficult. For example, one needs to know the size-frequency distribution for flares as well as the form of the electron distribution to derive the predicted limb fraction (e.g., Petrosian 1985). Furthermore, the results only represent an average for the flare sample. Different flaring regions are not likely to have identical geometries; nor are individual flares likely to have time-independent electron distributions.

Another, more direct, method for studying directivity in individual flares is the stereoscopic technique (Catalano and van Allen 1973). This method compares simultaneous observations made on two spacecraft that view the same flare from different directions. Kane et al. (1998) combined observations from the *Pioneer Venus Orbiter (PVO)* and *ISEE-3* satellites to produce stereoscopic flare observations of 39 flares in the energy range from 100 keV to 1 MeV. While the range of flux ratios measured by Kane et al. (1998) is consistent with the results of statistical studies (Vestrand et al. 1987), the deviations of the ratio from unity show no clear correlation with increasing difference in viewing angles. Later studies concluded that there was no clear evidence for directivity at hard X-ray energies (Kane et al. 1998; Li et al. 1994). Stereoscopic observations tend to suffer from cross-calibration issues between different instruments.

The high quality hard X-ray data from *RHESSI* have opened new opportunities and diagnostic techniques for the study of electron anisotropy in solar flares.

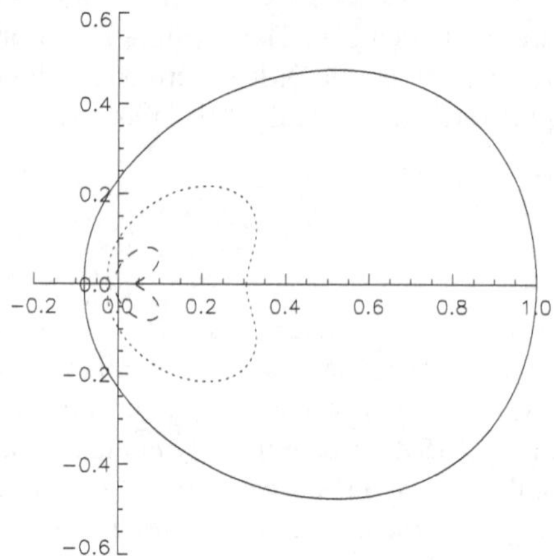

Fig. 16 Polar diagram of the bremsstrahlung cross-section (Gluckstern et al. 1953) for an electron of energy $E = 100$ keV and photon energies $\epsilon = 30$ keV (*solid line*), $\epsilon = 50$ keV (*dotted line*), and $\epsilon = 80$ keV (*dashed line*); the radial coordinate is proportional to the size of the cross-section and the angle from the x-axis corresponds to the angle between the incoming electron direction and the line to the observer. Note that at energies $E \gg \epsilon$, the cross-section peaks at $\theta = 0$, while for $E \simeq \epsilon$, the cross-section peaks at $\theta \simeq (30\text{–}40)^\circ$ (after Massone et al. 2004)

5.2 Anisotropy of X-ray Bremsstrahlung Emission

Using (1.2), it is straightforward to show that the emitted bremsstrahlung in the direction $\mathbf{\Omega}$ toward the observer can be written

$$I(\epsilon, \mathbf{\Omega}) = \frac{\bar{n} V}{4\pi R^2} \int_{\epsilon}^{\infty} \int_{\Omega'} \overline{F}(E, \mathbf{\Omega}') Q(\epsilon, E, \theta') \, dE \, d\mathbf{\Omega}', \tag{5.1}$$

where $Q(\epsilon, E, \theta')$ is the cross-section differential in photon energy ϵ and the angle θ' between the precollision electron direction $\mathbf{\Omega}'$ and the emitted photon direction $\mathbf{\Omega}$, summed over the polarization state of the emitted photon (Gluckstern et al. 1953). The bremsstrahlung cross-section is more angle-dependent for higher photon energies (Fig. 16), which is qualitatively consistent with the findings of the statistical analysis.

In (5.1) the mean source electron flux spectrum has been generalized to the form $\overline{F}(E, \mathbf{\Omega}')$, which takes into account the angular distribution of electrons. If we denote by (θ, ϕ) the polar coordinates of the electron velocity vector $\mathbf{\Omega}'$ relative to the mean direction of the electron velocity distribution (usually the same direction as the guiding magnetic field line), and assume azimuthal symmetry, i.e., no dependence on ϕ, so that $\overline{F}(E, \mathbf{\Omega}') \equiv f(E, \theta)$, and the angle of the mean direction of the electron velocity distribution is θ_0, (5.1) can be written:

$$I(\epsilon) = \frac{\bar{n} V}{4\pi R^2} \int_{E=\epsilon}^{\infty} \int_{\theta=0}^{\pi} \int_{\phi=0}^{2\pi} Q\left[\epsilon, E, \theta'(\theta, \phi; \theta_0)\right] f(E, \theta) \sin\theta \, d\phi \, d\theta \, dE, \tag{5.2}$$

where the directivity angle θ' between electron and photon directions is given by $\cos\theta' = \cos\theta \cos\theta_0 + \sin\theta \sin\theta_0 \cos\phi$. If we now define:

$$K_0(\epsilon, E, \theta) = \int_{\phi=0}^{2\pi} Q\left[\epsilon, E, \theta'(\theta, \phi; \theta_0)\right] d\phi, \tag{5.3}$$

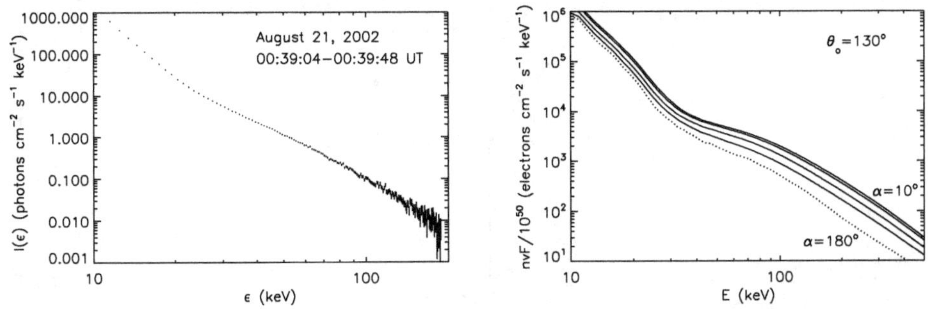

Fig. 17 *Left:* photon flux spectrum. *Right:* regularized electron flux spectra corresponding to $\theta_0 = 130°$ (angle between an observer and the electron beam direction) and (from top to bottom) $\alpha = 10°$, $30°$, $60°$, $90°$, $180°$. The choice of $\theta_0 = 130°$ is justified by the fact that this is the value corresponding to a vertically downward electron beam at the location of the selected flare (after Massone et al. 2004)

we can write

$$I(\epsilon) = \frac{\bar{n}V}{4\pi R^2} \int_{E=\epsilon}^{\infty} \int_{\theta=0}^{\pi} K_0(\epsilon, E, \theta) f(E, \theta) \sin\theta \, d\theta \, dE. \tag{5.4}$$

Inversion of (5.4) requires construction of the bivariate function $f(E,\theta)$ from knowledge of the (noisy) univariate function $I(\epsilon)$. The problem is significantly more tractable if we make the further assumption that $f(E,\theta)$ is separable in E and θ. We can then choose a particular form for the angular dependence of $f(E,\theta)$ and reconstruct the part of $f(E,\theta)$ that depends only on the electron energy (univariate problem in E) or, analogously, we can assume the E-dependence for $f(E,\theta)$ and recover the θ-dependence (univariate problem in θ).

In Massone et al. (2004), the θ-dependence of $f(E,\theta)$ is a prescribed (normalized) form $g(\theta)/\int_{\Omega'} g(\theta)\, d\boldsymbol{\Omega}'$. Specifically, it was assumed that at all energies the pre-collision electron velocities are uniformly distributed over a solid angle within a cone of half-angle α centered on a direction corresponding to a photon emission direction of θ_0. With such an assumption, (5.4) becomes

$$I(\epsilon) = \frac{\bar{n}V}{4\pi R^2} \int_{E=\epsilon}^{\infty} \overline{K_0}(\epsilon, E)\, \overline{F}(E)\, dE, \tag{5.5}$$

where

$$\overline{K_0}(\epsilon, E) = \frac{1}{2\pi(1-\cos\alpha)} \int_{\theta=0}^{\alpha} K_0(\epsilon, E, \theta) \sin\theta \, d\theta. \tag{5.6}$$

Equation (5.5) is formally identical to (2.4). Application of the usual Tikhonov regularization technique to (5.5) then yields the mean source energy spectrum $\overline{F}(E)$.

Massone et al. (2004) applied this analysis to a photon spectrum recorded by the *RHESSI* instrument from SOL2002-08-21T01:41 (M1.4) in the time interval 00:39:04-00:39:48 UT (left panel of Fig. 16). Values for the observation angle θ_0 ranging from 0° to 180°, and for the cone semi-angle α ranging from 10° to 180°, were considered. The results (right panel of Fig. 17) demonstrated clearly that the angular dependence cannot be neglected: use of the anisotropic cross-section yields electron spectra that are significantly different from the ones reconstructed by using the solid-angle-averaged cross-section (corresponding to $\alpha = 180°$). As the electron distribution becomes more anisotropic (decrease in α), the cross-section for

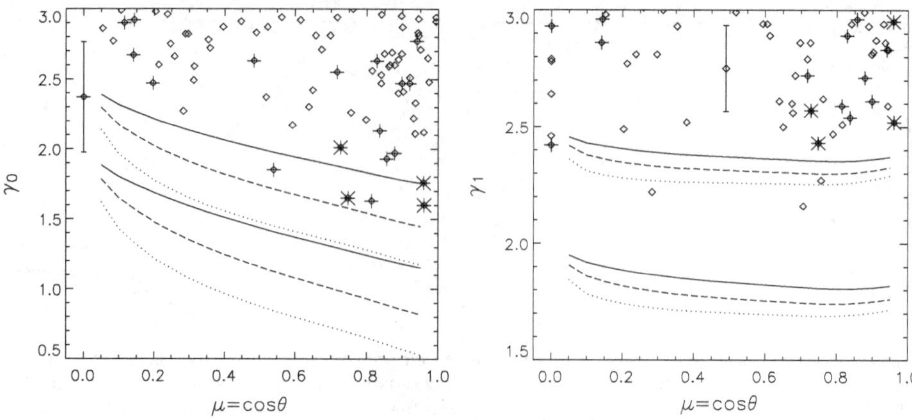

Fig. 18 Spectral index γ_0 (15–20 keV) and γ_1 (20–35 keV) versus cosine μ of heliocentric angle θ. *Vertical error bars* indicate average uncertainties on the values as determined from single power-law fits. *Lines* show the predicted dependency for single power-law primary spectra with spectral indices 2.0, 2.5 and for directivity $\alpha(\mu) = 1, 2, 4$ (*solid*, *dashed*, and *dotted lines* respectively). Flares with a dip in the mean electron distribution are denoted by *stars*; see also Sect. 4.5 (after Kašparová et al. 2007)

emission in the direction of the observer decreases for some parts of the cone and increases for others. The overall effect is a reduction in the cross-section, so that more electrons are required to produce the given photon spectrum. This effect is more pronounced at high energies (see Fig. 17), so the spectrum for small values of α is flatter than for $\alpha = 180°$ (isotropic distribution).

As the viewing angle θ_0 increases (source moves closer to the disk center), the cross-section for emission in direction of the observer in general decreases, especially at high values of E, and accordingly the required electron spectrum increases and also flattens.

5.3 Statistical Results on X-ray Anisotropy

As a result of the heliocentric angle dependence of albedo (Sect. 3), the shape of photon spectra should vary as a function of their position on the solar disk (Bai and Ramaty 1978; Kontar et al. 2006). A statistical analysis of *RHESSI* flares (Kašparová et al. 2007) demonstrates a clear center-to-limb variation of photon spectral indices in the 15–20 keV energy range and a weaker dependency in the 20–50 keV range. The observed spectral variations were found to be consistent with the predictions of albedo-induced spectral index changes (Fig. 18).

Because the number of albedo photons depends on the amount of downward directed primary emission, the characteristics of the albedo component can be used to get an estimate of the anisotropy of the primary hard X-ray emission. In this manner, Kašparová et al. (2007) obtained values for α (the ratio of downward-directed primary emission to observer-directed primary emission) in the 15–20 keV energy range; see Fig. 19. The values of directivity α range from 0.3 to 3; while they imply downward-collimated emission in some cases, overall they are also consistent with isotropic X-ray emission (see also Kontar and Brown 2006b). The model of Leach and Petrosian (1983, Fig. 4) predicts $\alpha \lesssim 3$ at 22 keV for disk-center events.

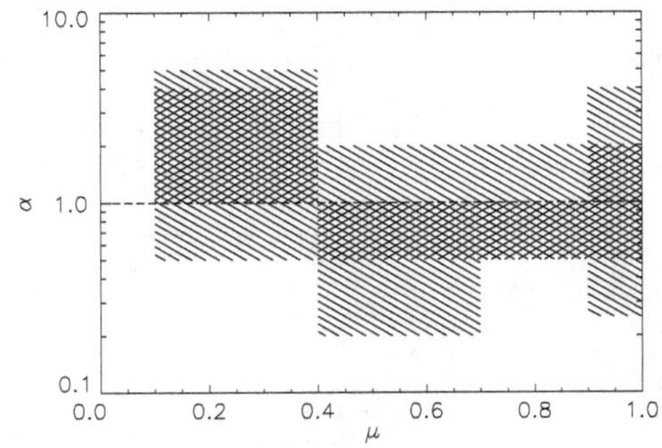

Fig. 19 Directivity at different heliocentric angle μ determined from the distributions of spectral index γ_0 in the 15–20 keV energy range at the limb and a given range of μ. *Cross-hatched areas* show the range of directivity values that are consistent (at the 0.05 significance level) with the hypothesis that the observed distribution at a given range of μ and the modeled distribution are drawn from the same parent distribution; *crossed areas* correspond to the 0.01 significance level. The *dashed line* shows the isotropic case, i.e., $\alpha(\mu) = 1$ (after Kašparová et al. 2007)

5.4 Albedo as a Probe of Electron Angular Distribution

The albedo spectral "contaminant" in fact offers very valuable insight into the anisotropy of the flare fast electron distribution. It does so by providing a view of the hard X-ray flare from behind, like a dentist's mirror. Moreover, the solar albedo "mirror" is spectrally distorting, so that its contribution to the overall spectrum can be distinguished. The observed spectrum in the observer's direction should contain an albedo "bump" feature, the strength of which is an indicator of the degree of downward beaming of the electron distribution. By use of this solar "mirror" we can achieve a degree of knowledge about the directionality of the primary photon distribution, and so the accelerated electron distribution, from single-spacecraft photon spectrometry (Kontar and Brown 2006b).

The required integration over solid angle in (5.1) can be approximated using a bi-directional representation. Invoking axial symmetry about a beam assumed to be perpendicular to the solar surface, we introduce the cross-sections

$$Q(\epsilon, E, \theta_0) = \frac{1}{\cos(\theta_0 - \Delta\theta) - \cos(\theta_0 + \Delta\theta)} \int_{\theta_0 - \Delta\theta}^{\theta_0 + \Delta\theta} Q(\epsilon, E, \theta') \sin(\theta')\, d\theta', \quad (5.7)$$

averaged over $[\theta_0 - \Delta\theta, \theta_0 + \Delta\theta]$ and centered at angle θ_0. We define $Q^F(\epsilon, E) \equiv Q(\epsilon, E, \theta_0 = \theta)$ and $Q^B(\epsilon, E) \equiv Q(\epsilon, E, \theta_0 = 180° - \theta)$, where θ is the heliocentric angle of the source. The electron spectrum $\overline{F}(E, \theta)$ is defined in a similar bi-directional approximation, through introduction of the quantities

$$\overline{F}_{u,d} = \frac{1}{\bar{n}V} \int F_{u,d}(E, \mathbf{r}) n(\mathbf{r})\, dV \quad (5.8)$$

for electrons propagating either upward (u) or downward (d) toward the scattering photosphere: $\overline{F}_u(E)$ and $\overline{F}_d(E)$ are the density-weighted volumetric mean flux spectra of electrons directed towards the observer (upward and downward, respectively), also averaged over $\Delta\theta$.

With these assumptions and definitions, the discretized hard X-ray spectrum $\mathbf{I}(\epsilon_i); i = 1, \ldots, N$, accounting both for the primary spectrum $\mathbf{I_o}$ and the reflected spectrum $\mathbf{I_r}$, can be

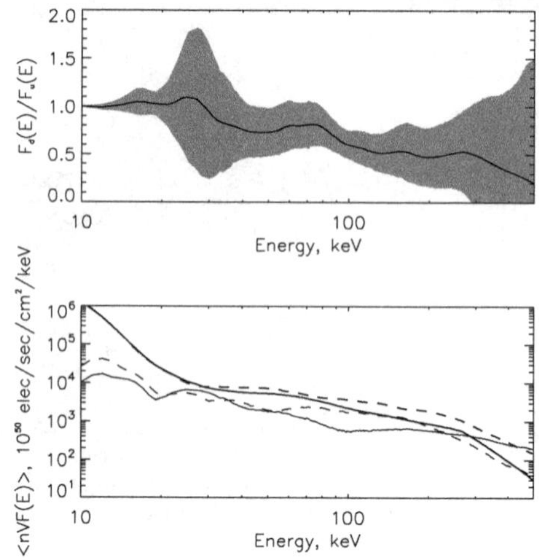

Fig. 20 *Lower panel:* The recovered mean electron flux spectra defined by (5.8) (*thick lines*) for SOL2002-08-20T08:25 (M3.4) (accumulation time interval 08:25:20-08:25:40 UT). Downward-directed $\overline{F}_d(E)$ (*solid line*) and observer-directed $\overline{F}_u(E)$ (*dashed line*) are shown, with corresponding 2σ errors (*thin lines*). *Upper panel:* electron anisotropy defined as $\overline{F}_d(E)/\overline{F}_u(E)$ with confidence values within the *shaded area* (after Kontar and Brown 2006b)

written (Kontar and Brown 2006b)

$$\mathbf{I} = \mathbf{I_o} + \mathbf{I_r} = \begin{pmatrix} \mathbf{Q^F} + \mathbf{G}(\mu)\mathbf{Q^B} & \mathbf{Q^B} + \mathbf{G}(\mu)\mathbf{Q^F} \end{pmatrix} \begin{pmatrix} \overline{\mathbf{F}}_\mathbf{u} \\ \overline{\mathbf{F}}_\mathbf{d} \end{pmatrix}, \tag{5.9}$$

where $\overline{\mathbf{F}}_{\mathbf{u,d}}(E_j)$; $j = 1, \ldots, M$ are the electron data vectors for downward- and upward-directed electrons, respectively and $\mathbf{Q}^{B,F}$ are matrix representations of the kernels of integral equations (5.7). The Green's matrices $\mathbf{G}(\mu)$ depend on the heliocentric angle of the source $\mu = \cos\theta$ and have been calculated in Kontar et al. (2006). Kontar and Brown (2006b) have solved the inverse problem (5.9) using first-order Tikhonov regularization method.

Figure 20 shows electron spectral solutions $(\overline{F}_u(E), \overline{F}_d(E))$ for the M3.4 flare SOL2002-08-20T08:25. The results are consistent with isotropy up to $E \approx 100$ keV, with some indication of upward anisotropy above 100 keV. These findings show a near-isotropic distribution (low electron directivity) of the mean electron spectrum of accelerated electrons in solar flares (Kontar and Brown 2006b; Holman et al. 2011, this issue). This strongly contradicts the models with purely collisional transport in solar flares (e.g., Brown 1972; Zharkova and Gordovskyy 2006; Karlický and Kašparová 2009; Holman et al. 2011, this issue).

5.5 X-ray Polarization and Electron Angular Distribution

5.5.1 Model Predictions

The difficulties of statistical and stereoscopic observations for measuring hard X-ray directivity evoke the need for a technique that can measure anisotropies for individual flares. Because the bremsstrahlung cross-section $Q(\epsilon, E, \theta')$ is in general polarization-dependent, polarization is a diagnostic that can, in principle, meet these requirements. Models of nonthermal (e.g., thick-target) hard X-ray production predict a clear and significant polarization signal with polarization levels $>10\%$ for beam-like distributions of electrons (Bai and Ramaty 1978; Brown 1972; Charikov et al. 1996; Emslie and Vlahos 1980; Langer and Petrosian 1977; Leach and Petrosian 1983; Zharkova et al. 1995) and an orientation parallel to the plane containing the guiding magnetic field direction and the direction to the observer. For vertical guiding fields, this orientation direction projects onto a radial line on the solar disk (Bai and Ramaty 1978; Leach and Petrosian 1983; Zharkova et al. 1995); for other orientations of the guiding field, other polarization vector orientations are possible (Emslie et al. 2008). For strong electron beaming, the polarization degree can reach values up to 60% at energies above 50 keV, or even higher for lower energies (Haug 1972).

All models predict a strong dependence between the observed value of polarization and the viewing angle. The highest polarization degrees are expected for large angles of view, when the line of sight is perpendicular to the magnetic field line. Thus, most theories predict higher polarization for flares located near the solar limb. The direction of the polarization depends on the energy: low energy hard X-rays are negatively polarized whereas, above 350 keV, the sign turns to positive (Bai and Ramaty 1978). Because collisions of beam electrons with ambient particles tend to isotropize the distribution, the highest polarizations (up to 85%) are expected at energies around 100 keV generated in the coronal portions of the top of the flare loop (Leach and Petrosian 1983). Photons observed from the footpoints (in the region of the dense chromosphere) would be polarized only to the level of around 20%. It should also be noted that even thermal models of the hard X-ray source predict a finite polarization of order a few percent, due to thermal gradients in the source (Emslie and Brown 1980). The thermal component, with its rather low polarization, tends to dominate the emission from all flares at energies below about 25 keV.

All these theoretical predictions, while clearly testable, could be criticized on the grounds that the modeling assumptions they contain may be oversimplistic. For example, each model to date assumes a single, simple magnetic field structure about which the emitting electrons spiral. It could be argued that any real flare, particularly one sufficiently large to produce a signal of sufficient strength to enable a polarization measurement, will in all probability contain a mix of structures that would average out any polarization signal present. However, hard X-ray imaging observations in the impulsive phase generally show a fairly simple geometry, consisting of two footpoint sources and perhaps a loop-top source (e.g., Masuda et al. 1995; Sakao et al. 1992). These observations suggest that simple magnetic structures are responsible for the energetic emissions and give support to the possibility that a statistically significant polarization signal could be produced in a large event.

As noted in Sect. 3, and in other places throughout this article, a substantial fraction of the observed hard X-ray flux is backscattered from the solar photosphere, as the so-called *photosphericalbedo*. The precise magnitude of this backscattered fraction depends, in part, on the polarization of the primary flux. Further, the reflected component will influence the degree of polarization of the total observed flux, since backscattering will tend to introduce polarization fractions of a few percent at energies below 100 keV (e.g., Bai and Ramaty 1978; Langer and Petrosian 1977). Direct imaging of this albedo patch would place a constraint on the contribution of such backscattered photons to the primary signal. Hudson et al. (2003) have also suggested that Compton scattering in the corona may lead to measurable polarization effects. Clearly, a simultaneous hard X-ray imaging capability (such as that provided by *RHESSI*) represents a major advantage for interpretation of a hard X-ray polarization measurement.

5.5.2 History of Observations

The history of observations of hard X-ray polarization from solar flares is a fascinating subject in its own. The first measurements of X-ray polarization from solar flares (at energies of ~15 keV) were made by Soviet experimenters using polarimeters aboard the *Intercosmos* satellites. Their polarimeters were made of a hexagonal Be-scatterer surrounded by six counters located in front of the Be-prism faces. Later versions of the instrument were mounted on a turnable drum to reduce systematical errors. In their initial study, Tindo et al. (1970) reported an average polarization of $P = 40\%$ ($\pm 20\%$) for three 1969 X-ray flares: SOL1969-10-20T10:50 (C9.0), SOL1969-10-23T05:15, and SOL1969-10-30T09:30 (M7.3). This study was followed by an analysis of two flares in 1970: SOL1970-10-24T05:41 (M6.4) and SOL1970-11-05T03:21 (X2.3) (Tindo et al. 1972a, 1972b) that showed polarizations of approximately 20% during the impulsive phase. These reports were met with considerable skepticism, on the grounds that they did not adequately allow for proper detector cross-calibration and had limited photon statistics (Brown et al. 1974). Subsequent observations with an instrument on the *OSO-7* satellite seemed to confirm the existence and magnitudes of the polarizations ($\sim$10%), but these data were compromised by in-flight gain shifts (Nakada et al. 1974). In a later study using a polarimeter on *Intercosmos* 11, Tindo et al. (1976) measured polarizations of only a few percent at $\sim$15 keV for two flares in July 1974. This small but finite polarization is consistent with the predictions for purely thermal emission that contains an admixture of polarized backscattered radiation (Bai and Ramaty 1978).

A decade later, a new polarimeter, designed to measure in the energy range from 5 keV to 20 keV (with about 1.5 keV energy resolution), was flown on the Space Shuttle *Columbia* (Tramiel et al. 1984). This design was based on metallic lithium scattering elements surrounded by gas-proportional counters. Contamination of the Li scattering elements invalidated the pre-flight instrument calibration. An in-flight calibration was performed using two flares near the center of the Sun, under the assumption of null polarization. The upper limits to the polarization derived from this calibration for 6 flares (C and M-classes) were in the range from 2.5% and 12.7% (99% confidence).

Recent measurements at energies below 100 keV have been performed with the SPR-N instrument on the *CORONAS-F* satellite (Zhitnik et al. 2006). The SPR-N instrument included a solar X-ray radiation monitor and a polarimeter capable of detecting signals in the energy ranges of 20–40, 40–60 and 60–100 keV. The polarization detector consisted of a hexahedral Be-scatterer and three pairs of CsI(Na) scintillation detectors located on the faces of the Be prism. With a total effective area ranging between $\sim$0.3 cm^2 at 20 keV to $\sim$1.5 cm^2 at 100 keV, it detected hard X-rays from more than 90 flares between 2001 and 2005. From a sample of 25 solar flares, one could determine the upper limits of the polarization degree from 8 to 40% (3σ). Only for the single case of SOL2003-10-29T20:49 (X10.0), located near Sun center, was a significant polarization level measured: its value increased from 50% (20–40 keV) to greater than 70% (60–100 keV). Although the same flare was observed by *RHESSI*, an independent polarization analysis was not possible because *RHESSI* was located at high magnetic latitude and was experiencing a high level of charged particle events (Suarez-Garcia et al. 2006).

5.5.3 RHESSI Polarization Measurements

RHESSI enables polarization measurements in a wide range of energies from 20 keV to 1 MeV. Its detection system of 9 cylindrical coaxial germanium detectors (Smith et al. 2002), coupled with a satellite rotation every 4 seconds, strongly reduces systematic effects in the polarization measurements. However, the *RHESSI* design is not optimized for studies of polarization and its small effective area for this kind of study, together with the high background contribution, have led to measurements with limited statistical significance.

For the energy range between 20 keV and 100 keV, polarization can be measured using the photons that are scattered into the rear segments of the Ge detectors by a (passive) Be

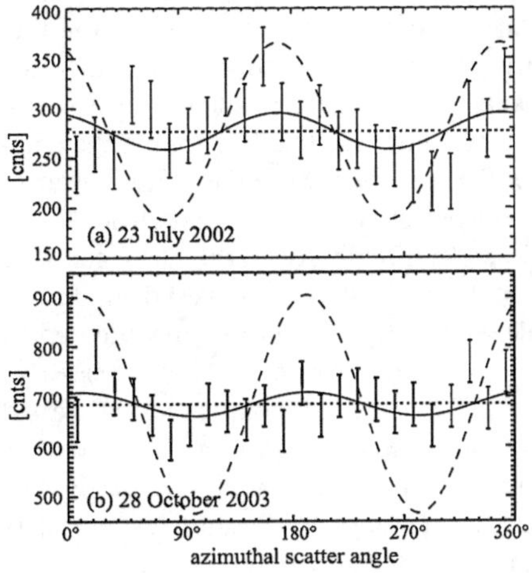

Fig. 21 Solar flare polarization data from Boggs et al. (2006) for photon energies between 200 keV and 1 MeV. The best fit curve is shown as a *solid line*. The case for 100% polarization is shown as a *dotted line*. The measured polarization for SOL2002-07-23T00:35 (X4.8) is 21% ± 9%. The measured polarization for SOL2003-10-28T11:10 (X17.2) is −11% ± 5%

scattering block that is placed between four of the germanium detectors within *RHESSI*'s spectrometer array (McConnell et al. 2002). At energies above ~100 keV, polarization measurements are performed using the so-called "coincidence mode": photons that Compton-scatter from one Ge detector to another are identified by a suitable coincidence timing window. In this way, one detector plays the role of an *active* scattering element, significantly reducing the background level.

McConnell et al. (2003) reported on a measurement of polarization covering the energy range from 20 keV up to 100 keV performed with the Be scattering block. The initial results from an analysis of SOL2002-07-23T00:35 (X4.8) showed evidence for polarization at a level of 15 ± 4%. The measured polarization angle (as measured counterclockwise from solar west) was $\Psi = 79^\circ$ ($\pm 5^\circ$), implying that the polarization vector is inclined ~64° with respect to the radial direction at the flare site. While the magnitude of the observed polarization is broadly consistent with the prediction of solar flare models which invoke the precipitation of a nonthermal electron beam into a dense chromospheric target, the orientation of the polarization vector is somewhat surprising. Emslie et al. (2008) suggest that the orientation of the polarization vector in this case could be explained by a tilt of the flaring loop with respect to the local vertical. Such a tilt is also consistent with gamma-ray line observations for this flare (Smith et al. 2003). Unfortunately, recent analysis of these data (McConnell et al. 2007) suggests that at least a part of the reported signal may be a result of systematic effects in the data analysis, underscoring the difficulty of making reliable polarization measurements.

Using the *RHESSI* coincidence mode, two results have been recently published for energies above 100 keV. The first one (see Fig. 21) describes the polarization measurements of two X-class solar flares, one located close to the solar limb and the other close to the center of the disk (Boggs et al. 2006). The energy range selected was between 200 keV and 1 MeV. The polarization degrees found were 21% ± 10% and −11% ± 5% in 1σ respectively. Assuming that the measured polarization was significant in each case, they measured a radial polarization direction for the flare close to the solar center, and an azimuthal direction for the flare near the limb. The levels of polarization, as well as their directions, are consistent with beamed electron distribution models (Bai and Ramaty 1978) and contrary to the results using the albedo method (Kontar and Brown 2006b; Kašparová et al. 2007). However, the level of inconsistency with more isotropic distributions is rather weak. Given that a report of GRB (gamma-ray burst) polarization by this same team has been met with considerable skepticism in the literature, one might also question these flare results. It should be noted, however, that the flare analysis differed significantly from the GRB analysis and has not been contested in the literature.

The second study (see Fig. 22) applies the coincidence method to the impulsive phase of seven solar flares in the 100 keV to 350 keV energy range (Suarez-Garcia et al. 2006). The flare sample consisted of six X-class (from X1.4 to X8.4) flares and one M7.0-class flare located either on the limb or in the outer part of the solar disc. Values for the polarization degree were 2% and 54%. The lowest degree of polarization obtained was found to be fully compatible with a 0% polarization measurement. The highest degree of polarization, which was the most statistically significant of the seven results obtained, was found to be 2.6 sigma away from 0. The angles of polarization Ψ were distributed between 35° and 85° independent of the flare location, contrary to both the results mentioned in Boggs et al. (2006), and to the expectation based on simple geometrical modeling. Additional attempts to correlate various parameters (e.g., polarization level, polarization angle, heliocentric angle, footpoint orientation, flare intensity) were also inconclusive. The results were compared with the theoretical predictions from Bai and Ramaty (1978) and Leach and Petrosian (1983) and with the 0% polarization hypothesis. The χ^2 analysis allowed only for rejection (90% of confidence) of one of the models from Leach and Petrosian (1983). In this model, predicting very high polarization values up to 85%, the magnetic field strength is constant along the loop and the electrons spiral at pitch angles close to 90°. Due to the statistical uncertainties, for the rest of the models the χ^2 values were very close to unity, making it impossible to distinguish between them. The polarization amplitudes measured by both Boggs et al. (2006) and Suarez-Garcia et al. (2006) are combined in Fig. 23 with diamonds and filled circles, respectively.

The results to date do not yet provide unambiguous evidence for solar flare polarization at hard X-ray energies. It is clear, however, that such data would constitute a unique probe into the electron acceleration process in solar flares, providing new constraints on theoretical models and allowing for more detailed studies of the acceleration processes and geometries.

6 The Electron Spatial Distribution

Hard X-ray imaging spectroscopy is a powerful tool with which to explore the underlying physics of particle acceleration and transport in solar flares, and has been a central component of the *RHESSI* concept since its beginnings.

6.1 Early Results

In its most basic form, imaging spectroscopy involves simply constructing and comparing two-dimensional (count) maps of the source for different energy bands. The earliest imaging of solar hard X-ray sources was carried out using the Hard X-ray Imaging Spectrometer (HXIS) (van Beek et al. 1980) on *SMM*. This instrument provided imaging information between 3.5 and 30 keV though the use of an array of subcollimator-defining grids; it had an angular resolution of 8 arcsec over a 160 arcsec field of view, and 32 arcsec over a wider, 444 arcsec field. Hoyng et al. (1981a) showed that the hard X-ray emission in SOL1980-04-07T01:07 (M4) was located "in two patches;" the patch with the harder spectrum "coincided with the brightest Hα emission." On the other hand, in SOL1980-04-10T09:23 (M4), the hard X-ray emission was "concentrated in a looplike structure, with the softer spectrum at the top of the loop and the harder spectrum in the legs, thus indicating preference for

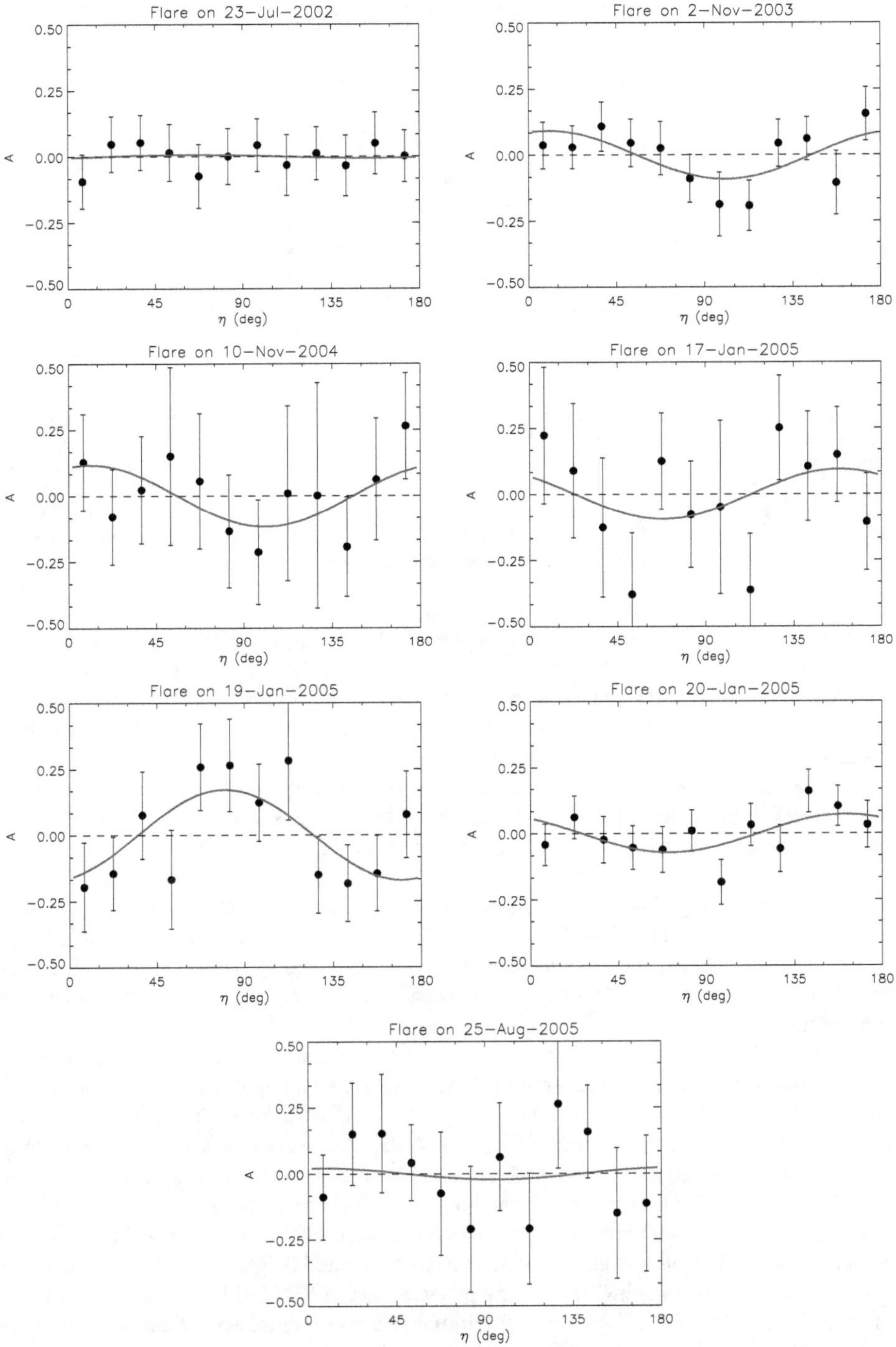

Fig. 22 Solar flare polarization data from Suarez-Garcia et al. (2006) for photon energies between 100 and 350 keV. The best fit curve is shown as a *solid line*. The best case for polarization is from SOL2005-01-19T08:22 (X1.3), where a polarization value of $54\% \pm 21\%$ was measured

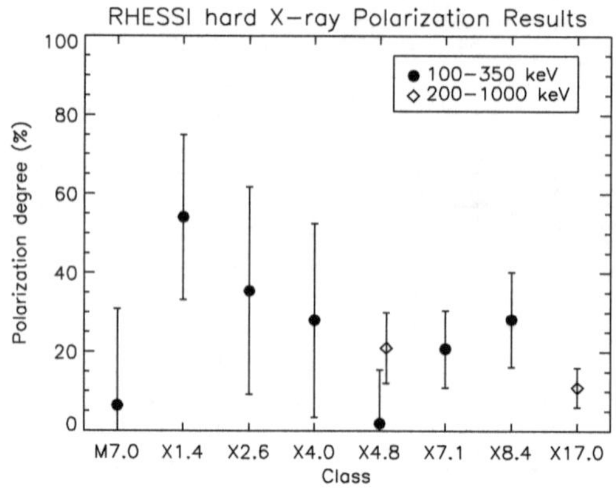

Fig. 23 Results on the absolute value of the degree of hard X-ray polarization above 100 keV achieved with *RHESSI* (1σ errors). The *diamonds* and the *filled circles* correspond to two independent measurements (see text for references). From low to high flare class (left to right) the flares represented are: SOL2005-08-25T04:40 (M6.4), SOL2005-01-19T08:22 (X1.3), SOL2004-11-10T02:13 (X2.5), SOL2005-01-17T09:52 (X3.8), SOL2002-07-23T00:35 (X4.8) (two measurements), SOL2005-01-20T07:01 (X7.1), SOL2003-11-02T17:25 (X8.3), and SOL2003-10-28T11:10 (X17.2). None of these results are significant at a level of greater than 3σ, suggesting that higher quality data are still needed

the thick-target model of hard X-ray production." Hoyng et al. (1981b) studied the hard X-ray emission from a large two-ribbon flare SOL1980-05-21T20:50 (X1) and concluded that the higher-energy (16–30 keV) emission originated in "separate locations of ∼8 arcsec width, coinciding in position with Hα flare kernels," while the softer (3.5–8 keV) emission originated from a "broader region in between."

Treating these observations as evidence for a hot coronal region (Krucker et al. 2008; Tomczak 2009) produced by the primary energy release process, and electron precipitation from this primary acceleration region into the dense chromosphere, Machado et al. (1982) studied the energetics of SOL1980-04-10T09:23 (M4) and concluded that "only a fraction of the total flare energy" was present in the accelerated electrons. Duijveman et al. (1982) continued the study of this event, in addition to SOL1980-05-21T20:50 (X1) and SOL1980-11-15T15:53 (X1). By comparing the energy in the accelerated electrons with an estimate of the thermal conductive flux out of the hot coronal acceleration region, they concluded that "a large fraction of the dissipated flare power has to go into electron acceleration."

The Solar X-ray Telescope (SXT) instrument (Takakura et al. 1983) on the *Hinotori* satellite used the rotating modulation collimator (RMC) technique with two pairs of rotating grids to make hard X-ray images (Oda et al. 1976); the FWHM angular resolution was ∼30 arcsec. Tsuneta et al. (1983, 1984) report observations of a near-limb flare, SOL1981-05-13T06:10 (X1.5), which exhibited a diffuse hard X-ray source situated some 40,000 km above the photosphere. Takakura et al. (1986) reported that a hard X-ray (20–40 keV range) image for the impulsive component of SOL1982-02-22T04:44 (M2.7) was an extended source elongated along the solar limb with a source height of 7,000 km. Tsuneta et al. (1983) also report imaging hard X-ray observations of SOL1981-07-20T14:41 (M5.5), for which "each hard X-ray source in the initial phase coincides with each Hα bright region," further evidence for a thick-target interpretation of the hard X-ray emission at such energies.

In contrast to the RMC technique used by *Hinotori* and *RHESSI*, the Hard X-ray Telescope (HXT) instrument on *Yohkoh* (Kosugi et al. 1991) used information from 64 pairs of occultation grids; each pair of grids yields a single Fourier component of the source. The FWHM resolution achievable was ∼8 arcsec. Matsushita et al. (1992), using observations of about a hundred flares observed with the *Yohkoh*/HXT, reported that, on average, the hard X-ray source height "decreased with increasing X-ray energy," consistent with the deeper penetration of higher-energy electrons (e.g., Brown and McClymont 1975; Emslie 1978, 1981). With nine RMC components, an angular resolution down to ∼ 2 arcsec, and high spectral resolution, the information available from *RHESSI* (Lin et al. 2002) is far superior for imaging spectroscopy studies.

6.2 Imaging Spectroscopy with *RHESSI*

One of the earliest results using *RHESSI* imaging spectroscopy data was by Aschwanden et al. (2002); see also Krucker and Lin (2002), who analyzed hard X-ray source height as a function of energy, improving the previous statistical analysis (Matsushita et al. 1992). With the superior data available from *RHESSI*, Aschwanden et al. (2002) were able to determine accurate source locations as a function of photon energy for the single event SOL2002-02-20T11:07 (C7.5), rather than a statistical average for an ensemble of events used by Matsushita et al. (1992). This was done by determining the centroid location of a circular Gaussian forward-fit to each of the two clearly resolved footpoints in this event. Then, using thick-target modeling of the expected centroid location vs. photon energy in a plane-parallel-stratified atmosphere model (Brown et al. 2002; Mrozek 2006; Kontar et al. 2008b), they were able to deduce a density vs. height structure for the atmosphere that was consistent with empirical models of flaring atmosphere density.

As mentioned at the outset of this section, imaging spectroscopy is, in principle, straightforwardly accomplished by constructing two-dimensional maps of the source at different energies (e.g., Krucker and Lin 2002; Saint-Hilaire and Benz 2002; Sui et al. 2002; Emslie et al. 2003; Ji et al. 2004; Alexander and Coyner 2006; Jin and Ding 2007; Alexander and Daou 2007; Liu et al. 2008; Dennis and Pernak 2009; Prato et al. 2009; Shao and Huang 2009; Petrosian and Chen 2010). These maps are produced by applying image processing algorithms (e.g., back-projection, CLEAN, Maximum Entropy or Pixon; for details see Hurford et al. 2002) to the temporally modulated fluxes from each of *RHESSI*'s detectors, in which spatial information on the source is encoded. Then,"interesting" regions in the field of view are selected and the intensity in such regions is determined in the map corresponding to each energy range. There results from this a set of intensity-versus-energy profiles, i.e., a count spectrum, for each feature. Using knowledge of the instrument response permits *photon* spectra for each feature to be determined. Finally, the corresponding spatially-resolved *electron* spectra are constructed through forward fitting or by applying regularized spectral inversion methods (e.g., Brown et al. 2006) to the spatially resolved photon spectra. Analysis of the variation of the electron spectrum throughout the target is a powerful diagnostic of the physical processes affecting the bremsstrahlung-producing electrons (see Emslie et al. 2001).

Emslie et al. (2003) performed such a "stacked-image" analysis for different energy bands via an imaging spectroscopy analysis of SOL2002-07-23T00:35 (X4.8). They identified four "interesting" features in the images (Fig. 24) and constructed a count spectrum for each (Fig. 25). They identified one of these features as a "coronal" source, and showed that its spectrum was consistent with a thermal source at a temperature of order 45×10^6 K. The spectra of the other three sources were more power-law-like in form. This, plus their location near the lower altitudes in the flaring structure, led Emslie et al. (2003) to identify these features as chromospheric footpoints.

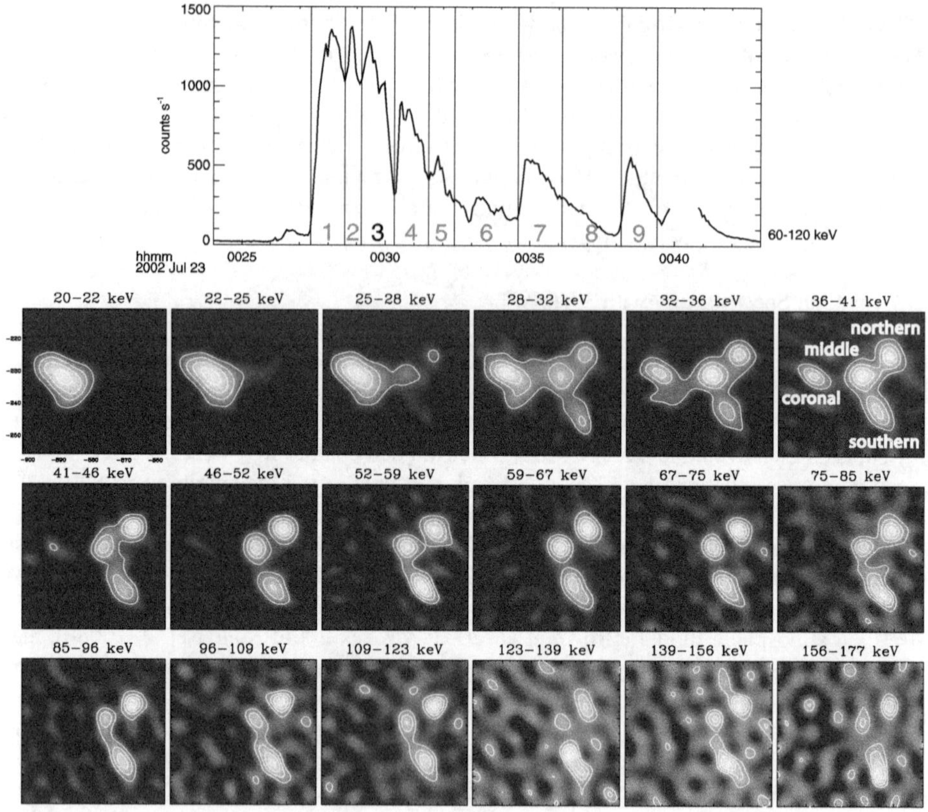

Fig. 24 *Top panel*: time profile for event. *Lower panels*: Images of SOL2002-07-23T00:35 (X4.8) in different count energy channels, for time interval 3 of the event (cf., flux profile in *top panel*). The four main regions in the event are labeled. Since this event was located near the East limb, *structures to the left* are higher in the atmosphere (after Emslie et al. 2003)

Interestingly, the spectra of two of these footpoints (labeled "North" and "South" in Fig. 24), while significantly different and varying with time, maintained a relatively constant *difference* throughout the event. Emslie et al. (2003) interpreted the similar time variation to a magnetic coupling of the two footpoints, and the systematic difference in spectral indices as due to differential spectral hardening associated with an asymmetric location of the electron acceleration region. From the magnitude ($\sim$0.3) of the spectral-index difference, they obtained an estimate of the differential column density between the acceleration region and each of the footpoints. More recently, however, Saint-Hilaire et al. (2008), in a statistical study of some 50 events, have pointed out that such a large intervening column density should in general (but, somewhat ironically, not necessarily in SOL2002-07-23T00:35) produce more emission in the legs of the loop than was observed by *RHESSI*; they conclude that electron precipitation in asymmetric magnetic field geometries is a more reasonable explanation for the observed footpoint spectral differences.

While the above imaging spectroscopy technique appears relatively straightforward to implement, there are a number of difficulties that warrant some commentary. First, for large events, pulse pileup (Datlowe 1976b, 1976a; Smith et al. 2002) is an issue. Pairs of low energy photons arriving nearly simultaneously are detected as a single energy count. Since

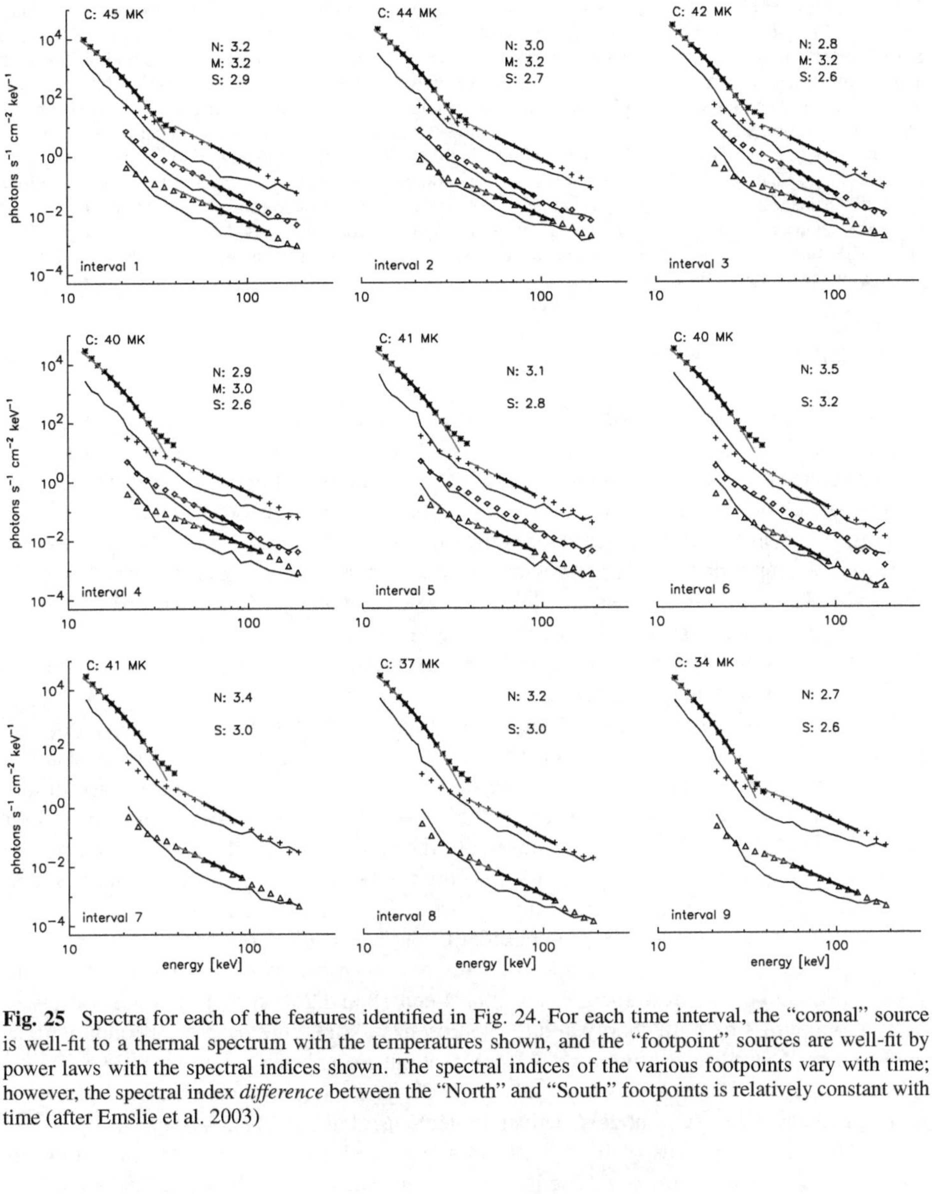

Fig. 25 Spectra for each of the features identified in Fig. 24. For each time interval, the "coronal" source is well-fit to a thermal spectrum with the temperatures shown, and the "footpoint" sources are well-fit by power laws with the spectral indices shown. The spectral indices of the various footpoints vary with time; however, the spectral index *difference* between the "North" and "South" footpoints is relatively constant with time (after Emslie et al. 2003)

the modulation of such counts corresponds to the low energy source, an image made at the higher energy returns a "ghost" image of the low energy source. Second, the dynamic range of *RHESSI* is such that features containing a few percent of the total flux cannot be reliably imaged, so that spectral information in the relevant energy ranges is not reliably recovered. For example, in Emslie et al. (2003) analysis of SOL2002-07-23T00:35, the spectrum of the "coronal" source was determined only at energies $\lesssim$40 keV, while the spectrum of the footpoints was determined only at energies $\gtrsim$30 keV. Only in the relatively narrow energy range from 30 to 40 keV could the spectra of all the features in the source be obtained.

Fig. 26 Construction of electron flux images from measured visibilities. Panel (**a**) shows the "traditional" ▶ approach to image reconstruction, in which the visibility information in count space from each of *RHESSI*'s nine detectors is used to construct an image based on hard X-ray counts. Note that there is no imposition of energy smoothing in this process: images in adjacent count energy channels ($q_i \pm \Delta q_i$) can exhibit significant differences due to count statistics. These differences would be further magnified if the count-based images were inverted to yield electron images. Panel (**b**) shows the innovative approach of Piana et al. (2007), in which the energy spectrum for each count visibility (*left panel*) is subjected to a regularized inversion procedure (see, e.g., Piana et al. 2003) to yield the corresponding (smooth) electron spectrum for each visibility. Once all the electron visibilities have been determined (*right panel*), they can be used to yield electron images using the same image reconstruction algorithms as used to produce count-based images (Panel (**c**)). These *electron flux images* are, by construction, necessarily smooth across electron energy ranges ($E_j \pm \Delta E_j$), and so are more amenable to further analysis

6.3 Visibilities and Imaging Spectroscopy

It is important to realize that, because of the RMC technique used by *RHESSI*, spatial information is encoded in the *RHESSI* data in a distinctive way, namely in rapid time variations of the detected counts in each of the *RHESSI* subcollimators. Schmahl et al. (2007) have developed a technique in which the observed temporal modulations produced are interpreted in terms of a set of *visibilities* (calibrated measurements of specific spatial Fourier components of the source distribution). As with image reconstruction in radio interferometry, the set of visibilities thus determined can then be used to infer the spatial properties of the X-ray source. It is important to note (1) that visibilities are a "first-order" product of the *RHESSI* data, and (2) that the data can be used to determine not only the values of the visibilities, but also their quantitative uncertainties.

Xu et al. (2008) have analyzed a set of ten events, each of which exhibits a rather simple, single-extended-source, geometry. They performed their imaging spectroscopy analysis by assuming a parametric form of the source structure (a seven-parameter curved elliptical Gaussian) and then forward fit not to the actual images themselves, but rather to the corresponding source *visibilities*. Because both the visibilities and their uncertainties were determined quantitatively, they were able to deduce not only the best-fit values of the source parameters, but also the uncertainties on the value of each parameter. Kontar et al. (2008b) have performed such an analysis with circular Gaussian fits (four parameters) of footpoints for the SOL2004-01-06T06:29 (M5.8) flare and have measured the sizes and heights of the hard X-ray sources in a few energy ranges between 18 and 250 keV. The height variations of footpoint emission with energy and the source size with height have been found with an unprecedented vertical resolution of ~150 km at chromospheric heights of 400–1500 km (cf., Brown et al. 2002; Aschwanden et al. 2002). The interpretation of these findings in terms of electron transport models is given by Holman et al. (2011, this issue).

One of the major goals of hard X-ray imaging spectroscopy is the determination not just of the source structure as a function of photon (or count) energy, and hence the variation of the hard X-ray spectrum throughout the source, but rather of the variation of the corresponding *electron* spectrum throughout the source. It is the spectral variation of this electron spectrum that provides the key insight into the physics of electron acceleration and propagation in solar flares.

Recognizing this, Piana et al. (2007) introduced a new approach to imaging spectroscopy which is optimized to the distinctive way in which spatial information is encoded in the *RHESSI* data. Although it is possible to use the count visibilities determined rather straightforwardly from the raw data to construct count images and then proceed to do imaging spectroscopy analysis in the "stacked image" manner described above, Piana et al. (2007) point out that such "traditional" imaging algorithms are completely ineffective in smoothing in

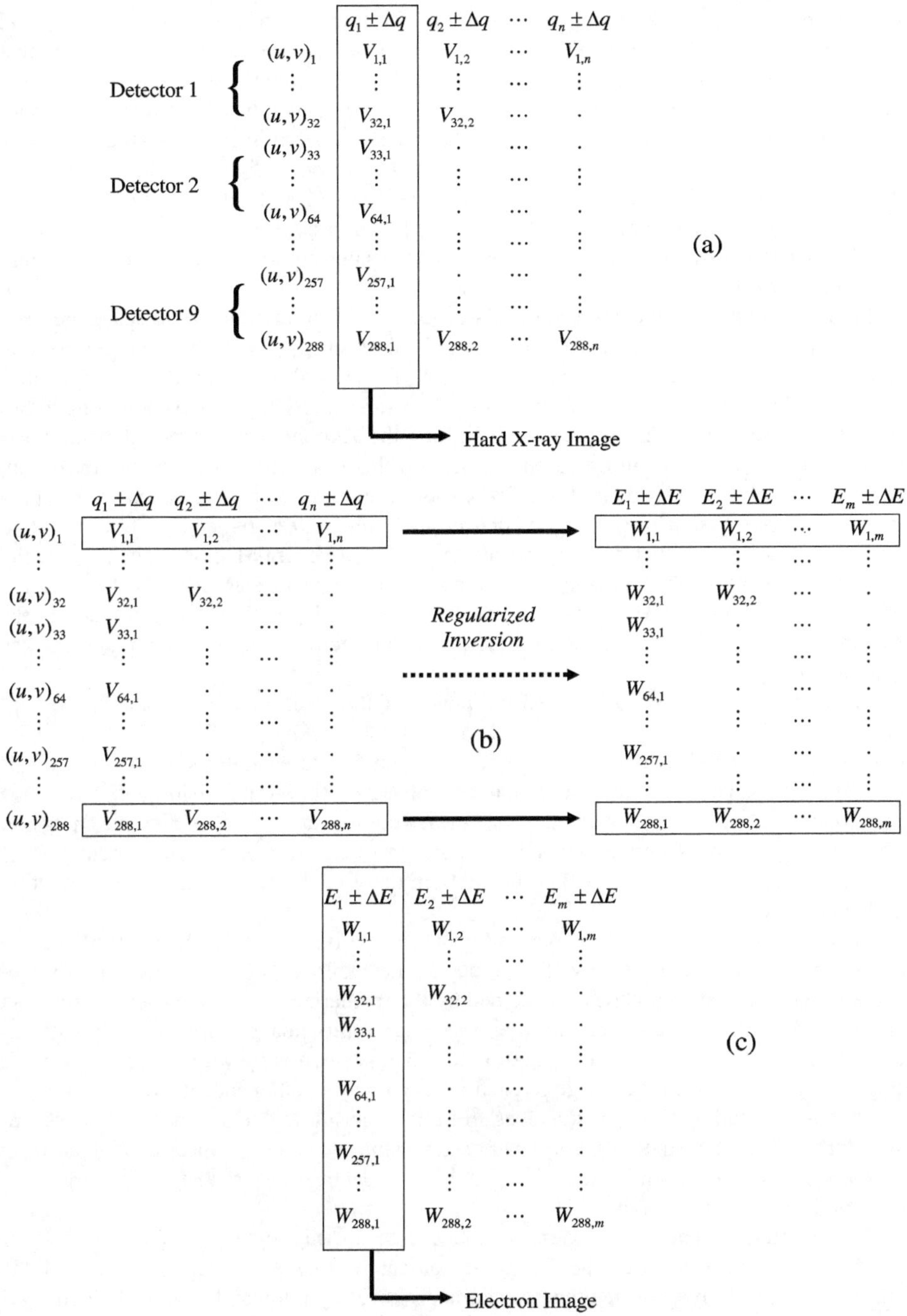

the energy direction, with the result that recovered images corresponding to adjacent energy bins can exhibit substantial differences. Further, *RHESSI*'s Fourier-component approach to imaging detects "patterns" of emission, rather than information in a "pixel-by-pixel" format, so that analysis of a particular sub-region of a source is affected by the signal (and noise) contained in all other features in the source. Piana et al. (2007) therefore argue that imaging spectroscopy analysis using *RHESSI* data is best accomplished through conversion of the temporal modulations in terms of *count visibilities* and subsequent analysis of these count visibilities to obtain information on the electron spectrum in the spatial-frequency (rather than spatial) domain. This leads to a set of *electron visibilities*, which contain all the information on the variation of the electron spectrum throughout the source (albeit as a function of spatial frequency, rather than position). If desired, such electron visibilities can then be used to construct *electron flux maps* using the same algorithms used to convert count visibilities into count maps.

Figure 26 illustrates the essence of this procedure. Visibilities in count space are used to determine the corresponding set of electron visibilities using the Tikhonov regularization technique (Tikhonov 1963) previously used (e.g., Piana et al. 2003) to ascertain spatially-integrated electron spectra $\overline{F}(E)$ from observations of spatially-integrated count (or photon) spectra $I(\epsilon)$ (see Sect. 4.2). Applied to visibilities, the Tikhonov regularization method permits the determination of information on electron fluxes at energies above the maximum count energy observed (Kontar et al. 2004). More importantly, it *forces smoothness in the inferred electron visibility spectra at each point in the spatial frequency domain* and so enhances real features that persist over a relatively wide energy band, while suppressing noise-related features that show up only over a narrow range of energies. The combination of visibility data and the Tikhonov regularization methodology allows the derivation of the most robust information on the spatial structure of the electron flux spectrum image, the key quantity of physical interest.

Piana et al. (2007) have illustrated the power of the electron visibility method by applying it to data obtained near the peak of SOL2002-02-20T11:07 (C7.5). Using visibilities from *RHESSI* RMCs 3 through 9, corresponding to spatial resolutions from $\sim 7''$ to $\sim 183''$, they construct the amplitude and phase of the count visibilities $V(u, v; q)$ (counts $\mathrm{cm}^{-2}\,\mathrm{s}^{-1}\,\mathrm{keV}^{-1}$) as a function of Fourier components (u, v) and count energy q. They then apply the Tikhonov inversion algorithm to construct the corresponding set of electron visibilities $W(u, v; E)$. By construction, these electron visibilities vary smoothly with electron energy E.

Figure 27 shows electron flux images for the illustrative SOL2002-02-20T11:07 (C7.5) event. They show evidence for two footpoints, connected by a “strand” of coronal flux. Because of the inherent smoothness demanded by the regularized algorithm used to construct electron visibilities from count visibilities, the electron flux images vary smoothly with energy. As pointed out by Piana et al. (2007), this contrasts markedly with the behavior in the count images, for which the image in each energy range is independent, so that statistical fluctuations, including those in sidelobes from neighboring features, result in a set of images that do *not* vary smoothly with energy. Inversion of the count spectra obtained from such images leads to amplification of such noise and electron spectra that exhibit large (and most probably) unphysical features.

Three different spatial subregions in the source are highlighted in Fig. 28. Two of these regions correspond to the footpoint sources visible at higher energies and the other one to similarly-sized regions located approximately midway between the two footpoints. The lower panel of Fig. 28 shows the areally-averaged[6] electron-flux spectra (elec-

[6]To get the total count spectrum for each region (counts $\mathrm{cm}^{-2}\,\mathrm{s}^{-1}\,\mathrm{keV}^{-1}$), simply multiply the areally-averaged spectrum by the area of that region, viz., $14.4 \times 14.4 = 207.36$ arcsec^2 (Footpoint 1), $22.8 \times 9 = 205.2$ arcsec^2 (Middle), and $14.4 \times 14.4 = 207.36$ arcsec^2 (Footpoint 2), respectively.

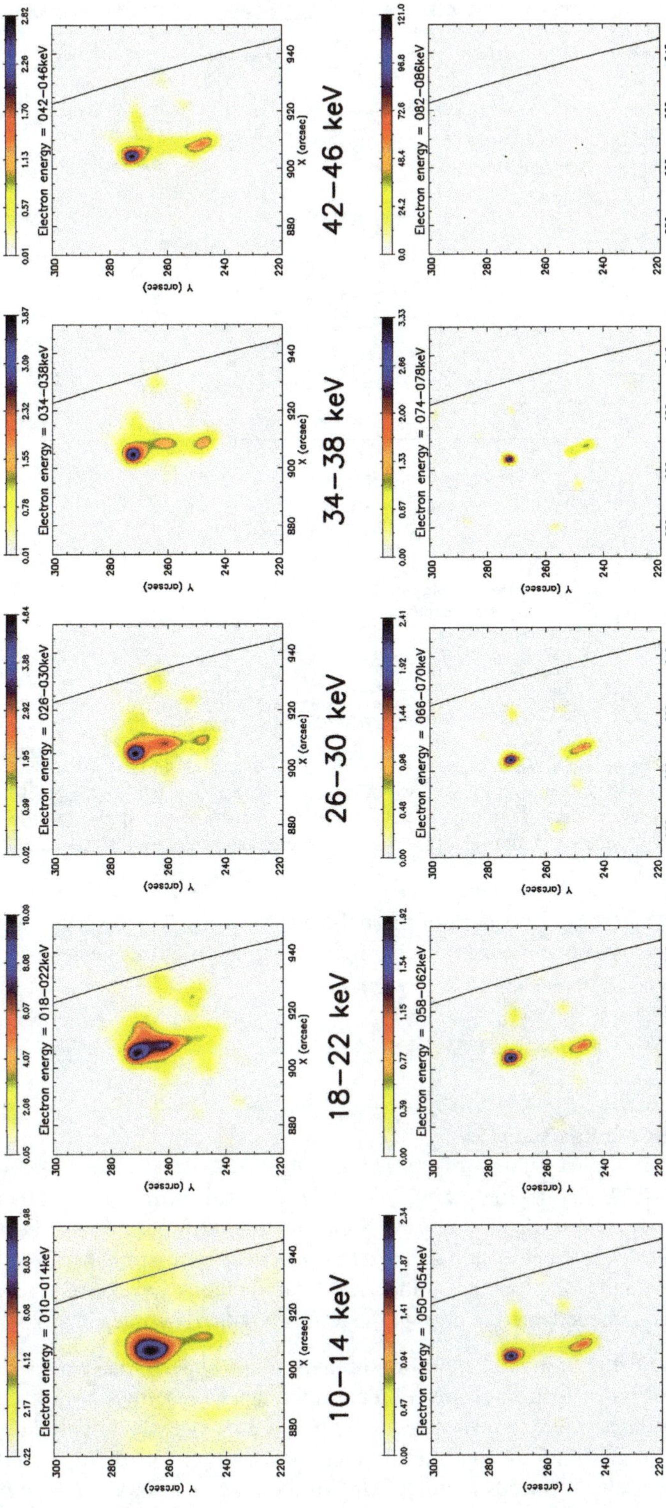

Fig. 27 Electron flux spectral images corresponding to the regularized electron flux spectral visibilities, obtained through application of Bong et al.'s (2006) MEM-NJIT algorithm (after Piana et al. 2007)

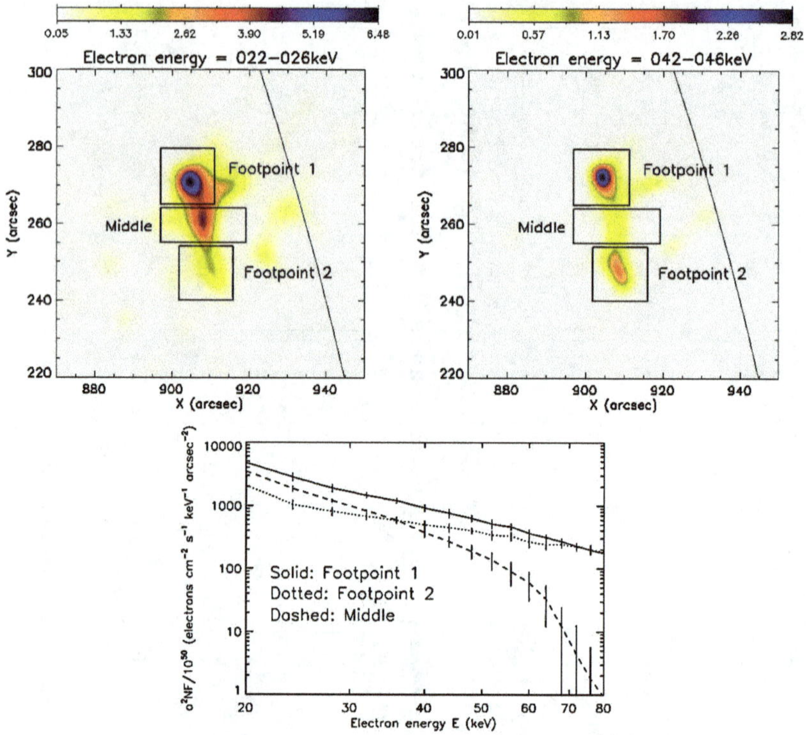

Fig. 28 *Top panels*: Electron images in the energy ranges 22–26 keV and 42–46 keV, respectively. Three sub-regions of interest are labeled on each image. Two of these correspond to bright footpoint-like sources and one to a region midway between the footpoints. *Bottom panel*: Areally-averaged electron flux spectra (electrons $\mathrm{cm}^{-2}\,\mathrm{s}^{-1}\,\mathrm{keV}^{-1}\,\mathrm{arcsec}^{-2}$) for each of the three sub-regions shown (after Piana et al. 2007)

trons $\mathrm{cm}^{-2}\,\mathrm{s}^{-1}\,\mathrm{keV}^{-1}\,\mathrm{arcsec}^{-2}$), for each of these three subregions. These spectra are sufficiently smooth that significant conclusions regarding the variation of the electron spectrum throughout the source can be made.

7 Summary

7.1 Hard X-ray Emission Processes

With the launch of *RHESSI*, not only have new types of hard X-ray data analysis become possible, but also the in-depth study of X-ray producing processes has been triggered (Sect. 2). The roles of electron-ion and electron-electron bremsstrahlung, free-bound electron-ion emission, and Compton backscatter of primary photons, have been highly scrutinized in view of the unprecedented quality of the *RHESSI* data. Examples include:

- As pointed out in Sect. 2.3, pure electron-ion bremsstrahlung spectra have a spectral index $\gamma \simeq \delta + 1$, for pure electron-electron bremsstrahlung the resulting X-ray spectrum has a significantly shallower photon spectrum, with $\gamma \simeq \delta$. Hence, the importance of the electron-electron bremsstrahlung contribution increases with photon energy and the enhanced emission per electron leads to a flattening of the photon spectrum $I(\epsilon)$ produced by a given $\overline{F}(E)$, or, equivalently, a steepening of the $\overline{F}(E)$ form required to produce a given $I(\epsilon)$.

- The recent work on the importance of long-neglected free-bound emission by energetic electrons (Sect. 2.2), with its emphasis on emission due to electron capture onto various ionization states of various elements, necessitates further work on the ionization structure of the solar atmosphere and its time dependence during flares.
- The solar atmosphere above an X-ray emitting region can be safely treated as an optically-thin medium, whereas the lower levels of the atmosphere are optically thick for X-rays. X-rays at energies below ∼11 keV are mostly photoelectrically absorbed in the photosphere, while Compton scattering dominates at the energies above. As a result, X-rays emitted downwards can be Compton-backscattered toward the observer; as discussed in Sects. 3.1 and 3.2, they play a major role in the spectral and imaging characteristics of the emission.

7.2 Electron Source Spectrum

The spatially-integrated photon spectrum is a key source of information about the mean electron flux spectrum in solar flares. High-energy-resolution data from *RHESSI* have permitted, for the first time, not only the extraction of basic parameters through forward-fitting the observed spectra (Sect. 4.1), but have also permitted the first reliable model-independent inversion of observed spectra (Sect. 4.2). Although the general form of a large number of nonthermal flare spectra can be adequately approximated by an isothermal Maxwellian for the low-energy component (Sect. 4.6), plus a broken power law for the high-energy component, *RHESSI* data have clearly demonstrated a wealth of features beyond such simple parametric models: the presence of high-energy and low-energy cutoffs (Sects. 4.3 and 4.5), spectral breaks (Sect. 4.4), and the presence of an albedo component leading to "dips" in the mean electron spectrum (Sect. 4.2). The inclusion of an isotropic albedo correction removes the need for low-energy cutoffs and, if low-energy cutoffs exist in the mean electron spectrum, they should be below ∼12 keV (Sects. 3.1 and 4.5).

7.3 Anisotropy

The electron-ion bremsstrahlung emission cross-section is generally anisotropic, with a dipole-like diagram at low energies (Sect. 5). The anisotropy or angular distribution of X-ray emitting electrons can be measured in a number of different ways, and various possibilities have been employed with *RHESSI* data. Parameter-free regularized electron flux spectra, reconstructed by assuming a parameterized form of the electron angular distribution, highlight the need to consider anisotropy in determining the true shape of the electron flux spectrum (Sect. 5.4). A statistical study of center-to-limb variations in the 15–50 keV energy range has shown that an anisotropy factor α, or the ratio of downward and upward directed fluxes, for hard X-ray emission that lies outside the range [0.2, 5] can be rejected with 99% confidence (Sect. 5.3).

The albedo portion of the observed spectrum can be rather effectively used to infer the mean electron flux in two directions simultaneously (Sect. 5.4). The reconstructions of the mean electron spectra in the downward and upward directions suggest that X-ray emitting electrons need not be significantly anisotropic in a broad range of energies from tens of keV to about 200 keV. This imposes significant challenges to solar flare models based on collisional transport; see Fletcher et al. (2011, this issue) and Holman et al. (2011, this issue) for details.

RHESSI has been used to attempt polarization measurements in two different energy bands using two different scattering processes. The results are, however, of limited statistical significance. If real, they would suggest that X-ray polarization above 50 keV is significant (with values up to 50% or more) and that high quality polarization measurements would provide significant constraints on particle acceleration models (Sect. 5.5). The *RHESSI* design is not optimized for studies of polarization, and attempts to measure it have been hampered by the small effective area and the high background contribution. Future progress will require a polarimetric instrument able to measure polarization with errors on the level of 1 to 2%. Such measurements would allow detailed studies of the electron beaming, quantifying its magnitude, and placing significant constraints on the acceleration geometry.

7.4 Spatial Variation of Electron Flux

Imaging spectroscopy results (Sect. 6) from *RHESSI* have (not surprisingly) revealed that the hard X-ray emission from solar flares is far from homogeneous: the centroid of high energy emission moves downward with higher energy, presumably due to the increased penetration of higher energy electrons; coronal sources have radically different spectra than chromospheric footpoints; and even systematic differences exist between different footpoint sources in the same event.

Since *RHESSI* provides spatial information on detected hard X-ray emission through a (Fourier-transform-based) rotating modulation collimator technique, it follows that the highest-fidelity spatial information is contained in the finite number of Fourier components ("visibilities") sampled. Both forward-fit and inversion techniques have been applied to *RHESSI* imaging spectroscopy data, with very positive results, including the first empirical estimates of the density and volume of the electron acceleration region, and the recovery of *electron flux maps* that, by construction, vary smoothly with energy and so provide valuable information on the variation of the emitting electron spectrum throughout the flare source.

Acknowledgements E.P.K., J.C.B. and P.C.V.M. acknowledge the support of a PPARC/STFC Rolling Grant, UC Berkeley NASA *RHESSI* Visitor funds (J.C.B.), a PPARC/STFCr UK Advanced Fellowship and Royal Society Conference Grant (E.P.K.) and a Dorothy Hodgkin Scholarship (P.C.V.M.). J.K. acknowledges support from Grant 205/06/P135 of the Grant Agency of the Czech Republic and the research plan AV0Z10030501 of the Astronomical Institute AS CR, v.v.i. A.G.E. was supported by NASA Grant NNX10AT78G and through a grant from the University of California-Berkeley. E.P.K., J.C.B., A.G.E., G.D.H., G.J.H., J.K., A.M.M., M.P. and M.P. have been supported in part by a grant from the International Space Science Institute (ISSI) in Bern, Switzerland. E.J.S. acknowledges a grant from NASA Goddard Space Flight Center to the University of Maryland and a Heliophysics GI grant 06-HGI06-15 from NASA HQ to NWRA for research in *RHESSI* albedo determination and applications. Financial support by the European Commission through the SOLAIRE Network (MTRN-CT-2006-035484) is gratefully acknowledged by E.P.K. and J.C.B. A.M.M., M.P. and M.P. acknowledge a grant by the Italian MIVR.

References

D. Alexander, A.J. Coyner, Astrophys. J. **640**, 505 (2006). doi:10.1086/500076

D. Alexander, A.G. Daou, Astrophys. J. **666**, 1268 (2007). doi:10.1086/520331

R.C. Alexander, J.C. Brown, Sol. Phys. **210**, 407 (2002). doi:10.1023/A:1022465615445

M.J. Aschwanden, *Physics of the Solar Corona. An Introduction with Problems and Solutions*, 2nd edn. (Pour la Science, Paris, 2005)

M.J. Aschwanden, J.C. Brown, E.P. Kontar, Sol. Phys. **210**, 383 (2002). doi:10.1023/A:1022472319619

T. Bai, Astrophys. J. **334**, 1049 (1988). doi:10.1086/166897

T. Bai, R. Ramaty, Astrophys. J. **219**, 705 (1978). doi:10.1086/155830

M. Bertero, C. DeMol, E.R. Pike, Inverse Probl. **1**, 301 (1985). doi:10.1088/0266-5611/1/4/004

S.E. Boggs, W. Coburn, E. Kalemci, Astrophys. J. **638**, 1129 (2006). arXiv:astro-ph/0510588, doi:10.1086/498930
S.V. Bogovalov, Y.D. Kotov, V.M. Zenchenko, G. Vedrenne, M. Niel, C. Barat, G. Chambon, R. Talon, Sov. Astron. Lett. **11**, 322 (1985)
S.-C. Bong, J. Lee, D.E. Gary, H.S. Yun, Astrophys. J. **636**, 1159 (2006). doi:10.1086/498225
J.C. Brown, Sol. Phys. **18**, 489 (1971)
J.C. Brown, Sol. Phys. **26**, 441 (1972)
J.C. Brown, in *Coronal Disturbances*, ed. by G.A. Newkirk. IAU Symposium, vol. 57 (1974), p. 395
J.C. Brown, M.J. Aschwanden, E.P. Kontar, Sol. Phys. **210**, 373 (2002). doi:10.1023/A:1022469402781
J.C. Brown, A.G. Emslie, Astrophys. J. **331**, 554 (1988). doi:10.1086/166581
J.C. Brown, A.G. Emslie, G.D. Holman, C.M. Johns-Krull, E.P. Kontar, R.P. Lin, A.M. Massone, M. Piana, Astrophys. J. **643**, 523 (2006). doi:10.1086/501497
J.C. Brown, A.G. Emslie, E.P. Kontar, Astrophys. J. Lett. **595**, L115 (2003). doi:10.1086/378169
J.C. Brown, P.C.V. Mallik, Astron. Astrophys. **481**, 507 (2008). arXiv:0706.2823, doi:10.1051/0004-6361:20078103
J.C. Brown, P.C.V. Mallik, Astrophys. J. Lett. **697**, L6 (2009). doi:10.1088/0004-637X/697/1/L6
J.C. Brown, P.C.V. Mallik, N.R. Badnell, Astron. Astrophys. **515**, C1+ (2010). doi:10.1051/0004-6361:20078103e
J.C. Brown, A.N. McClymont, Sol. Phys. **41**, 135 (1975)
J.C. Brown, A.N. McClymont, I.S. McLean, Nature **247**, 448 (1974)
J.C. Brown, R. Turkmani, E.P. Kontar, A.L. MacKinnon, L. Vlahos, Astron. Astrophys. **508**, 993 (2009). arXiv:0909.4243, doi:10.1051/0004-6361/200913145
J.C. Brown, H.F. van Beek, A.N. McClymont, Astron. Astrophys. **41**, 395 (1975)
A. Caspi, R.P. Lin, Astrophys. J. Lett. **725**, L161 (2010). doi:10.1088/2041-8205/725/2/L161
C.P. Catalano, J.A. van Allen, Astrophys. J. **185**, 335 (1973)
J.E. Charikov, A.B. Guzman, I.V. Kudryavtsev, Astron. Astrophys. **308**, 924 (1996)
A.J. Conway, J.C. Brown, B.A.C. Eves, E. Kontar, Astron. Astrophys. **407**, 725 (2003). doi:10.1051/0004-6361:20030897
I.J.D. Craig, J.C. Brown, Astron. Astrophys. **49**, 239 (1976)
I.J.D. Craig, J.C. Brown, *Inverse Problems in Astronomy: A Guide to Inversion Strategies for Remotely Sensed Data* (Adam Hilger, Ltd, Bristol, 1986), p. 159 (Research supported by SERC)
J.L. Culhane, Mon. Not. R. Astron. Soc. **144**, 375 (1969)
J.L. Culhane, L.W. Acton, Mon. Not. R. Astron. Soc. **151**, 141 (1970)
D.W. Datlowe, Space Sci. Instrum. **2**, 523 (1976a)
D.W. Datlowe, Space Sci. Instrum. **2**, 239 (1976b)
D.W. Datlowe, M.J. Elcan, H.S. Hudson, Sol. Phys. **39**, 155 (1974)
B.R. Dennis, Sol. Phys. **100**, 465 (1985)
B.R. Dennis, R.L. Pernak, Astrophys. J. **698**, 2131 (2009). doi:10.1088/0004-637X/698/2/2131
A. Duijveman, P. Hoyng, M.E. Machado, Sol. Phys. **81**, 137 (1982)
A.G. Emslie, Astrophys. J. **224**, 241 (1978). doi:10.1086/156371
A.G. Emslie, Astrophys. J. **245**, 711 (1981). doi:10.1086/158846
A.G. Emslie, Astrophys. J. Lett. **595**, L119 (2003). doi:10.1086/378168
A.G. Emslie, R.K. Barrett, J.C. Brown, Astrophys. J. **557**, 921 (2001). doi:10.1086/322261
A.G. Emslie, H.L. Bradsher, M.L. McConnell, Astrophys. J. **674**, 570 (2008). doi:10.1086/524983
A.G. Emslie, J.C. Brown, Astrophys. J. **237**, 1015 (1980). doi:10.1086/157947
A.G. Emslie, E.P. Kontar, S. Krucker, R.P. Lin, Astrophys. J. Lett. **595**, L107 (2003). doi:10.1086/378931
A.G. Emslie, D.F. Smith, Astrophys. J. **279**, 882 (1984). doi:10.1086/161959
A.G. Emslie, L. Vlahos, Astrophys. J. **242**, 359 (1980). doi:10.1086/158469
R. Falewicz, P. Rudawy, M. Siarkowski, Astron. Astrophys. **500**, 901 (2009). arXiv:0904.1588, doi:10.1051/0004-6361/200811364
F. Fárník, H. Hudson, T. Watanabe, Astron. Astrophys. **320**, 620 (1997)
L. Fletcher et al., Space Sci. Rev. (2011, this issue). doi:10.1007/s11214-010-9701-8
R.L. Gluckstern, M.H. Hull, G. Breit, Phys. Rev. **90**, 1026 (1953). doi:10.1103/PhysRev.90.1026
G. Han, Y. Li, W. Gan, Chin. Astron. Astrophys. **33**, 168 (2009). doi:10.1016/j.chinastron.2009.03.009
I.G. Hannah, E.P. Kontar, O.K. Sirenko, Astrophys. J. Lett. **707**, L45 (2009). arXiv:0911.0314, doi:10.1088/0004-637X/707/1/L45
I.G. Hannah et al., Space Sci. Rev. (2011, this issue). doi:10.1007/s11214-010-9705-4
E. Haug, Sol. Phys. **25**, 425 (1972)
E. Haug, Sol. Phys. **45**, 453 (1975)

E. Haug, Astron. Astrophys. **218**, 330 (1989)
E. Haug, Astron. Astrophys. **326**, 417 (1997)
D. Heristchi, Astrophys. J. **311**, 474 (1986). doi:10.1086/164787
G.D. Holman, Astrophys. J. **586**, 606 (2003). doi:10.1086/367554
G.D. Holman, M.J. Aschwanden, H. Aurass, M. Battaglia, P.C. Grigis, E.P. Kontar, W. Liu, G. Mann, P. Saint-Hilaire, E.J. Schmahl, V.V. Zharkova, Space Sci. Rev. (2011, this issue). doi:10.1007/s11214-010-9680-9
G.D. Holman, L. Sui, R.A. Schwartz, A.G. Emslie, Astrophys. J. Lett. **595**, L97 (2003). doi:10.1086/378488
P. Hoyng, A. Duijveman, A. Boelee, C. de Jager, M. Galama, R. Hoekstra, J. Imhof, H. Lafleur, M.E. Machado, R. Fryer, Astrophys. J. Lett. **244**, L153 (1981a). doi:10.1086/183501
P. Hoyng, A. Duijveman, M.E. Machado, D.M. Rust, Z. Svestka, A. Boelee, C. de Jager, K.T. Frost, H. Lafleur, G.M. Simnett, H.F. van Beek, B.E. Woodgate, Astrophys. J. Lett. **246**, L155+ (1981b). doi:10.1086/183574
G. Huang, Sol. Phys. **257**, 323 (2009). doi:10.1007/s11207-009-9381-2
H.S. Hudson, G.J. Hurford, J.C. Brown, Sol. Phys. **214**, 171 (2003)
G.J. Hurford, E.J. Schmahl, R.A. Schwartz, A.J. Conway, M.J. Aschwanden, A. Csillaghy, B.R. Dennis, C. Johns-Krull, S. Krucker, R.P. Lin, J. McTiernan, T.R. Metcalf, J. Sato, D.M. Smith, Sol. Phys. **210**, 61 (2002). doi:10.1023/A:1022436213688
H. Ji, H. Wang, E.J. Schmahl, J. Qiu, Y. Zhang, Astrophys. J. **605**, 938 (2004). doi:10.1086/382583
M. Jin, M.D. Ding, Astron. Astrophys. **471**, 705 (2007). doi:10.1051/0004-6361:20077202
C.M. Johns, R.P. Lin, Sol. Phys. **142**, 219 (1992a)
C.M. Johns, R.P. Lin, Sol. Phys. **137**, 121 (1992b)
S.R. Kane, K. Hurley, J.M. McTiernan, M. Boer, M. Niel, T. Kosugi, M. Yoshimori, Astrophys. J. **500**, 1003 (1998). doi:10.1086/305738
M. Karlický, J. Kašparová, Astron. Astrophys. **506**, 1437 (2009). arXiv:0909.0146, doi:10.1051/0004-6361/200912616
J. Kašparová, M. Karlický, E.P. Kontar, R.A. Schwartz, B.R. Dennis, Sol. Phys. **232**, 63 (2005). arXiv: astro-ph/0508636, doi:10.1007/s11207-005-1581-9
J. Kašparová, E.P. Kontar, J.C. Brown, Astron. Astrophys. **466**, 705 (2007). arXiv:astro-ph/0701871, doi:10.1051/0004-6361:20066689
H.W. Koch, J.W. Motz, Rev. Mod. Phys. **31**, 920 (1959). doi:10.1103/RevModPhys.31.920
E.P. Kontar, Sol. Phys. **202**, 131 (2001)
E.P. Kontar, J.C. Brown, Adv. Space Res. **38**, 945 (2006a). arXiv:astro-ph/0508418, doi:10.1016/j.asr.2005.09.029
E.P. Kontar, J.C. Brown, Astrophys. J. Lett. **653**, L149 (2006b). arXiv:astro-ph/0611170, doi:10.1086/510586
E.P. Kontar, J.C. Brown, A.G. Emslie, R.A. Schwartz, D.M. Smith, R.C. Alexander, Astrophys. J. Lett. **595**, L123 (2003). doi:10.1086/378172
E.P. Kontar, J.C. Brown, G.K. McArthur, Sol. Phys. **210**, 419 (2002). doi:10.1023/A:1022494318540
E.P. Kontar, E. Dickson, J. Kašparová, Sol. Phys. **252**, 139 (2008a). arXiv:0805.1470, doi:10.1007/s11207-008-9249-x
E.P. Kontar, A.G. Emslie, A.M. Massone, M. Piana, J.C. Brown, M. Prato, Astrophys. J. **670**, 857 (2007). arXiv:0707.4225, doi:10.1086/521977
E.P. Kontar, A.G. Emslie, M. Piana, A.M. Massone, J.C. Brown, Sol. Phys. **226**, 317 (2005). arXiv:astro-ph/0409691, doi:10.1007/s11207-005-7150-4
E.P. Kontar, I.G. Hannah, A.L. MacKinnon, Astron. Astrophys. **489**, L57 (2008b). arXiv:0808.3334, doi:10.1051/0004-6361:200810719
E.P. Kontar, N.L.S. Jeffrey, Astron. Astrophys. **513**, L2+ (2010). arXiv:1003.0884, doi:10.1051/0004-6361/201014066
E.P. Kontar, A.L. MacKinnon, R.A. Schwartz, J.C. Brown, Astron. Astrophys. **446**, 1157 (2006). arXiv: astro-ph/0510167, doi:10.1051/0004-6361:20053672
E.P. Kontar, M. Piana, A.M. Massone, A.G. Emslie, J.C. Brown, Sol. Phys. **225**, 293 (2004). arXiv:astro-ph/0409688, doi:10.1007/s11207-004-4140-x
A.A. Korchak, Sov. Astron. **11**, 258 (1967)
T. Kosugi, S. Masuda, K. Makishima, M. Inda, T. Murakami, T. Dotani, Y. Ogawara, T. Sakao, K. Kai, H. Nakajima, Sol. Phys. **136**, 17 (1991)
J. Kotoku, K. Makishima, Y. Matsumoto, M. Kohama, Y. Terada, T. Tamagawa, Publ. Astron. Soc. Jpn. **59**, 1161 (2007). arXiv:0708.0057
S. Krucker, M. Battaglia, P.J. Cargill, L. Fletcher, H.S. Hudson, A.L. MacKinnon, S. Masuda, L. Sui, M. Tomczak, A.L. Veronig, L. Vlahos, S.M. White, Astron. Astrophys. Rev. **8** (2008). doi:10.1007/s00159-008-0014-9
S. Krucker, R.P. Lin, Sol. Phys. **210**, 229 (2002). doi:10.1023/A:1022469902940

T.A. Kucera, P.J. Love, B.R. Dennis, G.D. Holman, R.A. Schwartz, D.M. Zarro, Astrophys. J. **466**, 1067 (1996). doi:10.1086/177576
E. Landi, G. Del Zanna, P.R. Young, K.P. Dere, H.E. Mason, M. Landini, Astrophys. J. Suppl. Ser. **162**, 261 (2006). doi:10.1086/498148
M. Landini, B.C. Monsignori Fossi, R. Pallavicini, Sol. Phys. **29**, 93 (1973)
S.H. Langer, V. Petrosian, Astrophys. J. **215**, 666 (1977)
J. Leach, V. Petrosian, Astrophys. J. **251**, 781 (1981). doi:10.1086/159521
J. Leach, V. Petrosian, Astrophys. J. **269**, 715 (1983). doi:10.1086/161081
P. Li, Astrophys. J. **443**, 855 (1995). doi:10.1086/175575
P. Li, K. Hurley, C. Barat, M. Niel, R. Talon, V. Kurt, Astrophys. J. **426**, 758 (1994). doi:10.1086/174112
R.P. Lin, B.R. Dennis, G.J. Hurford, D.M. Smith, A. Zehnder, P.R. Harvey, D.W. Curtis, D. Pankow, P. Turin, M. Bester, A. Csillaghy, M. Lewis, N. Madden, H.F. van Beek, M. Appleby, T. Raudorf, J. McTiernan, R. Ramaty, E. Schmahl, R. Schwartz, S. Krucker, R. Abiad, T. Quinn, P. Berg, M. Hashii, R. Sterling, R. Jackson, R. Pratt, R.D. Campbell, D. Malone, D. Landis, C.P. Barrington-Leigh, S. Slassi-Sennou, C. Cork, D. Clark, D. Amato, L. Orwig, R. Boyle, I.S. Banks, K. Shirey, A.K. Tolbert, D. Zarro, F. Snow, K. Thomsen, R. Henneck, A. Mchedlishvili, P. Ming, M. Fivian, J. Jordan, R. Wanner, J. Crubb, J. Preble, M. Matranga, A. Benz, H. Hudson, R.C. Canfield, G.D. Holman, C. Crannell, T. Kosugi, A.G. Emslie, N. Vilmer, J.C. Brown, C. Johns-Krull, M. Aschwanden, T. Metcalf, A. Conway, Sol. Phys. **210**, 3 (2002). doi:10.1023/A:1022428818870
R.P. Lin, R.A. Schwartz, Astrophys. J. **312**, 462 (1987). doi:10.1086/164891
W. Liu, V. Petrosian, B.R. Dennis, G.D. Holman, Astrophys. J. **693**, 847 (2009). arXiv:0805.1055, doi:10.1088/0004-637X/693/1/847
W. Liu, V. Petrosian, B.R. Dennis, Y.W. Jiang, Astrophys. J. **676**, 704 (2008). arXiv:0709.1963, doi:10.1086/527538
M.E. Machado, A. Duijveman, B.R. Dennis, Sol. Phys. **79**, 85 (1982)
P. Magdziarz, A.A. Zdziarski, Mon. Not. R. Astron. Soc. **273**, 837 (1995)
A.M. Massone, A.G. Emslie, E.P. Kontar, M. Piana, M. Prato, J.C. Brown, Astrophys. J. **613**, 1233 (2004). doi:10.1086/423127
A.M. Massone, M. Piana, A. Conway, B. Eves, Astron. Astrophys. **405**, 325 (2003). doi:10.1051/0004-6361:20030569
S. Masuda, T. Kosugi, H. Hara, T. Sakao, K. Shibata, S. Tsuneta, Publ. Astron. Soc. Jpn. **47**, 677 (1995)
K. Matsushita, S. Masuda, T. Kosugi, M. Inda, K. Yaji, Publ. Astron. Soc. Jpn. **44**, L89 (1992)
M.L. McConnell, J.M. Ryan, D.M. Smith, G.J. Hurford, M.D. Fivian, R.P. Lin, A.G. Emslie, W. Hajdas, in *American Astronomical Society Meeting Abstracts*, vol. 210 (2007), p. 93.01
M.L. McConnell, J.M. Ryan, D.M. Smith, R.P. Lin, A.G. Emslie, Sol. Phys. **210**, 125 (2002). arXiv:astro-ph/0209384, doi:10.1023/A:1022413708738
M.L. McConnell, D.M. Smith, A.G. Emslie, G.J. Hurford, R.P. Lin, J.M. Ryan, in *Bulletin of the American Astronomical Society*, vol. 35 (2003), p. 850
J.M. McTiernan, V. Petrosian, Astrophys. J. **359**, 524 (1990a). doi:10.1086/169084
J.M. McTiernan, V. Petrosian, Astrophys. J. **359**, 541 (1990b). doi:10.1086/169085
J.M. McTiernan, V. Petrosian, Astrophys. J. **379**, 381 (1991). doi:10.1086/170513
V.N. Mel'nik, V. Lapshin, E. Kontar, Sol. Phys. **184**, 353 (1999)
R. Morrison, D. McCammon, Astrophys. J. **270**, 119 (1983). doi:10.1086/161102
T. Mrozek, Adv. Space Res. **38**, 962 (2006). doi:10.1016/j.asr.2006.09.003
M.P. Nakada, W.M. Neupert, R.J. Thomas, Sol. Phys. **37**, 429 (1974)
N. Nitta, B.R. Dennis, A.L. Kiplinger, Astrophys. J. **353**, 313 (1990). doi:10.1086/168618
M. Oda, N. Muranaka, M. Matsuoka, S. Miyamoto, Y. Ogawara, Space Sci. Instrum. **2**, 141 (1976)
V. Petrosian, Astrophys. J. **299**, 987 (1985). doi:10.1086/163765
V. Petrosian, Q. Chen, Astrophys. J. Lett. **712**, L131 (2010). arXiv:1002.2673, doi:10.1088/2041-8205/712/2/L131
K.J.H. Phillips, C. Chifor, B.R. Dennis, Astrophys. J. **647**, 1480 (2006). arXiv:astro-ph/0607309, doi:10.1086/505518
K.J.H. Phillips, U. Feldman, E. Landi, *Ultraviolet and X-ray Spectroscopy of the Solar Atmosphere* (Cambridge University Press, Cambridge, 2008)
M. Piana, Astron. Astrophys. **288**, 949 (1994)
M. Piana, M. Bertero, Inverse Probl. **13**, 441 (1997)
M. Piana, J.C. Brown, A.M. Thompson, Sol. Phys. **156**, 315 (1995)
M. Piana, A.M. Massone, G.J. Hurford, M. Prato, A.G. Emslie, E.P. Kontar, R.A. Schwartz, Astrophys. J. **665**, 846 (2007). doi:10.1086/519518
M. Piana, A.M. Massone, E.P. Kontar, A.G. Emslie, J.C. Brown, R.A. Schwartz, Astrophys. J. Lett. **595**, L127 (2003). doi:10.1086/378171

G. Pizzichini, A. Spizzichino, G.R. Vespignani, Sol. Phys. **35**, 431 (1974)
M. Prato, A.G. Emslie, E.P. Kontar, A.M. Massone, M. Piana, Astrophys. J. **706**, 917 (2009). doi:10.1088/0004-637X/706/1/917
M. Prato, M. Piana, J.C. Brown, A.G. Emslie, E.P. Kontar, A.M. Massone, Sol. Phys. **237**, 61 (2006). doi:10.1007/s11207-006-0029-1
W.H. Press, S.A. Teukolsky, W.T. Vetterling, B.P. Flannery, *Numerical Recipes in FORTRAN. The Art of Scientific Computing*, 2nd edn. (Cambridge University Press, Cambridge, 1992)
J.-R. Roy, D.W. Datlowe, Sol. Phys. **40**, 165 (1975)
P. Saint-Hilaire, A.O. Benz, Sol. Phys. **210**, 287 (2002). arXiv:astro-ph/0210023, doi:10.1023/A:1022478300679
P. Saint-Hilaire, S. Krucker, R.P. Lin, Sol. Phys. **250**, 53 (2008). doi:10.1007/s11207-008-9193-9
T. Sakao, T. Kosugi, S. Masuda, M. Inda, K. Makishima, R.C. Canfield, H.S. Hudson, T.R. Metcalf, J.-P. Wuelser, L.W. Acton, Y. Ogawara, Publ. Astron. Soc. Jpn. **44**, L83 (1992)
N. Santangelo, H. Horstman, E. Horstman-Moretti, Sol. Phys. **29**, 143 (1973)
E.J. Schmahl, G.J. Hurford, Sol. Phys. **210**, 273 (2002). doi:10.1023/A:1022484822851
E.J. Schmahl, G.J. Hurford, Adv. Space Res. **32**, 2477 (2003). doi:10.1016/S0273-1177(03)00938-4
E.J. Schmahl, R.L. Pernak, G.J. Hurford, J. Lee, S. Bong, Sol. Phys. **240**, 241 (2007). doi:10.1007/s11207-007-0263-1
R.A. Schwartz, A. Csillaghy, A.K. Tolbert, G.J. Hurford, J. Mc Tiernan, D. Zarro, Sol. Phys. **210**, 165 (2002). doi:10.1023/A:1022444531435
C. Shao, G. Huang, Astrophys. J. **691**, 299 (2009). doi:10.1088/0004-637X/691/1/299
G.H. Share, R.J. Murphy, Astrophys. J. **485**, 409 (1997). doi:10.1086/304407
G.H. Share, R.J. Murphy, J. Kiener, N. de Séréville, Astrophys. J. **573**, 464 (2002). arXiv:astro-ph/0203215, doi:10.1086/340595
D.M. Smith, R.P. Lin, P. Turin, D.W. Curtis, J.H. Primbsch, R.D. Campbell, R. Abiad, P. Schroeder, C.P. Cork, E.L. Hull, D.A. Landis, N.W. Madden, D. Malone, R.H. Pehl, T. Raudorf, P. Sangsingkeow, R. Boyle, I.S. Banks, K. Shirey, R. Schwartz, Sol. Phys. **210**, 33 (2002). doi:10.1023/A:1022400716414
D.M. Smith, G.H. Share, R.J. Murphy, R.A. Schwartz, A.Y. Shih, R.P. Lin, Astrophys. J. Lett. **595**, L81 (2003). arXiv:astro-ph/0306292, doi:10.1086/378173
Y. Su, G.D. Holman, B.R. Dennis, A.K. Tolbert, R.A. Schwartz, Astrophys. J. **705**, 1584 (2009). doi:10.1088/0004-637X/705/2/1584
E. Suarez-Garcia, W. Hajdas, C. Wigger, K. Arzner, M. Güdel, A. Zehnder, P. Grigis, Sol. Phys. **239**, 149 (2006). arXiv:astro-ph/0609778, doi:10.1007/s11207-006-0268-1
L. Sui, G.D. Holman, B.R. Dennis, Astrophys. J. **626**, 1102 (2005). doi:10.1086/430086
L. Sui, G.D. Holman, B.R. Dennis, Astrophys. J. **670**, 862 (2007). doi:10.1086/522198
L. Sui, G.D. Holman, B.R. Dennis, S. Krucker, R.A. Schwartz, K. Tolbert, Sol. Phys. **210**, 245 (2002). doi:10.1023/A:1022417832283
T. Takakura, K. Tanaka, N. Nitta, K. Kai, K. Ohki, Sol. Phys. **107**, 109 (1986)
T. Takakura, S. Tsuneta, N. Nitta, K. Makishima, T. Murakami, Y. Ogawara, M. Oda, K. Ohki, S. Miyamoto, Astrophys. J. Lett. **270**, L83 (1983). doi:10.1086/184075
A.N. Tikhonov, Sov. Math. Dokl. **4**, 1035 (1963)
I.P. Tindo, V.D. Ivanov, S.L. Mandel'Stam, A.I. Shuryghin, Sol. Phys. **14**, 204 (1970)
I.P. Tindo, V.D. Ivanov, S.L. Mandel'Stam, A.I. Shuryghin, Sol. Phys. **24**, 429 (1972a)
I.P. Tindo, V.D. Ivanov, B. Valníček, M.A. Livshits, Sol. Phys. **27**, 426 (1972b)
I.P. Tindo, A.I. Shurygin, W. Steffen, Sol. Phys. **46**, 219 (1976)
F.F. Tomblin, Astrophys. J. **171**, 377 (1972)
M. Tomczak, Astron. Astrophys. **502**, 665 (2009). arXiv:0901.4227, doi:10.1051/0004-6361/200911732
L.J. Tramiel, R. Novick, G.A. Chanan, Astrophys. J. **280**, 440 (1984). doi:10.1086/162010
G. Trottet, N. Vilmer, C. Barat, A. Benz, A. Magun, A. Kuznetsov, R. Sunyaev, O. Terekhov, Astron. Astrophys. **334**, 1099 (1998)
S. Tsuneta, T. Takakura, N. Nitta, K. Makishima, T. Murakami, M. Oda, Y. Ogawara, I. Kondo, K. Ohki, K. Tanaka, Astrophys. J. **280**, 887 (1984). doi:10.1086/162064
S. Tsuneta, T. Takakura, N. Nitta, K. Makishima, T. Murakami, M. Oda, Y. Ogawara, K. Ohki, Sol. Phys. **86**, 313 (1983)
H.F. van Beek, P. Hoyng, B. Lafleur, G.M. Simnett, Sol. Phys. **65**, 39 (1980)
W.T. Vestrand, Sol. Phys. **118**, 95 (1988)
W.T. Vestrand, D.J. Forrest, E.L. Chupp, E. Rieger, G.H. Share, Astrophys. J. **322**, 1010 (1987). doi:10.1086/165796
W.T. Vestrand, D.J. Forrest, E. Rieger, in *International Cosmic Ray Conference*, vol. 3 (1991), p. 69

N. Vilmer, Astrophys. J. Suppl. Ser. **90**, 611 (1994). doi:10.1086/191882

Y. Xu, A.G. Emslie, G.J. Hurford, Astrophys. J. **673**, 576 (2008). doi:10.1086/524184

V.V. Zharkova, J.C. Brown, D.V. Syniavskii, Astron. Astrophys. **304**, 284 (1995)

V.V. Zharkova, M. Gordovskyy, Astrophys. J. **651**, 553 (2006). doi:10.1086/506423

V.V. Zharkova et al., Space Sci. Rev. (2011, this issue). doi:10.1007/s11214-011-9803-y

I.A. Zhitnik, S.V. Kuzin, A.M. Urnov, S.A. Bogachev, F.F. Goryaev, S.V. Shestov, Sol. Syst. Res. **40**, 272 (2006). doi:10.1134/S0038094606040022

Space Sci Rev (2011) 159:357–420
DOI 10.1007/s11214-011-9803-y

Recent Advances in Understanding Particle Acceleration Processes in Solar Flares

V.V. Zharkova · K. Arzner · A.O. Benz · P. Browning · C. Dauphin · A.G. Emslie · L. Fletcher · E.P. Kontar · G. Mann · M. Onofri · V. Petrosian · R. Turkmani · N. Vilmer · L. Vlahos

Received: 21 June 2010 / Accepted: 17 November 2010 / Published online: 26 July 2011

Abstract We review basic theoretical concepts in particle acceleration, with particular emphasis on processes likely to occur in regions of magnetic reconnection. Several new developments are discussed, including detailed studies of reconnection in three-dimensional magnetic field configurations (e.g., current sheets, collapsing traps, separatrix regions) and stochastic acceleration in a turbulent environment. Fluid, test-particle, and particle-in-cell

V.V. Zharkova (✉)
Department of Mathematics, University of Bradford, Bradford, UK
e-mail: v.v.zharkova@brad.ac.uk

K. Arzner · A.O. Benz
ETH, Zurich, Switzerland

P. Browning
Physics Department, University of Manchester, Manchester, UK

C. Dauphin · N. Vilmer
LESIA-Observatoire de Paris, CNRS, UPMC Univ Paris 06, Univ. Paris-Diderot, Paris, France

A.G. Emslie
Department of Physics and Astronomy, Western Kentucky University, Bowling Green, KY 42101, USA

L. Fletcher · E.P. Kontar
Department of Physics and Astronomy, University of Glasgow, Glasgow, G12 8QQ, UK

G. Mann
Astrophysikalischen Institut, Potsdam, Germany

M. Onofri · L. Vlahos
Dept. of Physics, University of Thessaloniki, Thessaloniki 54124, Greece

V. Petrosian
Center for Astrophysics, Stanford University, Palo Alto, USA

R. Turkmani
Imperial College, London, UK

approaches are used and results compared. While these studies show considerable promise in accounting for the various observational manifestations of solar flares, they are limited by a number of factors, mostly relating to available computational power. Not the least of these issues is the need to explicitly incorporate the electrodynamic feedback of the accelerated particles themselves on the environment in which they are accelerated. A brief prognosis for future advancement is offered.

Keywords Sun: flares · Sun: X-rays · Sun: acceleration · Sun: energetic particles

1 Introduction

It is by now well established that the primary source of the energy released in a solar flare is in stressed, current-carrying ("non-potential") magnetic fields. This magnetic energy is released through a *magnetic reconnection* process, in which magnetic field lines of opposite polarity change their topological connectivity and the field relaxes to a state of lower energy, thereby releasing energy in the form of accelerated particles, plasma heating and mass motion.

Magnetic reconnection can occur in several geometries. For example, closed magnetic loops embedded into the solar photosphere can collide in the upper solar atmosphere, resulting in a localized region of relatively high magnetic shear, as in the classical Sweet-Parker model (see, e.g., Sweet 1969). Alternatively, open magnetic field lines can collapse onto a closed helmet-like magnetic configuration. This leads to a region of high magnetic shear near the top of the closed structure—as evidenced by looptop sources of hard X-rays first observed by *Yohkoh* (Masuda et al. 1994)—and in the energization of successive field lines at greater and greater heights, which produces both coronal hard X-ray sources (Krucker et al. 2008a) and a concomitant increase in the separation of the associated hard X-ray and Hα footpoints (Kopp and Pneuman 1976). Magnetic reconnection can also be accompanied by dramatic large-scale changes in the magnetic field configuration, such as in the "breakout" model (Antiochos et al. 1999), in which a filament, previously restrained by the magnetic tension of a set of magnetic field lines exterior to it, is released into interplanetary space as a result of reconnection of the constraining field lines. Reconnection can also result in the formation of magnetohydrodynamic (MHD) shocks, which may be responsible for significant particle acceleration exterior to the region of magnetic reconnection itself.

Whatever the geometry, the fundamental process of magnetic reconnection somehow transfers a significant fraction (e.g., Miller et al. 1997) of the stored magnetic energy into accelerated charged particles, both electrons and ions. Some of the high-energy particles may be trapped in the corona by plasma turbulence (e.g., Petrosian and Liu 2004) and/or by magnetic mirroring (e.g., Melrose and White 1979; Leach 1984; Somov and Kosugi 1997), to produce hard X-ray (HXR) looptop sources (e.g., Masuda et al. 1994; Petrosian et al. 2002; Sui and Holman 2003; Jiang et al. 2006; Battaglia and Benz 2006; Liu 2006). Some particles may escape the acceleration region into interplanetary space and can contribute to solar energetic particle (SEP) events detected at 1 AU (e.g., Krucker et al. 2007a).

Understanding the conversion of magnetic energy into accelerated particles is one of the main goals of solar flare research. It also has attendant implications for our understanding of particle acceleration in other astrophysical sites and was indeed a key driver for the design of *Reuven Ramaty High-Energy Solar Spectroscopic Imager (RHESSI)*, which measures the diagnostic high-energy hard X-ray and gamma-ray radiation produced by the accelerated particles with unprecedented temporal, spectral and spatial resolution.

At a fundamental level, of course, charged particles are accelerated by an electric field. Differences in the large-scale structure and temporal evolution of the accelerating electric field allows us to discuss particle acceleration in terms of several basic scenarios (see Sect. 3). These include large-scale sub-Dreicer acceleration (see, for example Benka and Holman 1994); super-Dreicer acceleration in a relatively localized 2-D current sheet (e.g., Litvinenko and Somov 1993); and stochastic acceleration by MHD turbulence generated in fast plasma reconnection outflows (e.g., Miller et al. 1996; Pryadko and Petrosian 1997).

In Sect. 4, we report on recent work aimed at extending these basic mechanisms to more complex models. These include (1) first-order Fermi acceleration, either in a collapsing magnetic trap (Somov and Kosugi 1997) or in a fast termination shock (Tsuneta and Naito 1998; Aurass et al. 2007), or by a super-Dreicer electric field in a simple 3-D magnetic topology (Zharkova and Gordovskyy 2005b; Zharkova and Agapitov 2009); and (2) acceleration in more elaborate three-dimensional magnetic topologies derived from MHD simulations (Dalla and Browning 2006, 2008). We also discuss some recent suggestions of acceleration by large-scale Alfvén waves (Fletcher and Hudson 2008) and acceleration involving multiple reconnection regions or in a turbulent environment (Vlahos et al. 2004; Turkmani et al. 2005; Dauphin and Vilmer 2007).

In the latter portion of Sect. 4 we also consider both "local" and "global" aspects of the acceleration process. At the local level, accurate simulations of particle trajectories are needed to correctly infer the general properties of the acceleration. At the "global" level, we recognize that, in view of the large number of accelerated particles involved, the traditional "test-particle" approach, in which the electromagnetic environment in the acceleration region is *prescribed*, is not applicable (Sect. 4.7)—a self-consistent approach requires that we incorporate the electric and magnetic fields associated with the accelerated particles themselves in current sheets with and without tearing instabilities (Sects. 4.8 and 4.9).

In Sect. 5 and 6 we relate the results of the recent modeling efforts to flare observations, particularly from *RHESSI*, and we offer a brief prognosis for future models of the particle acceleration process.

2 Review of Pertinent Observations

2.1 Hard X-Ray Lightcurves and Spectra

2.1.1 Lightcurves

Spatially-integrated hard X-ray light curves obtained by *RHESSI* confirm many of the temporal features established by previous observations: sharp increases (bursts) of hard X-ray intensity over a relatively short ($\sim$0.5–5 s) timescale accompanied by a more slowly-varying component with a timescale up to a few tens of minutes (Holman et al. 2011). This suggests that electron acceleration is characterized by two fundamentally distinct timescales: a very short timescale associated with the actual *acceleration* of ambient electrons to high, bremsstrahlung-emitting energies and a much longer timescale associated with the *duration* of the acceleration process, a process that, as evidenced by steady hard X-ray emission for a substantial fraction of an hour or even longer, can continue to accelerate particles over such a sustained period.

At lower photon energies ε, where thermal bremsstrahlung dominates the total emission, the emission at higher photon energies is weighted more heavily by plasma at high temperatures T.

As a result, the decrease in conductive cooling time with temperature—$\tau \approx T^{-5/2}$ for collisionally-dominated conductive processes (see Spitzer 1962)—results in the emission at higher energies peaking sooner (Aschwanden 2007), and hence the light curve peaking progressively earlier with increase in photon energy. At higher energies, where *non-thermal* bremsstrahlung dominates, the relative timing of the emission at different energies depends on a balance between the reduced "time-of-flight" for higher energy electrons before they impact upon the thick target of the lower atmosphere (which tends to *advance* high energy emissions relative to low energy ones—Aschwanden and Schwartz 1996; Brown et al. 1998) and on the decrease in collision frequency with energy (which tends to *delay* high energy emissions relative to low energy ones—Aschwanden et al. 1997).

2.1.2 Photon and Electron Energy Spectra

Hard X-ray emission during flares typically exhibits a very steep spectrum at lower energies $\varepsilon \approx 10$ keV, indicative of a thermal process. It must be noted, however, that the frequent assumption (e.g., Holman et al. 2003) of an *iso*thermal plasma is neither expected on the basis of simple physics nor required by the observations (e.g., Brown 1974). Indeed, Aschwanden (2007) has shown that the assumption of an isothermal source is *inconsistent* with the observations of temporal variations of the hard X-ray spectrum in certain flares observed by *RHESSI*. Indeed, in most cases analysis of the variation of residuals (observed hard X-ray flux vs. flux calculated using an isothermal spectrum) with photon energy shows that an isothermal model *does* provide an excellent fit to the observed spectrum; generally a broad-band hard X-ray spectrometer shows the (unique) hottest component of a multithermal source. Nevertheless *RHESSI's* superior resolution and good photometric calibration have made it possible to derive a two-temperature source characterization from a single bremsstrahlung spectrum (e.g., Caspi and Lin 2010) in some cases.

In most powerful flares the photon spectra reveal double power laws below 300 keV, with the spectral index below ~30–60 keV being smaller (flatter) than that above ~30–60 keV by 2–4 units (Grigis and Benz 2006; Krucker et al. 2008a). For the most energetic flares, there can be discernible emission up to a few hundred MeV (for reviews, see Chupp and Ryan 2009; Vilmer et al. 2011). At photon energies ~500–1000 keV, the spectrum tends to flatten again, a flattening that has been, at least partially, attributed to the increasing contribution of electron-electron bremsstrahlung (Kontar et al. 2007). However, the contribution of electron-electron bremsstrahlung to this flattening does not exceed ~0.4, while the observed spectral hardening, as observed in a few events with *GRANAT*/PHEBUS, can be as high as 1.5–2 units at energies from 600 keV up to several tens of MeV (see, e.g., Trottet et al. 1998; Vilmer et al. 1999).

RHESSI observations reveal a relationship between spectral hardness and time. Values of the "local" spectral index $\gamma(\varepsilon) = -d\log I(\varepsilon)/d\log\varepsilon$ are generally lowest around the peak of the event—a so-called "soft-hard-soft" evolution originally noted by Parks and Winckler (1969). *RHESSI*'s superior spectral resolution shows that nearly every discernible intensity peak (which can number up to a dozen or more in a single flare) has a soft-hard-soft structure (Grigis and Benz 2004). Also the decrease in spectral index before, and increase in spectral index after, this time does not occur at the same rate for all photon energies ε (Kontar and MacKinnon 2005); see Fig. 1. Rather, a flare typically starts with a spectral index γ that is strongly dependent on energy, with the minimum value of γ showing a tendency to grow with time.

Non-thermal emission in the corona is identified in the impulsive phase by its softer spectrum (Mariska and McTiernan 1999; Petrosian et al. 2002), consistent with the small

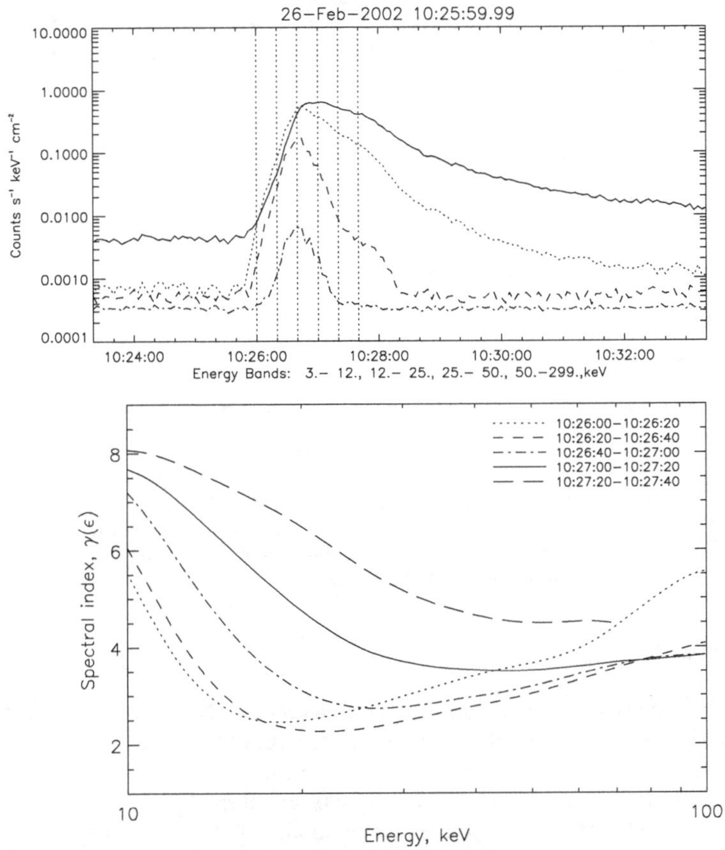

Fig. 1 *Upper panel*: Temporal variation (4-second cadence) of the count rates in seven front *RHESSI* segments for SOL2002-02-26T10:27 (C9.6) (10:27 UT). The *vertical lines* show five 20-second accumulation intervals for spectral analysis. *Lower panel*: Temporal variation of the energy-dependent photon spectral index $\gamma(\varepsilon) = -d\log I(\varepsilon)/d\log\varepsilon$. *Each line* corresponds to one of the time intervals in the upper plot. From Kontar and MacKinnon (2005)

column depth of the coronal part of the source. The absence of a significant amount of (energy-dependent) collisional losses in this relatively thin target should result in a spectrum two powers steeper than the target-averaged spectrum (Brown 1971; Hudson 1972; Datlowe and Lin 1973).

It should be noted that even if the accelerated electrons have a pure power-law energy spectrum $F_0(E_0) \approx E_0^{-\delta}$, for which a characteristic energy does not exist, characteristic energies associated with either electron transport and/or radiation physics may produce deviations from the power-law behavior in the observed spatially-integrated photon spectrum. Further, as elaborated upon by Kontar et al. (2011), anisotropy in the mean electron distribution, combined with the intrinsic directivity of the bremsstrahlung emission process, produces an anisotropic distribution of primary photons. Compton backscattering of photons from the photosphere ("X-ray albedo"; Langer and Petrosian 1977; Bai and Ramaty 1978) not only influences the observed photon spectrum, but also has a diagnostic potential for determining the electron angular distribution (Kontar and Brown 2006).

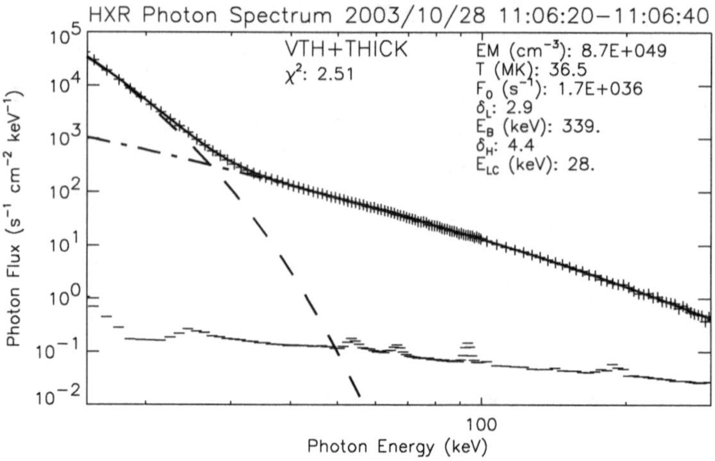

Fig. 2 *RHESSI* hard X-ray photon spectrum for SOL2003-10-28T11:10 (X17.2) (11:06:20-11:06:40 UT). The photon flux (*plus signs*) is fit with a thermal bremsstrahlung contribution from an isothermal plasma (*dashed curve*) plus a double power-law non-thermal electron flux component with a low-energy cutoff (*dash-dotted curve*). The best-fit parameters are shown in the plot

2.1.3 Electron Numbers

Analysis of *RHESSI* data (Brown et al. 2007; Dennis et al. 2007) has confirmed the important point that a significant fraction of the energy released in flares is deposited into accelerated particles. In addition, recent calorimetric analysis of the evolution of a radio spectrum from a dense flare (Bastian et al. 2007) reveals that a substantial fraction of the energy in the energetic electrons can be deposited into the coronal loop, rather than into the chromosphere; it is estimated that the energy deposited in the corona can approach about 30% of the estimated magnetic energy released in the whole flare.

As has been pointed out by several authors (e.g., Brown 1971; Tandberg-Hanssen and Emslie 1988), the total injected electron flux depends critically on the value of the low-energy cutoff E_c. A brief discussion of this issue is in order. Historically, a low-energy cutoff E_c was assumed simply in order to keep the injected power $\int_{E_c}^{\infty} E_0 F_0(E_0)\, dE_0$ finite (e.g., Brown 1971; Holman et al. 2003). To determine whether or not such a cut-off is actually *required* by observations, it is best to adopt a non-parametric approach to interpreting the photon spectrum $I(\varepsilon)$—i.e., to infer from $I(\varepsilon)$ what range of mean electron source functions $\overline{F}(E)$ (Brown et al. 2003) allow a statistically acceptable fit to $I(\varepsilon)$. Kašparová et al. (2005), in their analysis of SOL2002-08-20T08:25 (M3.4), have shown that a low-energy cutoff, or even a gap, in the mean source electron spectrum exists if the observed spectrum is considered as primary bremsstrahlung only; however, such cutoffs and/or gaps disappear if an albedo correction (for an isotropic primary source) is applied to the observed photon spectrum (see, e.g., Kontar et al. 2011). Further, Emslie (2003) has shown that, when allowance is made for warm-target effects in the electron energy-loss rate, the injected electron energy $\int_0^{\infty} E_0 F_0(E_0)\, dE_0$ corresponding to a pure power-law photon spectrum can be finite even if no low-energy cutoff exists.

Figure 2 shows the photon spectrum for the time interval indicated during SOL2003-10-28T11:10 (X17.2). Assuming that the non-thermal hard X-ray emission is thick-target bremsstrahlung (Brown 1971), one can use a forward-fitting method, including both thermal and nonthermal fit components, (e.g., Holman et al. 2003) to derive the injected non-thermal electron acceleration rate, differential in energy (electrons s^{-1} keV^{-1}). This method (see Kontar et al. 2011) also determines an observational upper limit to E_c (as the intersection of the thermal and nonthermal fit components), and hence a lower limit on the associated electron number and energy fluxes. Integration over energy provides the total electron acceleration rate. For this event, one obtains a total non-thermal electron production rate $F_e = 1.7 \times 10^{36}$ s^{-1} above $E_c = 28$ keV, corresponding to an injected power $P_e = 1.5 \times 10^{29}$ erg s^{-1} above the same energy. For a nominal footpoint area $S \lesssim 10^{18}$ cm^2, such a particle rate corresponds to an electron flux $F_e \gtrsim 10^{18}$ cm^{-2} s^{-1} and hence, since the average electron velocity $v \approx (c/3) \approx 10^{10}$ cm s^{-1}, to a beam density $n_b \gtrsim 10^8$ cm^{-3}. This is a substantial fraction of the ambient density $n \approx 10^{10}$ cm^{-3}.

The total electric *current* corresponding to such a particle acceleration rate, *if the acceleration is unidirectional*, is $I \approx 3 \times 10^{17}$ A, and the associated current density $j = I/S \approx 0.3$ A cm^{-2}. Such high current densities, especially since they are introduced over a timescale of seconds, produce unacceptably large inductive electric and magnetic fields, unless (1) the acceleration is near-isotropic, as in stochastic acceleration models (Sect. 3.4), (2) the source has very fine structure (Holman 1985), or (3) the beam current is effectively neutralized by a co-spatial return current formed by both the ambient and returning beam electrons (see Knight and Sturrock 1977; Emslie 1980; Brown and Bingham 1984; Larosa and Emslie 1989; van den Oord 1990; Zharkova and Gordovskyy 2006).

Finally, the total number of electrons in the acceleration region $\mathcal{N} = nV \approx 10^{37}$, a number which corresponds to only a few seconds of acceleration at the rates quoted above. Hence the issue of replenishment of the acceleration region (Emslie and Hénoux 1995) has to be accounted for by any acceleration and/or transport mechanism.

2.2 Lightcurves and Energy Spectra of γ-Rays

2.2.1 Gamma-Ray Lightcurves

Due to the limited sensitivity of current gamma-ray detectors, information regarding the time scales for ion acceleration is less stringent than that for energetic electrons (Sect. 2.1).

The prompt gamma-ray line emission is known to peak roughly simultaneously, within ± 1 s, of the hard X-ray continuum emission, based on a few events (e.g., Kane et al. 1986). This near-simultaneity of hard X-ray and gamma-ray emissions indicates that both ion acceleration to a few MeV, and interaction of the accelerated ions with the ambient atmosphere to produce nuclear lines (Vilmer et al. 2011), can occur within a timescale $\sim$1 s. However, in a few flares there *are* time delays between the appearance of hard X-rays and gamma-ray lines (see Sect. 2.2.6 in Vilmer et al. 2011); this is often interpreted in terms of differential transport of electrons and ions between the acceleration and emission sites.

RHESSI observations have also provided important information on the temporal evolution of prompt gamma-ray lines (Lin et al. 2003). Comparison with the temporal evolution of the X-ray flux at 150 keV shows that the temporal evolution of hard X-rays and gamma-rays are roughly similar, indicating a common origin of accelerated electrons and ions. However, for SOL2002-07-23T00:35 (X4.8), there is a small delay of around 10 s between the maxima of the X-ray and gamma-ray fluxes. This has been interpreted in terms of transport of energetic electrons and ions by Dauphin and Vilmer (2007), and is discussed in Vilmer et al. (2011).

At higher energies (above 10 MeV) several flares have been observed for hours after the impulsive phase and so pose a formidable constraint on the ion acceleration mechanism.

2.2.2 Energy Spectra and Abundances of Ions in Flares

As reviewed by Vilmer et al. (2011), interactions of energetic ions in the energy range from $\sim$1 MeV nucleon^{-1} to 100 MeV nucleon^{-1} produce an extensive spectrum of gamma-ray line emission. Narrow gamma-ray lines result from the bombardment of the ambient nuclei by accelerated protons and α-particles, while broad lines occur from the inverse reactions, in which accelerated heavy nuclei collide with ambient hydrogen or helium nuclei. De-excitation gamma-ray line spectra provide information on flare ions of energies above $\sim$2 MeV nucleon^{-1}. The shape of the ion distribution below this energy is unknown, but is comparably important for the evaluation of the total ion energy as is the low-energy cutoff to the evaluation of the total electron energy content (Sect. 2.1; see Emslie et al. 1997).

Quantitative gamma-ray line observations have now been obtained for more than twenty events. The accelerated ion spectra are found to extend as unbroken power laws down to at least 2 MeV nucleon^{-1}, if a reasonable ambient Ne/O abundance ratio is used (for details, see Vilmer et al. 2011). For the 19 flares observed by the *Solar Maximum Mission (SMM)* prior to the launch of *RHESSI*, the average spectral index was around 4.3; however, the events observed with *RHESSI* tend to have much harder slopes (Lin et al. 2003; Share and Murphy 2006).

Even though most of the energy contained in ions resides in protons and α-particles, the abundances of heavier accelerated ions provide crucial constraints for acceleration processes. Information on the abundances of such flare-accelerated ions was deduced for the *SMM*/GRS (Gamma-Ray Spectrometer) flares as well as for a few flares observed by *RHESSI*. A noticeable enhancement has been observed in the numbers of α-particles, as well as of accelerated ^{3}He isotopes, compared to the standard coronal abundances. Furthermore, accelerated heavy ions, such as Ne, Mg and Fe, are also generally found to be over-abundant with respect to normal coronal compositions, as pointed out by Vilmer et al. (2011). This poses another constraint on models for ion acceleration.

2.2.3 Ion Numbers

Since the first detection of solar flare gamma-ray lines in 1972 (Chupp et al. 1973), information on the energy content in ions has been obtained for more than 20 events. The analysis carried out using 19 events observed by *SMM*/GRS and a few events observed with *RHESSI* shows that the energy contained in >1 MeV ions ranges from 10^{29} to 10^{32} erg, comparable to the energy contained in sub-relativistic electrons. However, there are still large uncertainties in the determination of these quantities and a large dispersion of the ratio of electron ($\sim$20 keV) to ion ($\sim$1 MeV nucleon^{-1}) energy contents from one flare to another (for details, see Vilmer et al. 2011).

The issue of relative importance of electron and ion acceleration can also be addressed by comparing the production ratios of ions to quasi-relativistic electrons ($E \gtrsim 300$ keV). Such an analysis has been performed using both *SMM*/GRS (Forrest 1983; Chupp 1984; Chupp and Benz 1994) and *RHESSI* (Shih et al. 2009b) observations. Good correlations have been found between the total energy content in protons above 30 MeV and the total energy content in electrons above 300 keV, suggesting that high-energy electrons and ions are linked through a common acceleration processes. It should be noted, however, that in the statistical study by Shih et al. (2009a), a correlation between the number of protons with energies above 30 MeV, the electron number above 50 keV, and the soft X-ray *Geostationary Operational Environmental Satellite (GOES)* class exists, but only above a certain threshold in the level of ion production.

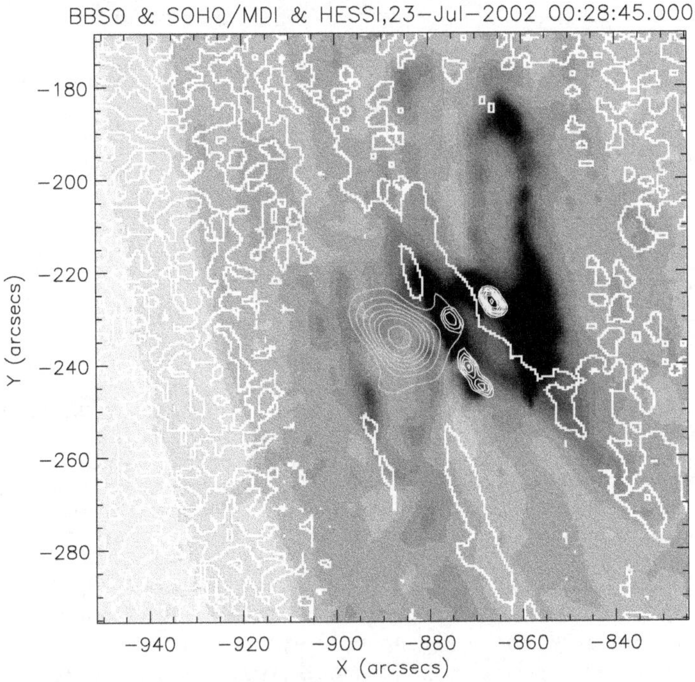

Fig. 3 *RHESSI* hard X-ray images of SOL2002-07-23T00:35 (X4.8) (*white thin contours*) taken at 00:28 UT, overlaid on Michelson Doppler Imager (MDI) neutral-line magnetograms (*white thick contours*) and (negative) Hα taken at 00:28:45 UT. One extended (coronal) and three compact (footpoint) hard X-ray sources are evident. From Zharkova et al. (2005). Reproduced by permission of American Geophysical Union

2.3 Geometry of Hard X-Ray and Gamma-Ray Sources

The imaging capabilities of *RHESSI*, combined with its high spectral resolution, have allowed us to resolve in detail coronal sources (e.g., Krucker et al. 2008a) and footpoints occurring in the same flare (see Fig. 3) and hence to study the acceleration processes that give rise to such sources.

The coronal source often appears before the main flare hard X-ray increase and the appearance of footpoints. In the impulsive phase, the coronal hard X-ray emission is, generally, well correlated in both time and spectrum with the footpoints (Emslie et al. 2003; Battaglia and Benz 2006). These observations suggest strong coupling between the corona and chromosphere during flares, a coupling that is presumably related to transport of accelerated particles from one region to the other.

2.3.1 Differences in Footpoint Spectral Indices

An interesting *RHESSI* observation is the approximate equality of the spectral indices in different footpoints of the same loop. Emslie et al. (2003) reported differences $\Delta\gamma \approx 0.3$–0.4 between the spectral indices of the two dominant footpoints in SOL2002-07-23T00:35 (X4.8). In a few smaller events analyzed by Battaglia and Benz (2006), $\Delta\gamma$ is even smaller and indeed is significant only in one out of five cases. However, other observations of X-class flares reveal a much stronger difference (up to 5) between the spectral indices of footpoints of the same loop (Takakura et al. 1995; Battaglia and Benz 2006; Krucker et al. 2007b).

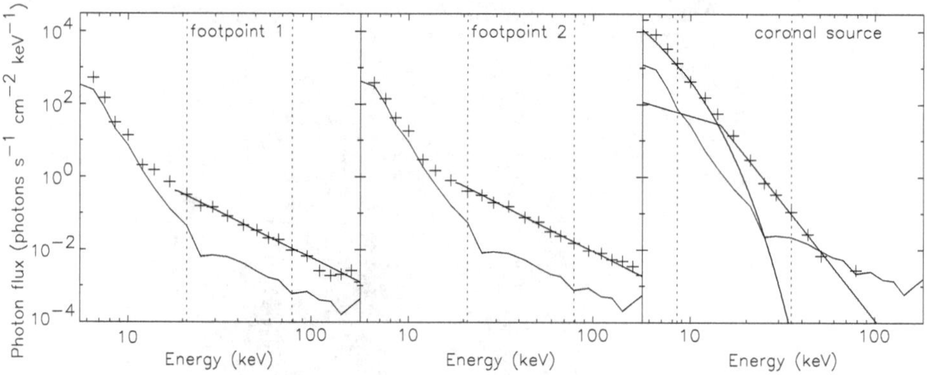

Fig. 4 Comparison of the spectra of the footpoint and coronal sources of SOL2005-07-13T14:49 (M5.0). *Left and middle panels:* Footpoint spectra. A power-law has been fit between the *dotted vertical lines*. *Right panel*: Spectrum of the coronal source. A power-law plus an isothermal spectrum has been fit between the *dotted vertical lines*. The *bold line* is the power-law fit in the *dashed interval* and a Maxwellian fit for the coronal source at low-energies. The *thin curve* is the background (mostly instrumental) determined at a location outside the enhanced flare emission. From Battaglia and Benz (2006)

More surprising than the occasional difference between the spectral indices of footpoints are the differences (or lack thereof) between the spectra of the coronal source and the footpoints. The spectral index difference $\Delta\gamma = 2.0$ predicted (Brown 1971) between thin- and (collision-dominated) thick-target sources is *not* observed in most cases (Holman et al. 2011). Differences $\Delta\gamma$ smaller than 2 could in principle be interpreted by invoking a coronal target that is intermediate between thin- and thick-target conditions; such spectral differences could also be partially accounted for by the transport effects of precipitating electrons, e.g., by the effect of a self-induced electric field (Zharkova and Gordovskyy 2006). However, $\Delta\gamma$ ranges from 0.59 to 3.68 (Grigis and Benz 2006; Krucker et al. 2007b), which includes values beyond the capability of simple models to explain. For example, Fig. 4 (after Battaglia and Benz 2006) displays the photon energy spectra of two footpoints and a coronal source, obtained for the same time interval during the flare SOL2005-07-13T14:49 (M5.0). The footpoint non-thermal power-law indices γ are near-identical (2.7 ± 0.1), while the spectral index of the non-thermal emission for the coronal source is substantially steeper ($\gamma = 5.6 \pm 0.1$).

2.3.2 Hard X-Ray and Gamma-Ray Sources

Gamma-ray images were first obtained by *RHESSI* for the flare SOL2002-07-23T00:35 (X4.8) (Hurford et al. 2003). They revealed the striking result that the hard X-ray sources are separated by a considerable distance ($\gtrsim 20''$) from the 2.223 MeV neutron-capture gamma-ray source. This presumably reflects differences in the acceleration and/or transport processes for ions and electrons, respectively (Vilmer et al. 2011). A similar, but smaller, separation was also detected for the "Halloween" flare SOL2003-10-28T11:10 (X17.2) (Hurford et al. 2006).

2.4 Hard X-Ray and Radio Emission

The information obtained through study of radio emission from solar flares complements that obtained from analysis of hard X-ray emission in several ways (White et al. 2011).

Radiation at centimeter and millimeter wavelengths is gyrosynchrotron emission originating from mildly relativistic electrons with energies above about ∼100 keV. Analysis of this emission hence permits an estimate of the ambient magnetic field strength, reported to be ∼200 G in the corona (White et al. 2011). Coronal sources have been observed to emit in centimeter radio emission even before X-ray footpoints appear (Asai et al. 2006), suggesting that the pre-flare heating may not be unrelated to acceleration in the main phase. Comparison of radio and hard X-ray emissions by Benz et al. (2006) also revealed, in some events, coherent radio emission *after the hard X-ray emission had stopped* (this phenomenon was, however, reported in only 5% of the flares studied). Gyrosynchrotron radio emission in flares in general correlates well with hard X-rays in *time* but not in *space* (Krucker et al. 2008b; White et al. 2011); in addition, the electron spectra deduced from the radio emission are typically harder than the electron spectra deduced from the hard X-ray emission. Flare emission at decimeter wavelengths concurrent with hard X-rays has also been found to originate from different locations: for example, Battaglia and Benz (2009) report the origin of associated decimetric spike sources separated by some 100″ from the coronal X-ray source.

Emission at meter and decimeter wavelengths originates not from individual particles, but from coherent radiation processes associated with various plasma waves. Benz et al. (2005) have studied the association of *RHESSI* hard X-rays with coherent radio emission in the meter and decimeter band in 201 flares with flare classes larger than C5. They found that coherent radio emission occurred before the onset of hard X-ray emission in 9% of the flares studied. Most of the radio emission was type III radiation at the plasma frequency and its first harmonic, emitted by a beam of electrons moving upward in the solar atmosphere into regions of progressively lower density. In a few cases, they also found a pulsating continuum, possibly caused by cyclotron maser emission. Both types of emission indicate the presence of electron acceleration in the pre-flare phase, which is often accompanied by X-ray emission from a thermal source in the corona (Krucker et al. 2008a; Fletcher et al. 2011).

Emission at higher frequencies, such as decimetric narrow-band spikes, pulsations, and stationary Type IV events, correlate more frequently with the hard X-ray flux and thus appear to be more directly related to the acceleration process (Arzner and Benz 2005). While there is a good *association* between coherent radio emission and hard X-rays, a strong *correlation* in the details of the time profile is less frequent, so that coherent emission is not a reliable proxy for the main flare energy release. This can be accounted for by invoking multiple reconnection sites connected by common field lines, along which accelerated particles propagate and serve as a trigger for acceleration in other regions at some distance (Benz et al. 2005).

2.5 Magnetic Field Changes Associated with the Flare

Sharp temporal increases of hard X-ray emission are often closely correlated in time with variations of the magnetic field measured on the photosphere (see, e.g., Kosovichev and Zharkova 2001; Sudol and Harvey 2005; Wang et al. 2005; Zharkova et al. 2005). For example, in the flare SOL2002-07-23T00:35 (X4.8), the magnetic flux change over the flare duration was about 1.2×10^{21} Mx, most of which was concentrated in the area spanned by the magnetic inversion line. Such magnetic field changes are irreversible: the magnetic field reaches a new steady state and does not return to its pre-flare value (Kosovichev and Zharkova 2001; Sudol and Harvey 2005; Zharkova et al. 2005). Cross-correlation analysis with a time lag between the temporal variations of magnetic field covering the neutral line and the hard X-ray light curve reveals a significant positive correlation with a time difference less than ∼100 s. This observation strengthens the theoretical notion that irreversible changes in the magnetic field are responsible for the initiation and development of the flare phenomenon.

3 Models of Particle Acceleration

In this section, we first review some basic concepts, such as acceleration by a direct electric field in a plasma environment (Sect. 3.1) and magnetic reconnection (Sect. 3.2), in which magnetic energy is released into various forms, including the acceleration of charged particles, both electrons and ions. Recognizing the strong observational evidence for the formation of current sheets (localized regions in which a strong magnetic shear is present) in solar flares, our initial discussion relates to traditional two-dimensional (2-D) current sheets, with subsequent elaboration to the three-dimensional (3-D) case (Sect. 3.3). Two likely products of the fundamental reconnection process are magnetohydrodynamic shocks and stochastic MHD turbulence, and particle acceleration by these agents is discussed in Sect. 3.4.

3.1 Basic Physics

As remarked in Sect. 1, the essential physics of particle acceleration is in principle rather straightforward—charged particles gain energy when subjected to an electric field in their rest frame. Such an electric field $\mathbf{E}$ may be a large-scale externally-imposed field, or a $\mathbf{v} \times \mathbf{B}$ field associated with particles crossing magnetic field lines, or a collective field associated with the environment in which the particle finds itself (e.g., a collisional Coulomb field or a field associated with a level of plasma wave energy). The richness of these various means of creating local electric fields is reflected in the richness of particle acceleration models in magnetized plasmas.

One of the most basic concepts is that of the *Dreicer field* (Chen 1974). Consider an electron subject to an externally-imposed large-scale electric field E, plus the frictional force due to collisions with stationary ambient particles. The equation of motion for such a particle is (in one dimension)

$$\frac{dv}{dt} = eE/m - f_c v, \tag{3.1}$$

where m and e are the electron mass and (absolute value of) charge, f_c is the collision frequency, and v is the velocity. Since for Coulomb collisions $f_c \sim v^{-3}$ (e.g., Chen 1974), we may write this as

$$m\frac{dv}{dt} = e\left(E - E_D\left[\frac{v_{\rm th}}{v}\right]^2\right), \tag{3.2}$$

where $v_{\rm th}$ is the electron thermal speed and we have defined the *Dreicer field* by

$$E_D = \frac{f_c(v_{\rm th})\, v_{\rm th}}{e}; \tag{3.3}$$

numerically, $E_D \simeq 10^{-8}\, n(\text{cm}^{-3})/T(\text{K})\ \text{V cm}^{-1}$.

If $E < E_D$, then only high-velocity particles ($v > v_{\rm crit} = v_{\rm th}[E_D/E]^{1/2}$) see a net positive force and gain velocity. This increased velocity reduces the drag force, leading to an even greater net acceleration and eventual runaway acceleration. In contrast, particles with $v < v_{\rm crit}$ suffer a net *retardation* force and are not accelerated; they remain part of the collisionally-redistributed Maxwellian core. For a Maxwellian distribution of velocities, the number density of accelerated particles $\int_{v_{\text{crit}}}^{\infty} f(v)\,dv$ (where $f(v)$ is the phase-space distribution function) is a (rapidly) increasing function of E. If $E > E_D$, all particles with velocity in excess of the thermal speed v_{th} are accelerated, and the efficient collisional re-population of the accelerated electrons thus leads effectively to acceleration of *all* the electrons in the distribution.

3.2 Magnetic Reconnection Models Associated with Flares

It is generally accepted that the energy release in solar flares occurs through a reconstruction of the magnetic field, caused by the change of connectivity of magnetic field lines during a magnetic *reconnection*. The electric field associated with this changing magnetic field and/or with the associated driven currents leads to particle acceleration.

Reconnection is a fundamental process defined by the magnetohydrodynamics of a magnetized plasma (see Priest and Forbes 2000). The magnetic diffusion equation (see, e.g., equation (3.91) of Tandberg-Hanssen and Emslie 1988) is

$$\frac{\partial \mathbf{B}}{\partial t} = \nabla \times (\mathbf{V} \times \mathbf{B}) + \frac{\eta c^2}{4\pi} \nabla^2 \mathbf{B}, \tag{3.4}$$

where c is the speed of light, $\mathbf{B}$ is the magnetic field, $\mathbf{v}$ the fluid velocity and η the resistivity. The order-of-magnitude ratio of the terms on the right side of this equation defines the *magnetic Reynolds number*

$$R_m = \frac{4\pi \ell V}{\eta c^2}; \tag{3.5}$$

it measures the ratio of the advective to diffusive contributions to the change in the magnetic field. For typical solar coronal values of η and with $\ell \approx 10^9$ cm, $R_m \approx 10^{14}$; for such high magnetic Reynolds numbers, the plasma is effectively "frozen-in" to the field and negligible change in the field topology, with its concomitant release of magnetic energy, can occur. A topological change in the magnetic field requires a breakdown in this ideal "frozen-in" flux condition; by (3.5), this can occur either in small-scale regions (high values of $\nabla^2 \mathbf{B}$) and/or in regions in which the resistivity η is anomalously enhanced. As a result, magnetic reconnection fundamentally occurs in narrow boundary layers called *diffusion regions*.

3.2.1 Basic 2-D MHD Theory of Magnetic Reconnection

In the simple neutral sheet geometry originally proposed by Sweet and Parker (Sweet 1969), oppositely-directed magnetic field lines $\mathbf{B}$ in close proximity follow a plasma inflow $\mathbf{V}_i$ oriented in the direction perpendicular to the field lines (note that evidence does exist for such reconnection geometries in flares—Aulanier et al. 2000; Fletcher et al. 2001; Sui and Holman 2003; Des Jardins et al. 2009). The increasing magnetic pressure in the localized region of field reversal is alleviated by reconnection in a small region of high $\nabla^2 \mathbf{B}$ near the plane of field reversal, which allows plasma outflows with velocity V_o toward the sides of the diffusion region.

The original analysis by Sweet (1969) showed that the reconnection rate V_i is given by

$$\frac{V_i}{V_A} = \frac{2}{\sqrt{R_m}}, \tag{3.6}$$

where V_A is the Alfvén velocity. Because of the $R_m^{-1/2}$ dependence on the inflow velocity, and hence on the rate of magnetic energy dissipation, such a neutral sheet scenario

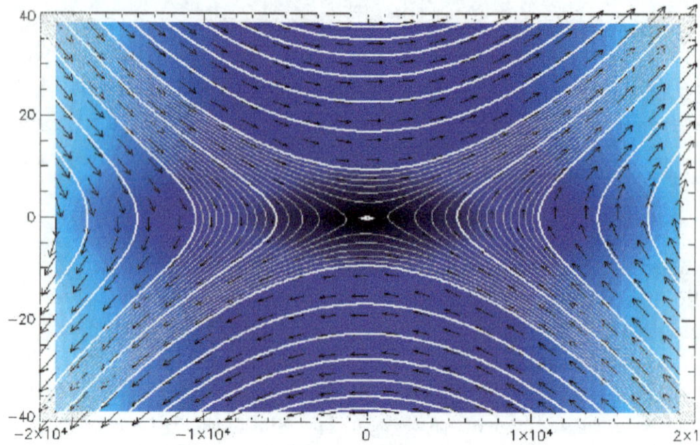

Fig. 5 Topology of the reconnecting magnetic field lines in the vicinity of an X-null point $B_y = \pm\alpha B_x$

describes a "slow" magnetic reconnection regime. In an effort to increase the reconnection rate, Petschek (1964) proposed a model involving the formation of shocks propagating outward from the diffusion region; in this scenario, the reconnection process occurs at a significantly faster rate, with $V_i \approx V_A$. Both these models were originally proposed with a classical value for the resistivity η and both resulted in reconnection rates too slow to explain the rapid energy release observed in flares. However, later realization that the resistivity in the reconnection region may be driven to anomalously high values through the presence of plasma instabilities in the current sheet (see, for example, Dahlburg et al. 1992; Bret 2009, and references therein). These include the lower hybrid drift instability leading to the Kelvin-Helmholtz instability (Landau and Lifshitz 1960; Chen et al. 1997; Lapenta et al. 2003), the kink instability (Pritchett et al. 1996; Zhu and Winglee 1996; Daughton 1999), the sausage instability (Büchner 1999) and magneto-acoustic waves (Schekochihin and Cowley 2007). This realization revived hopes that reconnection mechanisms could indeed be a viable candidate for the energy release in flares (Syrovatskii 1981; Priest 1982; Somov 2000). Questions remaining about the stability of current sheet geometries such as those envisioned by Sweet and Parker have been extensively investigated using a full kinetic approach (see Daughton et al. 2009, and references therein), and are discussed further in Sect. 4.8.

Another simple 2-D diffusion region geometry involves magnetic field lines of the form $B_y^2 = \alpha^2 B_x^2$. Such a form is associated with two types of neutral point: an O-type, elliptic topology ($\alpha^2 < 0$; see Sect. 4.9), and an X-type, hyperbolic topology ($\alpha^2 > 0$; see Sect. 3.3; Fig. 5).

3.2.2 Effect of the Hall Current on Reconnection Rates in 2-D Models

In the conductive plasma within the diffusive region where reconnection occurs, the current sheet tends to diffuse outwards at a slow rate with a timescale of $\tau_{\mathrm{dif}} = a^2\varpi$ where a is the current sheet half-width and ϖ is the magnetic diffusivity (Priest and Forbes 2000). This diffusion process ohmically converts magnetic energy into heat. As discussed in Sect. 3.3, this timescale is too slow to account for the fast reconnection rates observed in the impulsive phase of flares unless some other processes, such as resistive instabilities or Hall current effects, are invoked to enhance the reconnection process. The effects of the Hall current in

MHD models of the reconnection process are now discussed. Its effects in particle-in-cell (PIC) models will be discussed in Sect. 4.8, while the effect of resistive instabilities will be discussed in Sect. 4.9.

The basic Sweet-Parker 2-D reconnection model discussed above can produce more realistic reconnection rates if, instead of a resistive MHD model, one uses a two-fluid MHD model, which includes in the equation of motion a Hall current (Sonnerup 1979). Use of a two-fluid model means that the velocity vector $\mathbf{v}$ in (3.4) now represents the velocity $\mathbf{V}$ of the plasma as a whole, with an additional vector $\mathbf{V}_H$ representing the relative velocity between the electrons and the ions. The *Hall current* results from the fact that, inside the current sheet, the electrons and ions are magnetized to significantly different degrees—the spatial scales for effective magnetization is c/ω_{pe} for electrons and c/ω_{pi} for protons. (Here ω_{pe} and ω_{pi} are the respective plasma frequencies, which are in the ratio of the square root of the proton-to-electron mass ratio $\sqrt{m_p/m_e} \simeq 43$.)

At length scales between c/ω_{pe} and c/ω_{pi}, the motions of the electrons and ions are decoupled, and the current generated from the relative motion of electrons and ions can lead to the formation of whistler waves (Mandt et al. 1994a; Ma and Bhattacharjee 1996; Drake et al. 1997; Biskamp 1997). Inside such a region, the reconnection rate becomes insensitive to electron inertia, allowing the reconnection rates to be strongly increased to levels large enough to account for those measured in solar flares (Birn et al. 2001; Huba and Rudakov 2004). These results indicate a compelling need to consider a full kinetic approach for modeling magnetic reconnection processes in the diffusion region (for details, see Birn et al. 2001, and references therein); this will be discussed in more detail in Sects. 4.8 and 4.9.

3.2.3 Further Improvement of Reconnection Models

It must be recognized that magnetic reconnection is inherently a 3-D process, and that the addition of the third dimension inherently introduces fundamentally new physics. (As an elementary example, 2-D models may predict unrealistic energy gains since the induced electric field $\mathbf{E} = (\mathbf{V}/c) \times \mathbf{B}$ is infinite in extent in the invariant direction). In three dimensions, the magnetic field has a considerably more complicated structure (Fig. 6), involving *separatrices* and *separators* (surfaces separating different domains of magnetic connectivity, and lines of intersection of separatrix surfaces, respectively; see Priest and Démoulin 1995; Démoulin et al. 1996; Demoulin et al. 1996). As shown by Priest and Démoulin (1995), magnetic null points are no longer required for reconnection to take place, and, even when they do exist, magnetic nulls have a more complicated structure, typically involving a "fan" structure oriented around an axial "spine" (Priest et al. 1997).

In summary, reconnection can proceed in a variety of different magnetic field geometries and topologies (Priest and Forbes 2002). Furthermore, depending on the magnetic field configuration, reconnection can be accompanied by a variety of other physical processes such as generation of plasma waves, turbulence and shocks, each occurring on different temporal and spatial scales. These additional processes not only directly accelerate particles (e.g., in regions of enhanced magnetic field, which act as moving "mirrors"), but can also feed back on the magnetic field topology and hence the ongoing reconnection rate.

3.3 Particle Acceleration in a Reconnecting Current Sheet

Two of the most popular magnetic field topologies associated with solar flares involve the formation of a current sheet (Gorbachev and Somov 1989; Sui and Holman 2003), located either at the top of single helmet-like loop (Fig. 7) or at the intersection of two interacting

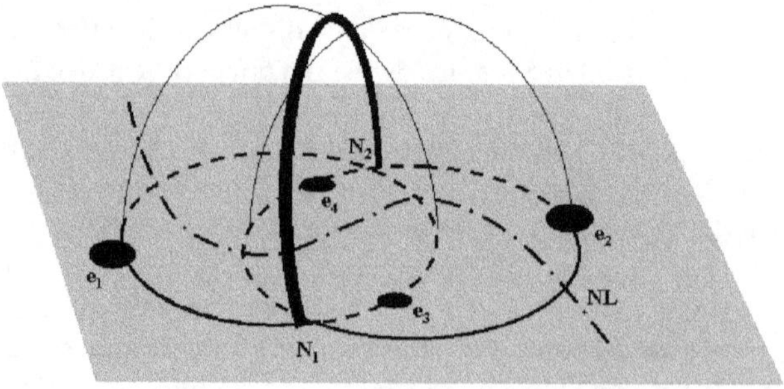

Fig. 6 The magnetic field geometry formed by four point-like sources of different magnetic polarity. The *dot-dashed line* labelled 'NL' is the neutral line. The *separatrix* surfaces are shown as *thin dome-like lines* connecting the photospheric sources. The intersection of these surfaces is the *separator*, shown as a *solid line*. From Titov and Hornig (2002)

loops (Fig. 8). Other more complicated topologies are essentially combinations of these two basic ones, and they are discussed in Sect. 4. In this section we therefore briefly overview the basic physics of particle acceleration in reconnecting current sheets.

Let us, then, consider the magnetic field topology as presented in the reconnecting current sheet (RCS) geometries illustrated in bottom plots of Figs. 7 and 8. The tangential magnetic field component $B_z(x)$ is approximated by the linear function

$$B_z(x) = B_0\left(-\frac{x}{d}\right), \tag{3.7}$$

where d is the RCS thickness (Martens and Young 1990; Litvinenko and Somov 1993). The width a of an RCS in the z-direction approaches 10^2–10^3 m (Somov 1992) and magnitudes of the other two magnetic field components B_x and B_y are assumed to be constant:

$$|B_x| = B_{x0}, \qquad |B_y| = B_{y0}. \tag{3.8}$$

Particles, effectively frozen into the field lines, are advected toward the RCS plane (the x-axis) by the flow field $\mathbf{V}_\mathrm{i}$, where they are accelerated by the $(\mathbf{V}_\mathrm{i}/c) \times \mathbf{B}$ electric field (Bulanov and Syrovatskii 1976; Litvinenko and Somov 1993; Litvinenko 1996b). As the reconnection process continues, ambient particles drift into the reconnecting current sheet; the rate of acceleration of particles is limited by the size of the thin diffusion region.

The induced electric field E_y is given by (Bulanov and Syrovatskii 1976; Martens and Young 1990; Litvinenko and Somov 1993; Zharkova and Gordovskyy 2004):

$$E_y = \frac{V_i B_{z0}}{c} - \frac{\eta c}{4\pi}\frac{\partial B_z}{\partial x}; \tag{3.9}$$

the first and second terms on the right hand side of (3.9) dominate outside and inside the diffusion region, respectively (Somov 2000). Priest (1982) has shown that the plasma inflow velocity V_i is smaller than the local Alfvén speed by about two orders of magnitude, i.e., $V_i \approx 0.01 V_A$. Hence the electric field is

$$E_y \approx 0.01 \frac{B_{z0}^2}{c\sqrt{4\pi n m_p}}, \tag{3.10}$$

where m_p is the proton mass. Let us assume the following physical conditions in the current sheet: temperature $T = 10^6$ K, density $n = 10^{10}$ cm^{-3}, and $B_{z0} \approx 100$ G. Then $V_A \approx 2 \times$

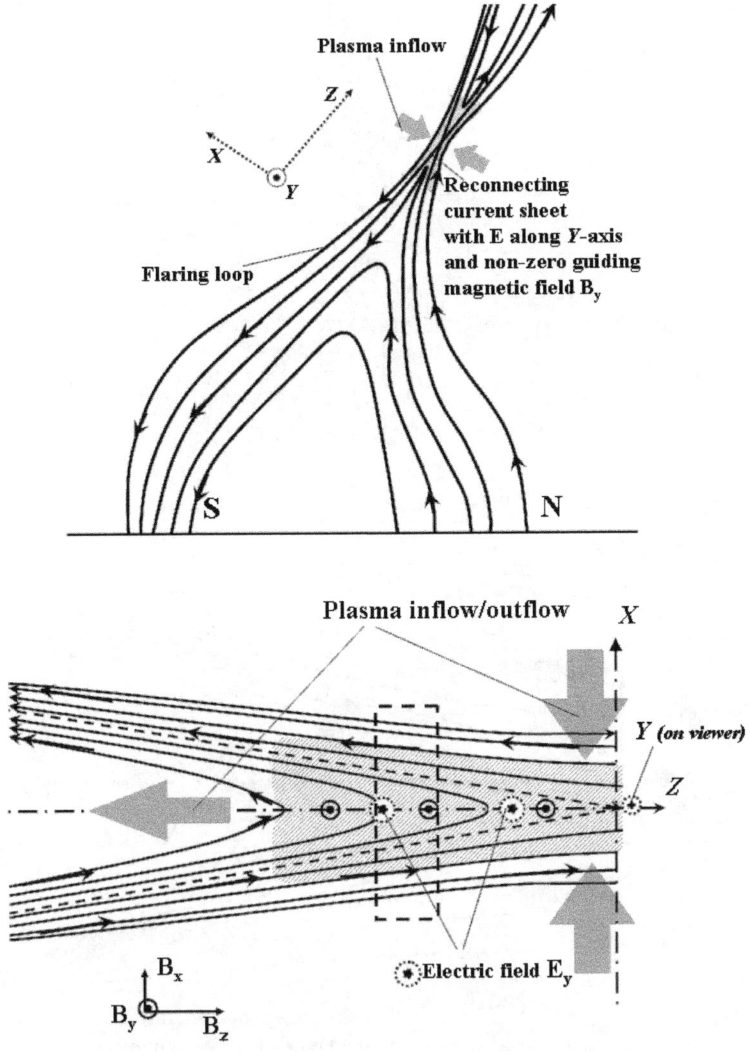

Fig. 7 *Top*: current-sheet location and *bottom* magnetic-field topology in a single helmet-type reconnecting non-neutral current sheet. The *large arrows* depict plasma inflows from the top and the bottom and outflows to the side. The x- and z-components of the magnetic field lie in the plane of the figure; a guiding magnetic-field component B_y and a drift electric field E_y are perpendicular to the plane

$10^6\ \mathrm{m\,s^{-1}}$ and the thermal velocities are about $10^5\ \mathrm{m\,s^{-1}}$ for protons and $10^7\ \mathrm{m\,s^{-1}}$ for electrons. Taking the initial inflow velocity V_i to be the proton thermal speed, we obtain $E_y \approx 2\ \mathrm{V\,cm^{-1}}$, which is roughly four orders of magnitude higher than the corresponding Dreicer field $\sim 10^{-4}\ \mathrm{V\,cm^{-1}}$ (Martens and Young 1990; Litvinenko 1996b, see Sect. 3.1). This electric field is directed along the longitudinal magnetic field component B_y, allowing efficient particle acceleration in the y-direction.

In these simulations the drifted, or reconnection, electric field is assumed to exist throughout the entire diffusion region, which in the MHD approach extends over some 10^2 m to 10^6 m (Bulanov and Syrovatskii 1976; Somov 1992; Oreshina and Somov 2000a).

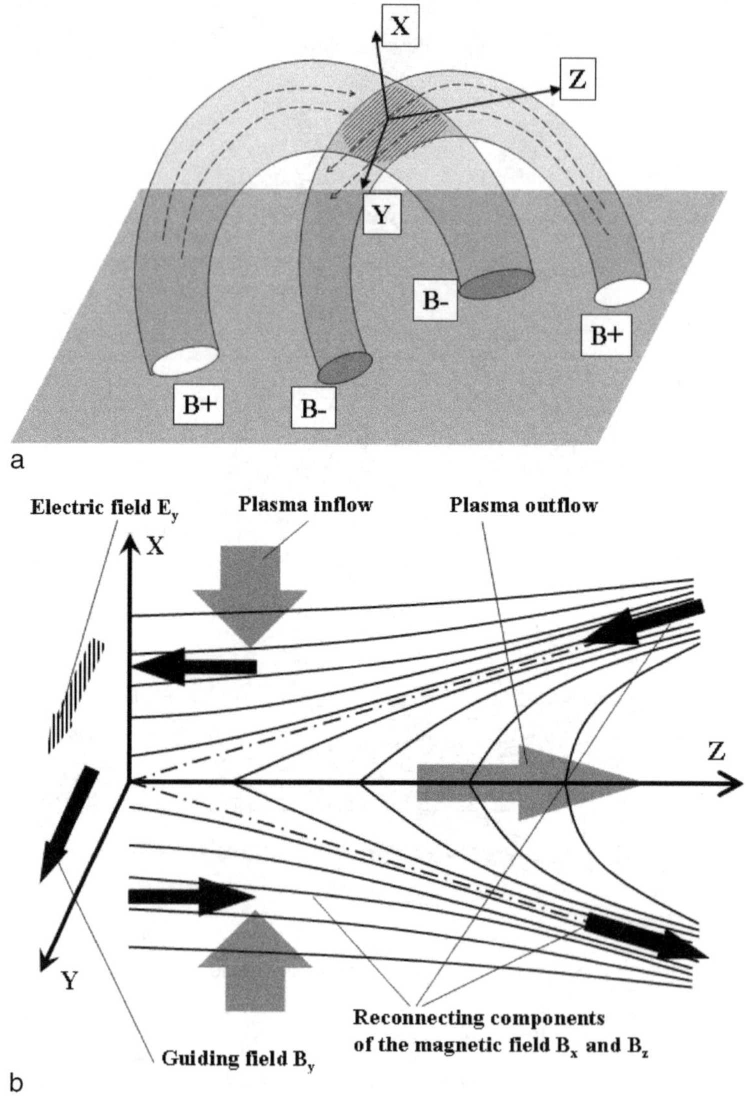

Fig. 8 *Top*: current sheet location, and *bottom:* magnetic-field topology in a reconnecting non-neutral current sheet formed by two interacting loops. The *large arrows* depict the plasma inflows. As for Fig. 7, the x- and z-components of the magnetic field lie in the plane of the figure; a guiding magnetic field component B_y and a drift electric field E_y are perpendicular to the plane

This assumption has been challenged by full kinetic PIC simulations (Furth et al. 1963; Drake and Lee 1977), which show the reconnection field to extend over a relatively small domain corresponding to the size of the electron diffusion region. Given the current limitations of PIC simulations (see Sect. 4.10), there is at present no definitive indication of which of the approaches (MHD or PIC) is more valid. In the remainder of this section we adopt the value provided by the MHD approach, recognizing that appropriate caveats must therefore be made.

The magnitudes of the magnetic field components B_x and B_y are chosen such that electrons remain magnetized in the vicinity of the midplane while protons are unmagnetized during the acceleration phase. In this case, the energy of the unmagnetized protons upon ejection can be estimated (see Litvinenko 1996b) as:

$$\varepsilon = 2m_\alpha c^2 \left(\frac{E_{y0}}{B_{x0}} \right)^2, \tag{3.11}$$

where m_α is the particle mass. However, since the electrons are magnetized, they follow magnetic field lines and gain an energy (Litvinenko 1996b):

$$\varepsilon = d \left| eE_{y0} \frac{B_{y0}}{B_{x0}} \right|, \tag{3.12}$$

where d is the current sheet thickness along the x-direction (see the bottom plot in Fig. 7). For RCSs with thicknesses from 1 m to 100 m, corresponding to proton Larmor radii in magnetic fields from 10^{-3} T to 2×10^{-2} T, electrons gain energies up to ~100 keV, comparable with those of bremsstrahlung-emitting electrons.

Although such a model does meet the challenge of (1) a sufficiently high reconnection rate and (2) sufficient energy gain by the accelerated electrons, the number of accelerated particles is fundamentally limited by the small width $2a$ of the current sheet. These deficiencies are addressed in the further development of this basic model to include three-dimensional considerations (Sect. 4.4), the use of multiple current sheets formed by the filamentation of the reconnection region (Sect. 4.6), and a full kinetic PIC approach to model portions of a 3-D current sheet (Sects. 4.8 and 4.9).

3.4 Particle Acceleration by Shocks and Turbulence

In this section, we review the basic physics involved in acceleration models that do not rely on a large-scale coherent electric field.

3.4.1 Shock Acceleration

Ever since the demonstration by several authors (Blandford and Ostriker 1978; Bell 1978; Cargill 1991; Giacalone and Jokipii 1996) that a very simple model of shock wave acceleration leads to a power-law spectrum of the accelerated particles (in good agreement with observations of galactic cosmic rays), shock acceleration has been frequently invoked in both space and astrophysical plasma contexts. However, this simple model, though very elegant, has a significant number of shortcomings. Inclusion of losses (Coulomb at low energies and synchrotron at high energies) or the influence of accelerated particles on the shock structure (see, e.g., Amato and Blasi 2005; Ellison 2005) can cause significant deviations from a simple power law.

Moreover, shocks are unable to accelerate low energy background particles efficiently; shock acceleration requires the injection of (fairly high energy) "seed" particles. Also, shock acceleration requires some scattering agents (most likely plasma waves or turbulence) to cause repeated passage of the particles through the shock front. In the case of quasi-parallel shocks, in which the magnetic field is nearly parallel to the normal of the shock front, the rate of energy gain is governed by the pitch-angle diffusion rate.

As pointed out by Wu (1983) and Leroy and Mangeney (1984), in the Earth's magnetosphere, many shocks can be quasi-perpendicular (i.e., the vector defining the normal to the shock is almost perpendicular to the shock velocity vector). In such a case the particles drift primarily along the surface of the shock and get accelerated more efficiently. However, acceleration in quasi-perpendicular shocks suffers from the fundamental limitation that injection of seed particles with very large velocities $v > u_{\rm sh}\xi$ (where $u_{\rm sh}$ is the shock speed and $\xi \gg 1$ is the ratio of parallel to perpendicular diffusion coefficients) is required (Amano and Hoshino 2010). Moreover, although there may be some *indirect* morphological evidence for the existence of such shocks in flares (see, e.g., Sui and Holman 2003), there is little *direct* evidence for the occurrence of these kinds of shocks near the top of flaring loop, where the acceleration seems to be taking place during the impulsive phase. Furthermore, such shocks typically do not appear in 3-D MHD simulations of the reconnection process.

In summary, many of the features that make acceleration of cosmic rays by shocks attractive are not directly applicable to the solar flare environment. As a result, this mechanism has, with some justification, received only limited attention in the literature. However, in Sect. 4.3.2 we do discuss the possibility of acceleration at fast magnetosonic shocks resulting from reconnection.

3.4.2 Stochastic Acceleration

A main contender for the particle acceleration mechanism in flares is stochastic acceleration by the magnetic field component of low-frequency magneto-acoustic waves (Miller et al. 1997; Schlickeiser and Miller 1998; Petrosian et al. 2006).

In the originally proposed Fermi acceleration process (Fermi 1949; Davis 1956), particles of velocity v scatter coherently off agents moving toward each other with a velocity $\pm u$. Since the velocity gain in a single collision is $2u$, the rate of velocity increase is $\Delta v/\Delta t \approx 2u/(L/v)$, where L is the separation of the scattering centers. Solving this, for the case in which a scatter occurs on a timescale significantly less than that required for L to decrease, shows that the energy grows exponentially with time: $E \propto e^{4ut/L}$. For particles scattering off *randomly* moving agents, both energy gains and losses are possible. However, the increased likelihood of a head-on (energy-increasing) collision leads to a general (slower than exponential) increase in energy, at a rate proportional to $(u/v)^2\, D_{\mu\mu}$, where $D_{\mu\mu}$ is the angular diffusion coefficient. This process is known as *second-order* Fermi acceleration.

In general, in a magnetized plasma characterized by a gyrofrequency Ω, a particle of velocity $\mathbf{v}$ will resonate with a wave of frequency ω and wavenumber $\mathbf{k}$ when

$$\omega = \mathbf{k}\cdot\mathbf{v} - \ell\Omega, \tag{3.13}$$

where ℓ is an integer. Particles with velocities near the Cherenkov resonance ($v \approx \omega/k_{\|}$) resonate with $\ell = 0$ waves. This process is known as *transit-time damping* (Hollweg 1974, and references therein). It acts to diffuse the particle distribution $f(\mathbf{p})$ in momentum space, described by the Fokker-Planck equation

$$\frac{\partial f(\mathbf{p})}{\partial t} = \left(\frac{1}{2}\sum_{i,j}\frac{\partial}{\partial p_i \partial p_j}D_{ij} - \sum_i \frac{\partial}{\partial p_i}F_i\right) f(\mathbf{p}) + Q(\mathbf{p}) - S(\mathbf{p})f(\mathbf{p}). \tag{3.14}$$

The diffusion coefficient D_{ij} and the secular coefficient F_i ($i = 1, 2$ correspond to energy and angular terms, respectively) contain the essential physics: the action of both the accelerating waves and decelerating effects (e.g., Coulomb collisions); the quantity $Q(\mathbf{p})$ corresponds to the injection of particle energy and the term $S(\mathbf{p})f(\mathbf{p})$ to escape.

For solar flares, the most likely sources for the energization term $Q(\mathbf{p})$ are plasma wave turbulence (Hamilton and Petrosian 1992; Pryadko and Petrosian 1997) and cascading MHD turbulence (Miller et al. 1996). In the first scenario, a presupposed level of plasma wave turbulence in the ambient plasma accelerates ambient particles of energy E stochastically with the rate D_{EE}/E^2. It has been shown (e.g., Hamilton and Petrosian 1992; Miller et al. 1996) that for flare conditions, plasma-wave turbulence can accelerate the background particles to the required high energies within the very short timescale derived from hard X-ray observations. More importantly, at low energies, and especially in strongly magnetized plasmas, the acceleration rate D_{EE}/E^2 of the ambient electrons may significantly exceed the scattering rate $D_{\mu\mu}$ for particular pitch angles and directions of propagation of this turbulence (Pryadko and Petrosian 1997,1998), leading to an efficient acceleration. However, high-frequency waves near the plasma frequency are problematic as drivers for stochastic acceleration in flares, because they would couple into decimeter radio waves and would be present at high levels in every flare, contrary to observations (Benz et al. 2005).

In the second scenario, involving cascading MHD turbulence, moving magnetic compressions associated with fast-mode MHD waves propagating at an angle to the magnetic field constitute the scattering centers. The wavelength of the induced MHD turbulence is assumed to cascade through an inertial range from large to very small, stopping at the point where wave dissipation becomes significant due to transit-time damping. Resonance with one wave will result in an energy change leading to resonance with a neighboring wave, and so on (Miller et al. 1996). Electrons can thus be accelerated all the way from thermal to relativistic energies through a series of overlapping resonances with low-amplitude fast-mode waves in a continuum broad-band spectrum.

In order to be efficiently accelerated, the test particle must have an initial speed that is greater than the speed of the magnetic compression, i.e., of order the Alfvén speed. For the pre-flare solar corona, *electrons* have speeds comparable to, or greater than, the Alfvén speed, and so are efficiently accelerated immediately. On the other hand, in a low-β plasma, the protons necessarily have an initial speed that is less than the Alfvén speed and so are not efficiently accelerated: the scattering centers converge before the test particle has suffered a sufficient number of energy-enhancing collisions. Efficient acceleration of ions therefore requires pre-acceleration to Alfvén speeds. In the model of Miller and Roberts (1995), this pre-acceleration is accomplished through (parallel-propagating) Alfvén waves, and it is essential to realize that *this process takes some time to accomplish*. During this pre-acceleration period, most of the energy is deposited in accelerated electrons; only after the protons are pre-accelerated to the Alfvén speed are they effectively accelerated—they then gain the bulk of the released energy by virtue of their greater mass (Miller and Roberts 1995). As explained by Emslie et al. (2004), this may be the explanation for the different locations of hard X-ray and gamma-ray sources in flares (see Sect. 2.3 and Hurford et al. 2003, 2006); efficient proton acceleration requires relatively long trapping times and so a larger magnetic structure, whereas electrons can be efficiently accelerated even in relatively small structures.

In practice, since there is a little difference *mathematically* between the plasma wave turbulence and shock mechanisms (Jones 1994), acceleration by both turbulence and shocks can be effectively combined. It should be also be noted that stochastic acceleration generally predicts an accelerated spectrum that is not a strict power-law, but rather steepens with increasing energy (e.g., Miller et al. 1996). However, a recent development in transit-time damping theory involves adding both the loss term $S(\mathbf{p})$ in (3.14) that allows electron escape, and a source term $Q(\mathbf{p})$ due to the return current replenishing the electrons in the acceleration region. These additions cause the solution to become stationary within less than one second under coronal conditions (Grigis and Benz 2006); furthermore, in the hard X-ray spectral region of observed non-thermal energies, the photon distribution is close to a power law, in agreement with observation. The predicted spectral index γ depends on the wave energy density and on the interaction time between the electrons and the waves. Additional trapping (e.g., by an electric potential) tends to harden the electron spectrum.

The main limitation of stochastic acceleration models at this juncture is the need to presuppose an *ad hoc* injection of the necessary waves (plasma or MHD). Furthermore, in the absence of pitch-angle scattering, the stochastic acceleration process generally leads to a decrease in particle pitch angles. This, in turn, reduces the efficiency of the acceleration process, since now only waves with very high parallel phase speeds, or those with pitch angles near 90°, can resonate with the particles being accelerated. Therefore, in order for significant stochastic acceleration to occur, the effective particle scattering rate (which is proportional to the collision frequency between particles and waves) must be greater than the frequency of Coulomb collisions between the charged particles. The latter condition imposes rather strict limitations on the growth rates of plasma waves or MHD turbulence, as appropriate.

4 Recent Theoretical Developments

In this section, we highlight some recent promising developments in our understanding of the physics of particle acceleration in the magnetized plasma environment associated with solar flares.

In Sect. 4.1, we turn our attention to some recent advances in stochastic acceleration models, namely the inclusion of (1) small-scale electric fields and (2) feedback of the accelerated particles on the turbulence level in the acceleration region. Some ideas on possible particle acceleration by large-scale Alfvén waves in a large-scale closed magnetic configuration are discussed in Sect. 4.2.

In Sect. 4.3, the evolution of the ubiquitous "helmet"-type magnetic field configuration observed in flaring regions is presented. We point out two likely acceleration regions in such a geometry: the collapse of previously-stretched field lines onto a closed magnetic loop (which leads to acceleration by both first-order Fermi and betatron processes, associated with the shrinking and compression of field line bundles, respectively), and the formation of "termination" shocks associated with high-speed jets, which accelerate particles through shock-drift acceleration.

In Sect. 4.4, we consider particle trajectories in a realistic 3-D current sheet geometry, considering, in particular, the asymmetric acceleration of electrons and ions and the resulting particle energy spectra, which turn out to be power laws. In Sect. 4.5, we discuss particle acceleration in 3-D magnetic field configurations, in which magnetic nulls have a fundamentally different topological structure than they do in a simple 2-D reconnection geometry. Both "fan" and "spine" reconnection modes are considered, and the resulting (rather hard power-law) spectra of the accelerated particles are presented.

In Sect. 4.6, we turn our attention to acceleration in *multiple* acceleration sites, considering cellular automaton (CA) models, MHD turbulent environments, and stochastic acceleration in a distribution of elementary current sheet sites.

In Sect. 4.7, we turn our attention to a critical aspect of all acceleration models—the need to recognize that, since a substantial fraction of the magnetic energy released in flares appears in the form of accelerated particles, these particles simply cannot be considered as an ensemble of test particles interacting with a *prescribed* electrodynamic environment. Rather, the electric and magnetic fields generated by the accelerated particles themselves are an essential element of this environment; the self-consistent inclusion of these fields represent a paramount consideration in acceleration models, without which unphysical and even paradoxical conclusions can result. We review the limitations of existing acceleration models in this context. In Sect. 4.8, we discuss recent attempts to include this electrodynamic feedback through the use of PIC models, for which the self-consistent electric and magnetic fields associated with the accelerated particles *do* form an essential element of the modeling. Finally, in Sect. 4.9, we briefly discuss a mechanism for particle acceleration that involves multiple small-scale closed magnetic "islands" formed by the tearing-mode instability in a reconnecting current sheet.

4.1 Stochastic Acceleration

As discussed in Sect. 2.1, the low value of the low-energy cutoff of the electron energy distribution suggested by *RHESSI* observations (Kontar et al. 2005; Saint-Hilaire and Benz 2005) has significant implications for the number of accelerated flare electrons. Indeed, for sufficiently low cutoff energies, the number of electrons required in a collisional thick-target model (Brown 1971) becomes so great that models in which the accelerated particles form a subset of the ambient distribution become untenable. The only viable mechanism of particle acceleration is then acceleration of the entire plasma within the acceleration region through a stochastic process that results in little or no net electrical current, to avoid the issues presented in Sect. 2.1.

As discussed in Sect. 3.4, high-frequency waves near the plasma frequency are problematic as drivers for stochastic acceleration, because they would couple into decimeter radio waves and would be present at high levels in every flare, contrary to observations (Benz et al. 2005). On the other hand, acceleration by transit-time damping do not suffer from this drawback—the driving waves have a frequency that is far below the plasma frequency.

4.1.1 Improvements to the MHD Stochastic Acceleration Model

The basic transit-time damping stochastic acceleration mechanism has been improved by adding low-frequency fluctuating electric fields parallel to the magnetic field (Grigis and Benz 2004, 2005). Such fields may originate from low-frequency and high-amplitude turbulence, such as kinetic Alfvén waves, as discussed in Sect. 5.1 (see also Miller et al. 1996; Pryadko and Petrosian 1997). They can accelerate and decelerate electrons (and protons), leading to a net diffusion in energy space (Arzner and Vlahos 2004). As for other models for stochastic acceleration, the authors assume an *ad hoc* distribution of turbulence. In this model, the non-thermal electron distributions in coronal sources "grow" out of the thermal population: the coronal source is initially purely thermal, then a soft non-thermal population develops, getting harder at the hard X-ray peak and softening toward the end of the emission.

As discussed in Sect. 2.1, such a "soft-hard-soft" behavior has been observed in a coronal source (Battaglia and Benz 2007). Indeed, Grigis and Benz (2004) report *RHESSI* observations suggesting the existence of a point in the spectrum at which the hard X-ray flux does not change with time during an event. Above this *pivot point* the spectral variation is consistent with the characteristic "soft-hard-soft" behavior of the photon spectrum. Note, however, that Zharkova and Gordovskyy (2006) have pointed out that such a "pivot-point" behavior in the variation of the hard X-ray spectrum with time could also be associated with the self-induced (return-current) electric field (e.g., Knight and Sturrock 1977; Emslie 1981) associated with an electron beam accelerated by the super-Dricer electric field in a current sheet and precipitating into a loop leg.

4.1.2 *Stochastic Acceleration with Particle Feedback*

A more self-consistent approach to stochastic acceleration was recently reported by Bykov and Fleishman (2009), who simulated electron acceleration by MHD waves, but included the effect on the MHD turbulence (spectrum and intensity) caused by the injection of high energy particle beams into the acceleration region. The initial MHD turbulence is prescribed, and is assumed to be produced in the form of transverse motions with a Gaussian spectrum on the scale $2\pi/k_0$. Through mode coupling, this turbulence produces a corresponding level of longitudinal turbulence in a system of a finite size R (where $Rk_0 > 1$). The model includes only large scale energy-conserving motions, with the kinetics of particles on smaller scale being determined by turbulent advection, valid only for electron energies up to $\sim$1 MeV (Bykov and Fleishman 2009); thus, the model can account for the feedback of the accelerated particles on the acceleration mechanism on large scales only. The phase-space diffusion coefficients D_{ij} in (3.14) are expressed in terms of the spectral functions that describe correlations between large scale turbulent motions (for details, see Bykov and Toptygin 1993).

The model assumes a continuous injection of monoenergetic particles (both electrons and protons) into the acceleration region. The initial phase of the acceleration is characterized by a linear growth regime which results in effective particle acceleration by the longitudinal large scale turbulent motions and thus leads to a spectral hardening. However, because the accelerated particles eventually accumulate a considerable fraction of the turbulent energy, the efficiency of the acceleration decreases; this predominantly affects the higher-energy particles, leading to a spectral softening.

4.2 Particle Acceleration by Large-Scale Alfvén Waves

In this model, reconnection is still at the heart of the magnetic energy release process, in that it allows the pre-flare field to reconfigure, releasing the energy needed to power the flare. However, motivated by *in situ* observations and theories for energy transport and electron acceleration in the Earth's magnetosphere during sub-storms (Stasiewicz et al. 2000), it is suggested (Fletcher and Hudson 2008) that the energy is transported through the corona not by particle beams but by Alfvén-wave pulses, launched in the primary reconnection site and carrying a large Poynting flux towards the chromosphere. The presence of such pulses has recently been verified in MHD simulations by Birn et al. (2009). They may play a direct role in accelerating the electrons either in the field-aligned electric fields that they generate in the corona, or through the magnetic energy that they transport to the chromosphere, leading to acceleration of electrons locally in the chromosphere by mechanisms as yet unspecified. A cartoon realization of this model is shown in Fig. 9.

To explain the rapid time variability observed in the flare time profiles, and also the simultaneity of hard X-ray footpoint sources within around 0.1 s, the wave pulses must move quickly—on the order of $0.1c$, which is significantly larger than the values usually quoted for the coronal Alfvén speed. However, such speeds *are* consistent with the properties of the core of an active region before a flare, where the density may be 10^9 cm^{-3} or less, and the magnetic field on the order of a few hundred Gauss (e.g., Brosius and White 2006. Linear wave pulses traveling at such speeds do not damp significantly *en route* through the corona.

Fletcher and Hudson (2008) calculate that for a coronal field of 500 G and an ambient density of 10^9 cm^{-3}, a magnetic field perturbation of 50 G is more than enough to deliver sufficient Poynting flux to the chromosphere to produce the flare energies observed. Of course, such a perturbation, on the order of a few to 10%, does stretch the linear approximation for Alfvén waves. An ideal, incompressible, non-linear Alfvén mode traveling in a

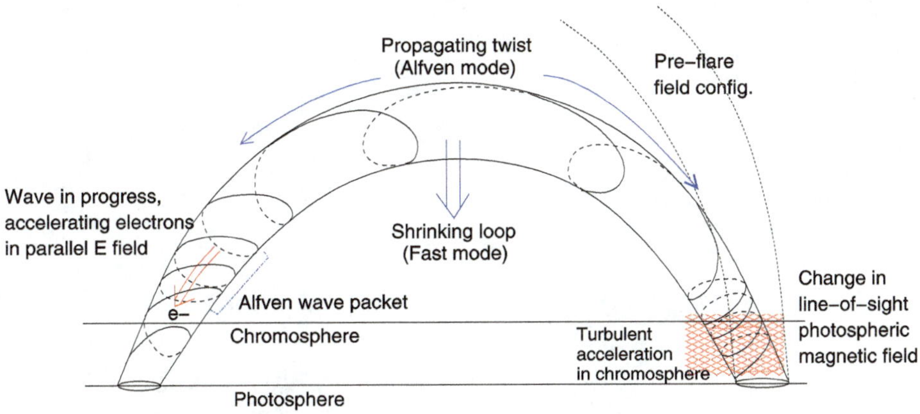

Fig. 9 The reconfiguring coronal field launches a torsional Alfvén wave pulse through the corona and into the chromosphere, as well as a fast-mode wave pulse. The Alfvén wave, which propagates in the inertial regime, can lead to electron acceleration in the corona. Any Alfvén wave energy entering the chromosphere can also lead to electron acceleration there. The wave will be partially reflected from the steep gradients in the chromosphere and then re-enter the corona (from Fletcher and Hudson 2008). Reproduced by permission of the AAS

completely uniform plasma does not steepen into a shock, or dissipate due to Landau damping (Barnes and Hollweg 1974), but the presence of density fluctuations, including those generated by the wave itself, can lead it to decay via a parametric instability into a forward-propagating ion-acoustic wave and a backward-propagating Alfvén wave (e.g. Goldstein 1978).

Analytical work by Jayanti and Hollweg (1993) suggests that, in the low-β coronal regime, the maximum growth rate of this instability for a wave with an amplitude of a few percent is a few times the wave frequency, so that this decay occurs for all but long-wavelength perturbations. However, in the low-β case, the Alfvén speed exceeds the sound speed, so the front of the Alfvén wave train (and presumably also the front of a simple Alfvén wave pulse) propagates ahead of the ion-acoustic wave produced, and is, therefore, not significantly affected by the instability. This has been demonstrated by Turkmani and Torkelsson (2003) in numerical simulations for roughly the parameter regime of interest. The process has not yet been fully explored in an environment where collisionless damping is important, though this may be the case for coronal depths in flares.

Because the Alfvén speed in such an environment is so high (higher than the electron thermal speed, such that an average electron lags behind the wave), the details of the wave propagation require two-fluid or even full kinetic simulations. Such simulations performed in the case of the magnetosphere (Kletzing 1994) indicate that a parallel electric field is generated by the wave pulse, in which ambient electrons can—if at the appropriate "injection" energy—be accelerated to twice the Alfvén speed. Similar considerations for solar coronal conditions indicated that modest non-thermal energies of a few tens of keV could be achieved in this manner. McClements and Fletcher (2009) have also recently confirmed this for solar plasma in the test-particle limit. This process is found to damp the wave energy, though it remains to be seen by how much.

On arriving at the steep gradient in Alfvén speed represented by the chromosphere, it is to be expected that this wave pulse would be at least partially reflected (Emslie and Sturrock 1982). However, any component of the wave energy entering the chromosphere could be

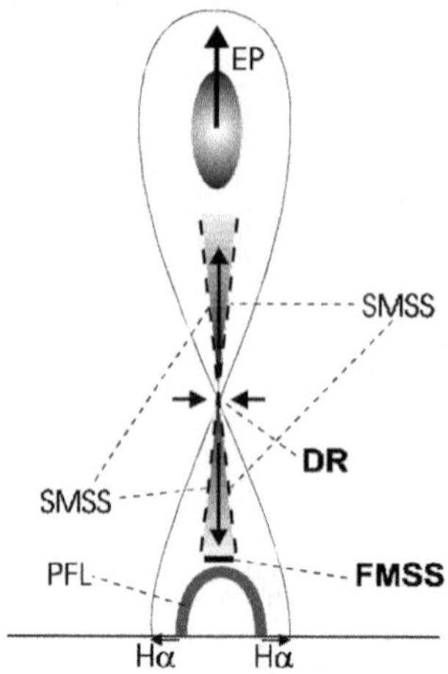

Fig. 10 Sketch of the flare scenario. EP: erupting prominence; DR: diffusion region; SMSS: slow-mode standing shock; FMSS: fast-mode standing (also called termination) shock; PFL: post flare loop. The *grey shaded areas* with the *arrows* show the outflow jets. From Mann et al. (2006)

locally damped—for example by ion-neutral coupling—leading to chromospheric heating. If one considers the wave pulse as imparting stress to the chromospheric magnetic field, then the multiple current sheets that this might generate could provide an environment for particle acceleration, as suggested by Turkmani et al. (2005). The timescales are also such that a perpendicular MHD cascade could be established in the reflected wave field, providing another opportunity for stochastic acceleration of the chromospheric electrons, if the acceleration timescale is less than the collisional loss timescale as discussed in Sect. 3.4.

Significant theoretical work is yet required to demonstrate theoretically whether any of these options are viable, e.g., whether any chromospheric electron acceleration mechanism can reproduce the timing delays observed between high and low energy hard X-rays (e.g., Aschwanden et al. 1996), which are normally interpreted as strong evidence for a coronal electron accelerator. It should also be noted that in the proposed model of energy *transport*, magnetic reconnection still plays a central role as the mechanism of energy *release*, and as the driver of the Alfvén waves. This means that concurrent electron acceleration in the corona can also take place by "traditional" means (e.g., by electric fields in current sheets or by turbulent acceleration by various MHD wave modes).

4.3 Electron Acceleration in Collapsing Current Sheets

Figure 10 depicts a helmet-type magnetic structure often observed in solar flares (see, for example, Sui and Holman 2003), in which a prominence is destabilized due to photospheric footpoint motions and rises upwards as an *eruptive prominence*. As the prominence rises, the underlying field lines are stretched, leading to the formation of a current sheet. We suppose that when the current exceeds a critical value, the resistivity is suddenly enhanced (say by plasma waves excited by various instabilities) and rapid magnetic reconnection begins in the diffusion region (DR), as shown earlier in Fig. 5.

MHD solutions for the large-scale magnetic configuration plotted in Fig. 10 allow us to investigate (see Sect. 3.2) the reconnection process inside the diffusion region, characterized by plasma inflows with velocity V_0 in the x-direction (from the sides in Fig. 10; from the top and bottom in Fig. 5) and outflowing with velocities V_1 in the z-direction (to the top and bottom of Fig. 10; to the left or right sides of Fig. 5). The outflows are generally very fast and create a rather complicated structure in the outer region at the RCS edge. The plasma confined on the newly reconnected field lines will start moving with the velocity V_1, both

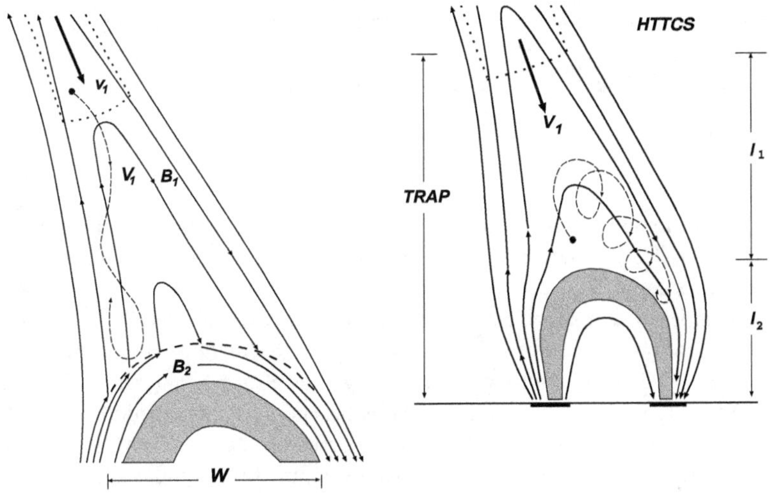

Fig. 11 Schematic models of collapsing magnetic field lines occurring in MHD models of magnetic reconnection with (*left panel*, *dashed line*) and without (*right panel*) a termination shock. From Somov and Kosugi (1997). HTTCS denotes "high temperature turbulent current sheet". Reproduced by permission of the AAS

upwards (not shown in the cartoon), and downwards, where it will encounter a magnetic obstacle in the form of the underlying loop. If the distance to this obstacle is large, a collapsing magnetic trap (see the bottom plot in Fig. 11) is formed. Moreover, the downward outflow is moving into a region of generally higher density and hence lower Alfvén speed. If a point is reached at which the outflow velocity V_1 exceeds the local magneto-acoustic wave velocity, then a fast MHD shock, termed a "termination shock," can appear ahead of this magnetic obstacle (Somov and Kosugi 1997), as shown by the dashed line in the upper panel of Fig. 11.

4.3.1 Acceleration in Collapsing Magnetic Traps

Somov and Kosugi (1997), Karlický and Kosugi (2004) and Bogachev and Somov (2005, 2007) have carried out simulations of the acceleration of electrons in collapsing magnetic traps. Particles with an initial Maxwellian distribution are injected at random points of a fragmented current sheet geometry obtained using a three-dimensional MHD simulation.

As a result of its motion from the reconnection region towards the chromosphere, the length L of the magnetic trap decreases, leading to an increase in energy of the trapped particles via a first-order Fermi process associated with conservation of the longitudinal invariant $\int p_\parallel\, ds$ (Benz 2002). Also, as a result of the *transverse* contraction of the magnetic trap, particles are accelerated by the betatron mechanism. This transverse contraction can be described by the ratio $b(t) = B(t)/B_0$ which changes from 1, when $B(0) = B_0$ to $b_m = B_m/B_0$, where B_m is the magnetic field in the magnetic mirrors. Thus the particle momenta $p_\parallel$ and $p_\perp$ change as follows:

$$p_\parallel = p_{0\parallel} \ell^{-1}, \tag{4.1}$$

where $\ell = (L/L_0)$, and

$$p_\perp = p_{0\perp} b^{1/2}. \tag{4.2}$$

Both the Fermi and betatron accelerations are accompanied by a change in particle pitch angle—a decrease (Fermi) or increase (betatron). When these two mechanisms act simultaneously, then the pitch angles of accelerated particle vary according to

$$\tan\alpha = \frac{p_\perp}{p_\parallel} = \ell b^{1/2} \tan\alpha_0, \tag{4.3}$$

which can result in an increase or decrease of pitch angle, according to the relative importance of the two effects. The particle kinetic energy increases in the magnetic trap until it falls into the loss cone.

Simulations show that particles with an initially Maxwellian distribution gain energies via both acceleration mechanisms; they retain a quasi-thermal distribution but attain much higher energies. The change of total kinetic energy has a form of the burst of particles with energies up to a few MeV. However, the magnetic energy is absorbed very rapidly by the electrons, so that a test-particle approach very rapidly ceases to be valid. A fuller treatment of this scenario therefore requires full kinetic simulations similar to those discussed in Sect. 4.8.

4.3.2 Acceleration by a Termination Shock

Due to the strong curvature of the magnetic field lines in the vicinity of the diffusion region in a helmet-type reconnection (see Fig. 10), the slowly inflowing plasma shoots away from the reconnection site as a hot jet. These oppositely-directed jets are embedded between a pair of slow magnetosonic shocks. If the speed of this jet is super-Alfvénic, a fast magnetosonic shock, also called a *termination shock*, is established, as shown previously in Fig. 11. The appearance of such shocks was predicted in the numerical simulations of Forbes (1986), Shibata et al. (1995) and Somov and Kosugi (1997).

Radio measurements (Aurass et al. 2002; White et al. 2011) of SOL2003-10-28T11:10 (X17.2) reveal the termination shock as a standing radio source with a (large) half-power source area of $A_s = 2.9 \times 10^{20}$ cm^2. A strong enhancement of the electromagnetic emission in the hard X- and gamma-ray range up to 10 MeV is simultaneously observed with the appearance of the radio signatures of the shock (Aurass and Mann 2004). These observations indicate that the termination shock could well be the source of the highly energetic electrons needed for the generation of the hard X- and gamma-ray radiation.

A fast magnetosonic shock formed during the MHD simulations of a reconnection process is accompanied by a compression of both the magnetic field and the density. Thus, it represents a moving magnetic mirror at which charged particles can be reflected and accelerated through a process termed *shock-drift acceleration*. This process can accelerate highly energetic electrons up to 10 MeV (Mann et al. 2006).

Generally, shock-drift acceleration represents reflections at the termination shocks. Analysis of such an acceleration is most straightforwardly performed in the *de Hoffmann-Teller frame*, the frame in which the electric field vanishes, so that the reflection takes place under conservation of both kinetic energy and magnetic moment. From such an analysis (Mann et al. 2006, 2009), one obtains the relationships between the electron velocities $\beta = v/c$ parallel and perpendicular to the upstream magnetic field before (index i) and after (index r) the reflection:

$$\beta_{r,\parallel} = \frac{2\beta_s - \beta_{i,\parallel}(1+\beta_s^2)}{1 - 2\beta_{i,\parallel}\beta_s + \beta_s^2} \tag{4.4}$$

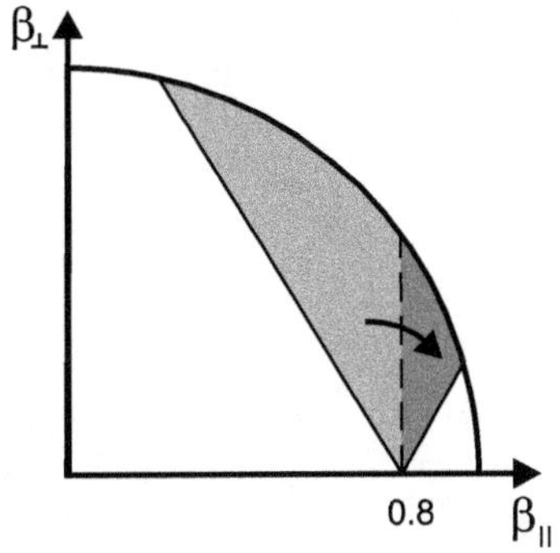

Fig. 12 Illustration of shock-drift acceleration in the $\beta_\perp$–$\beta_\parallel$ plane for a shock with $\beta_s = 0.8$ and $\alpha_{lc} = 45°$. All particles initially located in the light *grey area* are transformed into the *dark one*

and

$$\beta_{r,\perp} = \frac{(1-\beta_s^2)}{1-2\beta_{i,\parallel}\beta_s+\beta_s^2}\cdot\beta_{i,\perp}, \tag{4.5}$$

where $\beta_s = v_s \sec\theta/c$. Here v_s is the shock speed, c the velocity of light, and θ the angle between the shock normal and the upstream magnetic field. In addition, the reflection conditions:

$$\beta_{i,\parallel} \le \beta_s \tag{4.6}$$

and

$$\beta_{i,\perp} \ge \frac{\tan\alpha_{lc}}{\sqrt{1-\beta_s^2}}\cdot(\beta_s-\beta_{i,\parallel}) \tag{4.7}$$

must be satisfied in order for acceleration to occur. The loss-cone angle is defined by $\alpha_{lc} = \sin^{-1}[(B_{\mathrm{up}}/B_{\mathrm{down}})^{1/2}]$, where B_{up} and B_{down} denote the magnitude of the magnetic field in the up- and downstream region, respectively.

Shock-drift acceleration represents a transformation in the $\beta_\perp - \beta_\parallel$ plane as illustrated in Fig. 12. As seen in Fig. 12, the accelerated distribution function is simply a *shifted loss-cone distribution*. The flux of the accelerated electrons parallel to the upstream magnetic field is defined by

$$F_{acc,\parallel} = 2\pi N_0 c^4 \int_0^1 d\beta_\parallel \beta_\parallel \int_0^1 y d\beta_\perp \beta_\perp \cdot f_{acc}(\beta_\parallel, \beta_\perp), \tag{4.8}$$

and the differential flux is given by

$$j_{acc,\parallel} = \frac{dF_{acc,\parallel}}{dE}. \tag{4.9}$$

Figure 13 shows the differential fluxes obtained from the shock-drift acceleration model for various angles θ, assuming an initial distribution in the form of a Maxwellian with a temperature $T = 10$ MK. For comparison, the dotted line in Fig. 13 represents a pure Maxwellian distribution with the same temperature. One can see that shock-drift acceleration indeed provides a significantly enhanced component of energetic electrons, particularly for large values of the angle θ. These analytically-obtained results are qualitatively confirmed by numerical simulations performed for plasma conditions usually found at the Earth's bow shock (Krauss-Varban and Wu 1989; Krauss-Varban et al. 1989; Krauss-Varban and Burgess 1991).

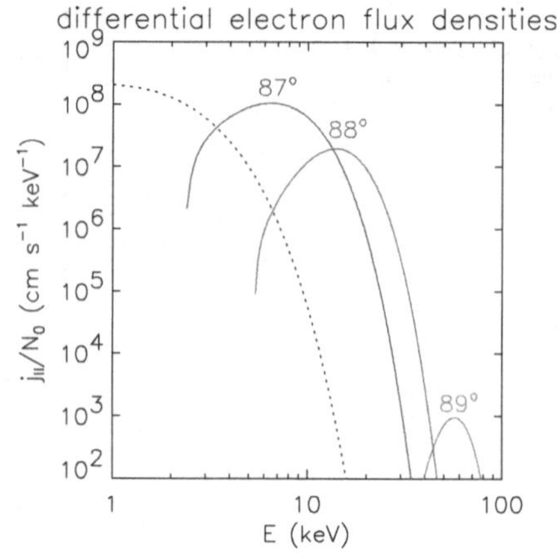

Fig. 13 Differential flux of electrons accelerated by shock-drift acceleration for various shock angles. The *dotted line* represents the flux of a pure Maxwellian distribution

In order to compare these results with observations, some basic parameters must be provided. Here, SOL2003-10-28T11:10 (X17.2) is considered as an example (Mann et al. 2009). In this event, the termination shock appears at 300 MHz for the harmonic emission. Matching harmonic emission to the observed 300 MHz frequency requires an electron number density of 2.8×10^8 cm^{-3}. According to a two-fold density model of the corona (Newkirk 1961), such a region is located ∼160 Mm above the photosphere, at which a magnetic field of ∼5 G is expected (Dulk and McLean 1978). These parameters lead to an Alfvén speed $V_A \approx 560$ km s^{-1}. For the temperature, a value of 40 MK is adopted, resulting in a thermal electron speed of 2.5×10^4 km s^{-1}.

Assuming a density jump of a factor of two across the termination shock, the Rankine-Hugoniot relations give a magnetic field $B_{\mathrm{down}}/B_{\mathrm{up}} = 2$ and an Alfvén-Mach number $M_A = 3.5$, corresponding to a shock speed of ∼2000 km s^{-1}. Due to the curvature of the magnetic field lines convected through the termination shock, there is a distribution of θ across the shock. Recognizing from Fig. 13 that the flux is very sensitive to θ, we model this curvature by a circle; the resulting flux of accelerated electrons can be integrated over θ, giving an accelerated electron production rate $F_e = 2 \times 10^{36}$ s^{-1} and a related power $P_e = 1 \times 10^{29}$ erg s^{-1} above 20 keV. This agrees well with the results inferred from *RHESSI* observations (Warmuth et al. 2007, 2009).

It must be stressed that a non-negligible part of the energy of the outflow jet is transferred into accelerated electrons. Thus, such a test-particle approach has its limitations: the termination shock acceleration process must be treated in a fully kinetic manner, in which the accelerated particles themselves are included in the conservation laws across the shock.

4.4 Particle Acceleration in a Single 3-D RCS with Complicated Magnetic Topology

4.4.1 Basic Equations

Following Sect. 3.3, we consider a highly-stressed current sheet oriented in the x-direction, with a lateral field

$$B_z(x) = B_0 \tanh\left(-\frac{x}{d}\right) \tag{4.10}$$

and a transverse magnetic field component B_x that is dependent on z according to Zharkova and Gordovskyy (2005a, 2005b):

$$B_x(z) = B_0\left(\frac{z}{a}\right)^{\alpha}, \tag{4.11}$$

where $\alpha > 0$ (we consider an illustrative value $\alpha = 1$). A weak, constant longitudinal component of the magnetic field B_y is also assumed to be present:

$$B_y = \beta B_0. \tag{4.12}$$

Particle trajectories are calculated from the relativistic equations of motion

$$\frac{d\mathbf{r}}{dt} = \frac{\mathbf{p}}{m_0\gamma} \tag{4.13}$$

$$\frac{d\mathbf{p}}{dt} = q\left(\mathbf{E} + \frac{1}{c}\frac{\mathbf{p}}{m_0\gamma} \times \mathbf{B}\right), \tag{4.14}$$

where t is time, $\mathbf{r}$ and $\mathbf{p}$ are the particle position and momentum vectors, $\mathbf{E}$ and $\mathbf{B}$ are the electric and magnetic field vectors, q and m_0 are the charge and rest mass of the particle, c is the speed of light, and $\gamma = \sqrt{(p/m_0c)^2 + 1}$ is the Lorentz factor. The integration time steps are chosen to be much smaller than the corresponding gyro-period, i.e., $dt \leq 0.1(m/q)B_z^{-1}$. For protons, this typically requires a time step $\lesssim 10^{-7}$ s, while for electrons the required time step is substantially shorter, $\lesssim 4 \times 10^{-11}$ s.

It should be noted that in the above description, the electric and magnetic field vectors $\mathbf{E}$ and $\mathbf{B}$ are *prescribed* quantities; the self-consistent electric and magnetic fields produced by the accelerated particles themselves are neglected. Such a test particle approach is valid for up to only $\sim 10^5$ test particles (Birdsall and Langdon 1985), and we shall return to this point in Sect. 4.7. It should also again be noted (see Sect. 3.3) that in this simulation the reconnection electric field is assumed to extend over the entire region covering the whole diffusion region. Thus, the caveats noted in this earlier section also apply here.

4.4.2 Particle Motion Inside a 3-D RCS

In the test-particle description used here, the electron trajectory and the direction of ejection are determined uniquely by its initial coordinate, velocity and the magnetic field geometry. Particles inflowing into a 3-D RCS are accelerated mostly near the lateral field reversal at the x-axis. In the presence of a non-zero B_y, electrons and protons are ejected in opposite directions from the midplane ($x = 0$); whether the particle is ejected to $x > 0$ or $x < 0$ depends on the sign of the particle charge, and the signs and magnitudes of B_x and B_y. If B_y is strong enough, then all protons (regardless of the side of the RCS they entered from) are ejected to the $x > 0$ semispace, while all electrons are ejected to the opposite semispace, as illustrated in Fig. 14 (Zharkova and Agapitov 2009). For particles of the same charge, there are two fundamental types of trajectories inside an RCS: particles that enter from and then are ejected into the *same* semispace are hereafter referred to as "bounced" particles, whereas particles that are ejected into the *opposite* semispace are referred to as "transit" particles.

Electron/proton asymmetry in the particle trajectories is dependent on the magnitude of the guiding field (Zharkova and Gordovskyy 2004). The asymmetry is *full* (equal to unity) if B_y is strong enough ($B_y/B_z \gtrsim 1.5 \times 10^{-2}$) that electrons are ejected into one semiplane and protons to the opposite one. If the guiding field is weak ($B_y/B_z \lesssim 10^{-6}$), electrons and

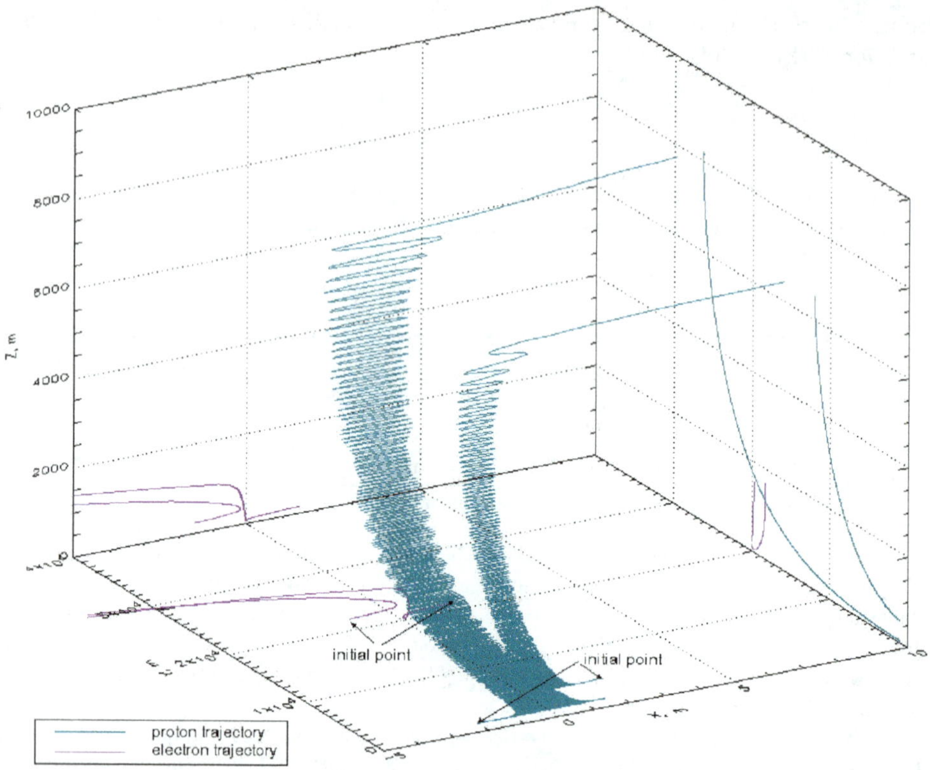

Fig. 14 Proton (*blue lines*) and electron trajectories (*purple lines*) in a 3-D RCS with a lateral magnetic field $B_x = 10$ G and an drift electric field $E_y = 100$ V m^{-1}. From Zharkova and Agapitov (2009)

protons can be ejected into the *same* semiplane (Zharkova and Gordovskyy 2004, 2005b). For a guiding field of intermediate magnitude, electrons and protons are ejected in a partially neutralized manner, with protons dominating in one semiplane and electrons in the other (Zharkova and Gordovskyy 2005b). The result of this separation of particle species is a polarization electric field appearing between the protons at the edge and electrons at the midplane of the RCS. Such a field was anticipated by Litvinenko and Somov (1993) and Oreshina and Somov (2000b), and has been investigated in some detail by Zharkova and Agapitov (2009).

The model produces power-law energy distributions with spectral indices $\delta = 1.5$ for protons and $\delta = 2$ for electrons. Such a result is confirmed by both analytic (Litvinenko 1996a) and numerical approaches (Zharkova and Gordovskyy 2005b; Wood and Neukirch 2005). For comparison, the particle energy spectra calculated for the model RCS of Oreshina and Somov (2000b) and presented in Fig. 15 show features beyond such simple power-law forms: there is a rapid rise in flux to a maximum at $E = E_m$; above this, the spectrum is approximately power-law. $E_m \simeq 10$–12 keV for protons ($\simeq$2–5 keV for electrons); these values increase slightly with the value of the parameter $\beta = B_y/B_z$. The energy E_m can be considered as the lower energy cutoff to the electron spectrum (see Section 2.1), and the values obtained here agree quite well with the values deduced from *RHESSI* observations (Holman et al. 2011).

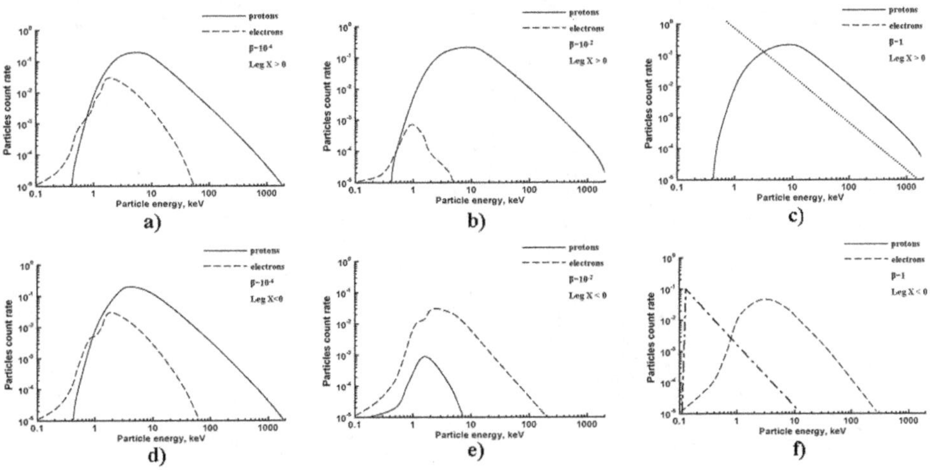

Fig. 15 Energy spectra of electrons (*dashed lines*) and protons (*solid lines*) ejected from the RCS. Spectra (**a**)–(**c**) are calculated for β ($= B_y/B_0$) equal to 10^{-4}, 10^{-2} and 1, respectively, and correspond to the semispace $x > 0$. Spectra (**d**)–(**f**) are calculated for the same values of β, but for the semispace $x < 0$. The estimated proton spectrum is presented as a dotted line in plot (**c**) and the estimated electron spectrum is shown as a *dash-dotted line* in plot (**f**). From Zharkova and Gordovskyy (2005b)

The spectral index δ for the power-law part of the spectrum ($E > E_m$) is found to vary slightly with the magnitude of the guiding field B_y. For a weak guiding field ($\beta = 10^{-4}$—Figs. 15, [a] and [d]), the particle trajectories are nearly symmetrical (Zharkova and Gordovskyy 2005b); the particles are ejected equally into both semispaces as approximately neutral beams. The particle energy spectra in both semispaces ($x < 0$ and $x > 0$) are very similar, with a sharp increase from zero to E_m, followed by a power-law with spectral indices $\delta \simeq 1.8$ for protons and $\delta \simeq 2.2$ for electrons. The power-law shapes extend up to about 100 keV (electrons) or 1 MeV (protons).

For a moderate guide field ($\beta = 10^{-2}$—Figs. 15 [b] and [e]), the protons and electrons are also ejected into both semispaces; however, the symmetry of their trajectories is partially destroyed. Hence, most protons are ejected into the semispace $x > 0$ with $\delta \simeq 1.7$, while a few low-energy protons are ejected into the semispace $x < 0$ with very soft, thermal-like, spectra ($\delta \simeq 4.8$). The opposite picture holds for electrons: the bulk are ejected into the semispace $x < 0$ with an energy spectrum $\delta \simeq 2.0$, while a smaller number of the low-energy electrons are ejected into the semispace $x > 0$ with a thermal-like spectrum having a maximum $E_m \approx 1$ keV.

Finally, if the guiding field is very strong ($\beta \simeq 1$—Figs.15, [c] and [f]), the particle trajectories are completely asymmetric. Particles are ejected completely separately from the RCS midplane into opposite semispaces: all protons are ejected into the semispace $x > 0$ while all electrons are ejected into the semispace $x < 0$. Their energy distributions show a sharp increase of particle number up to around a few keV followed by a power-law energy spectrum of index $\delta \simeq 1.5$ for protons and $\delta \simeq 1.8$ for electrons. (These spectral indices vary with the magnitude of a guiding field (Fig. 16) reflecting different reconnection scenarios and with different phases of magnetic reconnection, as discussed in Sect. 4.5.)

As shown in Fig. 17, during the reconnection process, the shape of the reconnecting current sheet changes from a non-compressed one with a symmetric X-type null point to a strongly-compressed shape with a well-defined elongation along the z-direction and a compression in the x-direction. The extent of this deformation depends on the value of the

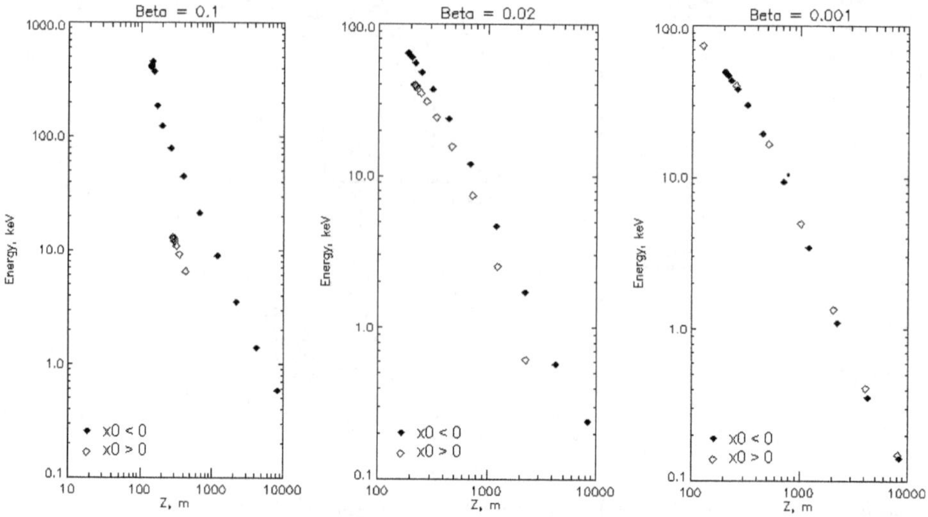

Fig. 16 The energy spectra resulting from the protons entering the current sheet from opposite sides, for different ratios $\beta = B_y/B_0$ and different magnitudes of the parameter α. From Zharkova and Agapitov (2009)

parameter α (Priest and Forbes 2000; Somov 2000) and possibly corresponds to different reconnection rates (Arber and Haynes 2006). At the reconnection onset ($\alpha = 0$) the diffusion region is located very near the null point. As the reconnection progresses (α increases from 0 to 1), the diffusion region becomes much wider in both x and z-directions. As α increases further, the reconnection rate begins to decrease (Zharkova and Agapitov 2009).

The spectral index of the accelerated particle energy spectra is associated with the parameter α and depends on the magnitude of the guiding field (Fig. 16). For an RCS with a density that increases exponentially with distance z and for a variation of density $n(z) \propto (z/a)^{\lambda}$, the following results are obtained (Zharkova and Agapitov 2009): for electrons in a weak (strong) guiding field B_y,

$$\delta_{e,\text{weak}} = 1 + \frac{2+\lambda}{2\alpha}; \qquad \delta_{e,\text{strong}} = \frac{1}{2} + \frac{1+\lambda}{2\alpha}, \tag{4.15}$$

while for protons in a guiding field of *any* strength,

$$\delta_p = 1 + \frac{1+\lambda}{2\alpha}. \tag{4.16}$$

In principle, therefore, by simultaneously measuring the spectral indices of electrons and protons in the same flare, one can deduce the magnetic field geometry under which these particles were accelerated (Zharkova and Gordovskyy 2005b; Zharkova and Agapitov 2009).

We again stress that all of these results have been arrived at using a test-particle approach, in which the feedback of the electromagnetic fields associated with the accelerated particles is not taken into account. Such a limitation is addressed in the PIC approach discussed in Sect. 4.8.

4.5 Particle Acceleration in 3-D MHD Models with Fan and Spine Reconnection

A further development in modeling particle acceleration in a reconnecting current sheet involves more advanced MHD models of fan and spine reconnection (Priest and Titov 1996).

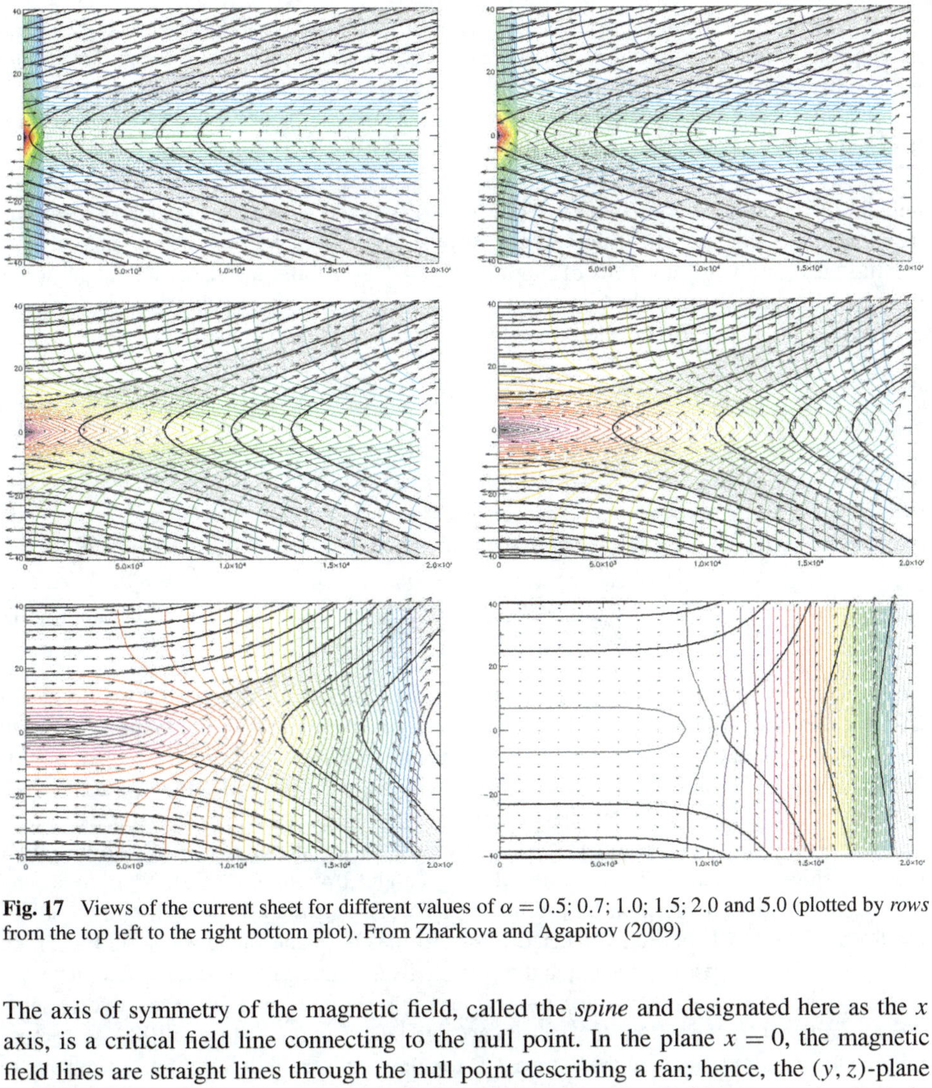

Fig. 17 Views of the current sheet for different values of $\alpha = 0.5; 0.7; 1.0; 1.5; 2.0$ and 5.0 (plotted by *rows* from the top left to the right bottom plot). From Zharkova and Agapitov (2009)

The axis of symmetry of the magnetic field, called the *spine* and designated here as the x axis, is a critical field line connecting to the null point. In the plane $x = 0$, the magnetic field lines are straight lines through the null point describing a fan; hence, the (y, z)-plane is termed the *fan plane*. The spine and the fan are the 3-D analogs of the separatrix planes in the classic 2-D X-point geometry. The field configuration near the potential magnetic nulls is given by (4.10)–(4.12).

Two regimes of reconnection at a 3-D magnetic null are possible: *spine reconnection* and *fan reconnection* (Lau and Finn 1990; Priest and Titov 1996). While the magnetic field configuration is the same in the two regimes, they are characterized by different plasma flow patterns and different electric fields. Spine reconnection has a current concentration along the critical spine field line, while fan reconnection has a current sheet in the fan plane.

For a potential null, the spine reconnection regime is characterized by plasma flows lying in planes containing the spine, while fan reconnection has non-planar flows that carry the magnetic field lines in a "swirling"-like motion symmetric about the fan plane. It should be noted that the electric field and flow pattern are non-axisymmetric in all cases. The model of Priest and Titov (1996) provides a useful analytical model for initial studies; it allows some of the key features of topology and geometry to be identified, and it is the closest 3-D analog of the widely-studied constant out-of-plane electric field used in 2-D studies. It describes only the outer ideal reconnection region and does not include acceleration due to the parallel electric field in the inner dissipation or resistive region; indeed, it breaks down very close to the spine/fan, as the electric fields formally diverge. However, since few particles will enter the small-volume dissipation regions, the results are likely to be appropriate for the bulk of the energy spectrum. This model has been used as a basis for particle acceleration at 3-D nulls in the regimes of both spine reconnection (Dalla and Browning 2005, 2006) and fan reconnection (Dalla and Browning 2008). Particle trajectories are obtained numerically by solving the relativistic equations of motion (see (4.14)). Again, a test-particle approach is used (for limitations, see Sect. 4.7).

The strength of the driving electric field is quantified by the parameter

$$\tilde{\mu} = \frac{v_\perp^2}{v_E^2(L)}, \tag{4.17}$$

where $v_\perp$ is the perpendicular speed associated with the gyro-motion, which may be equated to the thermal speed of the plasma. This can also be expressed as the dimensionless magnetic moment when speeds are normalized to the drift speed (Vekstein and Browning 1997):

$$\tilde{\mu} = \frac{m v_\perp^2 / 2B_0}{m v_E^2(L)/2B_0}, \tag{4.18}$$

where B_0 is the amplitude of the magnetic field. If $\tilde{\mu} \ll 1$, the electric drift speed at global length scales is strong compared with the thermal gyro-motion and we are in the *strong drift regime* corresponding to fast reconnection; in practice, this regime is reached when $\tilde{\mu} \approx 1$. In this case, particles may undergo significant acceleration (Dalla and Browning 2005) even in the outer "ideal" reconnection region, similar to those discussed in Sect. 4.4.

Typical individual trajectories in the fan reconnection are shown in Fig. 18, while similar plots for spine reconnection can be found in Dalla and Browning (2005). Note the strong dependence of the energy gain on the injection location, similar to those found for a single RCS in Sect. 4.4: particles injected closer to the fan plane (at lower latitudes β) are accelerated the most as they approach the strong-electric-field region near the fan plane.

Energy spectra for the spine and fan reconnection (Figs. 19 and 20, respectively) show that a significant fraction of protons can be accelerated to energies of up to 10 MeV, comparable to the values obtained in the simple 3-D topology presented in Sect. 4.4. As appears to be natural during any reconnection process, a generally power-law spectrum is obtained for the accelerated particles, but with a significant "bump" at higher energies, particularly for the spine reconnection, for which the bump arises from particles approaching close to the null itself.

This "bump-on-tail" distribution is similar to those reported for a single 3-D current sheet in Fig. 16 in Sect. 4.4 (see also Zharkova and Gordovskyy 2005c; Zharkova and Agapitov 2009) and is formed by the two populations of protons: "transit" and "bounced" ones, respectively. Note that the slopes of the spectra are similar for the spine and fan reconnection in the intermediate energy range, although fan reconnection is significantly less effective at generating high-energy particles.

Detailed information on both the spatial distributions of the accelerated of particles, and where these particles originate, are shown in Figs. 21 and 22, for spine and fan reconnection, respectively. These show the angular location of the particles at the initial (top plot) and the final (bottom plot) times. Each location is identified by its latitude β, i.e., the elevation

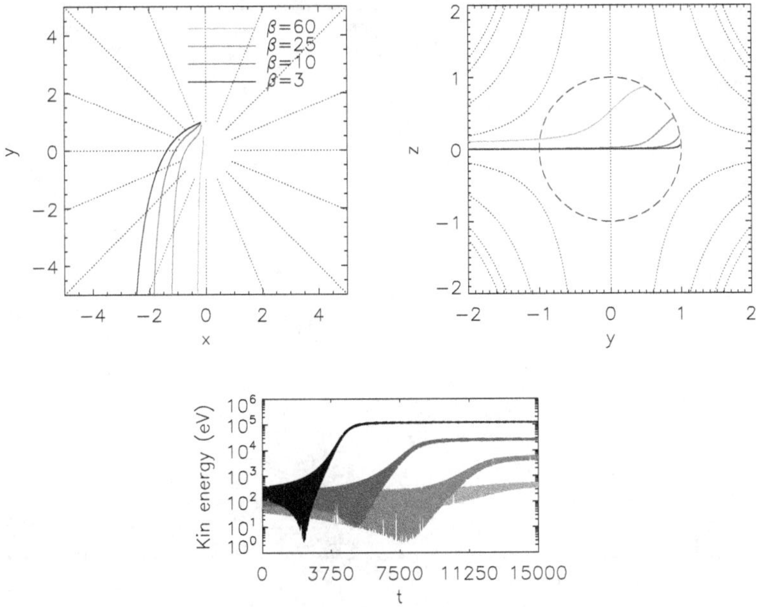

Fig. 18 Trajectories and energy gains in fan reconnection, for particles injected on the boundary at different latitudes β. From Dalla and Browning (2008)

Fig. 19 Energy spectra of particles in spine reconnection at successive times, showing evolution towards steady state. From Dalla and Browning (2006). Reproduced by permission of the AAS

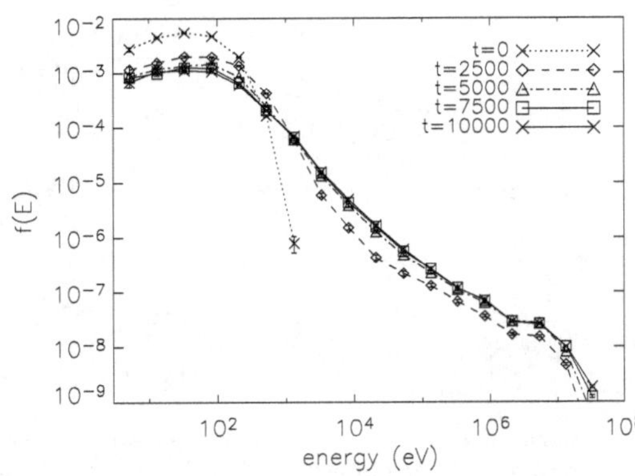

angle from the x–y plane, and longitude ϕ. Inflow regions are the top-right and bottom-left quadrants in the (ϕ, β) representation, but for clarity, only particles originating in the latter are shown (the other quadrant can be added by symmetry). Each point represents one particle and is color-coded according to the final particle energy.

In the case of spine reconnection, particles injected in the plane in which the electric field is weakest (longitude $\phi = 0$) actually have the greatest energy gain, although their acceleration times are longest (Dalla and Browning 2005). The high-energy particles which escape mostly emerge in jets along the spine. However, most of the highest-energy particles actually remain trapped in the magnetic field (Dalla and Browning 2006). The occurrence of these high-energy trapped particles may be an explanation for coronal hard X-ray sources (Krucker et al. 2008a).

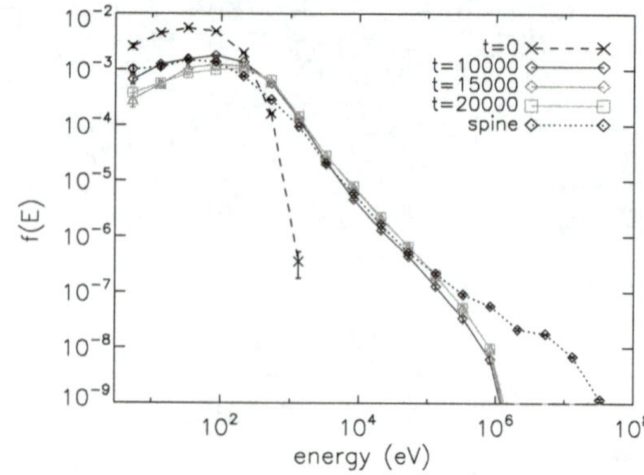

Fig. 20 Energy spectra of particles in fan reconnection at successive times, showing evolution towards steady state. From Dalla and Browning (2008)

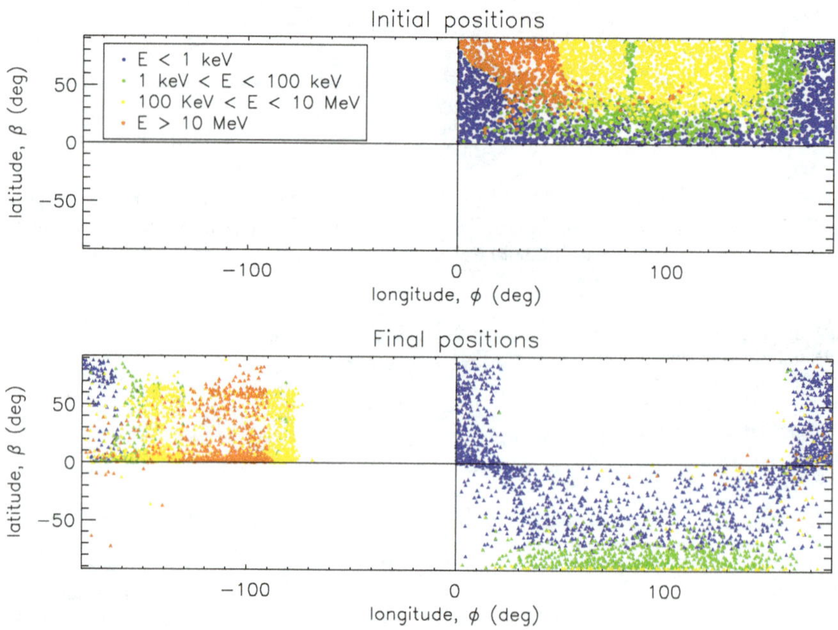

Fig. 21 Positions of particles at initial and final times for *spine* reconnection, color-coded according to their final energies (from Dalla and Browning 2006)

In fan reconnection, particles starting near the fan plane are those that gain the largest energy, as would be expected since the fan plane is the location where the electric field is largest. As one moves towards increasing latitudes, particles will reach the fan plane with greater difficulty and so gain less energy. Figure 22 shows that, at the final time, particles are confined to a range of longitudes $[-130°, -50°]$. This is a result of the tendency of the electric field drift in fan reconnection to push particles towards the $x = 0$ plane after they have passed the region near the null. Figure 22 also shows that the initial azimuthal locations near $\phi = 90°$ are the most favorable for particle acceleration since, near this plane, the transverse component of the electric field drift, pushing particles towards the fan plane, is largest. The fan reconnection regime is less efficient in seeding test particles to the region of strong electric field, compared with spine reconnection, and acceleration times are longer.

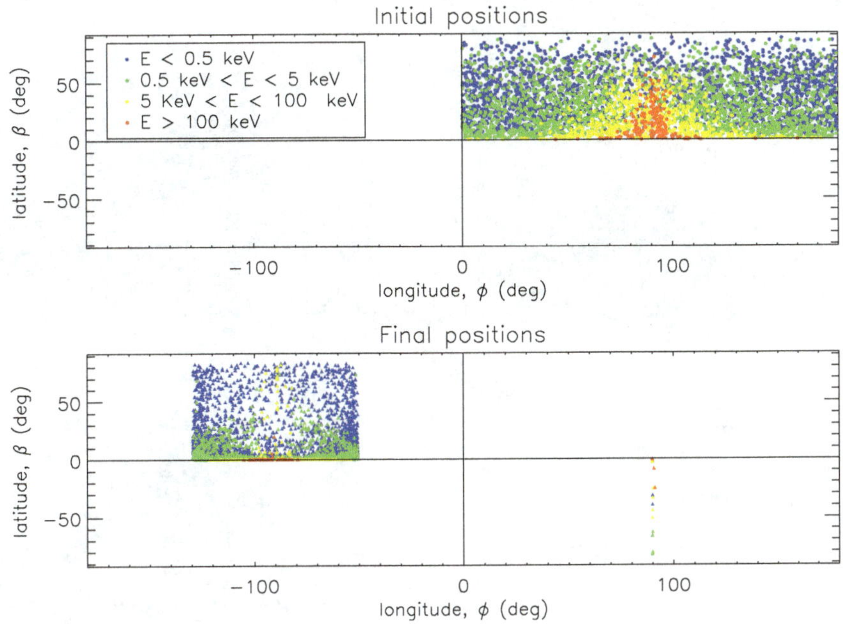

Fig. 22 Positions of particles at initial and final times for *fan* reconnection, color-coded according to their final energies. From Dalla and Browning (2008)

These results, obtained for a single-current-sheet reconnection model with a more complicated magnetic and electric field topology simulated by Priest and Titov (1996), advance further the conclusions derived from the basic 3-D reconnection models discussed in Sect. 4.4. This approach confirms that acceleration by a super-Dreicer electric field generated during magnetic reconnection is a viable mechanism, which naturally explains the energy input from magnetic field dissipation and its output in the form of power-law energy spectra of accelerated particles, with spectral indices related to the magnetic field geometry occurring during the reconnection. Apparently, this approach offers a good diagnostic tool for the magnetic field topology from the observed signatures of accelerated particles. We yet again note that the limitations of the test-particle approach (see Sect. 4.7) can restrict this diagnostic capability, and that a full kinetic approach to particle acceleration in such geometries is required.

4.6 Particle Acceleration in Multiple Current Sheets Formed by MHD Turbulence

Although particle acceleration in a single current sheet can, in principle, account for some flares or even flare ribbons, it is highly likely that flares are due to a number of unresolved magnetic tubes interacting with each other, so that more complex coronal geometries with multiple acceleration sites must be considered. One obvious benefit of such a scenario is that a particle can be accelerated at more than one location, perhaps enhancing the energization process.

4.6.1 Acceleration in Cellular Automata Models

The idea of multiple dissipation sites in flares is longstanding, but the first serious effort at modeling this was by Lu and Hamilton (1991), who took the ideas associated with self-

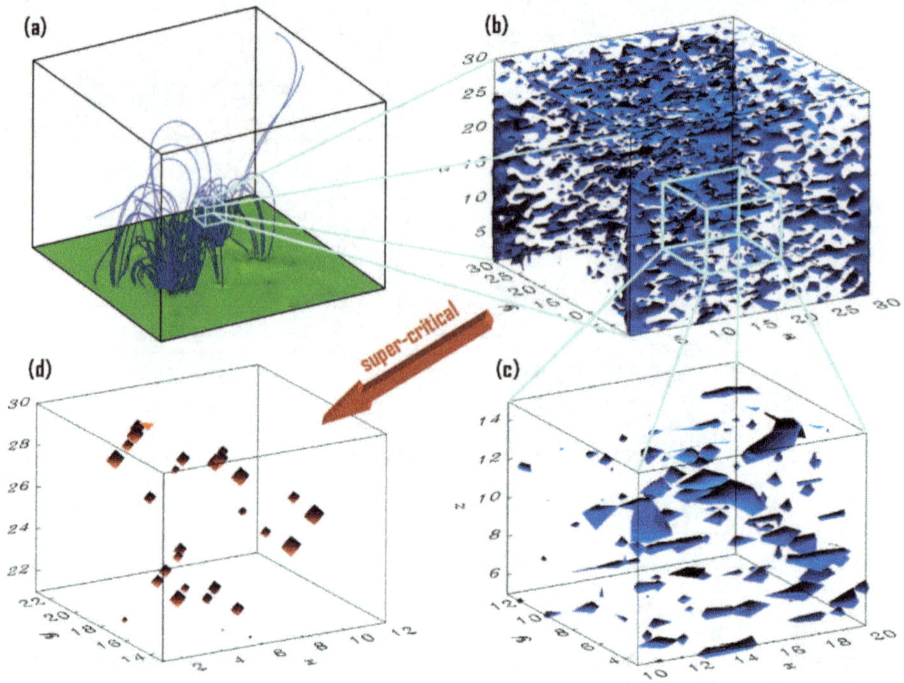

Fig. 23 The model of multiple current sheet formed by MHD turbulence in the CA model. From Vlahos et al. (2004). Reproduced by permission of the AAS

organized criticality (SOC) and applied them to the solar corona. They showed that if the right conditions were met, the triggering of dissipation at a single point could lead to a "spreading" of dissipation across a large coronal volume and that the distribution of event size as a function of energy followed a power law ($\mathcal{N}(\mathcal{E}) \propto \mathcal{E}^{-1.8}$) similar to that observed in flares (e.g., Crosby et al. 1993; see also Table 2 in Hannach et al. 2011, this issue). However, a fundamental question remains: can existing MHD models verify that the main rationale behind the SOC theory is valid?

Self-organized criticality models identify dissipation occurring at many spatially-separated regions as illustrated in Fig. 23, taken from Vlahos et al. (2004). The upper left panel shows a coronal field geometry, reconstructed from photospheric magnetograms, in a volume with characteristic dimensions of 10^9 cm. The current density in a localized region is shown in the upper right panel, and a snapshot of the locations where currents exceed the threshold for dissipation in the cellular automaton model (referred to as Unstable Current Sheets) is shown in the bottom right panel. Energy release in this model occurs at a number of locations of unstable current sheets (bottom left panel).

To assess particle acceleration in such a coronal geometry, Vlahos et al. (2004) have developed a model in which the particles move from one unstable current sheet to another unstable current sheet, gaining (or losing) energy at each and undergoing ballistic motion in a background magnetic field in between. This is illustrated in Fig. 24, where a particle first interacts with a site i, escapes along the magnetic field, encounters a site $i + 1$, escapes again, and so on. The energy gain thus occurs in many small increments, rather than at the single location of the current-sheet approach. By using the self-organized criticality hypothesis, one can identify the structure of the spatial distribution of the unstable current

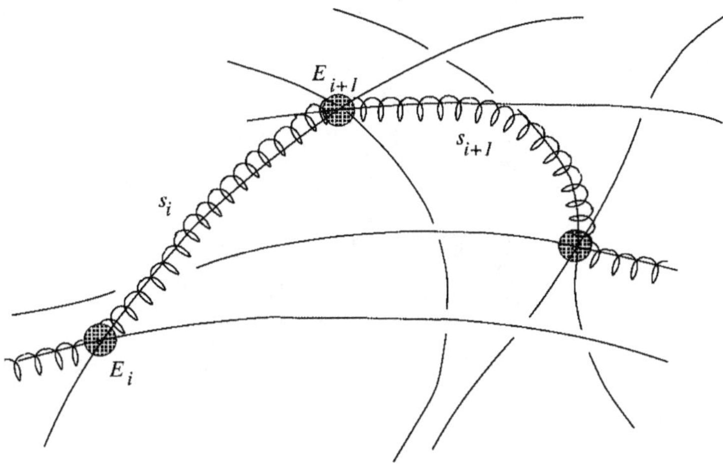

Fig. 24 Particle trajectories in the CA model. From Vlahos et al. (2004). Reproduced by permission of the AAS

sheets inside a complex active region which exhibits a specific fractal structure (McIntosh et al. 2002). Knowledge of the spatial structure of the unstable current sheets can help us to reconstruct the probability distribution of its spatial scales, and thus, the probability of a particle colliding with a given element of it (Isliker and Vlahos 2003). We do not yet understand the statistical properties of the "kernel"—the energy gain or loss for a particle colliding with an unstable current sheet—but this is relatively easy to study using three-dimensional MHD codes.

We can therefore start from a global reconstruction of the distribution of unstable current sheets and recover the kinetic properties of the particles, under the assumptions of the self-organized criticality theory. The acceleration is thus determined by three probability densities: one, defining the distance between a pair of accelerators; the second, the electric field strength at each accelerator; and a third, either the time spent in, or the length of, the unstable current sheet. The first two are assumed to be power laws, and the third is taken to be a Gaussian. Note that this model only addresses acceleration by direct electric fields, since that can be simply parameterized at the dissipation sites.

In the example shown in Fig. 23, the particle moves in a box of size 10^{10} cm and the distance between accelerators ranges from 10^4 to 10^{10} cm. The magnetic fields may be such that the particles become partially trapped and the trajectory becomes rather complex. The upper limit of the free travel distance is set to a value larger than the coronal volume in order to allow particles to leave the coronal region without undergoing any interaction with electric fields, besides the initial one. Note that for the maximum time $\Delta t = 1$ second for which the system is monitored, these escaping particles will move at most a distance $c\Delta t = 3 \times 10^{10}$ cm, which is of the order of the length of the coronal volume considered.

The electric field at each unstable current sheet varies between sub-Dreicer and 10^8 times Dreicer, and the acceleration time has a mean value of 2×10^{-3} s (Vlahos et al. 2004). A large number of test particles are tracked through this coronal geometry using the relativistic equation of motion described by (4.14), with acceleration taking place very rapidly ($\ll$1 s).

Figure 25 shows the distribution function of electrons at four different times. The initial distribution is a Maxwellian with a temperature of 100 eV (dash-dot line). One can see that

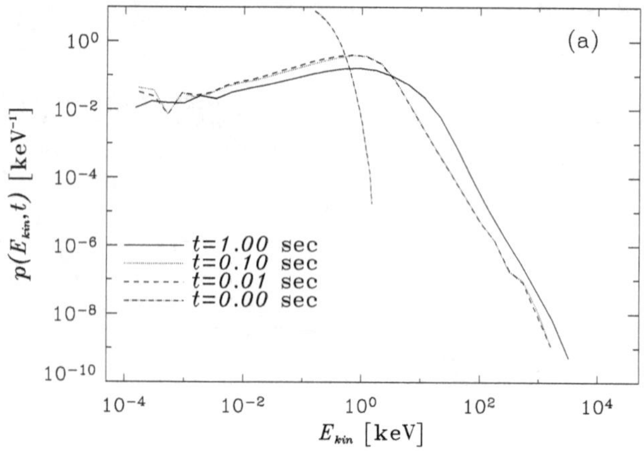

Fig. 25 Accelerated electron spectra in the CA model (Vlahos et al. 2004). Reproduced by permission of the AAS

(a) prompt acceleration to energies well in excess of 100 keV takes place, (b) just under half of the particles end up in the high energy tail and (c) the tail is a power law with slope $\delta \simeq 4$. When the test particles are protons, much longer acceleration times must be considered to allow for similar energies to be reached within the prescribed acceleration time.

Another approach based on the SOC models was proposed by Anastasiadis et al. (2004) to predict the kinetic energy distributions of electrons. The acceleration of electrons is still due to the presence of the randomly placed localized electric fields produced by the energy release process and simulated by the CA model. It is assumed that electrons interact successively with reconnecting current sheets with the electric field linked to the energy release time series. At each encounter, they randomly gain or lose energy. Electrons can be accelerated from a Maxwellian initial distribution up to a few MeV, and can be represented either by a power-law or an exponential distribution.

As the number of interactions increases, the electron energy distribution begins to diverge from a well defined power law, and an exponential tail develops (e.g., Trottet et al. 1998; Vilmer et al. 2003; Anastasiadis et al. 2004). Thick-target radiation from these energetic electrons was also computed, leading to hard X-ray spectra at low energies (10–100 keV) which are much flatter than usually observed. At higher energies (100–1000 keV), the X-ray spectra were consistent with the observations at these energies.

4.6.2 Acceleration in Three-Dimensional MHD Turbulence

Numerical experiments for test-particle acceleration in the turbulent magnetic and electric fields obtained from pseudo-spectral direct numerical solutions of the compressible three-dimensional MHD equations, with the Hall term included, have been carried out by Dmitruk et al. (2003) and Dmitruk and Matthaeus (2006). In both experiments electrons and protons were accelerated within very short times before particles left the simulation box (with length size of the order of a few turbulent correlation lengths). The momentum distribution functions very quickly develop long high-energy tails.

For electrons, there is no substantial difference found between the results with and without the Hall term in the MHD solution, while for protons significant differences are seen in the low-momentum part of the distributions. When a background uniform magnetic field (the guiding field) is added, the particle acceleration is anisotropic. Electrons develop large parallel momentum, while protons develop large perpendicular momentum similar to those reported for a single current sheet in Sect. 4.4.

4.6.3 Acceleration in Stochastic Current Sheets

Turkmani et al., Turkmani et al. (2005, 2006) have investigated the issue of particle acceleration in a 3-D coronal MHD model developed by Galsgaard and Nordlund (1996). The initial setting for the model is a straight stratified coronal loop with a magnetic field which extends between the two photospheric footpoints. The response of the loop to random photospheric motions is then followed. A turbulent cascade develops, leading to a formation of multiple current sheets in the corona. The resulting electric fields comprise both the inductive electric field and the resistive one, the latter appearing from the hybrid resistivity used in the model of Galsgaard and Roussev (2002). However, it is found that the impact of the (mainly perpendicular) inductive field on acceleration is weak. The distribution of current sheets changes in time with the turbulence development, giving rise to the possibility of very bursty particle acceleration events.

In order to study particle acceleration in this environment, the magnetic field topology deduced from the MHD simulation is assumed to be quasi-stationary, similar to that in Sect. 3.3. Then, test particles are followed through their motion in the electromagnetic fields associated with the multiple current sheets, using the relativistic equations of motion (see (4.14)). The model also allows particles to leave the loop once they hit the boundaries, thus limiting their energy gain. The simulation results show different types of particle dynamics, with particles both gaining and losing energy; however, not surprisingly, most leave the loop with net energy gain. About one quarter of the particles do not interact with the current sheet at all and thus gain no energy.

The results show that the resistive electric field, being a parallel field, is the principal means of accelerating particles. The acceleration is very fast and efficient, accelerating particles to relativistic energies in a very short time. The resulting distribution function of accelerated particles has a clear double-power-law structure, with a thermal component at low energies (Turkmani et al. 2006).

The MHD model currently in use produces particles with higher energies at somewhat shorter times compared to the Cellular Automata model discussed above (Vlahos et al. 2004). This is likely because of different choices in the basic parameters, especially the magnetic field strength and density. The same MHD model accelerates both electrons and protons without any need to adjust the electric fields, contrary to the CA-based model. The double power law does not arise in the CA model because of the possibility of particle loss in the MHD model. Both approaches show that fragmented current sheets in the corona are an reasonable scenario for particle acceleration.

Décamp and Malara (2006) performed a test-particle numerical experiment to simulate particle acceleration in the low-frequency turbulent environment generated by footpoint motions in a coronal loop. Only the effect of the resistive electric field E is retained, which is mainly parallel to the axial magnetic field. In the spatial frequency spectrum, the contribution of small scales is dominant. The spatial structure of E is obtained by a synthetic turbulence method (p-model), which allows intermittency with a power-law distribution to be reproduced. Spatial intermittency plays a key role in acceleration, enhancing both an extension of a power-law range, and an increase in the maximum energy.

By solving the relativistic equations of motion, the time evolution of the particle distribution can be determined. Electrons are accelerated to energies of the order of 50 keV in less than 0.3 s, and the final energy distribution also exhibits a power law. A correlation is found between heating events and particle acceleration that is qualitatively similar to that in the stochastic models discussed in Sect. 3.4.

4.6.4 *Limitations of MHD Turbulence Models*

With all their attractiveness in accounting for observed electron distributions in solar flares, models invoking MHD turbulence in an ensemble of current sheets nevertheless have a few essential problems.

First, MHD models involving multiple current sheets inevitably inherit the same size- and time-step problems as the PIC simulations discussed in Sect. 4.8, allowing us to simulate MHD turbulence and electron acceleration in only a very small region (about 1 m^3). It is then assumed, without justification, that the results from this small volume can be extended to the coronal scale.

Second, the basic model of a current sheet (discussed in Sect. 3.3) is still used for the energy gains by accelerated particles in these reconnecting current sheets, while more realistic 3-D models of the reconnecting current sheets (as discussed in Sects 4.4 and 4.5) show the energy gains to be significantly different.

Third, even if, as suggested above, multiple current sheets are formed in the whole flare volume, the diffusion regions in each have to be connected somewhere to reconnecting magnetic field lines. In a single current sheet model (see Sects. 3.3 and 4.4), and as also shown in the original magnetic field restoration from the magnetograms taken at the photosphere (Vlahos et al. 2004), these lines are embedded in the photosphere. The trajectory simulations presented in Sect. 4.4 show that the particle drifts follow the magnetic field lines of the main magnetic component and move towards the photosphere, not parallel to the solar surface. Therefore, in the proposed multi-current sheet models, the issue arises as to how the test particle can move across the magnetic field, in order to get to the another current sheet (see, in this context, Emslie and Hénoux 1995).

Last, but not least, a major problem arises from the fact that these simulations inherit the most common limitations related to the test-particle approach, as discussed in Sect. 4.7.

4.7 Limitations of the Test Particle Approach

As discussed above in Sects. 3.2 (basic reconnection in a single 2-D current sheet), 3.4 (acceleration by shocks and/or plasma turbulence), 4.4 and 4.5 (reconnection in 3-D geometries), and 4.6 (MHD models incorporating multiple reconnection sites), a considerable amount of simulation of the particle acceleration process involves a use of “test particles” moving in a prescribed electromagnetic field configuration. Such a test-particle approach is certainly useful at some level—for example, it allows us to investigate the individual trajectories of electrons and protons inside 3-D current sheets (Zharkova and Gordovskyy 2004; Dalla and Browning 2005; Wood and Neukirch 2005), and to obtain the aggregated energy distributions for a large volume of particles (Zharkova and Gordovskyy 2005a, 2005b; Dalla and Browning 2005; Browning and Dalla 2007), which can be compared with those inferred from observations. The approach can be used iteratively to extend the results of exact PIC simulations to a much larger region (see Sect. 4.8).

However, considerable caution must be exercised when applying a test-particle approach to situations in which the number of accelerated particles is sufficiently large (see Sect. 2.1.3). In such cases, the electric and magnetic fields associated with the accelerated particles themselves rapidly become comparable to the fields pre-supposed at the outset. Recall (Sect. 2.1) that a major flare can accelerate some 10^{37} electrons s^{-1} and that these electrons propagate within a magnetic flux tube of order 10^9 cm in radius. Attempting to model this as a single large-scale ejection leads to the well-known (e.g., Holman 1985; Emslie and Hénoux 1995) paradoxes that (1) the magnetic fields induced by the resulting current vastly exceed the pre-supposed field, and (2) that the electric fields resulting from both charge separation and induction effects associated with the accelerated particle stream (see Sects. 4.4 and 4.8) exceed plausible values by even more absurd margins. Specifically:

- Ampère's law, applied to a steady flow of 10^{37} electrons s^{-1} (current $I \approx 10^{18}$ amps), in a cylinder of cross-sectional radius $r \approx 10^7$ m gives a corresponding magnetic field strength $B \approx \mu_o I/2\pi r \approx 3 \times 10^4$ T $\approx 3 \times 10^8$ gauss. Using a propagation volume $V \approx 10^{21}$ m^3, the associated energy content $(B^2/2\mu_o)\,V \approx 10^{36}$ J $\approx 10^{43}$ ergs, at least ten orders of magnitude greater than the energy in the electron beam itself;
- The typical rise time for a hard X-ray burst in a solar flare is $\tau \approx 1$–10 s. To initiate a current I over a timescale τ in a volume of inductance $L \approx \mu_o \ell$ (where ℓ is the characteristic scale) requires a voltage $V \approx \mu_o \ell I/\tau$. Alternatively, the largest current that can appear when a voltage V is applied is $I \approx V\tau/\mu_o \ell$. With the characteristic voltage associated with the acceleration of hard-X-ray-producing electrons being $\sim$30 kV, and with solar loop sizes $\ell \approx 10^7$ m, this gives a maximum current $I \approx 10^4$ amps, some *fourteen* orders of magnitude less than the total current involved.

It is clear, therefore, that the self-consistent electromagnetic fields associated with electron acceleration rapidly become comparable to the fields pre-supposed to be present in the acceleration region; in such an environment, a test-particle approach breaks down severely. Given this, it is a clear imperative to take into account the self-consistent evolution of the electric and magnetic fields, *including the fields associated with the accelerated particles themselves*. Failure to do so can result in unrealistic numbers of accelerated particles predicted by acceleration models and even paradoxical results, such as the energy in accelerated particles exceeding the free energy in the pre-flare magnetic field configuration.

4.7.1 The Polarization Electric Field

As discussed in Sect. 4.4 (see Fig. 14), reconnection in a 3-D current sheet geometry leads to a separation of protons and electrons across the midplane. As shown in Fig. 26, such a charge separation results in a substantial (up to 100 V cm^{-1}) Hall-type polarization electric field across the midplane. As pointed out in Sect. 3.2, inclusion of such Hall-type electric fields can significantly *increase* the effectiveness of the whole magnetic reconnection process (Birn et al. 2001; Huba and Rudakov 2004; Drake et al. 2005); see also Sect. 4.9.

Electrons, because of their much smaller gyroradii, typically do not show much change in behavior in the presence of the polarization field. On the other hand, protons, with their larger gyroradii, gain slightly more energy when they move in the positive x-direction than in the negative x-direction. The net (nonlinear) result is that protons gain energy as a result of the polarization field, thus tending to make the proton energy spectrum flatter. A full investigation of the effect of the polarization electric field requires use of a PIC approach, as described in Sect. 4.8.

4.7.2 Turbulent Electric Fields

Various turbulent electric fields typically form inside a reconnecting current sheet—these are caused, for example, by the two-stream instability (Bret 2009) of the two electron beams that enter the current sheet from opposite sides (Zharkova and Agapitov 2009). Such electric fields, and their associated currents, typically oscillate near the plasma frequency and have an exponential growth rate Γ and an energy saturation level *PE* given by the analytical

Fig. 26 Polarization electric field induced by the separation of protons and electrons accelerated in a 3-D current sheet with $B_0 = 10$ G, at various distances z from the null point (from Zharkova and Agapitov 2009)

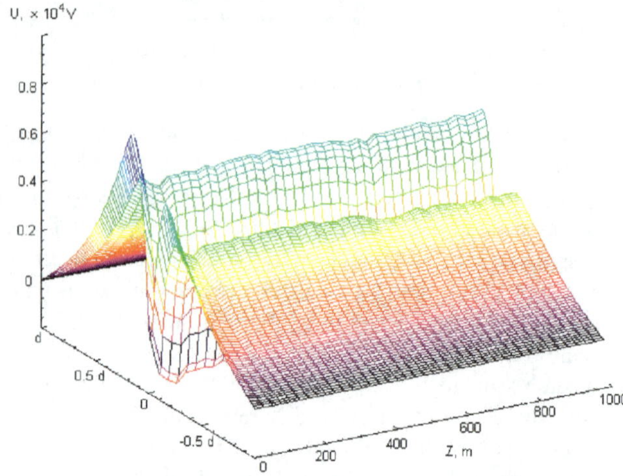

relations derived for the cold-beam plasma instability (Shapiro and Shevchenko 1988; Bret 2009):

$$\Gamma = \sqrt{1/32}^{-4/3}\Delta^{1/3}\omega_{pe}, \qquad PE = (\Delta/2)^{1/3}E_{\mathrm{beam}}, \tag{4.19}$$

where Δ is the ratio of the beam to plasma densities, and E_{beam} is the beam kinetic energy. The effect of the electric fields associated with such waves on the accelerated particle distributions was estimated by Zharkova and Agapitov (2009); this work confirmed the estimates of Bret (2009). Further progress requires full kinetic simulations.

4.8 Particle-in-Cell Simulation of Acceleration in a 3-D RCS

As has been pointed out repeatedly above, a full description of the particle acceleration process in solar flares requires that the electromagnetic fields and currents associated with the accelerated particles be self-consistently included. In this section, we describe early efforts to accommodate this crucial element of the modeling.

4.8.1 Description of the PIC Approach

In order to self-consistently include the electric and magnetic fields induced by the accelerated particles, one can use the 2D3V (PIC) simulation code (Verboncoeur et al. 1995). This PIC method solves the equations of motion for the plasma particles (see (4.14)), using values for the electric and magnetic fields which include both the pre-specified background fields **E** and **B** and the fields $\tilde{\mathbf{E}}$ and $\tilde{\mathbf{B}}$ associated with the accelerated particles, as calculated from Maxwell's equations.

In the 3-D simulations by Siversky and Zharkova (2009), the y-direction is chosen to be invariant. The system is periodic in the z-direction, so that a particle leaving the system through a boundary far from the reconnecting current sheet (see bottom of Fig. 8) reappears at the opposite boundary. The current sheet half-thickness is taken to be to 100 cm and the period along the x-direction is 2000 cm. In the PIC simulations discussed here, the computation region is extended well beyond the null point in order to explore the effect on the acceleration of different magnitudes of the transverse magnetic field. Plasma is continuously injected from the $x = \pm 1000$ cm boundaries of the simulation region.

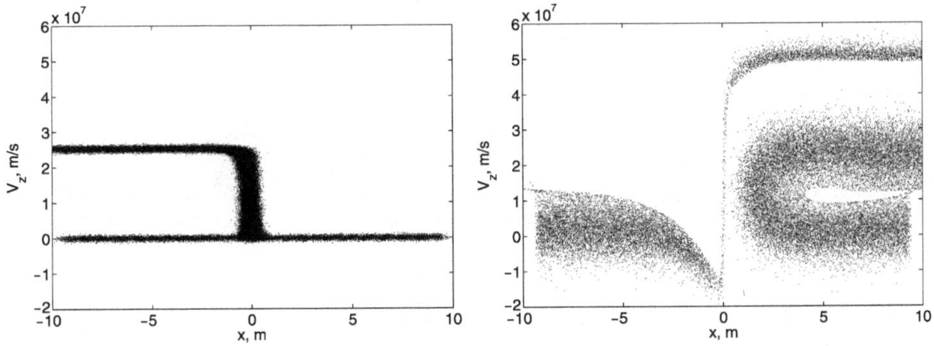

Fig. 27 PIC simulation snapshots of low-density ($n = 10^{10}$ m^{-3}) plasma particles on the x–V_z phase plane for protons (*left plot*) and electrons (*right plot*). The protons and the electrons enter the RCS from both sides and are ejected to opposite semiplanes either as a beam (transit particles) or as a flow (bounced particles); see the text for details. The current-sheet parameters are the same as in Sect. 4.4 (from Siversky and Zharkova 2009)

In order to avoid numerical instabilities in the PIC method, constraints on the steps in time and space must be satisfied (for details, see Siversky and Zharkova 2009). To satisfy these conditions while keeping the code running time reasonable, an artificially low plasma number density $n = 10^{10}$–10^{12} m^{-3} is used, and the proton-to-electron mass ratio is artificially reduced to $m_p/m_e = 100$. The spatial simulation grid has from 10 to 100 cells in the z direction and 100 cells in the x direction with $\Delta z = \lambda_D$ (the Debye length) and $\Delta x = \lambda_D/5$. There are ~100 particles per cell on average and the time step is 6×10^{-10} s.

4.8.2 PIC Simulation Results

The results of a simulation carried out with an extremely low number density $n = 10^4$ cm^{-3} are shown in Fig. 27 for both protons (left panel) and electrons (right panel). This low-density simulation was performed mainly to verify that the PIC code can reproduce the results obtained in test-particle simulations (see Sect. 4.4, Fig. 14).

Consistent with the test-particle results, the formation of separate beams of accelerated protons and electrons is clearly observed, and electrons and protons are ejected into the opposite semispaces with respect to the $x = 0$ midplane. The protons are largely unmagnetized inside the current sheet, so that the ejection velocity of the "bounced" and "transit" protons are almost the same. On the other hand, the electrons are much more magnetized, so that bounced electrons cannot reach the midplane and thus gain much less energy than transit electrons. In this particular case, however, the polarization electric field $\tilde{E}_x$ caused by the charge separation is rather weak (~20 V m^{-1}). It must be noted, however, that the electron skin depth for this simulation is about 5000 cm, which exceeds by a large factor the size (100 cm × 1000 cm) of the simulation region in both directions. The ion inertial length, which is also instrumental in controlling the electron/proton decoupling (Mandt et al. 1994b), is about 1000 cm, comparable to the size of the simulation region. Thus artificial decoupling of electrons and ion motions does result in this simulation.

Simulations for a higher density, $n = 10^6$ cm^{-3}, are shown in Fig. 28, for protons (left panel) and electrons (right panel). For this density, the electron skin depth is reduced to ~500 cm, which is within the size of the simulation region in either direction. Thus the decoupling of electron and ion motions is not an artifact of the size of the region compared to the skin depth. Although the assumed density is several orders of magnitude lower than in

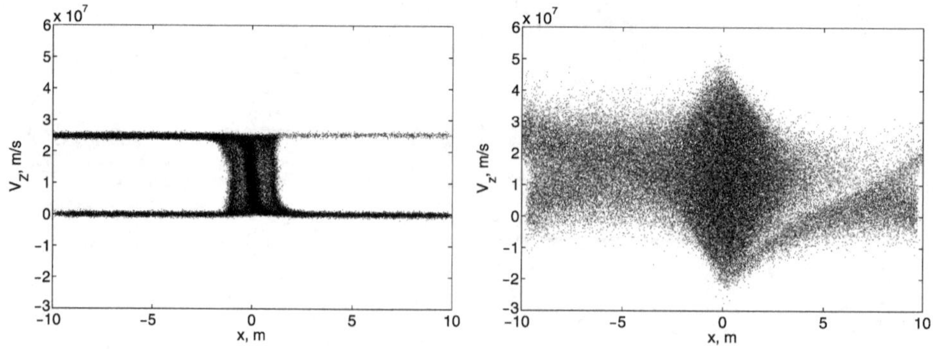

Fig. 28 Snapshots of higher-density plasma particles (10^{12} m^{-3}) on the x–V_z phase plane (PIC simulations) for protons (*left plot*) and electrons (*right plot*) entering the RCS from both sides. Protons are still ejected to the same semiplane, while the electrons form a cloud which experiences multiple ejections and returns to the acceleration region. As a result, the majority of the electrons ends up being ejected to the same semiplane as the protons, with only a small fraction of electrons being ejected into the opposite semiplane (see the text for details). The current sheet parameters are the same as in Sect. 4.4 (from Siversky and Zharkova 2009)

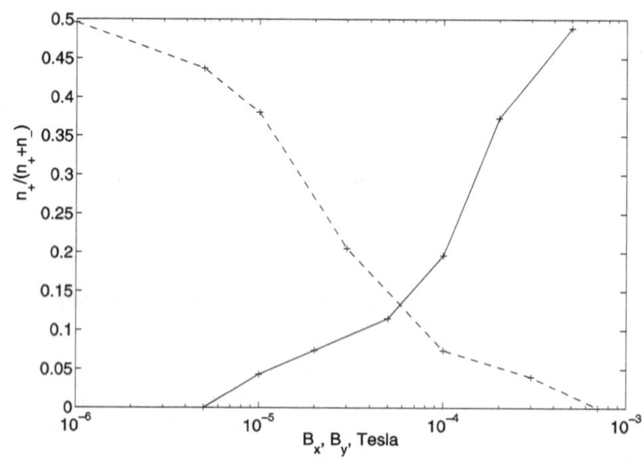

Fig. 29 The fraction of protons that are ejected into the $x > 0$ semispace, *Dashed line*: fraction vs. B_{y0} ($B_{x0} = 0.2$ G). *Solid line*: fraction vs. B_{x0} ($B_{y0} = 1$ G). All other parameters of the current sheet are the same as in Fig. 28

the solar corona (often taken to be 10^{10} cm^{-3}), already several differences between the test-particle and PIC approaches appear. The dynamics of the relatively unmagnetized protons are not substantially different: their trajectories are close to those obtained from the test-particle simulations (Fig. 14). They are still ejected mainly into one semispace ($x < 0$) with respect to the midplane, and their acceleration rate coincides with the theoretical magnitude given in Sect. 3.3.

In addition, however, a small number of protons is ejected to the $x > 0$ semispace, where electrons are normally ejected in test-particle simulations (Fig. 28, left panel). Figure 29 shows the fraction of the protons ejected into the $x > 0$ semispace as a function of B_{x0} and B_{y0}; the dependence on B_{y0} qualitatively coincides with that obtained by Zharkova and Gordovskyy (2004). For small B_{y0}, both PIC and test-particle simulations show that the protons are ejected symmetrically with respect to the midplane. However, in the PIC simulation, all the protons are ejected into the one semispace ($x < 0$) when $B_{y0} > 7$ G ($\approx B_{z0}$), while in the test-particle simulation (Zharkova and Gordovskyy 2004), the accelerated particles are fully

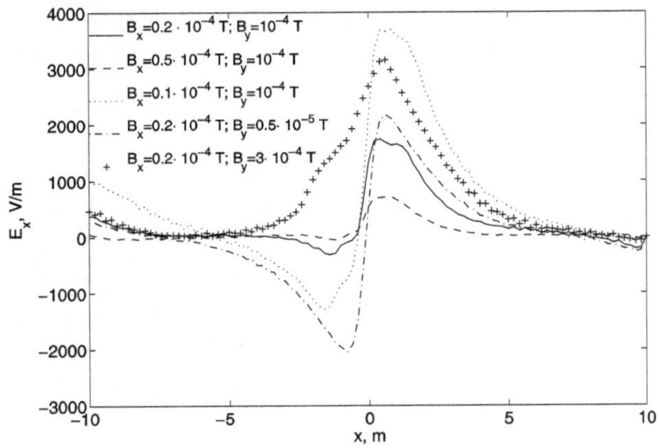

Fig. 30 Electric field $\tilde{E}_x$ induced by particles in the PIC simulations for different values of B_{x0} and B_{y0}. All other parameters of the current sheet are the same as in Fig. 28

separated if $B_{y0} > 1.5 \times 10^{-2} B_{z0}$. Thus, in PIC simulations, the particle trajectories have less asymmetry than those in the test-particle approach.

4.8.3 Polarization Electric Field Induced by Accelerated Particles

The PIC simulations show that the induced magnetic field $\tilde{\mathbf{B}}$ is typically much smaller than the background $\mathbf{B}$. On the other hand, inclusion of the induced *electric* field $\tilde{\mathbf{E}}$ is essential. Indeed, its absolute value is much larger (by an order of magnitude) than the field E_y induced by the reconnection itself (see Sects. 4.4 and 4.7.1).

Figure 30 shows the value of the charge separation electric field $\tilde{E}_x$, perpendicular to the current sheet, averaged over the z coordinate, for different values of B_{x0} and B_{y0}. The field becomes stronger when either B_{x0} decreases or B_{y0} increases. Also, when $B_{y0} \to 0$, the system becomes symmetric, and accordingly so does $\tilde{E}_x$. The distribution of the associated charge density $\rho(x) = (1/4\pi)\partial \tilde{E}_x/\partial x$ has been presented by Siversky and Zharkova (2009).

4.8.4 Particle Trajectories

As discussed above, proton trajectories do not change much in the presence of the polarization field, the only difference being that the "bounced" protons have somewhat larger orbits (Siversky and Zharkova 2009), leading to two smaller peaks at about ± 1 m in the charge density plot (Fig. 7 in Siversky and Zharkova 2009).

The trajectories of electrons are much more complicated. First, electrons which enter from the $x < 0$ semispace have dynamics similar to those of the "transit" electrons: they drift towards the midplane, become accelerated and are ejected to the $x > 0$ semispace. However, the polarization field $\tilde{E}_x(x)$, which extends well beyond the current sheet thickness and has a component parallel to the magnetic field, acts to decelerate the ejected electrons. For the chosen magnitudes of B_x and B_y, the majority of electrons cannot escape to the $x > 0$ semispace; instead, they are dragged back multiple times to the current sheet and after some time become indistinguishable from the electrons that entered from the $x > 0$ semispace.

The electrons that come from the $x > 0$ semispace demonstrate rather different dynamics compared with the case when $\tilde{E}_x = 0$. These electrons, which are "bounced" from the RCS in the absence of an $\tilde{E}_x$, can now reach its midplane. In the vicinity of the midplane, electrons become unmagnetized and oscillate with a gyro-frequency determined by the magnitude of B_y (Fig. 8 in Siversky and Zharkova 2009). After a large number of oscillations, the electron

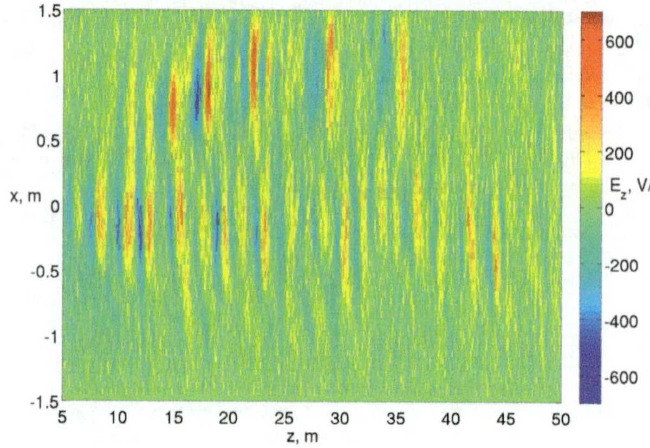

Fig. 31 Electric field $\tilde{E}_z$ induced by particles in PIC simulation ($B_{z0} = 10^{-3}T$, $B_{y0} = 10^{-4}T$, $B_{x0} = 4 \times 10^{-5}T$, $E_{y0} = 250\ \mathrm{V\,m^{-1}}$, $m_p/m_e = 10$, $n = 10^6\ \mathrm{cm^{-3}}$)

is ejected. If the electron's initial velocity is small, it can become quasi-trapped inside the RCS, similar to the situation in the vicinity of an X-type null point (Petkaki and MacKinnon 1997). Such electrons are accelerated on the midplane, ejected from it, then decelerated outside the RCS and return back to the midplane. This cycle is repeated until the electron finally gains enough energy to escape the RCS. Since the magnitude of the polarization field $\tilde{E}_x(x)$ is smaller at $x < 0$ than at $x > 0$ (see Fig. 30), it is easier for the electron to escape to the $x < 0$ semispace. Thus, eventually, *most of the electrons are ejected to the same semispace as protons*, contrary to the results of the test-particle simulations, while creating an electron cloud around an RCS before their joint ejection with protons. It seems that the asymmetry condition is shifted to much higher magnitudes of the guiding field than those found in the test-particle approach ($>0.01B_{z0}$, Zharkova and Gordovskyy 2004).

4.8.5 Energy Spectra

As shown by Siversky and Zharkova (2009), the ejected electrons, in general, form a wide, single-peaked energy distribution with the width being of the order of the mean energy. This is different from the test-particle simulations, in which two narrow-energy electron beams are formed. Consistent with Sect. 4.4, the mean electron energy still strongly depends on B_x, with lower B_x corresponding to a higher mean energy and a wider energy distribution.

4.8.6 Turbulent Electric Field

Another interesting effect observed in the PIC simulations for the selected physical conditions and proton-to-ion mass ratio is the excitation of plasma waves.

As can be seen from Fig. 31, the $\tilde{E}_z$ component of the induced electric field is structured with a characteristic length scale of $\lambda_{\mathrm{wave}} \approx 2$ m. This structure propagates in time in the positive z-direction at a speed $V_{\mathrm{wave}} \approx 1.3 \times 10^7\ \mathrm{m\,s^{-1}}$. This corresponds to an oscillation period $T_{\mathrm{wave}} \approx 1.5 \times 10^{-7}$ s, consistent with the plasma frequency for a density $n = 10^6\ \mathrm{cm^{-3}}$. Note, also, that the oscillating component of the excited wave $\tilde{E}_z$ is parallel to the direction of propagation which corresponds to the polarization of the Langmuir wave. The instability mechanism is evident from study of the electron velocity V_z. As shown in Fig. 32, the electrons have an unstable "bump-on-tail" distribution. The range of velocities V_z for which the derivative $\partial f/\partial V_z$ is positive is from 1.3×10^9 to $2 \times 10^9\ \mathrm{cm\,s^{-1}}$, corresponding to the phase velocity of the associated Langmuir wave V_{wave}.

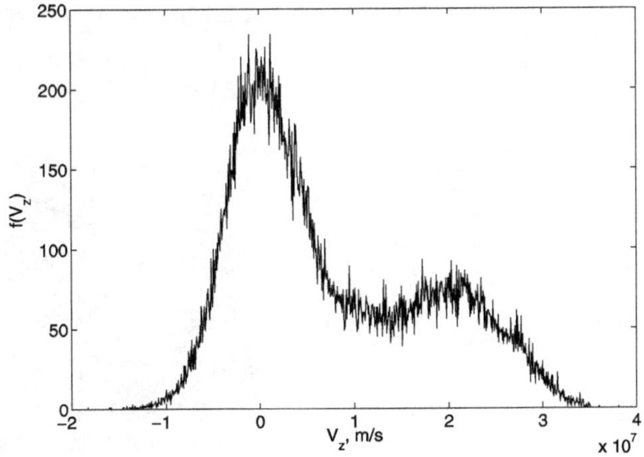

Fig. 32 Distribution function of electrons calculated for the region with turbulence plotted in Fig. 31

The use of plasma instabilities to produce anomalously high values of resistivity in the reconnection region have been studied by many authors. A comprehensive review of these results is beyond the scope of this article; however, a brief commentary is appropriate. Both 2-D and 2.5-D PIC simulations have revealed various types of instabilities to be present, from the lower hybrid drift instability (LHDI) leading to the Kelvin-Helmholtz instability (Landau and Lifshitz 1960; Chen et al. 1997; Lapenta et al. 2003) to kink (Pritchett et al. 1996; Zhu and Winglee 1996; Daughton 1999) and sausage (Büchner 1999) instabilities. However, full 3-D PIC simulations (with an artificially reduced speed of light $c = 15V_{Ae}$ and proton-to-electron mass ratio of 277.8; Zeiler et al. 2002), while retaining the lower hybrid drift instability, reveal no evidence of kink or sausage instabilities. This indicates that the appearance of various instability modes may be dependent on the details of the simulation used, and hence argues for an abundance of caution at this time in applying these results to actual flares.

4.9 Particle Acceleration in Collapsing Magnetic Islands

4.9.1 Tearing-Mode Instability in Current Sheets

As mentioned in Sect. 3.2, in reconnecting current sheets, diffusion converts magnetic energy into Ohmic heat, with a concomitant slow expansion of the current sheet at a rate proportional to the resistivity. Consequently, one way of enhancing the reconnection rate is to consider enhancement of the resistivity through resistive instabilities such as the tearing instability (Furth et al. 1963).

In a current sheet wide enough to meet the condition $\tau_{dif} \gg \tau_A$, where $\tau_A = d/V_A$ is the Alfvén traverse time, the tearing instability occurs on a timescale of $\tau_{dif}(\tau_A/\tau_{dif})^{\lambda}$, with the parameter λ in the range $0 < \lambda < 1$. The tearing instability occurs on wavelengths greater than the width of the current sheet, ($ka < 1$), with a growth rate (Priest and Forbes 2000):

$$\omega = \left[\tau_{dif}^3 \tau_A^2 (ka)^2\right]^{-1/5}, \tag{4.20}$$

where the wave number k satisfies the condition $(\tau_{dif}/\tau_A)^{1/4} < ka < 1$. The fastest growth is achieved for the longest wavelength (i.e., at $ka = 1$), while the smallest wavelength of the instability grows more slowly, on a timescale $\tau_{dif}^{3/5}\tau_A^{2/5}$. As first demonstrated by Furth et al.

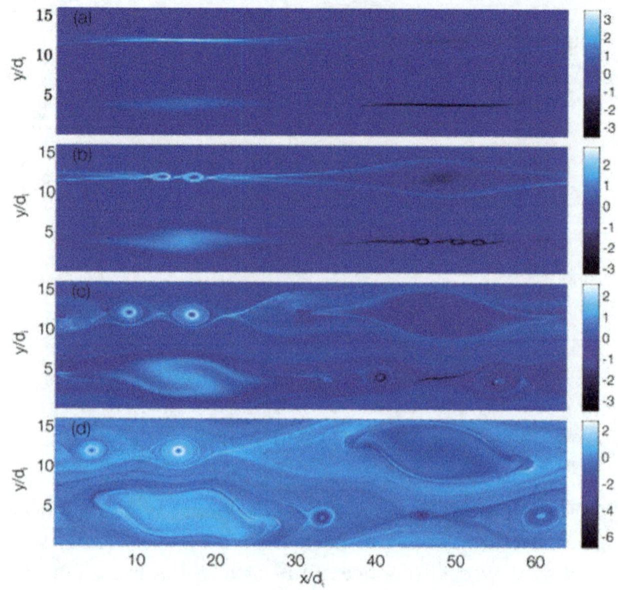

Fig. 33 Out-of-plane current density in 2.5-D PIC simulations of reconnection with a guiding field, for four consecutive time intervals, from top to bottom. The environment contains two current sheets and is characterized by the appearance of islands formed by the tearing instability. At the start, there are only (**a**) large islands, followed at later times (**b–d**) by the formation of smaller magnetic islands. Note that such islands do *not* appear in 2-D PIC simulations of the same environment without a guiding field. From Drake et al. (2006b). Reproduced by permission of American Geophysical Union

(1963) (see also Spicer 1977), tearing-mode instabilities generated during reconnection in a current sheet can produce closed magnetic "islands" with an elliptical O-type topology. Accordingly, large current sheets often observed in solar flares (e.g., see Sui and Holman 2003, and references therein) can, depending on their sizes and conditions, be subject to the formation of magnetic islands. For example, for a current sheet of size ~10,000 km (as reported by Sui and Holman 2003), embedded in a (coronal) medium with magnetic field $B \approx 100$ G and density $n = 10^{10}$ cm^{-3}, then the Alfvén velocity is about 2000 km s^{-1} and $\tau_A \approx 5$ sec. With a rather conservative collision frequency estimate 10^4 s^{-1}, $\tau_{dif} \approx 10^{15}$ sec. Substituting these values into (4.20) gives an island formation time in the range 10–1000 s.

4.9.2 Studying Particle Acceleration in Magnetic Islands Using a PIC Approach

In the past decade, the tearing-mode instability has been extensively investigated, both in 2.5-D current sheets (Drake et al. 1997; Shay and Drake 1998; Drake et al. 2005, 2006a, 2006b), and in a full 3-D kinetic PIC approach (Zeiler et al. 2002; Pritchett and Coroniti 2004; Karimabadi et al. 2007; Shay et al. 2007; Daughton et al. 2009; Drake et al. 2010).

Shay and Drake (1998) studied the dynamics of whistler waves generated inside the magnetic islands, using a code in which the electric and magnetic fields were calculated at each time step using the Poisson condition $\nabla \cdot E = 4\pi\rho$. The time step was normalized to the electron cyclotron time Ω_e^{-1}, calculated for a maximum initial magnetic field B_0, the spatial step normalized to the skin depth c/ω_{pe}, and the velocity normalized to the Alfvén velocity for electrons $V_{Ae} = B_0/\sqrt{4\pi n_0 m_e}$. The speed of light was artificially reduced to $5V_{Ae}$ in order to marginally resolve Debye radii. The simulation grid was 512×512 with 7 million particles involved. Small magnetic perturbations and associated currents were added as initial conditions to form seed magnetic islands centered around the two current layers, as shown in Fig. 33. Using such a formalism, Shay and Drake (1998) were able to resolve the electron dissipation region, despite the use of an artificially low ion-to-electron mass ratio of 200.

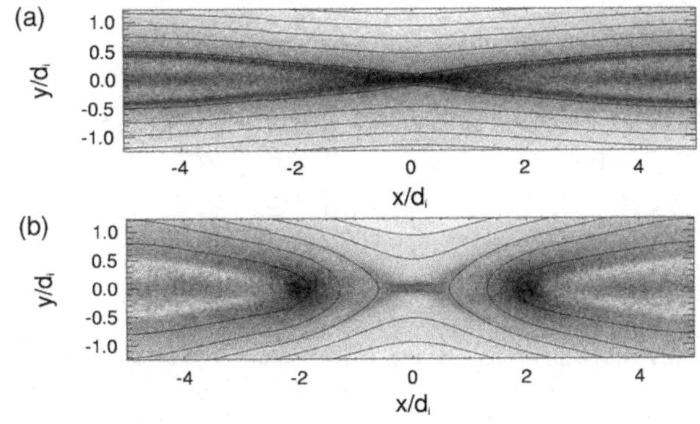

Fig. 34 Current density in PIC simulations for times (**a**) $t = 7.7/\omega_{ci}$ and (**b**) $t = 12.9/\omega_{ci}$. The *solid lines* correspond to magnetic flux lines. Adopted from Zeiler et al. (2002). Reproduced by permission of American Geophysical Union

A further modification of the PIC code was used to extend it to a 3-D model with 64 grid cells in the third direction, while increasing the proton-to-electron mass ratio to 277.8 and the speed of light to $c = 15V_{Ae}$ (Zeiler et al. 2002). Unlike a conventional Harris equilibrium, in which the variation of the density produces the equilibrium current, the density was taken to be nearly constant, with the current being produced by the $\mathbf{E} \times \mathbf{B}$ drift of the electrons. This resulted (see Fig. 34) in a "parallel" electric field occurring in the plane of the island current sheet (Drake et al. 1997; Shay and Drake 1998; Zeiler et al. 2002). Pritchett and Coroniti (2004) have further shown that in 3-D current sheets with a guiding magnetic field, the "parallel" electric field is formed only in the two quadrants $(+x, +y)$ and $(-x, -y)$. Electrons entering these sectors are accelerated by this "parallel" electric field; the other quadrants are characterized by an excess of ions, strongly reminiscent of the drift separation of electrons and protons in a 3-D single current sheet (Sects. 4.4 and 4.8).

Once formed, magnetic islands tend to shrink due to magnetic tension forces. Conservation of the longitudinal invariant leads to efficient first-order Fermi acceleration of particles tied to the island's field lines. As shown by Drake et al. (2006a), the process of island shrinking halts when the kinetic energy of the accelerated particles is sufficient to halt the further collapse due to magnetic tension forces; the electron energy gain is naturally a large fraction of the released magnetic energy. A sufficient number of electrons can be produced to account for a moderately-sized solar flare. The resultant energy spectra of electrons take the form of power laws, with harder spectra corresponding to lower values of the plasma β. For the low-β conditions appropriate to the solar corona, the model predicts accelerated spectra that are much harder ($\delta \approx 1.5$) than observed.

Further progress with 3-D PIC simulations of magnetic islands, using a larger 2560×2560 grid, revealed that the electron diffusion region formed by the electron flows occurs over a width of $5d_e$ (where d_e is the electron inertial length–Birn et al. 2001), while other simulations show an extent up to $20d_i$ (where d_i is the ion inertial length—Karimabadi et al. 2007; Shay et al. 2007); see Fig. 35. The electron diffusion region becomes elongated with a two-scale structure: an inner region (size D_e in Fig. 35) with a strong electron-drift-associated electric field (see (3.9)) and a wider diffusion region characterized by electron jets (Karimabadi et al. 2007; Shay et al. 2007). Shay et al. (2007) found that the electrons form super-Alfvénic outflow jets, which become decoupled from the magnetic field and extend to a large distance from the X-type null point, similar to the manner in which electron jets decouple from the protons in a test-particle approach to reconnection in a single 3-D current sheet (Sect. 4.4).

Recently, Daughton et al. (2009) showed that for collisional Sweet-Parker reconnection with Lundquist numbers S higher than 1000, the reconnection region becomes susceptible to

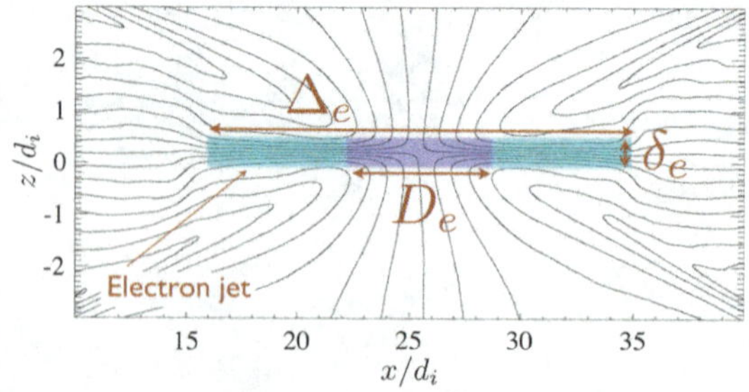

Fig. 35 The multi-scale structure of the electron diffusion region around an X-type null point of total length of Δ_e and thickness δ_e, as derived from 3-D PIC simulations. The *solid lines* show the electron trajectories for the time $t\omega_{ci} = 80$ (ω_{ci} is the ion cyclotron frequency). In the inner region of length D_e, there is a steady inflow of electrons and a strong out-of-plane current; the outer region is characterized by electron outflow jets. From Karimabadi et al. (2007). Reproduced by permission of American Geophysical Union

Fig. 36 Evolution of the current density over a region of extent $800d_i$ (where d_i is the ion inertial length), for different times t. White lines are the magnetic flux surfaces. The *bottom plot* presents a close-up of a region in which formation of new islands is occurring. From Daughton et al. (2009)

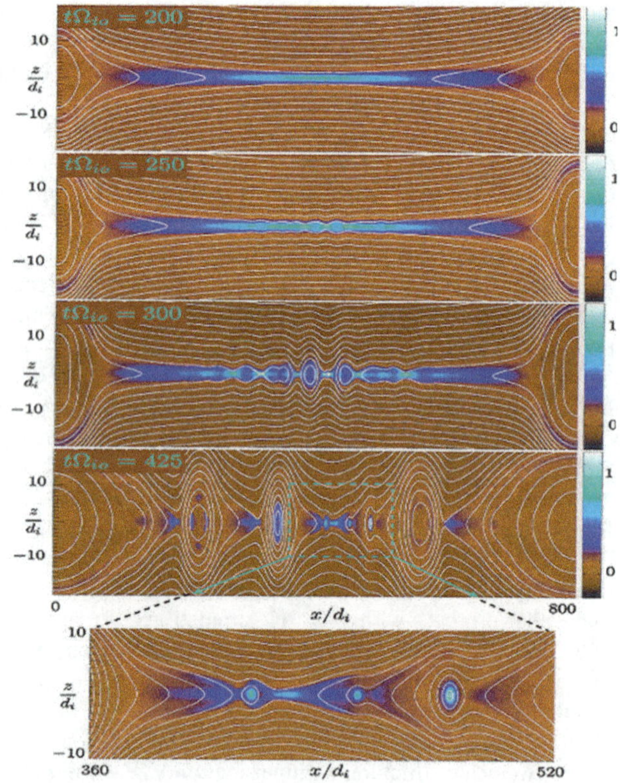

the tearing instability, leading to the formation of many magnetic islands with half-thickness approaching the ion inertial length d_i (see Fig. 36). This result shows that the current sheet size is a key factor in defining timescales and the role of different physical processes at different stages of the reconnection process. For example, the shrinking of a large current sheet during MHD reconnection is halted when it approaches a size conducive to the formation of magnetic islands.

4.10 Limitations of the PIC Approach

Many PIC simulations (Birn et al. 2001; Drake et al. 2006b; Tsiklauri and Haruki 2007; Shay et al. 2007; Karimabadi et al. 2007) have been concerned with the study of reconnection rates in the vicinity of a magnetic X-type null. Even for the very low densities used in these simulations, it is apparent that the feedback of the fields induced by accelerated particles can be substantial (see Sect. 4.7): it can significantly modify the particle trajectories and the energies gained during their acceleration.

However, because of computer power limitations, the size of the region in which the induced electric and magnetic field are simulated and coupled to the particle motion is necessarily rather small: $\sim 100 \times 50 \times 1000$ cm for 2-D simulations (Birn et al. 2001; Tsiklauri and Haruki 2007), or $4000 \times 40000 \times 86$ cm for the most recent 3-D simulations (Shay et al. 2007; Drake et al. 2010). These are orders of magnitude smaller than the 10^8–10^{10} cm sizes associated with flares (Masuda et al. 1994; Sui and Holman 2003; Jiang et al. 2006). Only in the PIC simulations discussed in Sect. 4.8 is the computation region extended well beyond the null point in order to explore the variations of acceleration process, at different magnitudes of the transverse magnetic field. Thus, *the narrowness of the simulation region* still imposes strong limitations on the number of particles being accelerated in PIC simulations.

Other puzzling features of the simulations relate to the origin of the electric fields in the models. One example is the "parallel" electric field reported in many 3-D simulations (Zeiler et al. 2002; Pritchett and Coroniti 2004; Drake et al. 2010) and produced either as a result of an advective electric field (see the first term in (3.9)) and/or by a gradient of the magnetic field component across the current sheet (see the second term in (3.9)). This "parallel" electric field is the electric field induced by accelerated electrons known as a self-induced (return-current) electric field associated with the propagation of the accelerated electrons, which can exceed the local Dreicer field by a factor of up to ~ 30 (Zharkova and Gordovskyy 2006). Although both the "parallel" electric field, or the self-induced (return current), electric field associated with the propagation of accelerated electrons can cause various interesting effects (e.g., the charge separation electric field can decelerate electrons and even return them back to the acceleration region), causality arguments would strongly suggest that neither field can be considered as the main driver of particle acceleration since the field is *formed by the accelerated particles themselves*.

A related problem is the difference in electron and ion gyration radii/periods (Birn et al. 2001). Because of this difference, within a timescale of $\lesssim 10^{-7}$ s, electrons can move towards the midplane, be accelerated and ejected, while the protons have only just started their gyrations in the RCS environment. In the simulations to date, the proton-to-electron mass ratio has been arbitrarily lowered to 100–200, an order of magnitude smaller than the actual value. Even the speed of light is conveniently reduced by some one or two orders of magnitude to allow a reduction in simulation times (Birn et al. 2001; Drake et al. 2006a). Such simulations, then, serve to highlight that by artificially adjusting fundamental physical constants (such as the electron mass or the speed of light) in order to achieve results with a reasonable amount of computational effort, certain features of the results can be altered significantly. Although Shay and Drake (1998) have shown that the *reconnection rate* may be indeed insensitive to the adopted values of both the electron mass and speed of light.

Issues such as the ratio of macro- to micro-scales can be resolved only by increasing the reconnection region size in the PIC simulations up to sizes comparable with observations. However, even with exponential increases in computing power, there are substantial doubts that one can achieve this goal within the next few decades. Possibly a hybrid approach, combining PIC simulations convoluted with test-particle extensions to greater source volumes, will allow us to make substantial progress in this area.

5 How Well Do Recent Theories Account for Observed Flare Characteristics?

5.1 Interrelation Between Acceleration and Transport

Before we address the relative merits of the various particle acceleration models, it is important to realize that many flare observations consist of radiation integrated over the source volume; the quantity observed involves integration over not only space but also quantities such as particle pitch angles. Therefore, in order to relate to observed features (Sect. 2), one must incorporate the effects of particle acceleration with the *transport* of the accelerated particles throughout the remainder of the flaring source. Although such an inclusion will in all probability obscure information on the acceleration process, it can also, more profitably, be used as a diagnostic of particle transport processes themselves.

As we have seen (Sects. 4.4, 4.5 and 4.6), accelerated electrons can attain quasi-relativistic energies within an extremely small timescale and can thus propagate over the extent of a solar flare of size $\sim 10^9$ cm within $\sim$0.01–0.1 s. The proton acceleration time is longer than that for electrons by up to two orders of magnitude (Sect. 4.4), so that protons still remain within the acceleration region while the electrons are ejected from it. Thus, particle transport mechanisms are essential for completing the electric circuit for particles escaping the acceleration region (Emslie and Hénoux 1995) and/or in establishing co-spatial return currents (Knight and Sturrock 1977; Emslie and Hénoux 1995; Zharkova and Gordovskyy 2006). Such return currents can be formed both by ambient electrons (Knight and Sturrock 1977; Nocera et al. 1985) and by relatively high-energy beam electrons (Zharkova and Gordovskyy 2006) that return to the corona with nearly the same velocities as the injected electrons. The self-induced electric fields created by the beam electrons and/or the return current have a structure very favorable for the generation of plasma turbulence (Karlický and Bárta 2008), which naturally creates the conditions required for stochastic acceleration of electrons to sub-relativistic energies as discussed in Sect. 3.4 (see also Miller et al. 1996; Pryadko and Petrosian 1997).

Another issue that can arise is consideration of *radiative* transport effects, such as radiative transfer of optically-thick lines and continua, and reflection (or albedo effects) of emission of hard X-rays from the photosphere.

5.2 Testing Acceleration Models Against Observational Constraints

Here we briefly summarize the extent to which the various models discussed in Sect. 4 have increased our ability to account for the characteristics (e.g., number, spectrum) of accelerated particles in a solar flare. In other words, can *any* hitherto-identified acceleration process produce the rate, number and energy spectrum of accelerated particles inferred from observations?

5.2.1 Stochastic Acceleration

Stochastic acceleration remains one of the most promising candidates to account for most of the characteristics deduced from hard X-ray, radio and gamma-ray observations. This mechanism does, however, require a pre-supposed level of either plasma or MHD turbulence, the basis for which has not yet been fully established. Further, high-frequency waves near the plasma frequency can be excluded as drivers for stochastic acceleration, since they would couple into a high level of decimetric radio emission in every flare, contrary to observations (Benz et al. 2005).

On the other hand, acceleration by transit-time damping *is* compatible with a frequent lack of radio emission because the frequency of the postulated MHD turbulence is far below the plasma frequency. The non-thermal electron distribution "grows" out of the thermal population, resulting in a "soft-hard-soft" evolution of the hard X-ray spectrum that is consistent with observations. Further, such "soft-hard-soft" behavior has been found in coronal sources, (Battaglia and Benz 2007) and is therefore inherent to the acceleration process and not to collisional transport effects. *RHESSI* observations showing the existence of a "pivot point" energy at which the differential electron flux does not vary with time, leading to the observed "soft-hard-soft" behavior, have been used (Grigis and Benz 2004) as evidence for a stochastic acceleration model.

Recent models of stochastic acceleration that involve the interaction of turbulence with high energy particle beams injected from elsewhere (Bykov and Fleishman 2009) can also account for some basic features of accelerated particles and their evolution with time. During the initial (linear) phase of the acceleration, effective particle acceleration by the longitudinal large scale turbulent motions, leading to visible spectral hardening. Then, because the accelerated particles eventually accumulate a considerable fraction of the turbulent energy, they deplete the energy in the turbulence, leading to a spectrum softening. Again, this is consistent with observations (Battaglia and Benz 2007) of "soft-hard-soft" spectral evolution in the main stage and even some hardening (soft-hard-harder behavior) during later stages (Krucker et al. 2007a).

5.2.2 Collapsing Current Sheets

Estimates of the electron energy gained in collapsing magnetic traps associated with reconnection in a helmet-type geometry, with and without a termination shock, reveal that there is a sufficient amount of energy released in accelerated particles during such a process. In addition, betatron acceleration, possibly coupled with first-order Fermi acceleration in collapsing current sheets, can account for the electron number and energies gained, although the energy spectrum of the accelerated particles resembles more the higher energy tail of a Maxwellian distribution. It remains to be established how the accelerated particles can move across the magnetic field of the trap in order to precipitate into the flare footpoints.

5.2.3 Acceleration in 3D Reconnecting Current Sheets

New MHD simulations of particle acceleration rates and spectra from collapsing current sheets have revised our views on the efficiency of particle acceleration inside a diffusion region, although the number of particles (both electrons and protons) that can be accelerated in a current sheet is limited by the volume encompassed by the current sheet (Holman 1985; Litvinenko 1996a, see Sect. 3.3). However, more complex 3-D current sheets and/or multiple current sheets may be reconcilable with this model (see Sects. 4.4 and 4.5). Acceleration of electrons in a 3-D RCS occurs mostly around the RCS midplane, over a very short timescale comparable with the growth rate of hard X-ray emission in solar flares. Acceleration of protons has a much longer timescale and results in a spread of protons in a location outside the midplane. The particle orbits in the RCS lead to a separation of electrons and protons into the opposite semiplanes, corresponding to ejection in opposite directions and into opposite loop legs. The separation of electrons and protons during acceleration in an RCS leads to the formation of a polarization electric field (induced by Hall currents), which sets the frame for the generation of strong plasma turbulence that can lead to further particle acceleration by stochastic processes. Single 3-D current sheets with guiding fields have been found, through both test-particle and PIC approaches, to be able to supply sufficient energy into both electrons and protons in order to account for *RHESSI* hard X-ray and gamma-ray observations.

Spine reconnection in 3-D MHD reconnection simulations can naturally produce two symmetric jets of energetic particles escaping along the spine, while the fan reconnection produces azimuthally localized ribbons of high energy particles in the fan plane. Both features, in principle, are often observed in flares (e.g., magnetic jets in SOL2000-07-14T10:24, X5.7, and and SOL2002-07-23T00:35, X4.8) and two-ribbon flares as discussed in Fletcher et al. (2011).

Another possibility involves tearing instabilities in current sheets, which leads to the formation of large numbers of magnetic islands, on which a large number of particles can be rapidly accelerated; however, the simulated electron energy spectra are much harder than inferred from hard X-ray observations.

Despite these qualified successes, there are two outstanding problems with this kind of acceleration model: (1) incorporation of the self-consistent electrodynamic fields associated with the accelerated particles; and (2) production of sufficient numbers of accelerated particles to match the observations. The latter can be marginally accounted for if one notes that strongly stressed 3-D current sheets, discussed in Sects. 4.4 and 4.5, are characterized by much larger diffusion regions than in simple reconnection models (Sect. 3.2).

5.2.4 Complex Reconnection Models

The other way to tackle the particle number problem involves multiple dissipation sites associated with self-organized criticality (SOC) that could lead to "spreading" of the dissipation across a large coronal volume (see Sect. 4.6). Particles move between the current sheets, either gaining or losing energy in each one, resulting in a systematic second-order gain in energy. Particle energy distributions from the simulations fit the observed spectra for electrons very well, although proton spectra have yet to be simulated.

However, a few essential issues do remain:

1. The models invoke an *ad hoc* distribution of RCS sizes and physical conditions in order to account for the number and spectra of the accelerated particles;
2. The basic model of a current sheet (discussed in Sect. 3.3) is used to derive the energy gains by accelerated particles in a component current sheet, while the actual gains are significantly different in more realistic 3-D models of reconnecting current sheets;
3. Even if multiple current sheets are formed in the whole volume of the corona, there remains the issue of where the reconnecting magnetic field lines are embedded and how the test particles can move across the strong magnetic field threading the reconnection volume in order to reach another current sheet;
4. Some of these models also suffer from problems associated with the neglect of the feedback of the electric and magnetic fields associated with the accelerated particles themselves;
5. Multiple RCS models simulated with a more realistic PIC approach are applicable to only a very small region (about 10^6 cm^3), and it is assumed without justification that the results can be scaled to flare volumes.

6 Summary

Over the past decade, and especially in the *RHESSI* era, significant progress has been made in advancing various particle acceleration models and assessing their ability to account for the observed features in solar flares. *RHESSI* observations have been instrumental in encouraging the further development and exploration of both "traditional" and "non-traditional" mechanisms, including more complicated magnetic field geometries and topologies, to enhanced MHD and PIC models of magnetic reconnection in a single current sheet, to multiple current sheets and stochastic acceleration by various plasma waves. Substantial progress has been made through the use of phenomenological estimations made in collapsing traps, termination shocks of Alfvén waves, and direct simulations using both test-particle and self-consistent PIC approaches.

Consideration of the full three-dimensionality of the magnetic reconnection process has been found to significantly enhance the acceleration efficiency in current-sheet diffusion regions. However, the question still remains whether such a mechanism can account for the very large number of electrons responsible for hard X-ray emission and whether consideration of the whole electric circuit formed by precipitating and returning beams in reconnecting loops can resolve the particle number problem. PIC simulations have demonstrated that the acceleration of substantial numbers of charged particles causes a number of important secondary effects: for example, polarization electric fields induced by separation of electrons and protons towards the midplane, return currents formed from precipitating and ambient electrons, and plasma turbulence related to beam instabilities.

Interactions of large-scale loops with strong magnetic fields can lead either to acceleration in collapsing traps by betatron and/or first-order Fermi mechanisms, or to acceleration in the termination shock formed by such the collapsing sheets (or by large-scale Alfvén waves produced during the reconnection). Interaction of smaller-scale magnetic fields can produce magnetic islands or multiple current sheets in which particles can be also accelerated in a cascade processes if the particle travels and gains energy from one feature or another.

The exact acceleration scenario will depend on the magnetic field topology and on the absolute magnitudes of physical parameters. Further, it is possible, or even likely, that all the acceleration processes discussed above play some role in solar flares and elsewhere in astrophysics. The strict demands imposed by the *RHESSI* observations have necessitated a thorough and critical review of all these models, leading to substantial new understanding along the way.

Acknowledgements AGE was supported by NASA Grant NNX10AT78G and by a grant from UC Berkeley. EPK acknowledge the support of a PPARC/STFC Rolling Grant, PPARC/STFC UK Advanced Fellowship and Royal Society Conference Grant and by the European Commission through the SOLAIRE Network (MTRN-CT-2006-035484). AGE and EPK were also supported in part by a grant from the International Space Science Institute (ISSI) in Bern, Switzerland. NV acknowledges support from Centre National d'Etudes Spatiales (CNES) and from the French program on Solar-Terrestrial Physics (PNST) of INSU/CNRS for the participation to the RHESSI project. VZ acknowledges the support of the Science Technology and Facility Council (STFC) project PP/E001246/1. The computational work for the test-particle and PIC simulations was carried out on the joint STFC and SFC (SRIF) funded cluster at the University of St Andrews, UK.

In addition to those noted in figure captions, we also acknowledge the following figures from A&A, reproduced with permission ©ESO: Figure 4—M. Battaglia and A.O. Benz, Astron. Astrophys. **456**, 751 (2006); Figures 10, 12, 13—G. Mann, H. Aurass, and A. Warmuth, Astron. Astrophys. **454**, 969 (2006); Figure 18—S. Dalla and P.K. Browning, Astron. Astrophys. **491**, 289 (2008); Figure 20—S. Dalla and P.K. Browning, Astron. Astrophys. **491**, 289 (2008); Figure 22—S. Dalla and P.K. Browning, Astron. Astrophys. **491**, 289 (2008).

The following figures are reproduced with kind permission from Springer Science+Business Media: Figure 1—E.P. Kontar and A.L. MacKinnon, Solar Phys. **227**, 299 (2005); Figure 8—V.V. Zharkova and M. Gordovskyy, Space Science Reviews **121**, 165 (2005b); Figure 15—V.V. Zharkova and M. Gordovskyy, Space Science Reviews **121**, 165 (2005b).

The following figure was reproduced with kind permission from Elsevier: Figure 6—V.S. Titov and G. Hornig, Advances in Space Research **29**, 1087 (2002).

The following figure was reproduced with kind permission from Wiley and Sons: Figure 7—V.V. Zharkova and M. Gordovskyy, Mon. Not. Roy. Astron. Soc. **356**, 1107 (2005a).

The following figure was reproduced with kind permission from APS: Figure 36—W. Daughton, V. Roytershteyn, B.J. Albright, H. Karimabadi, L. Yin, and K.J. Bowers, Physical Review Letters **103**(6), 065004 (2009), http://adsabs.harvard.edu/abs/2009PhRvL.103f5004D.

The following figures were reproduced with kind permission of Cambridge Univ. Press.: Figures 14, 16, 17, 26—V.V. Zharkova and O.V. Agapitov, Journal of Plasma Physics **75**, 159 (2009); Figures 27–32—T.V. Siversky and V.V. Zharkova, Journal of Plasma Physics, **75**, 619 (2009).

References

T. Amano, M. Hoshino, Phys. Rev. Lett. **104**(18), 181102 (2010)
E. Amato, P. Blasi, Mon. Not. R. Astron. Soc. **364**, L76 (2005)
A. Anastasiadis, C. Gontikakis, N. Vilmer, L. Vlahos, Astron. Astrophys. **422**, 323 (2004)
S.K. Antiochos, C.R. DeVore, J.A. Klimchuk, Astrophys. J. **510**, 485 (1999)
T.D. Arber, M. Haynes, Phys. Plasmas **13**(11), 112105 (2006)
K. Arzner, A.O. Benz, Sol. Phys. **231**, 117 (2005)
K. Arzner, L. Vlahos, Astrophys. J. Lett. **605**, L69 (2004)
A. Asai, H. Nakajima, M. Shimojo, S.M. White, H.S. Hudson, R.P. Lin, Publ. Astron. Soc. Jpn. **58**, L1 (2006)
M.J. Aschwanden, Astrophys. J. **661**, 1242 (2007)
M.J. Aschwanden, R.A. Schwartz, Astrophys. J. **464**, 974 (1996)
M.J. Aschwanden, M.J. Wills, H.S. Hudson, T. Kosugi, R.A. Schwartz, Astrophys. J. **468**, 398 (1996)
M.J. Aschwanden, R.M. Bynum, T. Kosugi, H.S. Hudson, R.A. Schwartz, Astrophys. J. **487**, 936 (1997)
G. Aulanier, E.E. DeLuca, S.K. Antiochos, R.A. McMullen, L. Golub, Astrophys. J. **540**, 1126 (2000)
H. Aurass, G. Mann, Astrophys. J. **615**, 526 (2004)
H. Aurass, G. Rausche, G. Mann, Astron. Astrophys. **471**, L37 (2007)
H. Aurass, B. Vršnak, G. Mann, Astron. Astrophys. **384**, 273 (2002)
T. Bai, R. Ramaty, Astrophys. J. **219**, 705 (1978)
A. Barnes, J.V. Hollweg, J. Geophys. Res. **79**, 2302 (1974)
T.S. Bastian, G.D. Fleishman, D.E. Gary, Astrophys. J. **666**, 1256 (2007)
M. Battaglia, A.O. Benz, Astron. Astrophys. **456**, 751 (2006)
M. Battaglia, A.O. Benz, Astron. Astrophys. **466**, 713 (2007)
M. Battaglia, A.O. Benz, Astron. Astrophys. **499**, L33 (2009)
T.E. Bell, Astronomy **6**, 6 (1978)
S.G. Benka, G.D. Holman, Astrophys. J. **435**, 469 (1994)
A. Benz (ed.), *Plasma Astrophysics*, 2nd edn. Astrophysics and Space Science Library, vol. 279 (2002)
A.O. Benz, P.C. Grigis, A. Csillaghy, P. Saint-Hilaire, Sol. Phys. **226**, 121 (2005)
A.O. Benz, H. Perret, P. Saint-Hilaire, P. Zlobec, Adv. Space Res. **38**, 951 (2006)
C.K. Birdsall, A.B. Langdon, in *Plasma Physics via Computer Simulation*, ed. by C.K. Birdsall, A.B. Langdon (McGraw-Hill, New York, 1985)
J. Birn, J.F. Drake, M.A. Shay, B.N. Rogers, R.E. Denton, M. Hesse, M. Kuznetsova, Z.W. Ma, A. Bhattacharjee, A. Otto, P.L. Pritchett, J. Geophys. Res. **106**, 3715 (2001)
J. Birn, L. Fletcher, M. Hesse, T. Neukirch, Astrophys. J. (2009)
D. Biskamp, *Nonlinear Magnetohydrodynamics* (1997)
R.D. Blandford, J.P. Ostriker, Astrophys. J. Lett. **221**, L29 (1978)
S.A. Bogachev, B.V. Somov, Astron. Lett. **31**, 537 (2005)
S.A. Bogachev, B.V. Somov, Astron. Lett. **33**, 54 (2007)
A. Bret, Astrophys. J. **699**, 990 (2009)
J.W. Brosius, S.M. White, Astrophys. J. Lett. **641**, L69 (2006)
J.C. Brown, Sol. Phys. **18**, 489 (1971)
J.C. Brown, in *Coronal Disturbances*, ed. by G.A. Newkirk. IAU Symposium, vol. 57 (1974), p. 395
J.C. Brown, R. Bingham, Astron. Astrophys. **131**, L11 (1984)
J.C. Brown, A.J. Conway, M.J. Aschwanden, Astrophys. J. **509**, 911 (1998)

J.C. Brown, A.G. Emslie, E.P. Kontar, Astrophys. J. Lett. **595**, L115 (2003)
J.C. Brown, E.P. Kontar, A.M. Veronig, in: *Lecture Notes in Physics*, vol. 725, ed. by K.-L. Klein, A.L. MacKinnon (Springer, Berlin, 2007), p. 65
P. Browning, S. Dalla, Mem. Soc. Astron. Ital. **78**, 255 (2007)
J. Büchner, Astrophys. Space Sci. **264**, 25 (1999)
S.V. Bulanov, S.I. Syrovatskii, *Motion of Charged Particles in the Vicinity of a Magnetic-Field Zero Line* (1976)
A.M. Bykov, G.D. Fleishman, Astrophys. J. Lett. **692**, L45 (2009)
A.M. Bykov, I.N. Toptygin, in *International Cosmic Ray Conference*, vol. 2 (1993), p. 267
P.J. Cargill, Astrophys. J. **376**, 771 (1991)
A. Caspi, R.P. Lin, Astrophys. J. Lett. **725**, L161 (2010)
F.F. Chen, *Introduction to Plasma Physics* (1974)
Q. Chen, A. Otto, L.C. Lee, J. Geophys. Res. **102**, 151 (1997)
E.L. Chupp, in *American Institute of Physics Conference Series*, vol. 115, ed. by S.E. Woosley (1984), p. 641
E.L. Chupp, O.A. Benz, Astrophys. J. Suppl. **90**, 511 (1994)
E.L. Chupp, D.J. Forrest, P.R. Higbie, A.N. Suri, C. Tsai, P.P. Dunphy, Nature **241**, 333 (1973)
E.L. Chupp, J.M. Ryan, Res. Astron. Astrophys. **9**, 11 (2009)
N.B. Crosby, M.J. Aschwanden, B.R. Dennis, Sol. Phys. **143**, 275 (1993)
R.B. Dahlburg, S.K. Antiochos, T.A. Zang, Phys. Fluids B **4**, 3902 (1992)
S. Dalla, P.K. Browning, Astron. Astrophys. **436**, 1103 (2005)
S. Dalla, P.K. Browning, Astrophys. J. Lett. **640**, L99 (2006)
S. Dalla, P.K. Browning, Astron. Astrophys. **491**, 289 (2008)
D.W. Datlowe, R.P. Lin, Sol. Phys. **32**, 459 (1973)
W. Daughton, Phys. Plasmas **6**, 1329 (1999)
W. Daughton, V. Roytershteyn, B.J. Albright, H. Karimabadi, L. Yin, K.J. Bowers, Phys. Rev. Lett. **103**(6), 065004 (2009)
C. Dauphin, N. Vilmer, Astron. Astrophys. **468**, 289 (2007)
L. Davis, Phys. Rev. **101**, 351 (1956)
N. Décamp, F. Malara, Astrophys. J. Lett. **637**, L61 (2006)
P. Demoulin, J.C. Hénoux, E.R. Priest, C.H. Mandrini, Astron. Astrophys. **308**, 643 (1996)
P. Démoulin, E.R. Priest, D.P. Lonie, J. Geophys. Res. **101**, 7631 (1996)
B.R. Dennis, H.S. Hudson, S. Krucker, in *Lecture Notes in Physics*, vol. 725, ed. by K.-L. Klein, A.L. MacKinnon (Springer, Berlin, 2007), p. 33
A. Des Jardins, R. Canfield, D. Longcope, C. Fordyce, S. Waitukaitis, Astrophys. J. **693**, 1628 (2009)
P. Dmitruk, W.H. Matthaeus, J. Geophys. Res. **111**, 12110 (2006)
P. Dmitruk, W.H. Matthaeus, N. Seenu, M.R. Brown, Astrophys. J. Lett. **597**, L81 (2003)
J.F. Drake, D. Biskamp, A. Zeiler, Geophys. Res. Lett. **24**, 2921 (1997)
J.F. Drake, Y.C. Lee, Phys. Rev. Lett. **39**, 453 (1977)
J.F. Drake, M.A. Shay, W. Thongthai, M. Swisdak, Phys. Rev. Lett. **94**(9), 095001 (2005)
J.F. Drake, M. Swisdak, H. Che, M.A. Shay, Nature **443**, 553 (2006a)
J.F. Drake, M. Swisdak, K.M. Schoeffler, B.N. Rogers, S. Kobayashi, Geophys. Res. Lett. **33**, 13105 (2006b)
J.F. Drake, M. Opher, M. Swisdak, J.N. Chamoun, Astrophys. J. **709**, 963 (2010)
G.A. Dulk, D.J. McLean, Sol. Phys. **57**, 279 (1978)
D.C. Ellison et al., in *International Cosmic Ray Conference*, vol. 3 (2005), p. 261
A.G. Emslie, Astrophys. J. **235**, 1055 (1980)
A.G. Emslie, Astrophys. J. **245**, 711 (1981)
A.G. Emslie, Astrophys. J. Lett. **595**, L119 (2003)
A.G. Emslie, P.A. Sturrock, Sol. Phys. **80**, 99 (1982)
A.G. Emslie, J.-C. Hénoux, Astrophys. J. **446**, 371 (1995)
A.G. Emslie, J.C. Brown, A.L. MacKinnon, Astrophys. J. **485**, 430 (1997)
A.G. Emslie, E.P. Kontar, S. Krucker, R.P. Lin, Astrophys. J. Lett. **595**, L107 (2003)
A.G. Emslie, J.A. Miller, J.C. Brown, Astrophys. J. Lett. **602**, L69 (2004)
E. Fermi, Phys. Rev. **75**, 1169 (1949)
L. Fletcher, H.S. Hudson, Astrophys. J. **675**, 1645 (2008)
L. Fletcher, T.R. Metcalf, D. Alexander, D.S. Brown, L.A. Ryder, Astrophys. J. **554**, 451 (2001)
L. Fletcher et al., Space Sci. Rev. (2011). doi:10.1007/s11214-010-9701-8
T.G. Forbes, Astrophys. J. **305**, 553 (1986)
D.J. Forrest, in: *Positron-Electron Pairs in Astrophysics*. American Institute of Physics Conference Series, vol. 101 (1983), p. 3
H.P. Furth, J. Killeen, M.N. Rosenbluth, Phys. Fluids **6**, 459 (1963)
K. Galsgaard, Å. Nordlund, Astrophys. Lett. Commun. **34**, 175 (1996)

K. Galsgaard, I. Roussev, Astron. Astrophys. **383**, 685 (2002)
J. Giacalone, J.R. Jokipii, J. Geophys. Res. **101**, 11095 (1996)
M.L. Goldstein, Astrophys. J. **219**, 700 (1978)
V.S. Gorbachev, B.V. Somov, Sov. Astron. **33**, 57 (1989)
P.C. Grigis, A.O. Benz, Astron. Astrophys. **426**, 1093 (2004)
P.C. Grigis, A.O. Benz, Astron. Astrophys. **434**, 1173 (2005)
P.C. Grigis, A.O. Benz, Astron. Astrophys. **458**, 641 (2006)
R.J. Hamilton, V. Petrosian, Astrophys. J. **398**, 350 (1992)
I. Hannach et al., Space Sci. Rev. (2011, this issue). doi:10.1007/s11214-010-9705-4
J.V. Hollweg, J. Geophys. Res. **79**, 1357 (1974)
G.D. Holman, Astrophys. J. **293**, 584 (1985)
G.D. Holman, L. Sui, R.A. Schwartz, A.G. Emslie, Astrophys. J. Lett. **595**, L97 (2003)
G. Holman et al., Space Sci. Rev. (2011). doi:10.1007/s11214-010-9680-9
J.D. Huba, L.I. Rudakov, Phys. Rev. Lett. **93**(17), 175003 (2004)
H.S. Hudson, Sol. Phys. **24**, 414 (1972)
G.J. Hurford, S. Krucker, R.P. Lin, R.A. Schwartz, G.H. Share, D.M. Smith, Astrophys. J. Lett. **644**, L93 (2006)
G.J. Hurford, R.A. Schwartz, S. Krucker, R.P. Lin, D.M. Smith, N. Vilmer, Astrophys. J. Lett. **595**, L77 (2003)
H. Isliker, L. Vlahos, Phys. Rev. E **67**(2), 026413 (2003)
V. Jayanti, J.V. Hollweg, J. Geophys. Res. **98**, 19049 (1993)
Y.W. Jiang, S. Liu, W. Liu, V. Petrosian, Astrophys. J. **638**, 1140 (2006)
T.W. Jones, Astrophys. J. Suppl. **90**, 969 (1994)
S.R. Kane, E.L. Chupp, D.J. Forrest, G.H. Share, E. Rieger, Astrophys. J. Lett. **300**, L95 (1986)
H. Karimabadi, W. Daughton, J. Scudder, Geophys. Res. Lett. **34**, 13104 (2007)
M. Karlický, M. Bárta, Sol. Phys. **247**, 335 (2008)
M. Karlický, T. Kosugi, Astron. Astrophys. **419**, 1159 (2004)
J. Kašparová, M. Karlický, E.P. Kontar, R.A. Schwartz, B.R. Dennis, Sol. Phys. **232**, 63 (2005)
C.A. Kletzing, J. Geophys. Res. **99**, 11095 (1994)
J.W. Knight, P.A. Sturrock, Astrophys. J. **218**, 306 (1977)
E.P. Kontar, J.C. Brown, Astrophys. J. Lett. **653**, L149 (2006)
E.P. Kontar, A.L. MacKinnon, Sol. Phys. **227**, 299 (2005)
E.P. Kontar, A.G. Emslie, M. Piana, A.M. Massone, J.C. Brown, Sol. Phys. **226**, 317 (2005)
E.P. Kontar, A.G. Emslie, A.M. Massone, M. Piana, J.C. Brown, M. Prato, Astrophys. J. **670**, 857 (2007)
E. Kontar et al., Space Sci. Rev. (2011). doi:10.1007/s11214-011-9804-x
R.A. Kopp, G.W. Pneuman, Sol. Phys. **50**, 85 (1976)
A.G. Kosovichev, V.V. Zharkova, Astrophys. J. Lett. **550**, L105 (2001)
D. Krauss-Varban, D. Burgess, J. Geophys. Res. **96**, 143 (1991)
D. Krauss-Varban, C.S. Wu, J. Geophys. Res. **94**, 15367 (1989)
D. Krauss-Varban, D. Burgess, C.S. Wu, J. Geophys. Res. **94**, 15089 (1989)
S. Krucker, E.P. Kontar, S. Christe, R.P. Lin, Astrophys. J. Lett. **663**, L109 (2007a)
S. Krucker, S.M. White, R.P. Lin, Astrophys. J. Lett. **669**, L49 (2007b)
S. Krucker, M. Battaglia, P.J. Cargill, L. Fletcher, H.S. Hudson, A.L. MacKinnon, S. Masuda, L. Sui, M. Tomczak, A.L. Veronig, L. Vlahos, S.M. White, Astron. Astrophys. Rev. **16**, 155 (2008a)
S. Krucker, P. Saint-Hilaire, S. Christe, S.M. White, A.D. Chavier, S.D. Bale, R.P. Lin, Astrophys. J. **681**, 644 (2008b)
L.D. Landau, E.M. Lifshitz, *Electrodynamics of Continuous Media* (1960)
S.H. Langer, V. Petrosian, Astrophys. J. **215**, 666 (1977)
G. Lapenta, J.U. Brackbill, W.S. Daughton, Phys. Plasmas **10**, 1577 (2003)
T.N. Larosa, A.G. Emslie, Sol. Phys. **120**, 343 (1989)
Y.-T. Lau, J.M. Finn, Astrophys. J. **350**, 672 (1990)
J. Leach, The impulsive hard X-rays from solar flares. Ph.D. thesis, Stanford University (1984)
M.M. Leroy, A. Mangeney, Ann. Geophys. **2**, 449 (1984)
R.P. Lin, S. Krucker, G.J. Hurford, D.M. Smith, H.S. Hudson, G.D. Holman, R.A. Schwartz, B.R. Dennis, G.H. Share, R.J. Murphy, A.G. Emslie, C. Johns-Krull, N. Vilmer, Astrophys. J. Lett. **595**, L69 (2003)
Y.E. Litvinenko, Sol. Phys. **167**, 321 (1996a)
Y.E. Litvinenko, Astrophys. J. **462**, 997 (1996b)
Y.E. Litvinenko, B.V. Somov, Sol. Phys. **146**, 127 (1993a)
Y.E. Litvinenko, B.V. Somov, Sol. Phys. **146**, 127 (1993b)
W. Liu, Characteristics of solar flare hard X-ray emissions: observations and models. Ph.D. thesis, Stanford University (2006)

E.T. Lu, R.J. Hamilton, Astrophys. J. Lett. **380**, L89 (1991)
Z.W. Ma, A. Bhattacharjee, Geophys. Res. Lett. **23**, 1673 (1996)
M.E. Mandt, R.E. Denton, J.F. Drake, Geophys. Res. Lett. **21**, 73 (1994a)
M.E. Mandt, R.E. Denton, J.F. Drake, Geophys. Res. Lett. **21**, 73 (1994b)
G. Mann, H. Aurass, A. Warmuth, Astron. Astrophys. **454**, 969 (2006)
G. Mann, A. Warmuth, H. Aurass, Astron. Astrophys. **494**, 669 (2009)
J.T. Mariska, J.M. McTiernan, Astrophys. J. **514**, 484 (1999)
P.C.H. Martens, A. Young, Astrophys. J. Suppl. **73**, 333 (1990)
S. Masuda, T. Kosugi, H. Hara, S. Tsuneta, Y. Ogawara, Nature **371**, 495 (1994)
K.G. McClements, L. Fletcher, Astrophys. J. **693**, 1494 (2009)
S.W. McIntosh, P. Charbonneau, T.J. Bogdan, H.-L. Liu, J.P. Norman, Phys. Rev. E **65**(4), 046125 (2002)
D.B. Melrose, S.M. White, Proc. - Astron. Soc. Aust. **3**, 369 (1979)
J.A. Miller, D.A. Roberts, Astrophys. J. **452**, 912 (1995)
J.A. Miller, T.N. Larosa, R.L. Moore, Astrophys. J. **461**, 445 (1996)
J.A. Miller, P.J. Cargill, A.G. Emslie, G.D. Holman, B.R. Dennis, T.N. LaRosa, R.M. Winglee, S.G. Benka, S. Tsuneta, J. Geophys. Res. **102**, 14631 (1997)
G.J. Newkirk, Astrophys. J. **133**, 983 (1961)
L. Nocera, I.I. Skrynnikov, B.V. Somov, Sol. Phys. **97**, 81 (1985)
A.V. Oreshina, B.V. Somov, Astron. Lett. **26**, 750 (2000a)
A.V. Oreshina, B.V. Somov, Astron. Lett. **26**, 750 (2000b)
G.K. Parks, J.R. Winckler, Astrophys. J. Lett. **155**, L117+ (1969)
P. Petkaki, A.L. MacKinnon, Sol. Phys. **172**, 279 (1997)
V. Petrosian, T.Q. Donaghy, J.M. McTiernan, Astrophys. J. **569**, 459 (2002)
V. Petrosian, S. Liu, Astrophys. J. **610**, 550 (2004)
V. Petrosian, H. Yan, A. Lazarian, Astrophys. J. **644**, 603 (2006)
H.E. Petschek, NASA Spec. Publ. **50**, 425 (1964)
E.R. Priest, *Solar Magneto-hydrodynamics* (1982)
E.R. Priest, P. Démoulin, J. Geophys. Res. **100**, 23443 (1995)
E. Priest, T. Forbes, *Magnetic Reconnection* (Cambridge University Press, Cambridge, 2000)
E.R. Priest, T.G. Forbes, Astron. Astrophys. Rev. **10**, 313 (2002)
E.R. Priest, V.S. Titov, Philos. Trans. R. Soc. Lond. Ser. A, Math. Phys. Sci. **354**, 2951 (1996)
E.R. Priest, T.N. Bungey, V.S. Titov, Geophys. Astrophys. Fluid Dyn. **84**, 127 (1997)
P.L. Pritchett, F.V. Coroniti, J. Geophys. Res. **109**, 1220 (2004)
P.L. Pritchett, F.V. Coroniti, V.K. Decyk, J. Geophys. Res. **101**, 27413 (1996)
J.M. Pryadko, V. Petrosian, Astrophys. J. **482**, 774 (1997)
J.M. Pryadko, V. Petrosian, Astrophys. J. **495**, 377 (1998)
P. Saint-Hilaire, A.O. Benz, Astron. Astrophys. **435**, 743 (2005)
A.A. Schekochihin, S.C. Cowley, *Turbulence and Magnetic Fields in Astrophysical Plasmas* (Springer, Berlin, 2007)
R. Schlickeiser, J.A. Miller, Astrophys. J. **492**, 352 (1998)
V.D. Shapiro, V.I. Shevchenko, Sov. Sci. Rev., E, Astrophys. Space Phys. Rev. **6**, 425 (1988)
G.H. Share, R.J. Murphy, Wash. DC Am. Geophys. Union. Geophys. Monogr. Ser. **165**, 177 (2006)
M.A. Shay, J.F. Drake, Geophys. Res. Lett. **25**, 3759 (1998)
M.A. Shay, J.F. Drake, M. Swisdak, Phys. Rev. Lett. **99**(15), 155002 (2007)
K. Shibata, S. Masuda, M. Shimojo, H. Hara, T. Yokoyama, S. Tsuneta, T. Kosugi, Y. Ogawara, Astrophys. J. Lett. **451**, L83+ (1995)
A.Y. Shih, R.P. Lin, G.J. Hurford, S.E. Boggs, A.C. Zoglauer, C.B. Wunderer, J.G. Sample, P. Turin, S. McBride, D.M. Smith, H. Tajima, P.N. Luke, M.S. Amman, Bull. Am. Astron. Soc. **41**, 436 (2009a)
A.Y. Shih, R.P. Lin, D.M. Smith, Astrophys. J. Lett. **698**, L152 (2009b)
T.V. Siversky, V.V. Zharkova, J. Plasma Phys. (2009)
B.V. Somov (ed.), *Physical Processes in Solar Flares*. Astrophysics and Space Science Library, vol. 172 (1992)
B.V. Somov (ed.), *Cosmic Plasma Physics*. Astrophysics and Space Science Library, vol. 251 (2000)
B.V. Somov, T. Kosugi, Astrophys. J. **485**, 859 (1997)
B.U.O. Sonnerup, in *Magnetospheric Boundary Layers*, ed. by B. Battrick, J. Mort, G. Haerendel, J. Ortner. ESA Special Publication, vol. 148 (1979), p. 395
D.S. Spicer, Sol. Phys. **53**, 305 (1977)
L. Spitzer, *Physics of Fully Ionized Gases* (1962)
K. Stasiewicz, P. Bellan, C. Chaston, C. Kletzing, R. Lysak, J. Maggs, O. Pokhotelov, C. Seyler, P. Shukla, L. Stenflo, A. Streltsov, J.-E. Wahlund, Space Sci. Rev. **92**, 423 (2000)
J.J. Sudol, J.W. Harvey, Astrophys. J. **635**, 647 (2005)

L. Sui, G.D. Holman, Astrophys. J. Lett. **596**, L251 (2003)
P.A. Sweet, Annu. Rev. Astron. Astrophys. **7**, 149 (1969)
S.I. Syrovatskii, in *Problems of Solar-Terrestrial Relationships* (Ylym, Askhabad, 1981), pp. 21–41
T. Takakura, T. Kosugi, T. Sakao, K. Makishima, M. Inda-Koide, S. Masuda, Publ. Astron. Soc. Jpn. **47**, 355 (1995)
E. Tandberg-Hanssen, A.G. Emslie, *The Physics of Solar Flares* (Cambridge University Press, Cambridge, 1988)
V.S. Titov, G. Hornig, Adv. Space Res. **29**, 1087 (2002)
G. Trottet, N. Vilmer, C. Barat, A. Benz, A. Magun, A. Kuznetsov, R. Sunyaev, O. Terekhov, Astron. Astrophys. **334**, 1099 (1998)
D. Tsiklauri, T. Haruki, Phys. Plasmas **14**(11), 112905 (2007)
S. Tsuneta, T. Naito, Astrophys. J. Lett. **495**, L67+ (1998)
R. Turkmani, U. Torkelsson, Astron. Astrophys. **409**, 813 (2003)
R. Turkmani, L. Vlahos, K. Galsgaard, P.J. Cargill, H. Isliker, Astrophys. J. Lett. **620**, L59 (2005)
R. Turkmani, P.J. Cargill, K. Galsgaard, L. Vlahos, H. Isliker, Astron. Astrophys. **449**, 749 (2006)
G.H.J. van den Oord, Astron. Astrophys. **234**, 496 (1990)
G.E. Vekstein, P.K. Browning, Phys. Plasmas **4**, 2261 (1997)
A.B. Verboncoeur, J.P. Langdon, N.T. Gladd, Comput. Phys. Commun. **87**, 199 (1995)
N. Vilmer, G. Trottet, C. Barat, R.A. Schwartz, S. Enome, A. Kuznetsov, R. Sunyaev, O. Terekhov, Astron. Astrophys. **342**, 575 (1999)
N. Vilmer, A.L. MacKinnon, G. Trottet, C. Barat, Astron. Astrophys. **412**, 865 (2003)
N. Vilmer et al., Space Sci. Rev. (2011). doi:10.1007/s11214-010-9728-x
L. Vlahos, H. Isliker, F. Lepreti, Astrophys. J. **608**, 540 (2004)
H. Wang, C. Liu, Y. Deng, H. Zhang, Astrophys. J. **627**, 1031 (2005)
A. Warmuth, G. Mann, H. Aurass, Cent. Eur. Astrophys. Bull. **31**, 135 (2007)
A. Warmuth, G. Mann, H. Aurass, Astron. Astrophys. **494**, 677 (2009)
S. White et al., Space Sci. Rev. (2011). doi:10.1007/s11214-010-9708-1
P. Wood, T. Neukirch, Sol. Phys. **226**, 73 (2005)
C.C. Wu, Geophys. Res. Lett. **10**, 545 (1983)
A. Zeiler, D. Biskamp, J.F. Drake, B.N. Rogers, M.A. Shay, M. Scholer, J. Geophys. Res. **107**, 1230 (2002)
V.V. Zharkova, O.V. Agapitov, J. Plasma Phys. **75**, 159 (2009)
V.V. Zharkova, M. Gordovskyy, Astrophys. J. **604**, 884 (2004)
V.V. Zharkova, M. Gordovskyy, Mon. Not. R. Astron. Soc. **356**, 1107 (2005a)
V.V. Zharkova, M. Gordovskyy, Space Sci. Rev. **121**, 165 (2005b)
V.V. Zharkova, M. Gordovskyy, Astron. Astrophys. **432**, 1033 (2005c)
V.V. Zharkova, M. Gordovskyy, Astrophys. J. **651**, 553 (2006)
V.V. Zharkova, S.I. Zharkov, S.S. Ipson, A.K. Benkhalil, J. Geophys. Res. **110**, 8104 (2005)
Z. Zhu, R.M. Winglee, J. Geophys. Res. **101**, 4885 (1996)

Space Sci Rev (2011) 159:421–445
DOI 10.1007/s11214-011-9801-0

Energy Release and Particle Acceleration in Flares: Summary and Future Prospects

R.P. Lin

Received: 15 February 2011 / Accepted: 10 May 2011 / Published online: 23 July 2011

Abstract *RHESSI* measurements relevant to the fundamental processes of energy release and particle acceleration in flares are summarized. *RHESSI's* precise measurements of hard X-ray continuum spectra enable model-independent deconvolution to obtain the parent electron spectrum. Taking into account the effects of albedo, these show that the low energy cut-off to the electron power-law spectrum is typically $\lesssim$tens of keV, confirming that the accelerated electrons contain a large fraction of the energy released in flares. *RHESSI* has detected a high coronal hard X-ray source that is filled with accelerated electrons whose energy density is comparable to the magnetic-field energy density. This suggests an efficient conversion of energy, previously stored in the magnetic field, into the bulk acceleration of electrons. A new, collisionless (Hall) magnetic reconnection process has been identified through theory and simulations, and directly observed in space and in the laboratory; it should occur in the solar corona as well, with a reconnection rate fast enough for the energy release in flares. The reconnection process could result in the formation of multiple elongated magnetic islands, that then collapse to bulk-accelerate the electrons, rapidly enough to produce the observed hard X-ray emissions. *RHESSI's* pioneering γ-ray line imaging of energetic ions, revealing footpoints straddling a flare loop arcade, has provided strong evidence that ion acceleration is also related to magnetic reconnection. Flare particle acceleration is shown to have a close relationship to impulsive Solar Energetic Particle (SEP) events observed in the interplanetary medium, and also to both fast coronal mass ejections and gradual SEP events. New instrumentation to provide the high sensitivity and wide dynamic range hard X-ray and γ-ray measurements, plus energetic neutral atom (ENA) imaging of SEPs above $\sim$2 $R_{\odot}$, will enable the next great leap forward in understanding particle acceleration and energy release is large solar eruptions—solar flares and associated fast coronal mass ejections (CMEs).

Keywords Sun: flares · Sun: X-rays · Sun: acceleration · Sun: energetic particles

R.P. Lin (✉)
Physics Department and Space Sciences Laboratory, UC Berkeley, Berkeley, CA 94720, USA
e-mail: rlin@ssl.berkeley.edu

R.P. Lin
School of Space Research, Kyung Hee University, Yongin, Gyeonggi 446-701, Korea

1 Introduction

Large solar flares are the most powerful explosions in the solar system, releasing up to 10^{32-33} ergs in 10^{2-3} s. They and their associated fast coronal mass ejections (CMEs) are the most energetic particle accelerators in the solar system, producing ions up to tens of GeV and electrons to hundreds of MeV. For flares, the accelerated particles often appear to contain the bulk of the total energy released in the flare, a remarkable efficiency that indicates that the particle-acceleration and energy-release processes are intimately related. Much of the observed flare impulsive-phase phenomena appear to be the result of the interaction of these accelerated particles with the ambient medium. Fast CMEs drive shock waves that accelerate solar energetic particles (SEPs) observed near 1 AU with efficiency of order 10%. Similar processes are believed to occur elsewhere in the universe, in stellar flares (e.g., Osten et al. 2007), magnetars (e.g., Hurley et al. 2005), young circumstellar disks, supernovae shock waves, etc. Solar flares and CMEs are the most accessible laboratories for understanding the fundamental physics of transient energy release and efficient particle acceleration in cosmic magnetized plasmas. Furthermore, these large solar eruptions produce the most extreme forms of space weather—the radiation hazard from the most intense SEP fluxes, and the disruption of the heliospheric plasma environment.

The first observation of a solar flare was made by Carrington (1859) in white light. The first evidence that the Sun could accelerate particles to high energies came from the detection of a ground-level event (GLE) in a cosmic-ray sensor, reported by Forbush (1946), who noted that it occurred near the time of a solar flare. Optical studies showed that flares typically occurred near sunspots, in regions of strong magnetic field, consistent with the release of energy stored in magnetic fields (see Giovanelli 1948). Swept-frequency radio observations provided evidence for the acceleration and escape of fast electrons (Wild et al. 1963) in flares. Soft X-ray (SXR) emission from a solar flare, indicating the presence of hot, $\sim 10^7$ K thermal plasma, was discovered by Chubb et al. (1957). Flare hard X-ray (HXR) emission was first reported by Peterson and Winckler (1958), and γ-ray line emission by Chupp et al. (1973); these radiations are generated through bremsstrahlung by energetic electrons and nuclear interactions of energetic ions, respectively, colliding with the ambient solar atmosphere. Since the cross-sections are known, and the solar atmosphere is optically thin to these energetic emissions, quantitative information about the parent electrons (see Kontar et al. 2011) and ions (see Vilmer et al. 2011), can be derived from the HXR and γ-ray line measurements.

For non-thermal particles with energy E much greater than kT (the average thermal energy of the ambient medium), the energy lost to Coulomb collisions is many orders of magnitude greater than the energy lost to HXRs and/or γ-rays; see Sect. 4 of Dennis et al. (2011, this issue). Assuming the particles lose all their energy to Coulomb collisions (collisional thick-target), the energy going into the source electrons and ions can be directly determined from the observed HXR and γ-ray line emissions (see Holman et al. 2011). In the idealized situation where the acceleration of the particles can be considered separate (e.g., in the tenuous corona) from their loss to collisions (in the dense chromosphere and photosphere), the spectra of the accelerated electrons and ions can be inferred from the observed HXR and γ-ray spectra, respectively. Typically, the flux spectra are fit to power-laws in energy, $F(E) = AE^{-\delta}$, where A and δ are constants; the total energy in electrons and in ions depends critically on how low in energy the power-law extends. This is the low-energy cutoff or roll-off parameter. For power laws extending down to $\sim$20 keV for the electrons, and down to $\sim$1 MeV for the ions, the accelerated electrons and ions each would contain a significant fraction, $\sim$10–50%, of the total energy released in flares (Lin and Hudson 1976). The rate of acceleration of tens-of-keV electrons required in a large flare would

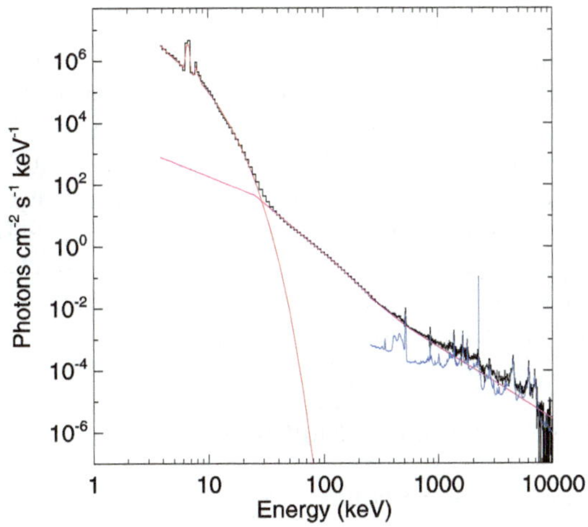

Fig. 1 *RHESSI* measurement of the energy spectrum of SOL2002-07-23 (X4.8) from 3 keV to 10 MeV. At energies below ~30 keV, the emission is dominated by thermal plasmas with temperatures up to ~40 MK, while accelerated electrons produce the spectrum detected above ~30 keV. Narrow and broad γ-ray line emissions produced by accelerated ions are observed from ~0.5 MeV to ~8 MeV (*blue line*). In the hard X-ray range, the *red line* shows the thermal component, and the magenta line the HXR power-law component (with cutoff)

be $\sim 10^{36}$ s^{-1}, an enormous number equivalent to an impossibly large current of $\sim 10^{17}$ A. This has led to models suggesting many filaments with oppositely directed currents, or with return currents (see Zharkova et al. 2011), or electron acceleration directly in the dense chromosphere (e.g., Fletcher and Hudson 2008).

With the poor energy resolution of early flare HXR measurements, the spectra could be fit to a thermal spectrum with $T \approx 10^{8-9}$ K (Crannell et al. 1978). Then $E \approx kT$, and Coulomb collisions would primarily exchange energy between electrons of comparable energy. In principle, there could be little or no net collisional loss, and the flare energy going into accelerated electrons could be orders of magnitude less than the total energy released in the flare. Flare HXR imaging has shown, however, that most of the HXR emission came from footpoints, co-spatial with chromospheric Hα brightenings and other low-temperature flare emissions, so $E \gg kT$ (see Fletcher et al. 2011). Furthermore, the first measurements with high spectral resolution showed that the hottest thermal plasmas had $T \leq 40$ MK (Lin et al. 1981), so the electrons producing HXR emission in coronal sources above ~20 keV are mostly non-thermal ($E \gtrsim 5kT$).

RHESSI, the *Reuven Ramaty High Energy Solar Spectroscopic Imager* mission (Lin et al. 2002) was designed to investigate how the Sun releases the energy for flares, presumably stored in the magnetic fields of the corona, and how electrons and ions are rapidly accelerated to high energies with such high efficiency. *RHESSI* provides high resolution imaging and spectroscopy of the X-rays and γ-rays that are emitted by the energetic electrons and ions in the flare. The measurements (Fig. 1) span almost four orders of magnitude in photon energy and more than 12 orders in flux, from the intense soft X-rays (SXRs) produced by hot thermal plasmas, through the HXRs emitted by accelerated electrons, to γ-ray line emission by accelerated ions. The cryogenically cooled germanium detectors (GeDs) provide uniquely high spectral resolution: ~1 keV FWHM (full width at half maximum) up to ≥200 keV, increasing to ~2 keV at the 511 keV positron-annihilation line and ~7 keV at 8 MeV, sufficient to resolve all the nuclear γ-ray lines except the 2.223 MeV neutron-capture deuterium line (≤0.1 keV FWHM intrinsic width). *RHESSI* provides HXR and γ-ray imaging (the first above 100 keV and the first ever for γ-ray lines) with the finest spatial resolution ever achieved (~2″ from 3 to ~100 keV, ~4″ up to ~300 keV, ~35″ up to 17 MeV), utilizing rotating modulation collimators (RMCs) that convert the spatial information into a temporal modulation of the count rates.

RHESSI observations span a broad range of solar flare (and quiet Sun) phenomena, described in the other chapters of this book. Here we summarize those results relevant to the fundamental processes of energy release and particle acceleration in flares, and synthesize them into a present understanding of the physics of flares. Some of the key questions addressed here are: How much energy is contained in the flare-accelerated electrons and ions? Where and how are the electrons and ions accelerated? Where and how is the energy released? What is the relationship of particle acceleration in flares related to the impulsive SEP events observed in the interplanetary medium? How are large flares related to fast CMEs and the acceleration of SEPs in large gradual events? Finally, we then discuss future research directions and prospects.

2 Flare Acceleration of Electrons

2.1 The Electron Spectrum

As discussed in Kontar et al. (2011), the observed flare HXR emission depends on the source electron spectrum, convolved with the bremsstrahlung cross-section, multiplied by the ambient density, and integrated over the line of sight. In the past, typically a source electron spectral shape was assumed, and then forward-fit to the HXR observations. Johns and Lin (1992) showed that the HXR continuum measurements can be directly deconvolved in a model-independent way to obtain the source electron spectrum (but crucially dependent on the statistics of the observations); they obtained source electron spectra for the HXR measurements of SOL1980-06-07T03:22 (M7.3), the first hard X-ray flare observed with high spectral resolution (Lin and Johns 1993).

For intense solar flares, *RHESSI* provides excellent statistics together with $\sim$1 keV FWHM resolution for the most precise HXR continuum measurements ever obtained from an astrophysical source. This has enabled routine, model-independent deconvolution of the HXR spectrum by powerful newly-developed mathematical techniques (e.g., regularized inversion; see Kontar et al. 2011) to obtain the spectrum of the source electrons. The observed HXR spectrum also depends on the angular distribution of the source electrons, and on the Compton scattering of the source HXRs by the solar photosphere, i.e., albedo. The latter can be significant, especially at deka-keV energies. An elegant Green's function approach to evaluating the albedo contribution that is independent of the primary spectrum has been developed (Kontar et al. 2006).

The number of electrons and the total energy they contain depends critically on the low-energy cutoff. *RHESSI's* spectral resolution is sufficient to resolve the steep high energy fall-off of hot flare thermal continuum (e-folding of $\sim$2 keV), allowing the precise determination of the energy above which the HXR emission must be non-thermal (see Holman et al. 2011). The newly-developed deconvolution methods have been applied (Kašparová et al. 2005) to flares with non-thermal HXR spectra that showed flattening at low energies (as expected for a cutoff or roll-off in the electron source spectrum spectra). For this initial sample of *RHESSI*-observed flares, the derived electron source spectra appeared to have a roll-off around 20–40 keV, but all these flares were located close to the solar disk center where the effects of albedo are strongest. After correcting for the expected albedo, all the derived source electron spectra extend in a power law with no roll-off down to where the hot flare thermal emission dominates, typically $\leq$20 keV and sometimes as low as $\sim$12 keV for these flares (Kontar et al. 2008). Low-energy cutoffs were also found for a number of other flares (Sui et al. 2005, 2007) in the $\sim$15–50 keV range, implying that the non-thermal electrons indeed must contain a large fraction of the energy released in many flares.

2.2 Early Phase

In the X4.8 flare SOL2002-07-23T00:35 (and several other large flares) *RHESSI* detected a weak HXR source high in the corona beginning ∼9 min before the impulsive phase, and prior to the detection of any footpoint HXR emission (Lin et al. 2002), or of any chromospheric counterpart in *TRACE* 195 Å, *SOHO*/MDI white light, or Hα (Krucker et al. 2003). The lack of footpoint emission indicates that the thermal source has its origin in the corona and does not come from chromospheric evaporation. Such a weak source could not have been detected by previous solar HXR instruments that had fixed windows optimized for the peak emission of large flares; *RHESSI* inserts attenuators to absorb low energy X-rays as the count rate increases, so its sensitivity is much higher when one or both of the attenuators are out. Both thermal (superhot) plasma and non-thermal electrons are detected (Fig. 2), with a much steeper (softer) energy spectrum than for electrons accelerated in the flare impulsive phase. As the pre-impulsive phase progresses, the coronal source appears to move downward, as expected when stored magnetic energy is being released by shortening of magnetic field lines. Some footpoint HXR emission appears later, but generally at much lower intensity than the coronal HXR source. In this phase only a small fraction of total flare energy is released. This pre-impulsive source may be due in part to the initial energization by magnetic reconnection process itself (to be discussed later).

2.3 Energy Release

During the impulsive phase, the rate $d\Phi/dt$ of magnetic flux Φ being reconnected in the corona can be inferred from the observed apparent velocity, v_{fp}, of the HXR footpoints (assumed to be magnetically connected to the photosphere just below; see Fig. 3) and the measured photospheric magnetic field, B_{fp}. The flux change $d\Phi/dt = v_c B_c a_c$ must be equal to $v_{fp} B_{fp} a_{fp}$, if the magnetic field is frozen to the plasma (see Fig. 4, left). Here a_{fp} is the footpoint width perpendicular to its motion, as given by the imaging, and v_c, B_c, and a_c are the velocity, magnetic field, and width in the corona respectively. Note that in three dimensions the convection electric field in the corona is $E_c = v_c B_c = v_{fp} B_{fp}(a_{fp}/a_c)$, as opposed to $E_c = v_c B_c = v_{fp} B_{fp}$ for the 2-D case. Figure 4 (right) shows that the HXR flux at 50 keV is roughly proportional to the measured reconnection rate $v_{fp} B_{fp} a_{fp}$ for the X10 flare SOL2003-10-29T20:49 (X10.0) (Krucker et al. 2005). The HXR flux is also roughly correlated with $v_{fp} B_{fp}^2$ (see Fig. 3.12 of Fletcher et al. 2011), consistent with models where a significant fraction of the magnetic energy released by reconnection goes to accelerating electrons. The correlation is better after ∼20:48 UT (red symbols in Fig. 4, right) when the geometry is close to the simplest 2-D reconnection model with just two HXR sources; earlier in the flare, the HXR emission is complex with at least two sources on each flare ribbon, and the scatter of points is much greater. For this flare the convection electric field $E_c = v_c B_c$ is as large as $6000(a_{fp}/a_c)\ \mathrm{V\,m^{-1}}$, and the reconnection rate reaches as high as $d\Phi/dt \sim 10^{18}\ \mathrm{Mx\,s^{-1}}$. Reconnection rates of the same order of magnitude are obtained for other large flares, together with similar rough linear correlation with the flare HXR fluxes. E_c is generally perpendicular to the magnetic field, and it should map down to the chromosphere, but if the reconnecting fields are not anti-parallel, there will be a component of E_c parallel to B in the reconnection region. These strong electric fields may be important for particle acceleration.

2.4 Energy Deposition

Where does the energy deposited by the non-thermal electrons into the solar atmosphere go? As discussed in Fletcher et al. (2011), energy deposition by energetic electrons at the

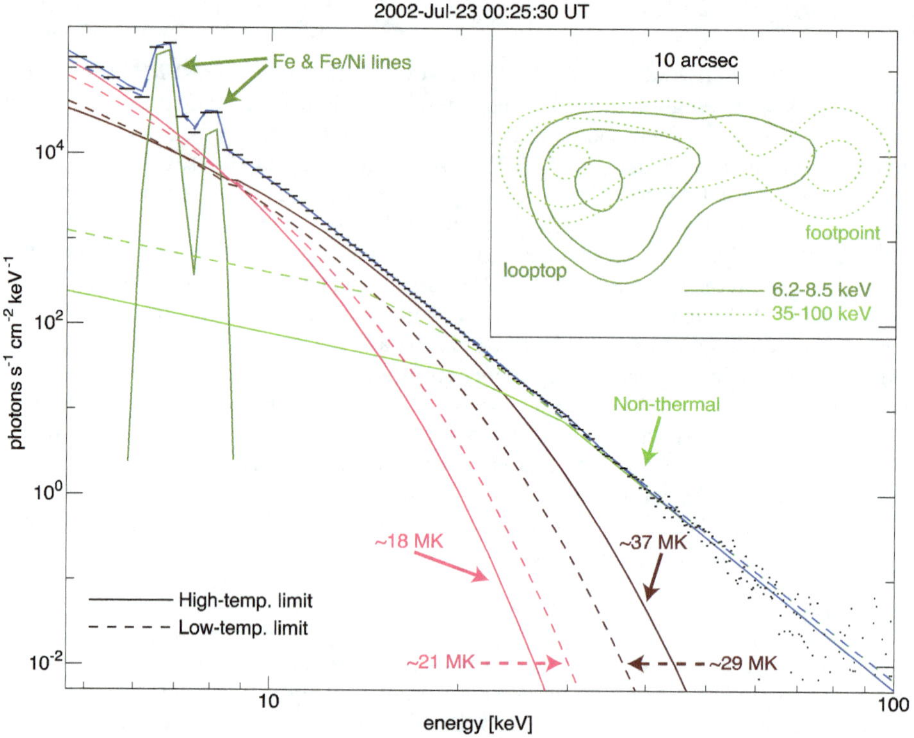

Fig. 2 Photon flux spectra (*black*) during the peak of the pre-impulsive phase of SOL2002-07-23 (~00:25:30 UT), with two acceptable model fits showing the upper (*solid*) and lower (*dashed*) temperature limits of the super-hot component (*brick*), as constrained by the Fe and Fe/Ni lines (*olive*). Cool isothermal (*magenta*) and non-thermal (*green*) components are also required. (*Inset*) 30%, 50%, and 90% contours of 6.2–8.5 keV (*black solid*) and 35–100 keV (*red dotted*) images at the same time. The peak non-thermal emission appears to be above the thermal looptop; the faint footpoint contains only ~16% of the total non-thermal flux within the 50% contour (from Caspi and Lin 2010)

top of the chromosphere produces hot thermal flare plasma through the process of chromospheric evaporation. Heating lower in the atmosphere is expected to result in very substantial amounts of radiation. The *SORCE* mission provided the first measurements precise enough to detect increases in total solar irradiance (TSI) for flares. For the very large X17 flare SOL2003-10-28, Woods et al. (2004) found an increase in TSI with a time profile that shows an impulsive component similar to the HXRs, as well as a gradual component similar to the soft X-rays. The maximum of the TSI increase ($\sim 10^{30}$ erg s^{-1}) is of the same order of magnitude as that expected from the energy deposited by the non-thermal electrons, and it occurs at the impulsive-component peak, prior to the SXR peak. Thus, at the present time, the observations are consistent with a large fraction of the total energy in a flare being released in the form of accelerated electrons (and ions) during the impulsive phase; these then heat the ambient atmosphere through collisions, evaporating chromospheric material to form the hot (above tens of MK) flare plasma, and heating the deeper solar atmosphere with the energy escaping as radiation. Most other impulsive phase flare phenomena (such as radio emission, etc.) can be explained as the consequence of the interaction of the accelerated particles with the ambient medium.

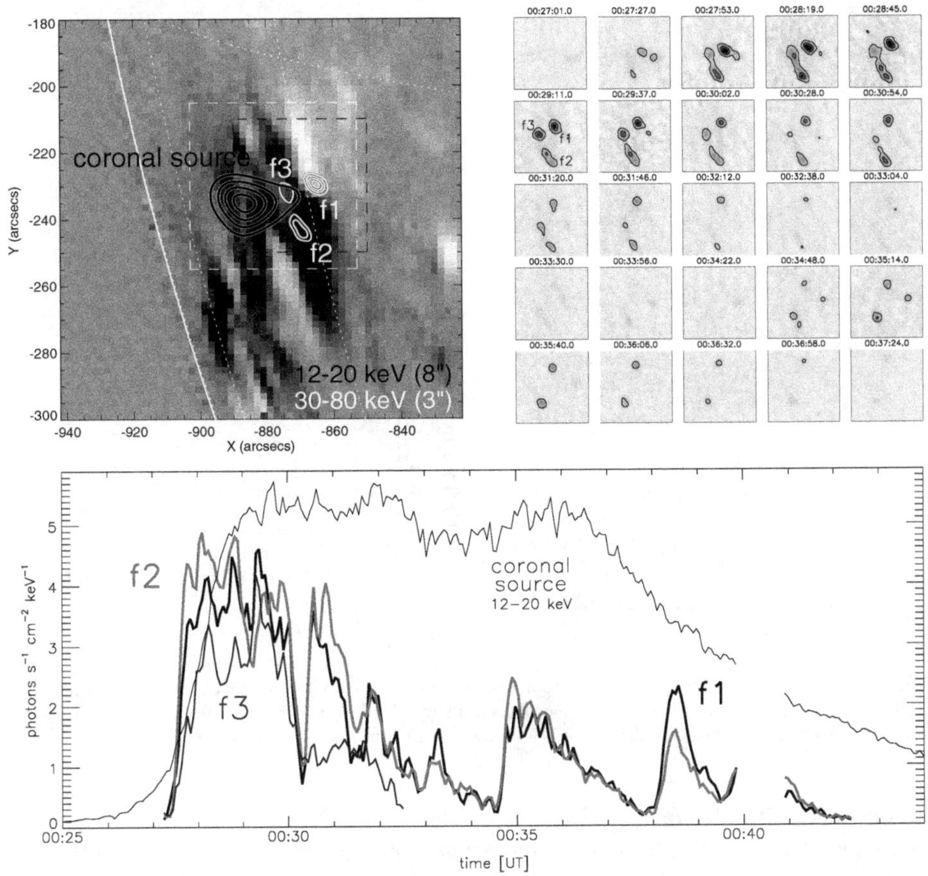

Fig. 3 Location and temporal variation of the different HXR sources observed in the γ-ray flare SOL2002-07-23 (X4.8). *Top left:* the three main 30–80 keV HXR footpoint sources (*white contours*) and the thermal plasma (12–20 keV, *black contours*), superimposed on the pre-flare magnetogram (MDI). *Top right:* 30–80 keV images over the area indicated by the *black dashed line* in the *top left panel*, taken with 26 s integration time, scaled to the maximum intensity of the time series. *Bottom:* flux-vs-time profiles of the three HXR (30–80 keV) footpoints, plus the coronal thermal source (*thin curve*, 12–20 keV flux divided by 1500) from images taken every 4 s (from Krucker et al. 2003)

The HXR spectra typically fit a double powerlaw with a downward break at an energy of tens of keV, reaching up to $\geq$100 keV in large flares. This break is often surprisingly sharp, but not unphysically so (Conway et al. 2003). As discussed in Holman et al. (2011), the break could be due to non-uniform ionization of the atmosphere (from the fully ionized corona to the neutral chromosphere) where the electrons lose their energy to collisions, or due to return-current energy loss or other wave-particle interactions, or due to the acceleration process itself. In impulsive solar electron events observed at $\sim$1 AU, the electron energy spectra are also observed to be a double powerlaw with a break (Krucker et al. 2007a).

RHESSI also provides Fourier-transform HXR imaging with the finest spatial resolution ever achieved ($\sim 2''$ from 3 to $\gtrsim$100 keV). Figure 3 (top right) shows a temporal sequence of HXR images for SOL2002-07-23T00:35 (X4.8), revealing three footpoints that are cospatial with flare Hα brightenings (Asai et al. 2004). This implies non-thermal electrons as an

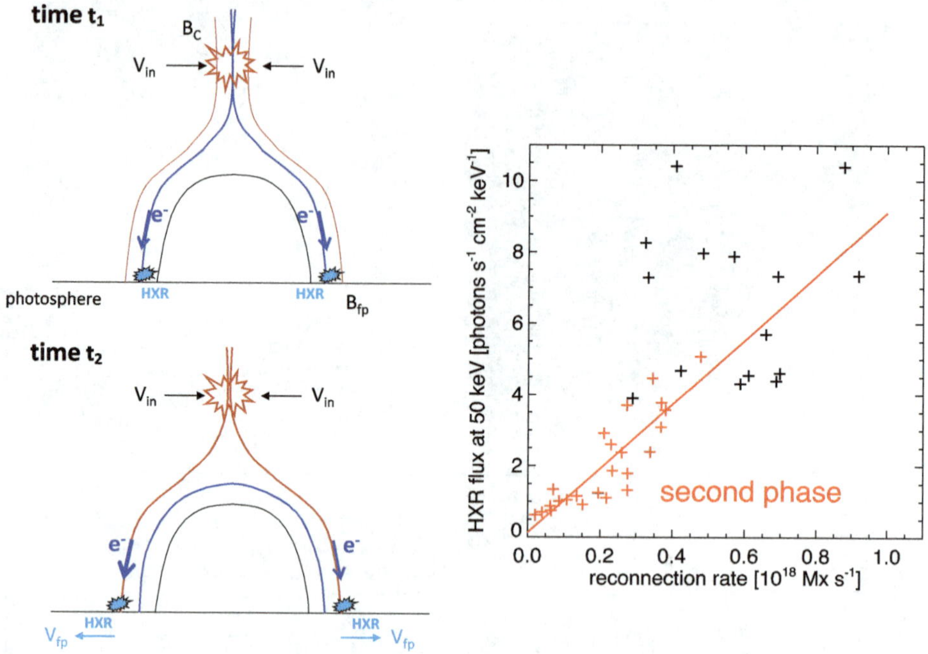

Fig. 4 *Left*: schematic of reconnecting field lines in the corona, showing the relationship to measurements in the chromosphere/photosphere. *Right*: the observed hard X-ray flux at 50 keV plotted versus the rate of magnetic flux being reconnected for SOL2003-10-29 (from Krucker et al. 2005)

energy source. The bottom panel shows that the HXR flux variations with time are generally the same for each of the three footpoints, with no obvious delays (within the few-second temporal resolution) of the peaks from one footpoint to another. Sakao et al. (2000) cross-correlated the HXR fluxes in paired footpoints of other flares observed by *Yohkoh*, and showed that the HXR emissions were simultaneous to within a few tenths of a second. These observations are consistent with the "standard" CSHKP[1] model of solar flares (e.g., Fletcher et al. 2011), where magnetic reconnection in roughly oppositely-directed magnetic fields above the closed loops forms a new loop (Fig. 4, left) in the corona and accelerates electrons which, because of their high speed ($\gtrsim 0.3c$), stream down the new loop and arrive at the two footpoints essentially simultaneously. This is a stringent constraint on models where the electron acceleration occurs in the chromosphere, as advocated by Fletcher and Hudson (2008); their Alfvén waves (proposed to transport the energy to the chromosphere) must have velocities comparable to $\sim 0.3c$.

2.5 Electron Acceleration Region

RHESSI's HXR imaging (and previous measurements) are generally consistent with the collisional thick-target model (see Sect. 7, Holman et al. 2011). Many lines of reasoning point to the corona as the location of the flare energy release and particle acceleration sites, and the coronal magnetic field as the source of the energy that powers flares. *RHESSI*

[1] Carmichael, Sturrock, Hirayama, Kopp & Pneuman.

sometimes detects two relatively weak coronal X-ray sources with opposing energy gradients (see Fig. 4.2, Fletcher et al. 2011), one above the other (Sui and Holman 2003; Liu et al. 2008), implying the energy release site lies between them at altitudes of ∼9–23 Mm, as expected for a current sheet formed between the top of flare loops and the coronal source. Using high-sensitivity HXR observations from the BATSE instrument on the *Compton Gamma-ray Observatory (CGO)*, Aschwanden et al. (1995) found a delay of low-energy HXRs for short-duration bursts, the delays being consistent with time-of-flight for the parent electrons from a high coronal source down to the chromospheric footpoints. Furthermore, several analyses of the HXR footpoint height as a function of energy show that the centroids of the emissions are at lower altitudes for higher energies, consistent with the parent electrons being injected from above the chromosphere (e.g., Aschwanden et al. 2002). In the well-known Masuda et al. (1994) flare, a HXR source was detected above the soft X-ray looptop. With the limited dynamic range of *Yohkoh*/HXT and *RHESSI*, however, weak coronal HXR sources are difficult to detect in the presence of the much brighter footpoint sources, but systematic studies of flares whose footpoints are occulted show that coronal HXR emission is commonly present (Krucker and Lin 2008).

More recently, *RHESSI* has detected an intense coronal HXR source located about 6 Mm above the soft X-ray flare loops (Fig. 5, middle and right upper panels) in SOL2007-12-31T01:11 (C8.3), where the footpoints were occulted, but with *STEREO B* providing unocculted EUV imaging of the whole flare (Krucker et al. 2010). This coronal HXR source shows several impulsive bursts at 20–50 keV (Fig. 5, left middle panel), with a power-law spectrum extending to >80 keV, with no detectable thermal emission (Fig. 5, lower right panel). Thus, essentially all the electrons in this source have been accelerated. The HXR imaging shows a source of volume $\sim 10^{27}$ cm^3, with a non-thermal, >16 keV electron density of $\sim 2 \times 10^9$ cm^{-3} for a total of $N(>16\ \text{keV}) \approx 2 \times 10^{36}$ electrons in the source (see Table 1, center column).

Microwave imaging at multiple wavelengths (from Nobeyama) shows a co-spatial source (Fig. 5, lower middle panel), consistent with gyrosynchrotron emission from the high-energy tail (electron energies of ∼1 MeV) of the same power-law electron distribution. The magnetic field in the source, estimated from the turnover in the radio spectrum, is ∼30–50 G, implying that the energy density of the non-thermal electrons in the source is comparable to that of the magnetic field, i.e., the plasma $\beta \approx 1$. Prior to the flare, no HXR emission was detected above background from this location; if we assume the density is the same as during the flare, the pre-flare plasma $\beta \approx 0.01$.

The impact of these observations on theoretical models of particle acceleration has been summarized in Zharkova et al. (2011). Particle-in-cell simulations of reconnection (Drake et al. 2006b) show that narrow current layers form at the X-line and produce secondary magnetic islands (Fig. 6b). Drake et al. (2006a) argue that in three dimensions the islands should be volume-filling, and that electrons should be efficiently accelerated, primarily by the contraction of the initially squashed magnetic islands (Fig. 6b), e.g, converting elongated magnetic fields to more potential fields. Some acceleration may also come from parallel electric fields in the reconnection process. Electrons circulating rapidly within the islands gain energy through a Fermi process, by reflecting off the ends of the islands as they move inward at the Alfvén speed; they interact with multiple islands to reach high energies and produce a powerlaw spectrum. This process operates for pre-event $\beta \ll 1$. The island contraction ceases when the energetic electron pressure approaches the local magnetic energy density, i.e., $\beta \sim 1$. Up to 60% of the released magnetic energy can be transferred to the electrons in the process.

As pointed out previously, a major issue is the supply of large numbers of electrons per second. The rate of loss of electrons from this source can be estimated by $N(>16\ \text{keV})/\tau_{\text{loss}}$,

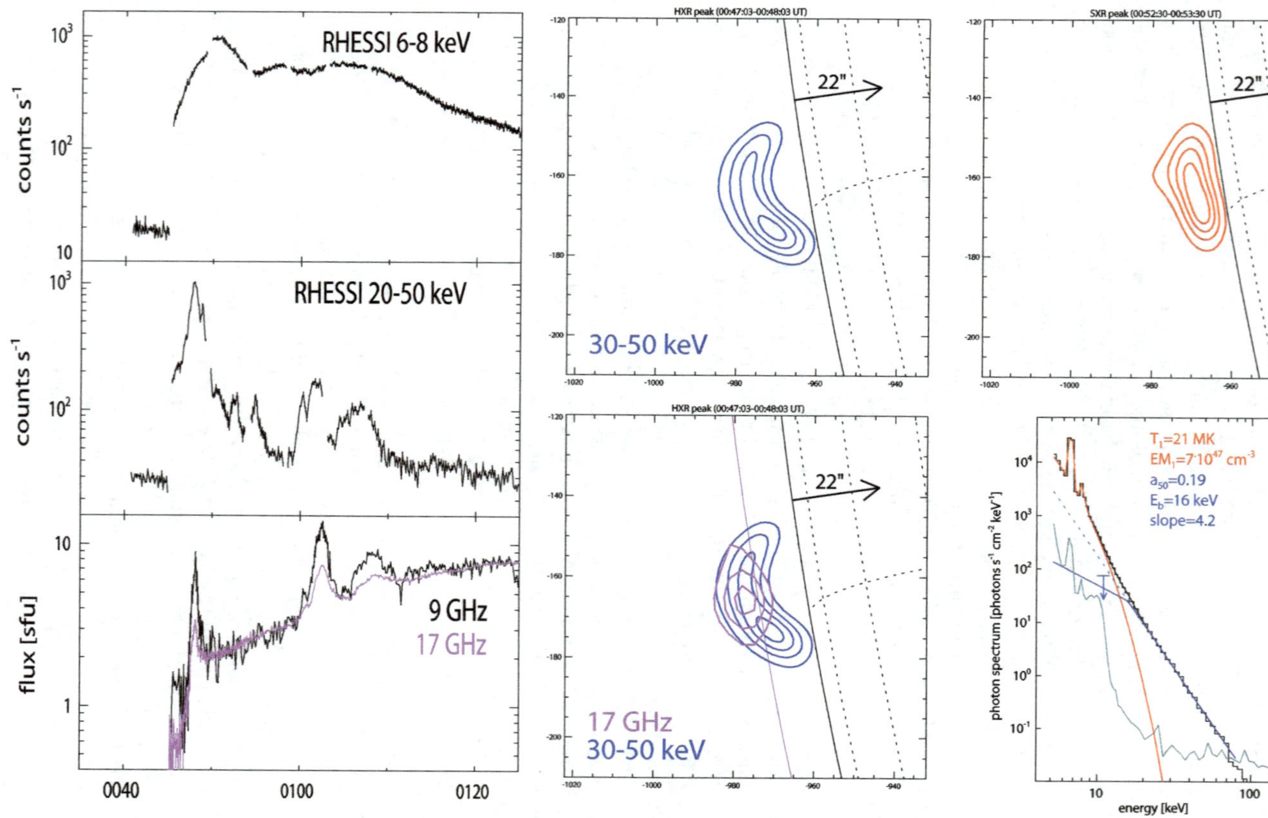

Fig. 5 HXR (*RHESSI*) and radio (Nobeyama) observations of coronal emission from the over-the-limb flare SOL2007-12-31T01:11 (C8.3). *Left*: flux vs. time profiles of thermal 6–8 keV X-rays (*top panel*), 20–50 keV HXRs (*middle panel*), 9 and 17 GHz radio (*bottom panel*). *Middle*: top, a 30–50 keV HXR image with limb indicated; *bottom*, a 17 GHz radio image (*purple*) with the radio limb indicated, superimposed on the HXR image. *Right*: top, Image of the soft X-ray source located below the HXR source; bottom, burst HXR energy spectrum (*blue*) with pre-burst SXR thermal spectrum (*red*)

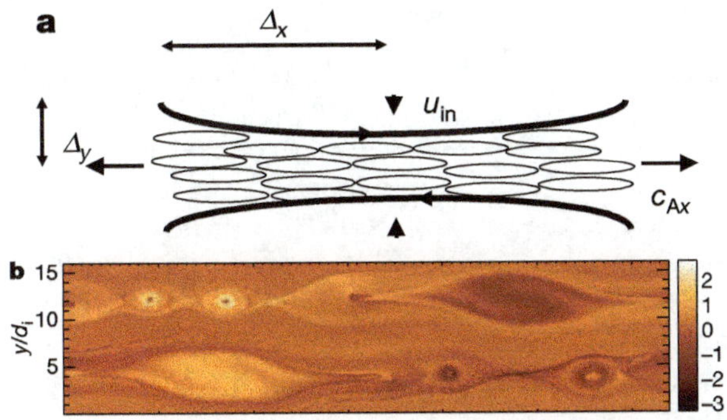

Fig. 6 (**a**) Diagram showing volume-filling islands expected around the field reversal region. (**b**) Particle-in-cell simulation of island formation during magnetic reconnection; shown here is the electron out-of-plane current at time $t = 20\ \Omega_{ci}^{-1}$ where Ω_{ci} is the ion cyclotron frequency (from Drake et al. 2006a)

and ranges from a maximum of 5×10^{37} electrons s^{-1} for $\tau_{loss} \approx 0.4$ s (the time for a 16-keV electron to traverse the source) to a minimum of $\sim 3 \times 10^{35}$ electrons s^{-1} for $\tau_{loss} \approx 20$ s (the fastest observed HXR decay time). Assuming the reconnection inflow speed is $\sim 0.1 v_A$ (discussed later), where v_A is the Alfvén speed, then the rate that electrons are brought into the reconnection region from both sides can be estimated as $dN/dt = 2 \times 0.1 n_e A v_A$, where A is the cross-sectional area of the inflow stream, assumed to arrive from both sides. For the parameters of this flare (Table 1), $v_A = 2.2 \times 10^8$ cm s^{-1}, and $dN/dt \approx 8 \times 10^{35}$ electrons s^{-1}. These numbers of electrons being convected into the acceleration region are sufficient to produce the footpoint HXR emission typical for this class (*GOES* M2) flare, provided the bulk of the electrons are accelerated, as suggested by these observations. Thus, the number of electrons and the rate that they can be accelerated may be sufficient, in principle, to produce the footpoint HXR emission in a large flare; this above-the-looptop source can be identified as the acceleration region for the flare.

Another above-the-looptop coronal source was observed in HXRs together with one footpoint source for one of the bursts in the event SOL2003-10-22T20:07 (M9.9) (Ishikawa et al. 2011). The energy spectrum of the electrons in the coronal source (inferred from the thin-target HXRs) has a power-law spectral index consistent with that required to produce the footpoint HXR (thick-target) emission as they lose their energy to collisions (see Table 1, right column). The rate (electrons s^{-1}) required to produce the footpoint HXR emission would empty the coronal source in about 4 s, compared to the 2 min duration of the burst. For comparison, in the Masuda flare the coronal source would be emptied in ~ 0.5 s compared to the ~ 2 min burst duration (Table 1, left column). These examples are consistent with the bulk acceleration of electrons in the corona through a mechanism involving magnetic reconnection, with an inflow speed of $\sim 0.1\ v_A$, but higher sensitivity and dynamic range are required to test whether these flares are unusual or whether this is common to all flares. The observed wide range of flares, however, suggest that perhaps a variety of possible mechanisms may be operating, ranging from energy release/particle acceleration in the coronal magnetic reconnection region, to particle acceleration in the chromosphere driven by Alfvén waves from reconnection region (as for the Earth's aurora; Fletcher and Hudson 2008).

3 Flare-Accelerated Ions

Information on the energy spectrum of the energetic ions is derived from ratios of γ-ray lines whose cross-sections have different energy thresholds. The handful of γ-ray lines that have been detected with sufficient statistics in flares provide information only for protons

Table 1 Coronal hard X-ray sources

Parameter	SOL1992-01-13 (Masuda flare)[a]	SOL2007-12-31	SOL2003-10-22 (peak 3)
Burst duration	~2 min	~2 min	~2 min
Fastest decay (1/e) time	~30 s	~20 s	~30 s
Energy range	~25–50 keV	~16–80 keV	~30–80 keV
Flux at 50 keV ph $(cm^2\,s\,keV)^{-1}$	~0.02	~0.2	~0.2
Height above photosphere	~20 Mm	~27 Mm	~25 Mm
Height above flare loop	~7 Mm	~6 Mm	~6 Mm
Length	~5 Mm	~29 Mm	~11 Mm
Width	~5 Mm	~6 Mm	~6 Mm
Volume[b]	$\sim 1 \times 10^{26}$ cm^3	$\sim 8 \times 10^{26}$ cm^3	$\sim 4 \times 10^{26}$ cm^3
Pre-flare ambient density[c]	low	low	low
Electron spectral index (thin-target hard X-rays)	~3–4.5	~3.7	~4.6
Non-thermal electron density	$\sim 2 \times 10^9$ cm^{-3}	$\sim 2 \times 10^9$ cm^{-3}	$\sim 1 \times 10^{10}$ cm^{-3}
Number of electrons >16 keV	$\sim 2 \times 10^{35}$	$\sim 2 \times 10^{36}$	$\sim 4 \times 10^{36}$
Energy, non-thermal electrons[d]	$>1 \times 10^{28}$ erg	$>1 \times 10^{29}$ erg	$>3 \times 10^{29}$ erg
Radio flux at 17 GHz	~600 SFU	~1.7 SFU	
Electron spectral index (radio)		~3.4	
Typical electron energy (17 GHz emission)		~1.2 MeV	
Magnetic field strength		~30–50 G	
Pre-flare plasma β[e]		~0.01	
Plasma β in HXR source		~1	
Footpoint X-ray flux at 50 keV	~0.1 ph $(cm^2\,s\,keV)^{-1}$		~0.1 ph $(cm^2\,s\,keV)^{-1}$
Footpoint area			12 Mm^2
Footpoint electron spectral index (thick target)	~4.2–5.0		~4.6
Footpoint electron loss rate (>16 keV)	$\sim 4 \times 10^{35}$ e$^-$ s^{-1}		$\sim 1 \times 10^{35}$ e$^-$ s^{-1}
Footpoint energy deposition rate (>16 keV)	$>2 \times 10^{28}$ erg s^{-1}		$>6 \times 10^{27}$ erg s^{-1}
Footpoint energy flux			$>5 \times 10^{10}$ erg $(cm^2\,s)^{-1}$

[a]Parameters taken from Masuda et al. (2000)

[b]Volume = length × width × width

[c]For the case that all electrons are accelerated, the pre-flare density is as given below

[d]Derived assuming that all electrons are accelerated

[e]Assumes a pre-flare temperature of 2 MK

above ~2.5 MeV. The inferred energy spectra of the accelerated ions, and the total energy contained in the ions, also depend on the composition and angular distributions of those ions; and on the composition, magnetic field, density, temperature, and scattering properties of the ambient medium—thus multi-parameter models must be utilized and the results can vary by up to an order of magnitude (see Sect. 2.1 of Vilmer et al. 2011). Assuming that the ion power-law spectra extend down to $\sim$1 MeV and the composition of the accelerated ions is the same as for impulsive SEPs events observed at 1 AU (i.e., with strong enhancements of heavy ions), the total energy contained in flare-accelerated ions is found to be comparable to the energy in flare-accelerated electrons above $\sim$20 keV and thus a substantial fraction of the total energy released in the flare (Ramaty et al. 1995; Ramaty and Mandzhavidze 2000; Lin et al. 2003).

RHESSI has provided a major breakthrough with the first imaging of energetic ions in flares (see Sect. 5 of Vilmer et al. 2011 for details), using the strong, narrow 2.223 MeV neutron-capture line that is produced primarily by >30 MeV protons. Five flares have been imaged to date (Hurford et al. 2003, 2006) and the γ-ray line emission is found to come entirely from compact unresolved ($\leq 35''$, *RHESSI's* γ-ray angular resolution) sources located in the flare region, with upper limits of order $\sim$10% for larger-scale diffuse emission. This clearly shows that the acceleration of the parent ions is predominantly flare-related and, for the most part, not due to acceleration by widespread shock waves (but see Vestrand and Forrest 1993), such as the fast CME-driven shocks that appear to accelerate the SEPs in large gradual SEP events observed near 1 AU.

For SOL2003-10-28T11:10 (X17), two γ-ray line sources are detected straddling the arcade of flare loops, strongly suggesting that the acceleration of ions is also associated with magnetic reconnection, similar to the acceleration of electrons. The two γ-ray line sources have about the same separation between footpoints and about the same orientation as the two >0.2 MeV HXR sources observed in the flare (contrary to expectations of stochastic acceleration models where ions are accelerated in larger loops; Emslie et al. 2004), but both γ-ray line sources are significantly displaced, by $\sim 15''$, from the HXR sources. Gradient and curvature drifts in a simple loop arcade field would produce a displacement much smaller than observed, although in the same direction. In the other four flares, two γ-ray line sources straddling the flare loops would not be resolvable, given the count statistics and flare loops widths. For SOL2002-07-23T00:35 (X4.8) the centroid of the γ-ray line source is displaced by $\sim 25''$ from the HXR source centroid. For two other flares, the γ-ray line sources appeared to be associated with one of the two HXR footpoints, although statistics for one flare were marginal; for the fifth flare the statistics were too poor to tell. These differences in spatial morphology between electrons and ions are surprising, given the otherwise close correlation between the 2.223 MeV line fluences and the >0.3 MeV electron bremsstrahlung fluences (discussed below). This may provide a clue to the acceleration and/or propagation of the two species, suggesting that electric fields may play a role (Litvinenko and Somov 1993; Zharkova and Gordovskyy 2004).

By comparing all the γ-ray flares detected by *RHESSI*, Shih et al. (2009b) showed that the fluence of the 2.223 MeV γ-ray line (produced by $\geq$30 MeV protons) is linearly proportional to the >0.3 MeV bremsstrahlung continuum fluence (produced by >0.3 MeV electrons), over more than three orders of magnitude in fluence, from the limit of detectability to the most intense flares (Fig. 2.18 of Vilmer et al. 2011). This strongly suggests that a single mechanism accelerates both $\geq$30 MeV protons and relativistic, >0.3 MeV electrons. A similarly close proportionality is observed between ion acceleration, given by either the 2.223 MeV line fluence, Fig. 7 (left), or the 4–8 MeV fluence, Fig. 7 (right), and tens-of-keV electron acceleration (either the fluence of 50 keV HXRs, shown in Fig. 7, right, or the peak *GOES* 1–8 Å soft X-ray flux, shown in Fig. 7, left; the peak soft X-ray flux is a proxy for the HXR fluence—i.e., the Neupert effect), but only for the most powerful ion-accelerating flares, i.e., those producing more than $\sim 2 \times 10^{31}$ protons $\geq$30 MeV. Less-powerful ion-accelerating flares, however, show a large uncorrelated excess acceleration of tens-of-keV

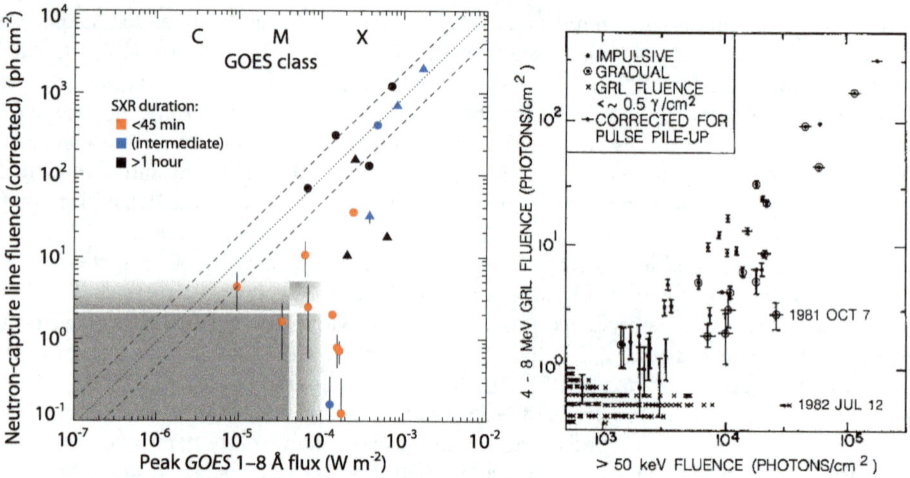

Fig. 7 *Left*: the 2.223 MeV neutron-capture line fluence (indicative of the number of $\geq$30 MeV ions) plotted versus the peak *GOES* soft X-ray flux (proportional to the non-thermal hard X-ray flux through the Neupert effect) for *RHESSI* flares. The *circles* (*triangles*) represent flares with complete (incomplete) coverage. The *dotted* and *dashed line* illustrate the direct proportionality observed for the flares with the largest line fluences. The *shaded area* has not been systematically searched (from Shih et al. 2009b). *Right:* the 4–8 MeV γ-ray line plus continuum fluence (indicative of the number of $\geq$10 MeV ions) measured by the GRS instrument on SMM, plotted versus the >50 keV hard X-ray fluence, showing direct proportionality between >50 keV fluence and γ-ray line fluence for the flares with the largest line fluences (from Cliver et al. 1994)

electrons, such that even the weakest detectable ion acceleration is associated with ~M-class or larger flares.

The observations thus are consistent with two acceleration processes: one that always accelerates both >30 MeV protons and >0.3 MeV electrons proportionally, and a second that accelerates the tens-of-keV electrons that heat the thermal flare plasma. These processes both occur in the flare impulsive phase, but the ion/relativistic electron acceleration may be dependent on substantial acceleration of tens-of-keV electrons (to *GOES* M class). For the flares that are the most powerful ion accelerators, however, the fraction of energy going to tens-of-keV electron acceleration reaches a constant minimum.

4 Collisionless Magnetic Reconnection and Flare Energy Release

The process of magnetic reconnection was first proposed by Giovanelli (1948) to explain the release of magnetic energy to power a solar flare (although his discussion was in terms of currents). Sweet (1969) proposed the "neutral point theory" for flares and Parker (1957) made the first consistent calculation of the magnetic reconnection rate for collisional resistivity. The Sweet-Parker reconnection rate, however, is many orders of magnitude too slow for the energy release in solar flares. In the last decade, significant progress has been made in the understanding of collisionless magnetic reconnection, through extensive theory and simulation work, i.e., the GEM challenge (Birn and Hesse 2001), *in situ* space measurements, and laboratory studies. In the collisionless case, the ions, with their large gyrodiameters, decouple from the magnetic field first, forming an ion diffusion region (Fig. 8a), while the electrons, with small gyrodiameters, are still tied to the magnetic field until they get much closer, eventually forming an electron diffusion region where they decouple from

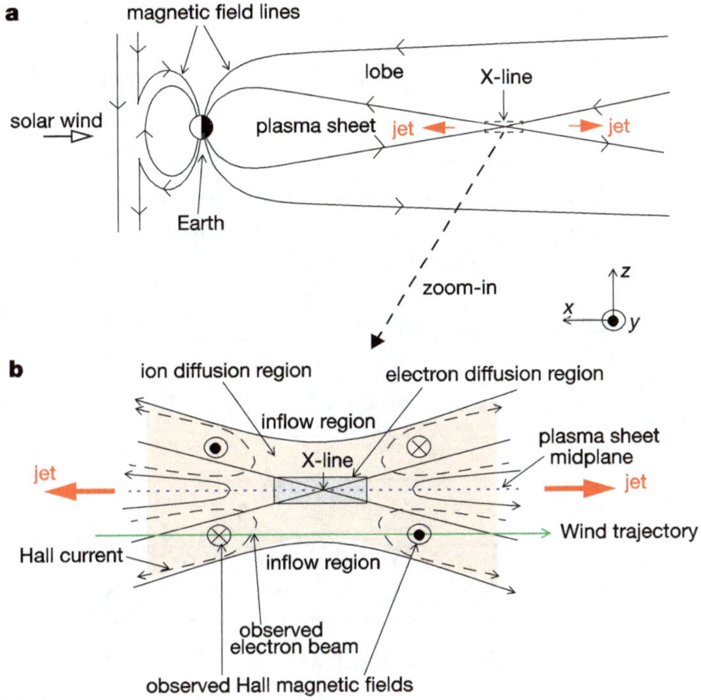

Fig. 8 (**a**) The geometry of the magnetic reconnection event observed in the Earth's distant magnetotail; (**b**) the ion diffusion region

the field; the different ion and electron motions generate a current that produces a characteristic quadrupolar Hall magnetic field (Fig. 8b). The Hall reconnection rate from theory and simulations (Birn and Hesse 2001) is expected to be of order 0.1 v_A, compared to $\sim 10^{-7} v_A$ for Sweet-Parker reconnection under solar coronal conditions (Cassak et al. 2006).

Direct evidence for Hall reconnection occurring in nature was first obtained from *in situ* plasma and field observations from the *WIND* spacecraft (see Fig. 9) in the distant magnetotail ($\sim$60 R_e) of the Earth (Øieroset et al. 2001). The Earth's magnetosphere/magnetotail (Fig. 8a) is similar in many respects to the standard picture of a solar flare. The inner magnetosphere has closed, nearly dipolar, magnetic loops, about the same size as the loops for a large solar flare. The solar wind drags the outer field lines of the magnetosphere into a magnetic tail where the field is roughly oppositely directed in the north and south tail lobes, with a current sheet in between. Magnetic energy is stored in this tail, whose field strength is much larger than that of a potential dipole field. Transient magnetic reconnection occurring in the magnetotail leads to the release of the stored energy to produce magnetospheric substorms (Angelopoulos et al. 2008). Figure 9, panel b, shows that the *WIND* spacecraft crossed from a 100–200 $\mathrm{km\,s^{-1}}$ Earthward-directed plasma jet directly to a region of $\sim$200 $\mathrm{km\,s^{-1}}$ tailward-directed jet, implying that the spacecraft remained in the reconnection layer, crossing from one outflow jet to the other, oppositely-directed outflow jet. The quadrupolar out-of-plane magnetic fields that are the signature of Hall reconnection are clearly evident (Fig. 9, panel d).

Soon afterwards, Hall reconnection was observed by the *POLAR* spacecraft near the subsolar point of the magnetosphere (Mozer et al. 2002), and the reconnection rate was

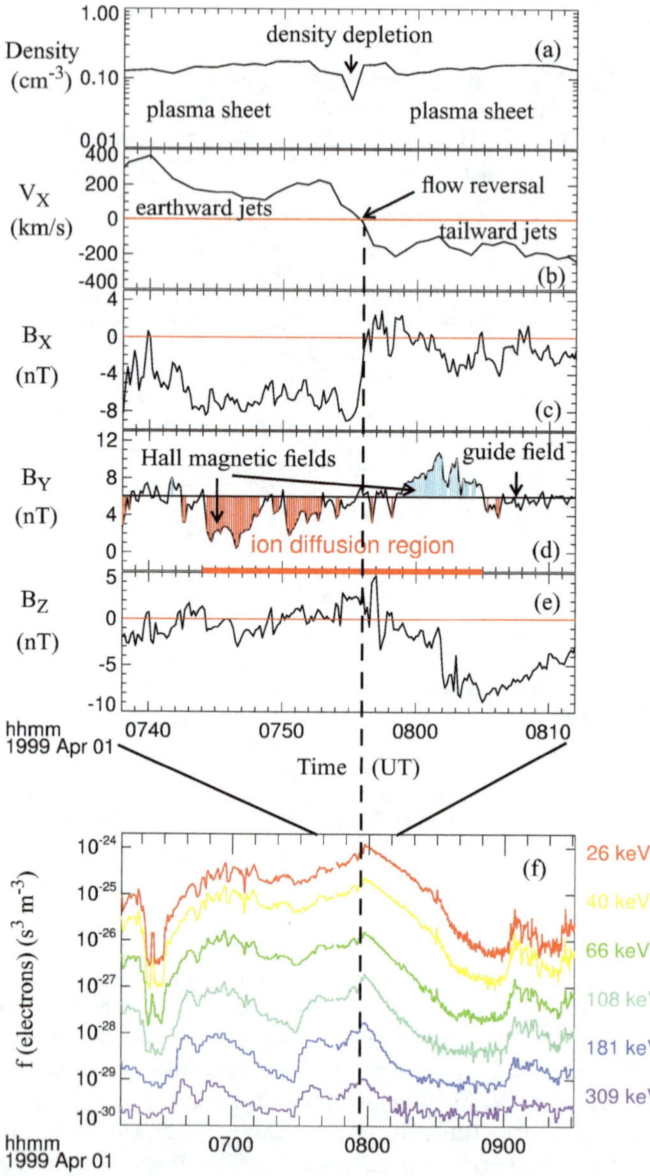

Fig. 9 Observations by the *WIND* spacecraft in the ion diffusion region (indicated by *red bar* in panel **d**) of a magnetic reconnection event (from Øieroset et al. 2001), identified from the large 'out-of-plane' Hall magnetic fields (B_y component, panel **d**) surrounding the flow (proton V_x in panel **b**) reversal. Panels **c**, **d**, & **e** show the three components of the magnetic field. Bipolar B_y field variations with polarities consistent with the Hall magnetic field pattern (see Fig. 8) were detected as the spacecraft crossed from Earthward side (*red shading*) to the tailward side (*blue shading*) of the flow reversal region. Panel **f** shows that the phase space densities for 20 to ~300 keV electrons peak in the ion diffusion region and decrease monotonically away from that region, with their spectrum softening with distance away

measured to be $\sim 0.02\ v_A$ by *in situ* electric-field measurements. Since then, many other *in situ* space measurements of Hall reconnection in the Earth's magnetosphere have been reported (see Eastwood et al. 2010, for a summary), and even detected at Mars (Halekas et al. 2009), indicating that this process commonly occurs in the collisionless plasmas found in nature. Remarkably, Hall reconnection has also been observed now in several laboratory plasma experiments (see the reviews by Yamada et al. 2010 and Zweibel and Yamada 2009) where conditions are completely different—density: $n \approx 0.1$ vs. $\sim 10^{14}\ \text{cm}^{-3}$, magnetic field: $B \approx 10^{-4}$ vs. ~ 100 G, temperature: $kT_e \approx 400$ vs. a few eV, scale size: ~ 100–1000 km vs. a few cm. This suggests that Hall reconnection is a fundamental process that can occur over a very wide range of plasma parameters, presumably including conditions in the solar corona. Laboratory measurements have also shown the rapid increase in reconnection rate as the plasma parameters are varied from collisional to collisionless (Ji et al. 1998).

The *WIND* spacecraft also observed intense fluxes of electrons up to ~ 300 keV energy (Fig. 9f) that peak in the ion diffusion region (Øieroset et al. 2002) of the distant magnetotail Hall reconnection event; the energetic >12 keV electron fluxes are the highest observed in the ~ 10 hour period that *WIND* was in the plasma sheet. The energetic electron fluxes decrease monotonically away from this region and their spectrum softens, strongly suggesting significant electron acceleration is occurring in the ion diffusion region. No increase in energetic, >30 keV ion fluxes, however, is observed.

Drake et al. (2006a) applied their model for electron acceleration by secondary islands from reconnection to this event—they predict a power-law spectral index of ~ 3.7 for the accelerated electron spectrum, remarkably close to the observed index of 3.8. More recently, for a reconnection event in the Earth's magnetotail observed *in situ* by the 4-spacecraft *Cluster* mission, Chen et al. (2008) showed that energetic electron bursts up to many tens of keV were closely correlated with multiple islands. Thus, electron acceleration appears to be associated with magnetic reconnection in the Earth's magnetotail, possibly with magnetic islands; we note, however, that other acceleration processes involving electric fields (Egedal et al. 2008) or coalescence of magnetic islands (Oka et al. 2010) also can explain these observations.

Cassak et al. (2005) noted that both Sweet-Parker and Hall reconnection are stable solutions for a given plasma regime. The slow rate of Sweet-Parker reconnection allows energy to be stored up in the magnetic field, and then a sudden transition to fast Hall reconnection can occur spontaneously when the dissipation region becomes thin enough—from convection of stronger magnetic fields into that region (Cassak et al. 2006)—thus resulting in a sudden rapid energy release. For the Sun, Longcope et al. (2005), in a study of *TRACE* observations of an active region reconnecting with a new flux loop, found evidence that energy was accumulated in the corona during a ~ 24 hour phase of slow reconnection (e.g., Sweet-Parker), followed by a phase of fast reconnection with an estimated rate of $\sim 0.05\ v_A$, presumably due to Hall reconnection in which the energy stored was released.

The magnetic reconnection region itself is likely to be very small, however, compared to the length of the elongated field lines in the standard flare picture or in the Earth's magnetotail, so only a very small fraction of the stored energy is released in the reconnection process itself. Some magnetic energy will be converted directly to particle energy through $P = |\mathbf{E}|I$, where $\mathbf{E} = \mathbf{V} \times \mathbf{B}$ is the convection electric field in the inflowing plasma. The current, I, can be obtained from Ampère's law. This goes into energizing the plasma, in principle increasing the average particle energy by $E_{\text{part}} = B^2/8\pi n$ (see, e.g., Lin et al. 1977). At present we cannot predict theoretically how much of this energy goes to thermal plasma and how much goes to accelerating non-thermal particles. The high coronal HXR source detected by *RHESSI* during the pre-impulsive phase of SOL2002-07-23T00:35 (X4.8) (and several other large flares) may be the result of energization by this initial reconnection (perhaps with some additional acceleration by the collapsing magnetic field); these observations show that both thermal (superhot) plasma and accelerated electrons are produced (Fig. 2). The primary effect of the magnetic-reconnection process, however, is to change the topology of the field from anti-parallel fields to highly elongated turbulence, loops or islands; then the bulk of the free energy in the magnetic field can be released as the turbulence or loops or islands relax to more potential configurations.

5 Connecting the Sun and the Heliosphere

5.1 Flares and Impulsive SEPs

The most common particle accelerations by the Sun (up to $\sim 10^4$ per year over the whole Sun near solar maximum) are impulsive solar ~1–100 keV electron events observed in the interplanetary medium (see Lin 1985; Wang et al. 2011). The SXR burst, when present, has a duration ≤ 1 hour, hence the term "impulsive." These events are accompanied by low-energy (~0.01–1 MeV/nucleon) ion emissions with large enhancements of ^{3}He (^{3}He/^{4}He ratio sometimes ≥ 1), heavy (e.g., Fe) and ultra-heavy (up to ~200 amu) ions, with high charge (e.g., Fe^{+20}) states (see Mason 2007, for a review). These events generally have relatively low particle fluxes, extend over ~30–60° in longitude, last for hours, and have an association (~99%) with type-III radio bursts observed in the ~10 kHz–14 MHz frequency range (Wang et al. 2011). The very low energies of impulsive event electrons (typically ~1 keV but sometimes reaching ~0.1 keV) suggests that the acceleration may be occurring high in the corona (Lin et al. 1995), thus these events are often referred to as "coronal flares" (Lin 1985). *RHESSI* (with attenuators out) has detected very weak 3–15 keV X-ray bursts in coincidence with type-III radio bursts (Christe et al. 2008), suggesting a coronal explosion. Associations with jets observed in EUV that occur close to a coronal hole boundary (Wang et al. 2006), and with fast (≥ 600 km s^{-1}) narrow ($\leq 20°$) CMEs (Kahler et al. 2001; Haggerty and Roelof 2002; Wang et al. 2011) have also been reported.

The e/p ratios defined as $J_e[0.5\text{ MeV}]/J_p[10\text{ MeV}]$ for impulsive events, where J_e and J_p are the electron and proton fluxes, respectively, are comparable to those measured for γ-ray line flares (Ramaty et al. 1993; Shih et al. 2009b). Furthermore, the detailed analysis of the γ-ray spectrum of the flare SOL1981-04-2727T09:45 (X5.5) suggests that the composition of the flare-accelerated heavy ions is enhanced, similar to the composition in impulsive SEP events (Ramaty et al. 1993, 1997; Mandzhavidze et al. 1997) and thus leading to the paradigm that these are due to flare acceleration (Reames 1995). Only ~25% of impulsive events, however, have an associated flare reported by *GOES* (although the association is much higher when compared with SXR bursts detected by *RHESSI*; see Krucker et al. 2007b), and only a very few events have been detected from γ-ray line flares; *RHESSI* has not detected γ-ray line emission from the flares associated with the impulsive events.

By assuming that the impulsive event electrons and ions at all energies travel the same path length L from the Sun to the observation site (typically near ~1 AU)—i.e., $L = v(E)(t(E) - t_0)$, both the injection time t_0 and L can be derived from the observed event onset times for particles with different velocities (Lin 1974). Surprisingly, for most (~80%) of the events, the derived injection times at energies above ~20 keV are delayed by ~10–30 minutes from the start of the associated type III burst (Krucker et al. 1999; Haggerty and Roelof 2002), while ~20% of the events are prompt (no delay). Interestingly, in the delayed events, the injection of lower-energy (∼1–10 keV) electrons begins before the start of the associated type III burst at the Sun (Wang et al. 2006). As these electrons escape from the Sun, their velocity dispersion results in a bump-on-tail distribution that generates Langmuir waves that in turn produce the associated type III radio burst. Consistent with this picture, intense Langmuir waves are detected *in situ* at ∼1 AU only when these ∼1–10 keV electrons arrive (Ergun et al. 1998).

The energy spectra of impulsive event electrons are typically double power laws with a downward break at tens of keV (Krucker et al. 2009b), very similar in shape to the non-thermal HXR spectra typically observed by *RHESSI* for flares. For prompt events, a direct comparison of *RHESSI* HXR spectrum with the escaping electron spectrum above ∼50 keV (above the break) shows that the power-law spectral indices are linearly correlated with $\gamma_x \approx \delta_e$ for prompt events (and uncorrelated for delayed events). This is clearly different from the $\gamma_x + 1 \approx \delta_e$ that would be expected if the electrons escaping to the interplanetary medium come from the single acceleration high in the corona, while most of the accelerated electrons go down to the chromosphere to produce the HXR emission in losing all their energy to collisions (thick target). The number of electrons escaping the Sun to produce the impulsive event is typically only of order ∼0.1% of the number required to produce the HXR burst.

RHESSI HXR imaging of the flares associated with impulsive events (Glesener et al. 2010) indicates that reconnection between open field lines and closed loops (interchange reconnection—see Heyvaerts et al. 1977; Shibata et al. 1996) is involved. This is consistent with a model where the electrons accelerated in this reconnection process that go upward escape to the interplanetary medium, while those going downward are trapped on closed field lines and are further accelerated as the field lines shrink (e.g., Karlický and Kosugi 2004) modifying the spectrum and increasing the number of electrons above 50 keV to produce the observed HXR emission. Thus, the escaping electrons (and perhaps the escaping ions) may be a sample of the particle energization in the flare magnetic reconnection process; note that the electron spectrum below ∼50 keV at ∼1 AU could also have been modified by wave-particle interactions in propagating to ∼1 AU (e.g., Reid and Kontar 2010).

5.2 Flares, CMEs, and Gradual SEP Events

The most powerful ion accelerators (sometimes ions reach 10 to ∼100 GeV, the most energetic particles accelerated in the solar system) are large gradual (associated flare SXR burst ≥1 hour duration) SEP events, that occur on average about once per month near solar maximum, and are dominated by intense fluxes of >10 MeV protons with e/p ratios about two orders of magnitude lower than the ratios for γ-ray line flares (Kallenrode et al. 1992; Cliver and Ling 2007; Shih et al. 2009b). These events have generally normal coronal abundances and ionization states (see Cliver 2009, for a review) extend over ∼100–180° in solar longitude, last for days, and are closely associated with fast CMEs and with type II radio bursts, suggesting that the SEPs are accelerated by shocks driven by fast CMEs and not by flares (Reames 1995). Fast CMEs are essentially always accompanied by simultaneous large flares; these solar eruptive events are the most powerful explosions and most energetic particle accelerators in the solar system, and they produce the most extreme space weather. The observed delays in gradual SEP event onset times (Kahler 1994), and the agreement of composition and temporal variations of the particle fluxes (Tylka and Lee 2006) with theoretical models indicate that diffusive shock acceleration (the same process that is believed to accelerate galactic cosmic rays in supernovae shock waves) at altitudes of ∼2–40 $R_\odot$ is responsible for gradual SEPs. For large, fast CMEs, the number of accelerated ions is sometimes significantly larger than the number required to produce the γ-ray line emission in the associated flare, and the total energy in the SEPs is typically of order ~10% of the total CME energy that is dominated by the kinetic energy of the fast ejecta (see Fig. 6.3 of Fletcher et al. 2011). Such a high efficiency is required for supernovae shock acceleration to produce galactic cosmic rays.

Recent studies show that acceleration profiles of fast CMEs are synchronous with, and closely resemble, the flare energy release as measured by the *RHESSI* hard X-ray flux, (Fig. 5.2 of Fletcher et al. 2011); this is consistent with the standard model for large solar eruptions of magnetic reconnection occurring in a current sheet behind the CME. *RHESSI* has detected a large (~1.5×10^5 km diameter) high coronal non-thermal HXR source that is expanding (at ~400 km s^{-1}) and moving outward at ~750 km s^{-1} behind a very fast (~2300 km s^{-1}) CME whose associated flare was ~40° behind the limb (Krucker et al. 2007b). The HXRs are emitted by non-thermal, >10 keV electrons (constituting ~10% of the total electron density) trapped in closed magnetic structures related to the CME. Furthermore, *RHESSI* has detected non-thermal HXR emission from every fast (>1500 km s^{-1}) "backside" CME where the associated flare site is 20° to 50° behind the limb.

Kiplinger (1995) found that flares with HXR burst spectra that evolved in time from soft to hard to harder (SHH, as opposed to the standard SHS behavior of most flares) are closely associated with high-energy SEP events observed in interplanetary space. In a statistical study of all *RHESSI* flares, Grayson et al. (2009) drew the same conclusion—they found that all *RHESSI* flares associated with an SEP event show SHH behavior, and none of the flares with SHS (normal) behavior are associated with an SEP event. *RHESSI* images show that the HXRs originate from footpoints during times with SHH behavior (Saldanha et al. 2008; Grigis and Benz 2008), and there is no abrupt change in footpoint motion at the onset of the hardening, indicating the same acceleration process taking place in the preceding impulsive phase appears to continue into the hardening phase. Furthermore, for a few flares that are well connected (W30–90), the energetic ion spectra inferred from the *RHESSI* γ-ray line observations appear similar to the spectra of SEPs observed near the Earth, suggesting the possibility that flare acceleration may contribute to gradual event SEPs on well-connected field lines. Thus, in large solar eruptive events there is evidence for a close physical connection between flares, fast CMEs, and SEP acceleration.

6 Summary

RHESSI's HXR and γ-ray line imaging and previous observations have provided strong evidence that flare electron and ion acceleration is related to magnetic reconnection, and *RHESSI's* spectroscopy and imaging have confirmed that in many flares most of the energy released is contained in the accelerated electrons and ions. Now a reconnection process fast enough for the release of energy in a flare—collisionless Hall reconnection—has been identified through theory and simulations, and directly observed in space and laboratory measurements to operate over a very wide range of plasma parameters. Presumably, magnetic energy is being stored in the corona by the motions of footpoints (or the emergence of new bipole from below the photosphere), when conditions are such that slow Sweet-Parker collisional reconnection applies. Then the flare begins with a spontaneous transition from slow Sweet-Parker collisional reconnection to fast collisionless Hall reconnection high in the corona, when the incoming magnetic fields thin the current sheet between regions of different magnetic direction sufficiently. This fast reconnection will start in one small region, but based on simulations, reconnection will rapidly develop at multiple different sites.

The high coronal HXR sources detected by *RHESSI* during the pre-impulsive phase of large flares (such as SOL2002-07-23) may be the result of energization by this initial reconnection (perhaps with some additional acceleration by the collapsing magnetic field); both thermal (superhot) plasma and accelerated electrons are produced. When the reconnection occurs with open field lines (interchange reconnection), this direct energization may provide the $\sim$1–10 keV electrons in impulsive events observed in the interplanetary medium. The upward reconnection jet may result in a fast narrow CME that then accelerates ions and further accelerates electrons higher up in the corona, leading to the observed delays (e.g., Wang et al. 2011).

The impulsive phase of the flare is characterized by the most intense and energetic non-thermal HXR continuum and γ-ray line emissions, mostly in the footpoints, implying that the strongest energy release and particle acceleration is occurring then, and that the energetic particles are precipitating into the chromosphere and photosphere. The high coronal HXR source observed in the Masuda flare above the soft X-ray looptops provided the first indication that the main flare energy release/particle acceleration occurred there. The occulted flare SOL2007-12-31T01:11 (C8.3) had a similar HXR coronal source above the looptops, but here *RHESSI's* high spectral resolution was able to show that this source was filled with energetic electrons with a power-law spectrum (with no thermal plasma detectable above background), indicating that most of the electrons in the source had been accelerated. The bulk of the electrons being convected into this region at a speed of $\sim 0.1\ v_{\mathrm{A}}$ (as expected for inflow to a Hall magnetic reconnection region) need to be accelerated to provide the rate of electrons per second required to produce the observed footpoint HXR emission.

Simultaneous multi-frequency microwave imaging showed that, in this source, the electron power-law spectrum extended to $\sim$1 MeV and the magnetic field is $\sim$30–50 G. The energy density of the accelerated electrons above $\sim$16 keV is then about equal to the magnetic energy density, i.e. $\beta \sim 1$. The upper limits on the X-ray emission from this region before the impulsive phase indicate that $\beta \sim 0.01$, implying a mechanism—such as that proposed by Drake et al. (2006a)—that very efficiently converts the stored magnetic energy, presumably in the form of elongated magnetic fields, to bulk acceleration of electrons. It is important for any theory of the acceleration of energetic particles in flares to show how they are related to magnetic reconnection, and how they can put the bulk of the energy released in a flare into energetic particles. In particular, acceleration mechanisms that rely on an intermediary such as waves, turbulence, or shocks to accelerate the particles, must show how a high overall efficiency—the efficiency for generating the intermediary, multiplied by the efficiency of the intermediary in accelerating the particles—can be achieved.

At present, we do not understand what triggers the sudden onset of the impulsive phase, or how such large numbers of electrons propagate down to the chromosphere. A possible clue is the abrupt change from downward motion of the coronal HXR source during the pre-impulsive phase of SOL2002-07-23, to upward motion at the onset of the impulsive phase. The downward motion suggests collapsing or shortening of magnetic field lines, which should result in further energization of particles in the source. Perhaps when the source gets low enough the bulk energization of electrons is triggered, and the accelerated electrons are then able to propagate to the footpoints to produce the observed HXR emission.

A single mechanism—also related to magnetic reconnection and occurring in the flare region during the impulsive phase—appears to accelerate both $>$30 MeV ions and relativistic, $>$0.3 MeV electrons proportionately. This mechanism, however, results in the spatial displacement of the accelerated ions from electrons; perhaps electric fields are involved. Significant acceleration of tens-of-keV electrons (enough to produce soft X-ray emission of *GOES* M-class or larger) appears to be a prerequisite to this acceleration, but when very large numbers of ions/relativistic electrons are accelerated, the acceleration of tens-of-keV electrons becomes proportional as well. At present, none of the current particle acceleration models (see Zharkova et al. 2011) provide a compelling explanation for these correlations between accelerated electrons and ions, and for their differences in spatial morphology.

Finally, the *RHESSI* observations show that large flares appeared to be closely connected to their associated fast CMEs and the acceleration of SEPs. At present we do not understand the physical mechanisms underlying this connection.

7 Future Prospects

Much higher sensitivity HXR imaging spectroscopy is clearly needed to study the temporal, spatial, and spectral evolution of the high coronal sources in typical flares, rather than just the ones for which that source is anomalously bright. The *RHESSI* observations show that this key region is where the energy release/electron acceleration appears to be happening. High sensitivity will also allow the detection of the HXR emission from the accelerated electrons as they travel down the legs of the loops to the chromosphere; such measurements will help to understand the propagation of these large numbers of electrons through the atmosphere. These measurements should also have much larger dynamic range than *RHESSI*, so the coronal sources can be observed simultaneously with the very bright footpoint sources. In the past decade, focusing optics have been developed for HXRs up to $\sim$80 keV, and now can provide angular resolutions of $\sim 7''$, fine enough for solar measurements. A Focusing Optics hard X-ray Spectrometer Imager (FOXSI) instrument is presently being developed for a rocket flight in late 2011 (Krucker et al. 2009a). A *FOXSI*-like instrument can provide much higher sensitivity and dynamic range than *RHESSI*.

Much more sensitive solar γ-ray line imaging with higher spatial resolution is required to follow the temporal, spatial, and spectral evolution of the ion footpoints, to make progress in understanding the acceleration of ions in solar flares. A Gamma-Ray Imaging Polarimeter for Solar flares (GRIPS) instrument that utilizes cooled germanium detectors with 3-D spatial resolution of 0.5 mm, together with a single rotating grid of novel design mounted $\sim$10 m away to provide the imaging spectroscopy, is presently being developed for a first balloon flight in 2012 (Shih et al. 2009a).

Now that $\sim$1.8–5 MeV energetic neutral atoms (ENAs) have been detected from an SEP event (Mewaldt et al. 2009) for the first time, it appears possible to image SEPs above $\sim 2\ R_\odot$—where they are presumably being accelerated by the fast CME shock wave—via the ENAs they produce through charge exchange. ENAs cannot be focused to form an image, but the *RHESSI* technique of Fourier-transform imaging using modulation methods can be applied.

The above measurements of the accelerated particles need to be combined with imaging measurements of the magnetic field plus plasma parameters—density, temperature, and flows—in the corona where the energy release/particle acceleration is taking place. The plasma parameters can be obtained with a combination of UV/EUV/soft X-ray imaging spectroscopy, while new techniques have been developed to measure the coronal magnetic fields through radio (e.g., FASR) and optical (CoSMO and ATST) imaging spectroscopy. Thus, the next solar maximum (estimated $\sim$2023) would be an ideal time to make the great leap forward in understanding the fundamental physics of transient energy release and efficient particle acceleration in large solar eruptions (and therefore in cosmic magnetized plasmas) especially since complementary measurements of the solar energetic particles escaping to space will likely be provided by the upcoming NASA *Solar Probe Plus* mission and ESA/NASA *Solar Orbiter* mission (which also provides coronagraphic and other imaging measurements) are planned to be going close to the Sun at the time of the next solar maximum. Planning for a *Solar Eruptive Events (SEE)* mission that provides the above-mentioned measurements is presently under way as part of the 2013 Heliophysics Decadal Survey.

Acknowledgements I'm pleased to acknowledge illuminating discussions and assistance from S. Krucker, H. Hudson, G. Emslie, A. Caspi, A. Shih, M. Øieroset, L. Wang, T. Phan, J. Drake, J. Egedal, and L. Glesener. This research was supported in part by NASA contract 5-98033 and by the WCU grant no. R31-10016 funded by the Korean Ministry of Education, Sciences, and Technology.

References

V. Angelopoulos, J.P. McFadden, D. Larson, C.W. Carlson, S.B. Mende, H. Frey, T. Phan, D.G. Sibeck, K. Glassmeier, U. Auster, E. Donovan, I.R. Mann, I.J. Rae, C.T. Russell, A. Runov, X. Zhou, L. Kepko, Science **321**, 931 (2008)
A. Asai, T. Yokoyama, M. Shimojo, S. Masuda, H. Kurokawa, K. Shibata, Astrophys. J. **611**, 557 (2004)
M.J. Aschwanden, A.O. Benz, B.R. Dennis, R.A. Schwartz, Astrophys. J. **455**, 347 (1995)
M.J. Aschwanden, J.C. Brown, E.P. Kontar, Sol. Phys. **210**, 383 (2002)
J. Birn, M. Hesse, J. Geophys. Res. **106**, 3737 (2001)
R.C. Carrington, Mon. Not. R. Astron. Soc. **20**, 13 (1859)
A. Caspi, R.P. Lin, Astrophys. J. Lett. **725**, L161 (2010)
P.A. Cassak, M.A. Shay, J.F. Drake, Phys. Rev. Lett. **95**(23), 235002 (2005)
P.A. Cassak, J.F. Drake, M.A. Shay, Astrophys. J. Lett. **644**, L145 (2006)
L. Chen, A. Bhattacharjee, P.A. Puhl-Quinn, H. Yang, N. Bessho, S. Imada, S. Mühlbachler, P.W. Daly, B. Lefebvre, Y. Khotyaintsev, A. Vaivads, A. Fazakerley, E. Georgescu, Nat. Phys. **4**, 19 (2008)
S. Christe, S. Krucker, R.P. Lin, Astrophys. J. Lett. **680**, L149 (2008)
T.A. Chubb, H. Friedman, R.W. Kreplin, J.E. Kupperian, Nature **179**, 861 (1957)
E.L. Chupp, D.J. Forrest, P.R. Higbie, A.N. Suri, C. Tsai, P.P. Dunphy, Nature **241**, 333 (1973)
E.W. Cliver, in *IAU Symposium*, vol. 257, ed. by N. Gopalswamy, D.F. Webb. (2009), p. 401
E.W. Cliver, A.G. Ling, Astrophys. J. **658**, 1349 (2007)
E.W. Cliver, N.B. Crosby, B.R. Dennis, Astrophys. J. **426**, 767 (1994)
A.J. Conway, J.C. Brown, B.A.C. Eves, E. Kontar, Astron. Astrophys. **407**, 725 (2003)
C.J. Crannell, K.J. Frost, J.L. Saba, C. Maetzler, K. Ohki, Astrophys. J. **223**, 620 (1978)
B.R. Dennis et al., Space Sci. Rev. (2011, this issue)
J.F. Drake, M. Swisdak, H. Che, M.A. Shay, Nature **443**, 553 (2006a)
J.F. Drake, M. Swisdak, K.M. Schoeffler, B.N. Rogers, S. Kobayashi, Geophys. Res. Lett. **33**, 13105 (2006b)
J.P. Eastwood, T.D. Phan, M. Øieroset, M.A. Shay, J. Geophys. Res. **115**, 8215 (2010)
J. Egedal, W. Fox, N. Katz, M. Porkolab, M. Øieroset, R.P. Lin, W. Daughton, J.F. Drake, J. Geophys. Res. **113**, 12207 (2008)
A.G. Emslie, J.A. Miller, J.C. Brown, Astrophys. J. Lett. **602**, L69 (2004)
R.E. Ergun, D. Larson, R.P. Lin, J.P. McFadden, C.W. Carlson, K.A. Anderson, L. Muschietti, M. McCarthy, G.K. Parks, H. Reme, J.M. Bosqued, C. D'Uston, T.R. Sanderson, K.P. Wenzel, M. Kaiser, R.P. Lepping, S.D. Bale, P. Kellogg, J. Bougeret, Astrophys. J. **503**, 435 (1998)
L. Fletcher, H.S. Hudson, Astrophys. J. **675**, 1645 (2008)
L. Fletcher et al., Space Sci. Rev. (2011)
S.E. Forbush, Phys. Rev. **70**, 771 (1946)
R.G. Giovanelli, Mon. Not. R. Astron. Soc. **108**, 163 (1948)
L. Glesener, S. Krucker, R.P. Lin, in *American Astronomical Society Meeting Abstracts #216*. Bulletin of the American Astronomical Society#320.06, vol. 41 (2010)
J.A. Grayson, S. Krucker, R.P. Lin, Astrophys. J. **707**, 1588 (2009)
P.C. Grigis, A.O. Benz, Astrophys. J. **683**, 1180 (2008)
D.K. Haggerty, E.C. Roelof, Astrophys. J. **579**, 841 (2002)
J.S. Halekas, J.P. Eastwood, D.A. Brain, T.D. Phan, M. Øieroset, R.P. Lin, J. Geophys. Res. **114**, 11204 (2009)
J. Heyvaerts, E.R. Priest, D.M. Rust, Astrophys. J. **216**, 123 (1977)
G. Holman et al., Space Sci. Rev. (2011)

G.J. Hurford, R.A. Schwartz, S. Krucker, R.P. Lin, D.M. Smith, N. Vilmer, Astrophys. J. Lett. **595**, L77 (2003)

G.J. Hurford, S. Krucker, R.P. Lin, R.A. Schwartz, G.H. Share, D.M. Smith, Astrophys. J. Lett. **644**, L93 (2006)

K. Hurley, S.E. Boggs, D.M. Smith, R.C. Duncan, R. Lin, A. Zoglauer, S. Krucker, G. Hurford, H. Hudson, C. Wigger, W. Hajdas, C. Thompson, I. Mitrofanov, A. Sanin, W. Boynton, C. Fellows, A. von Kienlin, G. Lichti, A. Rau, T. Cline, Nature **434**, 1098 (2005)

S. Ishikawa, S. Krucker, T. Takahashi, R.P. Lin, Astrophys. J. (2011)

H. Ji, M. Yamada, S. Hsu, R. Kulsrud, Phys. Rev. Lett. **80**, 3256 (1998)

C.M. Johns, R.P. Lin, Sol. Phys. **137**, 121 (1992)

S. Kahler, Astrophys. J. **428**, 837 (1994)

S.W. Kahler, D.V. Reames, N.R. Sheeley Jr., Astrophys. J. **562**, 558 (2001)

M. Kallenrode, E.W. Cliver, G. Wibberenz, Astrophys. J. **391**, 370 (1992)

M. Karlický, T. Kosugi, Astron. Astrophys. **419**, 1159 (2004)

J. Kašparová, M. Karlický, E.P. Kontar, R.A. Schwartz, B.R. Dennis, Sol. Phys. **232**, 63 (2005)

A.L. Kiplinger, Astrophys. J. **453**, 973 (1995)

E.P. Kontar, A.L. MacKinnon, R.A. Schwartz, J.C. Brown, Astron. Astrophys. **446**, 1157 (2006)

E.P. Kontar, E. Dickson, J. Kašparová, Sol. Phys. **252**, 139 (2008)

E. Kontar et al., Space Sci. Rev. (2011)

S. Krucker, R.P. Lin, Astrophys. J. **673**, 1181 (2008)

S. Krucker, D.E. Larson, R.P. Lin, B.J. Thompson, Astrophys. J. **519**, 864 (1999)

S. Krucker, G.J. Hurford, R.P. Lin, Astrophys. J. Lett. **595**, L103 (2003)

S. Krucker, M.D. Fivian, R.P. Lin, Adv. Space Res. **35**, 1707 (2005)

S. Krucker, E.P. Kontar, S. Christe, R.P. Lin, Astrophys. J. Lett. **663**, L109 (2007a)

S. Krucker, S.M. White, R.P. Lin, Astrophys. J. Lett. **669**, L49 (2007b)

S. Krucker, S. Christe, L. Glesener, S. McBride, P. Turin, D. Glaser, P. Saint-Hilaire, G. Delory, R.P. Lin, M. Gubarev, B. Ramsey, Y. Terada, S.-N. Ishikawa, M. Kokubun, S. Saito, T. Takahashi, S. Watanabe, K. Nakazawa, H. Tajima, S. Masuda, T. Minoshima, M. Shomojo, in *Society of Photo-Optical Instrumentation Engineers (SPIE) Conference Series*, vol. 7437 (2009a)

S. Krucker, P.H. Oakley, R.P. Lin, Astrophys. J. **691**, 806 (2009b)

S. Krucker, H.S. Hudson, L. Glesener, S.M. White, S. Masuda, J. Wuelser, R.P. Lin, Astrophys. J. **714**, 1108 (2010)

R.P. Lin, Space Sci. Rev. **16**, 189 (1974)

R.P. Lin, Sol. Phys. **100**, 537 (1985)

R.P. Lin, H.S. Hudson, Sol. Phys. **50**, 153 (1976)

R.P. Lin, C.M. Johns, Astrophys. J. Lett. **417**, L53 (1993)

R.P. Lin, K.A. Anderson, J.E. McCoy, C.T. Russell, J. Geophys. Res. **82**, 2761 (1977)

R.P. Lin, R.A. Schwartz, R.M. Pelling, K.C. Hurley, Astrophys. J. Lett. **251**, L109 (1981)

R.P. Lin, K.A. Anderson, S. Ashford, C. Carlson, D. Curtis, R. Ergun, D. Larson, J. McFadden, M. McCarthy, G.K. Parks, H. Rème, J.M. Bosqued, J. Coutelier, F. Cotin, C. D'Uston, K. Wenzel, T.R. Sanderson, J. Henrion, J.C. Ronnet, G. Paschmann, Space Sci. Rev. **71**, 125 (1995)

R.P. Lin, B.R. Dennis, G.J. Hurford, D.M. Smith, A. Zehnder, P.R. Harvey, D.W. Curtis, D. Pankow, P. Turin, M. Bester, A. Csillaghy, M. Lewis, N. Madden, H.F. van Beek, M. Appleby, T. Raudorf, J. McTiernan, R. Ramaty, E. Schmahl, R. Schwartz, S. Krucker, R. Abiad, T. Quinn, P. Berg, M. Hashii, R. Sterling, R. Jackson, R. Pratt, R.D. Campbell, D. Malone, D. Landis, C.P. Barrington-Leigh, S. Slassi-Sennou, C. Cork, D. Clark, D. Amato, L. Orwig, R. Boyle, I.S. Banks, K. Shirey, A.K. Tolbert, D. Zarro, F. Snow, K. Thomsen, R. Henneck, A. McHedlishvili, P. Ming, M. Fivian, J. Jordan, R. Wanner, J. Crubb, J. Preble, M. Matranga, A. Benz, H. Hudson, R.C. Canfield, G.D. Holman, C. Crannell, T. Kosugi, A.G. Emslie, N. Vilmer, J.C. Brown, C. Johns-Krull, M. Aschwanden, T. Metcalf, A. Conway, Sol. Phys. **210**, 3 (2002)

R.P. Lin, S. Krucker, G.J. Hurford, D.M. Smith, H.S. Hudson, G.D. Holman, R.A. Schwartz, B.R. Dennis, G.H. Share, R.J. Murphy, A.G. Emslie, C. Johns-Krull, N. Vilmer, Astrophys. J. Lett. **595**, L69 (2003)

Y.E. Litvinenko, B.V. Somov, Sol. Phys. **146**, 127 (1993)

W. Liu, V. Petrosian, B.R. Dennis, Y.W. Jiang, Astrophys. J. **676**, 704 (2008)

D.W. Longcope, D.E. McKenzie, J. Cirtain, J. Scott, Astrophys. J. **630**, 596 (2005)

N. Mandzhavidze, R. Ramaty, B. Kozlovsky, Astrophys. J. Lett. **489**, L99 (1997)

G.M. Mason, Space Sci. Rev. **130**, 231 (2007)

S. Masuda, T. Kosugi, H. Hara, S. Tsuneta, Y. Ogawara, Nature **371**, 495 (1994)

S. Masuda, J. Sato, T. Kosugi, T. Sakao, Adv. Space Res. **26**, 493 (2000)

R.A. Mewaldt, R.A. Leske, E.C. Stone, A.F. Barghouty, A.W. Labrador, C.M.S. Cohen, A.C. Cummings, A.J. Davis, T.T. von Rosenvinge, M.E. Wiedenbeck, Astrophys. J. Lett. **693**, L11 (2009)

F.S. Mozer, S.D. Bale, T.D. Phan, Phys. Rev. Lett. **89**(1), 015002 (2002)
M. Øieroset, T.D. Phan, M. Fujimoto, R.P. Lin, R.P. Lepping, Nature **412**, 414 (2001)
M. Øieroset, R.P. Lin, T.D. Phan, D.E. Larson, S.D. Bale, Phys. Rev. Lett. **89**(19), 195001 (2002)
M. Oka, T. Phan, S. Krucker, M. Fujimoto, I. Shinohara, Astrophys. J. **714**, 915 (2010)
R.A. Osten, S. Drake, J. Tueller, J. Cummings, M. Perri, A. Moretti, S. Covino, Astrophys. J. **654**, 1052 (2007)
E.N. Parker, J. Geophys. Res. **62**, 509 (1957)
L. Peterson, J.R. Winckler, Phys. Rev. Lett. **1**, 205 (1958)
R. Ramaty, N. Mandzhavidze, in *American Institute of Physics Conference Series*, vol. 522, ed. by S.S. Holt, W.W. Zhang (2000), p. 401
R. Ramaty, N. Mandzhavidze, B. Kozlovsky, J.G. Skibo, Adv. Space Res. **13**, 275 (1993)
R. Ramaty, N. Mandzhavidze, B. Kozlovsky, R.J. Murphy, Astrophys. J. Lett. **455**, L193 (1995)
R. Ramaty, N. Mandzhavidze, C. Barat, G. Trottet, Astrophys. J. **479**, 458 (1997)
D.V. Reames, Rev. Geophys. **33**, 585 (1995)
H.A.S. Reid, E.P. Kontar, Astrophys. J. **721**, 864 (2010)
T. Sakao, T. Kosugi, S. Masuda, J. Sato, Adv. Space Res. **26**, 497 (2000)
R. Saldanha, S. Krucker, R.P. Lin, Astrophys. J. **673**, 1169 (2008)
K. Shibata, T. Yokoyama, M. Shimojo, Adv. Space Res. **17**, 197 (1996)
A.Y. Shih, R.P. Lin, G.J. Hurford, S.E. Boggs, A.C. Zoglauer, C.B. Wunderer, J.G. Sample, P. Turin, S. McBride, D.M. Smith, H. Tajima, P.N. Luke, M.S. Amman, in *AAS/Solar Physics Division Meeting #40*, vol. 40 (2009a)
A.Y. Shih, R.P. Lin, D.M. Smith, Astrophys. J. Lett. **698**, L152 (2009b)
L. Sui, G.D. Holman, Astrophys. J. Lett. **596**, L251 (2003)
L. Sui, G.D. Holman, B.R. Dennis, Astrophys. J. **626**, 1102 (2005)
L. Sui, G.D. Holman, B.R. Dennis, Astrophys. J. **670**, 862 (2007)
P.A. Sweet, Annu. Rev. Astron. Astrophys. **7**, 149 (1969)
A.J. Tylka, M.A. Lee, Astrophys. J. **646**, 1319 (2006)
W.T. Vestrand, D.J. Forrest, Astrophys. J. Lett. **409**, L69 (1993)
N. Vilmer et al., Space Sci. Rev. (2011)
Y. Wang, M. Pick, G.M. Mason, Astrophys. J. **639**, 495 (2006)
L. Wang, R.P. Lin, S. Krucker, Astrophys. J. **727**, 121 (2011)
J.P. Wild, S.F. Smerd, A.A. Weiss, Annu. Rev. Astron. Astrophys. **1**, 291 (1963)
T.N. Woods, F.G. Eparvier, J. Fontenla, J. Harder, G. Kopp, W.E. McClintock, G. Rottman, B. Smiley, M. Snow, Geophys. Res. Lett. **31**, 10802 (2004)
M. Yamada, R. Kulsrud, H. Ji, Rev. Mod. Phys. **82**, 603 (2010)
V.V. Zharkova, M. Gordovskyy, Astrophys. J. **604**, 884 (2004)
V.V. Zharkova et al., Space Sci. Rev. (2011)
E.G. Zweibel, M. Yamada, Annu. Rev. Astron. Astrophys. **47**, 291 (2009)

Space Sci Rev (2011) 159:447–476
DOI 10.1007/s11214-011-9827-3

Index

Published online: 24 September 2011

B

I

J

K

L

P

Q

R

CPSIA information can be obtained
at www.ICGtesting.com
Printed in the USA
LVHW081445100319
610118LV00001B/2/P